T0140121

IFIP Advances in Information and Communication Technology 667

IFIP – The International Federation for Information Processing

IFIP was founded in 1960 under the auspices of UNESCO, following the first World Computer Congress held in Paris the previous year. A federation for societies working in information processing, IFIP's aim is two-fold: to support information processing in the countries of its members and to encourage technology transfer to developing nations. As its mission statement clearly states:

IFIP is the global non-profit federation of societies of ICT professionals that aims at achieving a worldwide professional and socially responsible development and application of information and communication technologies.

IFIP is a non-profit-making organization, run almost solely by 2500 volunteers. It operates through a number of technical committees and working groups, which organize events and publications. IFIP's events range from large international open conferences to working conferences and local seminars.

The flagship event is the IFIP World Computer Congress, at which both invited and contributed papers are presented. Contributed papers are rigorously refereed and the rejection rate is high.

As with the Congress, participation in the open conferences is open to all and papers may be invited or submitted. Again, submitted papers are stringently refereed.

The working conferences are structured differently. They are usually run by a working group and attendance is generally smaller and occasionally by invitation only. Their purpose is to create an atmosphere conducive to innovation and development. Refereeing is also rigorous and papers are subjected to extensive group discussion.

Publications arising from IFIP events vary. The papers presented at the IFIP World Computer Congress and at open conferences are published as conference proceedings, while the results of the working conferences are often published as collections of selected and edited papers.

IFIP distinguishes three types of institutional membership: Country Representative Members, Members at Large, and Associate Members. The type of organization that can apply for membership is a wide variety and includes national or international societies of individual computer scientists/ICT professionals, associations or federations of such societies, government institutions/government related organizations, national or international research institutes or consortia, universities, academies of sciences, companies, national or international associations or federations of companies.

More information about this series at https://link.springer.com/bookseries/6102

Frédéric Noël · Felix Nyffenegger ·
Louis Rivest · Abdelaziz Bouras (Eds.)

Product Lifecycle Management

PLM in Transition Times: The Place of Humans and Transformative Technologies

19th IFIP WG 5.1 International Conference, PLM 2022
Grenoble, France, July 10–13, 2022
Revised Selected Papers

Editors
Frédéric Noël
Laboratoire G-SCOP
Institut Polytechnique de Grenoble
Grenoble, France

Felix Nyffenegger
IPEK
OST Ostschweizer Fachhochschule
Rapperswil, Switzerland

Louis Rivest
École de Technologie Supérieure
Montreal, QC, Canada

Abdelaziz Bouras
Qatar University
Doha, Qatar

ISSN 1868-4238 ISSN 1868-422X (electronic)
IFIP Advances in Information and Communication Technology
ISBN 978-3-031-25184-9 ISBN 978-3-031-25182-5 (eBook)
https://doi.org/10.1007/978-3-031-25182-5

This Springer imprint is published by the registered company Springer Nature Switzerland AG
The registered company address is: Gewerbestrasse 11, 6330 Cham, Switzerland

Preface

Humanity is facing crises in many areas (health, environment, wars, social risks, etc.) which are reaching the end of the convergence cycle towards the globalisation model. These crises require us to think of the world in a transitional mode. Industrial systems for the production of goods and services are not spared this need for transition. The PLM conference series is completely dedicated to support production systems to meet emerging challenges. The concept of Industry 4.0 was introduced in Western countries to bridge the digital divide but has already been overtaken by the idea of a people-centred industry, which some call Industry 5.0.

In this context, it seemed reasonable in 2022 to approach product and service lifecycle management from the perspective of PLM in transition times, and particularly the place of humans and transformative technologies.

After the first wave of the COVID-19 crisis restricting our social life and freedom of movement, while information technology kept communication alive during the pandemic, PLM 2022 returned to the tradition of face-to-face conferences. This is not a step backwards. The place of people and social relations in the face of technology remains an imperative. For PLM, a conference is still a networking activity which leads to synergies between industrialists and academics, and while videoconferencing can maintain existing relationships it does not yet allow for the creation of new permanent contacts.

The international conference PLM 2022 was organized by the Université Grenoble Alpes and Grenoble-INP, at Grenoble, France on July 10-13, 2022. The conference started on July 11 and the Industry Day was held on July 13. The conference was attended by 120 researchers from 18 nations in three special sessions (Model-Based Systems Engineering for Human-Systems Integration; Advanced Visualization and Human Interaction for PLM; and Transformative Technologies in Production Systems) and an overall set of 18 scientific sessions plus one industrial session. In addition, PLM 2022 had the following academic keynotes, industrial keynotes and roundtables:

Academic Keynotes: Guy André Boy (CentraleSupélec and ESTIA), Gülgün Alpan (Grenoble Institute of Technology), Pierre David (Grenoble Institute of Technology), Gisela Lanza (wbk Institute of Production Science at KIT).

Industrial Keynotes: Pascal Hubert (ATLAS Program), Thierry Cormenier (Schneider Electric), Hugo Falgarone (CEO SkyReal).

Roundtables: "Technology-Organization-People Tryptic in Industry 4.0", moderator Thomas Reverdy (Grenoble INP), panelists Abdelaziz Bouras (Qatar University), Guy André Boy (Centrale Supélec and ESTIA), Bruno Vilain (ADTP); "Model-Based Systems Engineering", moderator Clément Fortin (Skoltech), panelists Eric Muller (MEWS Partners), Klaus-Dieter Thoben (Universität Bremen), Maxime Cassé (Inétum); "Transformative Technologies in Industry", moderator Lionel Roucoules (Arts & Métiers ParisTech), panelists Eric Muller (Mews Partners), Nicolas Maussang

(PETZL), Emmanuel Darlix (Poma), Jean Luc Conedera and Christophe Meunier (Caterpillar), Alex Laurens (Dassault Systèmes).

In the pre-conference programme, a PhD workshop was held, dedicated to young researchers, here again initiating new networking activities. Also, on each half-day of the conference, one keynote, either academic or industrial was followed by intense discussions with researchers and industry participants.

All submitted papers were reviewed by at least two reviewers. In total, 94 full papers were submitted, of which 67 were accepted to be presented at the conference and grouped into 18 thematic sections plus an industrial session. The 18 scientific sessions are reported in this book, split into 5 chapters balancing issues between organisation, processes and tools for PLM:

Organisation Issues: Knowledge Management; Business Models; Sustainability; End-to-End PLM.

Modelling Tools: Model-Based Systems Engineering; Geometric Modelling; Maturity Models; Digital Chain Process.

Transversal Tools: Artificial Intelligence; Advanced Visualization and Interaction; Machine Learning.

Product Development Issues: Design Methods; Building Design; Smart Products; New Product Development.

Manufacturing Issues: Sustainable Manufacturing; Lean Manufacturing; Models for Manufacturing.

This book is included in the IFIP Advances in Information and Communication Technology (AICT) series that publishes state-of-the-art results in the sciences and technologies of information and communication. In addition to this conference, the International Journal of Product Lifecycle Management (IJPLM) is the official journal of WG 5.1. Selected papers of this conference might be published as extended versions in this journal.

We would like to thank everyone who directly or indirectly contributed to making the PLM 2022 conference a success, particularly by making a collective effort to meet all of us in Grenoble.

January 2022

Frédéric Noël
Felix Nyffenegger
Louis Rivest
Abdelaziz Bouras

Organization

Conference Chairs

Alain Bernard	École Centrale de Nantes, France
Klaus-Dieter Thoben	BIBA, Germany

Program Chairs

Frederic Nöel	Grenoble Institute of Technology, France
Felix Nyffenegger	Rapperswil University of Applied Sciences, Switzerland

Local Organization Committee

Romain Pinquié	Grenoble Institute of Technology, France
Abdourahim Sylla	Grenoble Institute of Technology, France
Pierre David	Grenoble Institute of Technology, France
Gilles Foucault	Université Grenoble Alpes, France
Cédric Masclet	Université Grenoble Alpes, France
Guillaume Thomann	Grenoble Institute of Technology, France

Steering Committee

Chairs

Louis Rivest	University of Quebec, Canada
Osiris Canciglieri Junior	Pontifical Catholic University of Parana, Brazil

Members

Abdelaziz Bouras	Qatar University, Qatar
Anderson Luis Szejka	Pontifical Catholic University of Parana, Brazil
Balan Gurumoorthy	Indian Institute of Sciences Bangalore, India
Benoit Eynard	University of Technology of Compiegne, France
Clement Fortin	Skolkovo Institute of Science and Technology, Russia
Debashish Dutta	University of Michigan, USA
Eduardo Zancul	University of São Paulo, Brazil
Fernando Mas	University of Seville, Spain
José Rios	Polytechnic University of Madrid, Spain
Klaus-Dieter Thoben	University of Bremen, Germany
Miroslav Trajanovic	University of Nis, Serbia
Néjib Moalla	University of Lyon, France
Paolo Chiabert	Polytecnico di Torino, Italy

Ramy Harik	University of South Carolina, USA
Romeo Bandinelli	University of Florence, Italy
Sebti Foufou	Université de Bourgogne, France
Sergio Terzi	Politecnico di Milano, Italy

Doctoral Workshop Committee

Monica Rossi	Polytecnico di Torino, Italy
Yacine Ouzrout	University of Lyon, France

Scientific Committee

Abdelaziz Bouras	Qatar University, Qatar
Alain Bernard	Central School of Nantes, France
Anderson Luis Szejka	Pontifical Catholic University of Parana, Brazil
Alexander Smirnov	Russian Academy of Sciences, Russia
Alison McKay	University of Leeds, UK
Améziane Aoussat	Paris Institute of Science and Technology, France
Ângelo Marcio Oliveira Sant'Anna	Federal University of Bahia, Brazil
Antonio Batocchio	University of Campinas, Brazil
Balan Gurumoorthy	Indian Institute of Sciences Bangalore, India
Benoit Eynard	University of Technology of Compiegne, France
Carla Cristina Amodio Estorillio	Technological Federal University of Parana, Brazil
Chris McMahon	Technical University of Denmark, Denmark
Christophe Danjou	Polytechnique Montreal, Canada
Claudio Sassanelli	Politecnico di Bari, Italy
Clement Fortin	Skolkovo Institute of Science and Technology, Russia
Darli Rodrigues Vieira	Université du Québec à Trois-Rivières, Canada
Debashish Dutta	University of Michigan, USA
Detlef Gerhard	Ruhr University Bochum, Germany
Dimitris Kiritsis	Federal Institute of Technology in Lausanne, Switzerland
Eduardo de Freitas Rocha Loures	Pontifical Catholic University of Parana, Brazil
Eduardo Zancul	University of São Paulo, Brazil
Erastos Filos	European Commission, Belgium
Frédéric Demoly	University of Technology Belfort, Montbeliard, France
Frédéric Segonds	Paris Institute of Science and Technology, France
Fernando Antonio Forcelinni	Federal University of Santa Catarina, Brazil
Fernando Mas	University of Seville, Spain
George Huang	University of Hong Kong, China
Gianluca D'Antonio	Polytecnico di Torino, Italy

Ubiratã Tortato	Pontifical Catholic University of Parana, Brazil
Umberto Cugini	Politecnico di Milano, Italy
Vishal Singh	Indian Institute of Science, India
Wen Feng Lu	National University of Singapore, Singapore
Yaroslav Menshenin	Skolkovo Institute of Science and Technology, Russia
Young Won	Park University of Tokyo, Japan

Sponsors

Contents

Organisation: Knowledge Management, Business Models, Sustainability, End-to-End PLM

POSEIDON: A Graphical Editor for Item Selection Rules Within Feature Combination Rule Contexts

Daniel Bischoff[1(✉)], Wolfgang Küchlin[2], and Oliver Kopp[1]

[1] Mercedes-Benz AG, Böblingen, Germany
{daniel.bischoff,oliver.kopp}@mercedes-benz.com
[2] Universität Tübingen, Tübingen, Germany
wolfgang.kuechlin@uni-tuebingen.de

Abstract. A car line can offer more than 10^{100} variants, and for each customer order a concrete selection of features and parts needs to be done. The respective selection rules are interconnected and are subject to constraints imposed by different car lines. We address the problem of finding and fixing logical errors in these interconnected selection rules within different contexts of allowed feature combinations. Previous work has focused on text-based or matrix-like representations which presented challenges regarding cognitive complexity, size of the representation, and usability. We present an integrated visualization of the combined practical effects of item selection and feature combination rules. The implemented tool detects logical errors and supports user workflows to fix the data visually.

Keywords: Product configuration · Configuration systems · Configuration rules · Data quality · Visualization

1 Introduction

Customization of cars increases the space of available variants. The customer can select multiple equipment options (features). Based on this selection, the parts (mechanical, electronic, or software functions) to assemble the car have to be chosen. These parts are documented in the Bill-of-Materials (BoM). Typically, this is not a 1:1 feature/part relation, but there are complex rules in place [8]. Each rule specifies a Boolean condition on the chosen features and outputs whether the associated part has to be chosen. In this paper, we will refer to them as *item selection rules*. A general so called 150% BoM, which contains all possible parts of all possible variants, is converted into a specific 100% BoM by evaluating all rules. From this 100% BoM, a car can be produced.

The distance between the headlights of a car, for instance, might differ between a limousine and a convertible. When producing one of these cars, the

Published by Springer Nature Switzerland AG 2023
F. Noël et al. (Eds.): PLM 2022, IFIP AICT 667, pp. 3–14, 2023.
https://doi.org/10.1007/978-3-031-25182-5_1

same Electronic Control Unit (ECU) could still be used but needs to be configured. This ECU has to be informed which headlight distance is an accurate description of the ECU's context, i.e. the car. To derive that information from the equipment options, item selection rules for the configuration of the ECU software are defined.

In general, a customer can order from different product lines (e.g., gasoline or electric line). In certain product lines a combination of convertible with a sports package might be permitted, whereas other lines forbid this combination. The context constraining the allowed combinations is called *feature combination rule context.*

Item selection rules for identical parts in different contexts might be similar and are often used as the basis for new rules. Thus, item selection rules share similar parts but also differ in other parts. In general, a product line can also evolve over time. Thus, an item selection rule can be applicable in, and be subject to, multiple contexts. The main challenge is the creation of the initial rules and recontextualization for new contexts.

In this paper we address the challenges during initial creation and recontextualization by a) visualizing item selection rules, while b) being able to switch seamlessly between different contexts, and c) enabling interactions on the visual data itself. For that we start by presenting background and related work (Sect. 2). Then we provide a running example (Sect. 3). Afterwards, we discuss an algorithm to calculate the visualization (based on the given context; Sect. 4) and its implementation (Sect. 5). Subsequently, the implementation is evaluated and discussed (Sect. 6). Finally, we summarize our findings and provide an outlook on future work (Sect. 7).

2 Background and Related Work

In the automotive context, a product line defines which vehicle variants are possible. The term *feature combination rule context*, or *context* for short, entails the declaration of features and the constraints that make feature combinations valid or invalid. Different professional domains have different names for this concept. Examples are: High Level Configuration, Feature Tree, Product Overview, Vehicle Description Summary, and Model Description [2,7]. Automotive product lines have been encoded in Boolean logic at least since the 1990s s and formal validation methods based on Satisfiability Solving (SAT) have been used at least since 2000 [8], whereas feature trees were first transformed to Boolean logic three years later [9].

Each equipment option is represented as a Boolean variable indicating equipment option presence. Thus, a concrete vehicle is modeled as an assignment of the Boolean variables defined by the context which indicates presence or absence of equipment options. During assembly, concrete vehicles are built. The *item selection rules* define for which vehicle variants which of the items have to be selected. In the automotive context, examples of an item are a physical component to build a car, a software version, or a configuration parameter value.

A concrete vehicle is segmented into alternative bundles called *positions*, which need to be filled. For each position, there needs to be an item selected. Thus, each position lists respective <*item*, *rule*> pairs. In our BoMs, positions can contain roughly up to 30 alternative items, while up to 25 variables are used in some item selection rules.

For each position, there are quality criteria in place [8]: 1. Each item selection rule is statisfiable in its context (satisfiability). 2. For each variable assignment, there is at most one item hit (uniqueness). 3. For each variable assignment, there is at least one item hit (completeness). Identical data quality goals have also been found by Astesana et al. [2]. The quality criterion of satisfiability has been identified by Berndt et al. [3]: Multi-valued decision diagrams (MDDs) have been there for computation only. The quality criteria of uniqueness and completeness also have been identified by Voronov et al. [14].

Tidstam et al. [12] present an approach for item selection rules captured in matrices. One dimension captures the product families, the other dimension captures the items to be selected. An element in the matrix indicates whether an item is always selected, not selected, and sometimes selected. The case "sometimes" appears, because the context cannot be fully captured in the matrix. In contrast, our approach presents an explicit visualization of selections, the implications of a context, and the visualization is not bound to a matrix representation. Tidstam and Malmqvist [13] compared their matrix-based approach with a list-based approach. It turned out that the matrix-based approach was preferred by their target audience.

Shafiee et al. [11] conducted a survey on visual representation techniques for product configuration systems in industrial companies. Their result indicates that companies using visual knowledge representation techniques tend to have (i) higher quality of the product configuration system's knowledge base and (ii) higher quality with respect to communications with domain experts [11]. This supports our approach to use visual representations instead of text-based ones.

Amilhastre et al. [1] encoded context validity in a Constraint Satisfaction Problem (CSP), compiled the CSP into an automaton and represented it graphically. The automaton accepts valid feature selections, but, in contrast to our approach, does not encode which part selection follows, i.e. which and how many parts are involved in a potential collision.

POSEIDON is the first decision-diagram-based item selection rule editor, that has build-in quality checks and while allowing for seamless context switches.

3 Running Example

The example we will use throughout the rest of this paper centers around the configuration of an ECU that is used in two vehicle contexts called Electric (E-Series) and Gasoline (G-Series), respectively. We will focus on a single configuration value that represents the distance between the headlights called *headlight_distance_mm.*

Among the variables we use are *LIMOUSINE, KOMBI, COUPE,* and *CABRIO* to indicate body types and *FR, LU, BE, NL, DK, PL, CZ, AT, LI,* and *CH* to name distribution countries. The variables *SPORT*, *NIGHT*, and *ELEGANCE* are used to represent extras.

The context of E (γ_E) and G-Series (γ_G) is defined in Eq. (1) and Eq. (2), respectively. Here, pseudo Boolean constraints are used for brevity and can be transformed into pure Boolean constraints [5]. They are generally of the form $\Sigma(v) \leq 1$, which we refer to as an *at-most-one* (AMO) constraint, or $\Sigma(v) = 1$, an *exactly-one* (EXO) constraint. Some of these are shared between the contexts. We generally use the following operator precedence: $\neg, \wedge, \vee, \implies, \iff$ (from strongest to weakest binding).

$$
\begin{aligned}
\gamma_E = & [LIMOUSINE + KOMBI + COUPE + CABRIO = 1] \wedge \\
& [SPORT + NIGHT + ELEGANCE \leq 1] \wedge \\
& [FR + LU + BE + NL + DK + PL + CZ + AT + LI + CH \leq 1] \wedge \\
& ((LIMOUSINE \vee KOMBI) \implies \neg SPORT \wedge \neg NIGHT) \wedge \\
& (NL \implies NIGHT) \wedge \\
& (CABRIO \implies \neg DK \wedge \neg PL \wedge \neg CZ)
\end{aligned} \tag{1}
$$

$$
\begin{aligned}
\gamma_G = & [LIMOUSINE + KOMBI + COUPE + CABRIO = 1] \wedge \\
& [SPORT + NIGHT + ELEGANCE \leq 1] \wedge \\
& [FR + LU + BE + NL + DK + PL + CZ + AT + LI + CH \leq 1] \wedge \\
& (NL \implies \neg SPORT \wedge \neg ELEGANCE) \wedge \\
& (CABRIO \implies SPORT) \wedge \\
& \neg(KOMBI \wedge (BE \vee NE \vee LU) \wedge SPORT)
\end{aligned} \tag{2}
$$

Table 1. The position for `headlight_distance_mm` used in the example.

Item	Item selection rule ϕ
1650 mm	$CABRIO \vee COUPE \vee$ $SPORT \wedge (\neg KOMBI \wedge \neg LIMOUSINE \vee KOMBI \wedge$ $(\neg(BE \vee CZ \vee DK \vee FR \vee NL \vee PL) \vee LI \vee AT \vee CH \vee LU)) \vee$ $NIGHT \wedge (\neg LIMOUSINE \vee KOMBI)$
1700 mm	$KOMBI \wedge \neg NIGHT \wedge \neg SPORT \vee$ $NIGHT \wedge (LIMOUSINE \vee \neg CABRIO \wedge \neg COUPE \wedge \neg KOMBI)$
1750 mm	$LIMOUSINE \wedge \neg NIGHT \vee$ $ELEGANCE \wedge \neg(CABRIO \vee COUPE \vee KOMBI \vee LIMOUSINE) \vee$ $KOMBI \wedge SPORT \wedge (BE \vee CZ \vee DK \vee FR \vee NL \vee PL)$

The position configuring `headlight_distance_mm` is given in Table 1. To produce simpler diagrams, the rules used in this example initially meet all quality criteria (Sect. 2).

4 Visualizing Item Selection Rules

Algorithm 1 describes a procedure how to generate a decision diagram data structure based on a context, items, and item selection rules. This data structure can then be used as input for a graph layout algorithm. We will describe the most important aspects of Algorithm 1 in the following sections. The algorithm described here extends our previously published technique [4] by a) adding grouping of variables and b) the capability to visualize under different contexts.

Algorithm 1. Generate Visualization Data Structure

1: Given Context γ
2: Given Map of <Item i, Item Selection Rule ϕ_i > of length n
3: configVars $\leftarrow$ Variables that occur in $\{\phi_0, \dots, \phi_{n-1}\}$
4: $\gamma^* \leftarrow$ project(γ, configVars)
5: groups $\leftarrow$ group(γ^*, configVars)
6: $\psi \leftarrow \gamma^* \wedge \bigwedge_{i=0}^{n-1} (\phi_i \iff H_i)$
7: result $\leftarrow$ bdd(ψ, groups)
8: result $\leftarrow$ terminals(result)
9: result $\leftarrow$ mdd(result, groups)
10: result $\leftarrow$ labeling(result)
11: return result;

4.1 Projection

Projection aims to provide focus on the visualization in terms of the configuration variables: Variables not used in the configuration, but defined in the context should be removed. However, they still might have an effect on the configuration. Thus, their effects on valid combinations of the configuration variables are kept. To implement the projection, we use the model enumeration-based quantifier elimination as described and evaluated by Zengler and Küchlin [15]. In the running example, the contexts (γ_E and γ_G, Sect. 3) are already projected to the set of configuration variables for better readability.

4.2 Grouping

The grouping step (Line 5) partitions the configuration variables such that the variables in each group exclude each other. The context is the main source of potential groups, since it contains explicit at-most-one (AMO) and exactly-one (EXO) constraints. To generate more potential groups we implemented functionality which allows our users to propose groups, which we then validate against the context by checking whether *context* $\implies$ *AMO(proposal)* is a tautology. If the proposal turns out to be valid, we store it in our database, which realizes a user-defined group storage for each context. This shared and growing common knowledge is used during future groupings. In our example the groups are 1. the four body types, 2. the options SPORT, NIGHT and ELEGANCE, and 3. the distribution countries.

4.3 Constructing a BDD

The first step of constructing our visualization is to build a Binary Decision Diagram. For this purpose, we create a set of Boolean helper variables $H = \{H_0, \ldots, H_{n-1}\}$ representing the n items. Together with the n item selection rules $\phi = \{\phi_0, \ldots, \phi_{n-1}\}$ and the projected context γ^* they are used to create the core formula

$$\psi = \gamma^* \wedge \bigwedge_{i=0}^{n-1} (\phi_i \iff H_i). \tag{3}$$

We call H the set of *item variables*, in contrast to the configuration variables that occur in the selection rules ϕ.

The core formula ψ consists of two components – the context and a representation of the position. The construction of the position representation $\bigwedge_{i=0}^{n-1} (\phi_i \iff H_i)$ assures that no assignment of only the configuration variables can falsify this second part of ψ. There is always an otherwise unconstrained helper variable H_i to match the value of ϕ_i, thereby providing a trivial solution to all constraints. We can therefore conclude that any assignment of the configuration variables falsifying ψ represents a violation of the context constraints, and thus an invalid vehicle. As a result, the item selection rules determine the resulting item for each variant, while the context defines which variants are considered.

Subsequently, we compute a BDD for ψ. Thereby, the item variables H are always last in the BDD's ordering, while the configuration variables that belong to the same group are next to each other. For this purpose we adapted the sifting algorithm [10], which turned out best [6]. For our example this yields the result shown in Fig. 1 for the E-Series context.

4.4 Terminal Generation

To compress the visualization further, we traverse the graph top-down. Every path eventually assigns all configuration variables. At this point, a terminal node is generated that contains exactly those item variables H_i which must be true in order to reach the final $true node of the BDD. This path is unique at this point, because the core formula (3) constrains all H_i to be equivalent to their already fixed ϕ_i counterparts. For our example this yields Fig. 2.

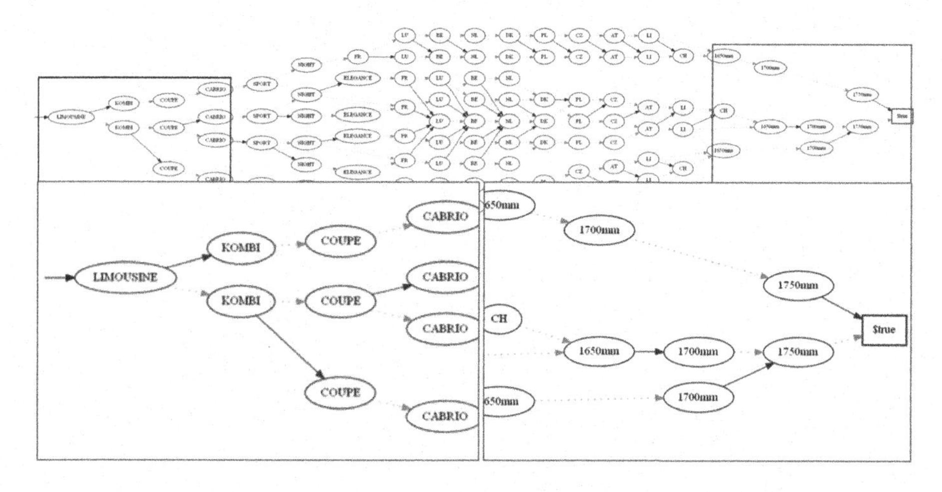

Fig. 1. The BDD for `headlight_distance_mm` in the E-Series context. The *$false* node is hidden with all its incoming edges for better readability.

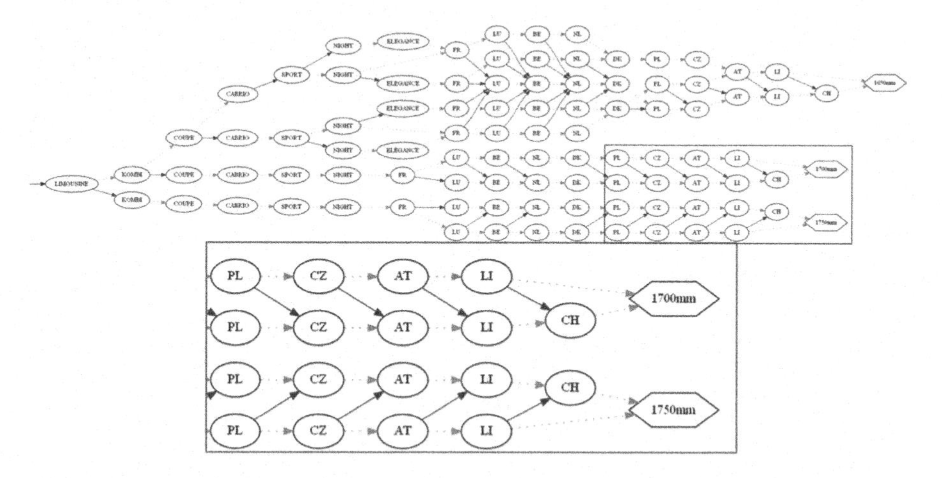

Fig. 2. The BDD for headlight_distance_mm in the E-Series context with terminals. The *$false* node is hidden with all its incoming edges for better readability.

4.5 Multi-decision Diagram Generation

We traverse the graph top-down. For every node that contains a configuration variable, we collect the variables of its children nodes recursively until we find the first variable of another group or a terminal. For those collected variables we form a *multi-decision node* representing the possible selections from the group. For our example this yields Fig. 3.

The multi-decision nodes replace the binary ones and have several outgoing edges. For each group member an outgoing edge exists, which may be shared with other members if they lead to the same result. One additional edge represents the *no selection* option for that group, where all variables are assigned *$false*. It

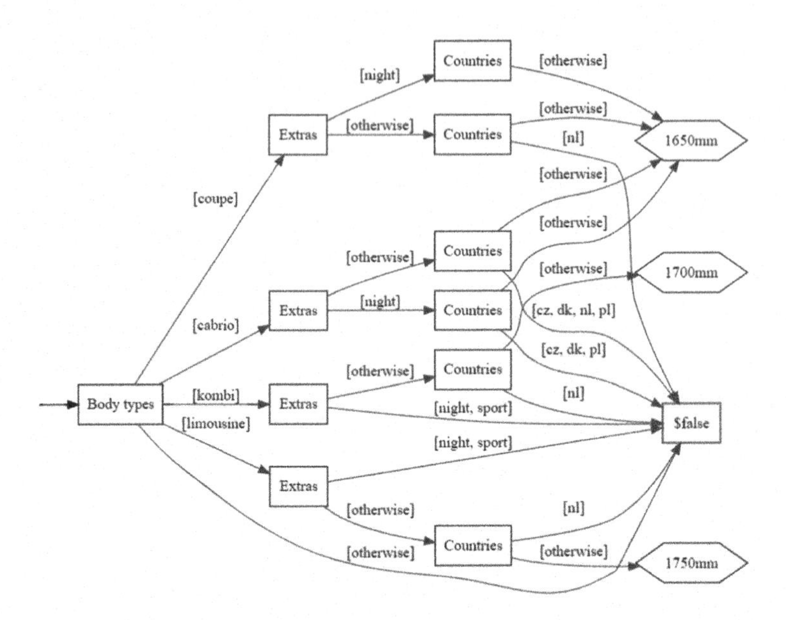

Fig. 3. The MDD for headlight_distance_mm in the E-Series context. The text in square brackets represents the decision in the previous MDD node.

may happen that *no selection* leads to the same item as choosing some member of the group, so that the edges may be shared.

The *no selection* edge is called *otherwise* in our figures, representing all options that are not explicitly noted on other outgoing edges of the same MDD node.

For example the topmost *Countries* node in Fig. 3 has a single outgoing edge *otherwise* which represents all possible countries or no country at all since they all result in the same item selection of 1650 mm. The *Countries* node right below has two outgoing edges, where NL leads to *$false* and therefore represents a configuration in contradiction with the context constraints and the *otherwise* edge representing all other or no country selection resulting in the 1650 mm item being chosen.

4.6 Label Creation and Propagation

We introduce labels to reduce the size of the graph. For this, we traverse the graph top-down. If a node has the *$false* node as a child, a *label* is generated to indicate that this assignment contradicts the context. The label contains the opposite variable assignment and is placed on all incoming edges of the node. The edge to the *$false* node is then removed (cf. Fig. 4). If a node has the same label on all its outgoing edges, the label is propagated up from all outgoing to all incoming edges. Labels on the same edge are combined using the $\wedge$ operator. This may result in nodes with only a single unlabeled outgoing edge, which are then removed. For our example this yields Fig. 5.

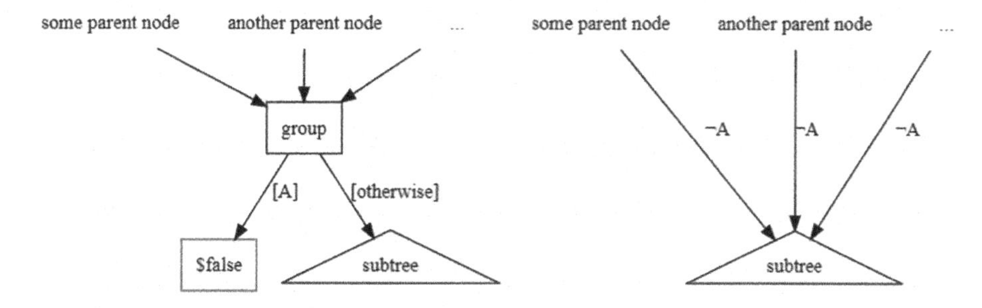

Fig. 4. Selecting A leads to a contradiction, thus A needs to be assigned *$false*. We force $\neg A$ by adding labels and (since the group now has a single unlabeled child) remove the decision node by rerouting its incoming edges to its valid subtree.

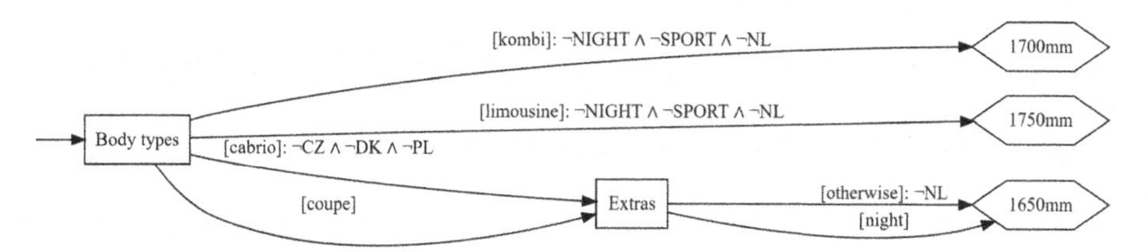

Fig. 5. The final MDD for headlight_distance_mm in the E-Series context with labels. The text in square brackets represents the decision in the previous MDD node.

4.7 Roundtrip

We allow users to manipulate the graphical representation, e.g., by introducing new nodes or redirecting edges. Effectively, this allows for a complete (re-) definition of item selection rules merely through the GUI. The resulting graph can then be transformed back into individual rules by reading off the possible necessary assignments to reach a terminal. E.g., we can read the resulting item selection rule for 1700 mm from Fig. 5, as $KOMBI \wedge \neg NIGHT \wedge \neg SPORT \wedge \neg NL$ for the top-most path. Hence, we can transform a position and a context into an MDD representation, let the user change the MDD visually, and transform the result back into the updated position for the used context.

5 Implementation

We implemented Algorithm 1 in a web-based tool which layouts the decision diagram and makes it editable. In order to demonstrate the use cases of our application and to be able to show how we deal with questionable data quality, we now change the position by adding another item 1800 mm with its selection rule $CABRIO \wedge NIGHT$. Since this case was already covered by the rule for 1650 mm, this leads to a data quality issue which we visualize to the user as shown in Fig. 6. The ambiguity highlighted in red indicates that there are vehicle configurations in this context that select multiple items. The path from the

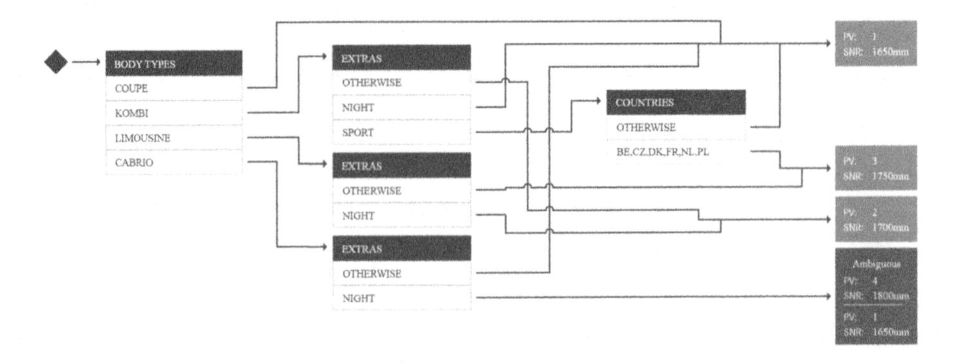

Fig. 6. A position containing an ambiguity in the context shared by E and G-Series.

diamond-shaped root node to the ambiguity identifies the vehicle(s) affected, in this case all *Cabrios* with the *Night* extra need correction.

Our user interface allows users to rearrange the ordering of groups in which case we simply skip automatic ordering in the algorithm. If a user thus has a certain ordering in mind they can manually force this ordering. This works when creating new positions as well as when working with preexisting positions. We implemented editing functionality which allows the user to work with the graph while keeping it valid until they want to retrieve the item selection rules (based on the roundtrip functionality described in Sect. 4.7). Therefore, a user can manipulate the graphical representation and retrieve the resulting rule set after the change.

6 Evaluation and Discussion

Using the implementation, we evaluated a model series from 2019 and took positions with 10 or more items resulting in 535 positions with sizes from 10 to 100 and a median of 17. Thereby, the runtime of the complete algorithm ranged from 0.8 s to 12.3 s, whereby the median runtime was 1.4 s. Within this, projection (Sect. 4.1) took 40% of the total runtime (in the median) with a minimum of 3% and a maximum 87%. As a consequence, we implemented caching for the projection: When a context had been projected to a given configuration variable set previously, the result is reused. At a cold start with caching activated, the runtime of the complete algorithm changed to the range of 0.02 s to 13 s, with a median of 1 s.

The scaleability primarily depends on the context. More precisely the number of combinations of the variables used in the item selection that are considered context-valid, is the primary driver of the visualization size. A decision diagram is always smaller than an analogous table, since inner nodes can be re-used.

The solution is applicable for other products and domains whenever the *context* and *item selection rules* are either written in or translatable to Boolean logic. This translation is possible for feature trees [9].

7 Conclusion and Outlook

We showed the theoretical foundation and a practical implementation of a tool for visualizing and editing item selection rules. It supports seamless context switches and direct visual data editing during initial creation and recontextualization. We showed that the tool can handle industry-grade sizes of item selection rules and contexts.

By allowing each branch to have its own ordering, we foresee that it is possible to reduce the size of the graph even further. Since the concrete impact on industry-sized item selection rules, especially when it comes to readability, is unclear, this investigation is our next research step.

Acknowledgements. This work has received funding by the BMWK-funded project SofDCar (19S21002D).

References

1. Amilhastre, J., Fargier, H., Marquis, P.: Consistency restoration and explanations in dynamic CSPS-application to configuration. Artif. Intell. **135**(1–2), 199–234 (2002)
2. Astesana, J.M., Cosserat, L., Fargier, H.: Constraint-based vehicle configuration: a case study. In: ICTAI. IEEE (2010)
3. Berndt, R., Bazan, P., Hielscher, K.S.: MDD-based verification of car manufacturing data. In: CIMSIM. IEEE (2011). https://doi.org/10.1109/CIMSim.2011.40
4. Bischoff, D., Küchlin, W.: Adapting binary decision diagrams for visualizing product configuration data. In: Eibl, M., Gaedke, M. (eds.) INFORMATIK 2017. GI e.V. (2017). https://doi.org/10.18420/in2017_149
5. Eén, N., Sörensson, N.: Translating pseudo-boolean constraints into sat. J. Satisfiability Boolean Model. Comput. **2**(1–4), 1–26 (2006)
6. Krawietzek, N.: Minimierung der graphischen Repräsentation von Coderegeln durch Variablensortierung. Master's thesis, SCMT, Filderstadt, Germany (2020)
7. Küchlin, W.: Logic and verification of product configuration in the automotive industry. In: Proof and Computation II, pp. 387–408. World Scientific (2021). https://doi.org/10.1142/9789811236488_0009
8. Küchlin, W., Sinz, C.: Proving consistency assertions for automotive product data management. J. Autom. Reason. **24**(1–2), 145–163 (2000)
9. Mannion, M., Camara, J.: Theorem proving for product line model verification. In: van der Linden, F.J. (ed.) PFE 2003. LNCS, vol. 3014, pp. 211–224. Springer, Heidelberg (2004). https://doi.org/10.1007/978-3-540-24667-1_16
10. Rudell, R.: Dynamic variable ordering for ordered binary decision diagrams. In: Kuehlmann, A. (ed.) The Best of ICCAD, pp. 51–63. Springer, Boston (2003). https://doi.org/10.1007/978-1-4615-0292-0_5
11. Shafiee, S., Kristjansdottir, K., Hvam, L., Felfernig, A., Myrodia, A.: Analysis of visual representation techniques for product configuration systems in industrial companies. In: IEEM. IEEE (2016). https://doi.org/10.1109/ieem.2016.7797985
12. Tidstam, A., Bligård, L.O., Ekstedt, F., Voronov, A., Åkesson, K., Malmqvist, J.: Development of industrial visualization tools for validation of vehicle configuration rules. In: TMCE (2012)

13. Tidstam, A., Malmqvist, J.: Comparison of configuration rule visualizations methods. In: Bernard, A., Rivest, L., Dutta, D. (eds.) PLM 2013. IAICT, vol. 409, pp. 550–559. Springer, Heidelberg (2013). https://doi.org/10.1007/978-3-642-41501-2_55
14. Voronov, A., Tidstam, A., Åkesson, K., Malmqvist, J.: Verification of item usage rules in product configuration. In: Rivest, L., Bouras, A., Louhichi, B. (eds.) PLM 2012. IAICT, vol. 388, pp. 182–191. Springer, Heidelberg (2012). https://doi.org/10.1007/978-3-642-35758-9_16
15. Zengler, C., Küchlin, W.: Boolean quantifier elimination for automotive configuration – a case study. In: Pecheur, C., Dierkes, M. (eds.) FMICS 2013. LNCS, vol. 8187, pp. 48–62. Springer, Heidelberg (2013). https://doi.org/10.1007/978-3-642-41010-9_4

Software Parts Classification for Agile and Efficient Product Life Cycle Management

Anmol Gaurav[1](✉), Bhanu Prakash Ila[2](✉), Naveen Mehata Kondamudi[1](✉), and Pinky Deshwal[1](✉)

[1] Business Consulting NPI/PLM Europe Manufacturing, Tata Consultancy Services, 1/G1, SIPCOT IT Park, Kanchipuram 600119, Tamil Nadu, India
{anmol.gaurav,naveen.kondamudi,pinky.deshwal}@tcs.com

[2] Business Consulting NPI/PLM Europe Manufacturing, Tata Consultancy Services, Initial Tower, 1 Terr. Bellini, 92800 Puteaux, France
bhanuprakash.i@tcs.com
http://www.tcs.com/

Abstract. The proliferation of software components and embedded systems in bundling of smart and innovative products and services has augmented the need of software classification in manufacturing context. Existing software classification standards are limited to information technology, networking, and mobile applications. To gain efficiency in product development through reusability and reduction in search and retrieval time for multi domain part structure development; software classification is imperative.

This paper discusses the different classification and nomenclature systems available for software, its benefits in Product Lifecycle Management (PLM) and other business areas, as well as limitation of existing PLM applications. This paper explains a novel, and contextual framework for classification of software components to enable strategic cataloguing for improved traceability, reusability, and Bill of Materials (BOM) management.

Keywords: Software part · Software part classification · Classification framework · Software classification standards · Classification for ALM

1 Introduction

Traditional business drivers are no longer sufficient to provide sustainability and growth to an organization. Globalization, pricing pressure, product complexity and competition are new drivers demanding organizations to substitute part of their existing portfolios with innovative products. Burgeoning of embedded systems in products to make it smarter, complaint with new regulatory requirements and experiential offerings for customers have raised the bar [1, 2]. Embedded software market is estimated to grow with CAGR of 7% and would create a value of more than $20 billion by 2027 [3]. Surge in its demand is being seen more in automotive industry due to increased regulatory pressure and advanced features requirements, and in healthcare industry due to COVID-19

Published by Springer Nature Switzerland AG 2023
F. Noël et al. (Eds.): PLM 2022, IFIP AICT 667, pp. 15–24, 2023.
https://doi.org/10.1007/978-3-031-25182-5_2

impact. PLM applications provide capabilities to define firmware and software as a part and features to link with their executables. But these are not sufficient for manufacturers who are looking to classify their software and firmware parts like hardware parts. There are international standards like UNSPC[1], ECLASS[2], GPC[3], CPV[4], ETIM[5] etc. which provide taxonomy for products and services [4, 5]. These standards are highly skewed towards the classification of hardware components. Adopting the current standard of four level hierarchical classification structure is not a fool-proof solution since hardware part can have one to many relationships with the corresponding software parts. Multiple relationships exist due to different supplier, versions, and specific features of the software. This paper covers investigation and analysis on existing global classification standards and has proposed a structured approach for software part classification.

Development of new software classifications standards and harmonized meta data attribute definition will significantly reduce search retrieval time, improve correctness, reliability, reusability and traceability and ease trade across countries [6–8]. For instance, when a designer is looking for a software component to meet a new product specification, the classification attributes would help to narrow down search and give relevant information from large repository of data. This would also help to avoid creation of duplicate parts. In another scenario, a field engineer would be able to reduce root cause analysis time and identify similar components when a warranty claim is reported.

2 Classification System

In this digital age, for a company to stay relevant and survive, depends on rate of innovation and its introduction through complex supply chain web [9]. Classification is essential to businesses to improve supplier sourcing strategy, optimize productions, for targeting buyers and reduce business transactions time.

Classification means logical grouping and ordering of large and expanding range of commodity types according to common characteristics. Existing international standards vary in terms of objectives, data model, granularity, breadth of category and adoption with respective to the region and refer Sect. 3 for more details. In general, all classification standards provide coded name, identification number and textual description in noun-modifier combination with different schema. Variety of part types, for example, hardware, electrical, mechanical, software etc. are categorized at different levels. The bar graph given in Fig. 1. shows distribution of categories of software parts at multiple hierarchy levels by 4 popular international standards.

[1] United Nations Standard Products and Services Code.

[2] ECLASS is a cross-industry product-data standard for the classification and unique description of products and services.

[3] Global Product Classification.

[4] Common Procurement Vocabulary.

[5] European Technical Information Model.

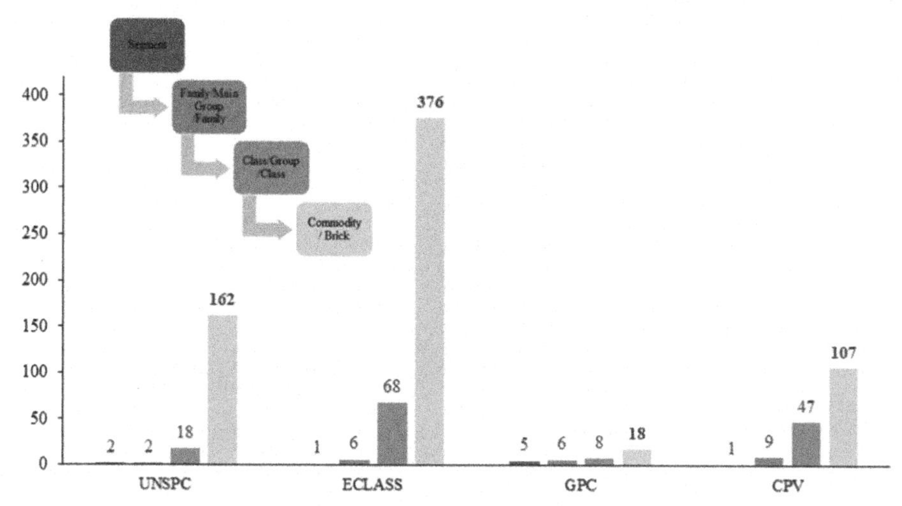

Fig. 1. Software classes at different levels for 4 global classification standards (Data is compiled from the codeset of each of the mentioned global standards).

2.1 Need for Software Classification System and Standardized Data

Business discontinuity due to COVID-19 has accelerated the process to have virtual collaborative environment to take design decisions quickly and need of IT systems with robust integrated data management capabilities. Design specific decisions get delayed due to variability in product data model definition and schema across enterprise level systems like PLM, ERP[6], MES[7], CRM[8] and SCM[9]. Uniform, classified & unambiguous data definition form the backbone of any enterprise-wide digital initiative. Manufacturing companies are looking for implementation of international standards for classification of different part types because proprietary standard of classification restricts an organization to collaborate with suppliers, partners, dealers, and retailers.

Available international standards do not fully define and classify different software parts created by R&D. On average, only 1% of total product classes (bottom of the hierarchy) are defined for software parts as shown in Fig. 2.

They provide rich taxonomy and attribute definition for electronic, hardware, electric and mechanical part types but not for software part types. Existing PLM systems only offer association between different part types as shown in an illustration (refer Fig. 3).

[6] Enterprise Resource Planning.

[7] Manufacturing Execution System.

[8] Customer Relationship Management.

[9] Supply Chain Management.

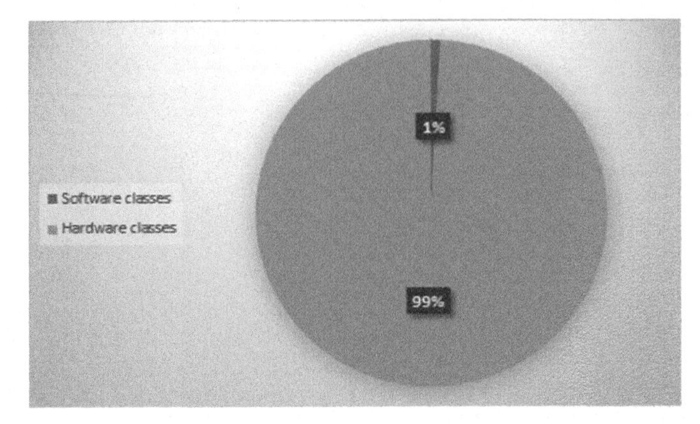

Fig. 2. Consolidated software vs hardware classes in current standards.

Only association feature of PLM system is not enough to fulfill the needs of engineering departments and associated areas. There is considerable dearth of software data definition and classification. More and more companies are adopting ALM[10]-PLM integration to strengthen the new product development to reduce time to market but its full potential is not being realized in the absence of standardized software attributes and taxonomy. In current scenario, software data is fragmented and residing in multiple systems with inaccurate, incomplete, and inconsistent information. So, it is imperative to have well-defined classification and nomenclature system for software and its adoption into extended enterprise systems.

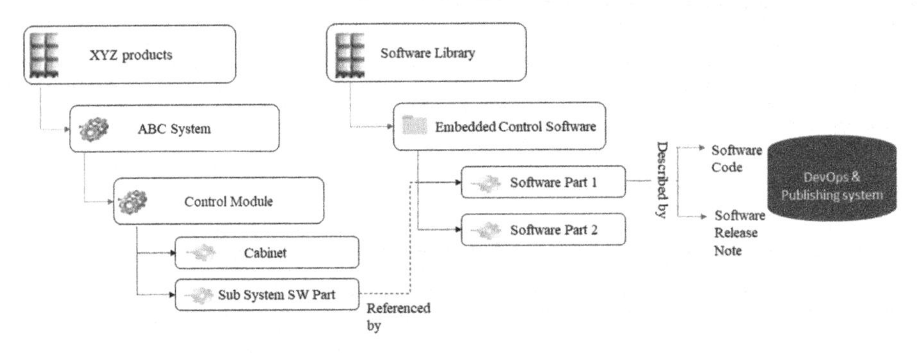

Fig. 3. Illustrative diagram showing software part type relationship with hardware part types.

2.2 Benefits of Classification System

Uniform, standard, and common engineering language for product characteristics has significant impact on metrices like engineering hours, procurement time, material costs

[10] Application Lifecycle Management.

etc. with error free data and less time-consuming processes [10, 11]. It declutters the process for searching of information from different solution providers, catalogues, on internet, by data conversion, and users from different field of engineering. Especially, software classification in digital era in context to hardware will significantly boost mass personalization and enhanced traceability and reuse-based software development [12].

3 Analysis of Existing International Standards

Table 1 shows the comparison of top 5 international standards categorizing software components at different levels, and it could act as reference map and further populated

Table 1. Eleven Criteria comparison of international classification standards

Criteria	ECLASS	UNSPC	GPC	CPV	ETIM
Standard Objective	Empower procurement, sales, engineering	Procurement	Buyers & manufacturers in cataloguing	Public procurement	Taxonomic classification of technical products
Classification or Product data dictionary	Product data dictionary	Classification	Product data dictionary	Classification	Product data dictionary
Schema (Classification level)	Four	Four	Four	Five	Two
Applicable Industry	Multi sector e.g., Automotive, construction, Electrical, Textile etc.	Multi sector e.g., electronic, Oil, Software, Energy, Healthcare etc.	Specific sectors such as apparel, consumer goods, etc.	Specific sectors e.g., Raw material, tourism, toys, Biotechnology, etc.	Specific sectors such as Electrotechnical, plumbing, shipbuilding etc.
Semantic definition and Attributes or feature availability	Partially available but rich in attributes	Not available	Available but number of attributes less compared to ECLASS	Not available	Partially available but rich in attributes
Language support	16	15	25	24	17
Geographic dominance	Germany & other European countries	US, Asia, Australia	Global	European countries	European countries
Adoption by companies	4000 +	2100 +	20L+	2.5L +	300+
Industry 4.0 (Digital Twin)	Provide library for machine readable characteristics	No references found	No references found	No references found	No references found
R&D specific features	ECLASS ADVANCED is mainly used in engineering and CAx areas.	No references found	No references found	No references found	Characteristics for 3D product data & Building information management
Licenses	Required	Free	Required	Required	Free

with more contextual criteria. Few standards like UNSPC are more focused on categorization of products and services rather than defining properties and associated values and units.

3.1 Classification Schematic Comparison for Software

ECLASS classification standard provides more software classes compared to other global standards (as mentioned in Sect. 2), but it lacks in quality and quantity as well. Newness in the name of software classes is created by appending information in parentheses signals lack of due diligence. Figure 4 elucidates variation in nomenclature and variety of category at different levels of software part classification schema.

Grouping of software are generic and based on how they are used instead of what they are. For example, nomenclature of software is based on application areas like telecommunication devices, office applications, engineering applications etc. There is a lack in variety, associated definition, and redundancy of software characteristics. The repetition of similar attributes also makes difficult to distinguish two disparate classes.

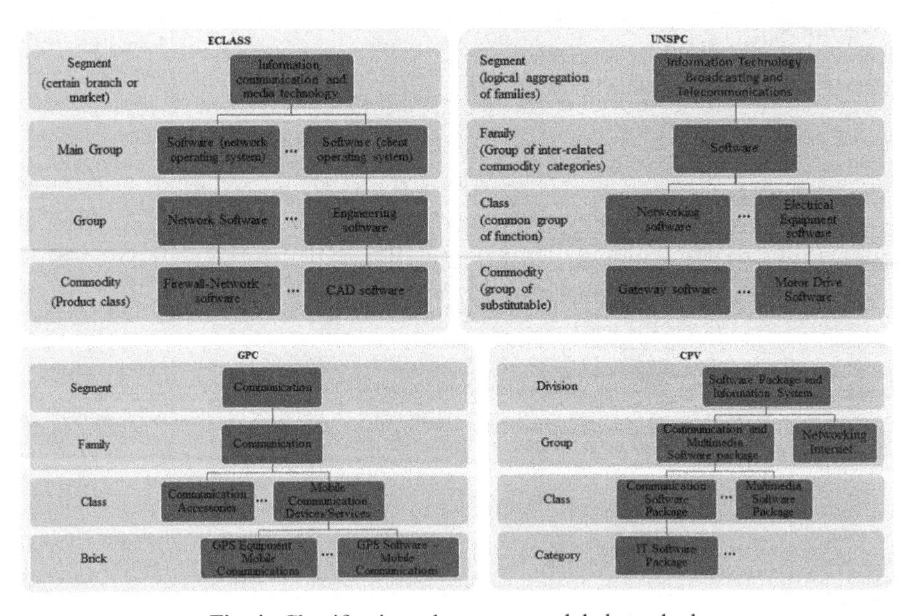

Fig. 4. Classification schema across global standards

4 Framework for Software Part Classification

Limitation posed by existing international standards urges for further exploration for new approaches to categorize software components. This necessity triggers pertinent questions as mentioned below:

- Are business needs for classification of hardware and software components same?
- What should be the approach? Should it be like current approaches of HW classification; and clustering software components in the same group of HW components?
- How should an organization classify software components? Should it adopt one international standard or custom approach to best serve the organization need?
- What should be the classification schema? Multi hierarchy vs flat taxonomy structure needs to be investigated.

The structured approach discussed in this section is based on our experience in implementation of software part classifications projects and can be guide for an organization to establish center of excellence for software part classification and disseminating knowledge across virtually integrated value chain. The success of enterprise-wide program depends upon strong governance structure and adoption of industry best practices for execution [13]. Setting up vision in line with organization objectives, audit frequency, hierarchical data governance structure and standardized business processes are key enablers for sustenance of such program. Framework shown in Fig. 5 presents end to end view of contextual approach with necessary processes, concepts, and tools for digital journey of software parts classification and their standardized data definition.

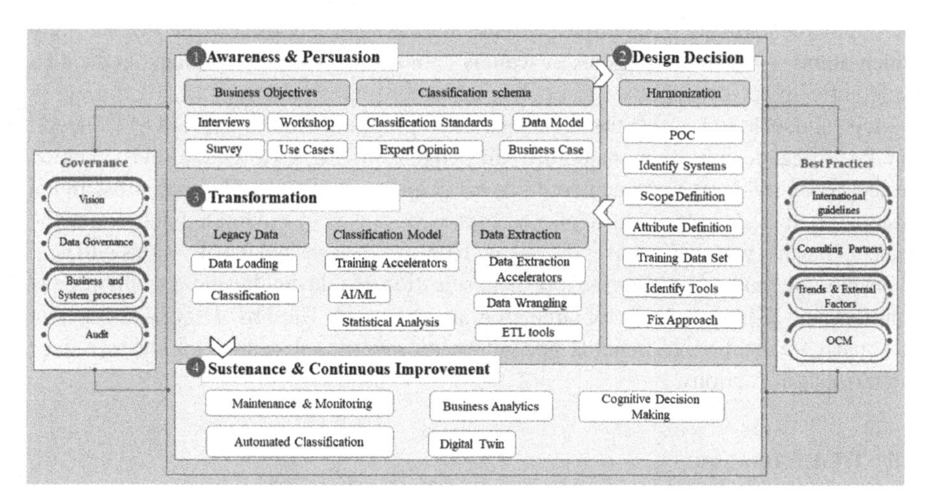

Fig. 5. Proposed framework for software part classification

4.1 Awareness and Persuasion

The digital journey of classification starts with dissemination of knowledge within enterprise and suppliers to create awareness. The potential adopters who could be at any level in their organization and acquired information from various communication channels. They should first enrich themselves with knowledge from primary and secondary research. Good understanding of the business use cases is essential to highlight the benefits and values that classification brings to the table.

The next step is by persuasion of top management or steering committee to approve business case for designing and implementation of standard coding & taxonomy system for products and services. A strong business case is prerequisite for buy-in from top management. Partnering with right consulting services to conduct interviews, workshops, and survey can trigger learning environment for employees to assimilate knowledge about global nomenclature and classification systems and will expedite overall implementation program. In depth, due diligence by evaluating pros and cons of existing schema with respect to various global schemas will be beneficial. Understanding existing internal processes and data model and analyzing it carefully with the support of experts are required to come up with multiple evaluation criteria. Identify industry best practices and latest trends while defining classification hierarchy.

4.2 Design Decision

Harmonization in classification of product and services and associated attributes definition across multiple systems are critical aspects for digital transformation journey of an organization. Standard coding system leads to part data synchronization, process flow integration and data standardization across company divisions, suppliers, and global locations. Identify and define SMART[11] metrices and set the base levels to be compared after implementation of standardized classification system. Choose few metrices initially which aligns with near term goals as well as with business key performance indicators s and perform benefit appraisal just before and after implementation.

It is advisable to kick off the implementation program with PoC (Proof of Concept) to take critical design decisions before full fledge rolling out. Doing part classification project using big bang approach could be risky and may create additional stress in the system. So, it is advisable to have phased approach for design and implementation of the project. Starting with PoC in an agile way with limited data set and few divisions will help to develop methods, frameworks, tools and selection of relevant business technological solutions and IT interventions for large program execution. The aim of PoC is to generate awareness, realization of benefits and return on investment demonstration to reduce organizational frictions.

4.3 Transformation

Tools and methods identified in previous steps will be leveraged for scaled agile roll out across divisions, product groups and suppliers. In general, organization product data resides on multiple systems with unique information. IT tools for data extraction automates the process for large volume of data. Based on the type of enterprise system and the situation, it is possible to select the partial or full data model. Extracted data can't be used as it is and needs to go through data wrangling process. The next step of transformation is the classification. Manual approach using statistical methods could suffice if the volumes are low otherwise it is advisable to go for automated classification solutions with quality assurance checks. Artificial Intelligence (AI) based automated classification approach involves training, tuning, model selection and testing using different types of

[11] Specific, Measurable, Achievable, Relevant, Time-bound.

data set. In AI based classification, predominantly, there are 2 steps. First is the fully automated classification step, where ML (Machine Learning) model parameters are generated based on the training data. Next one is the guided automation where some degree of manual intervention required for quality assurance. The final step of transformation is data loading, where classified data is updated into the enterprise systems. If consolidated item master or electronic catalogs already exists because of standardized processes and enterprise applications can skip few steps in transformation.

4.4 Sustenance and Continuous Improvement

Continuous evaluation, identification of new use cases, and automation of classification systems are necessary to stay agile, innovative, competitive and remove inefficiency in the system. To retain integrity of the data over time, regular maintenance based on the principles of information asset life cycle management is required [14]. Any modification must be reported through formal change request for analysis and evaluation by change control board of the data governance structure. Updating of existing data set or introduction of new category should follow operational guidelines of implementation. It should be open, flexible, easier to adopt and attract classification of new software-based products and services.

5 Summary

In the era of Industry 4.0, where each manufacturing company is looking to extract business value leveraging digital twins; it is imperative to have standardized product data across entire vertically integrated value chain. Harmonized product characteristics and classification not only simplify R&D processes but also streamline procurement, sales & marketing, material management and manufacturing processes. Common engineering language or semantic system facilitates automation of design works to create circuit diagrams, part lists, wiring lists, assembly diagram as well as automation of robotic manufacturing process. Since number and variety of embedded software and firmware are increasing, organization should look for new use cases to support adoption of software part classification. Each of the elements mentioned in the proposed framework needs to be adapted and contextualized as per company needs and could be utilized for classification of other part types as well.

References

1. Tomás, S.L., Damith, C.R., Bela, P., Duncan, M.: Taxonomy, technology and applications of smart objects. Inf. Syst. Front. **13**, 281–300 (2011)
2. Müllerburg, M.: Software intensive embedded systems. Inf. Softw. Technol. **41**(14) (1999)
3. GMI. https://www.gminsights.com/industry-analysis/embedded-software-market#:~:text=Industry%20Trends,to%20spur%20the%20industry%20growth. Accessed 05 Apr 2022
4. http://www.infoterm.info/pdf/members/strategic_documents/cen_ws_agreements/cwa_16138.pdf. Accessed 27 May 2022

5. Wolfgang, W., Peter, J.A.R., Laura, E.G.M.: Flexible classification standards for product data exchange. In: Handbook of Research on E-Business Standards and Protocols: Documents, Data and Advanced Web Technologies. IGI Global (2012)
6. ECLASS. https://eclass.eu/aktuelles/broschueren. Accessed 27 May 2022
7. Thomas, T., Sven, A., Christian, K., Ina, S., Gunter, S.: A classification and survey of analysis strategies for software product lines. ACM Comput. Surv. **47**(1), 45, Article no 6 (2014)
8. Alberto, P., Jose, D.R., John, S.W.: Beyond the information technology agreement: harmonization of standards and trade in electronics. World Econ. **33**(12), 1870–1897 (2010)
9. Elena, C., Orietta, M.: Survivor: the role of innovation in firms' survival. Res. Policy **35**(5), 626–641 (2006)
10. ECLASS. https://www.eclass.eu/en/news/videos.html. Accessed 04 Apr 2022
11. Granada Research white paper: Why coding and classifying products is critical to success in Electronic commerce. Published September 1998, Updated October 2001
12. Mahdi, N., Ebrahim, B., Weichang, D.: Non-functional Properties in Software Product Lines: A Taxonomy for Classification. Published in SEKE 2012. Computer Science, Business
13. UNSPC. https://unece.org/fileadmin/DAM/trade/agr/meetings/ge.11/2005/2005_i04_UNSPSC.pdf. Accessed 04 Apr 2022
14. Verdantis. https://www.verdantis.com/downloads/whitepapers/adopting-unspsc.pdf. Accessed 04 Apr 2022

General Methodology for the Generation and Dissemination of Manufacturing Knowledge: A Case Study with the Double Diamond AM Knowledge Approach

Francesco Serio[1], Emiliano Traini[1], Julia Barret[2], Frederic Parisot[3], Paolo Chiabert[1](✉), and Fréderic Segonds[2]

[1] Politecnico di Torino, Corso Duca degli Abruzzi, 24, 10129 Turin, TO, Italy
paolo.chiabert@polito.it
[2] Laboratoire de Conception de Produits et Innovation LCPI, Arts et Métiers Institute of Technology, HESAM University, 151 Boulevard de l'Hôpital, 75013 Paris, France
[3] AddUp, 13-33 rue verte, ZI de Ladoux, 63118 Cébazat, France

Abstract. The ever-increasing consolidation of industry 4.0 technologies and the imminent advent of the industry 5.0 paradigms makes it essential to use new methodologies for the generation and the transfer of knowledge in the manufacturing sector.

Considering the state of the art regarding pedagogy and Additive Manufacturing (AM) and starting from the need to create a unique tool to make the most of the potential of additive technology, a case-study based on the method called "Double Diamond AM Knowledge Approach" (D^2AMKA) is introduced with a deep discussion of the results obtained in the European project CAPT'N'SEE, managed by EIT manufacturing, which saw the collaboration of the Polytechnic of Turin, of the École Nationale Supérieure d'Arts et Métiers of Paris and of Add-Up. In extending the D^2AMKA, arose the need to create an information system to carry out the Product Lifecycle Management (PLM) for an AM process taking into account the differences with traditional processes. In order to satisfy this need will be shown how to apply to the case a previously created model to manage the information of a production process with a lean perspective.

To summarize this paper presents a general methodology to (i) capture knowledge needs in a specific manufacturing area and about a specific manufacturing sector, (ii) develop an e-learning path in that manufacturing sector with the collaboration of partners of that manufacturing area, and (iii) organize a journey in the name of training, dissemination, sharing and brainstorming.

Keywords: Additive manufacturing · Knowledge · Luxury industry · Industry 5.0

1 Introduction

"Given AM's recent introduction to volume production contexts, most engineers and manufacturing workers were not exposed to, or trained with, the principles or execution

Published by Springer Nature Switzerland AG 2023
F. Noël et al. (Eds.): PLM 2022, IFIP AICT 667, pp. 25–34, 2023.
https://doi.org/10.1007/978-3-031-25182-5_3

of AM during their formal education. It is likewise challenging and time consuming for universities to construct new degree programs and commensurate curriculum in AM." [1] Since their invention, Additive Manufacturing (AM) technologies are quickly spreading in many manufacturing sectors; however, in most cases a purely technical approach in integrating such technologies may not be sufficient to exploit the full potential. For this reason, sharing knowledge appears to be the unique way to change the perception of additive technologies and help companies in understanding all its potential. Because the AM technologies introduce a completely different approach to design, manufacturing and control of the finished product, the paper develops a methodology to improve the diffusion of AM awareness within the different phases of product and process development.

The case study presented focuses on the luxury sector because it suffers from a lack of multidisciplinary knowledge and it requires two types of specific knowledge: standard design methodologies and the revolution for process designers provided by additive subprocesses. In addition, the fear of investment and business model change is common in the sector, probably due to the inability to have an overview of the AM process and its benefits. Moreover, the intellectual property question, regarding the repeatability of such products and the digital content aspect, represent a real cognitive barrier for companies to guarantee their investment. Finally, many luxury brands are worried about the potential loss of capabilities and the consumers' retention that could be not attracted to this technology facing the old and traditional one.

In this paper, after the introduction, you will find the state-of-the-art section regarding the sharing knowledge methodologies. The third chapter introduces the design framework proposed to share knowledge as an implementation of a previous research work, while the fourth chapter reports a case study to validate the theoretical framework proposed in the luxury sector context.

2 State of the Art

This section summarizes the existing methods and tools used today by companies to speed up the process of sharing knowledge and its use in innovative processes in order to design disruptive products and services.

2.1 Knowledge Sharing Process

Because of the evolution of the various industrial sectors, there has also been an evolution of products and production processes. All this has led to an increase in the knowledge required to work and the need to pass this knowledge on to employees. The main approaches used nowadays to share knowledge through a company organizational structure are trainings, work and innovation labs, seminars, conferences and workshops (Fig. 1). From these approaches different learning methodologies for innovation were born such as agile methodologies, creativity, coaching and U theory (Fig. 1) [2–5].

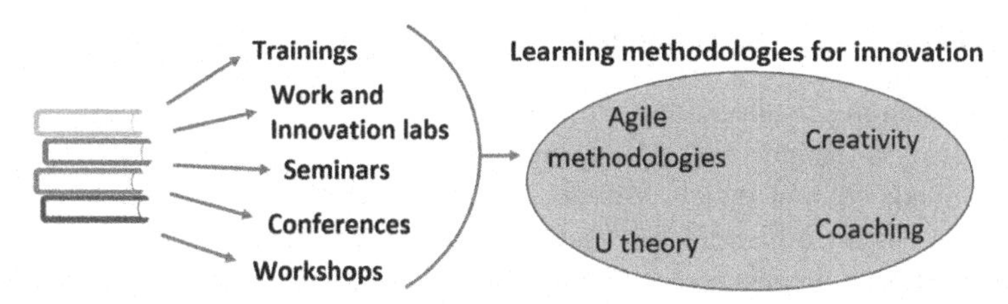

Fig. 1. Knowledge sharing approaches

Seminars, Conferences and Trainings. These three similar approaches of sharing knowledge are the most common ways adopted from companies to train their employees.

Seminars and conferences are generally considered the same; however, the only things in common between the two are that they both use audio-visual tools and that they involve people with the same educational background. In fact, seminars, often included in a conference of several days, have a shorter duration and can be classified as lectures to impart knowledge to the attendants that eventually receive a certificate of participation. The aim of the conference, on the other hand, is to allow the sharing of opinions and knowledge related to a given topic.

The other approach is represented by trainings, which always have the aim of training employees, but with a different form and duration compared to seminars and conferences. In fact, generally the trainings can last even months and are aimed at smaller groups of staff who need to acquire specific skills to be integrated into the activities daily carried out.

Workshops. This sharing knowledge approach could be considered a mix of conferences and seminars because it has a less rigid structure than a conference but at the same time, is more formal than a seminar. In addition, workshops tend to aggregate multidisciplinary attendants with the aim of teaching new skills or increase their awareness about not well-known topics.

Work and Innovation Labs. This approach of sharing knowledge it is a more recent way of training the staff of a company through the exchange of skills. In literature there are different taxonomies to identify work labs, but they could be summarized into three main typologies [4]: (I) Innovation Hub: sharing knowledge through distance learning networks, (II) innovation intermediary: innovative material shared in open access way, (III) Ecosystem attendants: local narrow networks and Fablabs.

Usually work labs are funded by public subsidies and in a small percentage even by private subsidies with the main objectives of educate, increase R&D skills, create social links and help entrepreneurs [4, 6, 7].

3 The Extended Framework D2AMKA Including the Nuggets and Journeys Design

This section aims to recall the Double Diamond AM Knowledge Approach [8] as an evolution of the double diamond methodology [9, 10] to help apprentices gain knowledge about the Additive Manufacturing process step by step.

3.1 The Double Diamond Methodology and the D^2AMKA

The double diamond methodology, officially invented by the British Design Council in 2005 [10], is a design thinking approach that consists in two phases represented by two diamonds each of which presents one phase of divergence and one of convergence (Fig. 2). The first diamond represents the problem defining process that is reached through two phases: the divergent one that aims to explore the problems from different perspectives and the convergent phase where the real problem is selected. The second diamond instead is the solution one, where the divergence is represented by the exploration and development of different solutions, prototypes and tests while the convergence is represented by the delivery of the final product.

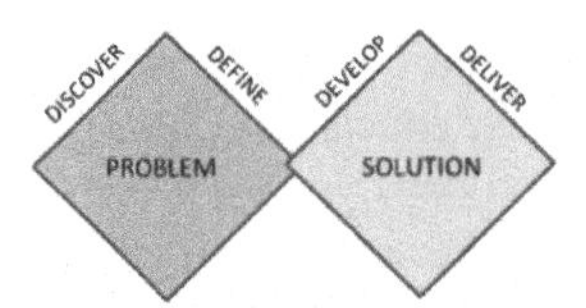

Fig. 2. The double diamond methodology

Getting inspired from the double diamond methodology we created the D^2AMKA (Fig. 3), in extended form Double Diamond AM Knowledge Approach [8], which helps learners to become aware of AM encouraging its diffusion. The D^2AMKA consists of two different learning timeframes corresponding to the two diamonds: the first one is an online collection of nuggets to address learning needs in the long term through a divergent thinking provided by videos and experts testimonies and a convergent thinking obtained from feedback quizzes. The second diamond instead focuses on few major topics treated in the nuggets with the divergence given by experts' presentations and the convergence by workshops experiences.

3.2 Nuggets Design and Objectives

This paragraph aims to explain all the actions undertaken to create the final nuggets. First, we decided the general topics related to AM that could be interesting for the potential learners and industrialists conceived as the target of the approach. The topics were selected thanks to the expertise of the project consortium members and academic partners who shared their knowledge on the AM process and on the luxury sector during some meetings held to define the contents to be shared. Secondly, we defined the learning path framework on which to fit each nugget. In particular, each learning path consists of a minimum of five different nuggets: (I) the connect nugget used as introduction or preface of the learning path that explains how it interacts with the other learning paths. No duration recommended for this kind nugget (II) the info nugget requires the participation of an expert who speaks about a specific topic strictly related to the whole learning path. Recommended duration from 5 to 15 min (III) the question nugget that follows each info nugget with a set of questions to help the development of a convergent thinking in each learner and to give a feedback on their enrollment in the knowledge process. Recommended duration from 2 to 5 min (IV) the task nugget, which involves the knowledge sharing of another expert belonging to a firm that has already had a practical experience in the info nugget topic treated. Maximum recommended duration of 45 min (V) the set up nugget that summarizes all the topics faced in the learning path to enhance the durability of the delivered knowledge but also increase the visibility of the project. No duration was recommended for these kind of nugget.

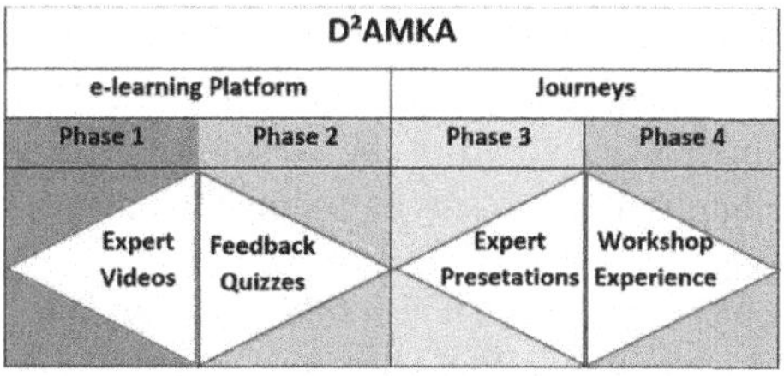

Fig. 3. The double diamond additive manufacturing knowledge approach

As third step, we started a call for experts by e-mail and by phone for the contacts in the research group network, 166, and by LinkedIn for the contacts outside the network, 187. These contacts were selected taking into account the correlation of their profession with the topics of the project, therefore the luxury sector, the AM sector, and in a small percentage the manufacturing industry in general. The final step was related to the definition of some Key Performance Indicators (KPIs) needed to assess the quality of the material created. In particular the KPIs selected are: (I) training and mentoring activities– 3 of institutions and 8 of courses, (II) 8 digital nuggets created related to creativity, innovation, design, part finishing, part control and Manufacturing Execution System technologies for AM in the luxury industry.

3.3 Journeys Design and Objectives

The training program has been divided into two "journey" programs, which integrate the same topics developed on the training platform by the nuggets. The first step taken was the careful definition of the time tables to be followed taking into account the nuggets produced, with the addition of some process demonstration, a fablab tour, an all-day workshop dedicated to creativity and brainstorming activities to easily involve participants. Subsequently we have selected the candidates based on their potential interest in the fields related to the project topics: luxury sector, AM and manufacturing in general. The following step was the call for participants sent to all the potential people interested by means of a registration form. With the same strategy applied in the call for experts for the nuggets, we used LinkedIn and the research group network as main source to target the right audience, delivering a high quality and tailored training. As done for the nuggets we created some KPIs to assess the quality of the journey: (I) 20 Participants in education and training, (II) Training and mentoring activities– 3 of institutions and 8 of courses, (III) 2 Educational Journeys related to creativity, AM and MES technologies applied to luxury industry with attendants active interaction.

4 Case Study: Spreading AM in the Luxury Industry

By carrying out research on the services available today online, there is an obvious lack of AM courses focused on the luxury industry, notwithstanding the compatibility of the latter with AM in terms of scales of production and of need to be in proximity of the client [12–14]. Furthermore, several major issues have been identified among the literature for this sector such as the unseen potential of AM by companies, the fear of the economic investment, [15–17] and the lack of awareness about the creative power of AM [14]. In addition there are doubt due to the need of adapted materials to be processed and the intellectual property question related to digital content ownership and repeatability [5, 12, 18]. Finally, many brands are afraid of the consumer point of view that could be disinterested in such technology because is in contrast with the old and traditional one that made the brand image.

In addition to the problems presented it becomes more and more necessary to manage the AM New Product Development (NPD) phase towards mass customization guaranteeing a comprehensive product information about intellectual properties and quality, and sharing at the same time knowledge related to Product Lifecycle Management (PLM) with particular attention to the design phase. This is mainly because, when approaching to AM, many companies do not know the differences in NPD between traditional and AM processes. If a company uses to develop new products by means of traditional "linear" processes, it will face many difficulties in applying the same scheme to AM. In fact, in subtractive manufacturing, the standardization of production techniques allows for greater linearity in the NPD process and so the changes made at each step do not challenge the choices made at previous stages, indeed features can be added to a part without calling the concept of the product into question. Conversely, in AM as soon as one of the parameters is modified, all the design choices must be reviewed because there are strong interdependencies between the way a part is produced, the material used to make it and the final properties. All the choices made during the design phases will potentially have an impact on surface quality and material integrity, and hence on the mechanical characteristics of the part.

This is why it is important to set up multidisciplinary working group, which meets at regular intervals, when creating a part in additive manufacturing until the optimum is obtained. Experts for the various stages must be brought together from the start of the project, so they can confer and make sure no aspect is forgotten.

4.1 Presentation of the Case Study

The case study presented in this work has been the foundation of the CAPT'N'SEE project and it gives a solid framework to achieve its objectives of offering an efficient and relevant training offer fostering a more creative, flexible, and smart use of AM in the manufacturing sector in general, and in the luxury sector in particular. The implementation of the project allowed, in turn, testing the methodology presented and enriching it with valuable feedback from a direct application.

CAPT'N'SEE (CAPTure aNd foStEr additive manufacturing knowlEdge for luxury industry) is a training program dedicated to professionals willing to enhance their expertise in the use of Additive Manufacturing technologies. The project has particularly treated the AM early design stages and the Manufacturing Execution Systems that allow the real-time control of processed parts. Starting from the expertise of its consortium members and gathering experienced industrial and academic partners, the project has initiated a dynamic shift in the way to work with AM in the luxury industry, which is in strong growth and where high value-added productions require enhanced creativity and high precision. This activity has received funding from the European Institute of Innovation and Technology (EIT).

The CAPT'N'SEE network associates two renowned academic partners and one skillful industrial: Arts & Metiers Institute of Technology (ENSAM) in France, Politecnico di Torino (Polito) in Italy, and AddUp which is operative mainly in France. AddUp is a joint venture born in 2016 from Michelin and Fives with the aim of develop additive metallic manufacturing solutions for industrials and supports them throughout their projects with services offered for a profitable experience.

The project outcome consisted firstly in the production of two series of nuggets filmed and edited and secondly in the organization of two educational journeys related to creativity, AM and MES technologies applied to luxury industry. The two journeys took place in Paris from 27 to 29 of September, under the supervision of ENSAM, and in Turin from 20 to 23 of October under the supervision of Polito.

4.2 Nuggets Produced

As required from the project KPIs, we developed eight learning paths related to AM, according the structure described in Sect. 3.2, in order to address most of the needs highlighted in the analysis reported as introduction of this section. We choose to enroll some AM experts of the luxury industry to share their knowledge on these topics and give a practical approach, showing all the potential of AM we wanted to highlight. In terms of content, the different activities have been structured as mix of videos and quizzes allowing the participant to diverge and converge their knowledge according to the first stage of the D^2AMKA. This knowledge is then further consolidated thanks to the Setup nugget, a summary sheet of the entire learning path, which narrow the learning experience to few important elements.

The eight learning paths created are: (I) **Multidisciplinary to learn and innovate:** Multidisciplinary working groups, Transform your company into a learning organization, Innovate trough collaborative projects; (II) **Stand out through AM:** Use value at the core of innovation, Adding value through AM; (III) **Innovation with AM as an economic asset:** Be able to consider AM as an alternative technology, The real economic benefits of adopting AM in your company; (IV) **Creativity and AM:** Creativity as a systematic process that has its place in the industrial company, Experiencing creativity in AM, Disseminate AM innovative culture in a globalized enterprise; (V) **Lean management:** AM and Human Resources: the hike in skills, Lean management of the AM process; (VI) **AM costing:** Knowing how to estimate the cost of a part, Installing an AM machine in an existing factory, Cost comparison between AM and traditional production processes; (VII) **Enterprise Information Systems:** Introduction to Computer Integrated Manufacturing (CIM), PLM and AM: key resources towards mass customization, Additive Manufacturing Resource Planning (MRP), Additive MES: managing the production to optimize the design; (VIII) **Artificial Intelligence:** Data generated during the lifecycle of an AM product, Generate information through an AM data architecture, Artificial Intelligence for a smart AM;

4.3 Journeys Outcome

As said previously we organized two different three-day journeys considering the study of industrial use-cases proposed by the experts in the network. The objectives of these journeys were to disseminate the learning contents and test the training program to support teams that deals with AM in developing strong skills in creativity and process management, with a specific focus on the luxury industry. The timeline encompassing each journey spans around three phases: journey design, journey execution and the post journey analyses. The journey design was already treated, so this section will describe the following two phases for each journey.

The first journey, managed by ENSAM, aimed to reconsidering the product development process from the early design stages in order to satisfy new creativity and multidisciplinary needs required by the AM processes. For this reason, the main activities concerned the enhancement and the promotion of the creativity allowed by AM thanks to the use of collaborative methods and tools. In this journey, we had 48 participants, 43 of which responded to a satisfaction survey we sent after the event declaring that they increased their knowledge of AM during the journey. The main topics in which attendees declared to have learned new things are innovation thanks to AM, creativity enabled by AM, multidisciplinary of AM projects.

The second journey, managed by Polito, was dedicated to the smart monitoring and control of the AM process or, more in detail to Manufacturing Execution Systems and finished part control solutions. Therefore, the main objective was to support attendants in developing strong skills in the management of the AM digital chain. For this reason the journey has been designed starting from the concept of AM Information Systems, continuing with the design and the execution of an additive process, and ending with the quality techniques and sustainability criteria implemented in AM. Moreover, we presented a model to manage the information of a production process with a lean perspective showing an appropriate methodology for specific use-cases [19]. Subsequently there was the execution of a simple open innovation experiment to engage the audience and share the open philosophy. In this second journey, we had 71 participants, with a daily attendance of 41 on the first day, 52 on the second day and 39 participants on the third day. To ensure continuous participation of the attendees, we organized a quiz with prizes to win a backpack offered by the sponsor for each day of the journey.

In order to allow the development of the convergent thinking and the birth of a common wisdom between attendees we scheduled Q&A sessions during the two seminars. This was helpful as well to measure the degree of knowledge of the participants on AM and its possibilities. For instance, we asked if it is possible to print gold or copper, which showed that 50% of the participants were unaware of the possibility to use gold or copper in AM.

Although the number of participants was quite high for both journeys, the maximum registration capacity given by the capacity of the classrooms was not reached. We therefore did not have to apply the selection process, based on the following priority levels: (I) professionals from the luxury sector, (II) professionals from the AM industry, (III) with equally relevant profiles, favored women. However if it had been asked to choose, with equally relevant profiles, we would have preferred women to balance the gender inequality. Moreover, during the advertising of these events, we were very careful to distribute equally to men and women all the material and the invitations. However a bit more of 2/3 of the total registered people were men.

In the end, the vast majority of the registered attendees for this event were professionals, followed by professors, academic researchers and students. An half of them were from the Additive Manufacturing sector, others worked in the luxury sector and the remaining were in other industries such as consultancy, energy and sustainability.

5 Conclusions and Future Works

By examining the KPIs required for the project assessment, we can state the successful ending of the same with the production of 8 learning paths and 2 journeys related to creativity, innovation, design, part finishing, part control and MES technologies for AM in the luxury industry. Thanks to this path, we have had the opportunity to understand the needs of the players in the luxury sector who want to introduce AM into their processes and above all to test the D^2AMKA for the first time. From the responses received from the participants, we can only be satisfied to find an increase in their general knowledge on the subject. However, it will be necessary in the future to question the participants about the investments made in relation to the increase in knowledge and to test the D^2AMKA in other contexts to have more confirmation of its effectiveness.

References

1. Peppler, K., et al.: Key principles for workforce upskilling via online learning: a learning analytics study of a professional course in additive manufacturing. arXiv (2020). https://doi.org/10.48550/arXiv.2008.06610
2. Matos, F., Vairinhos, V., Salavisa, I., Edvinsson, L., Massaro, M.: Knowledge, People, and Digital Transformation, 1st edn. Springer, Cham (2020)
3. Rias, A., Bouchard, C., Segonds, F., Abed, S.: Design for additive manufacturing: a creative approach. In: International Design Conference 16–19, Design 2016 (2016)
4. Seo-Zindy, R., Heeks, R.: Researching the emergence of 3D printing, makerspaces, hackerspaces and fablabs in the global south: a scoping review and research agenda on digital innovation and fabrication networks. Electron. J. Inf. Syst. Dev. Ctries. **80**(1), 1–24 (2017). https://doi.org/10.1002/j.1681-4835.2017.tb00589.x
5. Ropin, H., Pfleger-Landthaler, A., Irsa, W.: A FabLab as integrative part of a learning factory. Procedia Manuf. **45**(2019), 355–360 (2020). https://doi.org/10.1016/j.promfg.2020.04.033
6. Suire, R.: Innovating by bricolage: how do firms diversify through knowledge interactions with FabLabs? Reg. Stud. **53**(7), 939–950 (2019). https://doi.org/10.1080/00343404.2018.1522431
7. Catherine, F., Serikoff, G., Zacklad, M.: Le lab des labs. HAL science ouverte. Paris (2019)
8. Barret, J., et al.: CAPT'N SEE: a methodological proposal to capture and foster additive manufacturing knowledge for luxury industry. Econference. Belgrade, Serbia (2021)
9. Tschimmel, K.: Design thinking as an effective toolkit for innovation. In: XXIII ISPIM Conference: Action for Innovation: Innovating from Experience, pp. 1–20 (2012). https://doi.org/10.13140/2.1.2570.3361

10. Design Council: Eleven lessons: managing design in eleven global companies. Engineering, vol. 44, no. 272099 (2007)
11. Design Council. https://www.designcouncil.org.uk/. Accessed 22 Nov 2022
12. Cabigiosu, A.: Digitalization in the Luxury Fashion Industry, 1st edn. Palgrave Macmillan (2020)
13. LaFrenchFAB: FABulous project. https://www.fabulous.com.co/blog/infographie-marche-luxe-impression-3d/
14. Petit, J., Brosset, P., Bagnon, P.: Smart factories @ scale. Capgemini (2019). https://www.capgemini.com/wp-content/uploads/2019/11/Report-Smart-Factories.pdf
15. Godina, R., Ribeiro, I., Matos, F., Ferreira, B.T., Carvalho, H., Peças, P.: Impact assessment of additive manufacturing on sustainable business models in industry 4.0 context. Sustainability **12**(17), 1–21 (2020)
16. Ciulla, S., Benfratello, L.: The adoption of Additive Manufacturing in the dental prostheses industry and its impact on firm performance Academic Supervisor. Master's thesis, Politecnico di Torino (2018)
17. Savolainen, J., Collan, M.: How additive manufacturing technology changes business models? – review of literature. Additive Manuf. **32** (2020)
18. Carnot, I.: Cartographie des acteurs clés de la R&D en fabrication additive. Cetim (2017). https://www.instituts-carnot.eu/sites/default/files/images/CartoFabAdd_CETIM-filiere-Manufacturing-juil2017.pdf
19. Serio, F., Sordan, J.E., Chiabert, P.: The value stream hierarchical model: a practical tool to apply the lean thinking concepts at all the firms' levels. In: Junior, O.C., Noël, F., Rivest, L., Bouras, A. (eds.) Product Lifecycle Management. Green and Blue Technologies to Support Smart and Sustainable Organizations: PLM 2021, Part II, pp. 410–424. Springer, Cham (2022). https://doi.org/10.1007/978-3-030-94399-8_30

Structuring SMEs Collaborations Within a Cluster

Mélick Proulx[1](✉), Mickaël Gardoni[1], and Shadi Farha[2]

[1] École de Technologie Supérieure, Montréal, Québec H3C 1K3, Canada
melick.proulx.1@ens.etsmtl.ca, mickael.gardoni@etsmtl.ca
[2] APEX Precision, Saint-Lazare, Québec J7T 2B5, Canada
sfarha@apexprecision.ca

Abstract. Innovation is not limited to products and services; it is also used in a company's processes such as interactions with its environment, i.e., open innovation. According to our research, there are few approaches to effectively enable and support collaboration between manufacturing companies, even if they are in the same industrial cluster. We selected a platform to structure collaboration between companies, based on a business model framework. We tested this platform with two partners in the Quebec aerospace cluster that knew each other. To evaluate the collaboration, we identified the criteria to quantify collaboration effectiveness, such as the number of topics or projects discussed, the number of meetings created or the success rate of the knowledge exchange. Experimentation was carried out over five months, with detailed observations of collaboration's evolution. This platform initiated several collaborations between the two partners and increased the partners' technological knowledge maturity. However, several obstacles, such as difference of vocabulary and intellectual property exist and hinder communication. These could become troublesome when partners who don't know each other. Resolving these difficulties is the purpose of our future research.

Keywords: Collaboration · Platform · Knowledge exchange · Open innovation

1 Introduction

In today's environment, collaboration is a key aspect of business survival [5]. Collaboration can take several forms such as resources, manufacturing or business processes sharing [12]. Collaboration and open innovation can be beneficial and are already used in few industries such as technology, energy, and accommodation. However, the capacity of a firm to open up to innovation processes is determined by the firm itself and not by its operating market [22]. Manufacturing industry, specifically aerospace could benefit from more collaboration. One question emerges: how to structure collaboration between firms in the same cluster? Literature presents few tools to structure collaboration within the manufacturing industry even less in cluster-based sectors such as aerospace. According to our research, setting up a platform can foster collaboration [14]. Platforms and open innovation use innovation to create value and increase firms' competitiveness [14]. Few

Published by Springer Nature Switzerland AG 2023
F. Noël et al. (Eds.): PLM 2022, IFIP AICT 667, pp. 35–44, 2023.
https://doi.org/10.1007/978-3-031-25182-5_4

literatures present business exchange platform. Its objective is to support collaboration between manufacturing firms in the same cluster. First section will present state of the art on collaboratives models linked to open innovation. Second section will present the pre-experimentation with the platform and the industrial partner's choice, and the collaboration's criteria to assess the platform. Third section will present the methodology and the monitoring method, and the collaboration's criteria results. The last section will present points to improve, strengths and project benefits.

2 State of the Art

2.1 Collaboration Needs Between Companies Operating in the Same Cluster

Collaboration between manufacturing SMEs is difficult mainly because the competition minding is deeply embedded [15]. Competition's rules are changing, innovation cycles are getting shorter, managers must learn to collaborate more efficiently [9]. Firms strategics activities such as research and development and innovation are usually done within the firm itself, with minimal external intervention, this minding is outdated [10, 21]. Nowadays, human resources are volatile, firms cannot afford to innovate on their own [32]. SMEs facing a lack of resources make them more open to collaborate [32]. Collaboration between SMEs can increase their market competitiveness [24].

The project was conducted within the Quebec aerospace cluster, which regroups manufacturing SMEs. SMEs are more agile, innovate faster and react quicker to solve problems but they often lack resources [21]. Firms have reciprocal benefits operating in cluster, such as stimulate innovation by having suppliers, universities, and SMEs to collaborate creating new knowledge and sharing technologies infrastructures [8, 21, 26]. Easier access to resources and an efficient diffusion of the best practices are other benefits [26]. Group's ability to solve problem is better than the members alone [23]. Literature presented above suggests that collaboration is difficult between manufacturing firms but can bring major gains.

2.2 Collaboratives Models

In recent years, collaboratives models have been changing as the collaborative economy and business as a service model increased in these sectors: accommodation, transportation, and home supplies [30]. Lyft or Uber for transportation sector and SnapGood for household supplies are some examples [19, 31]. Collaborative economy's structure is to match suppliers and customers through a platform which make collaboration easier [19]. These examples can be used as starting point to support collaboration between firms. Successful platforms share online collaboration characteristics such as common goal's pursuit between participants [13]. Manufacturing sector's collaboration is not widespread. Formabilio is a collaboration example between manufacturing firms. One of its objectives is to link furniture manufacturing SMEs, designers, and customers [5]. Using this as a foundation, our first paper proposed a collaborative business model for same-clustered manufacturing firms. Designing this model, we took in consideration that more exchanges within the network, more the network have a success chance [2]. A key feature while creating our model was trust and reciprocity, two pillars to sharing rich information [1, 9, 15, 17]. Our model sits on four value propositions presented below. The first three values propositions will be addressed in this paper [27]:

1. Foster collaboration by a unique counter offering a business strategy.
2. Increase industrial and technological maturity of the members and cluster.
3. Fostering knowledge exchange between the members.
4. Respond to an entrepreneurial achievement need.

These value propositions can provide benefits to SMEs such as effective assets' use, development time and expenses reduction [4]. Literature suggests that collaboration is crucial to firms' survival, but only few articles present options to support it.

2.3 Open Innovation Materialized with a Digital Platform

Two types of knowledge exchange define open innovation: outside-in and inside-out. Outside-in is defined as: opening innovation processes to external firms. Inside-out is defined as: allocating developed knowledge, not used it internally, to the outside of the company thus others can use it [6]. Open innovation's future will be more collaborative and include more participants [6]. Literatures suggest that open innovation is less used in SMEs. SMEs face various challenges to acquire external knowledge such as organizational structure, culture, financial resource's availability, collaboration's costs, and intellectual property right management [12, 16, 28]. Long term, external technology acquisition will be a necessity for companies to remain competitive [22]. Literature also suggests that collaboration between firm can be improved using a platform [13, 19]. Platforms can facilitate collective ideas' development between partners utilizing external knowledge by giving a common interface to innovators to interact [16, 29]. A study of 254 platforms shows that digital platforms can be found under different categories: sharing, new and used item purchased, lease, donation and borrowing [13]. Platforms facilitate peer to peer business by sharing assets or services [19, 31]. Firms that own the platform play a vital role, such as coordinate partners and favoring value creation by sharing knowledge and resources [14]. Leader's platform position will be reinforced if the partners activities are complementary from one to another [14]. Both types of open innovation are experimented in this project. Literature suggests that platform can support collaboration between firms. The platform leader has an important role in its success. However, there is limited literatures on manufacturing firms that collaborate through a platform.

3 Pre-experimentation Phase

3.1 Selection of the Platform

Few platforms were available to support collaboration and implement our collaborative business model. The main feature is the use of tiles, called opportunity cards and they are used to support collaboration between the partners. Four types of opportunity card are available on the platform: wish, research, offer and information. Cards are also labeled according to categories: business, production, knowledge, or technological watch. The card's title and owner can be found on the card. Some platforms already available can be studied. The platform has to be flexible and must allow more execution and less administrative requirements. It must also facilitate communication between participants. The platform facilitates agile management. The evaluated platforms are an Excel file, a web portal development, a Trello board and a Smarsheet table. To select the more suitable platform for our application, we used six quality criteria. The criteria were adapted, initially they were used to evaluate e-learning platform [25].

1. User experience: the platform is easy to navigate.
2. Communication: the platform has email, chat and other communication options.
3. Availability: the platform is available for free and run with few plug-ins.
4. Content management: the platform makes it easy to track cards and their progress.
5. Administration: the platform makes it easy to manage members and their rights.
6. Accessibility: the platform can be consulted from any browser.

To these criteria, five-point Likert scale was added, strongly disagree had a value of 1 and strongly agree had a value of 5. Scores have been assigned by project's members. Each platform has their pros and cons: Excel table has few communication's options. Web portal will be costly to design and make it usable. Trello offers options to communicate but does not offer a user right management feature. Smartsheet is similar to Trello, but its navigation is more complicated. After the criteria analysis, Trello will be used for experimentation with the higher score of 23 on 30. Figure 1 present the different scores for each platform.

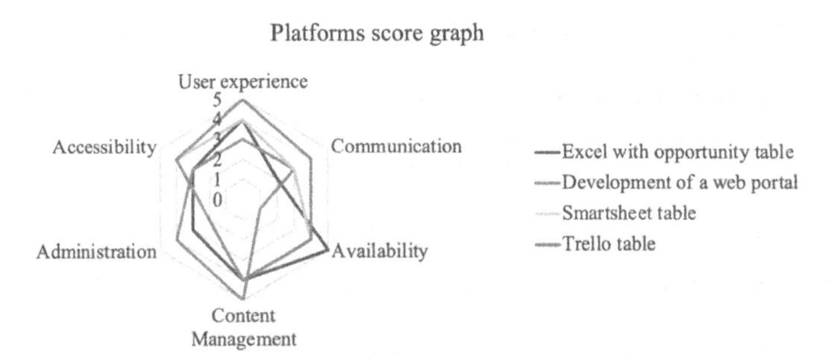

Fig. 1. Platforms score graph

3.2 Industrial Partners

The two Quebec based partners are Groupe Meloche Inc and APEX Precision. Groupe Meloche works in aerospace since 2008 and supplies structural and aircraft engine parts to Tier 1 companies. APEX Precision specializes in metal transformation and products assembly such as aerospace. The partners choice can influence collaboration's quality and efficiency. Partners were motivated to collaborate, to increase their maturity, to solve problems quicker and to perform knowledge exchanges.

3.3 Presentation of the Collaboration's Criteria

The collaboration's criteria were chosen to evaluate experimentation's success. Each of them will be presented and a summary table presenting minimum and maximum targets will follow. The number of collaborative projects created, or themes addressed quantify the interest level and allow to see if any discussions are initiated on a card. This criterion is independent of meetings created. Projects created can be discussed in follow-up meetings. The partners can have discussions on a card without having a knowledge exchange. This criterion's score will increase by one point if the partners have a discussion on a card. The number of meetings created quantifies the number of meetings between the partners generated by this project. The score will increase by one every time the partners have a meeting triggered by the project. Independent meetings won't be counted. The number of cards describes the number of cards added on the platform. This criterion's score will increase by one every time a card is added. The rate of a card knowledge exchange success represents the discussion about a specific card which lead to a knowledge exchange between partners. The knowledge exchange is qualified by a transfer of information from one partner to the other, which can be oral or written. The score will increase by one in case of knowledge exchange. The rate of a card success quantifies whether the knowledge exchange previously described has been beneficial by the receiving partner. In that case, this criterion's score will increase by one point. The waiting time of a card quantifies the time until a card is assigned to a partner. This criterion aims to determine whether partners are active and consult the platform. The attribution date minus the deposit date will be defined has the card's waiting time. Table 1 presents each criterion with minimum and maximum targets. The targets' criteria were established by the project's members during an iteration process.

Table 1. Collaboration's criteria minimum and maximum targets

Collaboration's criteria	Min. target	Max. target
Collaborative projects created or themes addressed	3	9
Number of meetings created	6	12
Number of cards	12	24
Rate of card knowledge exchange success	50%	75%
Rate of card success	25%	50%
Waiting time in days of a card	14	21

4 Experimentation Phase

4.1 Methodology and Monitoring

The experimentation phase will be performed with internal validation using a real case. The objective is to support collaboration between the partners presented above. Phase's duration was of five months. Meeting's frequency was established to one meeting every two or three weeks, but various factors influenced its frequency. More meetings were held at early, middle and closing stage. At least one partner's member was required every meeting. During meetings, the first point was validation criteria's review and monitoring graphs' analysis. Active card's review was also conducted, any development or additional information's need were discussed. At this stage, decision to involve technical people on a specific card was made. New cards were discussed to confirm an equal partners' comprehension. The fourth point was the blocking points, which could slow the progress. These blocking points were addressed, and an action plan was established. The improvement points were the penultimate meeting's point. They were divided into two categories: improvements applicable now or applicable in a future phase. The following points were the next steps, the questions, and the varia point.

4.2 Collaboration's Evaluation

Table 2 presents the experimentation's results by criterion with their minimum and maximum targets. The number of collaborative projects created, or themes addressed is beyond the maximum target. The criterion's evolution shows that it is directly related to the meetings between the partners. Information's exchanges took place mostly in the meetings, which shows that they are inevitable in knowledge exchange. The number of meetings created is between the targets. A meeting took place every two or three weeks. The meeting's frequency was decisive in the knowledge exchange between the partners. The more meetings were held in the project, the more knowledge exchange between the partners was happening. The number of cards is within targets. The cards put on the platform were considerably higher at the beginning. Both partners had many wishes, information to seek, to offer or to share at this stage. Card's addition on the platform is independent of having meetings. This means that the platform was used even though there was no meeting scheduled. The rate of card knowledge exchange success is a bit over the maximum. The knowledge exchanges began around week five. A correlation between knowledge exchange and cards' number was observed. When many new opportunity cards were created, knowledge exchange's level decreased until subsequent meeting, since it was performed during meetings. The rate of card success was evaluated only at the end of the experimentation phase to gather as much information as possible. The criterion's result is a bit over the maximum target. The last criterion evaluated is waiting time in days of a card which is beyond the maximum. This criterion is less suitable for the project. Only two partners participated in the project, every active cards were reviewed during the meeting, independently if the card was assigned or not. However, if the number of partners on the platform grows, this criterion will become more important.

Table 2. Collaboration evaluation's criteria results

Collaboration's criteria	Min. target	Max. target	Result
Collaborative projects created or themes addressed	3	9	20
Number of meetings created	6	12	10
Number of cards	12	24	24
Rate of card knowledge exchange success	50%	75%	83%
Rate of card success	25%	50%	53%
Waiting time in days of a card	14	21	21,58

5 Feedback on the Experimentation Phase

5.1 Improvements Points

To increase collaboration's quality, one aspect to improve is communication which is the cement of innovation [18]. Knowledge exchange happened because of the partners' interactions, and these are related to creativity and innovativeness [3]. The two main communication channels were emails and bi-monthly meetings. The vocabulary used by the partners was different which made the communication harder. Adding live chat, video calls, an expert's contact list, a term lexicon on the platform would be good options to improve collaboration and communication's effectiveness. In contrast, communication overuse may cause platform's participants creativity to decrease [18].

Intellectual property rights are another aspect to improve while using the platform. Separate intellectual property rights already owned from the partner from those which are created by the partners is important [1]. Patent can be used as a tool to promote collaboration between firms [7]. In this project, the exchanged information's was not sensible. Involving more partners on the platform to create new knowledge implies to exchange sensible information, causing intellectual property rights to become a major problem. Gassmann and Bader knowledge area for managing intellectual property right model would be a good foundation by presenting background, side ground and post ground knowledge areas throughout a collaboration timeframe [11]. Using Trello, it was impossible to set a confidentiality's level, so that the confidential cards were only seen by people who had access to it. In our case, confidentiality was not a problem since only two partners were involved. Again, involving more partners, confidentiality would become an issue. Having a list of the card's visitor would represent a good addition to track card's history. The addition of filters by partner, by projects or card confidentiality's level would be beneficial for collaboration.

The reciprocal trust between partner allows them to share their strengths and weaknesses. Again, more partners using the platform could create trust issues. An important aspect to keep in mind is that we want to support collaboration between firms through human interaction, which is a very important collaboration's factor. Setting up acceptance criteria for new platform partner can mitigate this point. The criteria could cover soft and hard skills to accept a partner. For larger companies, it may be more challenging to join this type of platform. Organizational structures will have to change and adapt to promote collaboration and make it more efficient [9]. Dedicating a person to support collaboration in their organization is an option. Their tasks would be to create connections between outside knowledge and to apply it inside and vice-versa. Some card's related functionalities can be improved to support collaboration between the partners through a platform. Adding more details in the card, such as a precise question or problem to be answered and make a better attachment's use would also be beneficial. These improvements will allow the partners to respond to an opportunity card more effectively and have a faster and more efficient exchange of knowledge. Making this platform scalable and increase partners' number will lead to exponential new knowledge creation. Consequently, the platform value will increase [14].

5.2 Strong Points

The platform's manager plays a vital role in its success and this project confirms it. The platform leader oversaw meetings planning, updated the platform and the cards and kept the participants in line with each other. As a result, more meetings took place with this project than without it. The meetings helped support collaboration. The two partners knew each other before, which helped to support collaboration. Partners belonging to the same industry and with similar issues also helped the collaboration. Their company's profile and their openness to collaborate, even if they are sometimes competitor, also helped supporting collaboration. Participant's personalities that collaborate must be compatible to have a successful collaboration. This project showed that the partners' interactions and their capacity to build trust between them directly impact the collaboration's result. The partner's and their technical expert's availability and exchanged information's level remain strong point. The platform's choice also seems to be a strong point. From the 6 criteria used to choose the platform, 4 of them helped support collaboration. On the other hand, communication and administration criteria slowed collaboration. In general, the platform was suitable to support the collaboration between firms.

5.3 Project Benefits

On the industrial side, this project made two same-clustered firms to collaborate. Through a platform, this project has shown that opportunities to collaborate were multiple and increased shared opportunities between partners. Knowledge exchanges' subject were various, business processes, manufacturing processes, human resources or market trend were among them. The partners also increased their businesses and manufacturing maturity faster by accessing to additional knowledge, which would have not been possible without taking part in this project. The firm's culture is usually an aspect that can slow down collaboration [9]. In this project, both partners promoted the same culture, which facilitate the collaboration [20]. On the academic side, this article demonstrated that is it possible for SMEs to use open innovation through a platform. Setting up a platform enhanced knowledge exchanges and increased the members' maturity. Collaborative business model shows that it is possible for firms operating in the same cluster to collaborate. This project reinforces literature on the fact that platform leader, culture, and partner's availability are key features to a successful collaboration. This paper suggests a platform to support the collaboration in the manufacturing sector between same-clustered firms. It contributes to the literature on collaboration between SMEs. It also suggests 6 criteria to determine collaboration success through a platform.

6 Conclusion

This project's objective was to demonstrate whether it is possible to structure collaboration between same-clustered firms. Indeed, we selected a platform which improved collaboration and communication and we evaluated them according to 6 criteria. Open innovation played a key role supporting collaboration between firms. Platform leader's implication, firms' culture, and availabilities have contributed to support collaboration. Partners' number and project's duration represent some limitations. This project was also limited by its operating environment, the Quebec aerospace cluster. In terms of future perspectives, including more firms from different size and looking at their interactions represents an opportunity. Developing tools to encourage communication and trust between partners would also be beneficial. Firms' collaboration effectiveness and compatibility would also be a helpful tool to develop. Intellectual property rights management between partners operating in the same cluster and collaborating is also a future perspective.

References

1. Attour, A., Ayerbe, C.: Le management amont et aval des droits de propriété intellectuelle au sein des écosystèmes-plateformes naissants. Systemes d'information management **20**(3), 47–96 (2015)
2. Berglund, H., Sandstrom, C.: Business model innovation from an open systems perspective: structural challenges and managerial solutions. Int. J. Prod. Dev. **18**(3/4), 274–285 (2013)
3. Bullinger, A.C., Neyer, A.K., Rass, M., Moeslein, K.M.: Community-based innovation contests: where competition meets cooperation. Creativity Innov. Manag. **19**(3), 290–303 (2010)
4. Cagnazzo, L., Tiacci, L., Rossi, V.: Knowledge management system in SMEs within stable enterprise networks. WSEAS Trans. Bus. Econ. **11**(2014), 155–174 (2014)
5. Carida, A., Colurcio, M., Melia, M.: Designing a collaborative business model for SMEs. Sinergi Ital. J. Manag. **33**(98), 233–253 (2015)
6. Chesbrough, H.: The future of open innovation: the future of open innovation is more extensive, more collaborative, and more engaged with a wider variety of participants. Res. Technol. Manag. **60**(1), 35–38 (2017)
7. Cohendet, P., Pénin, J.: Patents to exclude vs. include: rethinking the management of intellectual property rights in a knowledge-based economy. Technol. Innov. Manag. Rev. **1**(3) (2011)
8. Delgado, M., Porter, M.E., Stern, S.: Defining clusters of related industries. J. Econ. Geogr. **16**(1), 1–38 (2016)
9. Fawcett, S.E., Magnan, G.M., McCarter, M.W.: A three-stage implementation model forsupply chain collaboration. J. Bus. Logistics **29**(1), 93–112 (2008)
10. Gassmann, O.: Opening up the innovation process: towards an agenda. R & D Manag. **36**(3), 223–228 (2006)
11. Gassmann, O., Bader, M.A.: Intellectual property management in inter-firm R&D collaborations. Taiwan Acad. Manag. J. **6**(2), 217–236 (2006)

12. Gumilar, V., Zarnic, R., Selih, J.: Increasing competitiveness of the construction sector by adopting innovative clustering. Inzinerine Ekonomika-Eng. Econ. **22**(1), 41–49 (2011)
13. Hamari, J., Sjöklint, M., Ukkonen, A.: The sharing economy: why people participate in collaborative consumption. J. Am. Soc. Inf. Sci. **67**(9), 2047–2059 (2016)
14. Isckia, T.: Écosystèmes d'affaires, stratégies de plateforme et innovation ouverte: vers une approche intégrée de la dynamique d'innovation. Manag. Avenir **6**(46), 157–176 (2011)
15. Jarillo, J.C.: On strategic networks. Strateg. Manag. J. **9**, 31–41 (1988)
16. Kathan, W., Matzler, K., Füller, J., Hautz, J., Hutter, K.: Open innovation in SMEs: a case study of a regional open innovation platform. Probl. Perspect. Manag. **12**(1), 161–171 (2014)
17. Keast, R., Mandell, M.P., Brown, K., Woolcock, G.: Network structure: working differently and changing expectations. Public Adm. Rev. **64**(3), 363–371 (2004)
18. Kratzer, J., Leenders, O.T.A., Engelen, J.M.V.: Stimulating the potential: creative performance and communication in innovation teams. Creativity Innov. Manag. **13**(1), 63–71 (2004)
19. Laamanen, T., Pfeffer, J., Rong, K., Van de Ven, A.: Business models, ecosystems and society in the sharing economy. Acad. Manag. Discoveries **4**, 213–219 (2018)
20. Lan, W., Zhanglui, W.: Research on interactive learning, knowledge sharing and collective innovation in SMEs cluster. Int. J. Innov. Manag. Technol. **3**(1), 24–29 (2012)
21. Lee, S., Park, G., Yoon, B., Park, J.: Open innovation in SMEs – an intermediated network model. Res. Policy **39**(2010), 290–300 (2010)
22. Lichtenthaler, U.: Open innovation in practice: an analysis of strategic approaches to technology transactions. IEEE Trans. Eng. Manag. **55**(1), 148–157 (2008)
23. Maithili, A., Kumari, V., Rajamanickam, S.: An open innovation business model based on collective intelligence. Int. J. Mod. Eng. Res. **2**(2), 245–252 (2012)
24. Osanna, P.H., Durakbasa, M.N., Crisan, L., Bauer, J.M.: The management and exchange of knowledge and innovation in environments of collaborating small and medium sized enterprises. Official J. Int. Bus. Inf. Manag. Assoc. **I**(7), 130–136 (2009)
25. Pop, C.: Evaluation of E-learning platforms: a case study. Informatica Economică **16**(1/2012), 155–167 (2012)
26. Porter, M.: Clusters and regional competitiveness: recent learnings. In: International Conference on Technology Clusters, Montreal, Canada, vol. 7, pp. 2007–2013 (2003)
27. Proulx, M., Gardoni, M.: Methodology for designing a collaborative business model – case study aerospace cluster. In: Nyffenegger, F., Ríos, J., Rivest, L., Bouras, A. (eds.) PLM 2020. IAICT, vol. 594, pp. 387–401. Springer, Cham (2020). https://doi.org/10.1007/978-3-030-62807-9_31
28. Salvador, E., Montagna, F., Marcolin, F.: Clustering recent trends in the open innovation literature for SME strategy improvements. Int. J. Technol. Policy Manag. **13**(4), 354–376 (2013)
29. Simcoe, T.S., Graham, S.J., Feldman, M.P.: Competing on standards? Entrepreneurship, intellectual property, and platform technologies. J. Econ. Manag. Strategy **18**(3), 775–816 (2009)
30. Sundararajan, A.: From Zipcar to the sharing economy. Harvard Bus. Rev. (03) (2013)
31. Sundararajan, A.: Peer-to-peer businesses and the sharing (collaborative) economy: overview. Economic Effects and Regulatory Issues (2014)
32. Van de Vrande, V., De Jong, J.P., Vanhaverbeke, W., De Rochemont, M.: Open innovation in SMEs: trends, motives and management challenges. Technovation **29**(6–7), 423–437 (2009)

Data Analytics Capability Roadmap for PPO Business Models in Equipment Manufacturing Companies

Prasanna Kumar Kukkamalla(✉), Veli-Matti Uski, Olli Kuismanen, Hannu Kärkkäinen, and Karan Menon

Tampere University, Tampere, Finland
{prasanna.kukkamalla,veli-matti.uski,olli.kuismanen, hannu.karkkainen,karan.menon}@tuni.fi

Abstract. The objective of this paper is to extend the knowledge of data analytics capabilities and the development process required to implement pay-per-output/-outcome (PPO) business models. To achieve this objective, we conducted qualitative research to study two equipment manufacturing companies in their business model transformation, with a special interest in their data analytics capability development path. The findings were threefold; first we validated the data analytics capabilities necessary for PPO business models synthesised from literature with the selected case companies. Second, two additional data analytics capabilities, namely *capability to influence how performance is measured and analysed,* and *capability to simulate the solution's financial performance)* were identified. Finally, this study presents two different roadmaps on how data analytics capabilities have been developed. The study findings suggest that remote monitoring capability is among the most critical data analytics capability to initiate a PPO business model in the equipment manufacturing industry. This study contributes to business model literature, and information systems literature.

Keywords: Data analytics capabilities · Pay-per-outcome · Business model · Equipment manufacturing · Remote monitoring · Roadmap

1 Introduction

Supported by advancements in IoT and remote monitoring technologies, new data-driven business models such as pay-per-output/-outcome (PPO) business models are gaining interest among the equipment manufacturing industry (EMI) [1–3]. Despite these technological opportunities, many companies have failed in implementing PPO business models due to technological challenges [2, 4]. Since the revenue logic in PPO business model is based on equipment performance and the outcome(s) the equipment creates, the equipment is often complex and the performance (outcome) data often a combination of things, data analytics capabilities play a key role in success of this PPO business model.

F. Noël et al. (Eds.): PLM 2022, IFIP AICT 667, pp. 45–54, 2023.
https://doi.org/10.1007/978-3-031-25182-5_5

A critical step for succeeding in PPO business models is the identification, collection and utilization of data [5, 6] and therefore new data analytics capabilities are needed [7].

Extant literature have identified the criticality of these data analytics capabilities for manufacturing companies [8]. For example, [3] studied how manufacturing companies use data analytics capabilities to support their PPO business model strategies. However, existing literature have dismissed how these capabilities can actually be developed and formalized [9] and how the resources and drivers align during this digital transformation [10].

Therefore, this study focuses on understanding how two manufacturing companies have developed their data analytics capabilities to succeed in their PPO business models. Our research questions are:

RQ1: *What are the major data analytics capabilities required to implement pay-per-output/-outcome business model in equipment manufacturing companies?*
RQ2: *How equipment manufacturing companies can develop data analytics capabilities for pay-per-output/-outcome business models?*

To answer these questions, we investigated two equipment manufacturing companies which have successfully implemented pay-per-output/-outcome business models. We built two capability development roadmaps to illustrate the development process and factors associated with the process, since capability roadmaps are useful for understanding how technology and business drivers interrelate with each other [11].

2 Theoretical Background

The global awareness and interest towards data-driven business models among manufacturers have increased in recent years [12]. Pay-per-X business models are types of business models were the ownership of the machine is not usually transferring to the customer but the customer is paying based on use or outcome of the machine [13]. In the literature, pay-per-X business models are discussed in varying terms, such as outcome-based contracts [14], performance-based contracts [15] use- and result-oriented business models, [16] and use- and result-oriented product-service systems [17].

Pay-per-X business models can be divided to different archetypes based on monetizing logic. In a pay-per-use business model the customer is paying based on the use of the solution based on usage time (e.g. Rolls-Royce pay-per-operating hours) [18]. In a pay-per-output/-outcome (PPO) business models the customer pays for the achieved output, outcome or result [13]. Therefore, the payment depend on contractually set quality, performance or output levels [1, 13]. The feasibility of these different pay-per-X business models might vary depending on the industry, product type and company's existing capabilities [2, 19].

Data analytics capabilities are a central set of capabilities needed for implementing pay-per-X business model [3, 16, 18]. According to [7] data analytics capability is an ability to effectively deploy resources and skills to capture, store, and analyse data to improve company's competitive performance. While implementing PPO business models, the company requires: *Capability for remote monitoring:* Being able to remotely monitor how the solution is operating (near-real time). Remote monitoring technologies (RMT) are considered as an important enabler of integrated service solutions [3]. RMT could help improve after-sales services, enabling solutions, and build service ecosystems [20]. *Capability to convince the customer to share data:* Being able to gain access to solution data and any other relevant data, such as maintenance data [21]. *Capability to translate data into value:* Being able to clean, transform and model the data to discover useful information to support decision making [3, 21]. *Capability to ensure data privacy and security:* Being able to protect data from unauthorized use, disruption, deletion and corruption [3, 21]. *Capability to simulate equipment performance:* Being able to digitally simulate how the solution will perform in varying environments and situations [3].

Although the needed data analytics capabilities are identified in the literature, there are no studies focusing on how these data analytics capabilities have been developed in companies implementing PPO business models.

3 Methodology

Our study implemented a multiple case study of two equipment manufacturers which offer their products through PPO business model to understand what kind of data analytics capabilities they need and how they developed these analytics capabilities to enable the respective PPO business models. The qualitative case study method was selected since we are trying to understand a complex real-life topic [22] and since we are trying to gain deep understanding on the 'how' and 'why'-type of questions regarding their development [23]. The studied companies were purposefully selected [24] to be both offering a) originally standardized installable capital-heavy equipment through a traditional product-oriented business model, and both b) have later successfully implemented a PPO business model.

Case company A is operating in the business of industrial compressed air. The company was founded in 2010 to develop a transformative technical concept for compressed air and to scale it up on the global industrial markets. Already at the time of launching the first product on the market the concept of pay-per-output was defined to be a second go-to-market strategy, to enable faster adoption of the technology in the fairly conservative, risk-averse industrial markets. From the beginning the technical solution used in their products were very high-tech, creating the possibility and capability to make the products not only smart but also quite easily capable to be used in the PPO business model. The products rely on a vast amount of data to operate, and the control algorithm already uses a lot of the data for operation, the same data which is relevant for the PPO business. The company has two parallel business models: in the pay-per-output business model the company charges the customers based on cubic-meters of air the machine has produced. In the pay-per-outcome business model in addition to cubic-meters of air the company get paid as well based on how much energy it has saved.

Case company B has been manufacturing metal bar punching and bending machines since 1963. With over 50 years of experience, the company has become the worldwide market leader in the precision processing of flat materials such as conductor rails, bar stock or profiles. In 2019 the company started to also offer its punching machines through pay-per-outcome business model in parallel with the traditional sell and service business model. Currently the company enables charging the customer based on number of punches parallel to conventional sales channels.

The research team conducted multiple interviews with both company's representatives. All interviews were recorded and transcribed. Respondents were chosen for interviews based on their roles and responsibilities in the development of products, technology, especially with regard to their respective PPO business models, or knowledge in general. The interviews took place between February–March 2022. We used Phaal et al. [11] capability roadmap framework as a structure for the interviewees. The target was to understand in which order the companies have developed their data analytics capabilities. In addition, based on [25] definition of operational capabilities we asked which drivers, resources & skills, processes & activities have affected the development of these capabilities. We used a web-based tool to map out the timeline to ensure the sequence of events visualized in the result section is correct.

4 Results

This section presents major data analytics capabilities and development process to implement PPO business models in selected cases (See Table 1).

Table 1. Needed data analytics capabilities

Data analytics capability	Case A	Case B	Additional comments
Capability for remote monitoring	X	X	
Capability to convince the customer to share data	X	X	*Prerequisite for PPO business model but has not been an issue (Case A)*
Capability to translate data into value	X	X	
Capability to ensure data privacy and security	X	X	*Third party financial partner is responsible (Case B)*
Capability to simulate equipment technical performance	X	–	*Simulation is necessary to ensure equipment reliability (Case A)* *Not needed specifically for PPO business (Case B)*
Capability to influence how performance is measured and analysed	X	–	*New capability* *Third party financial partner is responsible (Case B)*
Capability to simulate equipment financial performance	X	X	*New capability*

4.1 Capabilities Required for PPO Business Model

Regarding the data analytics capabilities required to implement the PPO business model, both case companies agreed that data analytics capabilities presented during the workshop were relevant for the effective implementation of PPO business model except *capability to simulate equipment technical performance* for Case B. With respect to *capability for remote monitoring*, both case firms confirmed that this capability is essential to initiate the PPO business model. *"...it is essential capability to start with, absolutely because if you don't, i think you shouldn't consider pay-per-X or pay-per-punch or anything. We should all consider the big picture and part of this picture is obviously the remote monitoring. ...have to be done remotely..."* (Case B).

Capability to convince the customer to share data was considered as a prerequisite for PPO business model by Case B. The interviewee from Case A stated: *"we didn't need to convince any customer [...] that had decided to go for our technology, that can we access the data or not"* However, since the ownership of the equipment belongs to a third-party financing institute in case B, they expressed as not themselves needing this capability.

Capability to translate data into value was categorized as essential capability for PPO contracts. Interviewee from case A stated: *"...for example, billing [...] what can be used directly to translate some signals or measurement signals to for example, the power or the some other easily understandable measures [...] what most probably at the early stage needs to be developed by company or provider"* (Case A).

Regarding *capability to ensure data privacy and security*, case A stressed that: *"it can be a barrier in this kind of business if it's not very well handled"*. Case B also agreed that is essential for PPO business model, and they highlighted that their existing technology enables data security, so they don't need any additional skills to secure the data.

The companies' opinions from capability to simulate equipment performance from technical point-of-view were slightly separated. Case A had this capability on different levels (component, subsystem, system), and even stated that it is in their case a necessary capability: *"if I would do something different now, i would actually start from there. Already [...] in the in the very first stage to do kind of capability do for the system simulation, [...] before, that simulation capability, we actually [...] started with the testing, we tried to do before we got this kind of capability"*. For case B this capability wasn't deemed to have a connection with the PPO model especially.

On top of these, we identified two additional capabilities: *Capability to influence how performance is measured and analysed*, and *capability to simulate equipment financial performance*. This was brough up by both companies, that there is a need to show the financial outcome of the PPO contract for the customer. *"honest discussion to show really the real performance and the real... potential savings and whatever it is, is it power consumption, or to produce the heat or what's then needed. [...] to transfer that to [...] already reliable data"*.

Only case A highlighted the need of *capability to influence how performance is measured and analysed* (See Table 1). The interviewee stated that since PPO involves measurement of equipment performance, specific standards for measurement would be useful in their industry, currently there is variation. They also stressed that companies should be able to develop these standards, to be able to calculate the outcome of the equipment, and to be able to convince the customer of the outcome's validity.

Both case companies identified that even more than the *capability to simulate equipment performance* from technical point of view, they need the *capability to simulate equipment's financial performance*. According to the interviewees, since PPO contracts lock the customer in for a longer period, it is very essential to be able to show financial benefits to the customer and to understand their own costs in the long run. Regarding this, case B pointed out that this is critical capability but, in their case, the third party handle it.

4.2 Data Analytics Capabilities Developing Roadmap

Case A: Case A data analytics development roadmap is presented in Fig. 1. The development of data analytics capabilities was initiated already at the beginning when the development of the current technical concept was started. The reason for implementing remote monitoring technologies (RMT) was that the company's customer was located far away from company's headquarters, but the company still wanted to know how well the machines are performing. The company's investors specifically required this. The key resource for implementing RMT was several company experts, who had experience on these technologies from previous companies. The RMT was already implemented to company's equipment in the test laboratory, and therefore implementing it to first customer was easier. From the beginning the company decided to build RMT over mobile network which therefore didn't require connection through the customer's network. Due to this the interviewee stated that they didn't need to convince the customer about data security or privacy nor having the whole discussion about this.

Being a newcomer on the market, some customers had concerns about the technology readiness and the company size. To counter this, the company proposed to the customer a pay-per-output business model as a risk mitigation method. In the industry of Case A, the accuracy of output measurements is not traditionally very high. Therefore, Case A realized that they need new capabilities to influence how things are measured and analysed since the customers were getting confused by the manufacturers' mixed messaging. Throughout the development of the technology, the capabilities to translate data into value were developed and used.

As a newcomer in a traditional industry the company started early on to develop capability to simulate financial value of the machine so the customers would better understand the cost and benefits in long run. Soon after the idea of pay-per-outcome business model was introduced.

At a later phase some of the customers also wanted to integrate the machine into their own IT-systems. This required new connections between company's and customer's system and in this point data security issues were highlighted the first time. However, since the products were developed already using newest technologies, the company was capable to do this without additional investments.

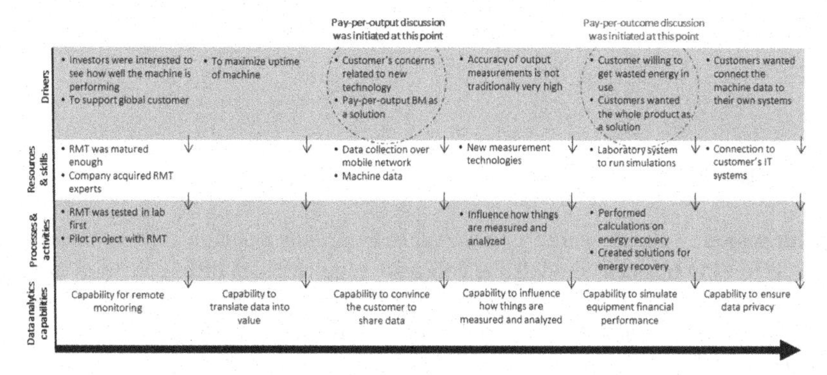

Fig. 1. Case A - Data analytics capability development roadmap

Case B: Our study found that Case B followed a different strategy to develop certain data analytics capabilities (See Fig. 2). The company decided to collaborate with a third-party financer and outsource some of the key activities to them. The company started to develop data analytics capabilities for its traditional business model in the 1990. In the beginning the company already translated data into value by solving customer issues using the data when physically in the customers premises. Always visiting customers to be able to troubleshoot the equipment wasn't efficient, so the company started to develop Capability for remote monitoring to be able to solve issues quicker and to save cost of travelling. To do this the company had to have Capability to convince customer to share data through their network but according to interviewee the benefits of the remote monitoring were in that time big enough that much convincing was not needed. Through the years the company develop their data analytics capabilities using several different technologies as the technology matured.

A few years ago, a third party approached the company and suggested the that they could finance and offer the company's machine through pay-per-output business model. According to the interviewee the third party had Capability to simulate equipment financial benefits and Capability to ensure data privacy and security. Therefore, the

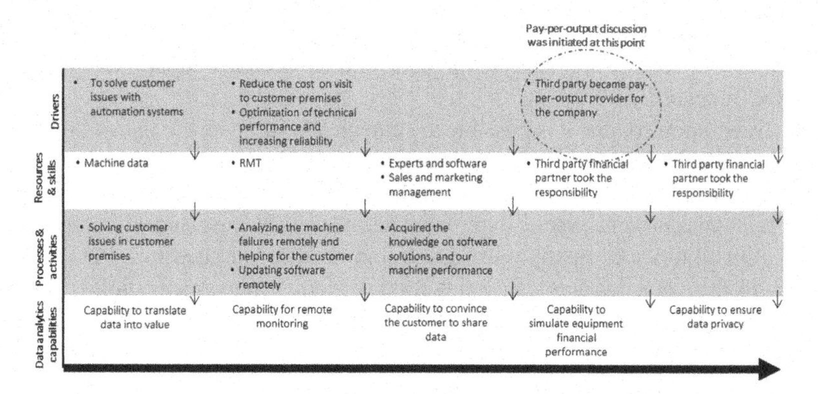

Fig. 2. Case B - Data analytics capability development roadmap

company did not need to develop additional capabilities to offer their pay-per-output business model.

5 Conclusions

The contribution of this research is three folds. First, our research shows that in order to succeed in PPO business models the company require seven different types of major data analytics capabilities, namely *Capability for remote monitoring, Capability to convince the customer to share data, Capability to translate data into value, Capability to ensure data privacy and security, Capability to simulate equipment technical performance, Capability to influence how performance is measured and analysed, Capability to simulate equipment financial performance.* We derived five data analytics capabilities from scientific literature and validated them. In addition, compared to earlier studies [3, 21] we identified two additional data analytics capabilities, *Capability to influence how performance is measured and analysed, Capability to simulate equipment financial performance.* Moreover, to authors' best knowledge, this is the first paper studying how data analytics capability development process can be developed for PPO business model implementation and to build roadmaps of data analytics capabilities development in equipment manufacturing companies. By illustrating two different case companies' data analytics capability roadmaps, this study has shown how companies can align key resources, and activities to develop required data analytics capabilities to implement PPO business model. We showed that to implement PPO business models, companies do not need have all the data analytics capabilities in advance.

Secondly, even though both case companies followed distinctive paths to develop data analytics capabilities, the need of remote monitoring capabilities were highlighted. Compared to earlier studies [3, 16, 18, 21] which have not made distinction between the priorities of data analytics capabilities and therefore have not studied in which order these could be developed, this study showed that remote monitoring capability was among the most critical data analytics capability to initiate PPO business model in EMI. Remote monitoring capability was an important prerequisite to other data analytics capabilities and the PPO business model. Both selected case companies have integrated RMT in their machines prior to implement PPO business model. This RMT integration has provided an opportunity to acquire and analyse the data. Almost all other capabilities were built on this RMT capability.

Thirdly, we showed, that it is possible to strategically mitigate the risk of developing some of the data analytics capabilities through a financing partner. This partner would take the responsibility of acquiring some of the needed capabilities.

This study makes some practical implications as well. We revealed what kind of data analytics capabilities a company must develop in order to succeed in a PPO business model. With this, practitioners can estimate how ready their organization is for PPO business model implementation and how to prioritize the resources. To succeed in PPO business model implementation, the company does not need to internally develop all the data analytics capabilities, they can adopt a strategy of outsourcing. PPO business model implementation is a continuous process, and data analytics capabilities can evolve along with the transformation. Roadmaps presented in this study could help managers how to integrate resources and skills to develop these capabilities.

The limitation of this study relates to the research theme and case selection criteria. Since the objective of this study is to identify major data analytics capabilities needed to implement PPO business models in equipment manufacturing companies, we did not focus on the data analytics capabilities implementation process, and how they work in a real-life situation. Therefore, the results might not be applicable to other industries with different kind of equipment. In future, we will validate these two data analytics capabilities by studying large sample. However, this offers interesting future research avenue to study development of data analytics capabilities in other industries. Secondly, we haven't focused on profitability of these PPO business models over time, so we can't evaluate the impact of these capabilities on the success of the PPO business model. Therefore, in future studies the development of profitability could be studied along with the data analytics capability development to find out does some capability have distinguishable effect on profitability.

Acknowledgement. This work was funded by Business Finland BF/Future Spaces.

References

1. Menon, K., Kärkkäinen, H., Mittal, S., Wuest, T.: Impact of IIoT based technologies on characteristic features and related options of nonownership business models. In: Fortin, C., Rivest, L., Bernard, A., Bouras, A. (eds.) PLM 2019. IAICT, vol. 565, pp. 302–312. Springer, Cham (2019). https://doi.org/10.1007/978-3-030-42250-9_29
2. Uski, V.M., et al.: Review of PPX business models: adaptability and feasibility of PPX models in the equipment manufacturing industry. In: Canciglieri Junior, O., Noël, F., Rivest, L., Bouras, A. (eds.) PLM 2021, vol. 639, pp. 358–372. Springer, Cham (2022). https://doi.org/10.1007/978-3-030-94335-6_26
3. Grubic, T., Peppard, J.: Servitized manufacturing firms competing through remote monitoring technology: an exploratory study. J. Manuf. Technol. Manag. **27**(2), 154–184 (2016)
4. Uuskoski, M., Menon, K., Kärkkäinen, H., Koskinen, K.: Perceived risks and benefits of advanced pay-per-use type of business models based on industry 4.0 enabled technologies in manufacturing companies. In: Chiabert, P., Bouras, A., Noël, F., Ríos, J. (eds.) PLM 2018. IAICT, vol. 540, pp. 498–507. Springer, Cham (2018). https://doi.org/10.1007/978-3-030-01614-2_46
5. Menon, K., Mittal, S., Kärkkäinen, H., Kuismanen, O.: Systematic steps towards concept design of pay–per-x business models: an exploratory research. In: Canciglieri Junior, O., Noël, F., Rivest, L., Bouras, A. (eds.) PLM 2021, vol. 640, pp. 386–397. Springer, Cham (2022). https://doi.org/10.1007/978-3-030-94399-8_28
6. Schroderus, J., Lasrado, L.A., Menon, K., Kärkkäinen, H.: Towards a pay-per-X maturity model for equipment manufacturing companies. Procedia Comput. Sci. **196**, 226–234 (2022)
7. Gupta, M., George, J.F.: Toward the development of a big data analytics capability. Inf. Manag. **53**(8), 1049–1064 (2016)
8. Manresa, A., Prester, J., Bikfalvi, A.: The role of servitization in the capabilities – performance path. CR **31**(3), 645–667 (2021)
9. Khanra, S., Dhir, A., Parida, V., Kohtamäki, M.: Servitization research: a review and bibliometric analysis of past achievements and future promises. J. Bus. Res. **131**, 151–166 (2021)

10. Agostini, L., Nosella, A.: Industry 4.0 and business models: a bibliometric literature review. BPMJ **27**(5), 1633–1655 (2021)
11. Phaal, R., Farrukh, C.J.P., Probert, D.R.: Technology roadmapping—a planning framework for evolution and revolution. Technol. Forecast. Soc. Chang. **71**(1–2), 5–26 (2004)
12. Baines, T., Ziaee Bigdeli, A., Bustinza, O.F., Shi, V.G., Baldwin, J., Ridgway, K.: Servitization: revisiting the state-of-the-art and research priorities. IJOPM **37**(2), 256–278 (2017)
13. Adrodegari, F., Alghisi, A., Ardolino, M., Saccani, N.: From ownership to service-oriented business models: a survey in capital goods companies and a PSS typology. Procedia CIRP **30**, 245–250 (2015)
14. Ng, I.C.L., Maull, R., Yip, N.: Outcome-based contracts as a driver for systems thinking and service-dominant logic in service science: evidence from the defence industry. Eur. Manag. J. **27**(6), 377–387 (2009)
15. Liinamaa, J., Viljanen, M., Hurmerinta, A., Ivanova-Gongne, M., Luotola, H., Gustafsson, M.: Performance-based and functional contracting in value-based solution selling. Ind. Mark. Manag. **59**, 37–49 (2016)
16. Möller, H., Shahnavaz, T.: Use-Oriented Business Model for Consumer Durables: an Exploratory Case Study on Business Model Capabilities (2020)
17. Yang, M., Smart, P., Kumar, M., Jolly, M., Evans, S.: Product-service systems business models for circular supply chains. Prod. Plan. Control **29**(6), 498–508 (2018)
18. Gebauer, H., Saul, C.J., Haldimann, M., Gustafsson, A.: Organizational capabilities for pay-per-use services in product-oriented companies. Int. J. Prod. Econ. **192**, 157–168 (2017)
19. Qi, Y., Mao, Z., Zhang, M., Guo, H.: Manufacturing practices and servitization: the role of mass customization and product innovation capabilities. Int. J. Prod. Econ. **228**, 107747 (2020)
20. Andreas Zolnowski, T.B., Schmitt, A.K.: Understanding the impact of remote service technology on service business models in manufacturing: from improving after-sales services to building service ecosystems. In: 19th European Conference on Information Systems (ECIS), p. 109 (2011)
21. Rösler, J., Friedli, T.: A capability model for equipment-as-a-service adoption in manufacturing companies. In: West, S., Meierhofer, J., Ganz, C. (eds.) Smart Services Summit. PI, pp. 59–71. Springer, Cham (2021). https://doi.org/10.1007/978-3-030-72090-2_6
22. Yin, R.K.: Case Study Research: Design and Methods, 3rd edn. Sage Publications, Thousand Oaks (2003)
23. Dyer, W.G., Wilkins, A.L.: Better Stories, Not Better Constructs, to Generate Better Theory: A Rejoinder to Eisenhardt, p. 8 (1991)
24. Eisenhardt, K.M., Graebner, M.E.: Theory building from cases: opportunities and challenges. AMJ **50**(1), 25–32 (2007)
25. Wu, S.J., Melnyk, S.A., Flynn, B.B.: Operational capabilities: the secret ingredient: operational capabilities. Decis. Sci. **41**(4), 721–754 (2010)

Systematic Literature Review About Sustainable Business Models and Industry 4.0

Grazielle Fatima Gomes Teixeira(✉) and Osiris Canciglieri Junior

Polytechnic School of the Pontifical Catholic University of Paraná – PUCPR, R. Imaculada Conceição, 1155 - Prado Velho, Curitiba, PR 80215-901, Brazil
grazielle.teixeira@pucpr.edu.br

Abstract. All over the world, Industry 4.0 (I4.0) and Sustainable Development (SD) have progressively gained the interest of scholars, politicians, and other parts of society. Besides being two of the most debated topics of the last decades, they also have overlaps between their independent research fields. Some examples are reductions of environmental impacts and improvements in production technologies. This integration of technologies and sustainable advances within an industrial context can enable a set of important competitiveness forces, which results can reflect in business improvement. However, the link between I4.0, SD, and business still needs a broader understanding. Basing on this perspective, this paper proposes an update of a systematic literature review to monitor the development of the topic and to check if there has been any progress, as well as to verify whether the findings are still valid. Results point to that there are seventeen research opportunities, showing the potential of I4.0 as an enabler of sustainable business models that changes the responsibilities of companies.

Keywords: Industry 4.0 · 4th industrial revolution · Sustainability · Business model · Systematic literature review

1 Introduction

Sustainable improvements aim to revolutionize the way that products are producing and using. In this line, novel technologies emerging under the Fourth Industrial Revolution or Industry 4.0 (I4.0) are creating new opportunities for solving sustainability-related needs [1, 2]. For example, I4.0 can help factories to improve sustainability efforts through real-time control of resource consumption, increased logistical efficiency, the extension of the product lifecycles by pro-active maintenance and remanufacturing [3]. Moreover, I4.0 can stimulate mass customization and product diversification strategies [4]. This brings new sides to customer-focused approaches because information systems (IS) can have a significant role to redesign products and business processes across the enterprise to turn its results more sustainable [2, 5, 6]. As a result, it can build or regenerate a cross-linked value chain of actors that compose a business activity [2].

This argumentation helps us to understand why the overlap between Sustainability and I4.0 has been one of the most important industrial debates in recent years [7–15]. But,

Published by Springer Nature Switzerland AG 2023
F. Noël et al. (Eds.): PLM 2022, IFIP AICT 667, pp. 55–65, 2023.
https://doi.org/10.1007/978-3-031-25182-5_6

although the I4.0 revolution can be described as a facilitator for sustainable development, this union remains underdeveloped in many aspects [11, 16–22]. One of these aspects is about Sustainable Business Models (SBM) into the I4.0 revolution [7, 8, 11, 13, 23].

There is still a lack of understanding about the impacts of I4.0 in many different issues of business to redesigning value chain configurations and adoption or creation of new business models focused more on sustainability goals [17, 24–27]. Sustainability is a concept that does not have a unique definition [28]. Due to this, its concrete implementation is considered difficult because there is a high degree of complexity concerning the depth and specifications of actions [11, 28]. Likewise, I4.0 is also a paradigm whose unique definition is not possible due to its wide range of approaches [8, 22, 29]. The business model concept is an abstract representation of the value flow and the workflow of an organizational unit [30].

As Stock *et al.* [11] and Dao [6] emphasized, there is still a lack of research on how I4.0 impacts the stabilization of business models according to sustainable development objectives. Some researchers add the concept of a maturity model to try to help understand the discussion, for example, Allais *et al.* [31], Gouvinhas *et al.* [32], Gaziulusoy [33], Romero and Molina [34], Murillo-Luna *et al.* [35]. The maturity models aim to develop and build knowledge in an evolutionary way so that a company can progress more solidly and securely. Based on this argument, it can be said that the literature is scarce on these conceptions. Teixeira *et al.* [2] presented a Systematic Literature Review (RSL) between May 2015 until May 2020. However, the investigation of this theme is still necessary and relevant for both academics and society. Due to this, the purpose of this article is to present the SLR update developed by Teixeira *et al.* [2]. In summary, this article addresses the set of issues:

(Q1) *What are the opportunities for the future research agenda?*
(Q2) *Where is this matter being investigated?*
(Q3) *What is the current situation in this field of research?*

To achieve the proposed objective, the article consists of four sections, the introduction of which is the first. In the second section, the review methodology is described. The results obtained to answer the set of questions are presented in Sect. 3 and analyzed in Sect. 4. Finally, the final considerations are presented in Sect. 5.

2 Review Methodology

Teixeira et al. [2] implemented a Systematic Literature Review (SLR) according to the procedures described by Seuring *et al.* [36], Durach *et al.* [37], and Tranfield *et al.* [38]. These procedures were reused to develop the update presented now. To keep rigorous of the result, an SLR must follow a set of steps. Durach *et al.* [37] describe six steps to build an SLR: (1) defining the research question (goal), (2) determining the required characteristics of primary studies (inclusion criteria), (3) retrieving a sample of potentially relevant literature (collect data), (4) selecting the pertinent literature (exclusion criteria), (5) synthesizing the literature (analyze data), and (6) reporting the results. Similar recommendations are made from Seuring *et al.* [36] who describe an SRL by four elements: (1) definition - who and what (goal), (2) boundaries and limitations – when and where (inclusion criteria), (3) variables and causalities – why and how (collect data), (4) predictions – could, should and would (analyze data).

Table 1. Keyword pairwise query.

1	TITLE-ABS-KEY ((sustainability AND business model AND industry 4.0))
2	TITLE-ABS-KEY ((Sustainability AND new business model AND industry 4.0))
3	TITLE-ABS-KEY ((Sustainability AND organisation improvement AND industry 4.0))
4	TITLE-ABS-KEY ((sustainability AND maturity model AND industry 4.0))
5	TITLE-ABS-KEY ((Sustainability AND maturity evaluation AND industry 4.0))
6	TITLE-ABS-KEY ((Sustainability AND dynamic capability AND industry 4.0))
7	TITLE-ABS-KEY ((Sustainability AND strateg* AND industry 4.0))
8	TITLE-ABS-KEY ((Sustainability AND competitiveness AND industry 4.0))
9	TITLE-ABS-KEY ((Sustainability AND competition AND industry 4.0))
10	TITLE-ABS-KEY ((Sustainability AND value creation AND industry 4.0))
11	TITLE-ABS-KEY ((Sustainability AND strategic planning AND industry 4.0))
12	TITLE-ABS-KEY ((Sustainability AND product AND development AND process AND industry 4.0))
13	TITLE-ABS-KEY ((Sustainability AND product AND development AND industry 4.0))
14	TITLE-ABS-KEY ((Sustainability AND product AND industry 4.0))
15	TITLE-ABS-KEY ((Sustainability AND continuous improvement AND development AND industry 4.0))

Moreover, to secure the validity and transparency of SLR, choose a database is also an important point. In this study, were selected specific databases: Scopus and Web of Science. Then, the second step was the selection of the inclusion and exclusion criteria. In this line, the review process considered only formal literature that was written full-texts in English and peer-review published papers.

Equal as Teixeira *et al.* [2] the choice of the time cut was based on evidence found in a previous analysis that the number of publications about SBM increased during the past two decades. However, a significant concentration of publications is observed between 2016 to now because the number of studies about I4.0 is more substantial during this same period [30]. In this way, the first review was limited to find papers published between May 2015 and May 2020. This second review was extended to find papers published until February 2021. The scope of this extension is to monitor the development of the topic and to check if there has been any progress, as well as to verify whether the findings are still valid. Then, was conducted a structured keyword search executed through a pairwise query, focusing on titles, abstracts, and keywords. The same keywords were used for this second review (Table 1).

3 Results

Before answering the questions listed, the results of the survey carried out in the two databases are presented. According to Table 2, the result of this extension of the SLR reached 279 articles. This represents an increase of approximately 81.17% concerning the previous number of the first SLR. Scopus was the database that brought the most results. Figure 1 illustrates which searches with the keywords obtained the highest and lowest results in each database.

With this, it is possible to observe that the combination of keywords 14 obtained better results in both databases and reviews. In the Scopus database, combinations of keywords with superior results was, respectively, 1, 2, 8, 13, 9, 12, and 4. The lowest results at Scopus were with keyword combinations 3, 5, 6, 10, 11 and 15. The combination of keywords 7 did not add new research.

In the Web of Science database, combinations of keywords with superior results was, respectively, 7, 6, 9, 12, and 13. The combinations of keywords 4 and 5 did not add new research. On the contrary, the lowest results were with keyword combinations 1, 2, 3, 8, 10, 11 and 15.

Of the total number of new articles found, only 19 articles met the criteria and objectives of the SLR. Therefore, adding the results of both reviews, 55 is the final number of articles that answer the listed questions. From the analysis of the content of these new contributions, the questions were answered according to their order.

Table 2. Different between the results in each review

Keyword pairwise query	1st Review (until 2020)		2nd Review (2020 - 2021)	
	Scopus	Web of Science	Scopus	Web of Science
1	8	4	29	5
2	4	2	14	3
3	1	1	3	1
4	1	0	6	0
5	0	0	2	0
6	2	0	5	4
7	17	3	17	14
8	7	1	15	3
9	9	2	16	5
10	8	4	10	5
11	0	0	2	1
12	12	3	19	6
13	20	3	28	6
14	33	7	42	14
15	1	1	3	1
Total in each database	123	31	211	68
Result	154		279	

4 Analysis

Started by **(Q1)** *What are the opportunities for future research agenda?* not all authors present directions for future research. However, in this second SLR three new opportunities for future investigation were identified: how to manage the transition, develop a scale to measure transition, and, how to make data analytics infrastructure. With these new notes, the total of appointments to future investigations is now seventeen (see Table 3).

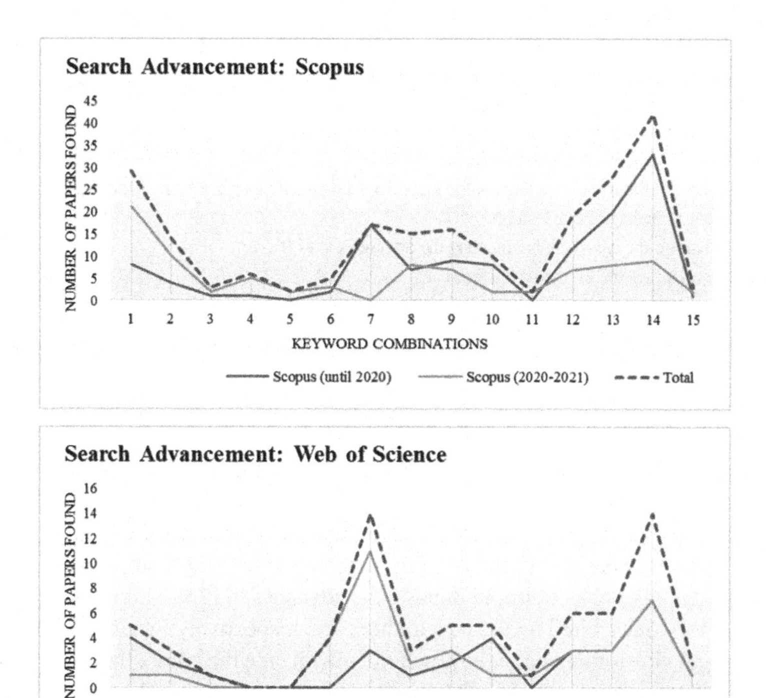

Fig. 1. Difference between the results.

Table 3. Opportunities for future research

N	%	Future research directions
1°	15,80%	Apply the study in large-scale
2°	12,00%	Analyze the different functions of the company in interaction with I4.0
3°	11,30%	How to set the TBL dmensions in the context of the I4.0
4°	9,00%	What is the impact of I4.0 on supply chain network design
5°	8,30%	How to keep job security stable
6°	7%	More research qualitative/quantitative
7°	6%	What is the impact of I4.0 on managing customer channels
8°	6%	How incorporating the I4.0 into theory about sustainable value propositions
9°	4%	How I4.0 will change the competition
10°	4%	How to measure cost-benefit analysis of I4.0 solutions
11°	3,01%	What are the policy-making efforts need to I4.0
12°	3,01%	What are the transformations in the identity of a manufacturing company promote by I4.0
13°	3,01%	How to manage the transition
14°	2,26%	What are the risks of implementing the advances in I4.0
15°	2,26%	Develop a scale to measure transition
16°	2,26%	How to make data analytics infrastructure
17°	1,50%	How to ensure data security

Comparing with the previous review it is possible to observe that the first three possibilities remained in the same order. However, the third was the most mentioned by the new studies. Among the possibilities of research that risen in the ranking are, respectively: how to keep job security stable, more research qualitative/quantitative, what are the transformations in the identity of a manufacturing company promote by I4.0, and, what are the risks of implementing the advances in I4.0. Those possibilities of research that were least cited by the new authors are, respectively: what is the impact of I4.0 on managing customer channels, how I4.0 will change the competition, and, how to measure cost-benefit analysis of I4.0 solutions. The remaining opportunities identified followed with the same percentage.

Data to answer **(Q2)** *Where this topic is being investigated?* reveal some changes (Fig. 2). First, the predominance of authors has now reverted to authors from Italy and later from Germany. In this update, contributions by authors from Germany were one of the smallest. Compared to the previous SLR, the countries with the largest contributions are, respectively, Spain, the USA, Hungary, and India. Also, 10 new countries were added to the list, respectively, in terms of the number of authors: France, Malaysia, Portugal, Romania, Slovenia, Bangladesh, Czech Republic, Hong Kong, Lebanon, and Pakistan. The increase in the number of countries demonstrates that the interest in investigating this field of research is increasingly global. Also, it is being discussed in countries with different levels of development.

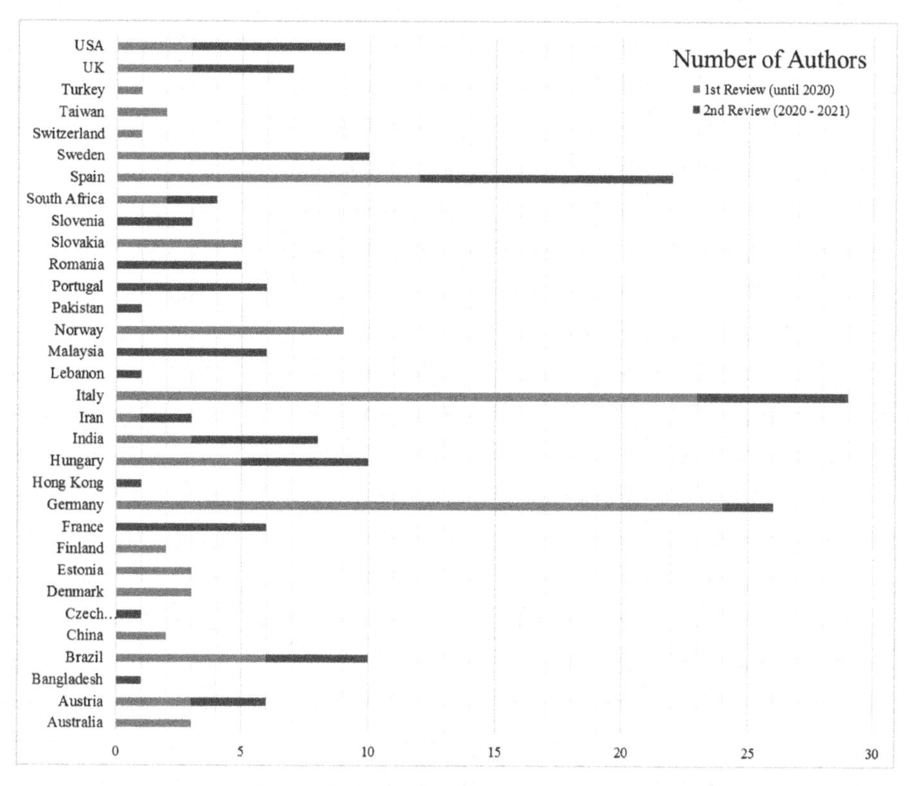

Fig. 2. Authors' affiliations countries.

According to the authors' affiliations, Fig. 3 shows which school each author is. In the first SLR, most of them were researchers in Business and Management and Industrial Engineering schools. This total number has not changed, but the number of new authors from the Industrial Engineering school is greater than those from the Business and Management school. The third school that added more authors was the School of Economics. In this second SLR, two new schools were added to the list: School of Mathematics and Computer Science. This result demonstrates that investigations are being developed by more diverse teams, expanding the multidisciplinary of discussion.

The chronologic data to answer **(Q3)** *What is the current status of this research field?* shows a growing trend, although there was a regression in 2019 (Fig. 4). The growth trend can be observed when comparing the numbers of 2020 in the first SLR with those of 2021. Even though it is in the first quarter of the year, the result is already higher. Also, the final result for 2020 is the highest in the entire sample. It is observed that 43,6% of total papers were published in the year 2015–2018 and 56,4% from 2019 until 2021.

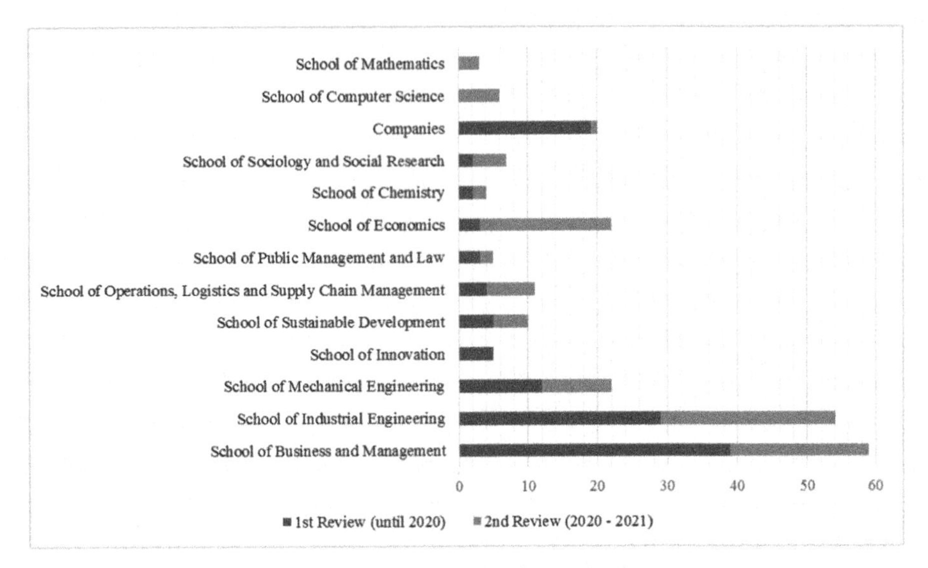

Fig. 3. Authors' affiliations school.

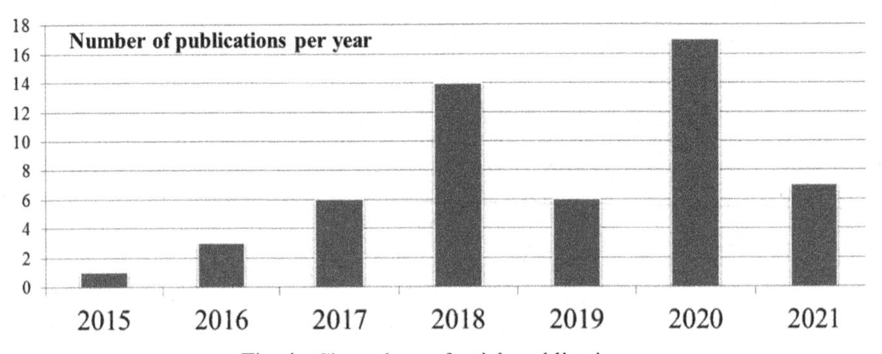

Fig. 4. Chronology of article publications.

5 Final Considerations

Results showed state of the art and a research agenda, allowing us to conclude that the topic is still legitimated but not consolidated because it is a phenomenon relatively new. Based on the relative scarce of investigations, the central importance of the study here is to be one-step further for the profound development of literature.

In addition, observing the notes for future investigations, it is possible to analyze that the study on the links between the I4.0 and the Sustainable Business Model is still at an early stage. Therefore, discussions on this topic restricted to market segments are opportunities for advancing the study. Another step forward in the discussion of the topic would be studies on the link between topics in countries with different development rates. As well as the observation of how topics could foster the creation of regulations and legislation in different industrial sectors. Regarding the new regulations, further studies could observe whether technological advances would be able to make global trade more equitable and fair.

Due to the qualitative nature of this study, our results had some limitations. The first limitation is related to the decision to not use statistical analysis against the limited number of papers analyzed. The second limitation is related to the selection of keywords because they may have other interesting articles outside the sample that have not been reviewed.

Acknowledgments. This study was financed in part by the Coordenação de Aperfeiçoamento de Pessoal de Nível Superior - Brasil (CAPES) - Finance Code 001.

References

1. Esmaeilian, B., Sarkis, J., Lewis, K., Behdad, S.: Blockchain for the future of sustainable supply chain management in Industry 4.0. Resour. Conserv. Recycl. **163**, 105064 (2020)
2. Teixeira, G.F.G., Junior, O.C., Szejka, A.L.: Future research agenda to understanding the sustainable business model in Industry 4.0. In: Rossit, D.A., Tohmé, F., Mejía Delgadillo, G. (eds.) ICPR-Americas 2020. CCIS, vol. 1407, pp. 357–371. Springer, Cham (2021). https://doi.org/10.1007/978-3-030-76307-7_27
3. Ghosh, D., Sant, T.G., Kuiti, M.R., Swami, S., Shankar, R.: Strategic decisions, competition and cost-sharing contract under Industry 4.0 and environmental considerations. Resour. Conserv. Recycl. **162**, 105057 (2020)
4. Fathi, M., Ghobakhloo, M.: Enabling mass customization and manufacturing sustainability in Industry 4.0 context: a novel heuristic algorithm for in-plant material supply optimization. Sustainability **12** (2020)
5. Butler, T.: Compliance with institutional imperatives on environmental sustainability: building theory on the role of green IS. J. Strat. Inf. Syst. **20**, 6–26 (2011)
6. Dao, V., Langella, I., Carbo, J.: From green to Sustainability: information technology and an integrated sustainability framework. J. Strat. Inf. Syst. **20**, 63–79 (2011)
7. García-Muiña, F.E., Medina-Salgado, M.S., Ferrari, A.M., Cucchi, M.: Sustainability transition in industry 4.0 and smart manufacturing with the triple-layered business model canvas. Sustainability **12**, 2364 (2020)
8. Rosa, P., Sassanelli, C., Urbinati, A., Chiaroni, D., Terzi, S.: Assessing relations between circular economy and Industry 4.0: a systematic literature review. Int. J. Prod. Res. **58**(6), 1662–1687 (2020). https://doi.org/10.1080/00207543.2019.1680896
9. Ghobakhloo, M.: Determinants of information and digital technology implementation for smart manufacturing. Int. J. Prod. Res. 1–22 (2019)
10. Kamble, S.S., Gunasekaran, A., Dhone, N.C.: Industry 4.0 and lean manufacturing practices for sustainable organisational performance in Indian manufacturing companies. Int. J. Prod. Res. 1–19 (2019)
11. Stock, T., Obenaus, M., Kunz, S., Kohl, H.: Industry 4.0 as enabler for a sustainable development: a qualitative assessment of its ecological and social potential. Process Saf. Environ. Prot. **118**, 254–267 (2018). https://doi.org/10.1016/j.psep.2018.06.026
12. Stock, T., Seliger, G.: Opportunities of sustainable manufacturing in Industry 4.0. In: Seliger, G., Kohl, H., Mallon, J. (eds.) 13th Global Conference on Sustainable Manufacturing - Decoupling Growth From Resource Use, Procedia CIRP, pp. 536–541 (2016)

13. Cornelis, J., de Man, J., Strandhagen, O.: An Industry 4.0 research agenda for sustainable business models. Procedia CIRP **63**, 721–726 (2017). https://doi.org/10.1016/j.procir.2017.03.315
14. Kiel, D., Müller, J.M., Arnold, C., Voigt, K.-I.I.: Sustainable industrial value creation: benefits and challenges of Industry 4.0. Int. J. Innov. Manag. **21**(08), 1740015 (2017)
15. Peukert, B., et al.: Addressing sustainability and flexibility in manufacturing via smart modular machine tool frames to support sustainable value creation. Procedia CIRP **29**, 514–519 (2015)
16. Bai, C., Dallasega, P., Orzes, G., Sarkis, J.: Industry 4.0 technologies assessment: a sustainability perspective. Int. J. Prod. Econ. **229**, 107776 (2020)
17. Felsberger, A., Qaiser, F.H., Choudhary, A., Reiner, G.: The impact of Industry 4.0 on the reconciliation of dynamic capabilities: evidence from the European manufacturing industries. Prod. Plan Control 1–24 (2020)
18. Ghobakhloo, M.: Industry 4.0, digitisation, and opportunities for sustainability. J. Clean. Prod. **252**, 119869 (2020)
19. Varela, L., Araújo, A., Ávila, P., Castro, H., Putnik, G.: Evaluation of the relation between lean manufacturing, Industry 4.0, and sustainability. Sustainability **11**, 1–19 (2019)
20. Beier, G., Niehoff, S., Xue, B.: More sustainability in industry through industrial internet of things? Appl. Sci. **8**(2), 219 (2018)
21. Kamble, S.S., Gunasekaran, A., Gawankar, S.A.: Sustainable Industry 4.0 framework: a systematic literature review identifying the current trends and future perspectives. Process Saf. Environ. Prot. **117**, 408–425 (2018)
22. Müller, J.M., Kiel, D., Voigt, K.-I.K.-I.: What drives the implementation of Industry 4.0? The role of opportunities and challenges in the context of sustainability. Sustainability **10**, 247 (2018)
23. Porter, M.E., Heppelmann, J.E.: How smart, connected products are transforming competition. Harv. Bus. Rev. **92**, 64 (2014)
24. Díaz-Chao, Á., Ficapal-Cusí, P., Torrent-Sellens, J.: Environmental assets, industry 4.0 technologies and firm performance in Spain: a dynamic capabilities path to reward sustainability. J. Clean. Prod. **281**, 125264 (2021). https://doi.org/10.1016/j.jclepro.2020.125264
25. Mattera, M., Gava, L.: Facing TBL with IoT: creating value and positively impacting business processes. Soc. Responsib. J. (2021)
26. Godina, R., Ribeiro, I., Matos, F., Ferreira, B.T., Carvalho, H., Peças, P.: Impact assessment of additive manufacturing on sustainable business models in industry 4.0 context. Sustainability **12**(17), 7066 (2020). https://doi.org/10.3390/su12177066
27. Lardo, A., Mancini, D., Paoloni, N., Russo, G.: The perspective of capability providers in creating a sustainable I4.0 environment. Manag. Decis. **58**(8), 1759–1777 (2020). https://doi.org/10.1108/MD-09-2019-1333
28. Teixeira, G.F.G., Canciglieri Junior, O.: How to make strategic planning for corporate sustainability? J. Clean. Prod. **230**, 1421–1431 (2019)
29. Piccarozzi, M., Aquilani, B., Gatti, C.: Industry 4.0 in management studies: a systematic literature review. Sustainability **10**(10), 3821 (2018). https://doi.org/10.3390/su10103821
30. Nosratabadi, S., Mosavi, A., Shamshirband, S., Zavadskas, E.K., Rakotonirainy, A., Chau, K.W.: Sustainable business models: a review. Sustainability **11**, 1–30 (2019)
31. Allais, R., Roucoules, L., Reyes, T.: Governance maturity grid: a transition method for integrating sustainability into companies? J. Clean. Prod. **140**, 213–226 (2017)
32. Gouvinhas, R.P., Reyes, T., Naveiro, R.M., Perry, N., Filho, E.R.: A proposed framework of sustainable self-evaluation maturity within companies: an exploratory study. Int. J. Interact. Des. Manuf. (IJIDeM) **10**(3), 319–327 (2016). https://doi.org/10.1007/s12008-016-0322-7
33. Gaziulusoy, A.I.: A critical review of approaches available for design and innovation teams through the perspective of sustainability science and system innovation therories. J. Clean. Prod. **107**, 366–377 (2015)

34. Romero, D., Molina, A.: Towards a sustainable development maturity model for green virtual enterprise breeding environments. IFAC Proc. Volumes **19**(3) (2014)
35. Murillo-Luna, J.L., Garcés-Ayerbe, C., Rivera-Torres, P.: Barriers to the adoption of proactive environmental strategies. J. Clean. Prod. **19**(13), 1417–1425 (2011)
36. Seuring, S., Yawar, S.A., Land, A., Khalid, R.U., Sauer, P.C.: The application of theory in literature reviews – illustrated with examples from supply chain management. Int. J. Oper. Prod. Manag. **41**(1), 1–20 (2020)
37. Durach, C., Kembro, J., Wieland, A.: A new paradigm for systematic literature reviews in supply chain management. J. Supply Chain Manag. **53**(4), 67–85 (2017)
38. Tranfield, D., Denyer, D., Smart, P.: Towards a methodology for developing evidence-informed management knowledge by means of systematic review. Br. Acad. Manag. **14**(3), 207–222 (2003)

Pay-Per-X Business Models for Equipment Manufacturing Companies: A Maturity Model

Joonas Schroderus[1], Sameer Mittal[1,2], Karan Menon[1](✉), Lester Allan Lasrado[3], and Hannu Kärkkäinen[1]

[1] Tampere University, Tampere, Finland
{sameer.mittal,karan.menon,hannu.karkkainen}@tuni.fi
[2] JK Lakshmipath University, Jaipur, India
sameer.mittal@jklu.edu.in
[3] Kristiania University College, Oslo, Norway
lesterallan.lasrado@kristiania.no

Abstract. Pay-per-X (PPX) business models are models where the ownership of the product is not transferred to the customer, but the customer has a right to use the product. The implementation of PPX business models can be time-taking, and complex because many company functions need to interact with each other, especially for equipment manufacturing industries (EMIs). The aim of this paper is to design a PPX maturity model for EMIs using a systematic maturity model development process. We present a PPX maturity model for EMIs with 7 dimensions, 19 sub-dimensions, 5 maturity levels, and relevant boundary conditions. This PPX maturity model is developed empirically with academic and industry experts from the maturity model and PPX perspective. The developed PPX maturity model will allow the EMIs to assess their current as-is situation in the most critical areas of PPX implementation and formulate a roadmap toward the implementation.

Keywords: Pay-per-X · Pay-per-use · Pay-per-outcome · Pay-per-output · Business models · Maturity model

1 Introduction

Pay-per-X (PPX) business models are models where the ownership of the product is not transferred to the customer, but the customer has a right to use the product. These PPX business models can be divided into pay-per-use, pay-per-output and pay-per-outcome business models [1, 2]. In pay-per-use business models, the customer pays for the time units of the machine used by the customer (e.g. per hour, per day) [2], whereas in pay-per-output business models the customer pays for the number of units the machine makes (e.g. per 100 units) [2]. In pay-per outcome business models, the focus is on achieving a specified outcomes or added value such as energy savings, rather than on a set of prescribed specifications [2].

The implementation of PPX business models can be time-taking, complex because many company functions need to interact with each other (for e.g. operations, analytics

Published by Springer Nature Switzerland AG 2023
F. Noël et al. (Eds.): PLM 2022, IFIP AICT 667, pp. 66–75, 2023.
https://doi.org/10.1007/978-3-031-25182-5_7

and sales need to work with each other to achieve PPX). As maturity models have been widely and quite successfully applied in implementations of different, complex systems, and strive for representing an evolution path towards the desired stages of maturity [3], they can also help with the implementation of business models [4]; see also [5]. Consequently, maturity models can help with the implementation of PPX business models, which are advanced and complex from both the technical and business-related implementation perspectives. However, despite the vast amount of maturity models in existence [3], maturity models in the implementation of PPX business models, and specifically in the context of business to business (B2B) equipment manufacturing industries (EMIs), present a research gap addressed in this paper. As PPX business models relate closely to concepts such as Industry 4.0, servitization, digitalization and product-service systems, the literature review of this research was consequently conducted by reviewing and analyzing maturity models in these relevant fields. Therefore, the aim of this paper is to design a PPX maturity model in the context of B2B EMIs.

The remainder of the paper is structured so, that first we introduce the main concepts, and explain what maturity and PPX models are. Second, we present the methodological background and choices. Then, through analysis making use of PPX-related literature and maturity models in relevant areas such as Industry 4.0 and digitalization, we present the preliminary maturity model and associated conditions. Lastly, we present our main conclusions and future work.

2 Theoretical Background

2.1 Pay-Per-X Business Models

PPX are the advanced servitization BMs that focus towards offering the use, output and outcome of the equipment offered by the EMIs to the customer [6].Therefore, ownership of equipment in case of PPX does not belong to the customer. PPX BMs are further divided into pay-per-use, pay-per-output, and pay-per-outcome BMs [6]. If the customer pays for a certain number of hours for which it has deployed the EMI's equipment, then it is referred to as pay-per-use BMs [2, 6]. Whereas, if the customer pays for the output units manufactured by deploying the equipment, we refer to it as pay-per-output BMs [6]. If the equipment is deployed to reach a desired outcome such as saving on costs and energy, then we refer to it as a pay-per-outcome BM [6]. EMIs can associate themselves with two major advantages while offering PPX BMs. Firstly, they can create more market segments and thus generate higher revenues [6–8], and secondly, lower the risks of offering expensive equipment by sharing risks with other players like a third party or a financial institute [9–11].

2.2 Maturity Models

Maturity can be defined as "the state of being complete, perfect or ready" [12]. Maturity models are understood as normative theories that should guide the audience towards some desired outcomes [13]. Maturity models are often represented as stage fixed level models, stage continuous level models or a matrix structure in form of focus area models

[14]. According to [15], researchers have used maturity models to facilitate (i) self-assessment or third-party assessment (also known as descriptive), (ii) benchmarking or comparison (comparative), and (iii) provide a roadmap for continuous improvement (prescriptive) [16, 17] with one main purpose being providing a common language to facilitate discussion among stakeholders and thus provide a structure for prioritizing actions [14, 17], which is also the aim of pay-per-X maturity models (PPX-MM). The five core components namely (i) maturity level, (ii) dimensions, (iii) boundary conditions, (iv) path to maturity and (v) assessment that constitutes a maturity model as used in our paper are the same as described in prior studies [17, 18].

2.3 Pay-Per-X Maturity Models

The advanced PPX business models can provide equipment manufacturing industries (EMIs) with new ways of earning in the globally saturated product-centric industries [19]. However, implementing PPX business models can be difficult. As EMIs can provide complex and highly customized solutions to their customers, finding new ways of earning can be challenging due to e.g. changing technologies, routines and business processes in general [20]. If companies lack understanding, these changes can have a negative effect on the performance of the equipment manufacturer [21], and potentially lead to e.g. difficulties in achieving expected returns from the new, service-oriented PPX business models [22]. Overall, the process of implementing business models is consequently still relatively underdeveloped [4, 23] and many business models fail during implementation [24]. Despite this fact, only little research has been done on standardized methods to assess and compare the maturity of business models [25].

Consequently, successful PPX business model implementation requires a systematic approach, that helps the equipment manufacturers to define the operational capabilities needed in the change process [26, 27]. For this, maturity models that have already been widely accepted and used for example in IT management [16], are also being recognized as prospective tools in other areas such as manufacturing and services [28] as well as more complex areas such as product-service systems [29]. Therefore, the argument in this research is that a maturity model can be developed to assess the PPX implementation readiness of EMIs, serving them as a starting point for assessing the companies' current as-is situation in the most critical areas needed in PPX business models. [30] Moreover, the maturity model provides a starting point for the development of a future roadmap towards the PPX business model implementation as well as overall helps the equipment manufacturers to define and reach the desired outcomes as efficiently as possible by providing them a common language within the company [3, 16, 29].

3 Methodology

The methodology deployed in the current research is based on action design research (ADR), developed by [31] (see Fig. 1). ADR combines the different action and design research approaches to emphasize the importance of both iteratively creating an artifact i.e., in this case the maturity model, as well as putting the artifact under development in the organizational context as well as theory. The process can be divided into 4 phases, which are the problem formulation phase where the problem and theory-based artifact are formulated; building, intervention, and evaluation (BIE) phase where the theory-based artifact is concurrently and iteratively developed in cooperation with the organizations; reflection and learning phase where the problem and artifact relevance are evaluated and finally the formalization of learning phase, where the outcome of the full process is generalized [31].

In the problem formulation phase of the ADR and development of the PPX-MM, the research focused on identifying and conceptualizing the research opportunity, formulating the research questions and more concretely, defining the scope of the PPX-MM. Once the scope was clear, the research included a literature review of both maturity model and PPX-related literature, to ensure there is relevance in the research and we can proceed with the development of a new PPX-MM. In terms of designing the maturity model, the research also utilized the overall maturity model development process created by [3, 12] maturity model design decision criteria helping to define the scope of the PPX-MM. Since scientific literature already consisted of a study [17] that followed ADR approach and focused on developing the dimensions and maturity levels of the PPX-MM. Therefore, we considered the seven dimensions and five maturity levels from an existing study [17].

The building, intervention, and evaluation phase of the development of the PPX-MM is where most of the actual design of the model occurred. However, as [31] state, this does not mean that the other stages are not overlapping or concurrent. The BIE cycle also started to move towards organizations: to assess the theory-based dimensions or the initial "artifact", the PPX-MM went through three series of expert interviews. The PPXBM was first taken to three academic experts in maturity models, and to follow that it was discussed with two academic PPX business model experts. Finally, the relevance of proposed PPX-MM was discussed into the organizational context, i.e., the partner companies and two end users of the PPX-MM. After the three rounds of expert and company workshops, the PPX-MM was again refined and developed within the research team according to findings from the expert and end user feedback, allowing a fully developed PPX-MM with dimensions, sub-dimensions, and unique level descriptions. With the help of the well-defined and validated dimensions, sub-dimensions, and maturity levels, we identified the boundary conditions with the help of a discussion between our research team.

However, even with the full PPX-MM, ADR emphasizes concurrent reflection and learning in the development of the artifact. Consequently, in addition to ensuring we reflected on the design of the PPX-MM against our formulated problem and research questions, we turned the maturity model into a web-based maturity assessment tool to test the validity of the fully developed PPX-MM. This tool was then used to once more test the whole maturity model in the organizational context through three partner companies in Finnish equipment manufacturing pilot company, ensuring we test the validity of the derived maturity model. However, the scope of the current paper is limited to the development of PPX-MM. Therefore, results of assessment, and the reflection and learning that happened with the assessment has not been shared in the current research. Similarly, the formalization of learning has not been done as it was out of the scope of the current paper. The ADR approach has limitation as the dimensions,

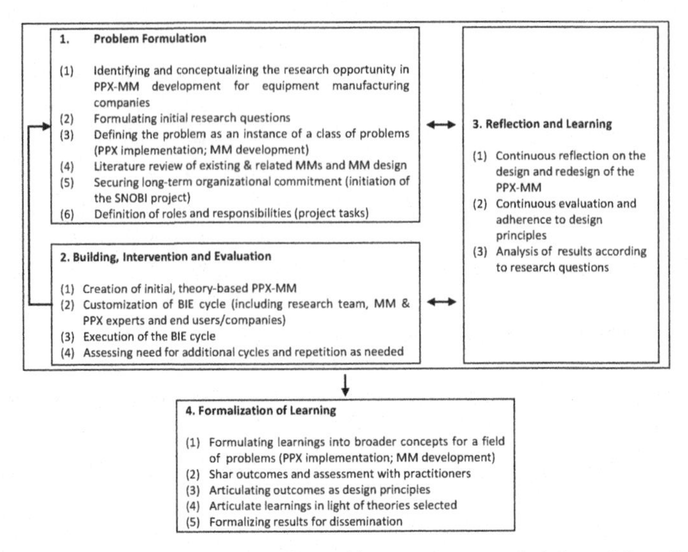

Fig. 1. Action design research followed in current research (adapted from [31])

and maturity levels can be defined in different ways from the perspective of different academic experts. However, since during the formalization and learning process the fully developed PPX-MM went through experts from Finnish equipment manufacturing pilot company, therefore, we were able to overcome this limitation.

4 Results

4.1 Pay-Per-X Maturity Model for Equipment Manufacturing Industries

In the current paper we develop and validate a full PPX-MM consisting of five maturity levels (initial, repeatable, defined, advanced, and optimized) and seven dimensions (organizational governance, strategy, risk management, competences & culture, product lifecycle processes, product & production technology, and data analytics). These maturity levels and dimensions were proposed in an existing study [17]. In the current research, our experts helped in validation of the maturity levels and the dimensions. Furthermore, with the help of several rounds of iteration of PPX-MM and validation of maturity levels and dimensions we were able to identify the boundary conditions. Additionally, we also assessed the entire PPX-MM with three Finnish equipment manufacturing industries. Below are the results of our study.

Organizational governance considers how the standards, rules and regulations are followed towards implementing PPX BMs, and who takes the responsibilities towards implementing the PPX BMs [17]. In the earlier study [17], organizational governance was suggested to have system, people, and data & information as its sub-dimensions. However, with the help of expert interviews we modified system governance to operational governance, as this sub-dimension deals with the operations performed in the organization. The boundary conditions corresponding to the organizational governance sub-dimensions show how standards and regulations are absent at initial level whereas at optimized level they are fully implemented across the organization (see Fig. 2).

Strategy deals with how the PPX related activities can be planned to align them towards vision of the company, and consists of business strategy, strategic alignment, and resource allocation as its sub-dimensions [17]. With the help of expert interviews, we added more clarifications to sub-dimensions. Business strategy is about how the overall business logic and plan is focused towards offering PPX. Whereas, resource allocation is how resources are efficiently allotted to offer PPX, and finally, strategic alignment is about how overall company strategy is focused towards offering PPX. The boundary conditions corresponding to the strategy sub-dimensions show any kind of strategic initiative is absent at initial level and finally, at optimized level the strategies are fully implemented and unified across the organization (see Fig. 2).

Risk management considers the activities and competences required to mitigate the risks related to offering PPX, its sub-dimensions are business risks, operational risks, and IT risks [17]. With the help of our expert suggestions, we identified that IT risks consists of broad category that can overlap with business and operational risks, therefore we changed IT risks to cybersecurity risks. The boundary conditions corresponding to the risk management sub-dimensions show how risk mitigation is absent at the initial level and finally, at optimized level the risks are being identified and mitigated (see Fig. 2).

Competences and culture dimension deals with the critical competences required for offering PPX BMs, e.g., co-creation (with customers), design of- process, product, and service engineering [17]. Whereas the culture part deals with the collaboration, sharing knowledge and attitude towards accepting PPX BMs in the organization [17]. With the help of our experts, we identified overlaps between leadership commitment (part of this dimension) and strategy dimensions. As a result, we excluded the leadership commitment that was earlier a part of this dimension. The boundary conditions corresponding to the competence sub-dimension suggests absence of PPX competences and no cooperation among the different units of a company at the initial level, whereas all PPX competences are present with support and cooperation at the optimized level (see Fig. 2).

Product lifecycle processes dimension deals with the process required at the beginning, middle, and end of the product lifecycle from the perspective of offering PPX BMs, e.g., product engineering, service design, sales, and logistics [17].

With the help of expert suggestions, we were able to redefine this dimension in terms of the processes involved in the three phases of a product lifecycle. Similarly, earlier we used tasks to define this dimension, however, after discussion with experts' tasks were changed to processes to make this dimension clearer. The boundary conditions corresponding to this dimension are related with absence of any PPX related beginning, middle-, and end- of product lifecycle processes at the initial level, whereas all the PPX related processes are present and continuously improved at the optimized level (see Fig. 2).

Product & production technology dimension consists of three sub-dimensions related to the technologies that help in optimizing the risks and benefits while offering PPX BMs, smart product & factory focus on the hardware and software technologies; connectivity deals with M2M and internet communication, and cloud deals with information access

Dimension	Subdimension	Maturity Level				
		1. Initial	2. Experimenting	3. Defined	4. Advanced	5. Optimized
Organizational Governance	Operational Governance	No PPX-specific operational governance.	Operational PPX architecture requirements identified with ad hoc implementation and development.	Necessary operational PPX architecture requirements are documented and related governance measures are standardized.	Operational PPX architecture requirements are defined and compliance is systematically monitored through related key performance indicators.	Operational PPX governance is integrated across company with best practices in place.
	People Governance	No PPX-specific roles or responsibilities related to PPX business model(s) defined.	Responsibilities related to PPX are identified with ad hoc implementation and development.	Necessary roles and responsibilities for PPX business model(s) are documented, defined and systematically governed.	PPX-related roles and responsibilities are defined with systematic performance monitoring through defined standards and key performance indicators.	Roles and responsibilities related to PPX are optimized and defined with respect to all company activities.
	Data & Information Governance	No set rules for PPX data & information governance.	PPX data & information governance requirements are identified with ad hoc implementation and development.	Necessary data governance requirements are documented and standardized, with data storage infrastructure defined in production.	Data & information governance requirements are defined, with compliance systematically monitored and developed through defined key performance indicators.	Data & information governance measures are optimized and integrated across company.
Strategy	Business Strategy	No defined business strategy for PPX business model(s).	Strategy for PPX business model(s) is experimental with ad hoc implementation and development.	Strategy for PPX business model(s) is defined and documented.	PPX is strategy is defined and continuously developed through defined key performance indicators.	PPX business strategy is fully developed and integral part of the corporate strategy.
	Resource Allocation	No plan for allocating resources towards PPX business model(s).	Basic PPX resource requirements are identified with ad hoc assignment.	Procedures for allocating resources towards PPX business model(s) are standardized, allowing systematic resource allocation for specific PPX activities.	PPX resource requirements are identified and documented across company, allowing systematic resource management and prioritization at an organizational level.	PPX resource allocation follows best practices and is optimized across company.
	Strategic Alignment	No strategic alignment between PPX and other strategic objectives.	Limited understanding of PPX and its relationship to other strategic objectives with ad hoc alignment practices.	Strategic understanding and objectives are shared between relevant business.	Strategic objectives are shared across company with compliance and performance monitored through common key performance indicators.	Full strategic alignment allowing optimization and development of common strategic goals across company.
Risk Management	Business Risks	No PPX-related business risk management.	PPX-related business risks are acknowledged with ad hoc management practices.	PPX-related business risk are documented, with systematic and defined risk management practices in place.	PPX-related business risk management is systematic and monitored, allowing predictive risk management.	PPX-related business risk management is proactive, with continuous improvement and optimization of risk management practices.
	Operational Risks	No PPX-related operational risk management.	PPX-related operational risks are acknowledged with ad hoc management practices.	PPX-related operational risk are documented, with systematic and defined risk management practices in place.	PPX-related operational risk management is systematic and monitored, allowing predictive risk management.	PPX-related operational risk management is proactive, with continuous improvement and optimization of risk management practices.
	Cybersecurity Risks	No PPX-related cybersecurity risk management.	PPX-related cybersecurity risks are acknowledged, with ad hoc management practices.	PPX-related cybersecurity risk are documented, with systematic and defined risk management practices in place.	PPX-related cybersecurity risk management is systematic and monitored, allowing predictive risk management.	PPX-related cybersecurity risk management is proactive, with continuous improvement and optimization of risk management practices.
Competences & Culture	Competences	No identified any PPX-related competences.	PPX-related competences are acknowledged with ad hoc acquisition.	Basic PPX-related competence requirements are defined and documented, allowing systematic competence acquisition.	PPX-related competences are acquired as well as developed systematically.	All PPX-related competences can be acquired and managed proactively.
	Culture	Culture is product-oriented, with no cooperation between different business units.	Organizational culture supports experimentation with limited & ad hoc cooperation between some business units.	Organizational culture supports innovation and is open towards PPX, with frequent collaboration between some business units.	Organizational culture is committed to PPX business model(s) with common incentives, with frequent collaboration across all related business units.	Organizational culture fully supports PPX, with complete trust and open communication at all organizational levels and relevant business units.
Product Lifecycle Processes	Beginning of Life Processes	No identified beginning of life processes for PPX business model(s).	PPX-related beginning of life processes are identified with ad hoc implementation.	PPX-related beginning of life processes are defined and systematically implemented for specific project(s).	PPX-related beginning of life processes are defined and implemented across company with systematic management through defined metrics.	PPX-related beginning of life processes are optimized and continuously improved across company.
	Middle of Life Processes	No identified middle of life processes for PPX business model(s).	PPX-related middle of life processes are identified with ad hoc implementation.	PPX-related middle of life processes are defined and systematically implemented for specific project(s).	PPX-related middle of life processes are defined and implemented across company with systematic management through defined metrics.	PPX-related middle of life processes are optimized and continuously improved across company.
	End of Life Processes	No identified end of life processes for PPX business model(s).	PPX-related end of life processes are identified with ad hoc implementation.	PPX-related end of life processes are defined and systematically implemented for specific project(s).	PPX-related end of life processes are defined and implemented across company, with systematic management through defined metrics.	PPX-related end of life processes are optimized and continuously improved across company.
Product & Production Technology	Smart Product & Factory	No machine data collection capabilities for PPX business model(s).	PPX data collection capabilities are tested in machine(s), allowing contract-specific, ad hoc data collection from customer(s).	PPX data collection technologies are standardized, with systematic data collection from customer machine.	PPX data collection capabilities is integrated in all machines, with performance monitored through defined key performance indicators.	Production technology fully supports data-based products for PPX, with performance optimized through cost minimization and efficiency.
	Connectivity	No connectivity between machines or production processes for PPX business model(s).	PPX product- and production-related connectivity technologies are experimental and non-standardized.	PPX product- and production-related connectivity technologies are standardized and we have access to customer(s)' machine.	PPX product- and production-related connectivity technologies are standardized and monitored through defined quality control measurements for development needs.	PPX product- and production-related connectivity technologies are optimized and continuously improved, allowing 2-way/remote connection and control of machines.
Data Analytics	Data Access	No access to PPX data.	PPX data is identified, but siloed and accessed manually & ad hoc.	PPX data is defined, enabling continous data flow and basic automation with online access.	PPX data is systematically accessed, with related key performance indicators defined and utilized in quality control.	All PPX data can be accessed, with cost efficient, high-performing and optimized best practices in place.
	Data Analysis	No PPX data analysis.	PPX data analysis is unstructured, allowing descriptive analysis and basic monitoring.	PPX data analysis capabilities are defined, enabling diagnostic analysis & recommendations and manual machine tuning.	PPX data analysis is systematic and predictive, with performance monitored through defined key performance indicators.	PPX data analysis is prescriptive/self-learning, with automation and self-adjusting capabilities.
	Data Utilization	PPX data not utilized in decision-making.	PPX data utilized for awareness purposes in basic reporting with ad hoc utilization in decision-making.	PPX data established as an asset and utilized to support decision-making.	PPX data utilzied broadly in the development of overall company strategy, with performance monitored through defined key performance indicators.	PPX data is considered as central to company strategy and operations development.

Fig. 2. Pay-per-X maturity model

applications, platforms, and databases [17]. With the help of expert suggestions, we identified overlaps between the cloud sub-dimension and data access (present in data analytics), therefore, the cloud sub-dimension was removed from this dimension. The boundary conditions corresponding to this dimension are related to absence of hardware, software, and connectivity from PPX perspective at the initial level, and fully supported hardware, software, and connectivity to offer PPX BMs at the optimized level (see Fig. 2).

Based on the functionality of the methods and tools the data analytics dimensions consist of three sub-dimensions, i.e., data collection, transformation & processing, visualization, and decision making [17]. With the help of expert interviews, we identified that data access, data analysis and data utilization are better terms to describe the sub-dimensions of data analytics. As a result, we renamed our sub-dimensions. The boundary conditions corresponding to this dimension are related with the absence of data access, analysis and utilization at the initial level, and full access, analysis, and utilization at the optimized level.

4.2 Pre-evaluation

We followed the recommendations suggested by previous studies [3], by pre-evaluation of our developed PPX-MM with the 3 criteria, i.e., i) comprehensiveness, ii) consistency, and iii) problem adequacy. During the evaluation of our PPX-MM 3 different academic experts were present from the area of MM and PPX, and additionally, three industry experts were from PPX were also present.

1. Comprehensiveness: Overall, experts found our PPX-MM to be very comprehensive, as it was able to cover various aspects that are required while offering PPX BMs. However, with the help of expert suggestions we divided product lifecycle processes into beginning-, middle-, and end of life processes. Similarly, leadership sub-dimension was removed as it overlapped with culture, and sub-dimension cloud was removed as it overlapped with data access.
2. Consistency: Overall, experts found our PPX-MM to be consistent. However, with the help of their suggestions we renamed various sub-dimensions, e.g., data collection became data access, and IT security became cybersecurity. Similarly, the sub-dimension system governance became operational governance.
3. Problem adequacy: Furthermore, we also iterated our PPX-MM several times leading to an improved version from understanding and application context, e.g., our description of boundary conditions was shortened, and reference levels were described with the help of quantitative brackets.

5 Discussion and Conclusions

The current research caters towards the need of guiding EMIs that plan to offer PPX BMs by developing a MM. We conducted focus group interviews with both academic (MM and PPX) and industry experts to check the evaluation criteria (i.e., comprehensiveness, consistency, and problem adequacy) and followed the meticulous ADR approach proposed in literature [3] to establish the comprehensive dimensions (and their sub-dimensions), and maturity levels available in literature [17]. The boundary conditions were identified with the help of scientific literature and were also validated and improved with the help of focus group interviews [3]. Thus, we were able to propose relevant and relatable boundary conditions in MM. Overall, we developed a PPX MM from the perspective of EMIs.

PPX BMs can help EMIs in generating more revenues by increasing their customer segments [6–8] and lowers the risks of offering PPX BMs [9–11]. However, the existing studies in scientific literature [32] do not cater towards this need. In the current research, we add to the descriptive and prescriptive knowledge on PPX for EMIs from the MM perspective, and contributed to the existing knowledge [25–27]. Overall, we offer the corresponding requirements (boundary conditions) for different dimensions and maturity models that an EMI should fulfil to offer PPX BMs. Additionally, we identified that offering PPX BMs, needs enablers from both technology and management perspective.

Furthermore, the process of developing PPX MM we were also able to develop an online PPX readiness assessment tool that manager of an EMI can deploy for assessing their readiness towards offering PPX BMs. We followed systematic ADR suggested by existing studies [3], and during the process of validating the MM we also used our tool for assessment; however, the results of the assessment are beyond the scope this article. Future research can also focus towards identifying the required technologies and management strategies that will help EMIs in meeting the requirements corresponding to various dimensions and maturity levels proposed in our PPX MM.

References

1. Menon, K.: Industrial internet enabled value creation for manufacturing companies: a data and information management perspective. Tampere University (2020)
2. Uski, V.-M., et al.: Review of PPX business models: adaptability and feasibility of PPX models in the equipment manufacturing industry. In: Canciglieri Junior, O., Noël, F., Rivest, L., Bouras, A. (eds.) PLM 2021. IFIP Advances in Information and Communication Technology, vol. 639, pp. 358–372. Springer, Cham (2022). https://doi.org/10.1007/978-3-030-94335-6_26
3. Becker, P.D.J., Knackstedt, D.R., Pöppelbuß, D.-W.I.J.: Developing maturity models for IT management. Bus. Inf. Syst. Eng. **1**, 213–222 (2009)
4. Poandl, E.M., Vorbach, S., Müller, C.: A maturity assessment for the business model of start-ups. In: ISPIM Conference Proceedings, pp. 1–9. The International Society for Professional Innovation Management (ISPIM), Manchester (2019)
5. Broekhuizen, T.L.J., Bakker, T., Postma, T.J.B.M.: Implementing new business models: what challenges lie ahead? Bus. Horiz. **61**, 555–566 (2018)
6. Baines, T., Lightfoot, H.W.: Servitization of the manufacturing firm: exploring the operations practices and technologies that deliver advanced services. Int. J. Oper. Prod. Manag. **34**, 2–35 (2014)
7. Oliva, R., Kallenberg, R.: Managing the transition from products to services. Int. J. Serv. Ind. Manag. **14**, 160–172 (2003)
8. Eggert, A., Hogreve, J., Ulaga, W., Muenkhoff, E.: Revenue and profit implications of industrial service strategies. J. Serv. Res. **17**, 23–39 (2014)
9. Porter, M.E., Heppelmann, J.E.: How smart, connected products are transforming companies. Harv. Bus. Rev. **93**, 96–114 (2015)
10. Smith, L., Maull, R., Ng, I.C.L.: Servitization and operations management: a service dominant-logic approach. Int. J. Oper. Prod. Manag. **34**, 242–269 (2014)
11. Erkoyuncu, J.A., Roy, R., Shehab, E., Cheruvu, K.: Understanding service uncertainties in industrial product–service system cost estimation. Int. J. Adv. Manuf. Technol. **52**, 1223–1238 (2011)
12. Mettler, T., Rohner, P., Winter, R.: Towards a Classification of maturity models in information systems. Presented at the (2010)

13. Damsgaard, J., Scheepers, R.: Managing the crises in intranet implementation: a stage model. Inf. Syst. J. **10**, 131–149 (2000)
14. Lasrado, L.A., Vatrapu, R., Andersen, K.: Maturity models development in is research: a literature review. Selected Papers of the IRIS, no. 6 (2015)
15. Lasrado, L.A.: Set-Theoretic approach to maturity models. Copenhagen Business School (CBS) (2018)
16. Poeppelbuss, J., Niehaves, B., Simons, A., Becker, J.: Maturity models in information systems research: literature search and analysis. Commun. Assoc. Inf. Syst. **29**, 27 (2011)
17. Schroderus, J., Lasrado, L.A., Menon, K., Kärkkäinen, H.: Towards a pay-per-X maturity model for equipment manufacturing companies. Proc. Comput. Sci. **196**, 226–234 (2022). https://doi.org/10.1016/j.procs.2021.12.009
18. Lasrado, L., Vatrapu, R., Andersen, K.N.: A set theoretical approach to maturity models: guidelines and demonstration. In: ICIS 2016 Proceedings (2016)
19. Kindström, D.: Towards a service-based business model – key aspects for future competitive advantage. Eur. Manag. J. **28**, 479–490 (2010)
20. Kohtamäki, M., Parida, V., Oghazi, P., Gebauer, H., Baines, T.: Digital servitization business models in ecosystems: a theory of the firm. J. Bus. Res. **104**, 380–392 (2019)
21. Zhang, W., Banerji, S.: Challenges of servitization: a systematic literature review. Ind. Mark. Manag. **65**, 217–227 (2017)
22. Gebauer, H., Fleisch, E., Friedli, T.: Overcoming the service paradox in manufacturing companies. Eur. Manag. J. **23**, 14–26 (2005)
23. Berends, H., Smits, A., Reymen, I., Podoynitsyna, K.: Learning while (re)configuring: Business model innovation processes in established firms. Strateg. Organ. **14**, 181–219 (2016)
24. Christensen, C.M., Bartman, T., van Bever, D.: The hard truth about business model innovation. MIT Sloan Manag. Rev. **58**, 31–40 (2016)
25. Mateu, J.M., March-Chorda, I.: Searching for better business models assessment methods. Manag. Decis. **54**, 2433–2446 (2016)
26. Teece, D.J.: Explicating dynamic capabilities: the nature and microfoundations of (sustainable) enterprise performance. Strateg. Manag. J. **28**, 1319–1350 (2007)
27. Menon, K., Mittal, S., Kärkkäinen, H., Kuismanen, O.: Systematic steps towards concept design of pay–per-X business models: an exploratory research. In: Canciglieri Junior, O., Noël, F., Rivest, L., Bouras, A. (eds.) PLM 2021. IFIP Advances in Information and Communication Technology, vol. 640, pp. 386–397. Springer, Cham (2022). https://doi.org/10.1007/978-3-030-94399-8_28
28. Wendler, R.: The maturity of maturity model research: a systematic mapping study. Inf. Softw. Technol. **54**, 1317–1339 (2012)
29. Neff, A.A., Hamel, F., Herz, T., Uebernickel, F., Brenner, W., vom Brocke, J.: Developing a maturity model for service systems in heavy equipment manufacturing enterprises. Inf. Manage. **51**, 895–911 (2014)
30. Silva, C., Ribeiro, P., Pinto, E.B., Monteiro, P.: Maturity model for collaborative R&D university-industry sustainable partnerships. Proc. Comput. Sci. **181**, 811–817 (2021)
31. Sein, M., Henfridsson, O., Purao, S., Rossi, M., Lindgren, R.: Action design research (2011)
32. Häckel, B., Huber, R., Stahl, B., Stöter, M.: Becoming a product-service system provider – a maturity model for manufacturers. In: Ahlemann, F., Schütte, R., Stieglitz, S. (eds.) WI 2021. LNISO, vol. 46, pp. 169–184. Springer, Cham (2021). https://doi.org/10.1007/978-3-030-86790-4_13

Supporting the Development of Circular Value Chains in the Automotive Sector Through an Information Sharing System: The TREASURE Project

Paolo Rosa(✉) and Sergio Terzi

Department of Management, Economics and Industrial Engineering, Politecnico di Milano, Piazza L. da Vinci 32, 20133 Milan, Italy
paolo1.rosa@polimi.it

Abstract. Circular Economy (CE) and Circular Value Chains (CVCs) are two topics commonly addressed in literature. However, there is a big research gap in terms of how to practically transform linear value chains into circular ones and very few application cases. The intent of this paper is to propose a way to support the creation of circular value chains through the implementation of an information sharing system in the automotive sector. To this aim, the TREASURE project will be taken into account as a reference example. A preliminary analysis of the context shows a high innovation potential related to the introduction of a similar technology linking car parts suppliers and carmakers with End-of-Life (EoL) actors.

Keywords: Circular value chain · Circular economy · Product lifecycle management · Information sharing · Automotive sector

1 Introduction

Car electronics is one of the most valuable sources of Critical Raw Materials (CRMs) in cars (Andersson et al. 2019; Restrepo et al. 2017). A modern medium-sized car can embed up to 15 electronic systems on average and luxury cars can reach up to 50 among microcomputers and electronic components (Wang and Chen 2011). A statistic of the BMW® Corporation has shown that these systems can account for more than 30% of total vehicle cost (and more than 50% in luxury cars) (Wang and Chen 2013). From 2000 onwards, electronics saw an increased penetration in the automotive sector (Restrepo et al. 2019). A recent report (MarketsandMarkets 2021) quantified the automotive microcontrollers market in about $989.2 Million in 2017, with a projection to $1,886.4 Million by 2022, at a CAGR of 13.78%. That said, remarkable is the historical lack of interest of car manufacturers (and the whole automotive sector) towards the recovery of these valuable components from End-of-Life Vehicles (ELVs). Arguably, the complex set of barriers (e.g. regulatory, governance-based, market, technological, cultural, societal,

Published by Springer Nature Switzerland AG 2023
F. Noël et al. (Eds.): PLM 2022, IFIP AICT 667, pp. 76–85, 2023.
https://doi.org/10.1007/978-3-031-25182-5_8

gender, etc.) result in difficulties for companies to implement Circular Economy (CE) by limiting its potential benefits. All these data show as the sectorial transition towards CE seems to be far from its completion, even if car manufacturers are investing big capitals trying to shift their business towards more sustainable mobility concepts. Especially at End-of-Life (EoL) stage, there are still many issues to be solved to functionally recover materials from cars (e.g. reuse recovered materials for the same purpose they were exploited originally) and the dependency from natural resources when producing new cars (e.g. electric/hybrid/fuel cell -powered) is still too high. This mandatory systemic transformation requires all companies/sectors to redefine products lifecycles since the beginning, by adopting CE principles already before designing them. Considering together the wide number of barriers impacting on the automotive sector and the limited collaboration among actors involved in traditional automotive value chains, the transition towards CE cannot be reached so easily. This issue is related (especially) to two elements. From one side, Beginning-of-Life (BoL) and EoL stages are still unconnected from an information sharing perspective. Data about materials embedded in cars are spread on a plethora of strictly protected databases accessible only by authorized actors. This way, even if data on materials embedded in cars are known since many years, no one can exploit them (e.g. to optimize ELV management processes). From another side, even if ELV management processes are active in Europe since the sixties, none of the actors involved in these processes is available to share their knowledge with car makers or car part suppliers, given their unavailability to collaborate. So, both car makers and car part suppliers cannot improve their design practices to make cars easier to disassemble and recycle. Considering all these issues, this paper wants to propose a way to support the creation of Circular Value Chains (CVCs) through the implementation of an information sharing system in the automotive sector. To this aim, the TREASURE project will be taken into account as a reference example. The paper is organized as follows. In Sect. 2, the theoretical background is presented. In Sect. 3, the research methodology is described. In Sect. 4, the expected findings are evidenced. In Sect. 5, concluding remarks and future research avenues are offered.

2 Theoretical Background

2.1 Current Automotive Material Loop

Current ELV recycling of steel and Al from vehicles unavoidably mix an increasing set of alloys which, in turn, leads to degradation of materials quality and forces a downgrading of recycled materials only in low-value applications (e.g. reuse recycled Al in engine basements), sometime outside the automotive sector (e.g. reuse recycled steel in buildings). Predictable technological changes (e.g. widespread adoption of electric vehicles) may lead to even worst performances. Therefore, a separation of different alloys (e.g. through the dismantling of some components) can become unavoidable in the next future to ensure an adequate recycling of these metals (Løvik et al. 2014; Ohno et al. 2017). In addition, the issues of environmental impacts and risks related to primary supply are not restricted to precious metals only. Often, critical/scarce metals have (in relation to the total vehicle's mass), disproportionally large environmental impacts and potential risks associated with their primary supply. Considering some studies on the lifecycle energy impacts of car electronics (Cassorla et al. 2017), total life cycle impacts of vehicle's Electronic Control Units (ECUs) are estimated to be 22.7 GJ of embodied energy and 654 kgCO2eq GHG emissions. That said, an average of 50 ECUs per vehicle represents 5% (or 8.5 GJ/vehicle) of the total vehicle manufacturing embodied energy. In case of an increase in the electronics content (i.e. 76 ECUs per vehicle), the embodied energy can increase by ~1.6 times. From a vehicle's lifecycle energy perspective, the ECU use phase contributes to ~63% of total lifecycle impacts. This way, the energy impacts of new automotive electronics recycling can be comparable to that of the automotive steel recycling and materials recycling together. However, the current ELV directive does not stimulate the recovery of these metals since the recycling targets are defined by the total mass. An interesting option can be focusing on thermodynamic rarity, an indicator measuring the downcycling of materials. This way, not only quantity, but also quality of metals becoming functionally lost in ELV recycling processes can be monitored (Ortego et al. 2018).Considering that the ELV directive (European Union 2000) aims to improve the "environmental performance of all of the economic operators involved in the lifecycle of vehicles", the recovery of critical/scarce metals should be motivated.

2.2 Current EU Knowledge Base on Automotive Critical/Valuable Raw Materials

Vehicle manufacturers do hold large amounts of data, specifically in the International Material Data System (IMDS), but these data are generally not accessible to others. The importance of vehicles and data from manufacturers was also highlighted in a past recommendation from the H2020 ProSUM (Downes et al. 2017) and ORAMA (Bide et al. 2017) projects. Again, the ELV directive states that: "In order to facilitate the dismantling and recovery, in particular recycling of ELVs, vehicle manufacturers should provide authorized treatment facilities with all requisite dismantling information, in particular for hazardous materials". This is currently ensured through the International Dismantling Information System (IDIS). However, similar provision of information from vehicle manufacturers (e.g. drawing on the IMDS data), can enable an obligation to dismantle certain parts based on their content of critical/scarce metals. The necessary data systems seem, to a large extent, to already be in place. Obviously, it's important to carefully consider a financing model and potential unintended consequences related with a mandatory dismantling of components, firstly the additional costs for dismantlers. Studies have been already made to investigate the magnitude of these costs (IEEP et al. 2010). Results state that these costs can probably not be fully compensated by increased incomes from material recycling, mainly because of current low economic value of recovered critical/scarce metals. In addition, there is a risk of disrupting the existing market for used spare parts. Setting material-specific recovery yield targets should be another option, particularly for those critical/scarce metals not mentioned in the current ELV directive. However, the compliance with such targets would need to be constantly monitored (e.g. by measuring the concentrations of some metals in various output fractions) and compared with data about ELV materials composition provided by manufacturers (e.g. by exploiting information from the IMDS).

3 Research Methodology

3.1 Digitalizing the Automotive Value Chain

One of the most important activities in TREASURE will be the physical and virtual data compilation for acquiring the initial knowledge and establishing recommendations in terms of: 1) eco-design; 2) dismantling, 3) recycling and 4) consumers involvement. In addition, this information will be the starting data of the platform, furtherly populated in the future by users themselves. The following Fig. 1 shows the logic followed within the TREASURE project. For each selected car component: i) it will be identified a disassembly procedure, ii) it will be assessed its recyclability in terms of material contents (and related metallurgical procedures) and iii) it will be identified a set of KPI to measure/monitor circularity performances.

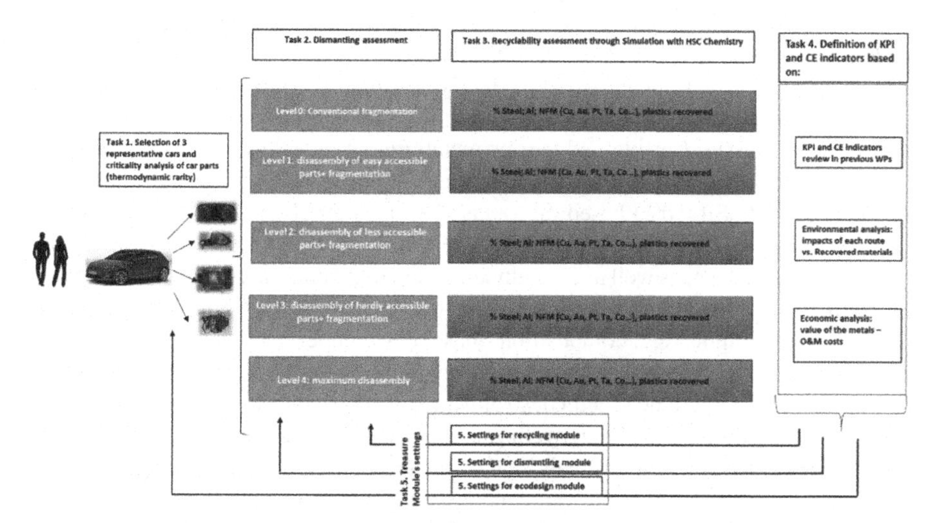

Fig. 1. Identification and selection of representative vehicles

Firstly, an identification and selection of three representative vehicles will be implemented. Such vehicles should represent different generations of cars covering as many different car parts and configurations as possible. Moreover, such car parts should be shared among as many vehicles from the same car maker as possible. Secondly, a thermodynamic criticality analysis will be undertaken. This analysis will allow to assess which car parts contain the most valuable metals from a physical point of view and classify them according to their scarcity in the earth's crust and the energy required to mine and refine them. To this aim, a specific database will be exploited. Such database includes material specifications of all car parts stemming from IMDS, a global data repository containing information on materials used by the automotive industry. Thirdly, disassembly levels will be established according to criticality of car parts and information from the specific database. The following Fig. 2 shows the different disassembly levels considered within the TREASURE project.

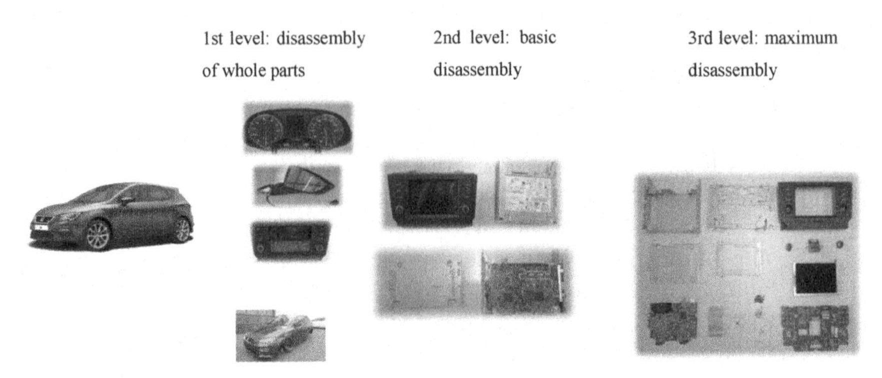

Fig. 2. Definition of the disassembly level of representative car parts (see Annex)

This way, three disassembly levels will be defined (see Fig. 2), together with a set of general recommendations (e.g. disassembly times and tools). Disassembled parts (for each level and for each vehicle), will be furtherly classified into different flows depending on their prevalent material composition. Fourthly, specific recycling routes consisting of physical processing/sorting and carrier metallurgical recycling infrastructures (e.g. pyro- and hydrometallurgical smelting and refining) will be assessed through simulation models (e.g. HSC Sim10®) based on a mass (but also energy/exergy) balance for all materials/metals/elements and compounds. Recycling/recovery rates will be available for whole parts/product as well as for individual elements/materials. This way, designers could link their decision to EoL stages. To that end, a detailed flowchart will be created, defining process inputs (e.g. composition, mass flow, temperature and pressure of sub-parts). Subsequently, the properties of each flow will be calculated by software. Fifthly, outflows of each unit will be specified, so closing the mass balance. This way, the behavior of all metals throughout the recycling process will be analyzed (together with the total quantity of recovered/lost metals) and translated in a set of recommendations about recyclability of car parts. These recommendations will also influence disassembly levels, disassembly activities and data generation/availability for disassembled parts. Different disassembly levels and approaches will be tested on optimized results from recycling and circularity (e.g. primaries required for dilution to produce alloys from recyclates). Finally, the so-constituted technological indicators will be integrated (together with CE assessment tools) in the TREASURE platform, so obtaining a complete circularity picture in the vehicle value chain. To this aim, different labels will be created depending on end users.

3.2 Designing, Developing and Integrating the TREASURE Platform

CE, by definition, involves all actors of a value chain, including BoL, MoL and EoL actors, processes and generating data. The TREASURE platform has the goal to provide the technical level of the project supporting data storage on which the different recommendation algorithms of the project run. Technically, the TREASURE platform relies on an open architecture in which the CE data about automotive electronics coming from different sources is stored in the TREASURE data lake and is combined with external data sources like public database (e.g. the Raw Materials Information System - RMIS). Data engineers will collaborate with end users to set up the correct data ingestion and data polishing and quality checks mechanism. Starting from initial findings, data scientists will be able to train algorithms and develop specific recommendation algorithm on top of the data lake. These recommendation algorithms are currently represented by an eco-design and circular (AI-based) advisory tool. However, they could be extended during and after the project depending on the TREASURE needs. The eco-design tool is aimed at helping car makers and car part manufacturers to identify the most critical (in terms of critical raw material contents) car parts in a vehicle, according to materials criticality levels and provide valuable recommendations based on improving the disassemblability and recyclability potential. The circular (AI-based) advisory tool is aimed at providing dismantlers and shredders with information regarding: i) critical and valuable car parts to be disassembled and ii) best recycling routes according to materials contained in car parts. In addition to these tools, a consumer involvement tool (based on a dedicated web platform) will provide a resource efficiency/CE label for awareness raising. The eco-design and circular (AI-based) tools will be fed through a specific database and will classify the different car parts of a vehicle according to their physical criticality (based on thermodynamic rarity). Based on composition and the recyclability insights, a set of recommendations (different for each end user) and a list of KPIs will be established for each car part in an automatic way. Information included in the algorithms will initially stem from the physical and virtual assessment carried out during initial activities, but it will be furtherly completed with feedbacks obtained from use cases. Even if this tool will be initially designed to work with a specific database, it could be adapted also to other automotive companies. The tool will be independent from internal systems, even if it will be fed with information from manufacturers. Confidentiality will be guaranteed

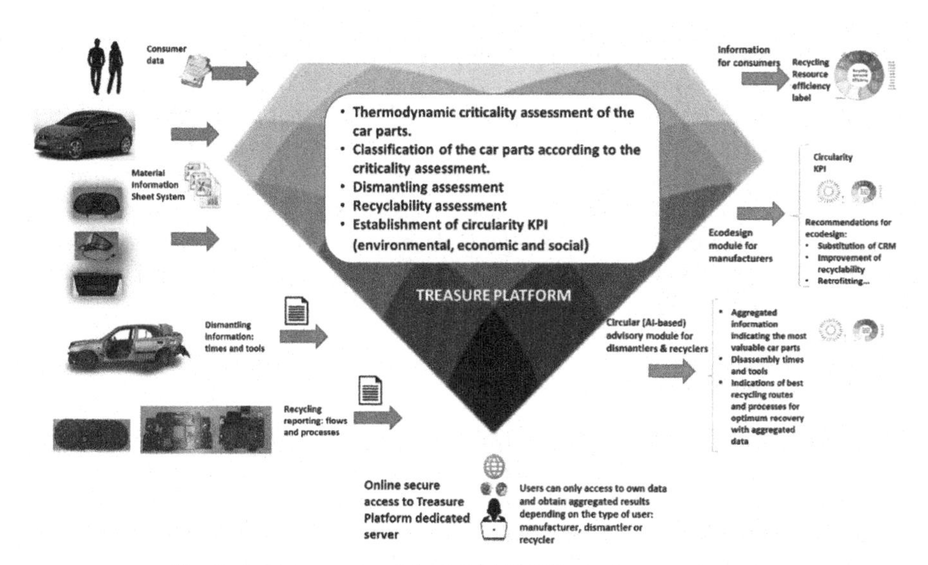

Fig. 3. Initial concept of the TREASURE solution (see Annex)

by a security access and by delivering only results with aggregated non-confidential data (see the previous figure) (Fig. 3).

4 Expected Findings

TREASURE wants to support the transition of the automotive sector towards CE trying to fill in the existing information gap among automotive actors, both at design and EoL stage, by both modifying the existing exploitation logics of international data spaces and trying to apply the concept of "cascade use" of resources (Kalverkamp et al. 2017). To this aim, a scenario analysis simulation tool dedicated to car electronics will be developed and tested with a set of dedicated demonstration actions. The, so implemented, scenario analysis simulation tool will have a multiple perspective. From one side, the TREASURE solution can assist both car parts suppliers and carmakers in assessing their design decisions in terms of circularity level, also considering the effects of their decisions on EoL processes (e.g. on car dismantlers and shredders operational performances and advanced metallurgical recycling processes). Vice versa, car dismantlers and shredders could benefit from the TREASURE solution by knowing about new design features of cars to be recycled in order to optimize their processes. Here, the TREASURE solution will exploit an already existing CE performance assessment methodology to measure and quantify CE-related performances through a set of dedicated KPIs. Specifically, economic KPIs will be identified through a Life Cycle Cost (LCC) analysis. Environmental KPIs will be identified through a Life Cycle Assessment (LCA), Material Flow Analysis (MFA) and Thermodynamic Rarity Assessment (an exergy-based analysis). Finally, social KPIs will be identified through a large-scale online ethnography of the CE movement in the automotive sector. From another side, the TREASURE project can act as an information hub, by exploiting data stored in the EU RMIS database. These data will be directly exploited by TREASURE to continuously monitor CE performances through a dedicated Circular Economy Performance Assessment (CEPA) methodology. About the EU RMIS database exploitation, an implicit hypothesis is that it will be updated with data both currently accessible by only re-known automotive actors (e.g. IDIS and IMDS) and stored in open generic databases (e.g. Eurostat, SCIP – Substances of Concern In Products). This way, data can be easily exploited also by Small and Medium Enterprises (SMEs) operating in other sectors in order to develop new businesses and value chains exploiting these data. The access to privately-owned DBs (e.g. IDIS and IMDS) will be processed according to the EU GDPR legislation.

4.1 Information Sharing System Logic

The TREASURE solution can be described as the sum of four main building blocks: 1) an eco-design tool, 2) a circular (AI-based) advisory tool, 3) a consumer involvement tool and 4) a web-based platform interacting with the previous tools and gathering information about valuable and critical raw materials embedded in cars by interconnecting with the RMIS. Even if all stakeholders will be interconnected through the platform, the information will be treated confidentially and that access to the platform by each stakeholder will be secure (Fig. 4).

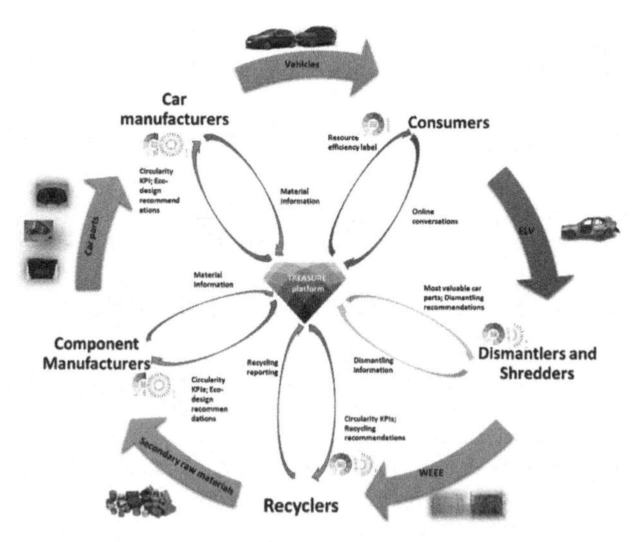

Fig. 4. Overall interaction of the TREASURE solution with different automotive value chain actors (see Annex)

When a new car electronic part must be designed, the eco-design tool will assist car parts suppliers and carmakers in selecting the most circular strategy basing on both general sustainability targets (e.g. substitute/reduce CRMs and/or plastics use) and a set of KPIs elaborated by the TREASURE platform quantifying positive and negative implications of CE on the whole automotive value chain. In order to calculate KPIs, an aggregation of data coming from both EoL actors (e.g. disassembly procedures, shredding issues, materials recovery strategies, etc.) and gathered from several databases (both private and public ones) will be executed by the platform. These KPIs will also ease the definition of design standards to facilitate disassembly and/or recyclability of cars/car parts. Vice versa, when a new car electronic model will enter the EoL stage, the AI-based advisory tool will assist car dismantler and car shredders in optimizing their processes from a CE perspective, by gathering information from the TREASURE platform. From one side, car dismantlers could gather information about 1) new disassembly procedures of specific vehicles/components (feeding from existing IMDS and IDIS platform, for instance), 2) critical car parts in a vehicle to be removed according to different perspectives (e.g. materials criticality, requirements from other EoL actors), 3) car parts circularity level (e.g. recyclability of materials, accessibility, disassemblability, hazardousness, etc.) and 4) new processes/equipments needed to recover valuable/critical raw materials from disassembled car parts. From another side, car shredders could be informed about 1) valuable materials to be separated and/or recovered before shipping materials to foundries, 2) new material separation/recovery techniques needed to recover valuable/critical raw materials from shredded cars. Finally, a consumer involvement tool will allow a bi-directional indirect interaction between automotive actors and final customers. From one side automotive actors could gather information from final customers about the impact of CE strategies on the market. From the other side, customers could increase their awareness about CE strategies adopted by automotive actors through a set of graphical indexes reporting the circularity level of their cars.

5 Conclusion

The ambition of TREASURE is offering to the European automotive sector interconnection among stakeholders of the value chain through an AI-based (secured-access) platform to foster communication for the proper implementation of CE practices and a set of new perspectives about innovative automotive components disassembly processes, materials recovery processes and secondary materials applications. This way, TREASURE wants to: i) guarantee a sustainable use of raw materials in the automotive sector, by reducing supply risks, ii) put in place CE practices in the automobile sector, acting as a showcase for the manufacturing industry, iii) offer a better environmental, economic and social performance of vehicles for users and stakeholders and iv) create new supply chains around ELV management, by focusing on the circular use of raw materials.

5.1 Contribution to Theory

- Define new information sharing channels in the automotive sector (both in onward and backward directions) through a secure access and ensuring confidentiality issues among stakeholders.
- Quantify positive and negative implications of CE in the automotive sector.
- Represent a set of success stories in three key value chains of the automotive industry (focusing on SMEs): 1) dismantlers/shredders, 2) recyclers and 3) manufacturers, practically demonstrating the benefits coming from the adoption of CE principles in the automotive sector.
- Integrate Key Enabling Technologies (KETs) for the efficient design of car electronics and subsequent disassembly and materials recovery.

5.2 Contribution to Practice

- Develop an information sharing system supporting the development of circular supply chains in the automotive sector.
- Close the material loop and reduce the dependency of the European automotive sector from raw materials supply.
- Perform a set of standardization and policy-related activities to make both industrial and politician actors aware about the current issues of the ELV management system.
- Increase the EU knowledge base on secondary raw materials.

References

Andersson, M., Ljunggren Söderman, M., Sandén, B.A.: Challenges of recycling multiple scarce metals: the case of Swedish ELV and WEEE recycling. Resour. Policy. **63**, 1–12 (2019). https://doi.org/10.1016/j.resourpol.2019.101403

Bide, T., et al.: Optimising quality of information in RAw MAterial data collection across Europe - deliverable 1.2: final analysis and recommendations for the improvement of statistical data collection methods in Europe for primary raw materials (2017)

Cassorla, P., Das, S., Armstrong, K., Cresko, J.: Life cycle energy impacts of automotive electronics. Smart Sustain. Manuf. Syst. **1**, 262–288 (2017). https://doi.org/10.1520/ssms20170009

Downes, S., et al.: Prospecting Secondary raw materials in the urban mine and mining wastes (ProSUM) recommendations report (2017)

European Union: Directive 2000/53/EC on end-of-life vehicles. Off. J. Eur. Union. 34–42 (2000)

IEEP, Ecologic, Arcadis, Umweltbundesamt, Bio Intelligence Service, VITO: Final report – supporting the thematic strategy on waste prevention and recycling (2010)

Kalverkamp, M., Pehlken, A., Wuest, T.: Cascade use and the management of product lifecycles. Sustain. **9**, 1–23 (2017). https://doi.org/10.3390/su9091540

Løvik, A.N., Modaresi, R., Müller, D.B.: Long-term strategies for increased recycling of automotive aluminum and its alloying elements. Environ. Sci. Technol. **48**, 4257–4265 (2014). https://doi.org/10.1021/es405604g

MarketsandMarkets: Automotive Microcontrollers Market by Application (Body Electronics, Chassis & Powertrain, Infotainment & Telematics, Safety & Security), Technology (ACC, Blind Spot Detection, Park Assist, TPMS), Vehicle, EV, Bit Size, Connectivity, and Region - Global F. https://www.marketsandmarkets.com/requestsampleNew.asp?id=162948952

Ohno, H., et al.: Optimal recycling of steel scrap and alloying elements: input-output based linear programming method with its application to end-of-life vehicles in Japan. Environ. Sci. Technol. **51**, 13086–13094 (2017). https://doi.org/10.1021/acs.est.7b04477

Ortego, A., Valero, A., Valero, A., Iglesias-Émbil, M.: Downcycling in automobile recycling process: a thermodynamic assessment. Resour. Conserv. Recycl. **136**, 24–32 (2018). https://doi.org/10.1016/j.resconrec.2018.04.006

Restrepo, E., Løvik, A.N., Wäger, P.A., Widmer, R., Lonka, R., Müller, D.B.: Stocks, flows, and distribution of critical metals in embedded electronics in passenger vehicles. Environ. Sci. Technol. **51**, 1129–1139 (2017). https://doi.org/10.1021/acs.est.6b05743

Restrepo, E., Løvik, A.N., Widmer, R., Wäger, P.A., Müller, D.B.: Historical penetration patterns of automobile electronic control systems and implications for critical raw materials recycling. Resources **8**, 1–20 (2019). https://doi.org/10.3390/resources8020058

Wang, J., Chen, M.: Recycling of electronic control units from end-of-life vehicles in China. JOM - J. Miner. Met. Mater. Soc. **63**, 42–47 (2011). https://doi.org/10.1007/s11837-011-0136-9

Wang, J., Chen, M.: Technology innovation of used automotive electronic control components recycling in China. Adv. Mater. Res. **610–613**, 2346–2349 (2013). https://doi.org/10.4028/www.scientific.net/AMR.610-613.2346

Requirements on and Selection of Data Storage Technologies for Life Cycle Assessment

Michael Ulbig[1], Simon Merschak[1(✉)], Peter Hehenberger[1], and Johann Bachler[2]

[1] Research Centre for Low Carbon Special Powertrain, University of Applied Sciences Upper Austria, Wels Campus, Stelzhamerstraße 23, 4600 Wels, Austria
simon.merschak@fh-wels.at

[2] AVL List GmbH, Hans-List-Platz 1, 8020 Graz, Austria

Abstract. The importance of a centralized data storage system for life cycle assessment (LCA) will be addressed in this paper. Further, the decision-making process for a suitable data storage system is discussed. LCA requires a lot of relevant data such as resource/material data, production process data and logistics data, originating from many different sources, which must be integrated. Therefore, data collection for LCA is quite difficult. In practice, relevant data for LCA is often not available or is uncertain and has therefore to be estimated or generalized. This implies less accuracy of the calculated carbon footprint. State of the Art research shows that the LCA data collection process can benefit from data engineering approaches. Key of these approaches is a suitable and efficient data storage system like a data warehouse or a data lake. Depending on the LCA use case, a data storage system can also benefit from the combination with other technologies such as big data and cloud computing. As a result, in this paper a criteria catalog is developed and presented. It can be used to evaluate and decide which data storage systems and additional technologies are recommended to store and process data for more efficient and more precise carbon footprint calculation in life cycle assessment.

Keywords: Carbon footprint · Life cycle assessment · Data engineering · Data storage technology

1 Introduction

One of the recent global concerns for politics and economy is the global climate change. This global climate change is caused by the emission of greenhouse gases (GHGs) like carbon dioxide [1]. With 84%, production and usage are responsible for the biggest part of energy related greenhouse gases emissions. The industrial sector makes 90% of the energy consumption of this part [2]. An important indicator for environmental performance of a product is the carbon footprint [1]. For the identification of the carbon footprint of mechatronic systems it is necessary to analyze the whole product life cycle [3]. Ecodesign is defined by the standard ISO/TR 14062 [7] as the integration of environmental aspects into product design and development. The approach is to have the

Published by Springer Nature Switzerland AG 2023
F. Noël et al. (Eds.): PLM 2022, IFIP AICT 667, pp. 86–95, 2023.
https://doi.org/10.1007/978-3-031-25182-5_9

environmental impacts of a product in mind during its entire life cycle while designing a product. A method for the evaluation of the environmental impacts on all stages of the product's lifecycle is called life cycle assessment (LCA). But the evaluation of the product carbon footprint is rather complex [3]. Especially the limited availability of environmental data across the entire life cycle is significant [4]. Another challenge is, that relevant data is often uncertain in the early design phase of a product [3]. A large number of LCA relevant data sources is already contained in IT-systems of companies like product lifecycle management (PLM) systems and enterprise resource planning (ERP) systems [5]. Other relevant data can be found in 3D CAD-models and technical drawings [6]. This variety of data sources and formats leads to a very time-consuming data acquisition process. It is important to face these problems and to design a system which enables the generation of a CO_2-report based on the available information. Information can be stored in internal systems of a company as well as in external data sources like LCA relevant databases. The objective of this publication is to supports the decision-making process for an appropriate data storage technology for the LCA use case. This approach is based on data engineering practices, which enable data-driven decision-making by collecting, transforming, and publishing data. The selection of an appropriate data storage system which should contain LCA relevant data is a first important step for the implementation of an LCA data pipeline. There are many different types of data storage systems but the most popular by literature are the concepts called data warehouse and data lake. The benefit of such an implemented system is, that the data acquisition process would be much faster and cost-efficient. The minimization of the human factor, which frequently results in mistakes, would also promise data of higher quality. In some large and dynamic productions with a high number of products like clothing - or industry 4.0 productions, where manual data inventory is not possible, such a system could be even an enabler. Furthermore, the data centralization would offer more possibilities to estimate missing values via interpolation or machine learning. But this presumes that the data storage system is used and implemented correctly. There is always the danger of malfunctions and technical issues. Additional, since the data is produced by different members of the supply chain, the system is still depending on their contribution. Overall, the system would make LCA more affordable, precise and offer new opportunities for other technologies and LCA approaches. But it must be actively maintained by IT and experts in the field of LCA to ensure overall quality.

2 Background and Related Research

2.1 Life Cycle Assessment

The International Organization for Standardization (ISO) has standardized the LCA method in its basics with ISO 14040 [4] and in detail with ISO 14044 [8]. According to ISO 14040, LCA addresses the environmental aspects and potential impact of the whole product life. This includes the raw material acquisition, production, use and disposal. The general environmental impact assessment requires the consideration of resource use, human health, and ecological consequences. In ISO 14044 LCA is parted into 4 different phases: goal and scope definition, inventory analysis, impact assessment and interpretation.

2.2 Data Engineering

A main task in data engineering is to provide a data infrastructure so that, at the final state, the required data is ready for further analysis like data science [6]. Part of this task is to select an appropriate storage system for the given use case. The data can be stored temporary in staging areas for example. In those, data usually stays only very shortly until it gets used and deleted. Other use cases require to store data in long-term archival storage to be stored for years [7]. Another part of the data infrastructure is a pipeline which extracts data from one system, transforms it and loads it to another system [6]. Recently, two aspects are very common. One is the handling of streaming data like in Internet of Things (IoT) use cases. The other one is handling a large volume of data [7].

2.3 Related Research

There has been research about facing the data acquisition problem of life cycle assessment before. Some projects suggest applying big data as well as cloud computing technologies as solutions. A case study by the German automotive industry came to the conclusion that the use of big data can provide considerable potential to gather and analyze product related data over the entire life cycle [8]. Another study concludes that environmental performance evaluations can profit from big data. One aspect is that big data can support the environmental supply chain. Integrated data resources in the big data era can help to evaluate environmental efficiencies and resource efficiencies in the industrial supply chain [9]. So far, there have been different approaches to the digital support of LCA. Especially in the beef supply chain, efforts have been made to collect and aggregate data over the supply chain via cloud technologies [10, 11].

2.4 Data Pipeline for Life Cycle Assessment

The whole process of data collection, storage and processing is shown in Fig. 1. The required data must be acquired from different LCA relevant data sources via an automated data integration process which centralizes the data. Two very common examples of this integration process are called Extract-Transform-Load (ETL) or Extract-Load-Transform (ELT). The data storage system collects all LCA relevant data automatically and provides them ready to use for LCA methods. The stored data can then be analyzed by different LCA methods. Finally, a carbon footprint report can be generated, and appropriate visualization methods can be used for the display of the assessment results.

3 Overview of Storage Technologies

In the literature, no umbrella term for the terms data warehouse and data lake could be found. Each of the systems mainly stores data and for this reason, the term data storage system is used as group term in this publication.

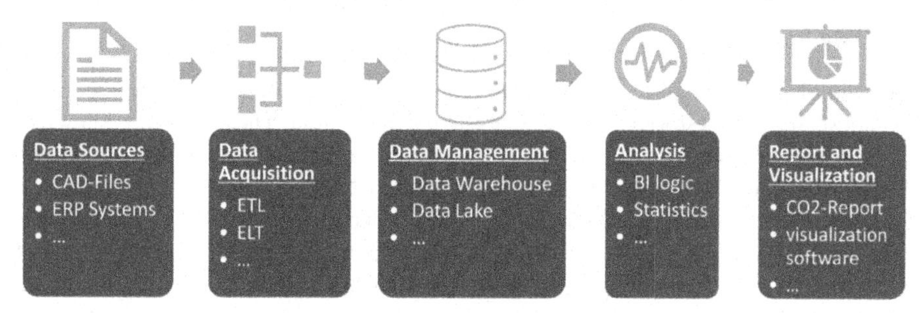

Fig. 1. LCA data pipeline

3.1 Data Warehouse

A data warehouse (DWH) is a logic, centralized data storage system which is separated from operative data storage systems. Ideally, it serves as a company-wide uniform and consistent database supplied by different types of data source systems.

The data flow is shown in Fig. 2. A data warehouse has four characteristics which are subject-orientation, integration, time-variance and non-volatile. Subject-orientation means, that the data management is designed for informing a decision maker about certain subjects. The decision maker should have direct access to the information about the subject. Typical subjects are time, location or product. A schema called Online Analytical Processing (OLAP) orientation can be applied to achieve subject-orientation. The key task of a data warehouse is the integration of data from different operative and external sources. It is important that the collected data must be consistent. Time-variance is the possibility to look at all the data over time. Nowadays also alternative time measures are considered like time transaction. In operative systems the data changes rather often. In contrast to that, the data in a data warehouse should not change after it is successfully integrated to create a history of the values [12].

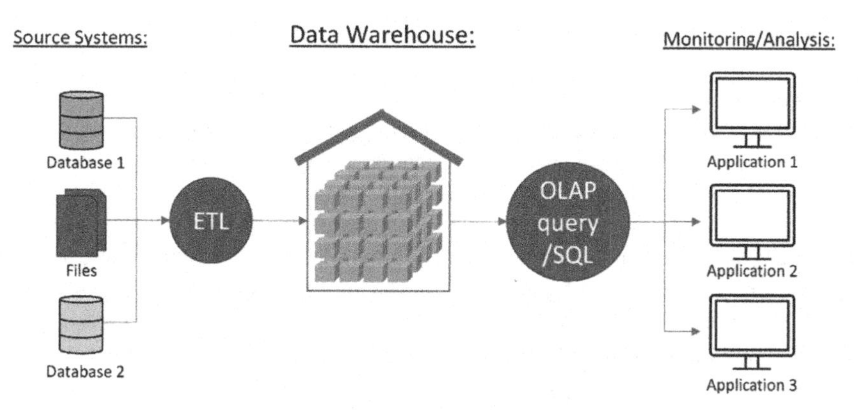

Fig. 2. Data flow for data warehouse

3.2 Data Lake

The main idea of a data lake is to store all relevant enterprise data. In contrast to a data warehouse, the data is stored in the data lake in the raw format. The data flow is shown in Fig. 3. A data lake supposes to include large amounts of data in various data structures. The data lake can be split into two layers. The landing layer contains identical copies of data from the sources systems. This layer is also called mirror layer. The landing layer is the base for the analytical layer. The analytical layer is very dynamic and consists of transformed data from the landing layer. The analytical layer then provides the data ready for business intelligence (BI) applications, analytical models and data visualization [13]. This data can then be accessed via, for example, Spark or structured query language (SQL).

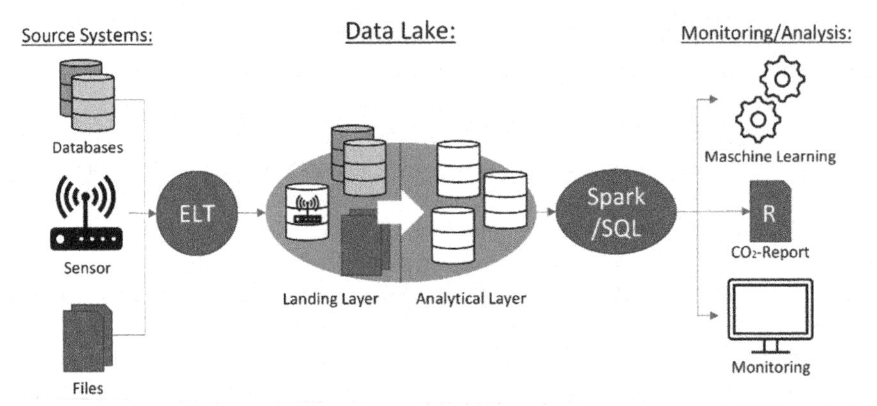

Fig. 3. Data flow for data lake

4 Derived Requirements

There is no standard criteria catalog for the evaluation of different data storage systems. For this reason, different aspects of data storage systems, which are discussed in literature, have been merged to create a criteria catalog. The first aspect which is considered, is the type of data which can be stored in the data storage system. Data can be structured (tables), semi-structured (xml, json) or unstructured (text, pictures, videos) [14]. The second and third aspect are the data in- and output. Looking at the input, there are three different ways how data can be ingested. Those are application, streaming and batch data [7]. For accessing the data storage system there are two methods. One option is to access the data individually and the other is to aggregate information over several datasets. Usual methods for aggregation are sums or medians. Connected to this is the topic of access control and storage duration of data [7]. Another topic which is also very relevant in production facilities, is performance and near-real time analytics. An example for near-real time analytics might be, that a Chief Marketing Officer needs up-to-date sales data of a new product for sales management [15]. The last two topics are big data and cloud computing. Both have become more and more important, and both have been used in LCA related projects as mentioned in Sect. 2.3. Both aspects are criteria from NIST standards [16, 17]. All relevant criteria are summarized in Table 1. The power consumption of a selected data storage system could also be an additional criterion, but since a solution like this would be integrated in already existing IT infrastructure, it will not be in the focus of this paper.

5 Evaluation and Discussion

5.1 Criteria Applied to Data Warehouse

The technology which is used to implement a data warehouse are relational databases. That means, that a data warehouse is made to store structured data and not semi- or unstructured data [18]. In classical data warehousing the data integration is performed in periodical intervals. The data will then be integrated via an ETL batch process. This process can also include application data from operational systems. Since the data gets integrated via an ETL process, it is not made for ingesting streaming data [12]. The data from a data warehouse can be accessed individually and via aggregation. If a data warehouse is implemented as a schema (OLAP), it is much more made to aggregate data then to access data individually. A data warehouse itself does not provide any access control, but there are extensions which allow such access control [12]. The same applies for the storage duration. A data warehouse by itself can be saved and archived via backups. Besides that, it can also just be stored like a normal relational database. In addition, a data warehouse provides a historicization concept called slowly changing dimension (SCD) [12]. The performance of a classical data warehouse is limited by the speed of the ETL process. So, it would not be possible to achieve near-real time analytics with a classical data warehouse. The challenges which big data is facing, cannot be solved with the use of a relational data warehouse [12]. Since a classical data warehouse would consist of a relational database and could not hold the volume of data which is usually used in big data. This argument can also be applied to the characteristic's velocity and variation. One addition which is rather common is a technique called slowly changing dimension (SCD). SCD has different strategies to historicize semantically or formally changing data which provides variability [12]. Since usually the technology which is used to implement a data warehouse are relational databases, it can be rather simply realized via cloud computing. Data warehouse solutions are hosted by cloud providers like Amazon Web Service (AWS) ® and IBM ® [19, 20].

5.2 Criteria Applied to Data Lake

A data lake stores structured, semi-structured and unstructured data [18]. In concept, the data lake is also able to include operational databases. This means that application data would be includable. The same applies for streaming data and batch data. A data lake has an additional serving layer which provides more efficient and comfortable access [12]. Since it is possible that the types of the stored data are very different as well as the storage technologies which would contain the data, the access is more complex than in a data warehouse. Individual access is possible, but one technology for individual access will be probably not sufficient. The same applies for aggregation as well. Access control can be implemented. The data governance should regulate the access to, and privacy of the data [13]. According to the characteristics, a data lake should have an information lifecycle management (ILM) strategy to manage the storage duration [13]. Since streaming data can be used and stored, it is possible to create near-real time analytics [12]. Considering that a data lake should be implemented on a scalable framework like Apache Hadoop, a data lake will be able to handle the characteristics of big data challenges [13]. Like the data warehouse a data lake is usually a repository and can be implemented locally or on a cloud platform. Many providers do offer such solutions like Google Cloud and AWS [21, 22].

5.3 Requirements from LCA

In general, it can be said that the decision criteria are dependent on the LCA use case. A company which produces clothing might fulfill criteria different than a company which produces cars or groceries. The most relevant data sources for LCA, at least for the production phase, are contained in PLM- and ERP-Systems [23]. Additionally, CAD-files [24] and descriptions of parts or orders in pure text form can also contain relevant information. So overall the LCA relevant data sources come in all structure types, which implies that a storage system for LCA use must be able to store all relevant data types as well.

The two most important data types to be ingested are application data, from PLM or ERP systems [23], and batch data like CAD-Files, tables or text files. Streaming data, like a constant stream of data from a sensor, is typically not used for LCA. An aggregation can be helpful when estimating the energy consumption of processes from sensor values, since the storage of a data stream needs a lot of memory. For the generation of an LCA report, the individual access to datasets and to aggregated data is both necessary. The necessity of access control depends highly on the LCA use case and company policies. For the purpose of a decision, it can be said that access control is not necessary by default. Usually, the LCA related data only needs to be available till the end of the LCA. To speed up future LCAs of similar products, it makes sense to store relevant data after that. On the other hand, the use of data which is too old, and not up to date, can be harmful as well.

LCA is a complex method with many aspects to consider in order to properly interpret the results and to draw reasonable conclusions [25]. For this reason, LCA can be usually seen as performance-uncritical task.

A high number of LCAs for multiple products or a LCA for a very complex product can lead to the necessity of big data solutions. One big data criterion which is relevant for LCA, is the high variation of data from different sources. But again, big data is not relevant for LCA by default. The same applies to cloud computing. The data collection and aggregation process of a product with a complex supply chain of multiple companies could benefit significantly from cloud computing. As shown in studies about the beef supply chain [10, 11]. That said, storing companies sensitive data at a third-party cloud provider could impose potential risks [26]. For this reason, there is still a lot of skepsis from companies to cloud computing.

5.4 Summary and Decision

To come to a decision, which data storage system is more suitable for the LCA use case, the hard criteria for LCA must be considered first. In the LCA use case, those are the types of structure. In general, both systems are suitable for storing data for LCA. Because the data sources of LCA are rather diverse and other use cases like LCA hotspot analysis and visualization might require the files in their raw format, a data lake would be a good choice. As shown in Table 1, a data warehouse can only store structured data, a data lake is made for storing all three types of data. All different data types can be extracted via an integration process and only structured and transformed data will be stored. Overall, as described above, the relevant data comes from a variety of sources, in all kinds of structure types and the amount can even extend to become a big data project. In all these cases, a data lake would be more suitable to store the data since it offers more flexibility than a data warehouse.

Table 1. Evaluation of data storage systems

Perspectives	Criteria	LCA (requirements)	Data Warehouse (possibilities)	Data Lake (possibilities)
Types of Structure	Structured	yes	yes	yes
	Semi-structured	(yes)	no	yes
	Unstructured	yes	no	yes
Ingest	Application data	yes	yes	yes
	Streaming data	no	no	yes
	Batch data	yes	yes	yes
Storage access	individual access	yes	simple	complex
	aggregation access	yes	simple	complex
	access controls	possible	yes	yes
	storage duration	middle	long	long
Perfor-mance	near-real time analytics	no	no	yes
Big Data	Volume	possible	no	yes
	Velocity	possible	no	yes
	Variation	yes	no	yes
	Variability	no	yes	yes
Cloud Computing	on-demand self-service	no	yes	yes
	broad network access	no	yes	yes
	resource pooling	no	yes	yes
	rapid elasticity	no	yes	yes
	measured service	no	yes	yes

6 Conclusions

Concluding, it has been found that there are many aspects to consider when choosing an appropriate data storage system for LCA. Even if both evaluated data storage systems are suitable, a data lake, thanks to its flexibility, is the better choice for the LCA use case. As a next step, a prototype of a data lake for LCA use will be implemented for further analysis. The goal is to design a data lake in a way to be appropriate for LCA. It is expected that the challenges will be the interface of the PLM and ERP systems as well as the evaluation of CAD files. Beside the data integration, there is the question how different data types will be treated and how to deal with missing information. The overall goal is to have a platform to test (semi-)automatic LCA. Other ideas like enhanced LCA with value estimation via machine learning for more exact LCA could be possible topics for further publications. In our vision, a data lake could be the base not only for the evaluation of products but could also support topics like LCA driven product design. A possible use case could be to compare different product designs. The system would get CAD geometry data and combine them with the material information of an ERP-system and the knowledge of a LCA databases to generate a CO_2-report.

Acknowledgments. The research has been applied for and was granted as COMET project under the guidance of the Austrian Research Promotion Agency FFG and is funded by the Federal Ministry for Transport, Innovation and Technology (BMVIT), the Federal Ministry for Digital and Economic Affairs (BMDW) and the provinces of Upper Austria and Styria.

References

1. He, B., Wang, J., Huang, S., Wang, Y.: Low-carbon product design for product life cycle. J. Eng. Design **26**, 321–339 (2015)
2. He, B., Wang, J., Deng, Z.: Cost-constrained low-carbon product design. Int. J. Adv. Manuf. Technol. **79**(9–12), 1821–1828 (2015). https://doi.org/10.1007/s00170-015-6947-z
3. Merschak, S., Hehenberger, P.: Ecodesign methods for mechatronic systems: a literature review and classification. In: 2019 20th International Conference on Research and Education in Mechatronics (REM). IEEE, Wels (2019)
4. DIN EN ISO 14040:2009-11, Environmental management - Life cycle assessment - Principles and framework (ISO 14040:2006)
5. DIN EN ISO 14044:2018-05, Environmental management - Life cycle assessment - Requirements and guidelines (ISO 14044:2006 + Amd 1:2017)
6. Crickard, P.: Data Engineering with Python. 1st edn. Packt Publishing (2020)
7. Sullivan, D.: Official Google Cloud Certified Professional Data Engineer Study Guide. Wiley, Indianapolis (2020)
8. Beier, G., Kiefer, J., Knopf, J.: Potentials of big data for corporate environmental management: a case study from the German automotive industry. J. Ind. Ecol. **26**, 336–349 (2020)
9. Song, M.-L., Fisher, R., Wang, J.-L., Cui, L.-B.: Environmental performance evaluation with big data: theories and methods. Ann. Oper. Res. **270**(1–2), 459–472 (2016). https://doi.org/10.1007/s10479-016-2158-8
10. Singh, A., Mishra, N., Ali, N., Shukla, S.I., Shankar, R.: Cloud computing technology: reducing carbon footprint in beef supply chain. Int. J. Prod. Econ. **164**, 462–471 (2015)

11. Singh, A., Kumari, S., Malekpoor, H., Mishra, N.: Big data cloud computing framework for low carbon supplier selection in the beef supply chain. J. Clean. Prod. **202**, 139–149 (2018)
12. Baars, H., Kemper, H.-G.: Business Intelligence & Analytics - Grundlagen und praktische Anwendungen: Ansätze der IT-basierten Entscheidungsunterstützung, 4th edn. Springer, Wiesbaden (2021)
13. Gupta, S., Giri, V.: Practical Enterprise Data Lake Insights: Handle Data-Driven Challenges in an Enterprise Big Data Lake. 1st edn. Apress®, Berkeley (2018)
14. Li, K.-C., Jiang, H., Zomaya, A. Y.: Big data Management and Processing. Taylor & Francis Group (2017)
15. Lakhe, B.: Practical Hadoop Migration: How to Integrate Your RDBMS with the Hadoop Ecosystem and Re-architect Relational Applications to NoSQL. Apress®, New York (2016)
16. NIST Big Data Public Working Group: NIST Big Data Interoperability Framework: Volume 1, Definitions, version 2. National Institute of Standards and Technology (2018)
17. Mell, P.M., Grance, T.: The NIST definition of cloud computing. National Institute of Standards and Technology, Gaithersburg (2011)
18. O'Leary, D.E.: Embedding AI and crowdsourcing in the big data lake. IEEE Intell. Syst. **29**(5), 70–73 (2014)
19. Data-Warehouse-Lösungen. https://www.ibm.com/de-de/analytics/data-warehouse. Accessed 05 Feb 2022
20. Was ist ein Data Warehouse? | Wichtige Konzepte | Amazon Web Services, https://aws.amazon.com/de/data-warehouse/. Accessed 05 Feb 2022
21. What is a Data Lake? Google Cloud. https://cloud.google.com/learn/what-is-a-data-lake. Accessed 05 Feb 2022
22. Data Lake | Implementierungen | AWS-Lösungen. https://aws.amazon.com/de/solutions/implementations/data-lake-solution/. Accessed 05 Feb 2022
23. Merschak, S., Hehenberger, P., Schmidt, S., Kirchberger, R.: Considerations of life cycle assessment and the estimate of carbon footprint of powertrains. SAE Technical Papers (2020)
24. Merschak, S., Hehenberger, P., Bachler, J., Kogler, A.: Data relevance and sources for carbon footprint calculation in powertrain production. In: Nyffenegger, F., Ríos, J., Rivest, L., Bouras, A. (eds.) PLM 2020. IAICT, vol. 594, pp. 203–214. Springer, Cham (2020). https://doi.org/10.1007/978-3-030-62807-9_17
25. Frischknecht, R.: Lehrbuch der Ökobilanzierung. Springer, Heidelberg (2020). https://doi.org/10.1007/978-3-662-54763-2
26. Fujita, H., Tuba, M., Sasaki, J.: Study on advantages and disadvantages of Cloud Computing - the advantages of telemetry applications in the cloud (2013)

Digital Twin Application to Energy Consumption Management in Production: A Literature Review

Daniele Perossa(✉), Roman Felipe Bastidas Santacruz, Roberto Rocca, and Luca Fumagalli

Politecnico di Milano, 20133 Milan, Italy
daniele.perossa@polimi.it

Abstract. Economic and environmental issues that translate into energy costs and contaminations in production are growingly attracting attention from several parts and actors. Therefore, Energy Consumption Management (ECM) is gaining ever higher importance within production environments. Industry 4.0 provides several opportunities to address these challenges. One of the technologies presenting the best potentialities is the Digital Twin (DT), which has been found able to promote ECM improvements related to production assets and processes in different ways. Nonetheless, in the academic literature has not been found an extensive review of DT application to ECM in manufacturing. Therefore, this paper proposes a systematic literature review to investigate the current state of the art of the applications, features and characteristics, and implementation strategies of DT applied to ECM in production contexts. Attention has been also paid to the human role inside the application of the DT technology to ECM and the interaction modalities between humans and the DT itself.

Keywords: Digital twin · Energy consumption management · Production · Industry 4.0 · Supervision and reconfiguration of complex industrial systems

1 Introduction

Industry 4.0 (I4.0) is revolutionising the manufacturing sector [1]. The result is an unprecedent interconnection between physical and digital environments [2] and great potentials in many different contexts. In manufacturing, environmental sustainability is one of the paradigm I4.0 can improve [3]. Energy Consumption (EC) and the related atmospheric emissions reduction is among the main I4.0 areas of application [4–6].

Within I4.0, Digital Twin (DT) is described as a key paradigm [7, 8]. DT provides an extremely high-fidelity virtual modelling, enabling process control and precise decision making [9]. Some authors investigate from a high level perspective the potentials of DT shop-floor in the context of Energy Consumption Management (ECM), discussing the related applications [10]. Others perform extensive reviews on DT application in manufacturing, mentioning ECM as one of the main areas of intervention, but a detailed investigation is missing [11, 12]. Even though the applicability of DT in ECM is considered a

Published by Springer Nature Switzerland AG 2023
F. Noël et al. (Eds.): PLM 2022, IFIP AICT 667, pp. 96–105, 2023.
https://doi.org/10.1007/978-3-031-25182-5_10

relevant topic, an extensive review of current applications, features and implementation modalities is missing. Thus, the main aim of this paper is to explore the state of the art of DT applied to ECM in production contexts.

The rest of the paper is structured as follows. In Sect. 2 the adopted Literature Review (LR) methodology is explained. In Sect. 3 the results of the LR are shown and discussed. In Sect. 4 the conclusions and possible future works are presented.

2 Methodology

In order to properly conduct a systematic LR and to identify the correct research topics, the CIMO (Context, Intervention, Mechanism, Outcome) framework was implemented [13]. For Context, the manufacturing sector was considered as reference, focusing exclusively on the production phase. For Intervention, DT application has been considered in this work. Since Mechanisms explains the relationship between interventions and outcomes and their circumstances of activity, the ECM has been considered. Finally, as an Outcome, in this case the effects of intervention are related to EC reduction. Aiming at covering all the relevant aspects inside the research framework, three main topics still not clear in literature were identified: (i) the identification of DT applications that support ECM in production; (ii) the identification of specific features typical of DTs implemented in ECM context in production; (iii) the identification of methodologies supporting DT implementation in the ECM in production. The recognition of these aspects allowed to define three Research Questions (RQs):

RQ1. What are the possible applications of DT to ECM in production?
RQ2. Are there specific features and architectures to adapt DT to the ECM context?
RQ3. Are there methodologies to support the DT implementation in ECM context?

The literature search was conducted on Scopus and Web of Science, as reported in Fig. 1. In the screening phase filters such as duplicates, language and unkown authors were applied, resulting in 244 papers. In the eligibility phase the 78 documents focused on ECM and production were identified. Lastly, the entire papers were analysed to identify the 37 of them relevant to answer the RQs.

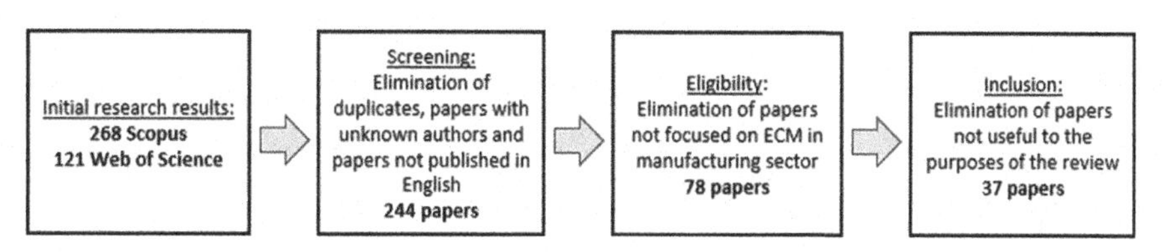

Fig. 1. Literature review methodology

The final papers were classified according to relevant criteria derived from the RQs:

- For RQ1, the interested area of ECM have been detected, i.e., the tasks that were described as performing by the DT to improve ECM.

- For RQ2, the description of the ECM DT features and the proposal of ECM DT architecture have been analysed.
- For RQ3, the description of implementation process, or structured implementation methodology have been considered.

All the analysed papers are presented and classified in Table 1 below. For each paper, in each column, has been inserted only the most underlined aspect in the document. Nonetheless, this does not mean, for each column, that the presented aspect was the only one discussed by the paper referring to that specific column.

Table 1. Synthesis of the analyzed papers

Paper	ECM applications	DT relevant features	Architecture proposal	Implementation
[14]	EM	DA&AI	N	N
[6]	EM	DA&AI	Y	N
[15]	EM	DA&AI	Y	PHI
[16]	EM	DA&AI	N	N
[17]	UO	VM&S	N	DI
[18]	EM	VM&S	N	DI
[19]	UO	–	Y	N
[20]	UO	HI	N	N
[21]	PEC	VM&S	N	PHI
[22]	AEW	DA&AI	N	SM
[23]	P&MO	DA&AI	Y	N
[24]	PEC	VM&S	Y	N
[25]	EPO	VM&S	N	N
[26]	EPO	VM&S	Y	N
[27]	Unspecified	–	N	PHI
[28]	P&MO	VM&S	Y	SM
[8]	P&MO	DAcq	N	DI
[29]	UO	–	Y	DI
[30]	AEW	DA&AI	Y	N
[31]	EPO	DA&AI	N	PHI
[32]	UO	–	N	PHI
[33]	UO	Dacq	Y	DI
[34]	EM	HI	N	DI

(*continued*)

Table 1. (*continued*)

Paper	ECM applications	DT relevant features	Architecture proposal	Implementation
[35]	P&MO	DA&AI	Y	DI
[36]	P&MO	_	Y	N
[37]	Unspecified	_	N	N
[38]	P&MO	VM&S	Y	DI
[39]	EPO	DA&AI	Y	DI
[40]	EM	Dacq	N	PHI
[41]	EM	DA&AI	Y	PHI
[42]	P&MO	DA&AI	Y	PHI
[43]	EM	HI	N	DI
[10]	EPO	DAcq	Y	PHI
[44]	EM	VM&S	Y	SM
[45]	EPO	DA&AI	N	DI
[46]	AEW	VM&S	N	N
[47]	EPO	VM&S	Y	SM

Energy Monitoring = EM; Unspecified Optimization = UO; Prediction of future Energy Consumption = PEC; Energy Parameters Optimization = EPO; Avoid Energy Waste = AEW; Process and Maintenance Optimization = P&MO
Data Analytics and Artificial Intelligence = DA&AI; Virtual Model and Simulation = VM&S; Human Interface = HI; Data Acquisition = DAcq
Yes = Y, No = N
Provides Hints for Implementation = PHI; Describes the Implementation = DI; Provides a Structured Methodology = SM

3 Results and Discussion

In this section the main results of the LR are described and discussed.

3.1 Applications of DT to ECM in Production

The main identified ECM applications are (i) Energy Monitoring (EM); (ii) Prediction of future Energy Consumption (PEC); (iii) Avoid Energy Waste (AEW); (iv) Process and Maintenance Optimization (P&MO) (v); Energy Parameters Optimization (EPO); (vi) Unspecified Optimization (UO).

PEC is related to the prediction of EC of assets, processes, or work cycles in future stages of the lifecycle or to compare alternative configurations. The main tool for PEC is simulation (e.g., [8, 21, 24, 28, 38, 42]). Simulation is also a key tool in the EC optimization through the production scheduling [28, 38, 42], which fell inside the P&MO cluster. Another interesting P&MO application is the optimization of machine states timing [8] to reduce EC. Maintenance process can also be a target of ECM optimization [23]. The analysis of the production process through a DT can lead also to AEW, by avoiding to use energy on not quality-compliant work in progress [22, 36], or by identifying energy consuming non value-added activities [28, 30]. A different approach concerns the possibility to perform the EPO, by directly controlling, through the DT, certain EC-related parameters [10, 25, 26, 35, 38, 39, 45, 47]. Finally, several cases did not specify how they wanted to optimize EC and fell under UO [17, 19, 20, 29, 32, 33]. As a final consideration, it is interesting to notice that all the analyzed applications do not substitute human-performed tasks: ECM DT enables the performing of new tasks, rather than allowing to perform better operations already human-performed. This is crucial, as it means that ECM DT would have a role of support and assistance to humans rather than substitution.

3.2 Features of DT for ECM

The main DT features identidied in the literature review are (i) Data Analytics and Artificial Intelligence (DA&AI); (ii) Virtual Model and Simulation (VM&S); (iii) Human Interface (HI); (iv) Data Acquisition (DAcq).

ECM-DT Features. *(i) DA&AI*. Heterogenous approaches can be observed according to situation, targets, and applications. Among the most notable cases, there are statistical approaches based on gaussian distribution [22, 39], mixed integer linear programming [24], and a hybrid model based on a self-adaptive population genetic algorithm and autoregressive moving average [31]. Such an heterogeneity is tackled by [16], underlining the absence of an overall accepted data exchange format. *(ii) VM&S*. It appears that many EC-related models are written in Matlab, for a Discrete Event Simulation (DES) approach [25, 40, 45–47]. There are exceptions, like [24] using Modelica for an Object Oriented Simulation, and [38] describing an integrated approach with elements of discrete event, dynamic system and agent based approaches. An important feature related to VM&S appears the integration between digital and physical entities. In [21] is used the sequential analysis technique "Page-Hinkley test" to let the virtual model cope with changes in the physical system. [25] describes an interactive mechanism where physical and digital spaces behave respectively as the client and the server. One last important feature about VM&S, can be the usage of an ontology to provide knowledge to the simulation [44]. *(iii) HI*. It is an aspect whose importance is enhanced by the crucial role it has in determining the humans-DT cooperation. Human intervention is crucial for many DT supported decision-making activities [48]. In [39] is used a Javascript Application Programming Interface (API) for rendering 2D and 3D interactive graphics. In [8] a Graphic User Interface (GUI) is built using Matlab. In [43] a GUI is programmed with Kivy, an open-source Python framework. In [34] is developed a mobile app for real-time visualization, which could have interesting consequences in the way humans relate to DT. No explicit mentioning was found about the usage of Augmented Reality with ECM DT, even though it can be a key element in the human-DT interaction. *(iv) DAcq*. A common feature is the usage of I4.0 technologies for the data sensing and transmission infrastructure [23, 29, 32, 34, 41, 42]. Another crucial point is the integration between DT and the field IT systems: [8] proposes a method for MES-integration and [33] for SCADA-integration.

ECM-DT Architectures. For what concerns architectures proposal, in the analyzed cases appears as quite recurrent the division in layers, usually three (in few cases four): one layer for DAcq, one for VM&S, and the last one for DA&AI (or applications) and HI. Sometimes, a further layer for data transmission is added between DAcq and VM&S [19, 23, 29, 30, 35, 42]. In [29] is also proposed to develop many separate DTs for the single physical elements, and then to link them all. Considering alternative typologies of DT architecture, in [10] the bottom layer presents synchronized physical and digital equipment, and above are placed ECM services. [39] presents an architecture focused on the transfer of data from PLCs to the web API. In [26] the data analytics module warns the operator, involved as a crucial component of the DT architecture, who triggers the simulation module. The architecture by [33] includes six modules (i.e., behaviour models repository, multi-physical models repository, parameters monitoring, process virtualization, forecast for process evolution prediction, planning management). [44] proposes a framework for a DT applied to an industrial robot split into physical and cyber space.

3.3 Strategies of Implementation of DT in ECM

Four of the analyzed works define a structured DT implementation and 11 describe the implementation. The descriptions usually focus on a specific implementation phase [29, 34, 43, 45]. [45] describes a process of design focused on the theoretical model and the results. In [29] the focus is set on the description of DT hardware implementation. [34, 43] center on the description of the DT software requirements and the hardware architecture of the solution. [44, 47] offer a complete description of the implementation strategy giving an idea of the path that companies could take to implement ECM DT. However, the DTs in both works are implemented on a cutting-edge industrial Robot which simplifies the process. To identify the high level implementation methodology of ECM DT, each document was analyzed and a common path of implementation was identified. Three main phases are counted (Fig. 2).

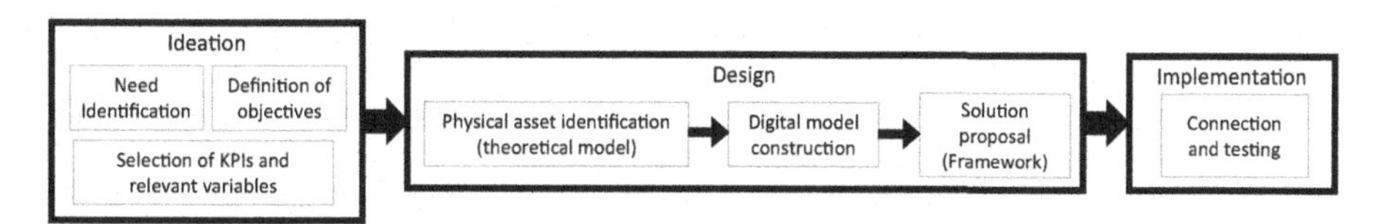

Fig. 2. Identified DT implementation methodology in ECM

The DT implementation requires a systematic approach to track the implementation process and the tasks to be executed in it [22]. In [28] is highlighted that one of the most relevant steps in the ECM DT implementation is the initial definition of KPIs and objectives to be achieved through the DT. Nevetheless, in most of the analysed implementation strategies, such definition varies depending on the interaction among such factors [8, 22, 28, 33, 35]. Specifically, in [8] the energy-Overall Equipment Effectiveness is defined to be the most relevant KPI to monitor. In [35] a dynamic sleep DT for ECM is proposed and modelled, and the KPIs and strategy to address are different from [8]. In some documents the way in which the definition of the objectives and KPIs for the DT were reached are not discussed [22, 29]. In summary, the way in which the implementation strategy varies in the different contexts of ECM is still fuzzy and can create confusion for companies interested in the DT implementation.

4 Conclusions and Future Work

This work proposed to investigate the state of the art of the of DT applied to ECM through a LR articulated into three RQs. The answer to RQ1 uncovered how the field of applications appears large, variegated, and full of potentiality. Furthermore, ECM DT clearly appears to be supporting humans rather than substituting them. RQ2 uncovered a great heterogeneity in the ways ECM DTs work and in the features that characterize their structuring. The only recurrent characteristic seems to be the modelling through Matlab and usage of DES, but without evidence of a correlation between this and the ECM objective of the DTs. Indeed, DT features seem to change according to specific targets, industries, and situations. Thus, a topic to be developed in future might be an analysis of the eventual correlation between DT features and ECM applications. There seems to be though a pattern towards the definition of a recurrent type of DT architecture, structured in layers, representing the sensing infrastructure, DA&AI and VM&S, and applications and HI. Finally, for RQ3, even though there are few implementation examples available and a common high-level implementation methodology has been identified, there is a lack of structured methodologies to support companies in the ECM DT implementation. This seems an issue related not just to ECM DT, but to generic DT implementation, and analysis and works about it should be performed. Finally, there are very few considerations about human role in ECM DT paradigm.

References

1. Da Xu, L., Xu, E.L., Li, L.: Industry 4.0: state of the art and future trends. Int. J. Prod. Res. **56**(8), 2941–2962 (2018). https://doi.org/10.1080/00207543.2018.1444806
2. Vaidya, S., Ambad, P., Bhosle, S.: Industry 4.0 - a Glimpse. Procedia Manuf. **20**, 233–238 (2018). https://doi.org/10.1016/j.promfg.2018.02.034
3. Stock, T., Seliger, G.: Opportunities of sustainable manufacturing in industry 4.0. Procedia CIRP **40**(lcc), 536–541 (2016). https://doi.org/10.1016/j.procir.2016.01.129
4. Liang, Y.C., Lu, X., Li, W.D., Wang, S.: Cyber physical system and big data enabled energy efficient machining optimisation. J. Clean. Prod. **187**, 46–62 (2018). https://doi.org/10.1016/j.jclepro.2018.03.149
5. Shrouf, F., Miragliotta, G.: Energy management based on internet of things: practices and framework for adoption in production management. J. Clean. Prod. **100**, 235–246 (2015). https://doi.org/10.1016/j.jclepro.2015.03.055
6. Ma, S., Zhang, Y., Liu, Y., Yang, H., Lv, J., Ren, S.: Data-driven sustainable intelligent manufacturing based on demand response for energy-intensive industries. J. Clean. Prod. **274**, 123155 (2020). https://doi.org/10.1016/j.jclepro.2020.123155
7. Biesinger, F., Meike, D., Kraß, B., Weyrich, M.: A digital twin for production planning based on cyber-physical systems: a case study for a cyber-physical system-based creation of a digital twin. Procedia CIRP **79**, 355–360 (2019). https://doi.org/10.1016/j.procir.2019.02.087

8. Rocca, R., Rosa, P., Sassanelli, C., Fumagalli, L., Terzi, S.: Integrating virtual reality and digital twin in circular economy practices: a laboratory application case. Sustain. **12**(6), 2286 (2020). https://doi.org/10.3390/su12062286
9. Kritzinger, W., Karner, M., Traar, G., Henjes, J., Sihn, W.: Digital twin in manufacturing: a categorical literature review and classification. IFAC-PapersOnLine **51**(11), 1016–1022 (2018). https://doi.org/10.1016/j.ifacol.2018.08.474
10. Zhang, M., Zuo, Y., Tao, F.: Equipment energy consumption management in applications. In: 2018 IEEE 15th International Conference on Networking, Sensing and Control, pp. 1–5 (2018)
11. Cimino, C., Negri, E., Fumagalli, L.: Review of digital twin applications in manufacturing. Comput. Ind. **113**, 103130 (2019). https://doi.org/10.1016/j.compind.2019.103130
12. Tao, F., Cheng, J., Qi, Q., Zhang, M., Zhang, H., Sui, F.: Digital twin-driven product design, manufacturing and service with big data. Int. J. Adv. Manuf. Technol. **94**(9–12), 3563–3576 (2017). https://doi.org/10.1007/s00170-017-0233-1
13. Denyer, D., Tranfield, D.: Producing a systematic review. In: The Sage Handbook of Organizational Research Methods, pp. 671–689. Sage Publications Ltd. (2009)
14. Peter, O.A., Anastasia, S.D., Muzalevskii, A.R.: The implementation of smart factory for product inspection and validation a step by step guide to the implementation of the virtual plant of a smart factory using digital twin. In: 2021 10th Mediterranean Conference on Embedded Computing, MECO 2021, pp. 7–10 (2021). https://doi.org/10.1109/MECO52532.2021.9460140
15. Tong, X., Liu, Q., Pi, S., Xiao, Y.: Real-time machining data application and service based on IMT digital twin. J. Intell. Manuf. **31**(5), 1113–1132 (2019). https://doi.org/10.1007/s10845-019-01500-0
16. Schmetz, A., et al.: Evaluation of industry 4.0 data formats for digital twin of optical components. Int. J. Precis. Eng. Manuf.-Green Technol. **7**(3), 573–584 (2020). https://doi.org/10.1007/s40684-020-00196-5
17. Arkouli, Z., Aivaliotis, P., Makris, S.: Towards accurate robot modelling of flexible robotic manipulators. Procedia CIRP **97**, 497–501 (2020). https://doi.org/10.1016/j.procir.2020.07.009
18. Soares, R.M., Câmara, M.M., Feital, T., Pinto, J.C.: Digital twin for monitoring of industrial multi-effect evaporation. Processes **7**(8), 1–14 (2019). https://doi.org/10.3390/PR7080537
19. Barni, A., Fontana, A., Menato, S., Sorlini, M., Canetta, L.: Exploiting the digital twin in the assessment and optimization of sustainability performances. In: 9th International Conference on Intelligent Systems 2018: Theory, Research and Innovation in Applications, IS 2018 - Proceedings, pp. 706–713 (2018). https://doi.org/10.1109/IS.2018.8710554
20. Pacaux-Lemoine, M.-P., Berdal, Q., Guérin, C., Rauffet, P., Chauvin, C., Trentesaux, D.: Designing human–system cooperation in industry 4.0 with cognitive work analysis: a first evaluation. Cogn. Technol. Work **24**(1), 93–111 (2021). https://doi.org/10.1007/s10111-021-00667-y
21. Bermeo-Ayerbe, M.A., Ocampo-Martinez, C., Diaz-Rozo, J.: Data-driven energy prediction modeling for both energy efficiency and maintenance in smart manufacturing systems. Energy **238**, 121691 (2022). https://doi.org/10.1016/j.energy.2021.121691
22. Trauer, J., Pfingstl, S., Finsterer, M., Zimmermann, M.: Improving production efficiency with a digital twin based on anomaly detection. Sustainability **13**(18), 10155 (2021). https://doi.org/10.3390/su131810155
23. Bányai, Á.: Energy consumption-based maintenance policy optimization. Energies **14**(18), 5674 (2021). https://doi.org/10.3390/en14185674
24. Kohne, T., Theisinger, L., Scherff, J., Weigold, M.: Correction to: data and optimization model of an industrial heat transfer station to increase energy flexibility. Energy Inform. **4**(3), 1–17 (2021). https://doi.org/10.1186/s42162-021-00179-z

25. Wang, J.F., Huang, Y.Q., Tang, D.L.: A digital twin simulator for real time energy saving control of serial manufacturing system. In: 2021 IEEE International Conference on Real-time Computing and Robotics, RCAR 2021, pp. 720–725 (2021). https://doi.org/10.1109/RCAR52367.2021.9517579
26. Pires, F., Ahmad, B., Moreira, A.P., Leitao, P.: Digital twin based what-if simulation for energy management. In: Proceedings of the 2021 4th IEEE International Conference on Industrial Cyber-Physical Systems, ICPS 2021, pp. 309–314 (2021). https://doi.org/10.1109/ICPS49255.2021.9468224
27. Teng, S.Y., Touš, M., Leong, W.D., How, B.S., Lam, H.L., Máša, V.: Recent advances on industrial data-driven energy savings: digital twins and infrastructures. Renew. Sustain. Energy Rev. **135**(February), 2021 (2020). https://doi.org/10.1016/j.rser.2020.110208
28. Park, K.T., Lee, D., Noh, S.D.: Operation procedures of a work-center-level digital twin for sustainable and smart manufacturing. Int. J. Precis. Eng. Manuf.-Green Technol. **7**(3), 791–814 (2020). https://doi.org/10.1007/s40684-020-00227-1
29. Wang, Y., Wang, S., Yang, B., Zhu, L., Liu, F.: Big data driven hierarchical digital twin predictive remanufacturing paradigm: architecture, control mechanism, application scenario and benefits. J. Clean. Prod. **248**, 119299 (2020). https://doi.org/10.1016/j.jclepro.2019.119299
30. Wang, W., Zhang, Y., Zhong, R.Y.: A proactive material handling method for CPS enabled shop-floor. Robot. Comput. Integr. Manuf. **61**(July 2019), 101849 (2020). https://doi.org/10.1016/j.rcim.2019.101849
31. Zhou, H., Yang, C., Sun, Y.: A collaborative optimization strategy for energy reduction in ironmaking digital twin. IEEE Access **8**, 177570–177579 (2020). https://doi.org/10.1109/ACCESS.2020.3027544
32. Senna, P.P., Almeida, A.H., Barros, A.C., Bessa, R.J., Azevedo, A.L.: Architecture model for a holistic and interoperable digital energy management platform. Procedia Manuf. **51**(2019), 1117–1124 (2020). https://doi.org/10.1016/j.promfg.2020.10.157
33. Cardin, O., et al.: Energy-aware resources in digital twin: the case of injection moulding machines. In: Borangiu, T., Trentesaux, D., Leitão, P., Giret Boggino, A., Botti, V. (eds.) SOHOMA 2019. SCI, vol. 853, pp. 183–194. Springer, Cham (2020). https://doi.org/10.1007/978-3-030-27477-1_14
34. Lima, F., Massote, A.A., Maia, R.F.: IoT energy retrofit and the connection of legacy machines inside the industry 4.0 concept. In: IECON Proceedings of the Industrial Electronics Conference, vol. 2019-Octob, pp. 5499–5504 (2019). https://doi.org/10.1109/IECON.2019.8927799
35. Wang, J., Huang, Y., Chang, Q., Li, S.: Event-driven online machine state decision for energy-efficient manufacturing system based on digital twin using max-plus algebra. Sustain. **11**(18), 5036 (2019). https://doi.org/10.3390/su11185036
36. Gupta, A., Basu, B.: Sustainable primary aluminium production: technology status and future opportunities. Trans. Indian Inst. Met. **72**(8), 2135–2150 (2019). https://doi.org/10.1007/s12666-019-01699-9
37. Wanner, J., Bahr, J., Full, J., Weeber, M., Birke, K.P., Sauer, A.: Technology assessment for digitalization in battery cell manufacturing. Procedia CIRP **99**, 520–525 (2021). https://doi.org/10.1016/j.procir.2021.03.110
38. Leiden, A., Herrmann, C., Thiede, S.: Cyber-physical production system approach for energy and resource efficient planning and operation of plating process chains. J. Clean. Prod. **280**, 125160 (2021). https://doi.org/10.1016/j.jclepro.2020.125160
39. Assad, F., Konstantinov, S., Ahmad, M.H., Rushforth, E.J., Harrison, R.: Utilising web-based digital twin to promote assembly line sustainability. In: Proceedings of the 2021 4th IEEE International Conference on Industrial Cyber-Physical Systems, ICPS 2021, pp. 381–386 (2021). https://doi.org/10.1109/ICPS49255.2021.9468209

40. Negri, E., Assiro, G., Caioli, L., Fumagalli, L.: A machine state-based digital twin development methodology. Proc. Summer Sch. Fr. Turco **1**, 34–40 (2019)
41. Kannan, K., Arunachalam, N.: A digital twin for grinding wheel: an information sharing platform for sustainable grinding process. J. Manuf. Sci. Eng. Trans. ASME **141**(2) (2019). doi: https://doi.org/10.1115/1.4042076
42. Park, K.T., Im, S.J., Kang, Y.S., Do Noh, S., Kang, Y.T., Yang, S.G.: Service-oriented platform for smart operation of dyeing and finishing industry. Int. J. Comput. Integr. Manuf. **32**(3), 307–326 (2019). https://doi.org/10.1080/0951192X.2019.1572225
43. Karanjkar, N., Joglekar, A., Mohanty, S., Prabhu, V., Raghunath, D., Sundaresan, R.: Digital twin for energy optimization in an SMT-PCB assembly line. In: Proceedings of the 2018 IEEE International Conference on Internet Things Intelligent Systems, IOTAIS 2018, pp. 85–89 (2019). https://doi.org/10.1109/IOTAIS.2018.8600830
44. Yan, K., Xu, W., Yao, B., Zhou, Z., Pham, D.T.: Digital twin-based energy modeling of industrial robots. In: Li, L., Hasegawa, K., Tanaka, S. (eds.) AsiaSim 2018. CCIS, vol. 946, pp. 333–348. Springer, Singapore (2018). https://doi.org/10.1007/978-981-13-2853-4_26
45. Zečević, N.: Energy intensification of steam methane reformer furnace in ammonia production by application of digital twin concept. Int. J. Sustain. Energy **41**(1), 12–28 (2022). https://doi.org/10.1080/14786451.2021.1893727
46. Karandaev, A.S., Gasiyarov, V.R., Radionov, A.A., Loginov, B.M.: Development of digital models of interconnected electrical profiles for rolling–drawing wire mills. Machines **9**(3), 1–28 (2021). https://doi.org/10.3390/machines9030054
47. Vatankhah Barenji, A., Liu, X., Guo, H., Li, Z.: A digital twin-driven approach towards smart manufacturing: reduced energy consumption for a robotic cell. Int. J. Comput. Integr. Manuf. **34**(7–8), 844–859 (2021). https://doi.org/10.1080/0951192X.2020.1775297
48. Lu, Y., Liu, C., Wang, K.I.K., Huang, H., Xu, X.: Digital twin-driven smart manufacturing: connotation, reference model, applications and research issues. Robot. Comput. Integr. Manuf. **61**(November), 101837 (2020). https://doi.org/10.1016/j.rcim.2019.101837

Product Lifecycle Management and Open Innovation in the Deep Tech Start-Ups Development

Bernardo Reisdorfer-Leite(✉), Marcelo Rudek, and Osiris Canciglieri Junior

Polytechnic School, Industrial and Systems Engineering Graduate Program (PPGEPS), Pontifícia Universidade Católica do Paraná (PUCPR), Curitiba, Brazil
{bernardo.leite,marcelo.rudek,osiris.canciglieri}@pucpr.br

Abstract. Deep tech start-ups are enterprises established on cutting-edge and radical technologies based on scientific discoveries (deep technologies), like advanced manufacturing and robotics, blockchain, agrotech and artificial intelligence. They start in university laboratories without a specific application and evolve from fundamental research to commercial application slower than start-ups based on off-line to on-line business model. With the increasing information exchange, complexity, and uncertainty among actors of innovation ecosystems and the world scientific community, it is required an approach that aims beyond the product development, covering the product lifecycle and understand the information flow among innovation ecosystem actors, i.e., how open innovation works across the boundaries of deep tech start-ups. This work performs a systematic literature review and proposes a conceptual model to understand how Product Lifecycle Management and Open Innovation support deep-tech start-ups in their development, seeking to encourage innovation ecosystems and members of the academic community towards the application of their research into disruptive solutions to complex issues.

Keywords: University spin-offs · Disruptive innovation · Science-based enterprises · Knowledge-based enterprises · Product innovation

1 Introduction

Product Lifecycle Management (PLM) represent a product centric lifecycle-oriented business model, supported by ICT, in which product data are shared among actors, processes and organisations in the distinct phases of the product lifecycle for achieving desired performances and sustainability for the product and related services [1]. As a technological solution, PLM establishes a set of tools and technologies that provide a shared platform for collaboration between product stakeholders and streamlines the flow of information throughout all stages of the product's life cycle [2].

As global competition intensifies and innovation becomes more risky, expensive, and necessary for survival, the concept of Open Innovation (OI) emerges as a novel strategy.

Published by Springer Nature Switzerland AG 2023
F. Noël et al. (Eds.): PLM 2022, IFIP AICT 667, pp. 106–115, 2023.
https://doi.org/10.1007/978-3-031-25182-5_11

According to Chesbrough & Bogers [3], OI is an innovation process distributed based on knowledge flows purposely managed across organisational boundaries, using monetary and non-monetary mechanisms in line with the business model of each organisation. These knowledge flows can involve knowledge flows to the focal organization (wing-leveraging external knowledge sources through internal processes), knowledge outputs from a focal organisation (leveraging internal knowledge through external marketing processes) or both (engaging external sources of knowledge and marketing activities).

As the innovation activity intensifies, Small and Medium-Sized Enterprises (SMEs) that generate innovation are being created worldwide exponentially, and the number is growing every day. In this universe, Start-ups play an essential role in the economy, differentiating themselves from other SMEs. Start-ups are companies created to evaluate business models developed around innovative ideas, typically proposed by several co-founders or members of the team [4]. Nevertheless, while the entire start-up market is growing, some parts are growing faster than others [5], the so-called Deep-tech start-ups (DTS), which are enterprises based on Deep-Technologies (Deep-techs).

Deep technology was sparked by Swati Chaturvedi, CEO of Propel(x) [6]. Deep techs are cutting-edge and disruptive technologies based on scientific discoveries and offer the potential to reinvent businesses for the better, maximizing impact, scale, profit, and social welfare [7]. Deep-techs usually originate in university laboratories, e.g., university spin-offs, which means that the technology itself is new and very sophisticated, it was necessarily undesigned with a concrete application or a specific market need [8], i.e., the business strategy of DTS is built round unique, differentiated, protected or difficult to reproduce, technological or scientific advances. Unlike software-based projects, where standard methods can be employed at any stage to rapidly develop a product, Deep-tech projects require frequently more time and specific knowledge. In addition, the core technology on which the start-up is based is extremely sophisticated and developed after years of fundamental research [8].

Start-ups concentrate efforts on product development [9]. A DTS, however, is slower and more ex-pensive than a digital start-up, for several reasons: 1) Strong research base: product development depends on fundamental research and/or advanced R&D, which requires support from a robust set of sophisticated skills, knowledge and infrastructure, in addition to prolonging the time to market products; 2) Heavy industrialization process: in addition to be supported by ICTs, most products in this field are hardware—typically based on advanced materials and resources—requiring highly developed industrial skills to acquire, manufacture and scale. These products are much more difficult to scale than products associated with the Internet and mobile technologies are used to; 4) Great investment needs: the infrastructure, skills and resources needed for a DTS require substantial financing capacity over an extended period; 5) Commercial application yet to be defined: the specifications of the final product can be undefined in the process. Blockchain developed as a specific technological solution for Bitcoin, for example, opened the door to an emerging financial market that its developers did not foresee. According to the Global Start-up Ecosystem report [5], the top four fastest growing sub-sectors of start-ups are: advanced manufacturing and robotics, blockchain, agrotech and new foods, and artificial intelligence, big data and analytics, that is, they depend completely on technological advances and tangible IP.

Most of this knowledge is generated by the R&D units (universities, Institutes, etc.), attempt to reduce the distance from academic environment to commercial application [10], and open innovation strategies, which perform a substantial role in the development of these new businesses, supporting the development of new products and services with a significant content of innovation that retain disruptive characteristics. Another relevant aspect represents the rise of the emerging need for organisations to become agile, not only their product development, which theme was highly associated with, but to respond quickly to the needs and desires of the market, that is, all the components of the socio-technical that the organisation and/or system is inserted must be agile. This characteristic contrasts with the development of deep-techs, which, as previously described, are based on scientific discoveries and have an indefinite time-to-market.

Founded on the concepts described above and performing systematic literature review, this work proposes to answer the following research question: ***"Product Lifecycle Management and Open Innovation approaches can support Deep-tech Start-ups?"***.

The subsequent sections of the paper are organised as follows. Section 2 presents the method. Section 3 presents the results and Sect. 4 discusses the elements and concludes the work.

2 Method

With the aim to answer the research question described in Sect. 1, we conducted a Systematic Literature Review (SLR) and content analysis of the literature on PLM and OI applied to the universe of DTS. To answer this question, we searched on two electronic data base platforms, i.e., Scopus and Web of Science. The time span selected was from 2011 until 2021. We exclusively selected articles published in journals and which language is English. The queries searched for DTS, PLM, OI and related terms on titles, abstracts, and keywords of the articles. The query that retrieved documents from DTSs and OI is *("deep tech*" OR "scientific discover*" OR "engineering innovation*" OR "technological innovation*" OR "science based start-up*" OR "university spinoff" OR "applied research") AND ("open innovation")*. The query that retrieved articles from DTSs and PLM is *("deep tech*" OR "scientific discover*" OR "engineering innovation*" OR "technological innovation*" OR "science based start-up*" OR "university spinoff" OR "applied research") AND ("product life cycle" OR "PLM" OR "product information" OR "product data" OR "beginning of life" OR "BOL" OR "middle of life" OR "MOL" OR "end of life" OR "EOL")*.

The two queries (DTS AND OI, and DTS and PLM) returned 478 articles that meet the search strategy. We removed 115 duplicates and retrieved the remaining 363 documents. The screening of titles, abstracts and keywords indicated ten articles and one article were included. We additionally searched for the articles published by the authors of the ten articles and included another seven articles which fulfil the criteria, totalizing seventeen articles to final selection.

3 Results

In this section, we present the findings from the systematic literature review about DTSs, PLM and OI, which can be combined into the schema in Fig. 1. The schema was conceived using inductive reasoning (drawing conclusions by going from the specific to the general).

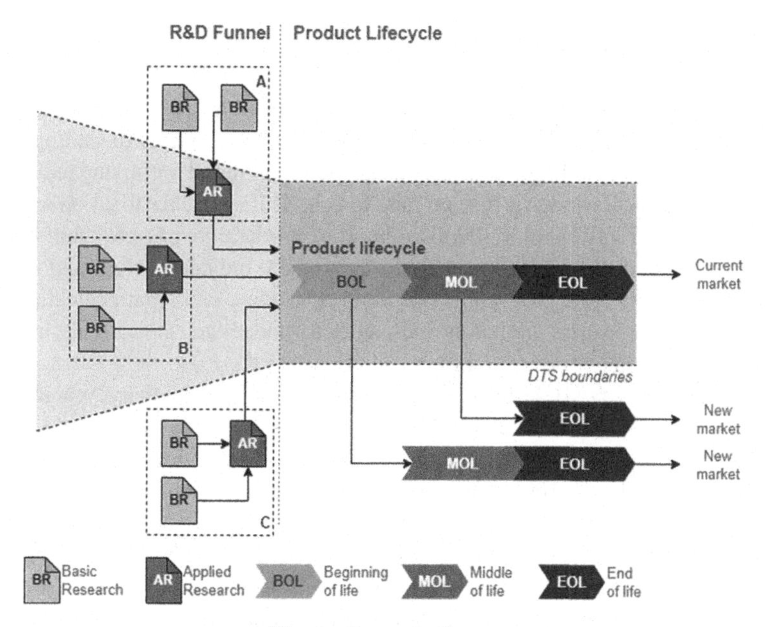

Fig. 1. General schema

The results indicate two clusters of studies: R&D-funnel and Product lifecycle. The cluster R&D-funnel is composed by studies that mostly use metric information studies (MIS) to harvest potential technologies and analyses whether technologies analysed mostly in scientific papers and patents are in the maturity level suitable to be translated into viable commercial products. Meanwhile, the other cluster, Product lifecycle deals with the challenges of translating research into a commercially viable product.

The section is presented in two subsections according to the findings. Subsection 3.1 describes the direct influence of OI on PLM applied to DTSs, while Subsect. 3.2 describes the indirect influence of OI on PLM regarding the DTs universe.

3.1 R&D Funnel

Most articles that contribute to the R&D funnel section massively use MIS to harvest and assess whether a technology has achieved its maturity to be developed and launched to the market, or to be commercially viable. The eight articles of this section and their contributions and applications are described below.

Lezama-Nicolás et al. [11] developed an approach to obtain the technology Readiness Level (TRL) from scientific articles, intellectual property documents, and news records, diminishing the level of bias in the analysis (compared to TRL) and avoiding the costs and drawbacks associated with assessing technological maturity through expert opinions. The proposed method could be employed to assess the maturity level of a technology or part of a technology which the DTS is willing to acquire rather than develop. Whilst Yeo et al. [12] proposes a bibliometric analysis of research papers that can be used as a proxy to measure the degree of uncertainty in a product's technological innovation. The method is applicable for "cherry-picking" promising products among vast numbers of candidate products.

Yeo et al. [13] developed a recommendation algorithm to suggest promising technologies to SMEs. The experimental method contains a distinct goal of finding forward technologies on the technology-value-chains and to recommend promising technologies to SMEs that need "knowledge arbitrage" and to help SMEs produce ideas on new R&D. Meanwhile, Su et al. [14] explore the likelihood of product commercialization and the speed of product commercialization with thirteen patent indicators classified into three perspectives of R&D strategies: collaboration, knowledge, and legal protection. R&D organizations can consume the patent indicators to understand how to attain product approval and accelerate product commercialization speed.

Su [15] discuss the possible correlation of patent characteristics that reveal significant variances at transitions with the change of patent characteristics over time and its strategic implications. The No. of Patent References, No. of Non-Patent References, No. of Foreign References, No. of IPCs, No. of UPCs, and litigated patent promises technology lifecycle indicators for detecting and forecasting technology lifecycles. The application of this work remains the possibility of studying technology lifecycle and uncovering technological development strategies with carefully selected patent characteristics.

Rodríguez-Salvador [16] propose a hybrid data model that combines the assessment of scientific publications and patent analysis production and is further supported by experts' feedback. The model uncovered the core areas of research of the analysed topic in his work. The proposed model helps (i) to study emerging technological innovations, (ii) to assess ongoing research trends and expose the knowledge landscape, and (iii) to better understand any emerging technologies that are currently coming onto the market.

Jun et al. [17] use patent documents as objective data to develop a model for vacant technology forecasting. The authors propose a matrix map that provides broad vacant technology areas. The model identifies the vacant technology areas of a given technology field by operating the results from both matrix map and K-medoids clustering based on support vector clustering. The proposed technology forecasting model can be applied to diverse technology fields, including R&D management, technology marketing and intellectual property management.

Wang et al. [18] shows that two contingency variables (absorptive capacity and technological uncertainty) produce a moderating impact on the relation between the external technology scouting of open search and a firm's open innovation generation. The industry dynamism provided a significant impact on the relation between external technology scouting and innovation generation in the intensely competitive high-technology environment.

3.2 Product Lifecycle

The second cluster deals mostly with product development in the product lifecycle and the impacts of open innovation. The nine articles selected to this section are described below.

Bahemia et al. [19] examine the effect of including eleven distinct types of external parties within an NPD project. The authors develop and evaluate a novel measurement scale that examines three dimensions of "openness": breadth, depth, and partner newness. The aggregation of the three dimensions might influence on the desired outcomes of the innovation project and the strength of patents. The study equally extends the extant supplier involvement in new product development literature to examine the effect of up to eleven types of external actor in NPD projects.

Bazan [20] designs a framework as a structured methodology based on the Stage-Gate® model, best practices in translational research, project management, new product development, business development, science of team science, and IP management. The proposed translational R&D model and methodology helps link university science and engineering research to commercial outcomes, i.e., to promote a seamless transition from research to business. The translational R&D model and methodology can be used to understand, describe, and implement their research as translational activities conducive to delivering their ideas to market. The framework is equally relevant to university researchers who might not intend to turn their early-stage innovations into businesses but advance the development of the innovations to the point where they become attractive for others, to embrace the challenge of developing the innovations farther for the market.

Also based on the Stage-Gate®, Bers et al. [21] builds new stages and new dimensions, partitioning the product–market focus of Stage-Gate® into four concurrent, interwoven tracks: market/societal, scientific/technological, business/organizational, and innovation ecosystem. The first represents a stage of strategy development, not around the company but around the innovation itself. The second new stage is designing organizations around the innovation. The other stages include a pre-discovery period and three post-launch periods, which provide for the extended experimentation and incremental improvement required of radical innovation. The model facilitates entry and exit of the multiple parties involved in radical innovations and proposes a technique (discrete real options analysis) that enables each party to assess the financial value of the innovation as it unfolds. The model can assist scholars and university managers in the developing of organizations (potentially spin-offs) around the innovation itself, in an inside-out approach, rather than an outside-in.

Moaniba et al. [22] contributes to the understanding of drug commercialization by constructing knowledge-based indicators that should allow researchers to quantitatively track and measure the effectiveness of the diversity of knowledge acquired in the contexts of countries involved in the co-invention (of the new drug), and the industrial and technological varieties. The study introduces a new indicator of the intensity of knowledge search by taking into consideration the geographic distance considered in the search for external knowledge, the significant effects of both the intensity of the knowledge sourced and the geographic distance considered in the search extend the existing understanding of the broader concept of innovation strategy.

Germann et al. [23] describes the different modalities of cross-fertilisation between basic university or publicly funded research institutions and the applied research and development activities within the pharmaceutical industry. The study states that innovation opportunities found in more narrow windows are easier to find, select, develop, and translate into the own organisation of the pharmaceutical company because of the closeness of the idea towards the receiving organisation and the scientific expertise. The study proposes the primary elements of innovative translational research from academic perspective and how to integrate it to a product development process of a big pharmaceutical company perspective.

Slavova [24] concludes university-industry interactions demonstrate a direct association with a firm's engagement in open systems of scientific knowledge exchange. University-industry alliances are conduits for transfer of industrial practices like entrepreneurial activities and patenting to academia. Firms with direct ties to universities are better-off at generating inventions of greater quality, greater impact, and more quickly.

Zhu et al. [25] demonstrate that OI breadth and OI depth positively affect NPD speed. Varying types of business models must be aligned with OI breadth and depth to better facilitate NPD speed. Additionally, OI breadth increases NPD speed through accessing diverse information sources, recombining complementary information, and capturing innovation opportunities, however, OI depth advances NPD speed through information utilization, such as information sharing and transfer. External information sources widely and deeply employed can provide innovative ideas and resources for gaining more innovation opportunities and thus speed up NPD processes. To regulate the NPD desired speed, managers can select the appropriate depth and breadth of OI in their operations.

Zou et al. [26] show that during the stages of growth and maturity, firms experience a more active trend of valuing, acquiring, assimilating, transforming, and exploiting external knowledge than in the introduction and decline stages of the product lifecycle. A firm's potential and realized absorptive capacities achieve their peak during the product lifecycle growth stage. The changing tendency of a firm's technology-innovation achievements is consistent with the product lifecycle curve, which means that technology-innovation achievements are driven typically by market demand of product lifecycle. The study suggests on how firms can adjust management policies to improve their absorptive capacity and technological innovation performance during different product lifecycle stages.

4 Discussion and Conclusion

We have selected to link PLM to OI in DTSs, because PLM provides an adequate understanding of the product lifecycle, from inception to disposal, and especially because PLM encompass and at the same time look beyond product development and product lifecycle from marketing's point of view. Since many organisations acquire competitive advantages using PLM to manage their product lifecycles, this affirmation could not be different from a start-up perspective. To narrow the gap of information to support PLM, we complete this study with OI, representing all the information flow to the development of a DTS. Therefore, we conducted a systematic literature review of PLM and OI applied to DTSs and the results provided scientific evidence to develop a preliminary conceptual model. From the selected articles, we divided them in two clusters. The first primarily addresses aspects related to MIS to harvest and assess science knowledge and intellectual properties that could be use in DTS. The latter provided a comprehensive vision of the surroundings of OI on PLM application in DTSs and the translation of fundamental research to a commercial product.

Start-ups in general lack of formal managerial structures due to its size and activities risks. To minimise adversary situations to their existence, start-ups rely on the ecosystems to evolve and then at some point of their lifecycle exit via IPO, M&A, etc. Therefore, start-ups are learning organisations that need information to expand and achieve their milestones. Depending on closed innovation does not represent an option when developing innovative technologies. From founder's perspective, learning remains an ongoing activity. The ecosystem equally represents valuable opportunities to develop collaborative networks, which helps to share the risks founders typically take when they decide to initiate a start-up. DTSs sustain an inside-out business and technology push strategy, their value proposition is built round the unique and hard to reproduce developed technology. Therefore, it is challenging for an organisation to generate all the critical knowledge required for the innovation process internally. Especially for start-ups that have minimal structures and sometimes inexistent structures, relying in support structures within innovation ecosystems. Summarising the reasons of DTSs success, it would be noteworthy to identify which factors could be more relevant to the DTS universe.

As exposed in previous sections of this work, information performs a significant role in DTSs. DTSs rely in maturity levels which can move forward with the adoption of external sources of information, amongst then collaboration, i.e., OI. Additionally, to perform the proper management of the product information of technological innovation being developed by DTSs, PLM provides solutions that DTSs faces when dealing with their product lifecycle. Absorbing external knowledge from scientific sources, using proper tools to manage the project, DTSs can excel the initial phases of their existence and increase the product maturity, therefore surpassing the valley of death.

We have not found studies covering end-of-life (EOL) stages of product lifecycle, and this is a yet to be discussed topic even among practitioners. It is comprehensive that most of the effort necessary to establish a radical technological solution must be in the initial stages of the product lifecycle. Also, some models found in SLR performed by this study suggests that customer and users play a substantial role in the technology development and the preliminary stages of product lifecycle. This affirmation is not entirely applicable to DTSs. Because the strategies to develop their products, DTSs exist to solve big Environmental and Social issues and their attractiveness to investments increases due to a change of society's value and eager for radical innovations, which is incorporating sustainable elements to its set of relevant values.

DTSs have naturally an advantage over other start-ups, the abundance of information surrounding their business that could help to reduce uncertainty related to their business application. Although few studies are being conducted about DTSs, they represent until now an increasing field of study. Nevertheless, the increasing number of investments in these companies, innovation governments policies, technological advances and the future depletion of "offline to online business model innovation start-ups", must push an increasing number of DTSs entrepreneurs to develop radical solutions to solve even more complex problems of the present and futures.

This work contributes to promote the foundations of "research to market" and academic entrepreneurship performed by Universities and Research Centres. It provides guidelines to players of innovation ecosystems develop their own frameworks to support radical innovations in DTSs using PLM and OI.

Acknowledgements. The authors acknowledge the financial support provided by the Pontifícia Universidade Católica do Paraná (PUCPR), the Industrial and Systems Engineering Graduate Program (PPGEPS), the Coordination for the Improvement of Higher Education Personnel (CAPES) and the National Council for Scientific and Technological Development (CNPq).

References

1. Terzi, S., Bouras, A., Dutta, D., Garetti, M., Kiritsis, D.: Product lifecycle management–from its history to its new role. Int. J. Prod. Lifecycle Manag. **4**(4), 360–389 (2010)
2. Ameri, F., Dutta, D.: Product lifecycle management: closing the knowledge loops. Comput.-Aided Design Appl. **2**(5), 577–590 (2005)
3. Chesbrough, H., Bogers, M.: Explicating open innovation: clarifying an emerging paradigm for understanding innovation. New Front. Open Innov. 3–28 (2014)
4. Salamzadeh, A., Kesim, H.K.: The enterprising communities and startup ecosystem in Iran. J. Enterp. Commun.: People Places Glob. Econ. **11**(4), 456–479 (2017)
5. Global Startup Ecosystem Report 2019. https://startupgenome.com/gser2019. Accessed 7 Apr 2022
6. What are deep technology startups and why are they good investments? https://www.propelx.com/blog/what-are-deep-technology-startups-and-why-are-they-good-investments/. Accessed 7 Apr 2022
7. Siegel, J., Krishnan, S.: Cultivating invisible impact with deep technology and creative destruction. J. Innov. Manag. **8**(3), 6–19 (2020)
8. How to build a succesful deep tech acceleration program. https://hello-tomorrow.org/wp-content/uploads/2019/11/How-to-build-a-succesful-deep-tech-acceleration-program-Hello-Tomorrow-Bpifrance-1.pdf. Accessed 7 Apr 2022
9. Reisdorfer-Leite, B., Marcos de Oliveira, M., Rudek, M., Szejka, A.L., Canciglieri Junior, O.: Startup definition proposal using product lifecycle management. In: Nyffenegger, F., Ríos, J., Rivest, L., Bouras, A. (eds.) PLM 2020. IAICT, vol. 594, pp. 426–435. Springer, Cham (2020). https://doi.org/10.1007/978-3-030-62807-9_34
10. de Oliveira, M.M., Reisdorfer-Leite, B., Rudek, M., Junior, O.C.: Evaluation of the impact of open innovation and acceleration programs on research and development performed by universities. In: 27th ISTE International Conference on Transdisciplinary Engineering, pp. 92–101. IOS Press, Amsterdam (2020)
11. Lezama-Nicolás, R., Rodríguez-Salvador, M., Río-Belver, R., Bildosola, I.: A bibliometric method for assessing technological maturity: the case of additive manufacturing. Scientometrics **117**(3), 1425–1452 (2018). https://doi.org/10.1007/s11192-018-2941-1
12. Yeo, W., Kim, S., Park, H., Kang, J.: A bibliometric method for measuring the degree of technological innovation. Technol. Forecast. Soc. Chang. **95**, 152–162 (2015)

13. Yeo, W., Kim, S., Coh, B.Y., Kang, J.: A quantitative approach to recommend promising technologies for SME innovation: a case study on knowledge arbitrage from LCD to solar cell. Scientometrics **96**(2), 589–604 (2013)
14. Su, H.N., Lin, Y.S.: How do patent-based measures inform product commercialization?—The case of the United States pharmaceutical industry. J. Eng. Tech. Manage. **50**, 24–38 (2018)
15. Su, H.N.: How to analyze technology lifecycle from the perspective of patent characteristics? The cases of DVDs and hard drives. R&D Manage. **48**(3), 308–319 (2018)
16. Rodríguez-Salvador, M., Rio-Belver, R.M., Garechana-Anacabe, G.: Scientometric and patentometric analyses to determine the knowledge landscape in innovative technologies: the case of 3D bioprinting. PLoS One **12**(6), e0180375 (2017)
17. Jun, S., Park, S.S., Jang, D.S.: Technology forecasting using matrix map and patent clustering. Ind. Manag. Data Syst. **115**(5), 786–807 (2012)
18. Wang, C.H., Quan, X.I.: The role of external technology scouting in Inbound open Innovation generation: evidence from high-technology industries. IEEE Trans. Eng. Manage. **68**(6), 1558–1569 (2019)
19. Bahemia, H., Squire, B., Cousins, P.: A multi-dimensional approach for managing open innovation in NPD. Int. J. Oper. Prod. Manag. **37**(10), 1366–1385 (2017)
20. Bazan, C.: "From lab bench to store shelves:" a translational research & development framework for linking university science and engineering research to commercial outcomes. J. Eng. Tech. Manage. **53**, 1–18 (2019)
21. Bers, J.A., Dismukes, J.P., Mehserle, D., Rowe, C.: Extending the stage-gate model to radical innovation-the accelerated radical innovation model. J. Knowl. Econ. **5**(4), 706–734 (2014)
22. Moaniba, I.M., Lee, P.C., Su, H.N.: How does external knowledge sourcing enhance product development? Evidence from drug commercialization. Technol. Soc. **63**, 101414 (2020)
23. Germann, P.G., Schuhmacher, A., Harrison, J., Law, R., Haug, K., Wong, G.: How to create innovation by building the translation bridge from basic research into medicinal drugs: an industrial perspective. Hum. Genom. **7**(1), 1–6 (2013)
24. Slavova, K.: When firms embrace science: university alliances and firm drug development pipeline. J. Prod. Innov. Manag. **39**(2), 265–279 (2021)
25. Zhu, X., Xiao, Z., Dong, M.C., Gu, J.: The fit between firms' open innovation and business model for new product development speed: a contingent perspective. Technovation **86**, 75–85 (2019)
26. Zou, B., Guo, F., Guo, J.: Absorptive capacity, technological innovation, and product life cycle: a system dynamics model. Springerplus **5**(1), 1–25 (2016). https://doi.org/10.1186/s40064-016-3328-5

Distributed Approach for Integration in Industrial Systems

Sylvain Lacroix[1,2](✉), Benoit Eynard[1](✉), Julien Le Duigou[1](✉), Xavier Godart[2](✉), and Christophe Danjou[3](✉)

[1] Universités de Technologie de Compiègne, Compiègne, France
{sylvain.lacroix,benoit.eynard,julien.le-duigou}@utc.fr
[2] Inetum, Meudon-la-Forêt, France
{sylvain.lacroix,xavier.godart}@inetum.com
[3] Polytechnique Montréal, Montréal, Canada
christophe.danjou@polymtl.ca

Abstract. Digital thread, that is interoperability between the different layers of companies' organization and throughout their processes, is a key target for Industrie 4.0 or Smart Manufacturing. The historical drivers of integration, PLM and the Automation Pyramid (ISA-95), were decisive to structure processes and organizations when no technical solution to reach global interoperability existed. However, now, thanks to the maturity of new technologies at every level (growing computing power in machines, efficient communication protocols, reliable heterarchical architecture), new solutions and architecture models are emerging. After a broad state of the art, this paper intents to discuss the opportunities brought by new architecture model such as RAMI 4.0, Service Oriented Architecture and Multi Agent Systems applied to industrial use cases, and the necessary questioning of the hierarchical integration approaches that are still prevalent.

Keywords: Integration · PLM · Operation management

1 Introduction

The digital transformation, that is now widespread in industrial companies, is a mean to increase quality of products and services through a full deployment of ICT (Information Technology Tools). One of digital transformation core concept is digital continuity or digital thread. This rests upon the idea of a channel to capture and share data across the company. There have been many works, theoretical and practical, in the last 30 years, supporting companies to deploy a digital thread through their organization. There are several axes framing digital thread. The focus can be on a PLM (Product Lifecycle Management) aspect, or on a more OM (Operations Management) aspect, depending on the source of the data. Yet the digital thread along these axes was limited by the interoperability between systems. Lately, new proposals addressing interoperability (communication standards, Service Oriented Architecture, Multi Agent Systems) provide solutions to

Published by Springer Nature Switzerland AG 2023
F. Noël et al. (Eds.): PLM 2022, IFIP AICT 667, pp. 116–125, 2023.
https://doi.org/10.1007/978-3-031-25182-5_12

foster digital thread in these existing frames. But they also enable new, more distributed, architecture, that are becoming interesting alternatives.

The present paper intents to question the current divergences between these axes and explore the possibilities for a convergence. Section 2 will describe the digital thread along three identified axis. Section 3 will explore recent proposals that may challenge this axes division. Section 4 will discuss the risen challenges and opportunities.

2 Integration Axis

Digital thread, and its integration, relies on the concept of interoperability. The term "integration" can have several meanings while addressing interoperability questions. This section will clarify this ambiguity but as well define interoperability.

Integration can refer to the concept of interoperability: "interoperability lies in the middle of an 'integration continuum' between compatibility and full integration" [1]. Furthermore, [2] proposes three integration axes: vertical, horizontal, and end to end. Vertical integration describes the intelligent cross-linking and digitalization across the different aggregation and hierarchical levels of a value creation module from manufacturing stations via manufacturing cells, lines and factories. Horizontal integration describes the cross-company and company-internal intelligent cross-linking throughout the value chain of a product life cycle and between value chains of adjoining product life cycles. PLM integration describes the intelligent cross-linking and digitalization throughout all phases of a product life cycle: from the raw material acquisition to manufacturing system, product use, and the product end of life. These three integration axes are the subject of the present section.

2.1 Integration Through PLM

"PLM can be broadly defined as a product centric – lifecycle-oriented business model, supported by ICT, in which product data are shared among actors, processes and organizations in the different phases of the product lifecycle for achieving desired performances and sustainability for the product and related services" [3]. It is critical to ensure a data continuity linking the different tools used for product design and manufacturing. For the product development cycle (including design, simulation, manufacturing and assembly), [4, 5] conclude that the two main ways to reach interoperability between the different systems are the use of standard formats, and of ontology models.

Ontologies are explicit specifications of a conceptualization [6]. A conceptualization is an abstract, simplified view of the world that we wish to represent for some purpose. Every knowledge base, knowledge-based system, or knowledge-level agent is committed to some conceptualization, explicitly or implicitly. In a literature review [7] presents several ontology based product models such as PRONTO, PSL, OntoPDM, PROMISE, OntoSTEP. These ontologies can be built from scratch or based on existing standards. [8], makes a proposal for using ontology to enable semantic interoperability between different PLM systems, that does not lead to any technology constraint or tool.

Then, the link between data from the design phase to the actual production is handled by the Product Data Management. "PDM is the system developed to efficiently manage the product-related data and entire data flow for the work process emerging from new product development or existing product modification based on the work of the R&D division. Therefore, in the manufacturing industry, the connection from PDM of product data, that is, of Bill of Material (BOM) data, to ERP is an essential requirement and also the most important problem that needs to be solved" [9]. The link between engineering data and manufacturing data is done through the transformation of the Engineering BOM (EBOM) in the Manufacturing BOM (MBOM) at the ERP level [10].

Predictably the Beginning Of Life (BOL) is the phase with the largest quantity of data, and for which more efforts have been made to reach interoperability. Regarding the latter stage of PLM, Middle and End of Life (MOL, EOL), there are less use cases because of the lack of data. It would be interesting to assess accurately what frameworks now exist for these stages of PLM, and how it can interact with OM.

2.2 Integration for Operation Management

While PLM is product centric, horizontal and vertical integration refer to the architecture of the company. This architecture, although handling product data, is the support of Operations Management: "OM is management of a systemic transformation process to convert a set of inputs into outputs". It rests on the ERP, as a central pillar, but also embraces other management systems (Supply chain management, customer relationship management., supplier relationship management., knowledge management.) [11].

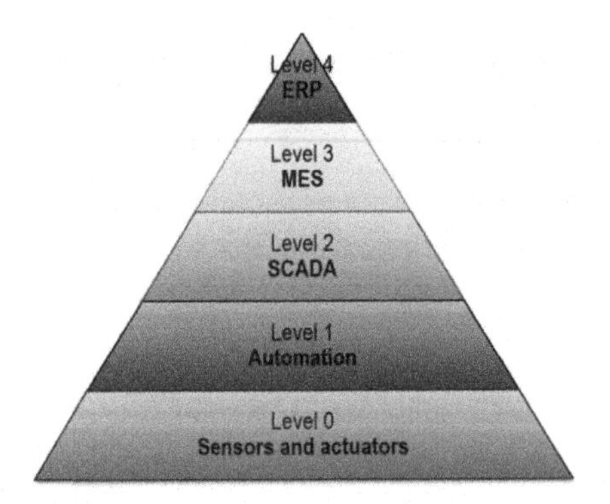

Fig. 1. Automation pyramid [12]

A breakdown in several business layers of the IT system has been proposed in the ISA-95 [12] (Fig. 1) framework: Level 0 and 1 are contained in the machines that compose the industrial system. There the integration issues are the prerogative of the Original Equipment Manufacturer (OEM).

Focusing on the 3 higher levels, (2:SCADA, 3:MES, 4:ERP) [13] proposes a new pyramid more consistent with the current integration questions (Fig. 2). The different machines are part of the field level and may be connected, although using distinct technologies (from legacy PLC to recent NC). The scope of the cell level is the orchestration of the different systems on the field level, in order to achieve the goals set at company level. MES and the human management of operations would occur at cell level; and ERP at company level.

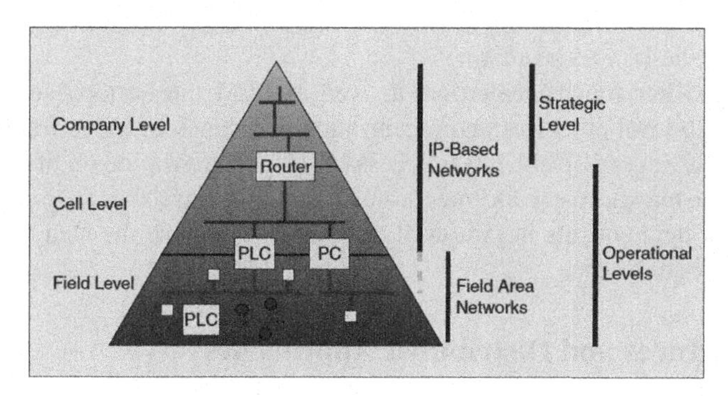

Fig. 2. Revised automation pyramid [13]

2.2.1 Integration Across Manufacturing Process (Horizontal)

Discussing the horizontal integration at field level is assessing the interoperability between the different machines. In the framework of [14], it can be achieved by "integration", "unification", or "federation". The integration approach means a standard is used by all the components. The current standard that is broadly spread is OPC UA, although MQTT is an emerging alternative [15]. Yet some systems that share more specific functions can have a more specific standard (AML, PackML). However it is common to see in a shopfloor successive generations of machine [16] that communicate and compute differently. In this case a semantic data model can oversee the communications, providing a translation between any component, suiting [14] definition of unification. Otherwise, interfaces can be set between components to manage semantic translations issues, realizing interoperability through federation. Asset Administrations Shell is an application of this strategy [17]. Notwithstanding these conceptually different approaches, it is usual to mix them to extend the scope of interoperability, to provide bridges between different standards or data models.

For cell level integration, MES solutions are usually unique for a given production system and do not need horizontal interoperability. Regarding company level, ERP interoperability in the context of extended enterprise is a challenge but it does not enter the scope of this paper.

2.2.2 Integration Across Business Layers (Vertical)

Data from machines is used strictly for follow up of production progress but an automated access at a sufficient rate enable more uses, such as monitoring for preventive maintenance [18] or near real time modelling for digital twin [19]. The main prerequisite of these use cases is the deployment horizontal integration with recent standards at field level (OPC UA or else). Regarding interoperability between ERP and MES layers, it is less of a challenge because both MES and ERP (Fig. 2) networks are IP based; there is no issue regarding technical interoperability. Still the question of semantic understanding stands. As a solution for this, MESA formalized a standard, B2MML (Business to Manufacturing Markup Language) that consists of a set of XML schemas written using the World Wide Web Consortium's XML Schema language (XSD) that implement the data models in the ISA-95 standard.

Vertical and horizontal integration, as well as PLM integration, were built on a partition of high-level processes and organization in distinct layers, to frame tools and methods to be use successively. It was necessary when there was no set of interoperable ICT solutions to manage the workflows throughout a company. Now, the progress of ICT tools and their deployments in industrial companies challenge the idea of integration following numbered steps.

3 Architectures and Distributed Approaches

3.1 Multidimensional Architecture Model

In a recent review of enterprise architecture model, [20] lists IIRA (2015), RAMI 4.0 (2015), IIVRA (2016) SITAM (2016), IBM industrie 4.0 (2017) and LAFSA (2019), that were created in the frame of Industrie 4.0. Amongst these 6 architectures, it is RAMI 4.0 and IIRA, that are the most popular and frequently mentioned. They both are supported by industrial consortia: German Electrical and Electronic Manufacturers' Association (ZVEI) for RAMI 4.0 and the Industrial Internet Consortium (IIC) for IIRA.

For instance, RAMI (Fig. 3) has three dimensions: business layers, life cycle value stream and hierarchy levels that create a multi-dimensional model allowing detailed partition of the enterprise layers.

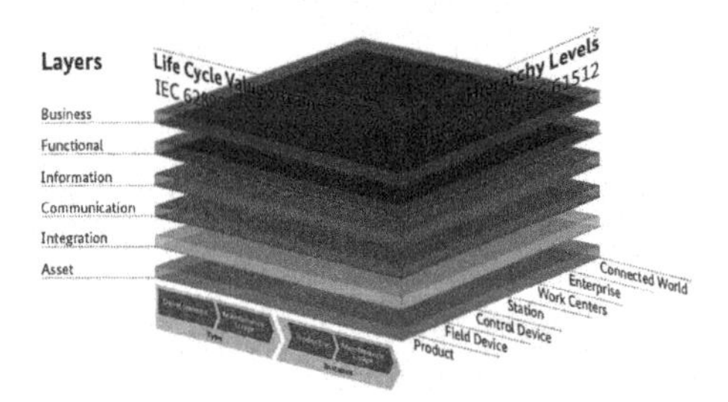

Fig. 3. RAMI 4.0 [21]

These reference architectures are remarkably different from the ISA95 proposal, and challenge the approach by integration (horizontal, vertical, PLM) because they intent to address all the different dimensions at once, and thus cannot be linearly described.

3.2 Service Oriented Architecture

Similarly, the service-oriented architecture is an architectural approach for software development that allows the recursive aggregation of services into new business processes and application. [22]. SOA takes business applications and breaks them down into individual functions and processes as services. Each existing services can be recomposited, reconstructed, and reused to new applications [23].

SOA addresses the requirements of loosely coupled, standards-based, and protocol independent distributed computing. Business operations running in an SOA comprise a number of invocations of these different components, often in an event-driven or asynchronous fashion that reflects the underlying business process needs. This functionality is provided by the Enterprise Service Bus (ESB) that is an integration platform that utilizes Web services standards to support a wide variety of communications patterns over multiple transport protocols and deliver value-added capabilities for SOA applications [24]. More practically, ESB is a three-party system with service providers on one side, service customers on the other side, and a service broker that orchestrate the exchanges.

Unfortunately, ESB is a solution only for software integration, and does not embrace industrial components. Yet, [25] and [26] offer an experimental extension of the ESB to the manufacturing operations, with a Manufacturing Service Bus (MSB). They explain that the traditional layers hierarchically ordered can be directly linked together with the MSB. Thus, there is more flexibility for the business processes to make interact directly any of its module notwithstanding any hierarchical difference. Other similar proposals based on ESB exist: Line Information System Architecture (LISA) [27] or a service bus gathering PLM, ERP and MES [28].

3.3 Multi Agent Systems in Manufacturing

[29] presents current and future industrials systems as Cyber Physical Production Systems (CPPS). CPPS are composed of industrial Cyber Physical Systems (CPS): collaborating computational entities which are in intensive connection with the surrounding physical world and its on-going processes, providing and using, at the same time, data-accessing and data-processing services available on the Internet. This new distribution of data-accessing and processing among the components of industrial systems challenges again the automation pyramid (Fig. 4): it is now possible to take decisions at a lower level for real time critical events.

To summarize, the latest enterprise architecture reference model are matrices with numerous interdependent partitions to be addressed and integrated. From the ICT perspective, SOA is now a mainstream solution to ensure interoperability in the IT layer of the companies in a *heterarchical* way (Fig. 4). Then, from the machines point of view, the concept of CPPS considers distributed computation power and new communication protocol ensuring direct interoperability at the lower level. All these approaches are consistent together but divergent compared to the linear and step by step integrations presented earlier.

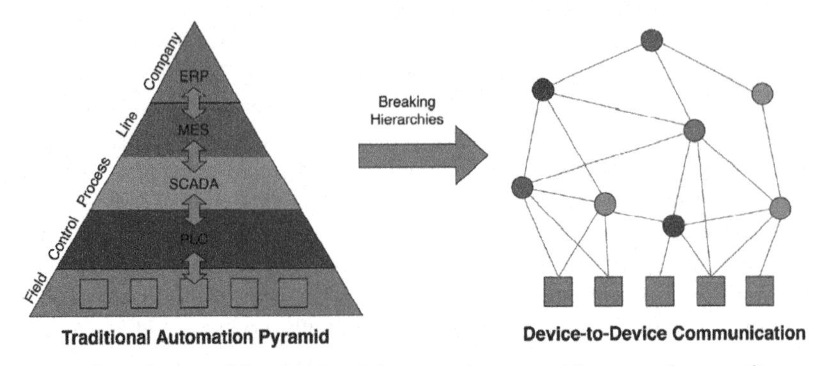

Fig. 4. Transition from traditional industrial automation pyramid to cross-layer entity-to-entity communication [20]

4 Discussion and Conclusion

Digital thread is a key challenge today because numerous software and data sources coexist and there is a high business value to have a consistent set of interoperable solution to capture and use data. Section 2 presented the major guidelines for product data and enterprise systems, namely PLM, and vertical and horizontal integration based on ISA-95 automation pyramid. Lastly Sect. 3 focused on newer approaches that lean on technologies recently available for industrial systems.

Although PLM and automation pyramid have structured industrials companies while ICT tools were not available and integrated, new technologies and concepts are now emerging that does not need so many layers and that provide more distributed abilities.

Firstly, the developments in this paper show a simplification of the issues about vertical integration. Originally, the automation pyramid [12] had five levels. Regarding industrial organizations, the integration effort from OEM has compressed in one "field level" the first 3 (0: sensors and actuators, 1: automation, 2: SCADA) [13]. In the meantime, the broad deployment of SOA in IT-based networks, through ESB, has flattened all hierarchy between software [24]. It then remains only two levels: the Information Technology (IT) layer (IP-based network) and the Operational Technology (OT) layer (Field area networks). Nowadays, the research efforts for vertical integration are thus mainly focusing on the IT/OT interface.

ESB extending to the manufacturing process seems a very promising SOA solution, using multi agent systems. Some experiments have been successful, but their scope is quite small. An interesting research axis would be to assess the scalability and the robustness of this concept at a company size. SOA could support the rising alternative to traditional hierarchical architectures: decentralized networks (Fig. 4) of CPS. These systems have by definition the ability to take decisions autonomously, thus providing indisputably quicker reaction time. The need for near real-time decisions, and for reactive and flexible systems [30] could become the issue that will trigger a reassessment of traditional architectures: "As a result rigid and centralized organizations may become obsolete" [29].

With the emergence of decentralized systems, also comes the possibility to set up local optimizers to deal with disturbances or unforeseen events. Yet the need for a global optimization remains to reach delivery dates and performance goals. To handle both, Dynamic Architectures have been proposed. [31] present a state of the art and the current challenges of hybrid control architectures, that include switching systems between a predictive mode (globally optimized) and a reactive mode (locally and temporarily optimized). These theoretical architectures are very complex; but if they reached a sufficient level of maturity, they could become mainstream thank to their unmatched industrial efficiency.

Lastly, the opposition between PLM integration and OM integration was seeming until the creation of new architecture reference models that support both axes. The link between data from the product (drawings, BOM, ...) and from the industrial system (OEE, machines static parameters, ...) are theoretically possible but a unified data model would have a very large scope. For instance [32] proposes compatible data models for product and process to link PLM and Equipment (machines) Lifecycle Management in aircraft manufacturing, resulting in an already very complex model.

To conclude, the historical PLM and OM architectures have been both designed and deployed to manage system and tools that were in no way interoperable. The recent technological advances, on software, hardware, and communication protocols, have unlocked new possibilities of integration, that bypass these architectures. The research axis emphasized in this paper are about the scalability of decentralized systems applied to manufacturing systems, SOA and dynamic architectures; and also, about the development of solutions handling both product data and operation management data overpassing the historical lack of awareness between PLM and Operation Management systems.

References

1. Tolk, A., Diallo, S.Y., Turnitsa, C.D.: Applying the levels of conceptual interoperability model in support of integratability, interoperability, and composability for system-of-systems engineering. Comput. Model. Simul. Eng. Fac. Publ. **27** (2007). https://digitalcommons.odu.edu/msve_fac_pubs/27/
2. Stock, T., Seliger, G.: Opportunities of sustainable manufacturing in industry 4.0. Procedia CIRP **40**, 536–541 (2016). https://doi.org/10.1016/j.procir.2016.01.129
3. Terzi, S., Bouras, A., Dutta, D., Garetti, M., Kiritsis, D.: Product lifecycle management - From its history to its new role. IJPLM **4**, 360 (2010)
4. Danjou, C.: Ingénierie de la chaîne numérique d'industrialisation : proposition d'un modèle d'interopérabilité pour la conception-fabrication intégrées. [PhD thesis]. Université de Technologie de Compiègne (2015)
5. Danjou, C., Le Duigou, J., Eynard, B.: Manufacturing knowledge management based on STEP-NC standard: a Closed-Loop Manufacturing approach. Int. J. Comput. Integr. Manuf. **30**(9), 995–1009 (2017)
6. Gruber, T.R.: A translation approach to portable ontology specifications. Knowl. Acquisition **5**, 199–220 (1993)
7. El Kadiri, S., Kiritsis, D.: Ontologies in the context of product lifecycle management: state of the art literature review. Int. J. Prod. Res. **53**, 5657–5668 (2015)
8. Sriti, M.F., Assouroko, I., Ducellier, G., Boutinaud, P., Eynard, B.: Ontology-based approach for product information exchange. Int. J. Prod. Lifecycle Manage. **8**(1), 1–23 (2015)
9. Lee, C., Leem, C.S., Hwang, I.: PDM and ERP integration methodology using digital manufacturing to support global manufacturing. Int. J. Adv. Manuf. Technol. **53**(1), 399–409 (2011)

10. Fortin, C., Huet, G.: Manufacturing process management: iterative synchronisation of engineering data with manufacturing realities. IJPD **4**, 280–295 (2007)
11. Bayraktar, E., Jothishankar, M.C., Tatoglu, E., Wu, T.: Evolution of operations management: past, present and future. Manage. Res. News **30**, 843–871 (2007)
12. American National Standards Institute. : ANSI/ISA-95.00.01–2010 (IEC 62264–1 Mod) Enterprise-Control System Integration - Part 1: Models and Terminology(2010)
13. Sauter, T.: The continuing evolution of integration in manufacturing automation. EEE Ind. Electron. Mag. **1**, 10–19 (2007)
14. International Standards Organization.: ISO 14258—Industrial automation systems—Concepts and rules for enterprise models(1998)
15. Profanter, S., Tekat, A., Dorofeev, K., Rickert, M., Knoll, A.: OPC UA versus ROS, DDS, and MQTT: performance evaluation of industry 4.0 protocols. In: 2019 IEEE International Conference on Industrial Technology, pp. 955–962. IEEE (2019)
16. Givehchi, O., Landsdorf, K., Simoens, P., Colombo, A.W.: Interoperability for industrial cyber-physical systems: an approach for legacy systems. IEEE Trans. Ind. Inf. **13**, 3370–3378 (2017)
17. Fuchs, J., Schmidt, J., Franke, J., Rehman, K., Sauer, M., Karnouskos, S.: I4.0-compliant integration of assets utilizing the asset administration shell. In: 2019 24th IEEE International Conference on Emerging Technologies and Factory Automation (ETFA), pp. 1243–1247 (2019)
18. García, M.V., Irisarri, E., Pérez, F., Estévez, E., Marcos, M.: An open CPPS automation architecture based on IEC-61499 over OPC-UA for flexible manufacturing in oil&gas industry. IFAC-PapersOnLine **50**, 1231–1238 (2017)
19. Stark, R., Kind, S., Neumeyer, S.: Innovations in digital modelling for next generation manufacturing system design. CIRP Ann. **66**, 169–172 (2017)
20. Nakagawa, E.Y., Antonino, P.O., Schnicke, F., Capilla, R., Kuhn, T., Liggesmeyer, P.: Industry 4.0 reference architectures: state of the art and future trends. Comput. Industr. Eng. **156**, 107241 (2021)
21. ZVEI. : Reference Architecture Model Industry 4.0. (2015)
22. Unger, T., Mietzner, R., Leymann, F.: Customer-defined service level agreements for composite applications. Enterp. Inf. Syst. **3**, 369–391 (2009)
23. Xu, L.D.: Enterprise systems: state-of-the-art and future trends. IEEE Trans. Ind. Inf. **7**, 630–640 (2011)
24. Papazoglou, M.P., van den Heuvel, W.-J.: Service oriented architectures: approaches, technologies and research issues. VLDB J. **16**, 389–415 (2007)
25. Morariu, C., Morariu, O., Borangiu, T., Raileanu, S.: Manufacturing service bus integration model for implementing highly flexible and scalable manufacturing systems. IFAC Proc. Vol. **45**, 1850–1855 (2012)
26. Schel, D., Henkel, C., Stock, D., Meyer, O., Rauhöft, G., Einberger, P., et al.: Manufacturing service bus: an implementation. Procedia CIRP **67**, 179–184 (2018)
27. Theorin, A., Bengtsson, K., Provost, J., Lieder, M., Johnsson, C., Lundholm, T., et al. : An event-driven manufacturing information system architecture for Industry 4.0. Int. J. Prod. Res. **55**, 1297–1311 (2017)
28. Trabesinger, S., Pichler, R., Schall, D., Gfrerer, R.: Connectivity as a prior challenge in establishing CPPS on basis of heterogeneous IT-software environments. Procedia Manuf. **31**, 370–376 (2019)
29. Monostori, L., Kádár, B., Bauernhansl, T., Kondoh, S., Kumara, S., Reinhart, G., et al.: Cyber-physical systems in manufacturing. CIRP Ann. **65**, 621–641 (2016)
30. Quintanilla, F.G., Cardin, O., Castagna, P.: Evolution of a flexible manufacturing system: from communicating to autonomous product. IFAC Proc. Vol. **45**, 710–715 (2012)

31. Cardin, O., Trentesaux, D., Thomas, A., Castagna, P., Berger, T., Bril El-Haouzi, H.: Coupling predictive scheduling and reactive control in manufacturing hybrid control architectures: state of the art and future challenges. J. Intell. Manuf. **28**, 1503–1517 (2017)
32. Belkadi, F., Troussier, N., Eynard, B., Bonjour, E.: Collaboration based on product lifecycles interoperability for extended enterprise. Int. J. Interact Des. Manuf. **4**, 169–179 (2010)

PLM Needs for Cosmetic and Fragrance Industries: A Preliminary Assessment

Francesca La Bella[1,2], Virginia Fani[1], Julien Le-Duigou[2], Romeo Bandinelli[1], Florent Ferrer[3], and Benoit Eynard[2](✉)

[1] Department of Industrial Engineering, University of Florence, Florence, Italy
[2] Université de Technologie de Compiègne, Compiègne, France
benoit.eynard@utc.fr
[3] Inetum, Toulouse, France

Abstract. Nowadays, issues such as shorter time-to-market, cost saving, and quality and compliance management are among the main concerns for the formulated industry. Thus, Product Lifecycle Management (PLM), already well-established in industries such as discrete manufacturing, is becoming increasingly popular, as a solution able to support industries operating in the formulated sector. On the market side, PLM vendors are offering increasingly verticalized solution to meet market demand. However, on the scientific side, limited research has been carried out about the use of PLM in pharmaceutical industry and very poor regarding the cosmetic and fragrance industries. With scientific research not keeping abreast of commercial development, the PLM concept is often misunderstood or not fully potential exploited by PLM developers. This research aims to assess actual and potential functionalities of cosmetic and fragrance industries PLM application. Findings are based on a scientific literature review about the use of PLM in pharmaceutical, cosmetic and fragrance industries, an assessment of cosmetic and fragrance regulation environment (limited to the EU market) and a study on PLM solution commercially available for the cosmetic and fragrance industries. Some well-established functionalities were found and some other potentially valuable PLM functionalities, which are currently missing or poor developed are identified and proposed. The need and the possibility for further development of such PLM application is determined.

Keywords: Product lifecycle management · Formulated product · Cosmetic industry · Fragrance industry · Enterprise information system

1 Introduction

The Product Lifecycle Management (PLM) is a business strategy for creating and sustaining a product-centric knowledge environment. As a technology solution, PLM is an integrator of tools and technologies that streamlines the flow of information through the various stages of the product lifecycle [1].

F. Noël et al. (Eds.): PLM 2022, IFIP AICT 667, pp. 126–135, 2023.
https://doi.org/10.1007/978-3-031-25182-5_13

Scientific research is addressing the PLM subject since more than two decades, discussing about its actual and potential value for the industry. On the commercial side, PLM applications, initially limited to some manufacturing industries, in the last years are spreading across several type of business, with ever-increasing verticalized solution addressing specific industry needs. However, especially in those sectors which are starting now to implement PLM, the development of commercial solution does not necessarily go hand in hand with scientific research and the PLM concept is often misunderstood or not fully potential exploited.

This study addresses the actual and potential use of PLM in cosmetic and fragrance industries, through a comparison with the pharmaceutical industry. The paper is organized as follows: Sect. 2 presents the research methodology; Sect. 3 defines pharmaceutical and then cosmetic and fragrance product lifecycle stages and in Sect. 4 a discussion of needed PLM functionalities is carried out. Section 5 concludes the paper.

2 Methodology

2.1 Scientific Literature Review

In order to identify how PLM can be used in the fragrance and cosmetics industries, a literature review was carried out, which consisted of two different phases.

Phase 1. For the first preliminary assess about the need for PLM in such kind of business, it was decided to broaden the research scope and include the pharmaceutical industry, leveraging on similarities between these sectors. Being the pharmaceuticals a more long-standing, developed, and regulated industry, such inclusion allowed to widen the literature results.

The mainly used searching engine was Scopus, with the integration of some results from Google Scholar and other indirect sources. The query that was at first used to search through papers' title, abstract and keywords is:

1. ("PLM" OR "Product Lifecycle Management" OR "Product Life Cycle Management") AND ("Pharmaceutical Industry" OR "Cosmetic Industry" OR "Fragrance Industry").

   Given the poor output result (see Table 1), especially regarding the cosmetics and fragrance industries, it was decided to further enlarge the research scope, including other keywords that could have been used to refer to the PLM concept or that can be anyhow related to it, with the aim to gain a general understanding of how, in such industries, data and all information related to the product are managed all along its lifecycle. Research queries are outlined below and obtained results are summarized in Table 1:
2. ("Product Data Management") AND ("Pharmaceutical Industry" OR "Cosmetic Industry" OR "Fragrance Industry")
3. ("Product Information Management") AND ("Pharmaceutical Industry" OR "Cosmetic Industry" OR "Fragrance Industry")
4. ("Laboratory Information Management System" OR "LIMS") AND ("Pharmaceutical Industry" OR "Cosmetic Industry" OR "Fragrance Industry")

Table 1. Research output (phase 1)

Query N°	Number of Scopus output documents	Number of documents form Google Scholar	Documents considered to be related to the research (relying on the abstract)	
			Number of documents dealing with pharmaceutical industry	Number of documents dealing with the cosmetics or with the fragrance industry
1	15	2	6	0
2	0	3	2	0
3	0	4	2	2
4	20	2	4	0

Such results allowed to gain a general understanding of the pharmaceutical industry and all challenges and issues related to the management of product data across its lifecycle, but still poor documentation was found dealing with the cosmetics and fragrance industries.

Phase 2. To try to better focus on the main research domain, the scope was narrowed back only to the fragrance and cosmetic industries and extended to every kind of IT systems and/or data management tool or method. Results from Google Scholar were not included, being considered not related to the research. The search query and related results are outlined below:

5 ("Electric Data Interchange" OR "Enterprise Resource Planning" OR "Material Requirement Planning" OR "Customer Relationship Management" OR "Supply Chain Management systems") AND ("Cosmetic Industry" OR "Fragrance Industry")
6 "Electronic Data Capture" AND ("Cosmetic Industry" OR "Fragrance Industry")
7 "Enterprise Content Management" AND ("Cosmetic Industry" OR "Fragrance Industry")
8 ("data warehouse" OR "data lake") AND ("Cosmetic Industry" OR "Fragrance Industry")
9 ("business intelligence software" OR "business intelligence systems" OR "business intelligence") AND ("Cosmetic Industry" OR "Fragrance Industry")
10 ("supporting information technology infrastructures" OR "supporting IT infrastructures" OR "supporting technology" OR "information infrastructure") AND ("Cosmetic Industry" OR "Fragrance Industry")
11 "Digital technologies" AND ("Cosmetic Industry" OR "Fragrance Industry")
12 "Information management system" AND ("Cosmetic Industry" OR "Fragrance Industry")
13 "Product Information File" AND ("Cosmetic Industry" OR "Fragrance Industry")
14 "Cosmetovigilance" AND ("Cosmetic Industry" OR "Fragrance Industry")

Table 2. Research output (phase 2)

Query N°	Number of Scopus output documents	Number of documents considered to be related to the research (relying on the abstract)
5	3	0
6	0	0
7	0	0
8	0	0
9	0	0
10	1	0
11	1	0
12	0	0
13	3	2
14	6	2

As shown, still poor results were obtained from the literature review. This led to the conclusion that, no significant scientific literature regarding the management of fragrance and cosmetic products information all along their lifecycle exists so far (Table 2).

2.2 Indirect Sources

To improve and try to enrich the results gained from scientific papers, some indirect sources were considerate. Standards and regulations for the cosmetics and fragrance sector have been reviewed, such as the European Regulation (EC) N° 1223/2009 [2] and the previous Directive 76/768/EC, as well as the trade association Cosmetics Europe guidelines [3].

Other source of information that were used are white papers from consulting companies regarding the use of PLM and the management of product information in such industry, other non-scientific articles, together with information from some PLM vendors reports and/or websites, such as Dassault's Centric Software[1], PLM Lascom[2] and Selerant PLM[3].

2.3 Summary of the Profile of Articles and References

Regarding the phase 1 of the scientific literature review, results allowed to gain a general understanding of the pharmaceutical industry. Covered topics are the pharmaceutical product life cycle, especially the new product development phase, the product data and

[1] https://www.centricsoftware.com.

[2] https://www.lascom.com.

[3] https://www.selerant.com.

information management, the challenges that pharmaceutical companies need to face and how PLM application or other IT systems can or could support them. Contributions are mainly case studies and review article, with also some original articles. In addition, two papers dealing with the cosmetic industry were found. Such papers, based on case studies, focus on the use of IT solution along the cosmetic value chain.

In the phase 2 of the scientific literature review, in which the scope is narrowed to the cosmetic and fragrance industry, results deal with the regulatory aspect and needed implications on cosmetic companies' organizational structure, especially regarding the management of product data before and after sale. Regarding the types of the references, these are communication from Cosmetics Europe[4], conference papers and reviews. In none of the above-mentioned references for the cosmetic and fragrance industry the PLM was mentioned.

Concerning the indirect sources, the study of standard and regulation that cosmetic companies must comply with allowed to deepen the results obtained in phase 2 of the scientific literature review, thus further understanding the cosmetic sector constraints and related needs. Regarding white papers from consulting companies, their main contribution is given by the fact that they represent the only reference dealing with the use of PLM in cosmetic and fragrance industry. Particularly, they give an overview of the commercially available PLM solution, their features, functionalities, and level of development of these. Such topics are further investigated analyzing some PLM vendors website. Finally, some insights from non-scientific sources are used to address the topic of the cosmetic and fragrance product life cycle, especially for the end of life.

3 The Product Lifecycle in Cosmetic and Fragrance Industries

To be able to understand actual and potential PLM application to the cosmetic and fragrance industries, it is key to understand which are the lifecycle stages of such products. As previous mentioned, poor scientific documentation was found regarding cosmetic and fragrance products lifecycle management. Therefore, such information was obtained through indirect sources and analogies with the pharmaceutical industry, leveraging on similarities between these industries in selling formulated products and dealing with governments safety regulation. Hence, prior to the presentation of cosmetic and fragrance products lifecycle, a description of pharmaceutical products lifecycle is provided. It is necessary to premise that, relying on the biographical documentation, the proposed lifecycle phases follow the marketing and commercialization point of view and issues related to manufacturing and product end of life are not fully considered.

In pharmaceutical industry the product lifecycle is divided into five phases [4] that has been reused to structure the proposal for cosmetic and fragrance industries (see below). Product development is a very long and complicated phase, with an average time required of 7, 2 years [4], only 2–3% of projects ending up as commercial products [5], and a very high pressure to launch new products into the market to face the branded competition [6]. In prevision of the approval of pharmaceutical products, it also needs a rigorous preclinical and clinical trials before the production scale-up from the laboratory

[4] https://cosmeticseurope.eu/.

to the pilot plant is made and the technology transfer process starts [7, 8, 9]. Both the technology transfer and the patent submission process need an extensive and complicated documentation. Commercialization phase consists in product grow and maturity and it is the phase in which there are the highest returns. During the growing phase, companies need to focus on increasing the market share, through promotion and marketing prices, but also through product availability and service development and improvement, to gain customer loyalty. To face the competition from generic products, it is also crucial to continuously assess new possibilities for patent protection, like for example, patent extension or litigations [4, 6, 10]. During the whole commercialization phase, companies need also to focus on pharmacovigilance[5], which tracks drug side effects. Last, during the decline phase, companies start withdrawing from the market variations of the product that are weak in their market position, abandoning the alternative channels, while the basic channel should be kept efficient [6]. Many different strategies can be applied to keep the product in the market [4, 10].

Through differences and similarities with drugs and pharmaceutical products, the lifecycle phases of cosmetic and fragrance product, can be outlined as below:

- **Product development phase**: The product development process requires different types of sequential and concurrent activities, in which several departments and external partners are involved. The product development phase is a very customer-centric activity, starting with a marketing brief, which outlines the new product requirements such as the product category, desired function, color, odor, packaging requirements, etc. Based on the inputs from marketing, available ingredients, and knowledge acquired from various publications and research papers, the formula research and development can start, together with the design of product packaging. While working on product formulation, each claim must be analyzed and concentration, allergens and chemical reaction must be evaluated to match the requirements and ensure product safety. Once the formula has been defined, raw material and packaging suppliers are selected, and the prototype or a trial production model is created [4, 5]. Unlike for pharmaceutical products, here the patentability aspect is not a primary issue, but still there is the need of a complete and extensive documentation for the registration phase.
- **Product registration**: unlike for pharmaceuticals, to enable product commercialization, companies do not have to deal with a long and complex product approval phase, but still, they must register the product on a centralized portal. There is the need to ensure product and raw material safety and compliance with the country of commercialization: several tests must be performed, such as microbiological, toxicology and stability tests. Other tests are also needed to prove any product claimed function. All such tests can be managed internally or with external laboratories. Then, for the product registration, all documents and information about the product that are required by countries legislation must be gathered [4]. In the European Community, according to the Cosmetics Regulation (EC) No. 1223/2009, these data must be collected in the Product Information File (PIF), containing information such as the cosmetic product

[5] The science and activities relating to the detection, assessment, understanding and prevention of adverse effects or any other medicine-related problem [11].

safety report[6], a description of the method of manufacturing, that need to comply to the Good Manufacturing Practice[7], and a proof of claimed effect. The PIF must be made accessible to the competent authority of the Member State where the file is located, shall be kept for a period of ten years, and updated as necessary [2].

- **Product introduction and launch**: After the last check of the final product, to ensure its conformity with the initial marketing brief, the product introduction phase can begin, the production can start, and the product can be commercialized [4]. To hit the product launch, so that it has the maximum impact, a detailed launch plan is needed, in which pricing, marketing and distribution plans need to be aligned [5], together with production and distribution plans.
- **Commercialization phase**: during this phase, the product is on the market and companies must keep the distribution pipelines loaded with products, while tracking every ingredient information, manufacturing details and batch number [6], for the previous mentioned safety reasons. Thus, they need to manage the whole supply chain, with chemical but also mineral and natural row material suppliers, often externalized manufacturing facilities and several distribution channels, including departmental stores, specialty stores, mass merchandisers and e-commerce. Adapting marketing and distribution strategies, through targeted and always up-to-date advertisements and promotions is vital. Moreover, during this phase, a high customer interaction is necessary for new product development, but also product updates and refreshes. Thus, companies usually manage after-sale services, including merchandise return and exchange policies, customer feedbacks, products warranties and satisfaction surveys [5, 7]. Just like for the pharmaceuticals, while the product is on the market, companies should also deal with regulation authorities check and with market and post-marketing surveillance issues. According to the above-mentioned European regulation, they must ensure post-marketing surveillance (cosmetovigilance[8]) through a good functioning post-market complaint system, enabling the reception, recording, and traceability of Undesirable Effects (UE) and Serious Undesirable Effects (SUE) reports. If needed, corrective action must be taken, and the cosmetic product safety report must be updated. [2, 3, 6].
- **Decline phase**: cosmetic and fragrance products reach the decline phase when they are displaced by new, more innovative products or when the demand for it falls, due to a change in consumer taste [8].

There are significant similarities between pharmaceutical industry and the fragrance and cosmetic industries, even if it is reasonable to assume that the management of a pharmaceutical product is a more critical business activity, given the more complicated and regulated environment and the higher product development cost and risk, due

[6] An expert report consisting of product and raw material safety information and showing that a safety assessment has been conducted. [3].

[7] Provided by the ISO 22716.

[8] The collection, evaluation and monitoring of spontaneous reports of undesirable events (serious and non-serious) observed during or after normal or reasonably foreseeable use of a cosmetic product [3].

to the higher degree innovation. Nevertheless, as pointed out, the cosmetics and fragrance industries still have significant needs related to product data and product related information flow management.

4 Preliminary Results and Discussion

As, previous mentioned, poor scientific documentation was found regarding the use of PLM in the cosmetic and fragrance industries and only few papers addressed the use of IT in such domains. Some example of already commonly used IT systems are supply chain management (SCM) systems, electric data interchange (EDI) and enterprise resource planning (ERP), together with customer relationship management (CRM) [7, 9].

Regarding PLM, the scientific literature review allowed to identify actual and potential PLM functionalities for pharmaceutical companies. These functionalities were classified according to different business area and then the same categorization was used to classify actual PLM functionalities for cosmetic and fragrance industries, that were deducted from PLM vendor websites and other indirect sources. Such business area that can benefit from PLM application are collaboration in the research and development activities, geographic regulations and strategies management, supply chain information flow, product portfolio management and quality management. Among them, well established PLM functionalities that were found are the ones for collaboration in product development, label and packaging management, portfolio management and supporting for regulatory and compliance issues [10]. Also, the formula management PLM functionalities are quite developed: being a formula development a very different activity from CAD modelling, dedicated tool are needed for example to compare formulation factors and possible formulation version against several targets, access to past formulation data and ingredient libraries, automatic label updating [10], as it is the case for Selerant and Centric PLM solutions. However, the research showed that, because of the lack of standardization in the pharmaceutical industry regarding product data modeling software and collaborative electronic data managed (like CAD software for the manufacturing industry), PLM products are still needed to evolve for managing formulated product design data [11]. It can be stated that this also applies to the cosmetic and fragrance industry.

Moreover, to sustain companies in facing the presented challenges, some PLM functionalities which are still missing, or poor developed in the commercially available offers were identified:

- **Distribution and sales functionalities**. There are loads of product-related information for sale and promotion, such as product versions, current price, and availability to be managed while dealing with the highly heterogeneity of cosmetic and fragrances distribution channels, requiring and producing even more various quantity and type of data, as explained by the PLM vendor Lascom. Also, knowledge coming from product sale is a valuable source of information for the next product development phase. Most of the times, these types of data are stored and managed using different IT systems, such as ERP, SCM or CRM, while an integrated functionality in PLM applications is still missing.

- **Government interaction functionalities**. Commercially available PLM capabilities for supporting cosmetic and fragrance products compliance mainly concern data retrieval and collection and automatic safety and regulatory document creation. However, given the complexity and the heterogeneity of the regulatory environment, some valuable PLM functionalities could be the ones enabling the automation of the interaction between companies and regulation authorities, for example with automatic document submission through a communication with a centralized government portal. It would be also even more valuable for the pharmaceutical industry. However, these kind of PLM functionalities would require a higher level of automation and standardization on the government side and issues about data exchange security may arise.
- **End of life functionalities**. At present, no commercially available PLM solution was found offering functionalities regarding product end of life. However, given product chemical composition and mixed material packaging, managing the cosmetic and fragrance product and packaging end of life is far from being a trivial issue [12]. Moreover, a product waste is not only an unused product in consumers hand, but also a formula tester, an unsold or returned product, expired or discontinued product in warehouses or on store shelves [13]. Many cosmetic and fragrance brands are starting to take responsibilities for the recycle/reuse of product they sell, for example switching to refillable product, either relying on consumer engagement or directly collecting empty containers [12, 14]. It is thus expected that this or similar action aiming to enhance companies' sustainability will be increasingly common in the foreseeable future and subsequently new product data management need will arise. This will pave the way to significant opportunities for further PLM development.

5 Conclusion

In the last years, the use of PLM, initially limited to some manufacturing industries, has been spreading across several type of business, with ever-increasing verticalized PLM solution and a lot of scientific research dealing with the PLM issue and discussing about its actual and potential applications. This study focuses on the use of PLM in the formulated industry, especially on the cosmetic and fragrance industries. Surprisingly poor scientific documentation was found dealing with such specific subject. Thus, obtained results are based on:

- Scientific literature review about the use of PLM in the pharmaceutical industry, that gave the theoretical foundation for the research topic.
- Market study about the commercially available PLM solution and features for the cosmetic and fragrance industries, relying on PLM vendor websites, consulting company reports and other indirect sources.
- Regulatory assessment of the cosmetic and fragrance industry market, that provided insight about cosmetic and fragrance companies' duties and constraints. For simplicity's sake, only the European regulations were considered.

This research has provided an overview about cosmetic and fragrance products and business peculiarities, in which regulation requirements, the importance of the packaging and multi-channel distribution are among the main distinctive features. The main available PLM functionalities supporting companies operating in such sector are identified. Some of theme, like collaboration in R&D activities, product data storage and retrieval, and portfolio management, are quite common also to the well-established PLM application, such the ones for the discrete manufacturing. On the other side, functionalities supporting the management of the highly diversified regulatory environment and tools for the formula, label and packaging management are more typical of the cosmetic and fragrance sectors.

Results show that, even if quite developed and with well-established functionalities, PLM application for cosmetic and fragrance industries still have room for growth to better meet specific needs. The need for a further step towards standardization to manage formulated product design data is identified and, undoubtedly, further research is needed to lead PLM application developers in exploiting all possible PLM functionalities that such industries can benefit.

References

1. Terzi, S., Bouras, A., Dutta, D., Garetti, M., Kiritsis, D.: Product lifecycle management - from its history to its new role. Int. J. Prod. Lifecycle Manag. **4**, 360 (2010)
2. Regulation (EC) No 1223/2009 of the European Parliament and of the Council of 30 November 2009 on cosmetic products, p.151
3. Renner, G., Audebert, F., Burfeindt, J., Calvet, B., Caratas-Perifan, M., Leal, M., et al.: Cosmetics Europe guidelines on the management of undesirable effects and reporting of serious undesirable effects from cosmetics in the European union. Cosmetics **4**, 1 (2017)
4. Melissandre: Cosmetic Product Life cycle: 6 Key Steps. CPG FB Cosmet Blog NPD Issues Softw (2017)
5. New Cosmetics Product Development. PharmaTutorn
6. Pauwels, M., Rogiers, V.: Safety evaluation of cosmetics in the EU. Toxicol. Lett. **151**, 7–17 (2004)
7. Wei, J., Suzuki, M., Guo, L.: How to improve business value in the e-cosmetic industry? IJSS **7**, 155–168 (2011)
8. Product life cycle of cosmeticsn
9. Nelson, T., Josephson, D.: Collaborative Software in the Flavor and Fragrance Industry. Perfumer Flavorist **27**, 4 (2002)
10. A Guide to PLM Providers for Formulated Packaged Goods Industries. Gartnern
11. Giridharan, M., Srinivasan, D.: Pharmaceutical product life cycle management-a comprehensive review. Int. J. Pharm. **11**, 9–12 (2021)
12. Unseen 2019: The ugly side of beauty waste. Mintlounge (2019)
13. Nast, C.: Beauty has a waste problem, and it's not packaging. Vogue Bus (2021)
14. cosmeticsdesign-europe.com.: Beauty and its waste: Brands need to 'add value' to justify extra efforts and costs, says expert. Cosmet-Eurn

Transiting Between Product Development and Production: How IT in Product Development and Manufacturing Integrate

Roland Stolt(✉) and Mohammad Arjomandi Rad

School of Engineering, Jonkoping University, Jonkoping, Sweden
roland.stolt@ju.se

Abstract. One crucial moment in product development is when all specifications of a product are ready, and the start of production is imminent. Companies very often have a milestone gate at this point where the development project is signed off and approved making sure that everything is ready and handing it over to the production department. This is a crucial moment involving checking a lot of information that has originated from different departments in the company. It can result in errors since the checking often is done manually and based on that staff check and report the results to a central function managing the transition. The situation is worsened by interoperability problems between different IT systems and databases used for different purposes in the company. In this paper it is examined how an automotive company currently is conducting the handover, gathering and checking all specifications before start of production. Risks of mistakes and unnecessary time loss are identified. In the paper it is discussed how the integration can be improved and the amount of manual work reduced. Results indicate that there is a potential for both saving time and reducing risk of error by automating parts of the handing over and integrating the IT systems and databases.

Keywords: Interoperability · Production · Stage-gate

1 Introduction

There are many different functions in a company that needs to work in close cooperation in product development projects. This is traditionally described as "concurrent engineering", See for example [1], meaning that the product and the production process are defined in parallel rather than in sequence to reduce the risk of having to re-work the design after completion to solve all producibility problems. With increasing complexity of products, this is becoming more important than ever. Defining a new product does not only involve the engineering and manufacturing departments but indeed most of the functions in the company. Not least the increased demands on the sustainability of products require that the supply chains are well understood already in the development phases. One example of a company function is the market and financial functions of the company that needs to be involved early in the development project to ensure the

Published by Springer Nature Switzerland AG 2023
F. Noël et al. (Eds.): PLM 2022, IFIP AICT 667, pp. 136–143, 2023.
https://doi.org/10.1007/978-3-031-25182-5_14

profitability of the intended product. After completing the product development project, the new product is going to be produced. During the development phase, the various functions likely worked in several and in many cases incompatible systems. To enable the production of the product, much product data and information needs to be defined. This is commonly done in the Enterprise Resource Planning system (ERP) or similar. The product structure of articles is often referred to as the manufacturing bill of materials (MBOM). Prior to the start of production, it needs to be defined to know which and how many of each article the product consists of. The expected sales volumes need to be entered so that the correct amount of material can be bought at the right point in time. Production also requires material handling in the company as well as external logistics services.

The purpose of this paper is to answer the question of how to the transition be-tween the development and the production phases can be done with as little time loss and risk of errors as possible. Currently, it is commonly done at a specific point in time when all the information produced in the development project is gathered. There is a great deal of manual information transfer introducing risks of errors. It is also stressful for staff to gather all information putting a lot of demands on the project managers. With an increased degree of integration between the IT systems used in the development phase with the ones used in production, the situation can perhaps be alleviated. In this paper, this has been investigated by examining how it is carried out in an industrial company in the automotive sub-supplier industry. It was found that the process is carried out according to a checklist requiring manual inputs from many different functions in the company. In the paper alternative ways of making the transition through the integration of IT systems are discussed.

2 Method

The current paper was written as a part of a larger research project. The paper is based on two sets of interviews. First, eleven 20 min long interviews that concerned the state of practice in digitalization in the studied company were conducted. These were recorded and transcribed. They were analyzed by writing a summery, yielding an understanding of the company state of practice and the level of digitalization. The first round of interviews was followed up with additional informal interviews with two company representatives with insight in the IT environment of the company. The result of the two interview rounds is accounted for in Sect. 4 of this paper.

It was understood that the transition between the development and the production phase is an area of concern for the company. The company also provided corporate documents describing the product development process and supportive documents pertaining to the transitioning between the product development and the production phases. The study resulted in an understanding of the company operations so that it could be investigated what future alternatives for information transfer could be possible for the company and assess their applicability in the organization.

3 Literature

Product Lifecycle Management (PLM) is the business activity of managing, in the most effective way, a company's products all across the lifecycle [2]. PLM is both the set of software to create and keep track of all the product data and a philosophy to include the complete life cycle when creating new products in a single environment. This environment can be described as containing software and databases needed to design and manufacture products. The IT environment can be referred to as the PLM system. It involves the traditional PDM (Product Data Management) for keeping track of the product documentation and defining the processes as well as the software needed to define the product throughout the life cycle. One of the key PLM ideas is to integrate the functionality as much as possible, perhaps into a single system. Understandably, including everything will result in a very large system, and to date, no such systems are commercially available. Perhaps future PLM systems will cater for all needs of a smaller company, to begin with. However, systems containing part of the functionality are still referred to as PLM systems.

The integration of different IT systems in the product realization process can be achieved by building single integrated environments for all the required functionalities. Examples of integrating the core systems of development and production are emerging. For instance, a novel system architecture is proposed by Madenas [3] for integrating PLM systems with cross supply chain maintenance information to ensure that previous failures will not reoccur with new product parts.

According to Avvaru et al. [4] the core systems are PLM, Manufacturing Execution Systems (MES), and ERP. The authors describe the integration of the functionality in a framework for interoperability and apply it to a one-of-a-kind product. This is a starting point for optimization of the processes, highlighting the most pressing interoperability needs. In another application by Prashanth and Venkataram [5], an ontology is proposed for cross communication between PLM and ERP. The paper identifies the interoperability needs and proposes an integration module based on their ontology. It uses a middleware Enterprise System integration (ESI). The connection between the ESI and the ERP is managed by a manual or custom process. Their application is limited to BOMs and Meta Data with a degree of manual or custom service ESI for transferring from ESI to ERP. The PLM system used is Windchill ®.

One problem with the integration is the different needs in the design and manufacturing phases. In the design phase, the system must be flexible and handle frequent changes. Further, the product structure that is defined is the engineering (EBOM). It is not the same as in manufacturing (MBOM). The latter has been put together from a functional perspective whereas the former keeps track of the manufacturing process, such as which of the parts are bought and which are manufactured in-house and at which facilities. It also keeps track of the assembly sequence. After the engineering phase is completed, the MBOM and a much other information need to be defined. It is a transition between the phases. Arthur [6] describe the phase transition between design and production in the case of naval architecture and shipbuilding. The paper describes a transition between design and manufacturing and explains that the PLM system widely has facilitated the process.

A PLM system can be used separately but with integrated data. Such integration can be achieved by making several systems using the same model or using a reference model which was the case in [7] for innovation management. The authors created an enhanced data model in conjunction with a reference process model to make strategic information flow more efficient between marketing and customer service teams.

Cloud PLM systems are other examples of integration that use integrated data in separate systems. Broadening the scope of the PLM systems require connected servers. This is challenging for companies why cloud solution has been identified as a way forward. In the possibilities of cloud solutions in aerospace are discussed. The paper concludes that the major challenges are performance, trust, and data security. The study was concentrated to an application in India. Model-Based Definition (MBD) means producing and documenting the product data in models. The same models are (as much as possible) used throughout the lifecycle. This makes it possible to define the information needed for the production in the same models. This is demonstrated by Rinos et al. [8] in a case where they show that at least the manufacturing information can be incorporated into a model. Thus, they eliminate the need for 2D drawing, widely simplifying the transition to production.

The integration of different IT systems in the product realization process can be achieved by a central knowledge base system (KBS) as a database connecting separate IT systems. To realize an integrated PLM system, the domain-specific products (ERP, CAx, etc.) need to be developed on a functional level, assuming they provide defined services that can be used by other services. However, most PLM implementations are carried out like a traditional IT project in product architecture form. To overcome this PLM systems integration hindrance, researchers propose a taxonomy of terminologies to help the semantics from different domains match and in this way remove the integration barriers [9]. Similar approach attempts to use existing PLM systems as a knowledge management tool to solve the semantic interoperability problem of heterogeneous data are described by Raza et al. [10]. The authors discuss that ontology developments will represent and capture the data and help the production process in manufacturing.

The integration of different IT systems in the product realization process can be achieved by extending the PLM footprint to all phases of the lifecycle. By relying on a system engineering process framework for contextualization of PLM standards, Moones et al. propose an extended interoperability approach for dynamic manufacturing networks [11]. This approach defines system boundaries that can be preserved between the business, applicative, and ICT layers in their manufacturing models. Moreover, the PLM concept has been extended to ePLM [12] by an electronic product code for tracing and tracking of the physical product after it is delivered to the user.

The literature survey has showed three different ways of integrating the IT environment.

1. The systems are integrated into a single environment
2. The systems are still separate, but the data is integrated via for example KBS and ontology.
3. The models are extended to incorporate all necessary information for the start of manufacturing.

What will lead to the best environment likely depend on the type of organization and the product. In this paper, it is reasoned on how this would apply at an automotive car accessory supplier.

4 Case

The case company manufacture and sell mainly automotive accessories for primarily private use. It includes transport solutions for private cars allowing transportation of equipment that is too large for transport inside the car such as skies, bicycles, and kayaks. The company sells its products world-wide, directly to consumers or as original equipment (OM) to car manufacturers. The products are considered high end and the company put much effort into design and testing for the best consumer experience. The company has been on the market for many years and has an annual turnover of about 0,8 billion Euros and a number of employees worldwide of roughly 3000 in 2020. The company has an in-house production and is especially prominent in sheet metal forming and thermoplastic injection molding. All product development is carried out in-house.

4.1 Organization

The development of a new product is initiated from the top management level. A product manager is appointed to oversee that the envisioned product is created. The product manager is responsible for specifying the product and ordering the organization to develop it. The project starts with appointing a project manager. It is the role of the project manager to plan and execute the project and to see that the different roles work together and that the project is conducted according to a time plan agreed with the product manager. The project manager plans the project in agreement with the different functions in the company. Making the time plan require frequent contacts with the various functions to establish the plan and oversee it being kept. Resources in the company are allocated for the project. This includes several people with different roles. Examples of roles are: production planning, purchasing, quality, mechanical design, finance, storage, customer service, and sales forecast responsible. These corresponds to the various functions that are needed to conduct the project. The project is conducted in five different stages according to an elaborate instruction: A- Market research, B- Concept, C - Design, D- production engineering and finally E - production. Each of the phases are followed by a project review. These are called gates A-E. It has been set-up much according to the well-known stage-gate model proposed by Cooper [13]. The project instruction specify what information must be available and what documents to prepare at each gate review so a decision can be taken to continue or terminate the project. If any information is lacking, then the project is not transferred to the next phase. After the production has ramped up (gate E) the project is closed, and the production is handed over to the production department.

4.2 IT Environment

During the first stages of the project A-C, a standardized catalogue structure is created on a server. The project is not yet defined in the ERP system of the company. CAD models, drawings and the engineering BOM structure are handled in a PLM system including work processes and versioning for the engineering part of the work. Other functions are working in their own systems. When transiting between phases C and D, preparations for pre-series production is made. This requires that the project results are transferred to ERP so that pre-series production can be carried out. Now, it must for example be possible to order components from suppliers and keep track of how many units have been produced and how many bought units to order. This require that the information that previously was in the engineering PLM system and on the catalogue structure on the server needs to be defined in ERP. One examples of information that needs to be entered in ERP is the final MBOM structure. All functions engaged in the product are expected to transfer project documentation into the ERP system. This includes for example that the product as well as the process has been approved. The process approval to ensure part quality is in automotive industry commonly known as PPAP (Production Part Approval Process). The process FMEA needs to be approved as well as the all testing needs to have been completed with approved results. There are several other documents such as the final financial calculations and the sales volume predictions that needs to be defined in the ERP systems. There is a requirement to check that the information entered in ERP is correct. Several people from different functions are responsible for making these checks. There are some tools that have been developed specifically for this purpose such as a viewer to check the structure created in ERP so that it can be compared to the product structures in PLM. However, much of the information is taken manually from the "development" server and is manually inserted in the ERP system as needed. This includes for example the predicted volumes, the list of suppliers and the agreements

Table 1. Items to be checked and entered in ERP.

Production Engineering	**Purchase**
Update MBOM	Define suppliers for puchased products
Create semi finished components	Calculate cost for purchased product and components
Define storage in production	Enter purchased items in ERP
Enter manufactiring lead times in ERP	Enter cost for purchased items
Verify product cost calculation	Connect agrements with suppliers
Create first manufacturing order	**Quality**
Verify manufacturing cost	Inspect first serial batch
Verify manufacturing lead times	Make sample orders at supplier
Customer service	**Production planning**
Assign customers item number to the product	Register sample orders for new manufacured items
Forecast	Create manual manufacturing order
Enter sales forcast	Relese first serial production order
Mechanical Design	
Check structures	
Finance	**Warehouse**
Specify logistics prize	Create packaging
Calculate product costs	Assign packaging to product item
Calculate std costs for prod facility	Send info logitics suppliers
Calculate distributed costs on sales facilities	Update sales and logistigs on manufacturing warehouse
Update status to production	

with the different suppliers. In the Table 1, the major items that are checked and placed in ERP are shown. Table 1 also shows which function in the company that is responsible for its completion.

The people responsible for entering the information report that they have completed it in an excel document on the server. However, this way of working is perceived as time consuming and ridged and is gradually falling into disuse. Instead, the company is working to replace it with another procedure. They are currently in the planning stages of this new procedure.

5 Discussion and Conclusion

The problem has been understood via corporate documents and interviews with the company. The transition between the different phases is a commonly occurring problem at many companies. All companies where product development and manufacturing undergo the phase transition need to handle this problem. It is currently accepted by companies that there are points in time, the so-called gates when the specified documents are put on the table and evaluated. However, there is a clear development towards making this transient more smooth and less prone to errors. Instead of relying on the project manager to drive all functions towards the gate, the functions could work toward a common database making it possible to follow the progress in real time towards completing the phase. One example of this is that, instead of at a point in time transferring the product structures into from PLM to ERP they could gradually be build up in the ERP system.

Three major ways have been suggested. One is that the PLM system is extended to include all life-cycle phases which indeed it one in the key ideas with PLM. All the applications would work towards a common database from which various views can be derived. The second line of development is that the various functions continue using "their" software that is specialized for each task. Then, an integration using for example ontologies needs to be made. The third way is defining the production phase in the same models used in the development phase. By doing so, the information would be gathered in a limited space, readily transferable to production.

For the organization studied the path forward is a gradual integration of the systems and to some extent a more homogeneous information model for integrating between the BOM:s in PLM and ERP. The option of a large PLM system encompassing both product development and production is not feasible due to the size of the company. Rather than storing the information on servers in many different and incompatible formats there should be a common database also allowing different views to be derived. For this the API:s of the different systems need to be used to allow each of the applications to operate on this common database.

6 Future Work

Finding and elaborating critical parts of the integration for the studied organization to gradually prepare the start of production is identified as the next step in this research. A test implementation on a limited product development to production case is planned at the case company.

References

1. Prasad, B.: Concurrent Engineering Fundamentals, Integrated product and process organization, vol. 1. Prentice Hall PTR, Upper Saddle River, N.J. (1996)
2. Stark, J.: Product Lifecycle Management: The Devil is in the Details. Springer International Publishing, Switzerland (2016)
3. Madenas, N., Tiwari, A., Turner, C.J., Peachey, S., Broome, S.: Improving root cause analysis through the integration of PLM systems with cross supply chain maintenance data. Int. J. Adv. Manuf. Technol. **84**(5–8), 1679–1695 (2015). https://doi.org/10.1007/s00170-015-7747-1
4. Avvaru, V.S., Bruno, G., Chiabert, P., Traini, E.: Integration of PLM, MES and ERP systems to optimize the engineering, production and business. In: Nyffenegger, F., Ríos, J., Rivest, L., Bouras, A. (eds.) PLM 2020. IAICT, vol. 594, pp. 70–82. Springer, Cham (2020). https://doi.org/10.1007/978-3-030-62807-9_7
5. Prashanth, B.N., Venkataram, R.: Development of modular integration framework between PLM and ERP systems. Mater. Today Proc. **4**, 2269–2278 (2017)
6. Arthur, S.: A PLM centric approach to the transition from design to manufacturing in first of class naval shipbuilding. In: Advances in Transdisciplinary Engineering (2017)
7. Löwer, M., Heller, J.E.: PLM reference model for integrated idea and innovation management. In: Fukuda, S., Bernard, A., Gurumoorthy, B., Bouras, A. (eds.) PLM 2014. IAICT, vol. 442, pp. 257–266. Springer, Heidelberg (2014). https://doi.org/10.1007/978-3-662-45937-9_26
8. Rinos, K., Kostis, N., Varitis, E., Vekis, V.: Implementation of model-based definition and product data management for the optimization of industrial collaboration and productivity. Procedia CIRP **100**, 355–360 (2021)
9. Fischer, J.W., Lammel, B., Hosenfeld, D., Bawachkar, D., Brinkmeier, B.: Do(PLM)Con: an instrument for systematic design of integrated PLM-architectures. In: Abramovici, M., Stark, R. (eds.) Smart Product Engineering. Lecture Notes in Production Engineering, pp. 211–220. Springer, Heidelberg (2013). https://doi.org/10.1007/978-3-642-30817-8_21
10. Raza, M.B., Kirkham, T., Harrison, R., Quentin, R.: Knowledge based flexible and integrated PLM system at ford. J. Inf. Syst. Manag. **1**(1) (2011)
11. Moones, E., et al.: Towards an extended interoperability systemic approach for dynamic manufacturing networks: role and assessment of PLM standards. In: Boulanger, F., Krob, D., Morel, G., Roussel, J.-C. (eds.) Complex Systems Design & Management, pp. 59–72. Springer, Cham (2015). https://doi.org/10.1007/978-3-319-11617-4_5
12. Riaz, A., et al.: Closing information loops with extended PLM. WSEAS Trans. Syst. **6**(2), 304–309 (2007)
13. Cooper, R.G., Edgett, S.J., Kleinschmidt, E.J.: Optimizing the stage-gate process: what best-practice companies do—II. Res. Technol. Manag. **45**(6), 43–49 (2002)

A Process-Oriented Approach for Shipbuilding Industrial Design Using Advanced PLM Tools

Lucia Recio Rubio[1,2], Amanda Martin Mariscal[1], Estela Peralta Alvarez[1], and Fernando Mas[1,2](✉)

[1] MyM Group, Cadiz, Spain
{lrecio,fmas}@us.es
[2] University of Sevilla, Sevilla, Spain

Abstract. The digitalization age is upon us; no one, not even the largest industries, is oblivious to the great advances in technology that are being achieved. Digital continuity, Industry 4.0, and Product Lifecycle Management (PLM) are currently major challenges for companies that cover the entire life cycle of a product, such as the aerospace or shipbuilding industry. The research aims to introduce functional and industrial product structures and process structure concepts into the shipbuilding industry. The aim of this paper is a proposal of a process-oriented approach to the organization of naval manufacturing engineering based on the generation of both a product structure for design and manufacturing, engineering Bill Of Material (eBOM) and manufacturing Bill Of Material (mBOM), and a process structure Bill Of Process (BOP) to ensure digital continuity using advanced PLM tools.

Keywords: PLM · Shipbuilding industry · Digital continuity · mBOM · eBOM · BOP · Industry 4.0

1 Introduction

Digital continuity, Industry 4.0, and Product Lifecycle Management (PLM) are currently major challenges for companies that cover the entire life cycle of a product, such as the aerospace, shipbuilding, and automotive industry.

Industry 4.0 is a production system that applies digital technologies to manufacturing and assembly and affects all activities related to manufacturing, from the industrial design of the product and the manufacturing and assembly processes to the planning of the product and the organization of work [1]. These digital technologies brought together under the umbrella of Industry 4.0 [2] contribute to digital continuity throughout the life cycle.

In particular, the design and manufacturing of vessels, changes in methods, processes, and enterprise information systems require new challenges for the management of product information. Industry 4.0 and PLM systems are a key digital continuity change within shipbuilding companies [3].

Published by Springer Nature Switzerland AG 2023
F. Noël et al. (Eds.): PLM 2022, IFIP AICT 667, pp. 144–152, 2023.
https://doi.org/10.1007/978-3-031-25182-5_15

The progressive implementation of Industry 4.0 and the digitalization tools in the shipbuilding industry justify the analysis of the functional and industrial design of the vessel. Also, some characteristics of the shipbuilding industry make it different from other industries applying Industry 4.0 and PLM. Shipbuilding is ordered-driven; a new design is made for each new vessel. In addition, the entire process, from contracting and design to construction, occurs simultaneously. Shipbuilding is a labor-intensive assembly industry; to produce a vessel, extensive labor, production costs, production time, products, and resources are required.

The aim of this work is to propose a process-oriented approach to the organization of the information handled by design engineering and manufacturing engineering, based on the generation of both a product structure for design and manufacturing, engineering Bill Of Material (eBOM) and manufacturing Bill Of Material (mBOM), and a process structure, Bill Of Processes (BOP), to ensure digital continuity using advanced PLM tools. The proposed product and process structures will facilitate collaborative engineering and concurrent definition of the functional and industrial design of the vessel.

2 Literature Review

Recently, there has been much research in the scientific literature on the conversion of the labor-intensive and experience-centered shipbuilding industry into a knowledge and technology-based industry making use of Industry 4.0 principles.

The term Industry 4.0 was introduced during the Hannover Fair in 2011 and was officially announced in 2013 by the German government as a strategic initiative to play a pioneering role in manufacturing industries [4]. It is a concept that describes the complex process of technological and organizational transformation of companies, the introduction of new business models, and the digitization of products and services [5].

The leaders in the shipbuilding industry, the United States Defense Advanced Research Projects Agency (DARPA) runs initiatives to develop a design system/environment that can reduce the cost of system design and development, Europe has led the design of shipbuilding system and naval/marine through projects funded by the European Community, and the Computerized Ship Design and Production System (CSDP) project led by the Korea Research Institute of Ships and Ocean Engineering (KRISO) aims to implement ship Computer Integrated Manufacturing (CIM) based technology for ship manufacturing systems [6].

Most of the research in [7–11] have focused its efforts on different approaches to generate a BOM in an automatic and efficient way, oriented to a block division to determine the best assembly sequence planning. This procedure can not represent a full BOP due to the tree nature of the BOM. An interesting approach in [12] attempts to understand the use, form and evolution of product structures and BOM concepts in shipbuilding with the aim of identifying equivalent notions in construction during the Beginning Of Life (BOL); however, the relationship between them throughout the lifecycle is unclear.

Digital continuity throughout the lifecycle, application of the digital mock-up of the vessel throughout the lifecycle, from contract to manufacturing using PLM tools, has been a research topic in recent years [13–15], fully supported by the major software vendors, such as Dassault Systèms [16] and Siemens [17], and the most relevant independent consultants, such as CIMdata [18] and ProSTEP [19]. Most of them rely on the use of the mBOM as BOP transferring this information to the Enterprise Resource Planning (ERP) to create the processes and relationships.

Several authors [6, 20, 21] have provided a most in-depth analysis of the entire product lifecycle from conceptual design to delivery, proposing the following production steps: part fabrication, part assembly, block assembly, block outfitting, painting, block erection, erection, launching, onboard outfitting, test & trials, and delivery.

3 Proposed Model

3.1 Complexity in Shipbuilding and Aerospace Industries

Vessels are complex products made of millions of single parts and components that must be designed, managed in bills of materials for manufacturing or purchasing, and assembled managing production planning and logistics. The shipbuilding industry differs from other typical manufacturing industries in the following characteristics.

- The vessels are truly diverse, difficult to standardize due to the user's requirements.
- Manufacturing begins while the functional design is not complete, so many engineering changes are expected during manufacturing and assembly.
- Shipbuilding is a labor-intensive industry, difficult to mechanize and automate.
- The accuracy required is high and the structure is complex, so it is difficult to standardize the manufacturing process.
- Vessels with different configurations are built at the same time.

The aerospace industry has similar complexity and shares some of the previous characteristics with the shipbuilding industry [22]. To highlight the similarities, Fig. 1, adapted from [6], shows how to fit the shipbuilding steps with the aerospace major component assembly lines and the aerospace Final Assembly Line (FAL).

The digital continuity and PLM methods and tools used by the aerospace industry [23] could be reused in the shipbuilding industry. An approach to managing "As design" and "As planned" product structures for the aerospace industry presented in [24, 25] can apply to vessel eBOM and mBOM product structures with some adaptations.

The shipbuilding industry does not use, in general, the joint concept. The representation of the joint definition is done directly into the product definition. The Bill Of Process (BOP) is used to represent the assembly processes.

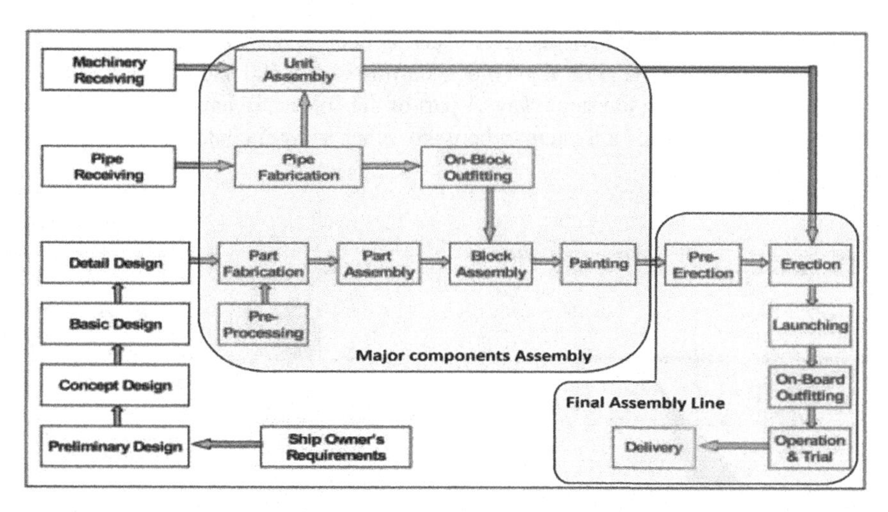

Fig. 1. Vessel design, manufacturing, and assembly steps. Adapted from [7]

3.2 Definitions of Product and Process Structures

Shipbuilding eBOM (engineering Bill Of Material) is a product structure oriented to represent the functional definition of the product: hull structure and mechanical, hydraulic, fuel, electrical and other systems, and equipment. It is a hierarchical system, the top is the ship, composed of various levels of intermediate products. Engineers design the ship and structure the eBOM based on requirements and customer needs. eBOM is usually represented by an abstract data type that reproduces a hierarchical tree structure with a root value and subtrees of children with a parent node. The last level of the subtrees, the leaves, contains 3D information.

Shipbuilding mBOM (manufacturing Bill Of Material) is a product structure oriented to represent the industrial definition of the product: physical products identified as an input or output of a manufacturing or assembly process, needed to program NC (Numerical Controlled) or automated machines and robots, moved and/or stored, involved in a process that needs to be simulated or to generate technical documentation for manufacturing, assembly, and verification or to have control of its configuration.

Shipbuilding BOP (Bill Of Processes) is a network of processes oriented to manufacture and assemble the products and their related components based on the mBOM and in accordance with the technological characteristics of the shipyard, the preparation of product parts, components, sections, and the requirements and conditions of the supply chain.

3.3 Proposed Use Case

The proposed use case is based on a typical Bulk-Carrier and a simple block decomposition of seven blocks represented in Fig. 2. Any other vessel can accommodate to the eBOM, mBOM and BOP structures definition despite the differences between the uppers level of the product structure: passengers, cargo, containers, Ro-Ro, tankers, tugs, yachts, or military vessels.

The Block 200, selected for the Use Case, is decomposed in Shells, Bulkheads, and Decks representing the Hull. Each block starting with B100 and ending with B400 should be decomposed in the same way. A part of the Bilge Discharge System running across the B200 stands for a System, where any other System could be represented in the same way.

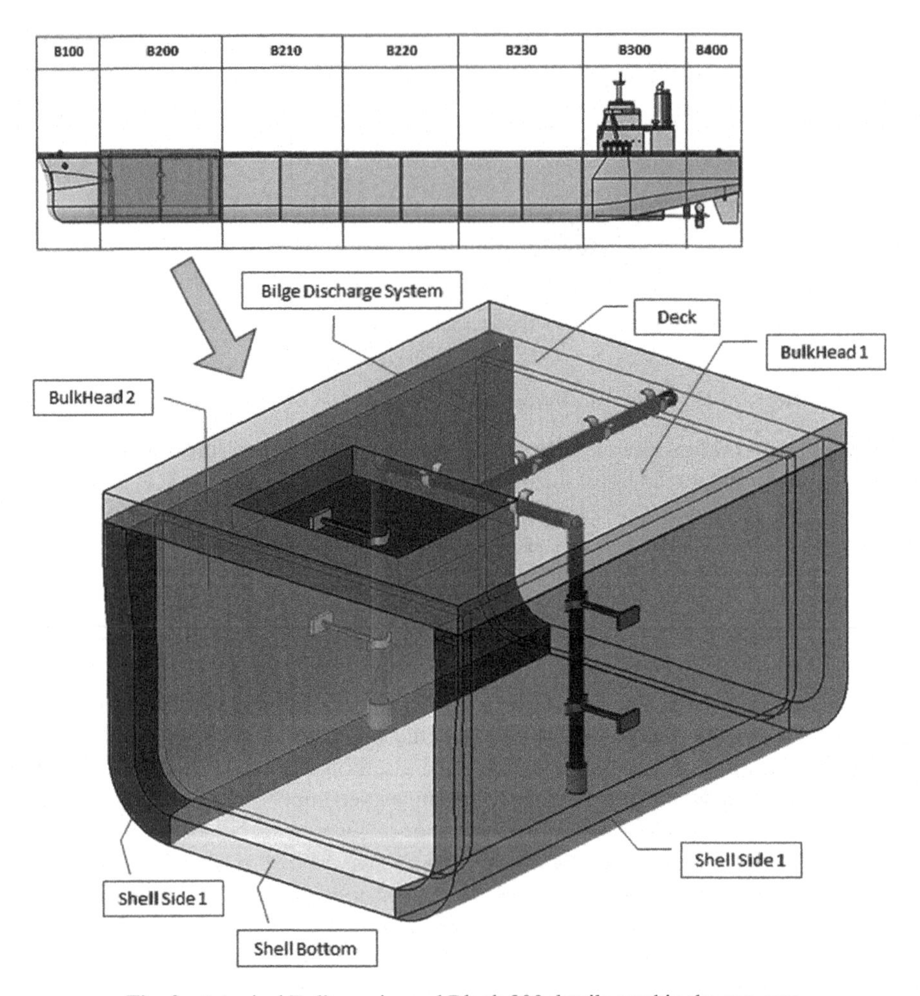

Fig. 2. A typical Bulk-carrier and Block 200 details used in the use case.

In summary, Block B200 is made up of panels: one bottom shell (bottom shell), two shell sides (Shell Side 1 and Shell Side 2), two bulkheads (Bulkhead 1 and Bulkhead 2) and one deck (Deck), represented with different colors and transparencies to improve visibility in Fig. 2. The bilge discharge system is represented with hull attachments in distinct colors to highlight the relationship with the panels.

3.4 Product and Process Structures

The eBOM product structure should represent the functional definition of the product, in terms of Hull and Systems, satisfying the customer requirements, standards, company and quality rules, etc. A proposed eBOM for B200 is shown in Fig. 3. To simplify the Use Case there is only one System represented in the B100, the bilge discharge system, represented as part of the water drained system as one of the multiple Systems onboard.

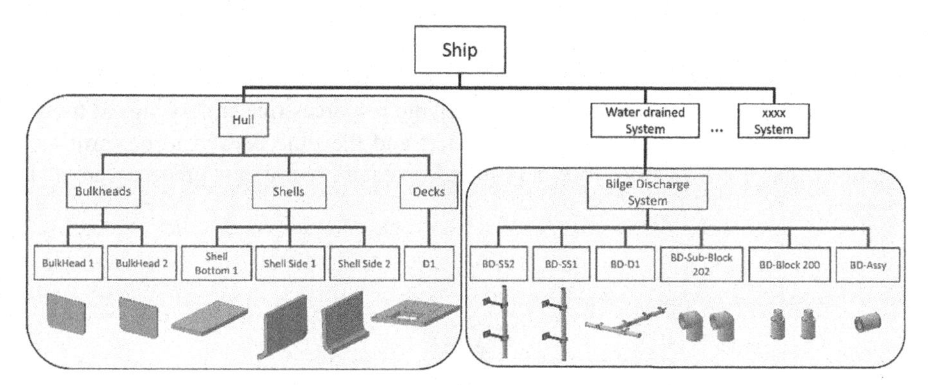

Fig. 3. Proposed eBOM for Bulk-Carrier Block 200.

The eBOM structure can be managed by a Product Data Management (PDM), which also stores multiples CAX files and other document files, allowing naval engineers to manage the vessel parts and systems from a functional view.

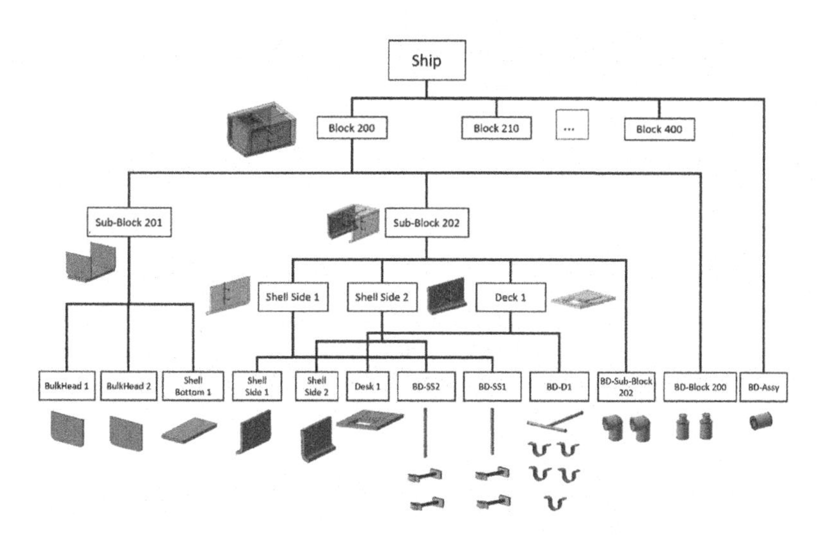

Fig. 4. Proposed mBOM for Bulk-Carrier Block 200.

The mBOM product structure should represent the industrial definition of the product, in terms building strategy, logistics system, resources availability, etc. A proposed mBOM for B200 is shown in Fig. 4. To simplify the Use Case, there is only one block represented and there is no mBOM structure for erection and further activities like onboard outfitting.

Both eBOM and mBOM structures are synchronized through the lower-level layer, the leaves of the structure, where the CAX files and documents reside. This is a shared layer between both structures.

The mBOM product structure is synchronized with the Bill Of Process. The BOP represents the manufacturing and assembly processes to build the vessel. In the Use Case the BOP represents the Block B200 processes in an industrial environment shown in the Fig. 5. Every process is characterized by de input (product represented in the mBOM), the output (product represented in the mBOM), the resources (machinery, jigs & tools, industrial means, workers, documentation, etc.) and the time needed to perform the operations. The BOP is a network, and therefore each process has predecessors and successors.

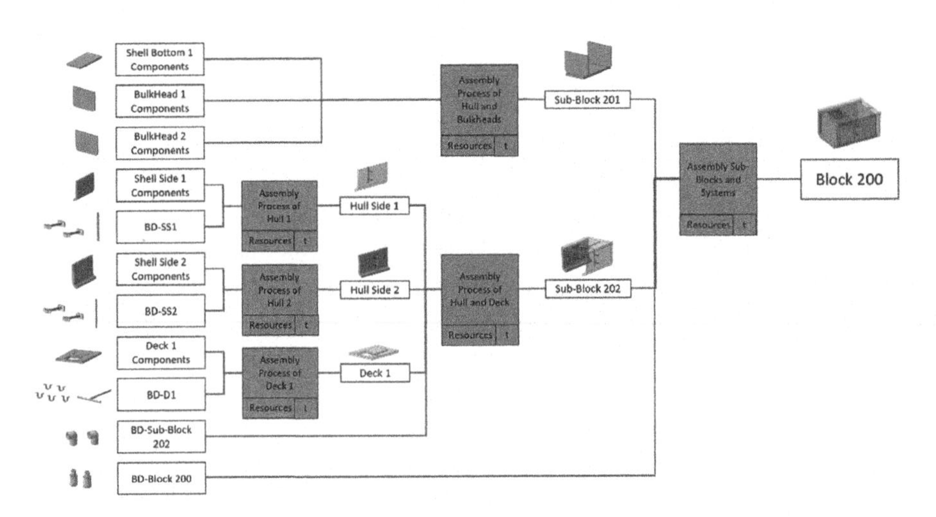

Fig. 5. Representative BOP at block level coordinated with the mBOM.

The combination of eBOM, mBOM and BOP let the naval engineers to perform a collaborative design process, maintaining an influence between the functional design and the industrial design and enabling other capabilities like virtual manufacturing, flow simulation trade-offs with scenarios, industrial digital twins, and other digital methods and tools.

4 Results and Conclusions

This paper presents a proposal to build vessel product and process structures in an analogous way to aerospace artifacts considering the similarities between both industries.

The approach is based on the definition of a unique product structure with two views and coordinated with the process network.

The first conclusion, derived from the literature review, is that the application to the shipbuilding industry of Industry 4.0 and PLM methods and tools is a relevant research topic.

The proposed Use Case has devised a method to maintain eBOM and mBOM coordinated through a common shared layer. Both product structures use the same components and data, and the upper structures of the eBOM and the mBOM represent the functional and the industrial representation of the product.

Consequently, eBOM, mBOM and BOP can be developed, maintained, and stored in a PDM. Commercial PLM software can be used to manage the technical information for the different vessel design skills: hull CFD (Computational Fluid Dynamics) analysis, stress and mechanical calculations, electrical and electronic system design and test, process and flow simulation, etc. BOP is fully coordinated with mBOM and represents the network of processes for the manufacturing and assembly of a block and further the full vessel.

Based on the mBOM and BOP, the digital continuity through the lifecycle is guaranteed with an interface connecting with the ERP systems, simulation capabilities, line balancing trade-offs and other tools.

Future studies should aim to extend the BOM and BOP to the erection, launching, onboard outfitting, test & trials, and delivery to cover the full shipyard process. In addition, we extend the research to the development and management of the "as built" configuration through the manufacturing and assembly processes.

Acknowledgements. The authors would like to recognize Maria Jose Garcia Rodriguez, M&M Group colleagues and Seville University colleagues for support during the development of this work and thank them for their contribution.

References

1. De Santis, S., Monetti, M.: Industry 4.0 for the future of manufacturing in the EU. Roma (2017)
2. PwC and GMIS, Industry 4.0: Building the Digital Industrial Enterprise. https://www.pwc.com/m1/en/publications/documents/middle-east-industry-4-0-survey.pdf. Accessed Mar 2022 (2016)
3. Gischner, B., Lazo, P., Richard, K., Wood, R.: Enhancing interoperability throughout the design and manufacturing process. J. Ship Prod. **22**(3), 172–183 (2006)
4. Xu, L.D., Xu, E.L., Li, L.: Industry 4.0: state of the art and future trends. Int. J. Prod. Res. **56**(8), 2941–2962 (2018). https://doi.org/10.1080/00207543.2018.1444806
5. Kagermann, H., Wahlster, W., Helbig, J., et al.: Recommendations for implementing the strategic initiative Industrie 4.0: Final report of the Industrie 4.0. Working Group: Industrie 4.0, National Academy of Science and Engineering: München, Germany (2013). http://forschungsunion.de/pdf/industrie_4_0_final_report.pdf
6. Kim, H., Lee, S.-S., Park, J.H., Lee, J.-G.: A model for a simulation-based shipbuilding system in a shipyard manufacturing process. Int. J. Comput. Integr. Manuf. **18**(6), 427–441 (2005). https://doi.org/10.1080/09511920500064789

7. Bruijn, D.P.: A model based approach to the automatic generation of block division plans: On the effective usefulness in ship production optimization algorithm. Ph.D. Dissertation (2017)
8. Choi, W.-S., Kim, D.-H., Nam, J.-H., Kim, M.-J., Son, Y.-B.: Estimating production metric for ship assembly based on geometric and production information of ship block model. J. Mar. Sci. Eng. **9**(1), 39 (2021). https://doi.org/10.3390/jmse9010039
9. Iwankowicz, R.R.: An efficient evolutionary method of assembly sequence planning for shipbuilding industry. Assem. Autom. **36**(1), 60–71 (2016). https://doi.org/10.1108/AA-02-2015-013
10. Basán, N.P., Achkar, V.G., Garcia del Valle, A., Mendez, C.A.: An effective continuous-time formulation for scheduling optimization in a shipbuilding assembly process. Lizandra Garcia Lupi Vergara; Iberoamerican J. Ind. Eng. **10**(20), 34–48 (2018)
11. Asok, K.A., Aoyama, K.: Module division planning considering uncertainties. J. Ship Prod. **25**, 153–160 (2009). https://doi.org/10.5957/jsp.2009.25.3.153
12. Boton, C., Rivest, L., Forgues, D., Jupp, J.R.: Comparison of shipbuilding and construction industries from the product structure standpoint. Int. J. Product Lifecycle Manag. **11**(3), 191–220 (2018)
13. Grau, M., et al.: Enabling the digital thread for shipbuilding and shipping. In: ICCAS 19, vol. 1 (2019)
14. Stanić, V., Hadjina, M., Fafandjel, N., Matulja, T.: Toward shipbuilding 4.0-an industry 4.0 changing the face of the shipbuilding industry. Brodogradnja **69**, 111–128 (2018). https://doi.org/10.21278/brod69307
15. Sender, J., Illgen, B., Flügge, W.: Digital design of shipbuilding networks. Procedia CIRP **79**, 540–545 (2019)
16. Dassault Systems. Accelerating Shipbuilding Business Transformation through Digital Continuity (2019). https://www.nsrp.org/wp-content/uploads/2019/03/08-Accelerating-Shipbuilding-Business-Transformation-through-Digital-Continuity.pdf. Accessed Mar 2022
17. Siemens. The digitalization of naval shipbuilding (2019). https://www.plm.automation.siemens.com/media/global/ko/The%20Digitalization%20of%20Naval%20Shipbuilding_tcm72-65038.pdf. Accessed Mar 2022
18. CIMdata. Shipbuilding Catalyst: Accelerating PLM Value for the Shipbuilding Industry (2014). https://www.cimdata.com/de/resources/complimentary-reports-research/commentaries/item/590-shipbuilding-catalyst-accelerating-plm-value-for-the-shipbuilding-industry-commentary. Accessed Mar 2022
19. ProSTEP, The Potential For Digitalization In The Shipbuilding Industry. https://prostep.us/wp-content/uploads/2019/03/PROSTEP-INC-Whitepaper-Shipbuilding-EN.pdf. Accessed Mar 2022
20. Kim, H., Lee, J., Park, J., Park, B., Jang, D.: Applying digital manufacturing technology to ship production and the maritime environment. Integr. Manuf. Syst. **13**(5), 295–305 (2002)
21. Lee, J.H., Kim, S.H., Lee, K.: Integration of evolutional BOMs for design of ship outfitting equipment. Comput. Aided Des. **44**(3), 253–273 (2012)
22. Mas, F., Menéndez, J.L., Oliva, M., Rios, J.: Collaborative engineering: an airbus case study. Procedia Eng. **63**, 336–345 (2013). https://doi.org/10.1016/j.proeng.2013.08.180
23. Mas, F., Rios, J., Gómez, A., Hernández, J.C.: Knowledge-based application to define aircraft final assembly lines at the industrialisation conceptual design phase. Int. J. Comput. Integr. Manuf. **29**(6), 677–691 (2016). https://doi.org/10.1080/0951192X.2015.1068453
24. Mas, F., Ríos, J., Menéndez, J.L., et al.: A process-oriented approach to modeling the conceptual design of aircraft assembly lines. Int. J. Adv. Manuf. Technol. **67**, 771–784 (2013). https://doi.org/10.1007/s00170-012-4521-5
25. Mas, F., Ríos, J., Menéndez, J.L., Hernández, J.C., Vizán, A.: Concurrent conceptual design of aero-structure assembly lines. In: 2008 IEEE International Technology Management Conference (ICE), pp. 1–8 (2008)

Modelling Tools: Model-Based Systems Engineering, Geometric Modelling, Maturity Models, Digital Chain Process

Model Signatures for Design and Usage of Simulation-Capable Model Networks in MBSE

Stephan Husung[1(✉)], Detlef Gerhard[2], Georg Jacobs[3], Julia Kowalski[3], Bernhard Rumpe[3], Klaus Zeman[4], and Thilo Zerwas[3]

[1] Technische Universität Ilmenau, Max-Planck-Ring 12, 98693 Ilmenau, Germany
stephan.husung@tu-ilmenau.de
[2] Ruhr-University Bochum, Universitaetsstr. 150, 44801 Bochum, Germany
[3] RWTH Aachen University, Eilfschornsteinstr. 18, 52062 Aachen, Germany
[4] Johannes Kepler University Linz, Altenbergerstraße 69, 4040 Linz, Austria

Abstract. Product development is characterised by numerous synthesis and analysis loops. Analysis provides information on the fulfillment of the required properties of the system under development. Analysis results therefore are an important basis for further synthesis steps. In the context of Model-Based Systems Engineering (MBSE), different types of simulation models play an important role. The overall system (product) can be broken down to system elements. For each system element a set of models has to be established, that provides the required degrees of fidelity, representations of system properties, flows, etc. reflecting the variety of modelling purposes. These models have to be integrated horizontally (same system level along the relevant flows of material, energy, and information) and vertically (aggregation from subsystem level to the overall system level, refinement in the opposite direction) to create a holistic model-based system representation. An important and challenging task is to identify and shape relevant subsystem models. In order to define an appropriate structure of these models, model developers may utilize criteria like selected properties of system elements and interrelations, their degree of detail or modelling assumptions. The relevant criteria have to be made transparent. For this purpose, the paper discusses the concept of model signatures that contain relevant meta information about each single model of all system elements, subsystems up to models of the overall system. This standardized meta information enables an identification and selection of those models and the decision on the necessary model integration. The concept is discussed on the basis of a roller bearing as an example out of an electro-mechanical drivetrain. A potential analysis provides information about the possible usage of the model signatures concept.

Keywords: MBSE · Product development · System models · Model signatures · Meta data · Model networks

Published by Springer Nature Switzerland AG 2023
F. Noël et al. (Eds.): PLM 2022, IFIP AICT 667, pp. 155–164, 2023.
https://doi.org/10.1007/978-3-031-25182-5_16

1 Introduction

Product development is characterised by numerous synthesis and analysis loops to reduce the delta between the required and as-is properties of a product [1, 2]. Analysis provides information on the fulfillment of the required properties of the product. Therefore, analysis results are an important basis for further synthesis steps. The analysis can be carried out by means of physical tests, calculations or simulations. In the context of model-based development, different types of simulation models are required.

For the development of mechatronic products, their description using Systems Theory is recommended [3]. Based on this, products are described via models of so-called mechatronic systems (systems are themselves models of the product). An essential technique to master the complexity of systems is their decomposition into system elements. Decomposition is, in principle, arbitrary, and different forms of decomposition can be chosen for a product depending on the objectives [4]. Decomposition is often driven by organizational structures or even better by focusing on specific (groups of) properties, thus leading to e.g. functional structures [5, 6]. In any case, the decomposition of a system into system elements results in hierarchically structured levels and leads to a desired modularity of a system, such that reuse of already existing system element libraries becomes feasible. Common system elements can then be reused in their given qualities for building up products as overall systems.

Decomposing a system constitutes a hierarchically structured **system architecture** which has to be defined at the beginning of the design phase [7]. Given the system architecture, the implemented system elements have to be **composed** accordingly, which works best, when a relevant set of properties is visible on the **system element interface and inside the system element**.

As representations of complex systems, models themselves exhibit complex structures, accordingly. Modelling works best, when the models are decomposed in accordance with a relevant (e.g. physical) structure of the system. Thus, system architecture typically coincides with the **model architecture**.

During the composition of a system model, for each system element an appropriate set of models has to be established, that provides the required degrees of fidelity, representations of system properties, flows, etc. reflecting the variety of modelling purposes. These models have to be integrated horizontally (same system level along the relevant flows of material, energy, and information) and vertically (aggregation from subsystem level to the overall system level, refinement in the opposite direction) to create a holistic model-based system representation. An important and challenging task is to identify and shape the relevant set of models. In order to define an appropriate structure of these models, model developers may utilize criteria like selected (groups of) properties of system elements and interrelations, their degree of detail or modelling assumptions. The relevant criteria have to be made transparent, especially in heterogeneous simulation environments. Naturally, models describing individual system elements may encapsulate internal parts as well. They should exhibit an explicit interface description, precisely specifying the ports and interactions of models during subsystem operation (also called runtime), the parameters of the model that allow to configure a model network for the system element before the simulation, and further semantic information about the model (describing objective of the model or model assumptions).

For this purpose, the concept of **model signatures** for the development, specification and compositional integration of models is introduced in this paper. The **objective** of the paper is to discuss the requirements for model signatures and first realisation concepts. First, the requirements and the state of the art are elaborated (cf. Sects. 2 and 3). Based on this, the concept of model signatures is explained (cf. Sect. 4), which is then illustrated by means of an example (cf. Sect. 5). The paper concludes with an analysis of potentials and further research questions (cf. Sect. 6).

2 Requirements for Model Signatures

The requirements for model signatures can be divided into requirements for supporting specific use cases, requirements for usability, and requirements for implementation. The relevant stakeholders for the model signatures are engineers who want to select the appropriate models for model networks (use case 1) and plan their interconnections and thus composition (use case 2). Necessary requirements for model signatures can be derived from these use cases:

For use case 1, engineers need relevant information on the model purpose as well as information on the specific model. This information includes the interfaces provided by the model to the outside and the flows available at the interfaces during operation (e.g., force, velocity, current, information etc.). Furthermore, it must be known which state parameters (e.g. temperature) the model exhibits during operation. For the flow variables and state parameters, it must be possible to distinguish whether they are static or dynamic during operation and, if dynamic, whether they are influenced from outside or controlled by the system element. In mathematical terms, a static parameter is just a constant value, while a dynamic parameter is a discrete or continuous function in a possible value range. Important for selecting a specific model is its level of idealization determined by model assumptions (e.g., linear stress-strain behaviour). Assumptions determine the scope of the model and necessarily lead to limitations in its validity.

For use case 2, the interconnectivity of models must be verified, quite like type checking in programming languages. As described in Sect. 1, it must be possible to network and link models horizontally and vertically. For the evaluation of the networking capability, it must be possible to provide relevant information via the model signature. For horizontal networking, the information about the interfaces and flows as well as the associated compatibility are required above all. Compatibility depends, among other things, on whether the flows fit together in terms of type, granularity, physical unit, direction, temporal behaviour, etc., as well as the modelling properties (e.g., data types and potentially given additional value ranges).

Regarding usability, it must be ensured that the relevant information is explicitly and unambiguously represented in the model signature.

Model signatures have to reflect (1) the interfaces of the system, sub-system, system element that they describe, and (2) the configuration and dimensioning parameters that engineers can decide upon during development. Furthermore, because of the need for reduction of complexity through abstraction, e.g., by omission of relevant flows (or their dynamics), and due to the need for discretization of continuous processes and distributed parameters, models exhibit restrictions in their connectivity. Such restrictions are typically not driven by the original system itself, but may be rooted in the design of the model, the underlying timing model, the form of simulation execution, etc. These forms of restrictions also have to be taken into consideration in model signatures.

The implementation of model signatures is subject to requirements regarding the definition of templates of model signatures, their instantiation and coupling with the executable simulation models. The implementation of these requirements is not described in concrete terms in this paper.

3 State of the Art

3.1 System Development

An established method for virtual development of mechatronic systems including their interacting subsystems is *Model-Based Systems Engineering (MBSE)* [6, 8, 9]. Thereby, the system to be developed is represented by a *system model* that is continuously both further specified (synthesis) and used for virtual behaviour verification (analysis) by all domains involved in the product development process. In order to control the complexity in the system model, the entire system is decomposed several times into *system elements*, so that a system architecture may span across several hierarchy levels (see e.g. Fig. 1) [6, 10]. Due to the encapsulation, these system elements have a handy complexity, interact with each other via interfaces and jointly represent the behaviour of the superordinate system.

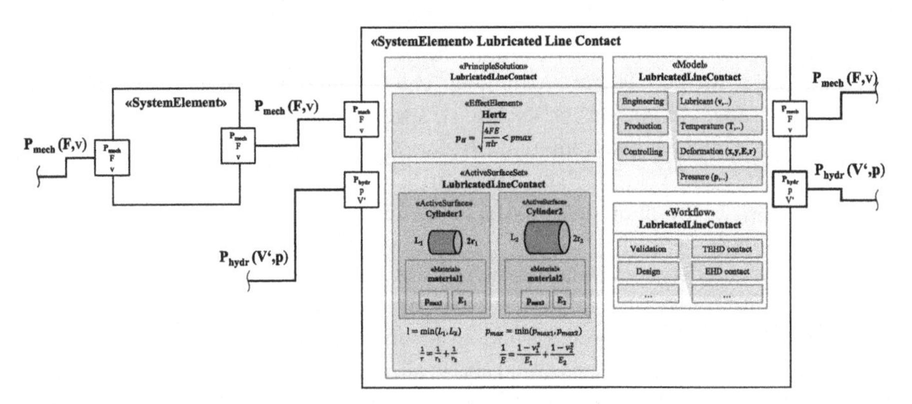

Fig. 1. Functional system architecture of system elements (according to [6, 11])

Using analysis based on defined characteristics of a product, its properties can be determined or – if the product does not yet exist – predicted [1]. In principle, analyses can be carried out using physical tests or simulation that can be based on proprietary and problem-specific simulation models. The reuse of models (product description [12] and simulation) is becoming increasingly important in order to reduce the effort required for model creation and to transfer existing knowledge between simulations. Standards such as FMI [13] have become established for the reuse of existing simulations, even in heterogeneous simulation environments.

3.2 Model Management

There is a variety of MBSE modelling tools to support engineers creating system models on different abstraction and fidelity levels for different purposes and views. The compatibility of those tools is limited; data exchange between different tools requires well-suited interfaces, model transformations and mappings. The limits in collaboration lead to a very restricted ability of composing and using model networks.

Most MBSE tools already enable some amount of analysis whereas deeper analysis or simulation requires software for simulation or custom analysis [14]. There are some approaches for integrating SysML and PLM systems for data and model management purposes. Heber and Groll [15] introduce a meta model to connect MBSE with PLM. Kirsch et al. [16] present an approach for SysML model management within PLM. PLM tools focus on data management in hierarchical (product) structures (BOM) but MBSE leads to data networks instead of hierarchies.

Wang [17] introduces an MBSE-Compliant Product Lifecycle Model Management (PLMM) as a methodology tailored to industrial application of MBSE, which implements a system of systems methodology in managing multitudinous models throughout a product lifecycle. One key aspect is the use of the SysML language as the unified top-level modelling method for building product meta models representing a set of specific product domain models (sub-models). The aim is to realize trans-phase and trans-domain model integration and synchronization, which tackle various challenges encountered in complex product developments.

Parrott and Spayd [18] focus in their research on the configuration management aspects of MBSE model management, in particular the change management of configuration controlled items. It was investigated, how changes to the models could be implemented with regard to how base content is affected and how the model versioning could be controlled. Hu et al. [19] propose a simulation models' meta model and ontology for their universal description. The meta model reflects the most essential features of simulation models without the consideration of a specific implementation method. Allen et al. [20] developed the Model-Driven Development Process (MDDP) methodology, which extends modelling to combine MBSE with Lean Information Management (LIM). The result is a fully consistent model for all relevant engineering items (EI) using only real-world semantics. Particular focus was put on model consistency and alignment to ISO 15288 to ensure that all information created during the process is automatically added to the model in its most appropriate form. In the context of mechatronic product development, Friedl et al. [21] propose the method of a Model Dependency Map (MDM) to disclose and describe the complex interrelations (e.g., inputs, outputs and their stakeholders) between different models of existing model networks. This approach may serve as a first step towards model signatures.

The management of a model network reflecting different abstraction and fidelity levels as well as the management of model transformations and mappings within the product development process is a challenge in PLM system environments. The model signature approach is a basis for tackling this issue on meta-model level.

4 Model Signature Approach

The requirements described in Sect. 2 illustrate that a formalised description is needed for the model signatures, which comprises the relevant information of a simulation model and can be read by the model user and, if necessary, be coupled without the specific executable simulation model (see Fig. 2). A model signature is necessary for each executable simulation model.

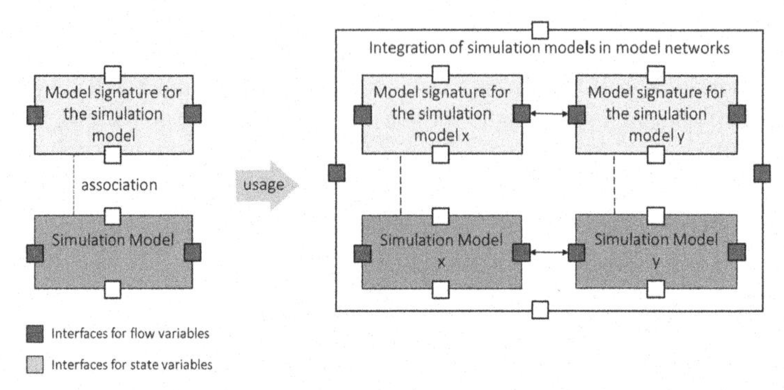

Fig. 2. Definition and usage of model signatures (with interfaces for variable exchange during runtime)

The analysis of different examples of simulation models (such as the lubricated line contact, see Sect. 5) shows that different types of simulation models can be found in the application. The possible types of simulation models are not considered in detail in this paper and are subject of further research activities. For the model signatures, this differentiation of the simulation models means that there can also be different classes of model signatures. A useful classification of model signatures supports the search for appropriate simulation models for specific applications.

If appropriate model signatures have been selected for specific use cases, their interconnectivity has to be ensured. Therefore, the model signatures must be represented in such a way that they can be used as a basis for checking whether the interfaces and flows or state variables are compatible in terms of content. Furthermore, due to the potential heterogeneous executable simulation models, it must be possible to ensure model compatibility on the basis of model signatures. Therefore, a model signature has to describe at least (1) the flows of the model of the system element for physical compatibility checking, and (2) the forms of encoding in the model, that unfortunately often also comprise tool specific properties such as data types, meshing parameters, time increments etc., for simulation compatibility.

In software development, object-orientation (OO) is a powerful concept to describe interfaces through their signatures and to encapsulate internal details. Domain Specific Languages (DSL) [22] tend to be defined from scratch and for a specific small purpose only. Composition, however, enforces them to provide explicit model signatures [23]. The concrete form of a model signature is rather dependent on the kind of model, but a common pattern of model signatures is an explicit naming of symbols that can be accessed or even influenced from outside [24]. Therefore, in this paper, the authors propose an approach based on semi-formal modelling languages such as SysML (see Fig. 3). They allow the representation of the relevant information in the model signatures as well as the interfaces including the flows or state variables when using the language in a clearly defined methodological form as discussed before. With such a sound use of SysML, semi-formal description languages for the model signatures can be used to determine their interconnectivity and various other models of consistency issues. In particular, it would allow to carry over the idea of a strong type system from programming to systems engineering models.

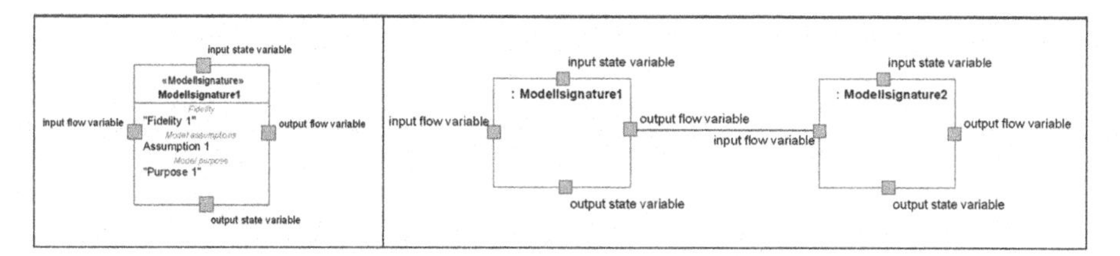

Fig. 3. Representation of model signatures using SysML

5 Example

The system element "lubricated line contact" as it appears in bearings and gear contacts, as well as the models belonging to it are well-known by literature. Therefore, it is selected as example for the intended model signature.

Figure 1 shows the system element "lubricated line contact" existing out of principle solution [11], a set of models and workflows [6]. The principle solution as a well-known element of design theory contains physical effects and acting surfaces [25]. The model set of the system element describes behaviours of the system element in different domains. Within this publication, we focus on the engineering domain, which includes models of analytical and numerical nature. With the intended model signature, the selection of appropriate models for specific purposes and fidelities is supported. Result is a case specific selection of models out of the system element's full model set. Those selected models are connected by workflows which allow the execution of the model chain.

The engineering models that are linked to the system element "lubricated line contact" are depicted in Fig. 4. They can be differentiated by purpose and fidelity [6]. For example, three models of different fidelity can be distinguished for the purpose "temperature calculation". In dependency of the required purpose of the system element (e.g. TEHD-calculation), a meaningful combination of multiple models as indicated by the black line has to be found. The decision on possible combinations of models needs deep insight into each model, which the authors aim to formalize by an appropriate model signature in order to support the compositional integration of models.

Figure 5 shows on the left hand side the parameters that are exchanged between the models when a TEHD calculation is executed. Obviously, state parameters, design

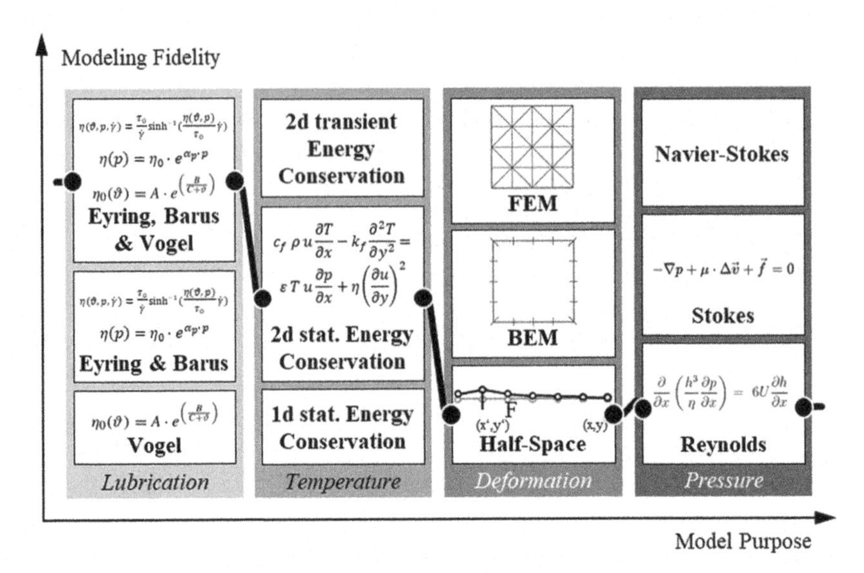

Fig. 4. Selection of engineering models for the scope "lubricated line contact"

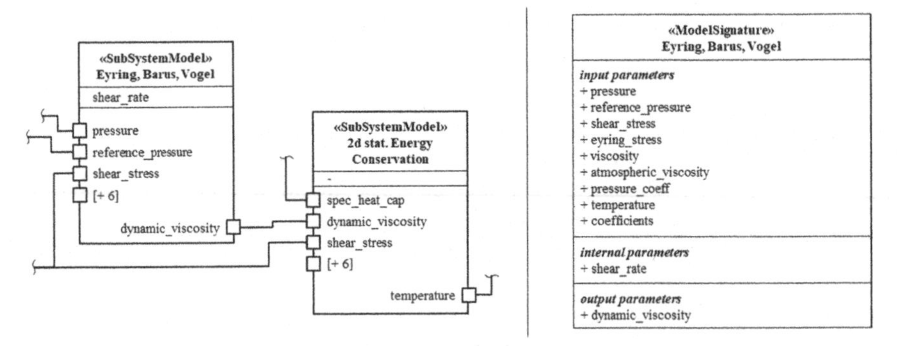

Fig. 5. Model chain of the lubrication model "Eyring, Barus, Vogel" and "2d stationary Energy Conservation" with their parameters (left side) as well as the proposed model signature of the lubrication model "Eyring, Barus, Vogel" (right side)

parameters and functional flows can occur as input, output or internal parameters. Each parameter can be of static or dynamic nature. A dynamic parameter can be driven externally or by the system element itself. The decision which parameter is input, output or internal depends on the purpose the model is used for. Especially, the lack of access to the instance of an internal parameter often leads to contradictions between the instances of the same parameter in other models. Models with inconsistent parameter instances are still a common problem in system modelling. Therefore, the model signature comprises input, output and internal parameters (Fig. 5, right hand side).

6 Relevant Research Questions for Further Steps

The discussions in this paper point out key messages and research questions derived therefrom:

- Model selection for analysis, and assurance of interconnectivity require explicit descriptions of model contents, model assumptions as well as modelling specifications in the form of model signatures in addition to executable models → derived research question: How must the model signatures (contents and formalization) be described (part of the paper)? Can the usual typing techniques known from software programming be adapted to these model signatures?
- The models of the system elements must be interconnected for composition. → derived research question: How is the interconnection of the models performed based on the model signatures? (research question is only partly addressed in this paper)
- In the context of today's modelling, different model types exist. For the different model types different model signatures are necessary → derived research question: Which model types exist and which model signatures are therefore necessary? Which technical requirements are necessary to implement the signatures (SysML or an extension)? (research question is only partly addressed in this paper)
- For compositional integration of models at different hierarchical levels, the signatures must reflect different hierarchical levels as well → derived research question: Which properties of the models have to be included in model signatures to ensure vertical compatibility and to enable hierarchical integration of models across different levels? (research question will be concretized in further publications)

The research questions will be investigated interdisciplinary in further activities.

References

1. Weber, C., Husung, S.: Virtualisation of product development/design - seen from design theory and methodology. In: 18th International Conference on Engineering Design (ICED 2011), pp. 226–235 (2011)
2. VDI: Entwicklung technischer Produkte und Systeme/Design of technical products and systems. Blatt 2 (VDI 2221:2019) (2019)
3. Ropohl, G.: Systemtechnik. Grundlagen und Anwendung, Hanser, München (1975)
4. Ariyo, O.O., Eckert, C.M., Clarkson, P.J.: Hierarchical decompositions for complex product representation. In: 10th International Design Conference, pp. 737–744 (2008)
5. Browning, T.R.: Applying the design structure matrix to system decomposition and integration problems: a review and new directions. IEEE Trans. Eng. Manag. **48**(3), 292–306 (2001). https://doi.org/10.1109/17.946528
6. Jacobs, G., Konrad, C., Berroth, J., Zerwas, T., Höpfner, G., Spütz, K.: Function-oriented model-based product development. In: Krause, D., Heyden, E. (eds.) Design Methodology for Future Products, pp. 243–263. Springer, Cham (2022). https://doi.org/10.1007/978-3-030-78368-6_13
7. VDI: Entwicklungsmethodik für mechatronische Systeme/Design methodology for mechatronic systems (VDI 2206:2004) (2004)

8. Rumpe, B.: Modeling with UML. Language, Concepts, Methods. Springer eBook Collection Computer Science. Springer, Cham (2016). https://doi.org/10.1007/978-3-319-33933-7
9. Friedenthal, S., Moore, A., Steiner, R.: A Practical Guide to SysML: The Systems Modeling Language, 3rd edn. The MK/OMG Press, Burlington (2015)
10. Husung, S., Weber, C., Mahboob, A.: Model-based systems engineering: a new way for function-driven product development. In: Krause, D., Heyden, E. (eds.) Design Methodology for Future Products, pp. 221–241. Springer, Cham (2022). https://doi.org/10.1007/978-3-030-78368-6_12
11. Zerwas, T., et al.: Mechanical concept development using principle solution models. In: IOP Conference Series: Materials Science and Engineering, p. 012001 (2021). https://doi.org/10.1088/1757-899X/1097/1/012001
12. Hick, H., Bajzek, M., Faustmann, C.: Definition of a system model for model-based development. SN Appl. Sci. **1**(9), 1–15 (2019). https://doi.org/10.1007/s42452-019-1069-0
13. Blochwitz, T., et al.: The functional mockup interface for tool independent exchange of simulation models. In: Proceedings of the 8th International Modelica Conference, pp. 105–114 (2011)
14. Bretz, L., Tschirner, C., Dumitrescu, R.: A concept for managing information in early stages of product engineering by integrating MBSE and workflow management systems. In: IEEE International Symposium on Systems Engineering (ISSE), pp. 1–8 (2016)
15. Heber, D.T., Groll, M.W.: A meta-model to connect model-based systems engineering with product data management by dint of the blockchain. In: IEEE International Conference on Intelligent Systems (IS), pp. 280–287 (2018). https://doi.org/10.1109/IS.2018.8710527
16. Kirsch, L., Müller, P., Eigner, M., Muggeo, C.: SysML-Modellverwaltung im PDM/PLM-Umfeld. In: Tag des Systems Engineering, pp. 333–342. Carl Hanser (2016)
17. Wang, C.: MBSE-compliant product lifecycle model management. In: 14th Annual Conference System of Systems Engineering (SoSE), pp. 248–253. IEEE (2019). https://doi.org/10.1109/SYSOSE.2019.8753869
18. Parrott, E.L., Spayd, L.C.: Configuration and data management of the NASA power and propulsion element MBSE model(s). In: 2020 IEEE Aerospace Conference, pp. 1–11. IEEE (2020). https://doi.org/10.1109/AERO47225.2020.9172375
19. Hu, C., Xu, C., Fan, G., Li, H., Song, D.: A simulation model design method for cloud-based simulation environment. Adv. Mech. Eng. **5**, 932684 (2013)
20. Allen, C., Di Maio, M., Kapos, G.-D., Klusmann, N.: MDDP: a pragmatic approach to managing complex and complicated MBSE models. In: IEEE International Symposium on Systems Engineering (ISSE), pp. 1–8 (2016). https://doi.org/10.1109/SysEng.2016.7753165
21. Friedl, M., Weingartner, L., Hehenberger, P., Scheidl, R.: Model dependency maps for transparent concurrent engineering processes. In: 14th Mechatronics Forum International Conference (Mechatronics 2014), pp. 614–621 (2014)
22. Fowler, M., Parsons, R.: Domain-Specific Languages. Addison-Wesley (2011)
23. Clark, T., van den Brand, M., Combemale, B., Rumpe, B.: Conceptual model of the globalization for domain-specific languages. In: Cheng, B.H.C., Combemale, B., France(†), R.B., Jézéquel, J.-M., Rumpe, B. (eds.) Globalizing Domain-Specific Languages. LNCS, vol. 9400, pp. 7–20. Springer, Cham (2015). https://doi.org/10.1007/978-3-319-26172-0_2
24. Butting, A., Hölldobler, K., Rumpe, B., Wortmann, A.: Compositional modelling languages with analytics and construction infrastructures based on object-oriented techniques—the MontiCore approach. In: Heinrich, R., Durán, F., Talcott, C., Zschaler, S. (eds) Composing Model-Based Analysis Tools, pp. 217–234. Springer, Cham (2021). https://doi.org/10.1007/978-3-030-81915-6_10
25. Koller, R.: Konstruktionslehre für den Maschinenbau. Grundlagen zur Neu- und Weiterentwicklung technischer Produkte mit Beispielen. Springer, Heidelberg (1994). https://doi.org/10.1007/978-3-662-08165-5

MBSE-PLM Integration: Initiatives and Future Outlook

Detlef Gerhard[1], Sophia Salas Cordero[2], Rob Vingerhoeds[2], Brendan P. Sullivan[3](✉), Monica Rossi[3], Yana Brovar[4], Yaroslav Menshenin[4], Clement Fortin[4], and Benoit Eynard[5]

[1] Ruhr-University Bochum, Universitaetsstr. 150, 44801 Bochum, Germany
[2] ISAE-SUPAERO, Université de Toulouse, Toulouse, France
sophia.salas-cordero@isae-supaero.fr
[3] Politecnico di Milano, Via Lambruschini 4B, 20156 Milan, Italy
Brendan.Sullivan@polimi.it
[4] Skolkovo Institute of Science and Technology, Bolshoy Boulevard 30, Bld. 1, Moscow 121205, Russia
[5] Université de Technologie de Compiègne, Rue du Dr Schweitzer, 60203 Compiègne Cedex, France

Abstract. The development of Cyber-Physical-Human Systems which are pervasive today as proposed by the Industry 4.0 vision, requires an efficient integration of the systems definition which is modelled within the Model Based System Engineering (MBSE) applications with the models representing the detailed design and analysis of these products, which are generally embedded in Product Life Cycle Management (PLM) systems. In this paper, we are presenting an overview of some of the important initiatives on this topic across the IFIP 5.1 community, and also projecting a future outlook based on promising new approaches to this emerging problem of MBSE and PLM integration.

Keywords: Systems Engineering · Model based systems engineering · Product life cycle management · Digital engineering · Cyber-physical-human systems

1 Introduction

Systems Engineering (SE) serves the purpose of managing systems complexity and reducing the uncertainty associated with the design process. SE appeared in the mid-twentieth century, when the system's complexity reached extreme levels in the aerospace industry. Since then, SE found its presence and utility not only in very large corporations and defence acquisition programs, but also in automotive, healthcare, energy, and other sectors of the economy. SE itself, as a discipline, has been extended towards different fields of knowledge. However, the increasing complexity of systems has created challenges in the life cycle management of projects, integration of diagrams, retention of knowledge and test planning.

The original version of this chapter was revised: The last name of the author has been corrected as "Salas Cordero". The correction to this chapter is available at
https://doi.org/10.1007/978-3-031-25182-5_67

Published by Springer Nature Switzerland AG 2023
F. Noël et al. (Eds.): PLM 2022, IFIP AICT 667, pp. 165–175, 2023.
https://doi.org/10.1007/978-3-031-25182-5_17

Outlined by IFIP5.1, there is a need for new approaches that "support systems that allow the information and data associated with products to be developed and sustained through the product life-cycle" [1]. This shift in development processes focus utilizes a more complex series of interfaces and necessitates a shift from previous dominant document-based approaches for the facilitation of communication, management of development risks, quality, process, and business productivity, as well as knowledge transfer. In response to this transition this paper presents a series of tools, methods and approaches that are enabled through an integrated model-based approach for the development and management of artifacts and reference models.

Potential advantages offered by MBSE and PLM integration, can include enhanced communications, reduced development risks, improved quality, increased productivity, and enhanced knowledge transfer, can be further scaled up, by increasing the level of automation throughout the system life cycle [2, 3]. In this way, the ability to leverage artifacts for decision making can be extended to larger parts of the development effort. However, both disciplines – MBSE and PLM – have grown to some extent independently, whereas we hypothesize that through better integration more value to our understanding and practice of product/system development is possible.

The paper is structured as follows. Section 2 reviews the trends related to digital engineering. To this purpose, fundamental concepts such as digital twin (DT), the V-Model [8] life cycle diagram and data continuity are reviewed. Section 3 then details the overview of existing initiatives and approaches in support of a MBSE-PLM integration. Section 4 presents a discussion, conclusions, and outlook.

2 Background

MBSE and PLM integration is a topic that keeps on gaining momentum. In the past years, there has been work on how to integrate MBSE and PLM throughout parametric models as in [4], also how to automate trade studies using MBSE and PLM [5], how MBSE and PLM industrial integration is a need for mission-critical systems [6], and incorporation of DT technology into MBSE [7].

2.1 Model Based Systems Engineering

The International Council on Systems Engineering (INCOSE) defines Model Based Systems Engineering (MBSE) as "the formalized application of modelling to support system requirements, design, analysis, verification, and validation activities beginning in the conceptual design phase and continuing throughout development and later life cycle" [9]. The focus on developing, managing, and controlling system models offers the potential to enhance product quality, enhance reuse of the system artifacts, and improve communications throughout all parts of the business and development team. Through the use of descriptive and analytical models that can be applied throughout the life cycle of a system, MBSE can reduce the time and cost to design, integrate and test a system, while simultaneously reducing risk and improving quality.

Through the Systems Modelling Language (SysML) it is possible to model numerous critical aspects required in the SE domain (structure, requirements, behaviour, parametric). SysML, which was developed as an extension to the Unified Modelling Language

(UML), is considered a standard modelling notation adopted in the context of MBSE and utilized in this paper [10, 11]. Additional MBSE solutions and languages such as Capella and the Object Process Methodology (OPM) can be applied/supplemented for SysML, and in the case of OPM even integrated with SysML.

2.2 Digital Engineering

INCOSE coined the term Digital Engineering [12] to characterize MBSE when complemented with simulation technologies. Digital Engineering supports an integrated model-based approach through the utilization of digital methodologies, tools, processes, and digital artifacts. Grieves and Vickers [13] define Digital Engineering as methods and tools to support the design of well-structured DTs, which are embodiments of both the systemic perspective and the product view. The main popularity of the DT paradigm came with the breakthrough of the development of big data analytics, simulation technologies, and Internet of Things. The DT is a set of virtual information constructs that fully describes all of the information required to define, describe, and produce a potential or actual physical manufactured product, including requirements, a fully annotated 3D model with geometric dimensioning and tolerancing (GD&T), material specifications, Product Manufacturing Information (PMI), etc. Rooted in SE, Digital Engineering [14] uses MBSE [15], to model the essential system characteristics including system requirements, structure, functions, and behaviour.

2.3 Product Life Cycle Management

Product Life Cycle Management (PLM) can be considered as a business strategy that focuses on the management of data, information, knowledge, and experience essential to creating and sustaining a product-centric knowledge environment throughout all passes of a system/product (beginning, middle and end of life) [16]. As such an integrated PLM environment enables collaboration between informed decision makers by combining and integrating various stakeholder perspectives of a product throughout its lifecycle. This collaborative information environment can be strengthened through the inclusion product data management (PDM) which ensures that the right information is received at the right time, in the right place through temporal product data evolution control (versioning, revision, bill of materials, etc.).

PLM vendors have gradually added MBSE functionalities into their solutions over the last decade, beginning with the linking of the system design and detailed design phases, however the integration requires substantial improvements to make it efficient. This is in part due to the still early development maturity of MBSE functionalities combined with the specificity of mechanical design functions. One should remember that PLM systems were developed to replace 2D drafting, which has been well standardized more than a century ago and is still used to legally define products in most industries. As an example, explicit time and state representations which effectively link space and time properties are absent in PLM systems but are fundamental for system, software and electronic design and thus are critical MBSE functionalities. Many approaches bring together multi-domain product development, including product requirements engineering, product architecture and system modelling, system simulation management, program planning, systems engineering, risk management and change management.

In addition to industrial PLM solutions, open-source frameworks such as TASTE, DocDokuPLM and Open Services for Lifecycle Collaboration (OSLC) (Sect. 3.2) are capable of integrating many aspects of MBSE and PLM. As an open-source tool-chain for embedded software development TASTE[1] was developed by the European Space Agency "to bring true, formal models-based Engineering into the way we develop space SW" [17], which supports modelling, model analysis, code generation & deployment, debugging & testing, and execution platforms [18]; DocDokuPLM[2], includes document management, product structure management, product configuration to manage alternatives, bill of materials (BOMs), process management change management, and a platform for data visualization and documents (Word, PDF, CAD...).

2.4 MBSE/PLM Integration

According to the basis and objectives of PLM, it can be seen that spanning from the first ideas, feasibility studies, through the actual development, the operation use and in the ultimately to the retirement of the product, different engineering skills come into play. Depending on the system/product type there can be mechanical engineering, electrical engineering, software engineering, etc. involved. Each of these engineering disciplines are "addressed using specific authoring tools, for example electrical/electronic CAD tools [19]. Building upon the expected advantages of MBSE development for the development of mechatronic products, the earlier phases should be able to rely on good MBSE tools, and then a seamless connection to the specific engineering authoring tools.

Ideally, a complete tool chain should cover all necessary engineering disciplines (e.g., mechanical, software, electronics and electrical), with seamless data continuity between tools, to guarantee upstream as well as downstream functionality. Which means that eventual decisions made in a specific authoring tool are not only cascaded down to the next tool, but also back up to the MBSE tool chain to enable to study specific impacts on other parts of the design.

Despite the initiatives mentioned throughout this paper, a number of obstacles hamper MBSE/PLM integration. Firstly, with the exception of tools commercialized by a single vendor, real data continuity is not guaranteed. This poses problems, as in industry, typically, different authoring tools, as well as data management tools are selected separately, and are partly based on historical observations. Moving from one authoring tool to another one, while maintaining a history on what is already in production, is very costly and time-consuming. Secondly, developing proficiency with a new tool represents is an important investment. It should be mentioned that to connect tools between them may require major adaptations to the tools. Additionally, extended enterprise leads to the need to connect tools between different companies. Such issues, if not addressed, can lead to the necessity to re-enter information that was already entered in other tools,

[1] https://gitrepos.estec.esa.int/taste/taste-setup.

[2] https://github.com/docdoku/docdoku-plm.

with the associated risks such as loss of coherency, introducing errors, etc. that could potentially lead to incoherent design decisions.

As a result of the above-mentioned issues, the complete value and potential of MBSE is partially diminished. In the light of absence of data continuity and harmonization, the use of MBSE often remains limited diagram drawing, leaving designers to transfer the necessary information to other tools.

3 Initiatives and Approaches for Life Cycle Collaboration

There are initiatives in a number of countries covering various approaches to support life cycle model collaboration and MBSE/PLM integration. We present an overview of important projects and initiatives which aim to foster collaboration in the field and increase model-based management, control, and analysis.

INCOSE has a number of initiatives advancing SE theory and practice, one of them is the working group "Digital Engineering Information Exchange" (DEIX WG). Its focus is on the digital artifacts (DA). DA is a digital form of information content that a digital engineering ecosystem produces and consumes by generally following the SE life cycle's process areas as defined in ISO 15288 [20]. DA provide "data for alternative views to visualize, communicate, and deliver data, information, and knowledge to stakeholders", in order to make informed and evidence-based decisions.

3.1 Business Process Modelling Notation

Through the efforts of the Object Management Group (OMG) the Business Process Modelling and Notation (BPMN) has been introduced to support the formalization of the Business Process Model (BPM) layer that offers a common unified visual language capable of defining the interactions amongst and between processes and organizations that make systems at the context and/or operational scenario analysis level [21]. Collaboration between BPMN during the concept of operations (CONOPS) phase provides businesses with the comprehensive capability of understanding development and business procedures through a graphical notation and give organizations the ability to communicate these procedures in a standard manner. Furthermore, life cycle model collaboration can enhance the performance, collaboration and business transactions between people and business entities necessary for a product-centric knowledge environment.

Integration of MBSE and BPMN is capable of increasing life cycle capabilities through increased abstraction and automation, ensuring that relevant artifacts, functions, and elements align with stakeholder needs [10, 11]. The technical system model (MBSE) and organizational model (BPM), enable improved complexity management, communication, and process controllability throughout the entire [22]. The automation layer through this abstraction is able to reflect on different BPMN models (Processes/Orchestration, Choreographies, and Collaborations) that connect distinct processes within the organization, facilitating the conversion and coordination of workflows. Through this graphical notation BPMN is capable of providing a comprehensive view to the system or business models in relation to one another.

3.2 Open Services for Lifecycle Collaboration

Through open formal modelling languages and standards, such as Modelica [23], Functional Mock-Up Interface (FMI) [24], Structure and System Parametrization (SSP) [25], etc., a growing interest in open-source tools that allow collaboration and shared access to information appeared. Modelica, as well as Acumen [26] or Bloqqi [27] are widely used, open-source and mature industrial language. The Modelica Association complements its solutions with machine learning frameworks [28] and continuously improve support for the FMI, enabling models exchange between tools.

Another initiative in line with the purpose of INCOSE towards tools integration is OSLC [29]. OSLC consists of specifications designed for different integration scenarios to make vendor-independent heterogeneous tool integration easier. The standard is built on web standards for communication and ontology definitions. The OSLC specifications consist of an OSLC Core specification, and a set of Domain specifications. The OSLC Core specification defines basic concepts and rules for integration methodologies and ensures consistency among different domain specifications. In turn, OSLC domain specification focuses on specific life cycle topics such as requirements management, change management, configuration management, architecture management, etc. [30]. It has also been recently proposed to improve OSLC through the CONEXUS tool, using category theory as a basis for database integration [31].

3.3 System Architecture Definition

Every system or product has an architecture which defines fundamental relationships between and within its elements during the high-level design phase. The system architecture is defined iteratively, in a series of decision-making activities [32] in which stakeholders, system architects, project managers and designers play a crucial and integrative role. Focusing on the upper-left side of the V-Model, the architecture includes feasibility study/concept exploration, concept of operations, system requirements, and high-level design. Nowadays, the system architecture definition tends to be realized through model-based approaches.

System architecture is "the embodiment of concept, the allocation of physical/informatic function to the elements of form, and the definition of relationships among the elements and with the surrounding context" [33]. Concept is a critical entity, as it rationalizes the architecture and maps function to form. Additionally, the presence of relationships amongst elements, and interfaces between elements, can be established according to subsystems, or departments and units within the organization.

3.4 Detailed Domain Specific Digital Engineering

Detailed engineering within different domains (bottom of V-model) requires the consistent use of standards throughout different digitalization fields and levels, to ensure the exchange of information in cross-company based cooperation. Customers and suppliers require digital representations (models) of components and systems, for geometry and product structure. JT (ISO/DIS 14306:2017) and STEP AP242 (ISO 10303–242:2020) are the most advanced standards in this context. As companies move from traditional paper-based workflows to MBSE, STEP AP 242, is the most commonly used file format for CAD interoperability with downstream departments, suppliers, subcontractors, and customers. STEP AP 242 includes PMI such as GD&T, BOM and other meta information, as well as references for part or assembly design and measurement. It comprises all the features of AP203 (ISO 10303–203:2005) and AP214 (ISO 10303–214:2010), and additionally contains features such as semantic 3D PMI, 3D shape quality and 3D design with parametric/geometric constraints. Therefore, it is particularly suited for the exchange of product and assembly structure data with external references to geometry files (regardless of file format). Based on merging of standards from PLM (different STEP Application protocols) and SE (ISO/IEC 15288, ISO/IEC/IEEE 12207, ISO 10303 - STEP AP233 (ISO 10303–233:2012) and new AP243 (ISO 10303–243:2021), an added value proposal for DT standardization is under development, also associated with the Product Lifecycle Management ATLAS Program[3].

3.5 Verification and Validation

The development of increasingly complex products and their functions requires the use of MBSE methods and advanced simulation and testing capabilities for a variety of applications. Simulation-based decision making and release of complex systems in collaborative development scenarios between partners is gaining significant importance in industry. The current trend shows that the utilization of simulations exceeds physical tests for verification and validation purposes due to the cost and general flexibility of the solution. The quality of the simulation results and their traceability throughout the entire development process is essential.

While multiple options exist, OpenModelica[4] was identified as one solution capable of integrated large-scale modelling, simulation, optimization, model-based analysis [34, 35]. In order to achieve interoperability among different behaviour modeling tools, the FMI standard defines the Functional Mock-Up Units (FMU) as a container and an interface to exchange dynamic models or to perform co-simulation or scheduled execution of simulation models. When exchanged, FMU metadata and FMU structural information are exchanged as well. The ProSTEP iViP Simulation data management (Sim PDM) recommendation [36] provides integration guidelines of simulation data in PLM environments and defines communication processes between simulation data management systems (SDM system) and CAE systems. Additionally, ProSTEP iViP "Smart Systems Engineering Behaviour Model Exchange" [37] describes a comprehensive and representative spectrum of the exchange of behaviour models particularly for the use case of joint SE development network between the contracting entity and the supplier. As a complementary development, the Simulation Model Meta Data (SMMD) specification, under development, defines a data file format as a consignment note to FMUs to integrate FMU exchange processes into PLM [38]. The Automated Functional Data FDX recommendation describes a data model and format for the standardized and traceable exchange of functional data and relevant, associated metadata.

[3] ATLAS is related also to the new ISO 23247-2021 or IEC 62832:2020 for the specific smart manufacturing field https://plmatlas.com/.

[4] https://github.com/OpenModelica.

4 Discussion, Conclusion and Outlook

The importance of the traceability and reproducibility of models and simulations, as a major task in the development process of increasingly complex products, requires the integration of MBSE and PLM as seamless as possible; since the importance of decisions derived from modelling and simulation is also growing.

The basis for a system architecture definition is set by requirements. Requirements information can be exchanged through the transmission of Requirements Exchange Format (ReqIF)-compliant XML documents at a systems level for general architecture definition and requirements management [39]. Collaboration between partners is enhanced by the benefits of applying such methods across organizational boundaries.

As discussed in Sect. 3.4, core PLM issues with respect to geometry, shape and part definition and digital thread are supported by AP242. It covers the entire development process from the detailed design stage until the end of production development. This well aligned standard integrates with existing solutions for life cycle support, with minimal migration disruption. It can additionally replace paper-based processes with 3D master models since it contains information normally found in technical drawings, according to a semantic PMI. Product data and non-geometric metadata are represented by AP242, as well as simulation. It enables the description of kinematic structures (e.g., connections, articulations, pairings, movements). Links from the kinematic structure to CAD parts can be established by external references.

While the benefits of BPMN and MBSE integration have strong practical value for PLM, it should be acknowledged that there remains at present a comparatively large modelling effort required for MBSE and BPMN. The integrated knowledge and process information facilitated through this integration will allow for a broader and more comprehensive view of all operations/activities, facilitating enhanced decision making. The expected improved integration of diagrams through intelligent operations will be fundamental to changing how businesses go about their entire engineering processes [40].

INCOSE defines some avenues for the future of SE: it is a model-based environment, with an increasing emphasis on AI powered by "large data sets and expert's domain knowledge" [39]. In this way, there will be a further development of systems engineering methods and tools based on established science and mathematics.

In MBSE, data are quantitative and qualitative variables that characterize an artifact (e.g., product or service) to be designed in terms of its specific requirement, constraint, functionality, behaviour, structure, etc. Data acquisition and processing is applicable to many activities in product development, such as interpretation of customer opinions, market information, customer needs and competitive benchmarking, identification of dependencies. Data in the context of data driven engineering is characterized by high volume and variety, which leads to the requirement for special tools and procedures. Data science provides approaches, algorithms, and technologies to gain knowledge, understanding and intelligence based on data. The product life cycle can be looped back, products already produced and sold provide data that can be used to design the next generation of products. The concept of DT, encompasses information, models, and data from all phases of the life cycle, is an important enabler for Data-Driven Engineering.

The next steps for integrating MBSE and PLM in a common digital backbone will certainly be involving DT and its various applications. According to [3] and [4] key enabling technologies are needed to be associated and merged to provide a real and effective solution to support DT. In the classical architecture of DT based on Virtual models, Physical entities, and Services, MBSE and PLM clearly allow the consistent storage, management, and integrity of DT data. Then, one of the biggest issues under process by academics and industrialists is the connection and integration with physical assets and service operations. Some developments with Internet of Things, Cyber-Physical Systems [11], and Servitization are promising but still require scaling before it will be possible to transform research concepts into viable enterprise and business solutions.

References

1. IFIP5.1 - International Federation for Information Processing. Accessed 19 May 2022. http://www.ifip-wg51.org/topics.php
2. Friedenthal, S.: Future directions for product lifecycle management (PLM) and model-based systems engineering (MBSE). Whitepaper (2020). Accessed 18 May 2022
3. Schuler, R., Kaufmann, U., et al.: 10 theses about MBSE and PLM - challenges and benefits of model based engineering (MBE). PLM4MBSE working group position paper (2015)
4. Katzwinkel, T., Löwer, M.: MBSE-integrated parametric working surfaces as part of a PLM design approach. In: Proceedings of the Design Society: International Conference on Engineering Design (2019)
5. Chadzynski, P.Z., et al.: Enhancing automated trade studies using MBSE SysML and PLM. In: INCOSE International Symposium, vol. 28, pp. 1626 1635 (2018)
6. Pavalkis, S.: Towards industrial integration of MBSE into PLM for mission-critical systems. In: INCOSE International Symposium (2016)
7. Madni, A.M., Madni, C.C., Lucero, S.D.: Leveraging digital twin technology in model-based systems engineering. Systems, **7**(1), 7 (2019). https://doi.org/10.3390/systems7010007
8. Fairley, D., et al.: System life cycle process models: V, systems engineering book of knowledge (SEBoK) (2014). https://www.sebokwiki.org/wiki/System_Life_Cycle_Process_Models:_V Accessed 15 Mar. 2022)
9. INCOSE Technical Operations. Systems Engineering Vision 2020. International Council on Systems Engineering, Seattle, WA, INCOSE-TP-2004–004–02 (2007)
10. Cwikła, G., Gwiazda, A., Banas, W., Monica, Z., Foit, K.: Analysis of the possibility of SysML and BPMN application in formal data acquisition system description. In: IOP Conference Series Materials Science and Engineering, vol. 227 (2017)
11. Lu, J., Ma, J., Zheng, X., et al.: Design ontology supporting model-based systems engineering formalisms. IEEE Syst. J. **16**, 5465–5476 (2021)
12. Giachetti, R., et al.: Digital engineering, systems engineering book of knowledge (2018). https://www.sebokwiki.org/wiki/Digital_Engineering. Accessed 15 March 2022
13. Grieves, M., et al.: Digital twin: mitigating unpredictable, undesirable emergent behaviour in complex systems. In: Kahlen, F.J., Flumerfelt, S., Alves, A. (eds.) Transdisciplinary Perspectives on Complex Systems, pp. 85–113. Springer International Publishing, Cham (2017). https://doi.org/10.1007/978-3-319-38756-7_4
14. Tsui, R., Davis, D. Sahlin, J.: Digital engineering models of complex systems using model-based systems engineering (MBSE) from enterprise architecture (EA) to systems of systems (SoS) architectures & systems development life cycle (SDLC). In: INCOSE International Symposium, vol. 28, no. 1, pp. 760–776 (2018)

15. Peterson, T.A.: Systems engineering: transforming digital transformation. In: INCOSE International Symposium, vol. 29, no. 1, pp. 434–447. https://onlinelibrary.wiley.com/doi/abs/10.1002/j.2334-5837.2019.00613.x (2019)
16. Terzi, S., Bouras, A., Dutta, D., Kiritsis, D.: Product lifecycle management - From its history to its new role. Int. J. Prod. Lifecycle Manag. **4**, 360 (2010)
17. ESA / ESTEC / TEC-SWT 2020: TASTE (2005). https://taste.tools/. Accessed 07 Aug 2021
18. Perrotin, M.: TASTE: an open-source tool-chain for embedded software development. https://taste.tools/. Accessed 07 Aug 2021
19. Vingerhoeds, R.A., et al.: Educational challenges for cyber-physical systems modelling. In: 12th International Conference on Modelling, Optimization and Simulation (2018)
20. INCOSE Systems Engineering Handbook: A Guide for System Life Cycle Processes and Activities, 4th edn. Wiley, INCOSE (2015)
21. OMG: Systems Modeling Language,Ver. 1.6, Object Management Group (2018). https://www.omg.org/spec/SysML/
22. Konrad, C., et al.: Enabling complexity management through merging business process modeling with MBSE. Procedia CIRP, **84**, 451–456 (2019)
23. Modelica Association. Modelica-A Unified Object-Oriented Language for Physical Systems Modeling-Language Specification Version 3.4. https://www.modelica.org/documents/ModelicaSpec34.pdf. Accessed 20 May 2022
24. MODELISAR. Functional Mock-Up Interface, Version 2.0. Interface Specification (2017). https://fmi-standard.org/downloads/. Accessed 20 May 2022
25. Modelica Association. System Structure and Parameterization, Version 1.0. https://ssp-standard.org. Accessed 20 May 2022
26. Zeng, Y., et al.: Modeling electromechanical aspects of cyber-physical systems. J. Softw. Eng. Robot. **7**, 100–119 (2016)
27. Fors, N., et al.: Modular feature-based block diagram programming. In: Proceedings of the 2016 ACM International Symposium on New Ideas, New Paradigms, and Reflections on Programming and Software, Onward! Amsterdam, The Netherlands, 2–4 November 2016, pp. 57–73 (2016)
28. Tensorflow.org. An end-to-end open source machine learning platform (2019). https:// www.tensorflow.org/. Accessed 05 Apr 2022
29. OCSL: OSLC core specification version 2.0. Open Services for Lifecycle Collaboration (2010)
30. El-khoury, J.: An Introduction to OSLC and Linked Data, Lecture. Stockholm, Sweden (2015). https://open-services.net/resources/video-20151118-jad-ecs/ Accessed Mar 2022
31. CONEXUS, Category Theory for Semantic Data Interoperability: Exchange and Consolidate Data for Agile PLM at Scale (2021)
32. Menshenin Y., et al.: Model-based system architecting and decision-making. In: Handbook of Model-Based Systems Engineering Madni, A.M., Augustine, N., Sievers M. (eds.), pp. 1–42. Springer, Cham (2022). https://doi.org/10.1007/978-3-030-27486-3_17-1
33. Crawley, E., Cameron, B., Selva, D.: System Architecture: Strategy and Product Development for Complex Systems. Prentice Hall Press, Boston (2015)
34. Otter, M.: Modelica Overview, Modelica Association, Presentation (2018). https://modelica.org/education/educational-material/lecture-material/english/ModelicaOverview.pdf/view.html. Accessed March 2022
35. Fritzson, P., et al.: The OpenModelica integrated environment for modeling, simulation, and model-based development. Model. Ident. Control **41**, 241–295 (2020)
36. Prostep ivip/VDA: Simulation Data Management - Integration of Simulation and Computation in a PDM Environment (SimPDM), Version 2.0 (2008)
37. Prostep ivip; SmartSE Recommendation V 2 - Smart Systems Engineering - Simulation Model Exchange, prostep ivip Association. PSI I 1. Darmstadt, Germany (2018)

38. OMG 2016 Requirements Interchange Format. https://www.omg.org/spec/ReqIF
39. Geissen, M.: Traceability of simulation tasks – building blocks for simulation-based decision-making. ProductData J. 20–23. (2020) ISSN 1436–0403
40. McDermott, T.A., Blackburn, M.R., Beling, P.A.: Artificial intelligence and future of systems engineering. In: Lawless, W.F., Mittu, R., Sofge, D.A., Shortell, T., McDermott, T.A. (eds.) Systems Engineering and Artificial Intelligence, pp. 47–59. Springer, Cham (2021). https://doi.org/10.1007/978-3-030-77283-3_3

An Open Benchmark Exercise for Model-Based Design Reviews

Victor Romero(✉), Romain Pinquié, and Frédéric Noël

Univ. Grenoble Alpes, CNRS, Grenoble, INP, G-SCOP, Grenoble, France
{victor.romero,romain.pinquie,frederic.noel}@grenoble-inp.fr

Abstract. The design of engineered systems is a process punctuated by design reviews pursuing various goals such as sharing a common understanding of intermediate representations, making sure that the design meets the requirements, and making informed decisions about the future of the project. In a model-based design approach, an issue of the design reviews is the difficulty to establish a common ground since participants have to understand various models that serve as intermediate representations to communicate during meetings. This observation motivates new technology-mediated design situations with no clear evidence of progress. Indeed, a wide variety of design review environments exist; however, there have been a very limited number of comparisons of these environments to date. This is mainly due to the lack of benchmark problems to evaluate candidate design review environments claiming to facilitate the understanding of model-centric designs. This paper proposes an open-science benchmark exercise that includes the definition of the pursued goals, the measures of performance, the sources of a telescope model-centric design from three viewpoints in line with the universal Function-Behaviour-Structure (FBS) ontology, and a systematic experimental protocol. This benchmark problem should enable anyone to provide objective evidence of progress regarding new environments for reviewing model-centric designs.

Keywords: Model-based design · Design review · Grounding · Benchmark problem

1 Introduction

The increasing complexity of designed systems pushed engineers to give up the document-based design and replace it with a model-based design approach (a.k.a. model-based systems engineering, model-centric design, model-based engineering, integrated model-centric engineering, model-centric engineering, etc.). If the use of models improves design meetings [1], the stakeholders involved in a new product development project still must meet in design reviews.

The establishment of common ground is crucial communication activity to ensure good design [2–4]. Grounding in communication is to "*coordinate on content without*

Published by Springer Nature Switzerland AG 2023
F. Noël et al. (Eds.): PLM 2022, IFIP AICT 667, pp. 176–185, 2023.
https://doi.org/10.1007/978-3-031-25182-5_18

assuming a vast amount of shared information or common ground that is, mutual knowledge, mutual beliefs, and mutual assumptions" [5]. In engineering design, as stated by Burkhardt [6], *"The establishment of common ground is a collaborative process in which the co-designers mutually establish what they know so that design activities can proceed. Grounding is linked to sharing of information through the representation of the environment and the artefact, the dialogue, and the supposed "pre-existing" shared knowledge. This activity ensures inter-comprehension and construction of shared (or compatible) representations of the current state of the problem, solutions, plans, design rules and more general design knowledge."*. Successive design reviews serve to update the common ground moment by moment and this is why we will focus on the common undersdanting of the design of engineered systems.

Today, the most common practice to coordinate content consists in sharing screenshots of models in a slide deck [7], which makes the understanding of models and the identification of relationships between the models difficult [7, 8]. Research studies try to improve design reviews by proposing new mediation technologies, but they show a lack of rigour in the validation phase which is also complained about in engineering design [9, 10]. Indeed, the validation of these claims remains weak since most of them demonstrate that the solution is functional without objective evidence of progress [11–16]. In some cases, experimentations are performed to demonstrate the efficiency of the solution for a specific activity [17–19]. Despite empirical validation, results remain incomparable because of a missing common benchmark problem.

After a literature review of the virtual environments supporting design reviews, Sect. 3 introduces a benchmark problem for evaluating a candidate for model-centric design review environments.

2 Literature Review

As stated in the previous section, with PowerPoint-like presentations, stakeholders of design projects face difficulties during design reviews to share and understand the models and their relationships [7, 8].

One way among others to pass by this issue is to visualise these models in specific environments. Several studies propose tools for the design review process in various domains such as building architecture [11], civil engineering [12], industrial engineering design [13, 14], but the most popular applications concern mechanical geometric definition and space allocation reviews [15–19]. These environments are desktop-based human-machine interfaces through which the users can interact with the models [11] or augmented reality devices [13] for a natural capture of the interactions with the environment. However, the majority of the papers concentrate on virtual reality devices [12], [14–19] for conducting an immersive design review in a collaborative mode [16] with 3D model interactions to move, scale, and rotate the object [14, 16, 17].

If these studies claim to improve design reviews, the validation of contributions remains weak. Most of them present a demonstrator [11, 13, 14, 16], sometimes completed by observation of a few subjects [12, 15], but only a small number of studies scientifically validate their finding by quantitative and objective experimentations [17–19]. If presenting a demonstrator is the first step to finding new interactions and visualisation techniques, all the more when targeted end-users experience the demonstrator, the coexistence of too many alternatives hampers the selection of the suitable design review solutions in academia and industry. In engineering design research, a company should rapidly find the leading edge of a desired field of knowledge, that is, what works, for whom, and under which circumstances. However, by proposing experimentations to validate the usefulness of a proposed environment, we are only halfway up because an objective comparison requires evaluating candidate solutions on the same benchmark problems.

3 A Benchmark Proposal

To improve the quality of validation, the authors propose the first version of an open benchmark problem to evaluate environments supporting model-centric design reviews.

3.1 Introduction

This benchmark problem in engineering design aims at evaluating model-centric design review environments, especially for the activity of grounding that consists in coordinating content by sharing intermediate representations to ensure inter-comprehension of the current design situation.

3.2 Goals and Constraints

All candidate solutions evaluated with this benchmark shall intend to improve the user's understanding of a model-based design including three main goals:

- To maximize the understanding of a model-based design, especially the top-down systemic decomposition (vertical traceability) and the relationships linking design objects across different views (horizontal traceability),
- To maximize the usability, and
- To minimize the cognitive load.

Anyone can reuse the go-to telescope dataset that will be introduced later or create a new one under the following constraints:

- The pursued goal is to facilitate the understanding of a model-based design during design reviews with an emphasis on vertical and horizontal traceability,
- The design concerns a multi-engineering system (not only mechanical),
- The design must be model-based (not document-based),
- The data must contain at least three different views (functional, logical, behavioural, structural, safety, quality, manufacturing, etc.).
- The data to be reviewed include engineering data (requirements, functional architectures, geometric definitions, mechanical kinematics, failure modes, system dynamic simulations, reliability block diagrams, etc.) and potentially project data (market studies, financial expectations, Gantt planning, etc.).

3.3 Dataset

To measure the efficiency of new candidate solutions, the authors propose a systematic experiment with a telescope as a use case. The model-based design of a simplified fully automated go-to telescope serves as an initial dataset to perform the systematic experiment, but it can be completed with new datasets that will help us to conduct a meta-analysis for drawing more generic conclusions.

The design review takes place during the conceptual design phase of the new product development process, and three models already exist: a system architecture model, a preliminary geometric definition CAD model, and a behaviour model.

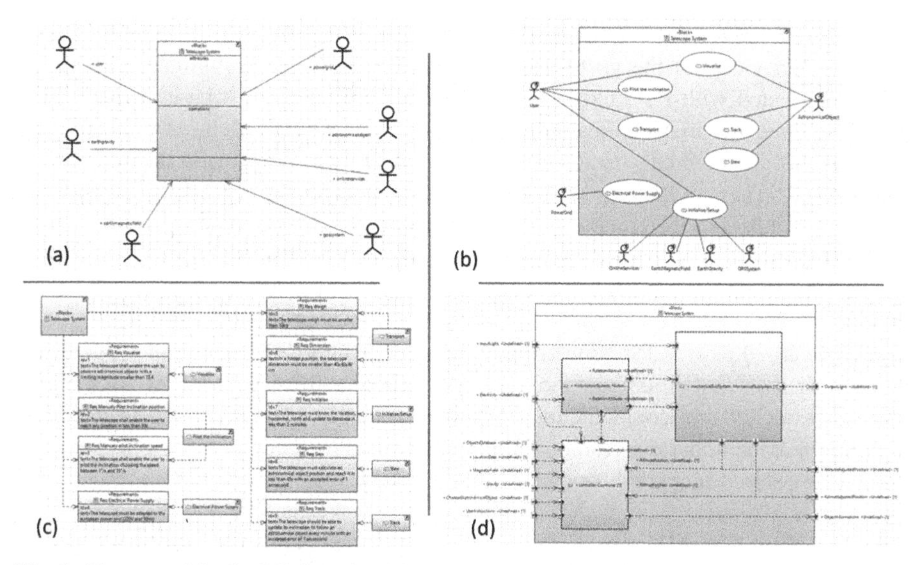

Fig. 1. Extracts of the SysML-based architecture model of the telescope. All these models can be find in the git page (https://github.com/GIS-S-mart/Benchmark-6_model-based-design-reviews)

We assume that the design process starts with the development of the architecture model in SysML with Papyrus as the editing environment. We begin by defining the environment of the telescope including human and non-human stakeholders that interact (directly) with or influence (indirectly) the system-of-interest (Block Definition Diagram Fig. 1a). Then we define the services (external functions) that the telescope must provide to the stakeholders and the constraints imposed by the environment (Use Case Diagram Fig. 1b). By adding performance intervals to the services, we get the telescope system requirements (Requirement Diagram Fig. 1c). Once the requirements are validated, we make the first design choice at the system level by decomposing the system into sub-systems (Internal Bloc Diagram Fig. 1d). Thus, the telescope is decomposed into a mechanical assembly sub-system, a controller sub-system, and a DC motor sub-system. Then, for each sub-system, we recursively follow the same process by defining its stakeholders, functions, and requirements. Finally, we decompose again these sub-systems into sub-sub-systems until we consider that the system is an elementary component that is acquired according to a buy, make or reuse strategy.

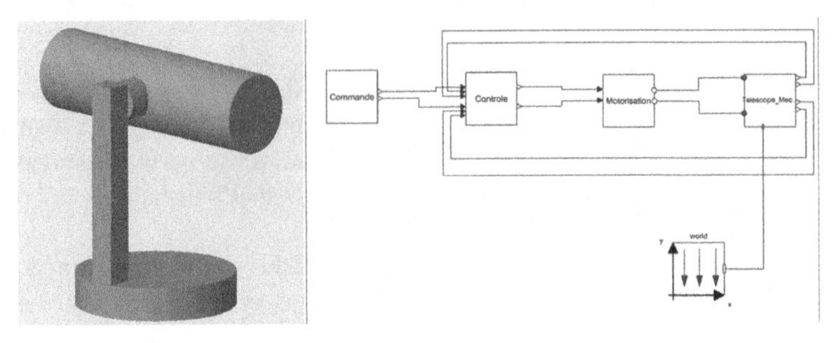

Fig. 2. Preliminary structural (left) and behavioural (right) design of the go-to telescope

Two other views of the go-to telescope correspond to the preliminary geometric definition created with PTC Creo (Fig. 2, left) and a first physics-based behavioural model implemented in Modelica with 3D Experience Dynamic Behaviour Modelling (Fig. 2 right). To facilitate data exchange and integration we provide the sources in standardized STEP or OBJ and Modelica formats, respectively.

Figure 3 illustrates the graph-oriented data model with objects, object properties, and relationships giving an overview of the interdependencies between the three views in line with the universal Function-Behaviour-Structure (FBS) ontology: function (architecture model), behaviour (behaviour model), and structure (CAD model). This data model should be reused as it is for comparing competing design review environments, even with new datasets that differ from the telescope, so as to avoid any biases due to changes in the data model that could facilitate the achievement of the tasks and lead to the comparison of incomparable results. The parsing of data is achieved with ad-hoc parsers before being stored in a graph-oriented Neo4j database.

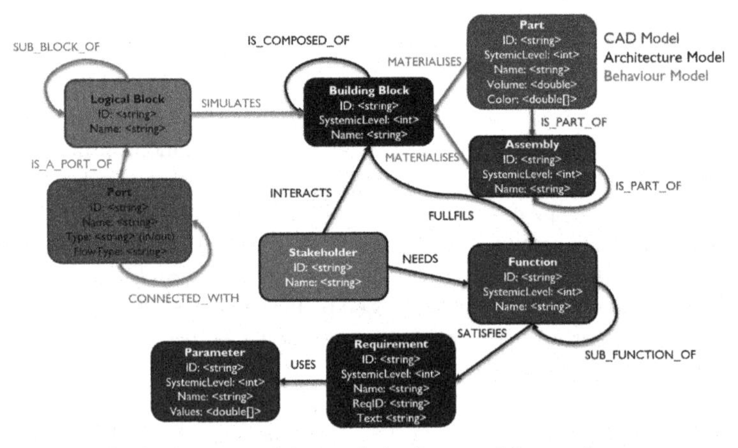

Fig. 3. Structure of the graph database used for our dataset

3.4 Measures of Performance

Remember that the candidate solutions that will be evaluated on this benchmark problem pursue three goals: 1) to maximize the understanding of a model-based design, especially the top-down systemic decomposition and the digital threads across different views; 2) to maximize the usability, and 3) to minimize the cognitive load. Four measures of performance serve to show how well a solution accomplishes these goals.

- ***To maximize the understanding of a model-based design***

Since the major difficulty during design reviews concerns grounding with a shared understanding of models [7], the questions concentrate on (in)direct relationships between design objects captured in different models and at various systemic levels.

There are two sets of questions, built then to force the user to find a link between elements of different models. For the first set, participants can answer using the design review environment to find the answer to questions 1 to 3. For the second set, participants have 5 min to explore the data and, after closing the environment, they must answer questions 4 to 6 based on what they remember or what they have understood about the system under design.

Questions *with* access to the design review environment:

1. What is the ID of the requirements satisfied by the assembly (from the CAD model) "TelescopeAssembly"?
2. What is the name of the functions needed by the stakeholder "PowerGrid"?
3. What is the name of the stakeholder which needs the requirement "Req Manually pilot inclination position" to be met?

To avoid answering all the questions in a row, they are displayed one by one.

Questions *without* access to the design review environment:

1. What are the stakeholders of the telescope?
2. What are the sub-logical blocks (of the behaviour model) of the telescope?
3. What are the stakeholders of the sub-system "MechanicalSubSystem" (in the architecture model)?

These two sets of questions depend on the telescope dataset. If someone decides to share a new dataset to generalize conclusions based on a meta-analysis, then it would be necessary to adapt the questions by replacing the instances of assemblies (e.g. TelescopeAssembly), stakeholders (e.g. PowerGrid), requirements (e.g. Req Manually pilot inclination position), etc. with instances of the new dataset without fundamentally changing the questions.

Three measures of performance serve to evaluate the level of understanding of the reviewed model-based design:

1. The time spent by each participant to answer the first set of questions 1 to 3. This criterion helps to compare the efficiency (the shorter the better) of exploring data with other design review environments.

2. For each set of questions, the total number of wrong answers.
3. The perceived level of confidence of the user in answering questions 1 to 6. A "confidence index" between 1 (not confident) and 7 (absolutely confident) indicates respectively if participants are sure that an answer is right or wrong. The measure of confidence helps to spot random answers.

- ***To maximize the usability***

Although advanced solutions (e.g. eyes tracking) could have been used to evaluate the usability of the candidate model-centric design review environment, the well-known System Usability Scale (SUS) questionnaire [20] is prefered for widespread adoption. The standardized SUS questionnaire gives a usability score between 0 (awful) and 100 (excellent). The SUS questionnaire is independent of the task and can consequently be used to compare candidate environments without any change.

- ***To minimize the cognitive load***

The evaluation of the cognitive load relies on the Task Load Index (TLX) measured with the NASA/TLX questionnaire [21]. The workload score varies between 0 (low cognitive load) and 100 (high cognitive load). This test is dependent on the task to achieve and therefore supports the comparison of results obtained with the same questions and dataset.

3.5 Participants

The evaluation of the usability with the SUS questionnaire requires 14 participants to have significant results [22], whereas experiments show that 12 subjects may already be significant [23]. Therefore, to be confident in the validity of the results, we select at least 15 subjects and, if possible, with varying profiles (position, age, sex, etc.), especially regarding their level of expertise in Model-Based Design ranging from novice to experts with various domain of expertise (mechanics, systems engineering, etc.).

3.6 Systematic Experimental Protocol

The systematic process to experiment is composed of three activities:

1. *Answering a demographic questionnaire*

To analyse the impact of participants' profiles on the results, e.g., a virtual reality expert may overestimate usability compared to an expert in systems engineering. Thus, participants must answer a demographic questionnaire when starting the experiment.

2. *Training*

Candidate model-based design review environments will often be new to users. Our goal is not to evaluate the intuitiveness of the solution but its efficiency. Therefore, a training phase is mandatory to reduce experimental bias. During this phase, participants have unlimited time to learn how to use the environment and get familiar with the visual and interactive metaphors for exploring data. The training relies on the review of an inverted pendulum system which also includes subsystems (a controller sub-assembly, a mechanical sub-assembly and a DC motor sub-assembly) with three viewpoints (function, behaviour and structure), that is, compliant with the data model (Fig. 3). Thus, it is relatively similar to the go-to telescope used for the experiment. During this phase, participants can ask any question about the tool and the engineering data. To validate the training, the participant must rightly answer three questions that are relatively similar to the ones asked during the experiment:

1. What is the name of the requirements associated with the stakeholder "Gravity"?
2. What is the name of the logical block (from the behaviour model) that satisfies the requirement "ReqPilotMotor"?
3. What are the functions fulfilled by the assembly (of the CAD model) "INVERTED-PENDULUM"?

3. *Experiment*

Once the participants pass the training questions, the experiment can start with the model-centric design review of the go-to telescope. The time is limited to 20 min. Indeed, by experience, depending on the complexity of the environment, 30 to 45 min are necessary to accomplish the training phase. The total time for the experiment is about 1 h, which is acceptable for convincing the targeted participants.

The first phase of the experiment is the evaluation of how well participants retrieved linked data with the model-based design review environment. In practice, participants read the first set of questions 1 to 3 and try to find the answers. During this phase, the participant can navigate back and forth between the environment and the questions for 20 min, and he does not receive any external feedback such as answers to questions arising during the task. This part of the experiment evaluates both the understanding of the model-based design without requiring memory capabilities and the usefulness of the tool for data exploration. To conclude this first phase, participants estimate a confidence index for each question.

Once the participant considers having answered the questions 1 to 3 or when the 20 min time limit is over, participants have 5 min more in the environment to better understand the model-based design. At the end of the 5 min, participants close the environment before answering the second set of questions 4 to 6 based on their understanding of the system. They evaluate the confidence index too. For this second phase, the time is neither limited nor tracked as we want to evaluate their understanding of the design without considering memorization abilities. To conclude the experiment, participants answer the SUS and NASA/TLX questionnaires.

4 Conclusion

Several research studies propose environments which claim to facilitate model-based design reviews, but the lack of focus on validation and common benchmark problems hampers clear evidence of progress. This paper proposes an open-science benchmark exercise that includes the definition of the pursued goals, measures of performance, sources of a telescope model-centric design from three viewpoints in line with the universal Function-Behaviour-Structure (FBS) ontology, and a systematic experimental protocol to compare such environments. In open access as a GitHub repository[1] of the open-science S.mart platform [24], this first version of the benchmark problem is open to community-based improvements (addition, removal or change of goals, criteria, datasets, etc.) for enhancing the quality of our research. Future works could focus on the development of new academic and industrial datasets and a meta-analysis to draw more general conclusions on the added value of such mediated design situations.

Acknowledgements. This work was supported by the French National Research Agency under the Investments for the Future Programs (PIA) with the grant ANR-21-ESRE-0030 (CONTINUUM project) and by the LabEx Persyval Mine'VR Project.

References

1. McDermott, T., Hutchison, N., Salado, A., Henderson, K., Clifford, M.: Benchmarking the benefits and current maturity of model-based systems engineering across the enterprise: Results of the MBSE maturity survey. in *Systems Engineering Research Center (SERC)* (2020)
2. D'Astous, P., Détienne, F., Visser, W., Robillard, P.N.: Changing our view on design evaluation meetings methodology: a study of software technical review meetings. Des. Stud. **25**(6), 625–655 (2004). https://doi.org/10.1016/j.destud.2003.12.002
3. Olson, G., Olson, J., Carter, M., Storrosten, M.: Small group design meetings: an analysis of collaboration. Hum.-Comput. Interact. **7**(4), 347–374 (1992). https://doi.org/10.1207/s15327051hci0704_1
4. Stempfle, J., Badke-Schaub, P.: Thinking in design teams - an analysis of team communication. Des. Stud. **23**(5), 473–496 (2002). https://doi.org/10.1016/S0142-694X(02)00004-2
5. Clark, H.H., Brennan, S.E.: Grounding in communication. In: Resnick, L.B., Levine, J.M., Teasley, S.D. (eds.) Perspectives on Socially Shared Cognition. American Psychological Association. Washington, pp. 127–149 (1991). https://doi.org/10.1037/10096-006
6. Burkhardt, J.-M., Détienne, F., Hébert, A.-M., Perron, L., Safin, S., Leclercq, P.: An approach to assess the quality of collaboration in technology-mediated design situations. In: Proceedings of ECCE 2009 European Conference on Cognitive. Ergonomics, pp. 8 (2009)
7. Pinquié, R., Romero, V., Noel, F.: Survey of model-based design reviews: practices & challenges? Proc. Des. Soc. **2**, 1945–1954 (2022). https://doi.org/10.1017/pds.2022.197
8. Baduel, R., Chami, M., Bruel, J.-M., Ober, I.: SysML models verification and validation in an industrial context: challenges and experimentation. In: Pierantonio, A., Trujillo, S. (eds.) ECMFA 2018. LNCS, vol. 10890, pp. 132–146. Springer, Cham (2018). https://doi.org/10.1007/978-3-319-92997-2_9
9. Barth, A., Caillaud, E., Rose, B.: How to validate research in engineering design. In: Proceedings of the 18th International Conference on Engineering Design, Copenhagen, p.11 (2011)

[1] https://github.com/GIS-S-mart/Benchmark-6_model-based-design-reviews.

10. Vermaas, P.E.: Design theories, models and their testing: on the scientific status of design research. In: Chakrabarti, A., Blessing, L.T.M. (eds.) An Anthology of Theories and Models of Design, pp. 47–66. Springer, London (2014). https://doi.org/10.1007/978-1-4471-6338-1_2
11. Shiratuddin, M.F. Thabet, W.: Information-rich virtual environment (VE) for design review, p. 9 (2007)
12. van den Berg, M., Hartmann, T., de Graaf, R.: Supporting design reviews with pre-meeting virtual reality environments. J. Inf. Technol. Constr. **22**, 305–321 (2017)
13. Verlinden, J., Horváth, I., Nam, T.-J.: Recording augmented reality experiences to capture design reviews. Int. J. Interact. Des. Manuf. IJIDeM **3**(3), 189–200 (2009). https://doi.org/10.1007/s12008-009-0074-8
14. Romero, V., Pinquié, R., Noël, F.: An immersive virtual environment for reviewing model-centric designs. Proc. Des. Soc. **1**, 447–456 (2021). https://doi.org/10.1017/pds.2021.45
15. Aromaa, S., Leino, S.-P., Viitaniemi, J., Jokinen, L., Kiviranta, S.: Benefits of the use of virtual environments in product design review meeting. In: Proceedings of DESIGN 2012, the 12th International Design Conference, p. 10. Dubrovnik, Croatia (2012)
16. Adwernat, S., Wolf, M., Gerhard, D.: Optimizing the design review process for cyber-physical systems using virtual reality. Procedia CIRP **91**, 710–715 (2020). https://doi.org/10.1016/j.procir.2020.03.115
17. Wolfartsberger, J.: Analyzing the potential of Virtual Reality for engineering design review. Autom. Constr. **104**, 27–37 (2019). https://doi.org/10.1016/j.autcon.2019.03.018
18. Freeman, I.J., Salmon, J.L., Coburn, J.Q.: CAD integration in virtual reality design reviews for improved engineering model interaction. In: Volume 11: Systems, Design, and Complexity, p. V011T15A006 Phoenix, Arizona, USA (2016). https://doi.org/10.1115/IMECE2016-66948
19. de Casenave, L., Lugo, J.E.: Effects of immersion on virtual reality prototype design reviews of mechanical assemblies. In: Volume 7: 30th International Conference on Design Theory and Methodology, p. V007T06A044. Quebec City, Quebec, Canada (2018). https://doi.org/10.1115/DETC2018-85542
20. Brooke, J.: SUS - A quick and dirty usability scale. In: Jordan, P.W., Thomas, B., Weerdmeester, B.A., McClelland, I.L., (eds.) Usability Evaluation in Industry, p. 8 Taylor & Francis (1996)
21. Hart, S.G., Staveland, L.E.: Development of NASA-TLX (Task Load Index): Results of Empirical and Theoretical Research. In: Advances in Psychology, vol. 52, , pp. 139–183. Elsevier (1988). https://doi.org/10.1016/S0166-4115(08)62386-9
22. Tullis, T.S., Stetson, J.N.: A comparison of questionnaires for assessing website usability, p. 12 (2004)
23. Guest, G., Bunce, A., Johnson, L.: How many interviews are enough?: an experiment with data saturation and variability. Field Methods **18**(1), 59–82 (2006). https://doi.org/10.1177/1525822X05279903
24. Pinquié, R., Le Duigou, J., Grimal, L., Roucoules, L.: An open science platform for benchmarking engineering design research. In: Presented at the 32nd Cirp Design Conference 2022 (2022)

Integrating Computational Design Support in Model-Based Systems Engineering Using Model Transformations

Eugen Rigger[1(✉)], Simon Rädler[2], and Tino Stankovic[3]

[1] Zumtobel Lighting GmbH, Schweizerstrasse 30, 6851 Dornbirn, Austria
eugen.rigger@zumtobelgroup.com

[2] Business Informatics Group, TU Vienna, Favoritenstraße 9-11/194-3, 1040 Vienna, Austria

[3] Department of Mechanical and Process Engineering Chair in Engineering, Design and Computing, CLA F 21.2 Tannenstrasse 3, 8092 Zürich, Switzerland

Abstract. Model-based systems engineering (MBSE) combines the rigor of systems engineering with formal models to support communication in multidisciplinary engineering. With industrial adoption of MBSE, the maturity of modeling environments supporting MBSE increased. Still, efficient means to integrate computational design methods in MBSE are missing. Here, we present a method that enables systems engineers to directly integrate computational methods for solving design tasks. The method relies on established semantics of the systems modeling language (SysML) and therefore can be directly integrated with existing system models so to avoid redundant knowledge formalizations for computational methods. Next, model transformations are applied to generate the mathematical model based on the relevant parts of the system model. These temporary models are used to solve the design task and generate output that is fed back to the system model. Therefore, the proposed method contributes by relying on a single and comprehensible knowledge formalization understandable to engineers. Further, it enables systems engineers to formalize design tasks for automated reasoning themselves by bundling the complexity of the mathematical modeling within the model transformations. An industrial case for designing sealing elements for piping is used to illustrate the potential of the proposed approach. Future work needs to further elaborate on automated selection of appropriate mathematical methods as well as computational support for the identification of opportunities for integration of computational design methods readily while developing a system model.

Keywords: Model-based systems engineering · SysML · Computational design method · Design automation

Published by Springer Nature Switzerland AG 2023
F. Noël et al. (Eds.): PLM 2022, IFIP AICT 667, pp. 186–195, 2023.
https://doi.org/10.1007/978-3-031-25182-5_19

1 Introduction

Systems engineering and in particular model-based systems engineering (MBSE) are design methodologies integrating different engineering disciplines in product development, promising increased design performance for the development of complex systems [4]. Establishing a central formalization of system characteristics and explicit modeling of dependencies within the system are core to MBSE [3]. In this respect, dedicated languages for MBSE exist, with the Systems Modeling Language (SysML) being an established standard for graphical modeling of systems [7]. From the inception of SysML, validation of systems using the formalized dependencies in the systems, particularly parametric constraints, are a core point of interest [9] and tools enabling evaluation of parametric relations using SysML models are currently commercially available. In this context, current state of MBSE practice focuses on the evaluation of a limited number of design alternatives [1]. Methodologies to integrate MBSE and set-based design for systematic design space exploration have recently been conceptually introduced [15]. Yet, the integration of computational design support and systems modeling that goes beyond parametric evaluation is cumbersome and causes redundant knowledge formalizations: Either, the semantics of the computational methods need to be integrated to the system model extending the modeling syntax and requiring fundamental insight to the computational method [14], or, the computational model is developed from scratch directly within the environment of the computational method guided by the information captured in the system model [16].

Here, we present a method for seamless integration of computational design support while developing a system, solely relying on established syntax of the standardized modeling language SysML and model transformations [2] from SysML to the computational formalization. The utilization of the computational design support does not require expert knowledge regarding the mathematical modeling or implementation specifics of the underlying computational method. For the type of computational design support, we focus on methods from the domain of design automation [12]. Thereby, this work contributes by proposing a means for direct integration of computational design methods with engineering design. A novel approach regarding the formalization of design tasks strengthening the role of the SysML as unified meta-model in systems engineering is presented. In this work, the definition of critical system parameters of pressure pipe sealing mechanisms is used as an example to illustrate the proposed concept for conceptual design of a mechanical subsystem.

In the following, first, the relevant background regarding systems modeling, computational design methods from the field design automation, and model transformations are introduced. Next, the principles for integration of computational methods in a systems engineering process are described and illustrated by a case study. In the discussion section, the observations from the case study are critically discussed as well as the scaleability of the approach for design automation tasks along the product lifecycle. The paper closes with concluding remarks.

2 Background

In the following, first, processes and methods for model-based systems engineering are reflected. In this context, the application of SysML in the context of computational design methods is assessed. Next, principles of computational design methods from the research field design automation are reviewed so to derive requirements regarding the semantics for design automation task formalization. Finally, the concept of model transformations is introduced.

2.1 Model-Based Systems Engineering

MBSE centers around a formal system model that captures system characteristics in a neutral domain-independent manner. In practice, several processes supporting MBSE can be identified [3] and industrial standards have been established such as VDI 2206 that follows the v-model [17]. Similarly, standardized modeling languages such as SysML exist [8]. SysML provides the semantic foundation for documentation of system requirements, behavior, structure, and parametric relations.

2.2 Computational Design Methods for Systems Engineering

Design automation aims to increase efficiency and effectiveness in design by the application of computational methods for reasoning with formalized knowledge. Regarding the actual formalization of a design automation task, conventional approaches rely on dedicated knowledge formalization for design automation [16]. Efforts towards using standardized languages such as SysML exist. However, they rely on extensions of SysML tailored for the targeted computational method, for example [14]. With the aim to derive a basis for a more unified representation independent of a specific computational method, [12] analyzed existing design automation methods with respect to the required types of knowledge to describe a design automation task fully. They derived multiple classes of design automation tasks that can be distinguished based on a unique combination of the required knowledge regarding input and goals of a design automation task as well as the generated output knowledge. Building upon this work, [10] introduced a method for modeling a product configuration problem solely relying on SysML syntax. Yet, a detailed elaboration on the actual processing of the SysML model to yield a computational formalization is missing.

2.3 Model Transformations

Models can be defined as machine-readable artifacts, enabling the representation of a relevant aspect of interest [2]. Model transformations allow deriving a specific viewpoint on an artifact to make use of these machine-readable artifacts and can be classified as model-to-model and model-to-code/text transformations [2]. Model transformations are typically applied to meta-models. In model-to-model, the mapping between the input and output meta-model is given by different conditions and operations [13]. The purpose of model-to-model transformation are bringing two systems together, derive a specific viewpoint or align models etc.

In model-to-text transformations, the input meta-model is depicted and mapped to an output, given as blocks of text, filled with input properties and combined in a logical order. Similarly, text-to-model transformations can be defined. Model transformations are supported by different transformation engines, like ATL [5] or Epsilon [6].

Based on the gaps identified in the preceding sections, the research gaps addressed in this work concerns the seamless integration of computational design methods within systems engineering.

3 Method for Integration of Computational Support in Model-Based Systems Engineering

In this section, we introduce the proposed method to show how computational design methods and model-based systems engineering can be seamlessly integrated. First, the overall principle is highlighted showing the interplay of a generic model-based systems engineering process, the system model and model transformations. Next, SysML modeling semantics are introduced so to define the context for integration of computational support. Finally, the details of applied model transformations are presented.

3.1 Linking the System Model and Computational Design Methods

Figure 1 shows the overall concept for the usage of computational design methods in model-based systems engineering: the system model is gradually refined and elaborated on during the design process. A design task can be described by the input, output and goals [12]. These design tasks can be mapped to design automation tasks supported by available computational design methods [11]. Once a design automation task is identified, relevant parts of the system model can be extracted and used to formalize the design automation task. The specific, modeling semantics are detailed in Subsect. 3.2. The yielded subset of the system model can then be used to generate the mathematical model using model transformations as described in Subsect. 3.3 and, finally, the obtained results are fed back to the system model integrating the output generated by applying the computational design method on the task formalization. It has to be noted, that the generated mathematical model solely serves the purpose of generating a new design and can be withdrawn after usage.

3.2 SysML Modeling Guidelines for Computational Support

To comprehensively introduce the required modeling semantics, first, the formalization of input knowledge is presented. Next, the semantics for the formalization of the goals of a task are specified.

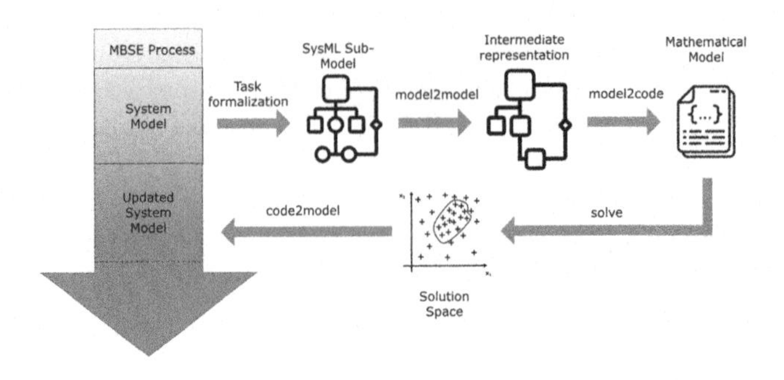

Fig. 1. Process overview

Formalizing Input of a Design Task. First, the product architecture of the model needs to be formalized. Depending on the maturity of the systems model, this can refer to a functional, logical or physical architecture. SysML Block Definition Diagrams are used to depict hierarchical structures of blocks using part-of associations. Associations' multiplicities can be used to depict degrees of freedom in the model, e.g. a multiplicity 1..4 defines that one to four instances of a specific component can be defined. Additionally, value properties are used to describe specific properties of a block, e.g. the diameter of a wheel. If a numeric value property is instantiated, but not assigned any value this property is considered a variable when generating the mathematical model. When it comes to the selection of types of sub-components/-systems, abstract blocks can be defined in SysML indicating that a specific instance of this block needs to be first selected. Using specialization relations, the value properties of the abstract block can be propagated to all possible variants. Using parametric diagrams relations among blocks can be defined so to evaluate a system's performance. In particular, value and part properties can be linked. In case parametric relations between the properties exist, constraint blocks and the related constraints and constraint properties can be used to formalize more complex dependencies. The constraint itself can be a regular mathematical expression following JAVA syntax but could also be an external simulation model [9]. Finally, to conclude the specification of input of a task definition, specific values need to be assigned to selected value and part properties in order to narrow down the solution space. In this respect, a block named "input" needs to be defined as well as corresponding properties. These properties can then be linked to the system's properties using parametric diagrams.

Formalizing Goals of a Design Task. Similar to the formalization of constraints for evaluation of the system, performance constraints can be formalized using parametric diagrams and constraint blocks, e.g. the weight of a system must not exceed a threshold value. When it comes to finding optimal solutions, the objective function can be defined simply by using the SysML stereotype "objectivefunction" upon a specific constraint block.

3.3 Model Transformations Relying on SysML

As depicted in Fig. 1, multiple transformations are required to transform the task formalization to a mathematical model and then migrate the yielded results back to the system model. First, a model-to-model transformation is performed generating an intermediate representation that is designed to efficiently provide all information required for deriving the mathematical model using a model-to-code transformation. In contrary to the SysML model, the intermediate model captures only the artifacts that are relevant for the mathematical model. Thus, the size of the model can be reduced to increase efficiency of transformations. For example, the artifacts of the intermediate model are enriched so to contain explicit references to children, parents etc. Thereby, unidirectional access to all relevant information is warranted required for the subsequent model-to-text transformation. As a representation of the mathematical model, various mathematical modeling languages targeted at optimization can be used such as AMPL or MiniZinc. These languages follow similar structures. In a first step, all parameters and variables of the mathematical model are created in plain text. Thereby, iterative looping through the intermediate model searching for variable and parameter declaration is performed. Following this, constraints are generated and constraint parameters are replaced by variable and parameter definitions defined in the previous step. Finally, the objective function is defined and a an appropriate solver is identified based on problem characteristics, such as types of constraints, variables etc. Figure 8 illustrates the concept of applied model transformations.

4 Case Study

In this section a case study is presented for a use case from oil&gas industries. Specifically, the design task to pre-dimension the sealing of a pressurized pipe is addressed that can be considered a crucial step in the early stages of the system development. In the following, the SysML model as well as the resulting MiniZinc model are introduced.

4.1 SysML Model

Figure 2 shows a schematic of an o-ring sealing system considered here as use case. Figure 3 shows the product architecture of the use case. The calculation model, which is the core of the concept, is shown on top. It consists of two sub-components, the Pipe and the O-Ring. Each component contains value properties describing the components. The italic font applied for naming the O-Ring indicates that the block is abstract. Figure 4 shows an excerpt of possible O-Rings assigning specific values to the properties inherited from the abstract block.

Figure 5 shows how input values of a block can be semantically linked to system properties using a parametric diagram . To evaluate specific designs, a constraint to calculate the smallest possible Groove_Diameter given the

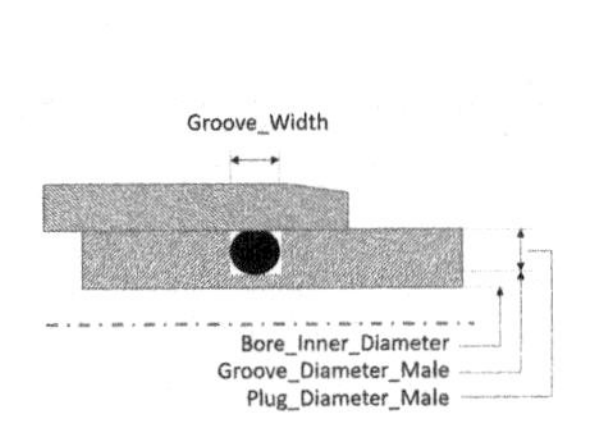

Fig. 2. Schematic of use case

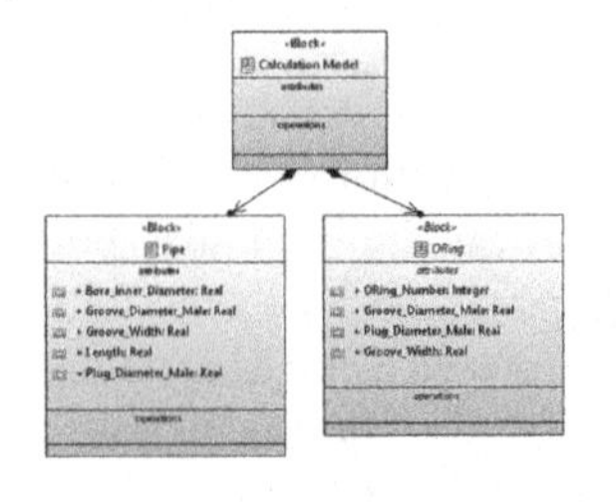

Fig. 3. Product architecture

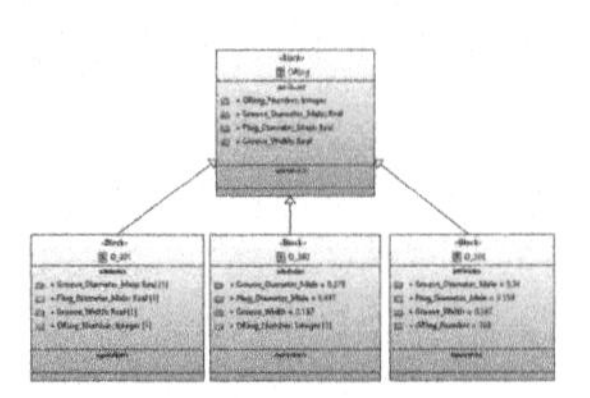

Fig. 4. O-Rings

Bore_Inner_Diameter, tensile and burst rating as well as corresponding safety factors. Figure 6 depicts the constraint block linked to the system model using a parametric diagram. The objective function of the design task is shown in Fig. 7. It states that an O-Ring needs to be selected that is as close as possible to the smallest possible Groove_Diameter. An additional functional constraint is added to the model stating that the pipe's groove diameter needs to be smaller than the O-Ring's diameter in order to enable physically meaningful solutions, only.

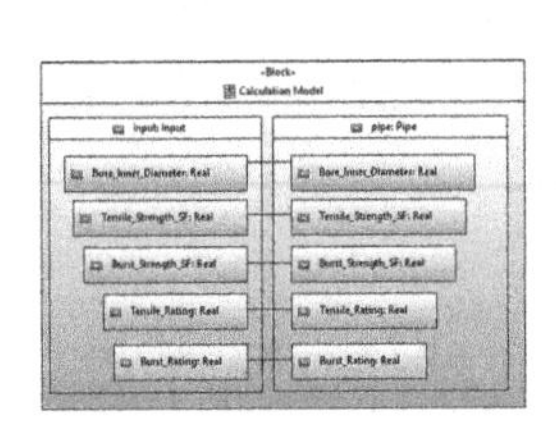

Fig. 5. Connecting input values to the model

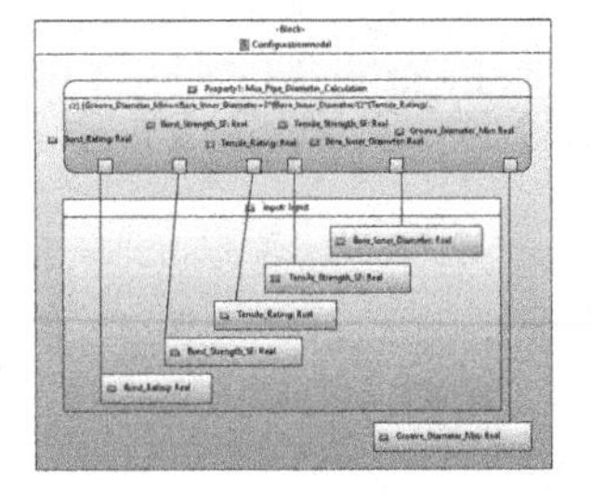

Fig. 6. Constraint block integrated using a parametric diagram

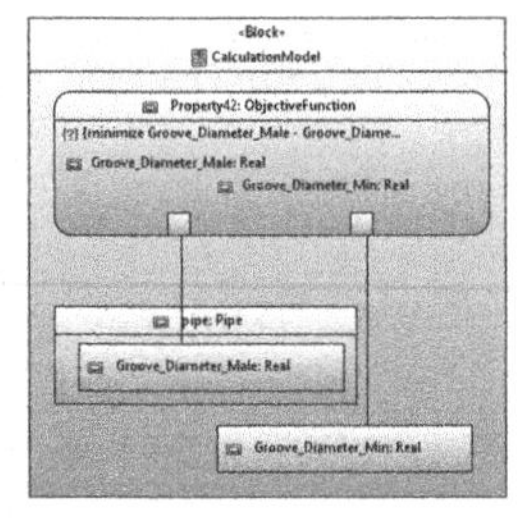

Fig. 7. Objective function

4.2 Model Transformations

The Java-based model transformation family of Eclipse Epsilon is used to transform the SysML 1.6 model from Eclipse Papyrus 6.0 into the intermediate model and further to a MiniZinc model. In this respect, the model-to-model transformation apply the Epsilon Transformation Language (ETL). For each instance of an entity in SysML such as a block, constraint block or connections, one or multiple transformation rules are applied. In ETL, *guards* can be defined so that a specific transformation is only applied for the right model artifacts.

Based on the intermediate model, the model-to-text transformation is applied using the Epsilon Generation Language (EGL) yielding a MiniZinc model.

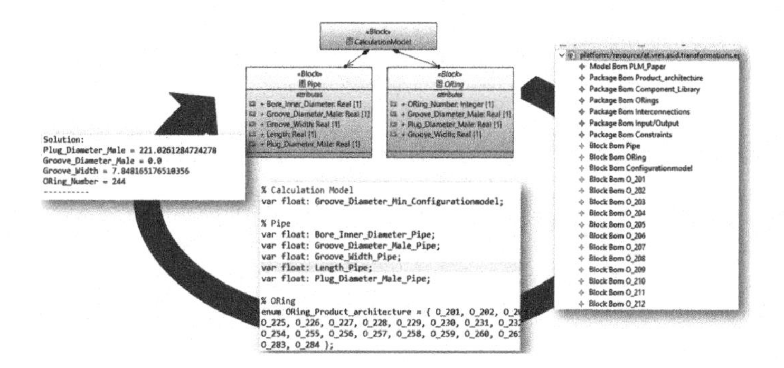

Fig. 8. Transformation Round-Trip

Within the EGL, text and loops are implemented. To enable execution of generated model, a specific solver needs to be selected, e.g. COIN-BC[1] solver.

5 Discussion

In the following, the proposed method is critically discussed from two perspectives: First, the formalization of the mathematical model using SysML without the creation of new stereotypes, and, second, the model transformations applied to integrate computational design methods and SysML models.

5.1 Mathematical Modeling Using SysML

By applying model transformations, the proposed method integrates design task formalizations directly within the system model. Therefore, the role of the system model as a single source for system modeling is strengthened by avoiding redundant formalizations as required by existing approaches from the domain of design automation. A major benefit of our proposed method is that already existing product knowledge captured by the system model can be directly reused preventing errors in formalizations and saving time. By strictly separating systems modeling and mathematical modeling using the transformations, systems engineers are enabled to integrated computational design methods in their work. Nevertheless, the systems engineer needs to be knowledgeable about the opportunities for the application of computational design methods. Methods exist [11], yet, more rigorous support is desirable, for example using pattern recognition upon the system model to identify design tasks that can be potentially supported using computational design methods. Future work needs to focus upon facilitating design task modeling for systems engineering by providing means to assess design task formalizations. For instance, network analysis of constraints and models could support early debugging of faulty relations in the design task formalization. Additionally, future work needs to investigate how machine learning

[1] https://github.com/coin-or/Cbc.

models can be formalized using SysML so to enable an even broader application of computational design methods.

5.2 Model Transformation and Information Backflow

The presented method relies on an intermediate model that acts as a comprehensive source of information for derivation of a mathematical model. For the depicted use case, a MiniZinc model was generated. However, any other target domain can be addressed. Yet, it needs to be taken into account that generating the mathematical model requires transforming an object-oriented model to a declarative representation. Hence, developing these transformations can be considered a challenge. For example, combining multiplicities and abstract blocks means that a number of decision variables corresponding to the degrees of freedom of the multiplicities times the available instance of the abstract block need to be defined. Therefore, complex mathematical models can be yielded for allegedly simple design task formalizations in SysML. In this respect, future work needs to elaborate on automatically selecting appropriate solvers for a given design task formalization. A challenge to be considered is that the solver's performance depend on the mathematical modeling. In this regard, automatic selection of appropriate algorithms and improvements based on the selection could enhance the results. Additionally, some solvers can describe a solution space or generate multiple solutions. Backpropagation of results should then enable to represent all generated variants in SysML.

6 Conclusion

This paper presents a method enabling seamless integration of computational design methods and model-based systems engineering. Based on a case study addressing the pre-dimensioning of a mechanical subsystem, it is shown how semantics of the systems modeling language SysML can be used to formalize a design automation task directly within the system model avoiding redundant formalization of knowledge. Further, the applicability of model transformations to transform a SysML model to a mathematical model is shown. Hence, seamless integration of computational design support is enabled. Future work will elaborate on the actual identification of design tasks as well as further elaborating on the optimization of generated mathematical models.

Acknowledgements. This work has been partially supported and funded by the Austrian Research Promotion Agency (FFG) via the "Austrian Competence Center for Digital Production" (CDP) no. 881843, the K2 centre InTribology, no. 872176.

References

1. Beihoff, B., et al.: A world in motion - systems engineering vision 2025
2. Brambilla, M., Cabot, J., Wimmer, M.: Model-driven software engineering in practice: Second edition 3(1), 1–207. https://doi.org/10.2200/S00751ED2V01Y201701SWE004
3. Estefan, J.A.: Survey of model-based systems engineering (MBSE) methodologies. Incose MBSE Focus Group **25**, 1–70 (2007)
4. Huldt, T., Stenius, I.: State-of-practice survey of model-based systems engineering. vol. 22(2), pp. 134–145. https://doi.org/10.1002/sys.21466,65
5. Jouault, F., Allilaire, F., Bézivin, J., Kurtev, I.: ATL: a model transformation tool. Sci. Comput. Program. **72**(1), 31–39 (2008). https://doi.org/10.1016/j.scico.2007.08.002
6. Kolovos, D.S., Paige, R.F., Polack, F.A.C.: The epsilon transformation language. In: Vallecillo, A., Gray, J., Pierantonio, A. (eds.) ICMT 2008. LNCS, vol. 5063, pp. 46–60. Springer, Cham (2008)
7. OMG: OMG SysML. http://www.omgsysml.org/
8. OMG: SYM SysML-modelica transformation. https://www.omg.org/spec/SyM/1.0/
9. Peak, R.S., Burkhart, R.M., Friedenthal, S.A., Wilson, M.W., Bajaj, M., Kim, I.: Simulation-based design using SysML part 1: a parametrics primer. In: INCOSE International Symposium, vol. 17, pp. 1516–1535. Wiley Online Library (2007). http://onlinelibrary.wiley.com/doi/10.1002/j.2334-5837.2007.tb02964.x/abstract
10. Rigger, E., Fleisch, R., Stankovic, T.: Facilitating configuration model formalization based on systems engineering. In: Proceedings of the Workshop on Configuration (ConfWS'21)
11. Rigger, E., Shea, K., Stanković, T.: Method for identification and integration of design automation tasks in industrial contexts. Adv. Eng. Inform. **52**, 101558 (2002). https://doi.org/10.1016/j.aei.2022.101558
12. Rigger, E., Stanković, T., Shea, K.: Task categorization for identification of design automation opportunities. https://doi.org/10.1080/09544828.2018.1448927
13. Sendall, S., Kozaczynski, W.: Model transformation: the heart and soul of model-driven software development. IEEE Softw. **20**(5), 42–45 (2003). https://doi.org/10.1109/MS.2003.1231150
14. Shah, A.A., Paredis, C.J.J., Burkhart, R., Schaefer, D.: Combining mathematical programming and SysML for automated component sizing of hydraulic systems. vol. 12(4), p. 041006. https://doi.org/10.1115/1.400776400009
15. Specking, E., Parnell, G., Pohl, E., Buchanan, R.: Early design space exploration with model-based system engineering and set-based design. Systems **6**(4), 45 (2018). https://doi.org/10.3390/systems6040045
16. Stjepandić, J., Verhagen, W.J.C., Liese, H., Bermell-Garcia, P.: Knowledge-based engineering. In: Stjepandić, J., Wognum, N., Verhagen, W.J.C. (eds.) Concurrent Engineering in the 21st Century, pp. 255–286. Springer, Cham (2015) https://doi.org/10.1007/978-3-319-13776-610
17. VDI-Verlag: VDI-guideline 2206: Design methodology for mechatronic systems

Knowledge Reuse of CAD Data in Parallel Development of Multiple Wiring Harness Variants

Kevin Eder[1,2](✉), Onur Tas[1], Jonas Neckenich[1], Roland Winter[1], Uwe Zielbauer[1], and Kristin Paetzold[2]

[1] Mercedes-Benz AG, HPC X973, 71059 Sindelfingen, Germany
kevin.eder@mercedes-benz.com

[2] Technical University Dresden, George-Bähr-Straße 3c, 01069 Dresden, Germany
kristin.paetzold@tu-dresden.de

Abstract. The high dependency of a wiring harness on its surrounding components requires changes to be made almost daily during development. To ensure operability, these adjustments have to be carried out in some cases even after the data freeze of a development phase. Since these changes may also be relevant for subsequent development phases, redundant design efforts are the consequence. One way to increase the efficiency of these processes is to reuse knowledge. This paper reviews the state of the art of knowledge reuse in engineering design and in software development. Furthermore, an overview of the structure of a wiring harness, as well as the knowledge contained in the corresponding CAD data is presented. Thereupon, all scenarios are specified in which knowledge reuse in the wiring harness development could be possible. Subsequently, a methodology for complete, partial and assisted reuse is specified enabling the transfer of gained knowledge into other CAD data sets of wiring harnesses. In addition, the conditions that have to be met for the application of each reuse method are presented. Finally, the requirements needed for the implementation of the shown methodology into a CAD system are specified.

Keywords: Knowledge reuse · Knowledge transfer · Knowledge management · Knowledge-based engineering · Computer aided design (CAD)

1 Introduction

The high complexity of wiring harnesses and the currently low level of automation cause a high production effort [1]. Consequently, early design freezes are required to ensure on-time assembly. In addition subsequent changes to ensure operability, as well as parallel development of various product variants on one platform, result in overlapping timelines. Since each development phase of any product variant sets new requirements [2], changes may only be valid from specific phases and dates. This causes the creation of variants for individual subareas of the wiring harness. Due to the modular and tree-like

Published by Springer Nature Switzerland AG 2023
F. Noël et al. (Eds.): PLM 2022, IFIP AICT 667, pp. 196–205, 2023.
https://doi.org/10.1007/978-3-031-25182-5_20

structure of the wiring harness other areas, can still remain the same (see Fig. 1). Since newly gained knowledge and the corresponding adjustments to one CAD model can also be relevant for other models that are being developed in parallel, redundant maintenance efforts are the result [3]. This ensures that variants can be kept as similar as possible, but also avoids the repetition of errors and therefore reducing costs [3, 4].

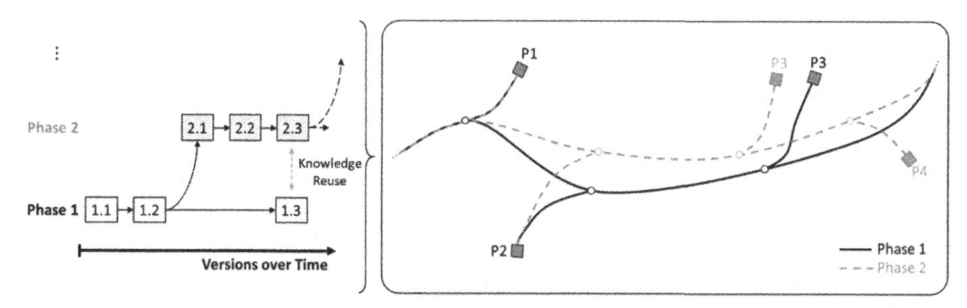

Fig. 1. Variants of wiring harnesses during parallel development

Since changes have to be done almost daily and the complexity of the wiring harness itself, the execution of changes require a particular high effort [5, 6]. Engineers in wiring harness development spend almost 80% of their time for change management, which is almost three times what engineers in other fields spend on average with approximately 25% of their time [7, 8]. Therefore, reducing the time for redundant design changes by reusing gained knowledge to automate individual design tasks offers a high potential for increasing the efficiency of those changes [3, 9]. To address this, Eder et al. [3] present an approach for a knowledge management system. In Order to advance the knowledge reuse stage of that approach, the following questions have to be answered:

- What are scenarios for knowledge reuse of CAD data in wiring harness development?
- How can knowledge be integrated in CAD data sets of wiring harnesses?
- What conditions must apply for the reuse of knowledge?

This paper is structured as follows: Sect. 2 reviews the state of the art of knowledge reuse in engineering design and in software development. Furthermore, it presents an overview of knowledge of wiring harnesses in CAD. Section 3 shows different reuse scenarios of CAD data in wiring harness development and a corresponding reuse methodology. Section 4 validates the presented methodology. Section 5 concludes and gives an outlook on future research.

2 State of the Art

Although knowledge reuse is a desire of any company, its implementation is difficult and the actual benefits are hard to determine [4, 10]. Knowledge can either be explicit in form of data or it can be tacit, which is based on personal experience and therefore difficult to express [11]. To reuse explicit knowledge, it has to be made available by the interpretation of relevant information tailored to its use case [12]. Since native CAD files are encrypted, all relevant information have to be exported via API (application programming interface) into a universal machine readable format like the XML-scheme [13]. In addition, it must be defined under which constraints a reuse of knowledge is possible and how it can be integrated [14].

2.1 Knowledge Reuse in Engineering Design

Knowledge reuse is based on a detailed understanding of the structure and internal logic of a product [10]. One massive challenge of knowledge reuse is, the more constraints and information are associated to the knowledge, the more unique the use case is, decreasing the reusability [4]. For example, the different boundary conditions make the transfer of knowledge between product families more difficult than within product families. In addition, reusability decreases the higher the product complexity [10]. To counter these challenges, the reuse of knowledge can vary over degree of completeness [15] and level of abstraction [4]. Instead of always reusing the whole knowledge, only individual aspects of the existing knowledge are reused, which reduces the applicable constraints. By increasing the level of abstraction, the focus on individual information rises. In general, the higher the degree of abstraction, the more flexible and reusable the existing knowledge becomes. But therefore, it is crucial to preserve the design intent, which requires a fundamental design knowledge and awareness of existing design to ensure a correct integration [15]. According to Lundin et al. [10], reusing knowledge with a high level of abstraction is more likely to result in a conceptual solution proposal, whereas a low level of abstraction is more likely to result in a direct implementation. Chen et al. [14] suggest, using individual reuse mechanisms for functions, geometry, features and interfaces. This should enable designers to decide for themselves what information interests them. Then a tool should assist the designer with the integration of the selected knowledge and thus avoiding that designers have to recreate the knowledge from scratch [14].

2.2 Knowledge Reuse in Software Development

For a long time, it is state of the art in software development to work on a program in parallel using a version control system (VCS) [16]. Programmer can create a local temporary variant, a so called branch, of the master data set at any time. This allows all desired changes to be made without external influences. Subsequently, all changes can be transferred to the master data set. The changes made are automatically detected by the VCS based on the specified merge algorithm and can be transferred where possible. If the program code has been changed since the local branch has been created, a conflict may occur. Depending on the merge algorithm, e.g. semantically or structural, these can be solved automatically in some cases [17]. Since each algorithm can only cover certain scenarios, conflicts often have to be solved by the respective developer manually [18]. A fully automated conflict resolving is generally considered to be not safe in practice, since errors can be caused during this process. Rather, it is considered useful to assist the developer in resolving the conflict [19].

2.3 Structure and Knowledge of a Wiring Harness in CAD

A wiring harness is composed of individual segments. Each segment consists of a bundle of individual wires, which can be built up modularly by a connector, a wire protection, fixings and accessories, such as labels [20]. In general a segment is represented with a maximum bundle diameter in order to reserve installation space in the vehicle. Following the top-down principle a wiring harness CAD assembly is designed around one central part containing all segments, wire protections and accessories. In addition parts like connectors and fixings are added from a PDM system and positioned as desired [21]. Furthermore, each connector is allocated to its specific electro logical function via unique reference. Due to the modular structure and the clearly defined linkage of each element a wiring harness assembly can be fully described by the knowledge about the part data, the associated information like the connector references, their positioning, as well as the routing path of each segment [20]. Eder et al. [20] show how the explicit knowledge about changes made in on CAD model can be formalized. Therefore, all types of changes that can be made to a CAD model of a wiring harness are characterized. In addition, a methodology is presented on how to automatically detect these changes. To increase reusability, all changes in a CAD model are analyzed and clustered according to their intentions automatically. Furthermore, these clusters are stored in a Diff-XML file, containing all information about the state before and after the change was made. In addition this information is enhanced by automatic generated change descriptions. According to Altner et al. [22] the usage of standardized description improves the engineering change management.

3 Methodology for Knowledge Reuse in Wiring Harness Development

In order to reuse knowledge in different CAD data, all scenarios are identified in which elements and attributes can be similar and thus potentially be transferred. These scenarios are used to develop the following methods. In addition, all requirements for application and implementation of these methods are specified.

3.1 Scenarios for Knowledge Reuse

In the development of wiring harnesses, there are three scenarios that can be considered for knowledge reuse. One of these is the development of several phases in parallel. During this process, variants arise due to different requirements of the individual development phases and changes after the design freeze. However, the resulting individual variants can still have some of the same elements and areas. When changes are made to one data set, these can also be relevant for other data sets [3]. Therefore, the more product variants share the same wiring harness, the more data variants can be in development at the same time and consequently be considered for knowledge reuse. Another opportunity for the reuse of knowledge are alternative installation paths of one particular wiring harness. Since a wiring harness is a flexible component, the same product can be installed in many different ways. The need for these can be caused, for example, by different environmental conditions or various equipment variants in the vehicle. Consequently, each alternative installation path requires an individual CAD model. Since alternative installation paths correspond to the same product, it is necessary that all part data, the associated elements, as well as the segment structure and length remain the same in all corresponding CAD models. This requires that all changes to one model of the wiring harness must be transferred to all the other models.

The third scenario is based upon product variants that are built on partially the same bodywork areas, such as different vehicle derivatives and steering types. For example, changes to body holes and repositioning of components can cause adjustments to fixings and connectors in several CAD models. An overview of which elements and attributes are relevant for knowledge reuse in each scenario is shown in Fig. 2.

Knowledge Reuse / Element	Parallel Development Phases				Alternative Installation Path				Same Bodywork Area			
	Part Data	Infor-mation	Position	Routing	Part Data	Infor-mation	Position	Routing	Part Data	Infor-mation	Position	Routing
Connector	✓	✓	✓		✓	✓	✓		✓	✓	✓	
Fixing	✓	✓	✓		✓	✓	✓		✓	✓	✓	
Accessory	✓	✓	✓		✓	✓	✓		X	X	X	
Wire Protection	✓	✓	✓		✓	✓	✓		X	X	X	
Segment	✓	✓		✓	O	O		✓ Length O	X	X		X

✓ = Relevant O = Only If Identical Routing X = Not Relevant

Fig. 2. Reusable elements and attributes of wiring harnesses in possible development scenarios

3.2 Knowledge Reuse Methods

CAD models that are developed in parallel usually differ in degree of similarity. Thus, the shared knowledge can cover entire areas or only individual elements and attributes. These variations lead to the fact that changes cannot be transferred completely in each case. In order to utilize the maximum potential of the acquired knowledge, it becomes necessary to reuse the knowledge in varying degrees of completeness and abstraction, as described in the state of the art. Based on the modular and tree-like structure of the wiring harness, the three categories complete reuse, partial reuse and assisted reuse are defined. Furthermore, all constraints and matching rules are specified to determine if knowledge reuse is possible and which method has to be used.

Complete Reuse: All elements and attributes of the changes made in one CAD model are transferred into another model (see Fig. 3). The basic requirement for this type of transfer is that all elements and attributes of the affected area are identical in both data sets. This transfer can be done automatically without any additional user input. As such, it offers the greatest time saving of the three reuse methods. In addition, multiple data sets can be kept as similar as possible, giving a higher potential for knowledge reuse and especially complete reuse in the future.

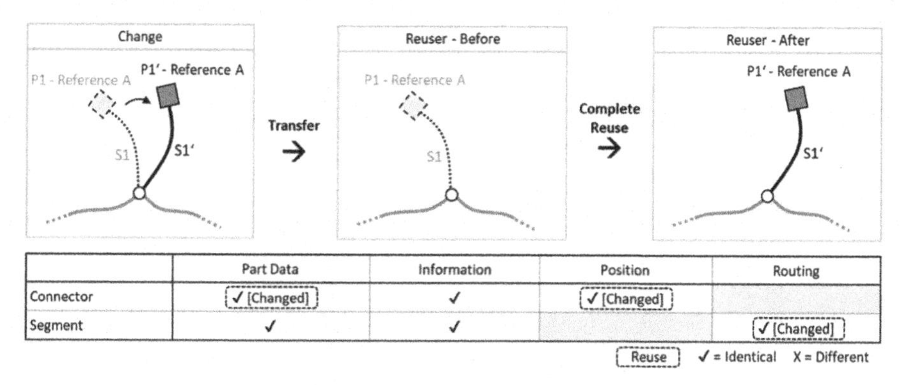

	Part Data	Information	Position	Routing
Connector	✓ [Changed]	✓	✓ [Changed]	
Segment	✓	✓		✓ [Changed]

Fig. 3. Principle of complete reuse

Partial Reuse: Only selected elements and attributes of the changes made in one CAD model are transferred into another model (see Fig. 4). Therefore, specific matching and transfer rules have to be met. For example, if the part data and the information of a connector are identical, but the position differs, the connector can be replaced by reusing the gained knowledge automatically without any additional user input.

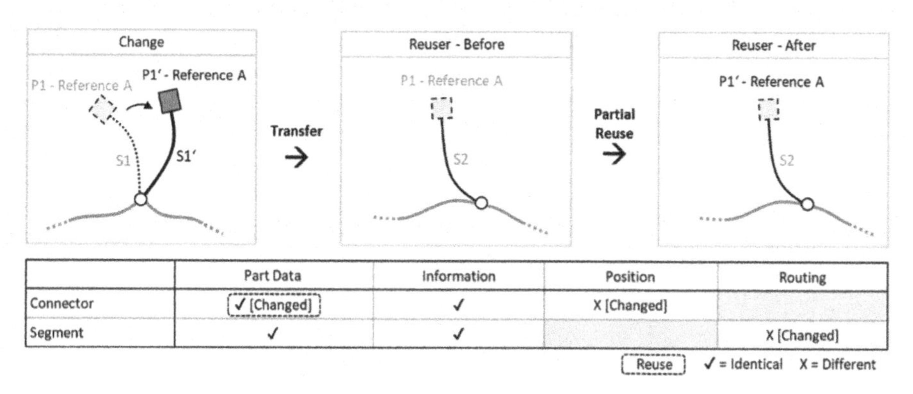

	Part Data	Information	Position	Routing
Connector	✓ [Changed]	✓	X [Changed]	
Segment	✓	✓		X [Changed]

Fig. 4. Principle of partial reuse

Assisted Reuse: Only selected elements and attributes of the changes made in one CAD model are transferred into another model. Thereby, a conflict arises between the source data and the destination data. This occurs, for example, when the position of one element has to be transferred to another data set whose segment path differs from the source data set (see Fig. 5). Consequently, there is no defined routing with the geometric boundary conditions of the designated data set and the position data to be reused. In order to solve the conflict and to be able to carry out the reuse process, the missing information must be supplemented by the user with support from the system. In analogy to software development, a fully automated conflict solution is considered to be too complex and prone to error. One reason for this is the lack of information about the environment in the available data, so collisions with the surroundings cannot be ruled out.

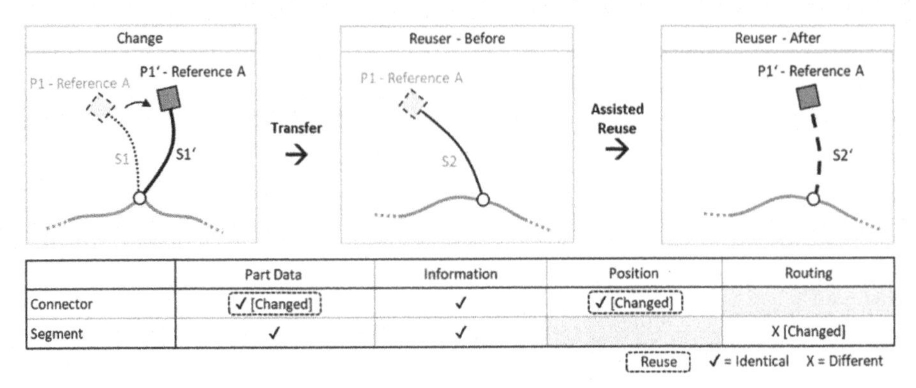

	Part Data	Information	Position	Routing
Connector	✓ [Changed]	✓	✓ [Changed]	
Segment	✓	✓		X [Changed]

Fig. 5. Principle of assisted reuse

3.3 Requirements for Implementation

The approach to integrate knowledge into other CAD models is to modify them in the CAD system using the information of a Diff-XML file. Therefore, the XML file should either be provided manually or should be made available to the CAD system by an appropriate knowledge management system via API. The system has to search for the elements and attributes that match the initial state of the Diff-XML file. Depending on the matched data, the specific reuse rules must be applied in order to determine all possible reuse options and thus the reuse method to be carried out in each case. Subsequently, all changes that can be integrated have to be offered to the user in a selection window using corresponding change descriptions. Based on the user input, the system will then execute the selected changes automatically. The Functions that need to be performed by the CAD system in order to reproduce all the change scenarios are listed in Fig. 6.

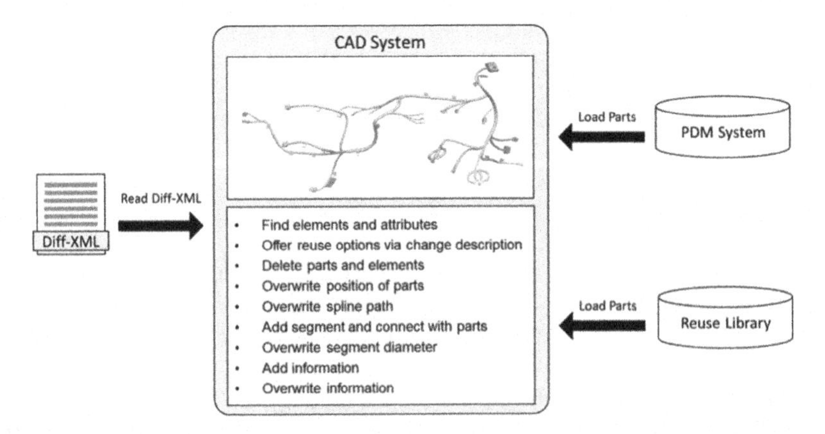

Fig. 6. Requirements for the implementation of the knowledge reuse methods

Changes for which assisted reuse is required must be offered to the user in a feature box that allows the user to enter the missing information. For example, if a new connector has to be added, the user is asked to define the branch-off point from an existing segment.

4 Validation

In Order to validate our methodology, several CAD models were modified in Siemens NX by using the possible operations add, delete, replace and change. Before and after the changes were carried out, XML files were generated containing all necessary information about the wiring harness CAD model. Based on these, the corresponding Diff-XML files were created. A prototype add-on for Siemens NX was used to integrate the contained knowledge of a Diff-XML file into a wiring harness CAD model. All implementable changes were displayed in a user interface and could be selected by the designer. To validate the complete reuse method, all changes were recreated with the previous versions of the manually modified CAD models. Then, each manually adjusted data set was compared to the ones created by knowledge reuse to ensure that there were no deviations. Furthermore, individual elements and attributes could be transferred and thus the technological feasibility of the partial reuse and assisted reuse could be shown. Figure 7 presents an example of how the knowledge reuse process implemented into a CAD system could look like.

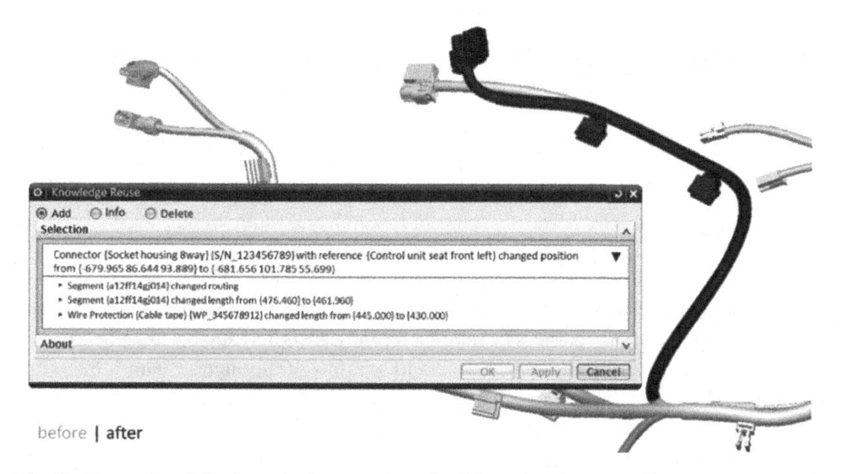

Fig. 7. Example of the knowledge reuse methodology implemented into a CAD system

Nevertheless, the tests also revealed optimization potentials. If, for example, a new connector was added to a CAD model, this connector could be theoretically integrated into almost all the other data models with assisted reuse. Thus, to prevent designers from being offered too many irrelevant change options, an additional relevance check might be useful. In addition, minor deviations of individual elements and attributes led to the fact that knowledge could only be implemented with assistance. A tolerance in the matching of the attributes might improve the efficiency.

5 Conclusion and Outlook

The increasing demand for high quality data during parallel development of multiple wiring harness variants requires individual changes to be carried out redundantly in multiple CAD models. In this paper, it was presented how these efforts can be reduced by reusing knowledge. First, the different use cases that occur in wiring harness development were specified. Based on these, a methodology for complete reuse and partial reuse were introduced, enabling the transfer of changes made to one CAD model into another one in varying degree of completeness automatically. In order to be able to integrate knowledge in situations where information are missing or conflicts occur, a method for assisted reuse was presented. Furthermore, it was specified that complete reuse can be applied, if whole areas are identical between two data sets. Partial reuse and assisted reuse, on the other hand, can be used when certain elements and attributes match. In addition, the requirements that have to be met in order to implement the reuse method into a CAD system were presented. For future work, these methods should be combined with a corresponding knowledge management system which is also responsible for the generation and transfer of the knowledge via Diff-XML files. In addition, the optimization potential identified should be investigated in order to increase the usability and efficiency of the reuse process.

References

1. Trommnau, J., Kühnle, J., Siegert, J., et al.: Overview of the state of the art in the production process of automotive wire harnesses, current research and future trends. Procedia CIRP **81**, 387–392 (2019). https://doi.org/10.1016/j.procir.2019.03.067
2. Schichtel, M.: Produktdatenmodellierung in der Praxis, 1st edn. Hanser, München (2002)
3. Eder, K., Tas, O., Zielbauer, U., et al.: A knowledge management approach to support concurrent engineering in wiring harness development. In: Canciglieri Junior, O., Noël, F., Rivest, L., et al. (eds.) PLM 2021. IFIPAICT, vol. 640, pp. 68–79. Springer, Cham (2022). https://doi.org/10.1007/978-3-030-94399-8_6
4. Busby, J.S.: The problem with design reuse: an investigation into outcomes and antecedents. J. Eng. Des. **10**, 277–296 (1999). https://doi.org/10.1080/095448299261335
5. Smith, N.: Enhancing automotive wire harness manufacturing through digital continuity. In: SAE 2015 World Congress & Exhibition (2015). https://doi.org/10.4271/2015-01-0238
6. Kuhn, M., Nguyen, H.: The future of harness development and manufacturing - Results from an expert case study. Bordnetze im Automobil - 7. Internationaler Fachkongress 2019 (2019)
7. Albers, A., Altner, M.M., et al.: Evaluation of engineering changes based on variations from the model of PGE - product generation engineering in an automotive wiring harness. Proc. Des. Soc. **2**, 303–312 (2022). https://doi.org/10.1017/pds.2022.32
8. Langer, S., Wilberg, J., Maier, A., et al.: Änderungsmanagement-Report 2012: Studienergebnisse zu Ursachen und Auswirkungen, aktuellen Praktiken, Herausforderungen und Strategien in Deutschland (2012)
9. Bracewell, R., Wallace, K., Moss, M., et al.: Capturing design rationale. Comput. Aided Des. **41**, 173–186 (2009). https://doi.org/10.1016/j.cad.2008.10.005
10. Lundin, M., Lejon, E., Dagman, A., et al.: Efficient design module capture and representation for product family reuse. J. Comput. Inf. Sci. Eng. **17** (2017) https://doi.org/10.1115/1.4035673

11. Nonaka, I., Konno, N.: The concept of "Ba": building a foundation for knowledge creation. Calif. Manage. Rev. **40**, 40–54 (1998). https://doi.org/10.2307/41165942
12. Chandrasegaran, S.K., Ramani, K., Sriram, R.D., et al.: The evolution, challenges, and future of knowledge representation in product design systems. Comput. Aided Des. **45**, 204–228 (2013). https://doi.org/10.1016/j.cad.2012.08.006
13. Colombo, G., Pugliese, D., Klein, P., et al.: A study for neutral format to exchange and reuse engineering knowledge in KBE applications. In: 2014 International Conference on Engineering, Technology and Innovation (ICE), pp. 1–10 (2014). https://doi.org/10.1109/ICE.2014.6871565
14. Chen, X., Gao, S., Yang, Y., et al.: Multi-level assembly model for top-down design of mechanical products. Comput. Aided Des. **44**, 1033–1048 (2012). https://doi.org/10.1016/j.cad.2010.12.008
15. Bergmann, R.: Experience management: foundations, development methodology, and Internet-based applications. Hot Topics, **2432**, 93–139 Springer, Berlin, New York (2002)
16. Ruparelia, N.B.: The history of version control. SIGSOFT Softw Eng Notes **35**, 5–9 (2010). https://doi.org/10.1145/1668862.1668876
17. Mens, T.: A state-of-the-art survey on software merging. IIEEE Trans Software Eng **28**, 449–462 (2002). https://doi.org/10.1109/TSE.2002.1000449
18. Menezes, J.W., Trindade, B., Pimentel, J.F., et al.: What causes merge conflicts? SBES 2020. In: Proceedings of the 34th Brazilian Symposium on Software Engineering, pp. 203–212 (2020). https://doi.org/10.1145/3422392.3422440
19. Shen, B., Zhang, W., Yu, A., et al.: SoManyConflicts: resolve many merge conflicts interactively and systematically. In: 36th IEEE/ACM International Conference on Automated Software Engineering (ASE), pp. 1291–1295 (2021). https://doi.org/10.1109/ASE51524.2021.9678937
20. Eder, K., Herzog, W., Altner, M., et al.: Knowledge documentation based on automatic identification and clustering of change intentions in CAD data of wiring harnesses. Proc. Des. Soc. **2**, 683–692 (2022). https://doi.org/10.1017/pds.2022.70
21. Neckenich, J., Zielbauer, U., Winter, R., et al.: An integrated approach for an extended assembly-oriented design of automotive wiring harness using 3D Master Models. In: DS 84: Proceedings of the DESIGN 2016 14th International Design Conference (2016)
22. Altner, M., Redinger, H., Valeh, B., et al.: Improving engineering change management by introducing a standardised description for engineering changes for the automotive wiring harness. In: Procedia of the 32nd CIRP Design Conference (2022)

Generative Engineering and Design – A Comparison of Different Approaches to Utilize Artificial Intelligence in CAD Software Tools

Detlef Gerhard, Timo Köring(✉), and Matthias Neges

Institute for Product and Service Engineering, Digital Engineering Chair, Ruhr University Bochum, Universitaetsstr. 150, 44801 Bochum, Germany
{detlef.gerhard,timo.koering,matthias.neges}@rub.de

Abstract. There are new CAD software tools emerging that will likely pave the way to next generation engineering design. They are labelled Generative Engineering, Generative Design, Algorithm Based Design, Simulation Based Design or similar. Those new tools have in common that they make intensive use of Artificial Intelligence methods and extensive computing power. The focus of these algorithm-driven approaches to developing products and finding solutions is not the explicit creation of geometry but rather the definition of constraints, boundary conditions, rules and procedures that allow the computation of feasible solutions including the implicit generation of geometric models. One major idea is to initially open up wide solution spaces that are goal-oriented based on given requirements and provide engineers with all relevant information to perform trade-off studies and create new and innovative solutions instead of perpetuating existing solutions. The new tools are often associated with topology optimization and generative manufacturing, but the concepts go far beyond and lead to a complete workflow in the creation of products.

The paper systematically analyses different software tools and shows that Generative Design and Engineering is interpreted and implemented differently by various software vendors, all pursuing different goals. In addition to the potentials, the paper mainly shows the current limitations of the implemented approaches in the different CAD tools. The focus is not only on the design phase, but also on how the tools take into account different aspects such as the automation of the entire development process, cost evaluation and manufacturing processes.

Keywords: Generative design · Generative engineering · Topology optimization · Artificial intelligence

1 Introduction

Generative Design (GD), Generative Engineering, or Algorithm Based Design are terms for the next generation of engineering design which combines human capabilities with

Published by Springer Nature Switzerland AG 2023
F. Noël et al. (Eds.): PLM 2022, IFIP AICT 667, pp. 206–215, 2023.
https://doi.org/10.1007/978-3-031-25182-5_21

artificial intelligence, algorithms, and computing power[1]. Especially the technological progress in cloud computing opens completely new ways of creating and evaluate designs through the extends computing power. Furthermore, the manufacturing technique generative manufacturing also removes many constraints from the design of manufacturing and extends the design of freedom to the engineer markedly [1]. The resulting possibilities of manufacturing nature-based and lattice structures requires new ways of construction. Lattice structures are for example body centered cubic or honeycombs structures. Simultaneously, shorter development times and higher requirements lead to first best solutions, often based on existing solutions, instead of exploring the complete solution space to find the best possible solutions. To integrate all these new possibilities and to tackle the challenges a new approach for the development process is given by the approach of GD.

The definition and understanding of GD vary and is interpreted and implemented slightly differently by different CAD tool vendors. One possible cause is that GD is used in a variety of different disciplines such as architecture, art, and engineering [2]. All approaches in the field of engineering have in common that algorithms generate and evaluate various designs or alternative solutions based on user defined boundary conditions [3]. The aim of GD is not to replace engineers but to shift the ability from modeling the solution to the definition of the boundary conditions and the evaluation of the solution. Hereby, the aim is to deliberately open up the possibility that established patterns of thought are broken down and solutions do not only emerge as a gradual further development of already existing solutions. This allows a wider range of options to consider and make the solution more suitable and effective. This paper presents the state of the art in GD in distinction to topology optimization, nature-based and lattice structure. Different GD interpretations and embodiments of the approach in CAD software tools for structural components are presented as well as a comparison of implementation inside three different software tools while identifying strengths and limitations. Therefore, a schematic workflow along the product development process is presented and investigated. Although there are already GD approaches for fluid paths and temperature applications for example, this paper only refers to structural components.

2 State of the Art in Generative Design

GD is an approach to autonomously generate and evaluate designs based on algorithms, that consider user specified boundary conditions like materials, manufacturing processes, loads and supports. Instead of the traditional approach where one concept is selected and evaluate through manual iterations, GD allows to generate and evaluate multiple design at the same time automatically and compare them [4]. As a result, in addition to the traditional approach, a significantly larger solution space is explored and compared. Particularly, the explicit generation of geometry and shape of parts is not any longer the foreground of engineering work, but the definition and characterization of the design problem itself. Geometry is then created automatically and in an implicit way. The approach also takes further stages like cost evaluation into account and allows engineers to evaluate a whole concept at an early stage of development [4]. This methodology

[1] For the sake of simplicity, we will only use the abbreviation GD in the rest of the text.

is not new, there are already first attempts during the late 1970s, but for a large-scale application, further developed algorithms and primarily high computing power were necessary [5]. The technological progress in cloud-based computing performance and machine learning opens new opportunities and increases the possibilities which lead to a new generation of CAD Software tools from different vendors. But as already mentioned the definition and implementation is not set and other terms like topology optimization, nature-based design and lattices structures generation for additive manufacturing are often associated or equated with GD [6].

Topology optimization is limited to the topology aspect and extracts a shape depending on the boundary conditions from a start shape based on a particular optimization algorithm. The goal of this approach is to reach a volume or mass reduction by complete focusing on the shape. As a conclusion of the necessary starting shape, material can only be subtracted from this shape. In contrast, GD is generating a form, can add material to the design space, evaluate different materials at the same time and create new shapes independent from an existing shape. Therefore, the options with GD are much larger than with topology optimization. Nonetheless, topology optimization algorithms can be part of a GD functionality to create a shape. The goals of GD are not limited to volume and mass-reduction and can also consider costs, materials, displacement, manufacturing processes and similar. Besides topology optimization, GD is also particularly excellent suited for biomimetic designs as well as the automated and algorithm-based generation of lattice structures for parts that are manufactured using additive manufacturing technologies. These structures are complex and time consuming to create with traditional modelling. In summary, topology optimization, bionics and lattice structures are like tools in a toolbox of GD methodology [1].

3 General Analysis of GD Approaches

3.1 Implicit Geometry Creation Workflow

The schematic workflow for a GD process for structural component contains is shown in Fig. 1.

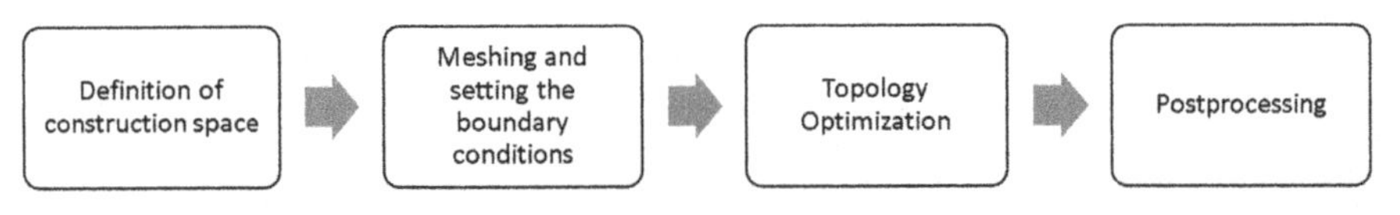

Fig. 1. Schematic workflow for GD process

Definition of the Construction Space. At the beginning of each geometry creation workflow, the definition of the construction space is necessary. This construction space contains not only a spatial area in which geometry can be created or changed (design space DS) but also the definition of spatial areas which have to be kept unchanged (non-design space NDS, retained geometries) or which are given obstacles. Initial geometry models can be imported from different CAD systems or created inside the software. For the definition of the construction space, there are two extreme options with intermediate levels possible. In the "shell-based" option the design space represents the complete construction space (like a shell). In this case the definition of obstacles is not necessary and areas outside this space would not be considered. In this, the non-design space is for geometries which are functional necessary like for instance holes with supports for screws. For the definition of the non-design space a selection through surfaces of the design space or required 3D-models are possible. The other extreme option "restraint geometries" is just the creation of retained geometries (like non-design space) and obstacles. A starting shape could also be created and is recommended if several retained geometries should be interconnected across obstacles, but it does not restrict the construction space.

Meshing and Setting the Boundary Conditions. The next step is to prepare the model for FEM analysis with meshing and defining the boundary conditions like forces, supports and material. Also, the constraints and objectives for the optimization are set. Since the objectives and constraints can be used as both, they are considered together.

For meshing, the range is between "low influence" and "customizable with parameters". In the "low influence" approach the options are limited to the resolution or similar. The "customizable with parameters" option offers many different mesh options as well as for the subsequent preparation, enhancements, and connection of multiple meshes. Mesh type (e.g., triangle or quad), mesh size and other settings can be individualized. Moreover, the option to import or integrate the mesh generation from a different tool is possible.

Potential loads are gravity, forces, moments, and pressure. Therefore, integrable load steps are linear static, buckling and modal. For the constraints or objectives are volume fraction, structural compliance, stress, and displacement as well as the consideration of one direction for extrusion and symmetry with different restrictions possible. Moreover, passive (retain) regions, buckling and modal frequency constraints or overhang constraints for additive manufacturing, can be included. To set the values for the various constraints a manual input or the automation with an import from Excel for example is possible. Another different is the support of manufacturing processes. Due to the consider of one direction for extrusion the tools are designed for additive manufacturing. Possible supported manufacturing processes are beside additive manufacturing, die casting, 2-axis cutting und milling while considering different manufacturing settings like tool properties.

Topology Optimization. Topology optimization for exploration of alternative results includes more methods than the classical topology optimization and describes the improvement of the topology with further methods. One "basic" alternative is to implement the traditional approach of topology optimization with local computing power and provide a block which used the model, variables, objectives, constraints, and a limitation of iterations can as inputs. Also, the integration of further settings like "boundary penalty" to prefer material on the boundaries or material inside, or an option to see all iterations with a different threshold is possible. The threshold defines how much material should be removed and has a major influence on the result. Even if the constraints are fulfilled, the final result does not necessarily meet the requirements depending on the threshold. Another"advanced" alternative is to use experimental solvers or different approaches like the level-set-approach in cloud computing to explore alternative designs and solutions. With respect to the different options, the number of solutions can vary between one solution from the traditional topology optimization and multiple solution depending on the number of materials or manufacturing processes.

Besides topology optimization, another aspect in GD software tools is the generation of a field-driven design, which means for example the adaption of a shell thickness due to the stress in a special area. If the stress in a certain area higher, the shell is denser than in an area with less stress.

Postprocessing. After the topology optimization, the result needs to be prepared for manufacturing and evaluated in relation to the goal achievement. Depending on the results of the topology optimization, the results could be rough and need to be postprocessed. This can be done either automatically using individual algorithms with voxel and smoothing or manually. The next postprocessing step is the fulfilment check of load requirements. For this, different simulations like static FE, nonlinear-static FE, modal frequency, and buckling-analysis could be implemented or exported to a specialized tool. The analysis of key performance indicators (KPI) can be automatically evaluated and compared to the target values or with a manual process for given KPI. For KPI analyses, a black box or an individual approach can be implemented. In the black box approach, the data is supplied by an external tool and there is no option to customize the data. The individual approach allows input data inside a tool for customizable to company-specific data or offers mathematical operations to create own workflows for the calculation. To prepare results for manufacturing, the tools can be integrated or depend on special postprocessing software. The number of supported manufacturing processes varies in different tools.

3.2 Solution Space Exploration and Trade-off Studies

With respect to solution space exploration and the capability to perform automized and assisted trade-off studies GD approach can be divided into two main categories.

The first category **"Assisted CAD modelling"** is an adaption to the traditional CAD approach with the GD module integrated as a workspace in the same way as e.g., simulation module. The GD module user interface is designed like a traditional CAD tool with a ribbon menu for modelling, design constraints and similar as well as a viewer for representing the geometry. The approach is geometry driven and provides a user friendly and familiar environment for the engineer. The configuration options are limited, and the approach is like a black box in terms of algorithms which reduces error-proneness on the one hand side but also limits the comprehensibility on how a solution space for a given problem is created. An impact and inspection of the several steps inside the software is strongly restricted and an automation of the whole process is not possible. The exploration of the solution space and the possibility to perform trade-off studies are implicitly integrated, and the adjustment of the process is not possible.

The second category is **"Functional CAD programming environment"**. Here, the user interface is different from traditional CAD and consists of a low code programming environment. Instead of predefined workflows and CAD functions, the exploration of the solution space and the functionalities are individual arrangeable and can be automized. In this approach, the integration of further tools inside the software or through an im- and export is strongly supported. The engineer is in complete control of all steps, i.e., it is a white box approach. A user needs knowledge about the generation and application of workflows which could lead to mistakes. Nonetheless, an advantage of this approach is the customizable parametric driven generation, which allows to change settings fast and especially explore many different designs through a design of experiments study. This can be used for design automation by only changing a few boundary conditions and is powerful with respect to solution space explorations and trade-off studies.

4 GD Interpretation of Selected Software Tools

The vendors of the different CAD software tools describe GD in various ways. To understand the implementations inside the tools, in this chapter the GD approaches of three software tools are presented.

4.1 Autodesk Fusion 360

Autodesk describes GD as an approach for simultaneously generation of multiple solutions based on real-world goals using cloud computing. The generation is not limited to special manufacturing processes like additive manufacturing and can also take traditional processes like milling into account. For this GD approach, a human operator, artificial intelligence algorithms, cloud computing power and a strong link to manufacturing and process awareness are required. Moreover, Autodesk describes the characteristics of GD as the capacity to solve more complex problems with designs which are physics based and mimics nature as well as with generative design factors in constraints such as load, weight, and cost. The fact that Autodesk implemented GD as a module inside Fusion 360 allows the operator to easily transfer results from GD to tools such as CAD, CAM and FEM. In addition to GD for structural component, Fusion 360 also integrates a preview of GD for fluid paths [4, 7, 8].

4.2 nTopology

nTopology (nTop) is a stand-alone solution for GD and uses an approach of implicit modelling in contrast to traditional explicit modelling (e. g. B-Rep, CSG, Direct) that allows to model geometries with implicit equation in a fast and robust way, especially for lattice structures and organic shapes. For nTop, GD "is a goal-oriented and simulation-driven design methodology that uses software and computational algorithms to generate high-performance geometry based on user-defined engineering requirements". Therefore, nTop defines three key components of GD, which a GD software combines: geometry generation, design analysis & evaluation and automated iteration loops. GD enables engineers to use the full of range of possibilities from additive manufacturing. In comparison to topology optimization, this GD approach allows to consider technical and non-technical requirements like costs. The advantages of nTop approach are the development of new designs (with no inspiration from previous), high performance products and acceleration of product development. Disadvantages arise from the limitation of the constraints GD can handle and the fact that the quality of the result is limited to the quality of the workflow/simulation. nTop is also not limited to structural components and integrates also GD for thermal applications [9].

4.3 ELISE

ELISE describes the product development process as an incremental process with many manual iterations due to different departments, software tools and data types. To improve this process, engineers should be able to use computers to develop optimal products automatically. ELISE can be interpreted as a low-code enabled CAD integration platform that combines several commercial engineering tools in a stable, all-encompassing software. In ELISE the algorithm-based design is defined as a technical DNA (workflow) which automatically generates shape and geometry, based on the boundary conditions. These boundary conditions contain clear design rules, limits, and target values. The workflow integrates and evaluates all different processes and tools like CAD design, FE-Simulation, and calculations inside and is completely customizable. The advantages of this approach are that changing boundary conditions automatically lead to new parts. There is no longer an individually creation of each component necessary and the DNA can be reused partially or completely. By this approach, the new degree of design freedom provided through additive manufacturing is captured and enabled well [10].

4.4 Assessment of the Modules

Table 1 shows an assessment and a comparison of the two categories.

The **assisted CAD-modelling** approach shows how to integrate GD as a module inside the existing solution while providing the traditional CAD-user interface. The generation of the structure is like topology optimization with focusing on a design for manufacturing and include a selection of further features like cost analysis and comparison of options. The simultaneous generated results are due to the cloud computing support and different solver involvement manifold and more nature based. But the options to configure individual solutions are restricted, the workflow cannot be completely automated, and the approach is limited customizable due to only changing boundary conditions and the resolution of the mesh.

In contrast, the new vendors follow with **Functional CAD programming environment** a similar concept for GD, although the used term is different. Both software solutions rely on the individual generation of reusable workflows for the automation of the entire process and allow to change many parameters. But the visualization options for comparison the results of different parameters as a trade-off study are limited. The approaches of both are not limited to generate a shape like with topology optimization, they are more holistic and individual depending on the circumstances. The engineer has all options to variate the whole process by creating an own workflow and can track all steps to create the design from the construction space over topology optimization to evaluation. The focus of both platforms is on additive manufacturing although not

Table 1. Assessment of different approaches to features

	Assisted CAD-modelling (e. g., Fusion 360)	Functional CAD programming environment (e. g., nTop/ELISE)
Geometry Definition		
Definition of construction space	Restraint geometries	Shell-based
Mesh generation for topology optimization	Low influence	Customizable with parameters
Considerable loads	Gravity, forces, moments, pressure	
Considerable load steps	Linear, in preview: modal, buckling	Linear
Constraints/objective for optimization	Volume fraction, structural compliance, stress, passive regions, extrusion direction	
	In preview: displacement, modal, symmetry	Displacement, symmetry
Manufacturing processes	Additive, die casting, 2-axis cutting, milling	Focused on design for additive manufacturing
Topology Optimization	Advanced (cloud computing)	Basic, field-driven (local)
Postprocessing	Manual	Automatic
Topology reconstruction	Not necessary	Voxel, smooth
Evaluation for fulfilment of the requirements	Static FE, modal frequency, buckling-analysis	
	Nonlinear-static FE	Im- and export for external tools
KPIs evaluation	Black box, automatically for technical, manual for non-technical	Individual, automatically through mathematical operators
Preparing for the manufacturing processes	Not inside GD	For additive
Further Aspects		
Design automation	No	Individual
Lattice structures	No	Yes
Design of experiments	No	Yes

limited to it. Differences are in the provided load steps and constraints. ELISE allows modal and buckling load steps and therefore constraints, while nTop allows moreover an overhang constraint with focus on additive manufacturing. For the evaluation of the KPI, ELISE has besides the mathematical operator's unique blocks for some KPI to evaluate them automatically.

An advantage of the implicit modeling platform nTop against the open platform ELISE is that nearly all features are integrated inside the platform and no other license or software is necessary. The only exception is the design of experiments in nTop. Against this, ELISE provides many features, but it also relies on partners for integration of different tools like lattice structures. Moreover, notable is the necessary license for the basic function of topology optimization. But besides the typical workflow, the new vendors provide further features like field-driven design, lattice structures.

Figure 2 shows the differences in the definition of the construction space for the three presented software tools and the generated results after the development process. All results show a reduction of the used material while take over the preserved areas and meet the stress requirements. Especially the experimental result from Fusion 360 shows a design that is clearly different from the traditional approach compared to the other.

Further advantages of GD are shown in the example of a seat bracket from General Motors and Autodesk where the GD solution is 40% lighter but also 20% stronger and eight parts are consolidate into one [11].

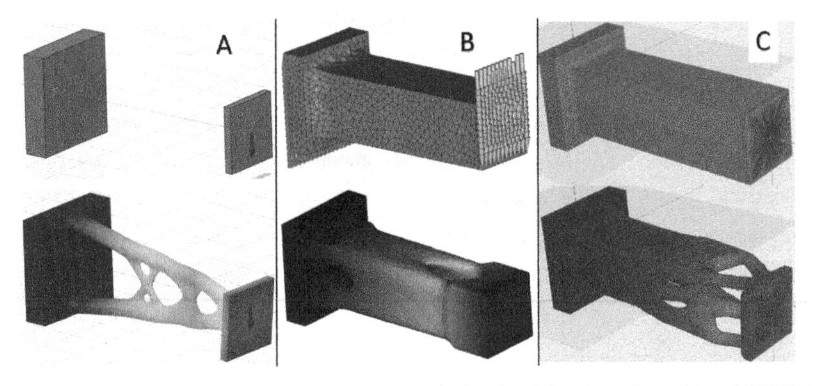

Fig. 2. Example of construction space and results in Fusion 360 (A), nTop (B) and ELISE (C)

5 Conclusion and Outlook

This paper shows a systematic overview and a comparison of different approaches to implement GD inside CAD software tools. For all mentioned features along the product development process, the possibilities for integration are manifold and differs in most cases between an individual and a black box approach as well as a manual against an automatic process. The implementation of the features inside three selected software tools were analyzed. The results show that the implementation can be divided into an "Assisted CAD-modelling" and "Functional CAD programming environment" approach while the embodiment was always slightly different and not completely equal. Each software tools have its strength and limitations in different aspects. In this paper, the focus was on the features for the generation of a single part, but no consideration was given to the generation of assemblies. Also, the reproduction and variation of results by similar boundary conditions as trade-off study was not analyzed. Moreover, the results need to be validated regarding building options and fulfillment of the requirements. Also, it needs to be considered that all software tools are still under development. Therefore, it is necessary to evaluation the functions of these tools in future and compare them if new functions are available.

References

1. Briard, T., Segonds, F., Zamariola, N.: G-DfAM: a methodological proposal of generative design for additive manufacturing in the automotive industry. Int. J. Interact. Des. Manuf. (IJIDeM) **14**(3), 875–886 (2020). https://doi.org/10.1007/s12008-020-00669-6
2. Jaisawal, R., Agrawal, V.: Generative Design Method (GDM) – a state of art. In: IOP Conference Series: Materials Science and Engineering, vol. 1104 (2021)
3. Pollák, M., Kočiško, M., Dobránsky, J.: Analysis of software solutions for creating models by a generative design approach. In: IOP Conference Series: Materials Science and Engineering, vol. 1199 (2021)
4. Harvard Business Review. https://hbr.org/resources/pdfs/comm/autodesk/The.Next.Wave.of.Intelligent.Design.Automation.pdf. Accessed 28 Mar 2022
5. Caetano, I., Santos, L., Leitão, A.: Computational design in architecture: defining parametric, generative, and algorithmic design. Front. Architectural Res. **9**(2), 287–300 (2020)
6. Babel, N., Metzger, M.: Untersuchung künstlicher Intelligenz im Bereich der Konstruktion mit Generativer Design Software. Hochschule für Angewandte Wissenschaften Landshut (2021)
7. Autodesk - Generative Design for Manufacturing. https://d1.awsstatic.com/partner-network/partner_marketing_web_team/Manufacturing-partner-solutions_Autodesk_brochure.pdf. Accessed 28 Mar 2022
8. Autodesk, KUKA: Deciphering Industry 4.0 – Part IV Generative Design (2018)
9. Generative Design: The Engineering Guide | nTopology. https://ntopology.com/generative-design-guide/. Accessed 08 Mar 2022
10. ELISE Portal. https://portal.elise.de/#/docs/2012250159/2012250203. Accessed 18 Mar 2022
11. Autodesk News - How GM and Autodesk are using generative design for vehicles of the future. https://adsknews.autodesk.com/news/gm-autodesk-using-generative-design-vehicles-future. Accessed 25 May 2022

Finite Element Mesh Generation for Nano-scale Modeling of Tilted Columnar Thin Films for Numerical Simulation

Noé Watiez[1](✉), Aurélien Besnard[1], Pavel Moskovkin[2], Ruding Lou[3], José Outeiro[1], Hélène Birembaux[1], and Stéphane Lucas[2]

[1] Arts et Métiers Institute of Technology, LaBoMaP, 71250 Cluny, France
noe.watiez@ensam.eu
[2] University of Namur, 5000 Namur, Belgium
[3] Arts et Métiers Institute of Technology, LISPEN, 71100 Chalon-sur-Saône, France

Abstract. Glancing Angle Deposition (GLAD) is a technique used in Physical Vapor Deposition (PVD) to prepare thin films with specific properties. During the deposition process, a tilt is introduced between the sputtered atoms flux and the normal of the substrate surface. By shadowing effect, this induces tilted nano-columns that affect the properties of the coating. To predict these properties, several existing tools simulate the different steps of the PVD deposition. First, a simulation of the sputtering of atoms from a metallic target is made, followed by the computation of the atoms transportation from the target to the substrate. Finally, the growth of the film is computed. All these simulations use point based representation to represent the deposited atoms. However, such representation is not suited for classical finite elements analysis (FEA).

In this paper, a methodology for generating FEA meshes from the points produced by film growth simulation is presented. Two major scientific challenges are overcome. Firstly, how to segment the "film" point cloud into a collection of individual "columns" and secondly, how to generate the meshes of the columns that are approximately represented by points. The point cloud segmentation is computed through neighbourhood notion. The mesh is obtained as an implicit surface by the marching cubes algorithm and smoothed by a Humphrey's class Laplacian algorithm. Numerical simulations based on the generated FEA meshes will be conducted using Abaqus FEA software.

Keywords: PVD · GLAD · Columns · FEM · Simulation · Point cloud · Meshes

1 Introduction

Physical Vapor Deposition (PVD) is a process used to coat parts with thin layers ranging from a few hundreds of nanometers to a few micrometers in thickness. It has a large scale of application due to a high degree of freedom in the composition of the layer as well as an easy application in the industry [1]. It is used in several domains, ranging

Published by Springer Nature Switzerland AG 2023
F. Noël et al. (Eds.): PLM 2022, IFIP AICT 667, pp. 216–226, 2023.
https://doi.org/10.1007/978-3-031-25182-5_22

from mechanical application (such as coated milling drills) [2], optics [3], electronics [4] and in several other fields.

The deposition is carried out at an atomic scale in a vacuum chamber, hence *in situ* data is sparse [5]. Most of the information comes from post-deposition characterization of the coating [1]. In this case, simulation is a key to evaluate the macroscopic properties from several micro-scale interaction through laws of physics.

Many simulation software are based on the Monte-Carlo algorithm, and, in particular Kinetic Monte-Carlo (KMC). Different successive codes are mandatory to conduct such study. For example, SRIM [6] or TRYDYN [7] can be used to characterize the sputtering from a target, SIMTRA [8] computes the transportation of matter in the chamber, NASCAM [9, 10] or SIMUL3D [11] simulate the growth of the columnar film. This list is not exhaustive, however it presents the three type of software used in PVD simulation (sputtering, transport, growth). However, no software currently support the mechanical simulation of thin films based on a simulated structure. The aim of this study is to address this lack and produce such simulation.

This paper presents a pilot study of FE meshes generation from point based film growth simulation result which will be used in Finite Element Method (FEM) simulation of the mechanical properties.

2 Technological background

2.1 Presentation of the GLAD Process

As its name implies, PVD process uses the physical phenomenon in order to turn the solid material into a vapor, which is then deposited onto the part. The phenomenon used in this study is called "Magnetron Sputtering".

Figure 1 presents the PVD process: in a deposition chamber, a plasma is condensed on the surface of a metal object, called target, which leads to the erosion of the surface and the sputtering of the eroded matter. Then, the sputtered matter is transported through the chamber toward a substrate, resulting in the growth of the film.

This study focuses particularly on the GLancing Angle Deposition process (GLAD) in which an angle is formed between the flux of sputtered matter and the substrate in order to constrain the growth. The major effect is the apparition of an inclined porous columnar structure due to the shadowing effect as presented by Hawkeye *et al.* [12]. It is then possible to nanostructure the film by controlling the deposition angle and thus vary its properties. This effect has already been documented for a variety of applications, such as anti-reflect optical films [13], magnetic films [14] or mechanical applications [15].

2.2 Simulation Steps

The current steps of the simulation is the successive simulation of the sputtering, the transport and the growth of the film [16]. In this study, the sputtering step is conducted with SRIM [5] and SIMTRA [8]. The former estimates the atom flux resulting from the ion and the target in both angle and energy in a localized manner (a few nm^2 surface of the target). Whereas the latter uses those angular and energetic distribution as well as a racetrack in order to create the emission overall the target.

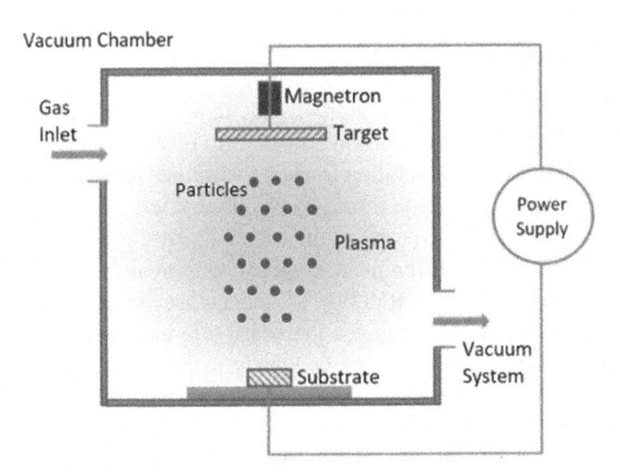

Fig. 1. Outline of the PVD process [2]

SIMTRA then proceeds to the computation of the transportation in the chamber, considering the collisions with the atoms present, in this case argon at a pressure of 2.2 $\times$ 10^{-3} mbar, and returns a file containing the final position, angle and energy of any atom landing on the target.

Finally, NASCAM [8, 9] simulates the growth from the transportation result given by SIMTRA. Several transportations can be used together to emulate multi-target deposition as well as a reactive atmosphere. It results into a point clouds on a cubic or hexagonal grid with information about the composition and the morphology. The three software are based on the KMC algorithm.

This paper presents a post-processing workflow made to prepare the results from the previously stated simulation for mechanical calculation by finite element model.

3 Post-processing

The construction of a mesh from the simulated PVD growth consists of several steps. The overall workflow of the process is presented in Fig. 2. The state of the input and output data is also provided between each two successive steps. The following parts elaborates on the column separation algorithm as well as the steps followed in order to produce a smooth mesh from the point cloud computed for each column morphology in the initial simulation step. The mesh assembly consists of simply adding all sub-meshes for each column to a global coating mesh.

3.1 Columnar Separation

As previously stated, the GLAD technique induces a columnar morphology, which in turn causes variations of the film properties. Hence, in order to compute the film properties, it is necessary to study the columns. This is the aim of the first part of the post-processing algorithm. NASCAM exports a point cloud where each point represents an atom and

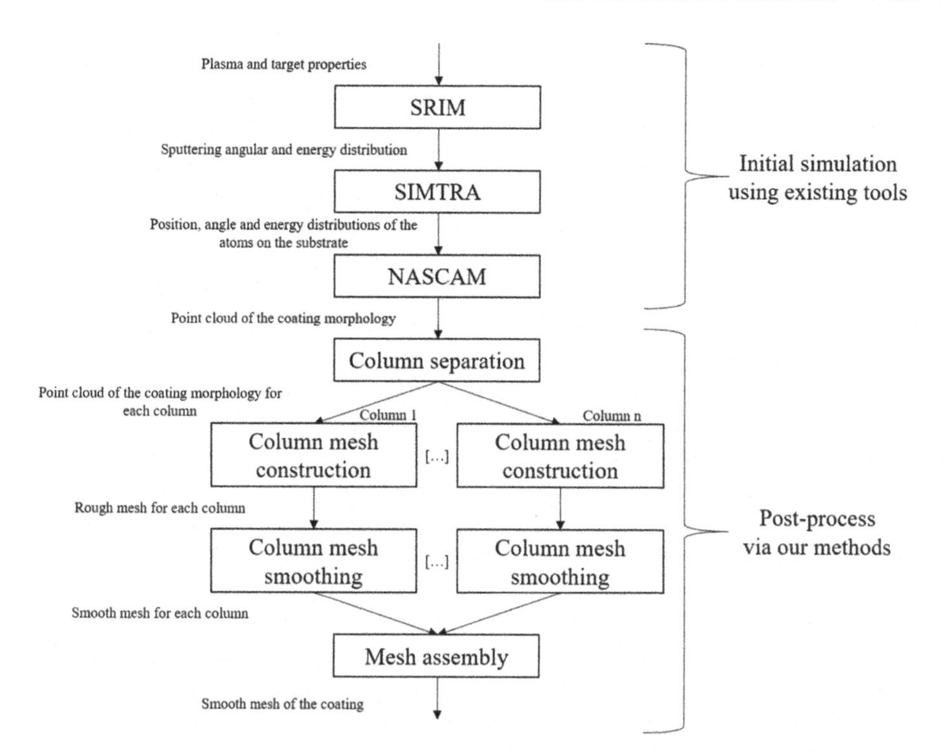

Fig. 2. Workflow of the simulation from the PVD process to the construction of the mesh

possesses the name of the chemical specie in addition to the xyz coordinates. Unfortunately, no information about the point clustering according to the columns is provided. Therefore, a method has to be designed to segment this global point cloud into a set of columns.

The designed method is based on the growth of a column and the chronological flux of atoms. Figure 3 illustrates three principal cases when an atom arrives on the substrate. If the atom arrives on the substrate without any neighbor, a new column is created (Fig. 3a). Else, the atom is added to one of the column from its neighbors. If there is only one neighbor atom, the newly arriving atom will belong to the same column of this unique neighbor (Fig. 3b). If the newly arriving atoms is in contact to several neighbors belonging to different columns, the one of highest weight is assigned to the newly arriving atom (Fig. 3c).

Figure 4 is a logic graph presenting the major steps of the column separation algorithm.

All neighbors are considered by relative attraction, which means with a decay along the distance to the point. For calculation purpose and according to the work on a cubic grid, only the points contained in the cube of three units edges centered on the newly added point are considered as neighbors as presented in Fig. 5.

In this figure the neighboring positions to any point on the grid, here the orange point with a halo are filled with virtual points. A color is given to each in accordance with the distance from this point to the central point. Positions on the faces (pink) are at the

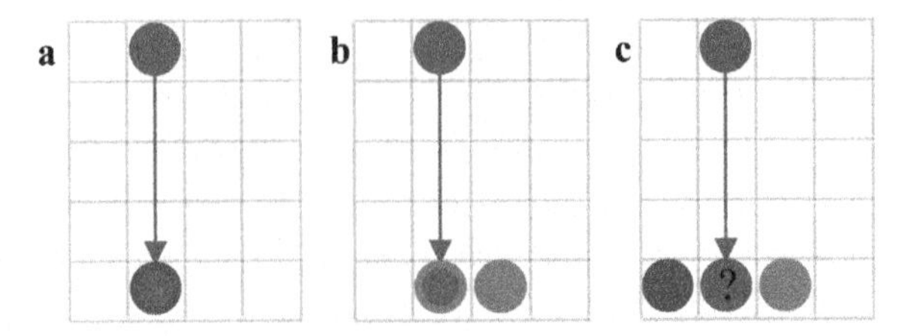

Fig. 3. Schematic representation of the attribution of an atom depending on the neighboring structure. (a) No neighbors, (b) A single neighbor, (c) several neighbors

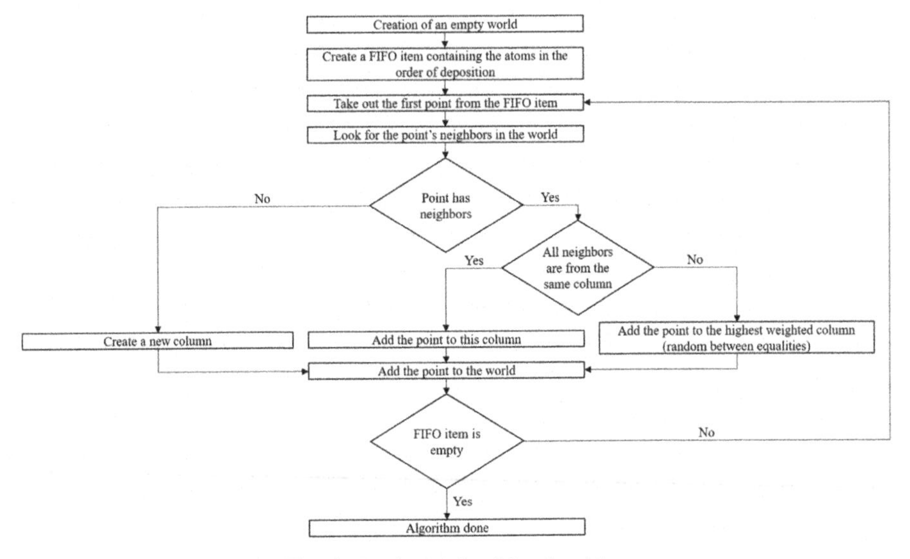

Fig. 4. Logic graph of the algorithm

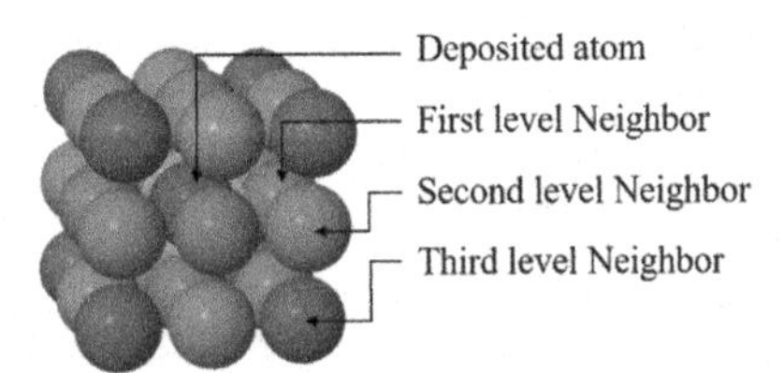

Fig. 5. Layout of the neighborhood of an atom in the algorithm

distance of 1 unit, on the edges (green) at $\sqrt{2}$ units and on the vertices (brown) at $\sqrt{3}$ units. In a first approximation, the parameters used by Besnard [17] on Simul3D will be kept for further simulation. It estimates the weight of any point to be linear with its inverted distance to the middle of the cube.

In order to develop the algorithm, it is necessary to know the structure around every point at the time of deposition. A new option implemented in NASCAM provides the deposition order and the successive position of each point. By browsing through this file, it is possible to have the structure at any time of the deposition.

NASCAM works in a prismatic world. To reduce computational costs, this world is periodic: its boundary planes along both the x and the y axis are considered adjacent. This has to be taken into account in the separation of columns as one can overlap over different boundaries of the world, mostly in highly tilted deposition (Fig. 6).

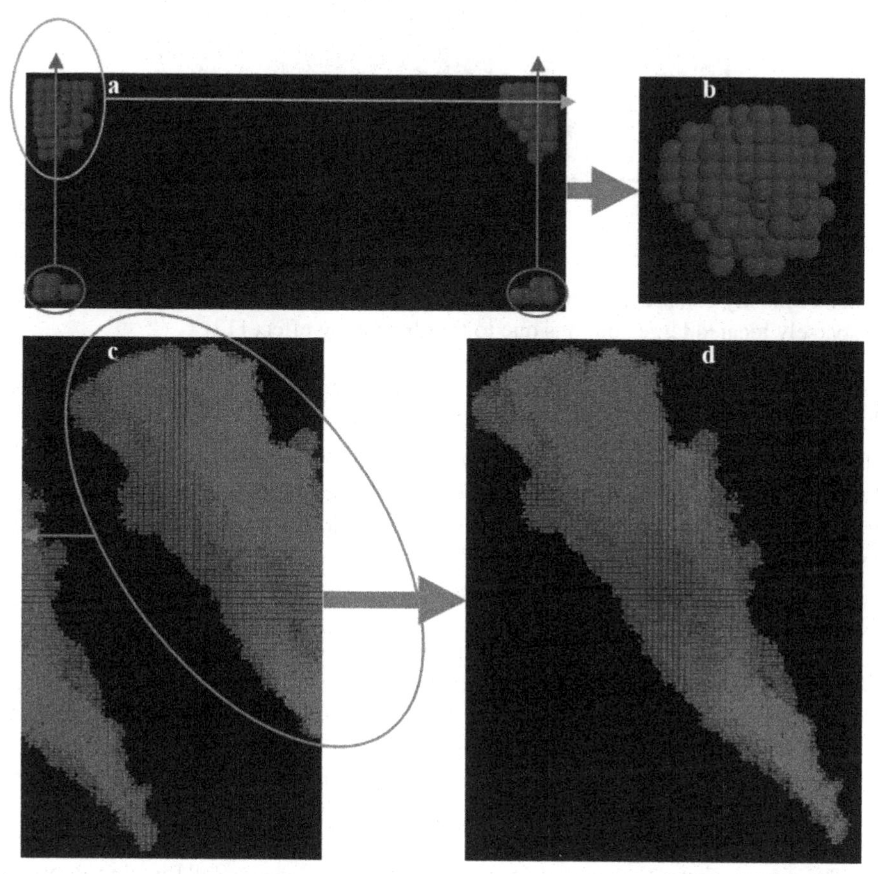

Fig. 6. Top view of a small column on a periodic world along with the steps needed for the deperiodicization (a) and of the same column in a non periodic world. Cross section of a large column in a periodic (c) and non-periodic (d) world

Figure 6a presents the effect of the periodicity on the top view of a single small column and Fig. 6b show the shape of the same column in a non-periodic world. Figure 6c and d present the same situation for a large column in cross section view.

The results of the column separation algorithm are presented on Fig. 7.

Figure 7a presents the raw structure exported by NASCAM, while Fig. 7b presents the segmented structure where columns are identified and displayed in different colors. It can be observed that, in accordance with the literature [13], the lower part of the

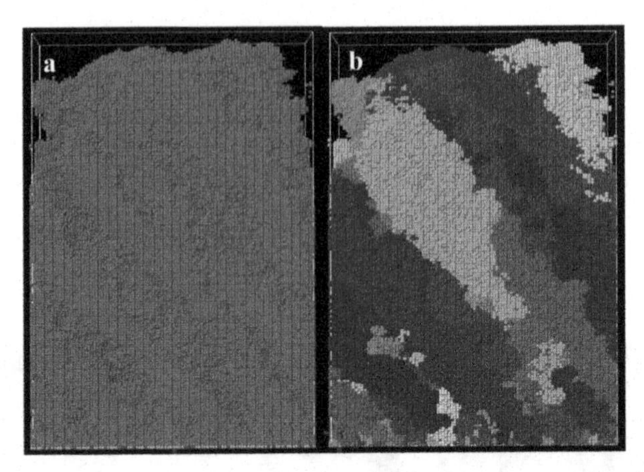

Fig. 7. Comparison of raw data from Nascam (a) and columnar separation algorithm (b)

structure mostly consists of densely packed small columns and the higher part are made of sparsely located large columns due to the shadowing effect [12].

3.2 Mesh Construction

The point cloud could not be used as it is for classic FEM mechanical calculation. Although point based mechanical simulations exist in the literature [18], it has been preferred for this study to work with FEM simulation software, which requires the construction of a mesh.

Concerning this point, several choices were made:

- Ignoring the internal porosities of the columns, which would be considered as computation artefacts from the probabilistic approach.
- Creating a mesh on an implicit surface set so that the surfaces created around adjacent points from different columns are in contact.
- Smoothing the mesh to improve the computability and to reduce the serrated patterns induced by the voxelization used in NASCAM.

The suppression of the internal porosities is made through the Ball Pivoting algorithm [19] in which a ball of given radius rolls on the edges, creating a triangle surface when it is in contact with three points. The created surface is then voxelized [20]. From these two steps, a new point cloud is obtained which presents no porosity nor holes smaller than the ball radius. Figure 8b presents a slice of the point cloud corresponding to the dark blue column on the data exported from NASCAM (Fig. 8a) with two added cavities, one on the surface and one in the bulk. Figure 8c corresponds to the slice of the point cloud obtained after applying a voxelization algorithm to the surface created by Ball Pivoting and filling the voxelized data. These two steps reduce the impact of surface cavities depending on the size of the ball chosen for the algorithm and guaranty the density of the columns.

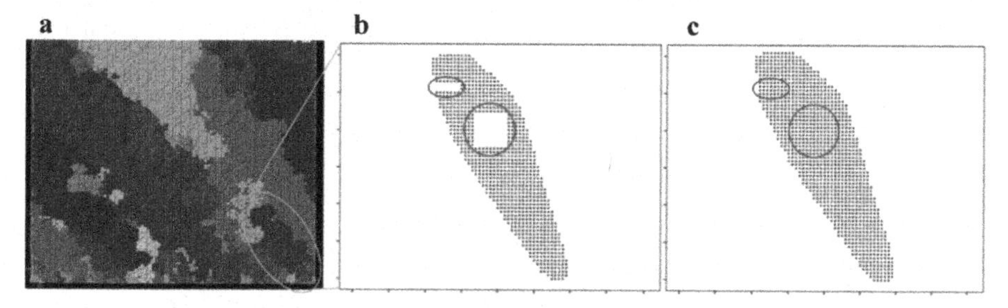

Fig. 8. Raw (a) and densified (b) point clouds representing a column

The implicit surface mesh is then constructed through the Marching Cubes algorithm [21, 22] applied to the voxelized point cloud. A cube is moved through the point cloud adding edges and surfaces by comparing the position of the points in the cube to a library of all possibilities. Thanks to this structure, this algorithm is fast and reliable on dense point clouds but very prone to errors in sparse ones. It is an efficient solution given the previous assumption to consider columns as dense entities.

Due to the fact that Marching Cube algorithm uses predetermined structure on a grid, the range of possible angles between surfaces are limited, sometime resulting in sharp angles, ill-suited for mechanical calculation. The Fig. 9a presents a slice of the surface constructed by Marching Cubes from the column highlighted in Fig. 8a. As expected, the grid parameters are obvious and the structure is very serrated along oblique angles.

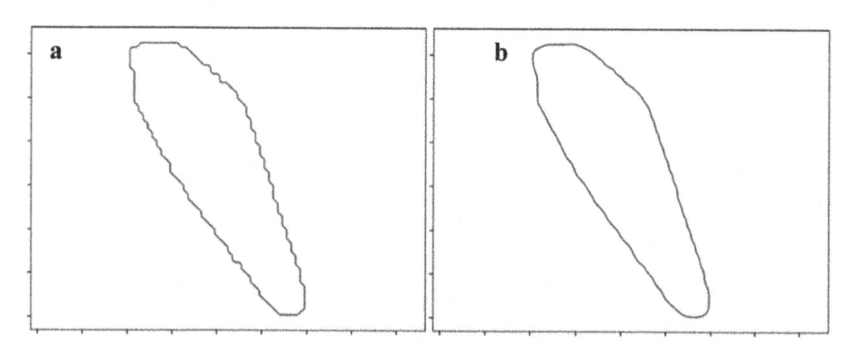

Fig. 9. Slice of the mesh created by the Marching Cubes algorithm from the dense point cloud presented in 7c corresponding to the column highlighted in 7a, before smoothing (a) and after smoothing (b)

Consequently, a smoothing is applied to reduce the impact of the grid. The Humphrey's Class Laplacian Smoothing [23] is used in order to produce a smooth mesh while at the same time limiting the volume shrinkage of the bulk. Nevertheless, a reduction of the global volume is still to be expected in the sharp boundaries. Volume measurement points to a loss of approximately 3% in volume. The smoothing has very little impact on the bulk.

As can be seen on Fig. 9b, the structure is mostly conserved with only a few rounder angles or straighter oblique edges replacing the serrated ones.

The 3D rendering of the column's mesh on the software Blender is presented in Fig. 10.

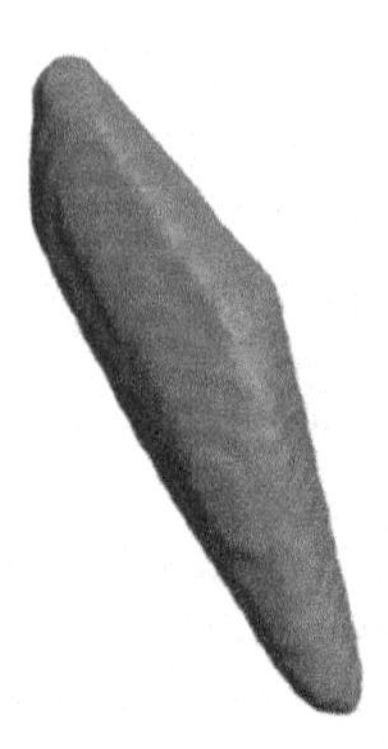

Fig. 10. 3D mesh of a simulated PVD column

At this point, it is possible to import the resulting mesh into a FEM calculator to compute the mechanical properties but the real interest comes from the possibility to create an assembly from all the columns of a given deposition simulation and run the calculation for the whole coating.

Figure 11 presents a side by side an SEM (Scanning Electron Microscopy) image of a coating and the assembly of the columns from the simulated world shown on Fig. 7 created by the application of the proposed 3D reconstruction method. The reconstructed structure is similar to the SEM images of a similar coating, both on the measured column tilt and on the shape of the columns, which involve multiple thin columns at the base and a few large columns at the top.

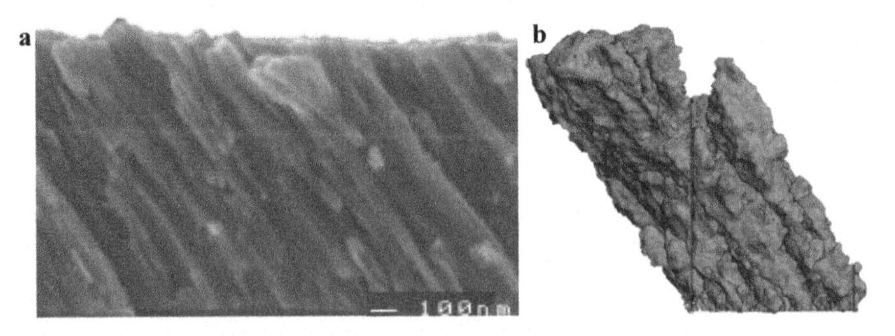

Fig. 11. (a) SEM image of a GLAD coating (b) 3D mesh of a simulated PVD world

4 Conclusion and Perspectives

This study presents a method devised from the numerical simulation of thin film growth to the construction of the FE mesh. This method has been successfully conducted in a two step-approach. The first step is a segmentation of the film in a set of columns while the second step consist in the creation of surface meshes for each column. The 3D representation of the current meshes shows consistency with experimental data obtained by SEM microscopy on both the tilt and the shape of the columns.

The process will be perfected in a batch of experimental and numerical investigations of the morphological (SEM) and mechanical (nano-indentation tests) properties of thin films.

The parameters expected to be reviewed this way are:

- The weight of the face, edge and vertex atoms in the definition of the attribution of a new deposited atom to the existing columns. The parameter presently used is a reduction proportional to the inverted distance as suggested by Besnard's previous work [17].
- The size of the mesh unit, which is currently based on the grid from NASCAM.
- The number of successive smoothing which may be increased or decreased depending on the computational capabilities of the machine.

Another potential upgrade factor for this study is the risk of collision between columns during the assembly part due to the shape modification from the smoothing.

Although the smoothing mostly induces a reduction of the shape in the boundaries, it may result in local expansions. In the case of this happening in the boundary between adjacent columns, it could lead to collisions and failure of the simulations, an additional part could be included at the assembly step to resolve the issue in those situations.

References

1. Martin, P.M.: Handbook of Deposition Technologies for Films and Coatings, third edition. Cambridge, Elsevier, p. 936 (2010). ISBN 978–0–8155–2031–3
2. Baptista, A.: On the physical vapour deposition (PVD): evolution of magnetron sputtering processes for industrial applications. Procedia Manuf. **17**, 746–757 (2018). https://doi.org/10.1016/j.promfg.2018.10.125
3. Stoessel, C.H.: Optical Thin Films and Coatings (Second Edition), Woodhead Publishing, p. 860 (2018). ISBN 978–0–08–102073–9
4. Ma, Y.: Materials and structure engineering by magnetron sputtering for advanced lithium batteries. Energy Storage Mater. **39**, 203–224 (2021). https://doi.org/10.1016/j.ensm.2021.04.012
5. Bobzin, K.: Enhanced PVD process control by online substrate temperature measurement. Surf. Coat. Technol. **354**, 383–389 (2018). https://doi.org/10.1016/j.surfcoat.2018.07.096
6. Ziegler, J.F.: "SRIM-2003", Nuclear Instruments and Methods in Physics Research B, vol. 219–220, pp. 1027–1036 (2004). https://doi.org/10.1016/j.nimb.2004.01.208
7. Möller, W.: Tridyn-binary collision simulation of atomic collisions and dynamic composition changes in solids. Comput. Phys. Commun. **51**, 355–368 (1998). https://doi.org/10.1016/0010-4655(88)90148-8
8. Aeken, K.V.: The metal flux from a rotating cylindrical magnetron: a Monte Carlo simulation. J. Phy. D: Appl. Phys. **41**, 205307 (2008). https://doi.org/10.1088/0022-3727/41/20/205307
9. Lucas, S., Moskovkin, P.: Simulation at high temperature of atomic deposition, islands coalescence, Ostwald and inverse Ostwald ripening with a general simple kinetic Monte Carlo code. Thin Solids Films **518**, 5355–5361 (2010). https://doi.org/10.1016/j.tsf.2010.04.064

10. Moskovkin, P., Lucas, S.: Computer simulations of the early stage growth of Ge clusters at elevated temperatures, on patterned Si substrate using the kinetic Monte Carlo method. Thin Solids Films **536**, 313–317 (2013). https://doi.org/10.1016/j.tsf.2013.03.031
11. Besnard, A., Martin, N., Carpentier, L.: Three-dimensional growth simulation: a study of substrate oriented films. Mater. Sci. Eng. **12**, 012011 (2010). https://doi.org/10.1088/1757-899X/12/1/012011
12. Hawkeye, M.M.: Glancing Angle Deposition of Thin Films: Engineering the Nanoscale, p. 320 Wiley, Hoboken (2013). ISBN 978–1–118–84756–5
13. Lacroix, B.: Nanostructure and physical properties control of indium tin oxide films prepared at room temperature through ion beam sputtering deposition at oblique angles. J. Phys. Chem. C **123**, 14036–14046 (2019). https://doi.org/10.1021/acs.jpcc.9b02885
14. Potočnik, J.: The influence of thickness on magnetic properties of nanostructured nickel thin films obtained by GLAD technique. Mater. Res. Bull. **84**, 455–461 (2016). https://doi.org/10.1016/j.materresbull.2016.08.044
15. Lintymer, J.: Influence of zigzag microstructure on mechanical and electrical properties of chromium multilayered thin films. Surf. Coat. Technol. **180–181**, 26–32 (2004). https://doi.org/10.1016/j.surfcoat.2003.10.027
16. Siad, A. : Etude numérique et expérimentale de la croissance de couches minces déposées par pulvérisation réactive (2016). tel-01418118
17. Besnard, A.: Three-dimensional growth simulation: A study of substrate oriented films. In: IOP Conference Series: Materials Science and Engineering, vol. 12 (2010). https://doi.org/10.1088/1757-899X/12/1/012011
18. Kudela, L.: Direct structural analysis of domains defined by point clouds. Comput. Methods Appl. Mech. Eng. **358**, 112581 (2019). https://doi.org/10.1016/j.cma.2019.112581
19. Bernardini, F.: The ball-pivoting algorithm for surface reconstruction. IEEE Trans. Vis. Comput. Graph. **5**, 349–359 (1999). https://doi.org/10.1109/2945.817351
20. Lai, S., Cheng, F.(: Voxelization of free-form solids represented by catmull-clark subdivision surfaces. In: Kim, M.-S., Shimada, K. (eds.) GMP 2006. LNCS, vol. 4077, pp. 595–601. Springer, Heidelberg (2006). https://doi.org/10.1007/11802914_45
21. Lorensen, W.: Marching cubes: a high resolution 3D surface construction algorithm. Comput. Graph. **21**, 163–169 (1987). https://doi.org/10.1145/37402.37422
22. Lewiner, T.: Efficient implementation of marching cubes' cases with topological guarantees. J. Graph. Tools **8**, 1–15 (2003). https://doi.org/10.1080/10867651.2003.10487582
23. Vollmer, J.: Improved laplacian smoothing of noisy surface meshes. Comput. Graph. Forum **18**, 131–138 (1999). https://doi.org/10.1111/1467-8659.00334

Geometric Coherence of a Digital Twin: A Discussion

Abdelhadi Lammini(✉), Romain Pinquié, and Gilles Foucault

Univ. Grenoble Alpes, CNRS, Grenoble INP, G-SCOP, 38000 Grenoble, France
abdelhadi.lammini@grenoble-inp.fr

Abstract. The topic of the digital twin has been widely investigated recently. However, few works superficially address its geometrical aspect. This paper attempts to clarify what it means to maintain the geometric coherence of a digital twin based on a taxonomy of geometric modifications. Each type of geometric change and existing capture solutions are reviewed. To conclude, the review encourages us to adopt an agile human-centric approach supported by extended reality technologies as a complement to automated supervision processes.

Keywords: Digital twin · Geometry · Coherence · Geometric change · Tracking

1 Introduction

The digital twin concept was introduced as the ideal Product Lifecycle Management (PLM) [1]. A digital twin is strongly intertwined with a real twin in a cyber-physical system. If the digital twin is misaligned with the physical twin, then the cyber-physical system is incapable of providing the services expected by the stakeholders. Although there is a bewildering array of studies related to the alignment of the behavioral properties between the digital and physical twins, proposals for maintaining the coherence of structural properties are virtually absent. For instance, the interface between the physical and digital twins in [2] uses the Industrial Internet of Things (IIoT) to update the behavioral properties without considering geometric changes.

This paper identifies the concept of geometric coherency of a cyber-physical system whose properties emerge from the intertwinement of a physical twin and digital twin. The first section concentrates on the definition of the concepts of the digital twin, digital twin coherence, and geometric coherence based on a literature review. The third section introduces a taxonomy of geometric modifications (GM) to be captured to maintain geometric coherency. The taxonomy supports the definition of each GM, the review of existing solutions to capture GM, and their limits according to some PLM phases.

2 Literature Review

The concept of the digital twin is much more influenced by ever-evolving technologies and business incentives than rational scientific knowledge. From an academic perspective, some definitions focus on the multi-physics and multi-scale characteristics [3],

Published by Springer Nature Switzerland AG 2023
F. Noël et al. (Eds.): PLM 2022, IFIP AICT 667, pp. 227–236, 2023.
https://doi.org/10.1007/978-3-031-25182-5_23

whereas others concentrate on the level of data integration and the real-time constraint [4]. The analysis of 23 papers on digital twins identified through a systematic literature review since 2010 on Google Scholar led us to a list of 10 key features. Figure 1 illustrates the number of papers containing each feature. A majority of papers consider a digital as an as-built representation of a real system and both are bi-directionally synchronized in near real-time. However, results show that only 13% of studied papers claimed the importance of geometry coherence as a key feature that will be discussed in this paper.

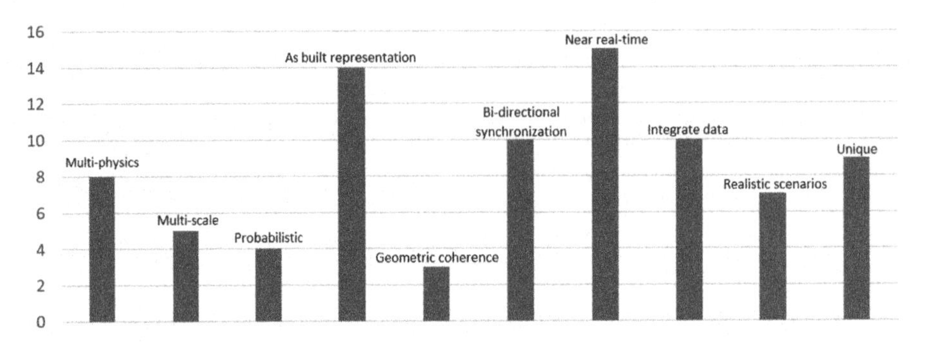

Fig. 1. Key features of 23 digital twin definitions since 2010

2.1 Digital Twin Coherence

Although the synchronization between the digital and physical twins is crucial, maintaining their coherence is more about finding the right level of fidelity for the pursued goals. However, studies ignore the geometric coherence and focus only on the behavior by developing data-based digital twins for predictive maintenance [5] and manufacturing [6]. The problem with data-based digital twins is that the geometry is either overlooked or considered perfect whereas dynamic behavioral properties' fidelity depends on it. Without capturing geometric changes, one cannot pretend to have a high-fidelity digital twin. For instance, by evaluating three scenarios, [5] shows that missing teeth in mechanical gears impact the wheel speed.

2.2 Digital Twin Geometric Coherence

Keeping the geometry coherent requires capturing geometric modifications before updating the geometry of the digital twin. To capture a geometric change, various tracking technologies exist (Image recognition, 3D scan, Infrared, Magnetic sensors, etc.). As far as the author is concerned, very few papers propose solutions to track geometric modifications of a digital twin [8, 9]. Scanning and reverse engineering solutions [7] provide a new geometric model, but they require a time-consuming process that does not need rapid synchronization.

In the next section, we propose to clarify the concept of geometric coherence by developing a taxonomy of geometric modifications and a review of tracking solutions.

3 Approach and Methodology

Despite the increasing research studies on digital twins, we still miss a formal definition of coherence, especially geometric coherence.

3.1 Digital Twin Coherence

Although the coherence of a digital twin is anecdotally discussed, Fig. 2 clarifies it by refining the service "Maintain the coherence" into solution-neutral technical functions:

- Acquire data: monitoring of the structural and behavioral properties of the digital and physical twins.
- Detect deviation: processing of measured data to identify a gap between the properties of the digital and physical twins.
- Make a corrective decision: once the deviation is detected and the geometric modification type is known, a corrective decision has to be made accordingly.
- Make a corrective action: once the corrective decision is made, corrective action should be executed by the digital twin, the physical twin, or both.

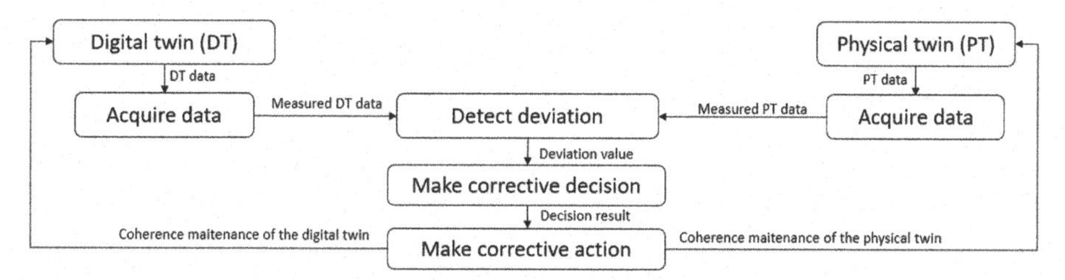

Fig. 2. Process for maintaining the coherence of a digital twin

The solutions to implement each function depend on the type of geometric modification and the context (business strategy, constraints, resources…). The next section will introduce a taxonomy of geometric modifications and the existing tracking solutions.

3.2 Digital Twin Geometric Coherence

Geometric coherence can be defined as the level of similarity between the geometry of the physical twin and the geometry of the digital twin. Having a high score of similarity requires capturing changes on the real twin and to update the digital twin or *vice-versa*. The proposed taxonomy (Fig. 3), which is inspired by the manipulation of a bill-of-materials, classifies the geometric modifications (GM) into three categories: 1) modification of the position and orientation of a part or an assembly (GM1), 2) addition, deletion, or replacement of a part or an assembly (GM2), 3) modification, addition or removal of material on a part (GM3).

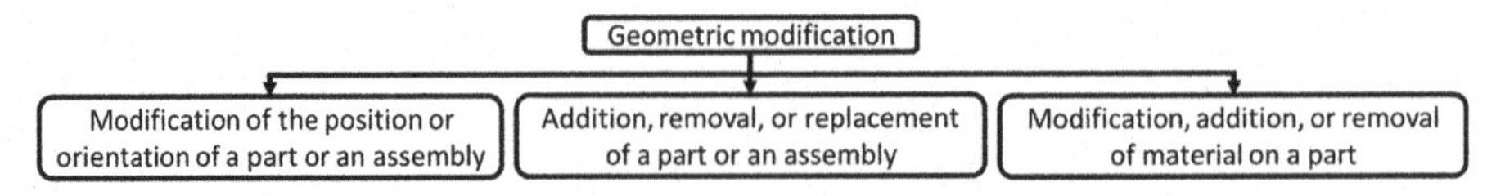

Fig. 3. Taxonomy of geometric modifications

a) Modification of the position and orientation of a part or an assembly

i) Definition

The position and orientation are two parameters to locate a rigid body in space, or relative to its parent. They can be presented either by a transformation matrix, or a position and quaternion vectors. Maintaining the geometric coherence (Fig. 2) when a position or orientation is modified consists of acquiring the new positions and orientations of the components that make up the digital and physical twins before calculating the deviation value and making a corrective decision to execute a corrective action on the real twin or the digital twin or both.

ii) Review of existing capture solutions

Getting an indoor position and orientation tracking system remains a bottleneck. Existing technologies to capture changes in position and orientation can be classified into three categories: 1) IIoT, 2) image processing, and 3) 3D scanning.

Concerning IIoT, [8] created a digital twin of a factory. A function of the digital twin is to detect the position of a forklift and assets using Bluetooth sensors. Despite the accuracy of the sensors (1 m), the time spent looking for the assets is reduced by 99.79% (from 960 min to 2 min).

Regarding image processing techniques, [9] developed a deep learning algorithm to estimate the orientation of a physical twin based on camera input. The estimated orientation value can serve to update the orientation of the digital twin with an error below 20%. Planned future work will add a depth camera to estimate the position too. Likewise, [10] developed two image-based measurement systems (an image-based distance measurement system and a parallel lines distance measurement system) to locate a mobile robot in an indoor environment by calculating its coordinates. The measurement errors range from 2.24 cm to 12.37 cm which could be improved by using a high-resolution camera.

Finally, [11] developed a non-contact measurement method using HoloLens. The HoloLens depth camera scans the area in 3D, and then the user's gaze and gestures select the measurement points that serve to calculate the distance, the area, and the volume between the points with a level of precision of 1 cm.

iii) Limits of existing capture solutions

[8] equipped multiple assets with sensors. The size of the assets is an important parameter, as tracking a small object (e.g. screw) can't be achieved here, since the attachment of the sensor to the object cannot be achieved.

In [9] results were accurate, but the authors face three main issues. First, full or partial symmetrical objects impact the pose estimation process. Second, the shape of some objects appears similar even if they are seen from different angles which causes the same impact on the process. Third, the objects with a dark light-absorbing surface engender a high error while estimating the rotation angles.

In [10] results depend on the camera quality as well as the lighting. In addition, the use of sensors to track the position is relative to the required precision level. For example, Bluetooth sensors can track assets in a factory with an approximate accuracy of 1 m [8] and GPS has a tracking accuracy between 20 and 30 cm [22].

Finally, the first version of the HoloLens offers a position calculation method with an error of about 1 cm [11], which is more precise than Bluetooth or GPS but it is still not enough in some situations such as the capture of tool wearing during CNC machining. Generally, the accuracy of 3D scans depends on the scanner's nature, however, the highest the precision level is, the longer the processing time.

b) Addition, removal, or replacement of a part or an assembly

i) Definition

There are multiple reasons for adding (e.g. adding new functionality to the system), changing (e.g. replacing a degraded component), or removing (e.g. removal of a functionality of the system) a component or an assembly of a physical twin. Consequently, the digital twin has to be updated. In some cases, the change occurs on the digital twin first leading to the update of the physical twin. In both cases, the process for maintaining the geometric coherence (Fig. 1) consists in capturing the bill of materials of the physical twin to compare it with the digital twin before making a corrective decision to take corrective action or not for adding, removing, or replacing a part or an assembly in the digital twin.

ii) Review of existing capture solutions

Recognizing and capturing the physical twin components is one way to control the presence or absence of objects, and it is mainly done by 3D scanning or image processing from video captures. For example, the Yolov5 object detection algorithm recognizes objects based on camera input [8] and mobile laser-scanned point clouds detect, capture and extract road objects [12]. Model-based comparison is one approach to detect 3D changes based on the conversion of the point cloud into digital surface models (DSMs) that are compared to detect deviations. Change detection relies on the classification of objects. For example, [13] detects the change areas by comparing two DSMs coming from multi-temporal LiDAR data using classification to estimate the percentage of pixels that have been converted from ground to building and vegetation.

As part of the deviation detection process, [14] has compared laser scan point-cloud data to CAD data as an inspection of industrial plant systems. Combing the inspection process tool with the recognition and extraction capabilities previously mentioned can provide a system to capture a real object to be compared with the digital twin CAD model to indicate the differences between the digital and physical twins. There also exists a set of common solutions to capture the presence of an object such as LED-based photoelectric sensors that emit light.

As far as the author is concerned, no studies have considered object replacement detection.

iii) Limits of existing capture solutions

The accuracy and the time process of the model-based comparison approach depend on the scan precision as well as the classification process. Thus, employing scanning technology to capture objects, then comparing them to a CAD model is a heavy and time-consuming process. Moreover, when small objects are drowned in a large scene, the detection and extraction process is challenging [12]. The utilization of LED-based photoelectric sensors is limited in the detection context, but it can be used to indicate in near real-time the removal of a fixed component.

c) Modification, addition, or removal of material on a part

i) Definition

A component undergoes several material changes. Some are made on purpose (e.g. additive manufacturing) to adapt the functionality of the component to a specific situation, others are usually undesired and are due to external factors (e.g. corrosion, cracking, creeping, etc.). Shape modifications are of three types [14]:

- Formative: "*The desired shape is acquired by application of pressure to a body of raw material, examples: forging, bending, casting, injection molding, the compaction of green bodies in conventional powder metallurgy or ceramic processing, etc.*"
- Subtractive: "*The desired shape is acquired by selective removal of material, examples: milling, turning, drilling, EDM, etc.*"
- Additive: "*The desired shape is acquired by successive addition of material*"

  By analogy, the undesired material changes can be brought into three categories:
- Undesired material modification: the undesired shape is formed by an external effort applied to the component body, resulting in a material modification (e.g. crack and delamination. The shape modification is either plastic or elastic).
- Undesired material removal: The undesired shape is acquired by external efforts, resulting in material removal (e.g. corrosion, erosion).
- Undesired material addition: The undesired shape is acquired by external factors, resulting in the addition of material (e.g. dental calculus).

ii) Review of existing capture solutions

Whatever the type of geometric deformation (formative, subtractive, or additive), 3D change detection based on a second multi-temporal point cloud approach is the point-based comparison based [15]. It uses LiDAR data and an adaptative threshold to highlight a red zone in the point cloud indicating the modified zone. The accuracy of this zone depends on three factors: registration errors, thresholds, and a neighboring number of 3D change detection. Experiments for evaluating the influence of each factor show that the adaptive threshold approach results have higher accuracy (95%) as the number of neighboring points increases and when the registration error is better than 0.041 m. The overall performance of change detection is measured by four variables: correctness, completeness, quality, and F1 measure.

To detect geometric modifications due to formative shaping, one can use a sensor to capture in 3D the shape of cables after bending in near real-time [23]. Moreover, a review of 61 deep learning-based crack detection approaches points out that semantic segmentation-based provides high accuracy (from 79.99% to 96.79%) in terms of crack detection and localization [16]. Finally, the demonstration of the linear relationship between crack propagation and the electromagnetic signal coming out of a sensor provides a low-cost solution for crack detection in structures [17]. The sensor was able to track small changes in the structure i.e. cracks widths below 1 mm efficiently.

To detect geometric modifications due to the removal of material, one can exploit the visual attribute of corrosion: colors with a tracking algorithm based on images, or textures with deep learning algorithms [18]. Rayleigh ultrasonic wavefield by laser probing was used to detect a non-visible drilled hole on a subsurface of aluminum plate [19]. The measurement accuracy is not examined under different circumstances, but the inspected hole of 0.5 mm in diameter and 5 mm in depth was successfully identified.

Finally, to capture geometric changes due to the addition of material, deep learning solutions recognize the overlap welding, that is, the bulge of weld metal beyond the root that looks like a circle that extends out of the component body [19]. Vibration tests can evaluate the limescale buildup on the walls of the pipes, the status of the pipe section, and the rate of deposition [20].

This review shows the extensive use of 3D scans, deep and machine learning algorithms as well as non-destructive techniques for material change detection.

iii) Limits of existing capture solutions

Detecting changes in a point cloud density with a model-based comparison threshold approach has high data accuracy [15] but it is a complex and time-consuming process. Regarding the tracking of geometric changes of cables and beams, the maximum length of the sensor is 115 cm. The machine learning recognition and detection algorithms showed an important accuracy result, but it is limited to the outside visible side of the component, unlike the method such as non-destructive tests which can capture changes inside a body. Moreover, the trained algorithms work mostly for the formative, subtractive, and additive shaping as well as the common undesired material changes where a supervision system is installed, but they are not able to detect an undesired change. In other words, maintaining the geometric coherence of a digital twin requires training the learning algorithms with massive datasets representing all kinds of changes to avoid overfitting.

4 Discussion

The literature review points out the lack of interest in maintaining the geometric coherence of a digital twin. We assume that this is due to the difficulties of automatically capturing geometric changes compared to the capture and update of behavioral properties with IIoT technologies.

Tracking the geometric modifications of a system may be fundamental in different PLM phases. For instance, considering the geometric deviations during the maintenance phase would improve the quality of the digital twin's maintenance predictions, as "*In addition to monitoring the deviations between collected data and expected values, a digital twin can interpret collected data from other perspectives. Comparison between digital twin simulated data and collected data can help determine the failure mode. Digital twin provides a high-fidelity accurate model and keeps updating through the product lifecycle.*" [21]. Having an up-to-date geometric definition of a digital twin also helps to improve the design of multibody system dynamics, especially control algorithms. It is also tremendously important regarding the dismantling phase in critical fields like the nuclear energy industry or aerospace to have a clear understanding of the latest product configuration that will facilitate the recycling process.

A complex system can contain so many components (e.g. sensors, actuators, gears, etc.) that tracking all geometric changes is important but idealistic. IIoT can track the positions and orientations of instrumented objects at different scales (factory, machine, etc.) as well as some specific deformations such as crack and wear detection. However, having an up-to-date digital twin means setting up sensors on all parts or at least knowing in advance all the critical parts to monitor. Sometimes components are not easily accessible, reaching them requires deep investigation inside the system, which makes image capturing and scanning difficult. Image processing techniques also rely upon a supervision system to integrate within an operational environment which can be inappropriate. Furthermore, the trained algorithm may potentially fail to recognize uncommon geometric changes (e.g. buckling) because of the insufficient volume of training and testing data to cover all potential scenarios. Regarding the 3D scan solutions, the capture of geometric deformations is highly accurate but the reverse engineering process is cumbersome.

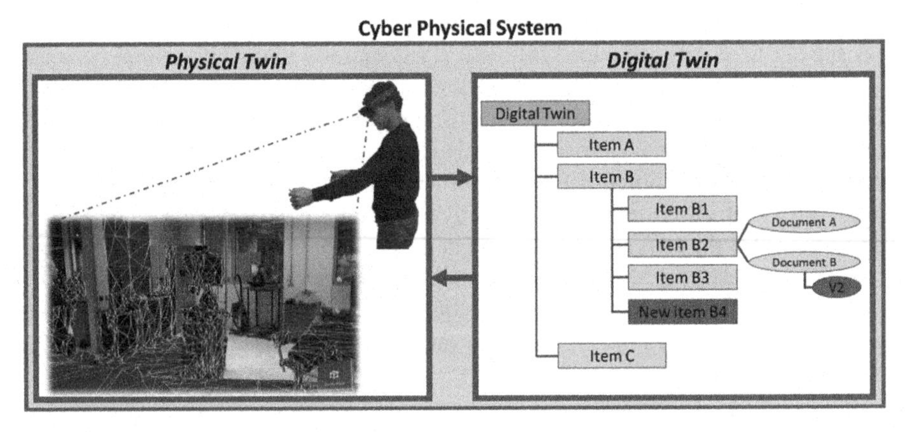

Fig. 4. Human-in-the-loop maintenance of geometric coherence of a digital twin

The automatic supervision and update of geometric changes are still, and will always remain, partial. This observation encourages us to promote an agile human-centric approach supported by extended reality technologies as a complement to automated supervision processes. The human ability to detect and communicate common and uncommon deviations remains crucial to maintain an up-to-date digital twin. As a complement to automated supervision processes, the review encourages us to investigate an agile human-in-the-loop approach supported by extended reality technologies that enable us to capture geometric changes and synchronize them with the bill-of-materials stored in a PLM software (Fig. 4).

5 Conclusion

The main chara of a digital twin is its synarecteristicronization with a physical twin. Studies aiming at maintaining the coherence between both twins concentrate on the behavioral properties using IIoT without considering components' geometry modifications, and components add/removal in the whole product structure. This paper attempts to clarify the concept of geometric coherence with a taxonomy of modifications. A review of geometric deviations capturing technologies shows that existing solutions require heavy automatic supervision systems (full instrumentation with sensors or camera tracking with image processing) which concentrate on intended geometric changes. Finally, we discuss the human role in the process by arguing that augmented reality, especially spatial mapping, synchronized with PLM software could bring a benefit to automatic supervision systems.

Acknowledgment. This work was supported by the French National Research Agency, grants ANR-20-CE10–0002 (Coherence4D project) and ANR-21-ESRE-0030 (CONTINUUM project).

References

1. Grieves, M.: Product Lifecycle Management. McGraw-Hill, Nova Iorque (2006)
2. Julien, N., Martin, E.: How to characterize a digital twin: a usage-driven classification. IFAC-PapersOnLine **54**(1), 894–899 (2021)
3. Aivaliotis, P., Georgoulias, K., Arkouli, Z., Makris, S.: Methodology for enabling digital twin using advanced physics-based modelling in predictive maintenance. Procedia CIRP **81**, 417–422 (2019). https://doi.org/10.1016/J.PROCIR.2019.03.072
4. Kritzinger, W., Karner, M., Traar, G., Henjes, J., Sihn, W.: Digital twin in manufacturing: a categorical literature review and classification. IFAC-PapersOnLine **51**(11), 1016–1022 (2018). https://doi.org/10.1016/j.ifacol.2018.08.474
5. Magargle, R., et al.: A simulation-based digital twin for model-driven health monitoring and predictive maintenance of an automotive braking system. In: Proceedings of the 12th International Modelica Conference, Prague, Czech Republic, 15–17 May 2017, no 132, pp. 35-46 (2017). https://doi.org/10.3384/ecp1713235
6. Ghosh, A.K., Ullah, A.S., Kubo, A.: Hidden markov model-based digital twin construction for futuristic manufacturing systems. AI EDAM **33**(3), 317–331 (2019). https://doi.org/10.1017/S089006041900012X
7. ALNuaimi, M.S., Alkaabi, E.D., Ziout, A.: The use of reverse engineering in a comparison between engineered and solid hardwood. Proc. ICETIT (2019)
8. Vijayakumar, K., Dhanasekaran, C., Pugazhenthi, R., Sivaganesan, S.: Digital twin for factory system simulation. Int. J. Recent Technol. Eng. **8**(1), 63–68 (2019)
9. Sundby, T., Graham, J.M., Rasheed, A., Tabib, M., San, O.: Geometric change detection digital twins. Digital **1**(2), 111–129 (2021). https://doi.org/10.3390/digital1020009
10. Li, I.-H., Chen, M.-C., Wang, W.-Y., Su, S.-F., Lai, T.-W.: Mobile robot self-localization system using single webcam distance measurement technology in indoor environments. Sensors **14**(2), 2089–2109 (2014). https://doi.org/10.3390/s140202089
11. Huang, J., Yang, B., Chen, J.: A non-contact measurement method based on HoloLens. Int. J. Performability Eng. **14**(1), 144 (2018). https://doi.org/10.23940/ijpe.18.01.p15.144150

12. Ma, L., Li, Y., Li, J., Wang, C., Wang, R., Chapman, M.A.: Mobile laser scanned point clouds for road object detection and extraction: a review. Remote Sens. **10**(10), 1531 (2018). https://doi.org/10.3390/rs10101531
13. Chaabouni-Chouayakh, H., Krauss, T., d'Angelo, P., Reinartz, P.: 3D change detection inside urban areas using different digital surface models (2010)
14. Nguyen, C.H.P., Choi, Y.: Parametric comparing for local inspection of industrial plants by using as-built model acquired from laser scan data. Comput. Aided Des. Appl. **15**(2), 238–246 (2018). https://doi.org/10.1080/16864360.2017.1375675
15. Liu, D., Li, D., Wang, M., Wang, Z.: 3D change detection using adaptive thresholds based on local point cloud density. ISPRS Int. J. Geo-Inf. **10**(3), 127 (2021). https://doi.org/10.3390/ijgi10030127
16. Hamishebahar, Y., Guan, H., So, S., Jo, J.: A comprehensive review of deep learning based crack detection approaches. Appl. Sci. **12**(3), 1374 (2022). https://doi.org/10.3390/app12031374
17. Gkantou, M., Muradov, M., Kamaris, G.S., Hashim, K., Atherton, W., Kot, P.: Novel electromagnetic sensors embedded in reinforced concrete beams for crack detection. Sensors **19**(23), 5175 (2019). https://doi.org/10.3390/s19235175
18. Matthaiou, A., Papalambrou, G., Samuelides, M.S.: Corrosion detection with computer vision and deep learning. In: Developments in the Analysis and Design of Marine Structures: Proceedings of the 8th International Conference on Marine Structures (MARSTRUCT 2021, 7-9 June 2021, Trondheim, Norway), p. 289 (2021)
19. Imano, K.: Detection of drilled hole on subsurface of aluminum plate with rayleigh ultrasonic wave field by laser probing. Sensors Mater. **32**(2), 791–797 (2020). https://doi.org/10.18494/SAM.2020.2659
20. Kocsis, D.: Modeling and vibration analysis of lime scale deposition in geothermal pipe. https://doi.org/10.30638/EEMJ.2014.315
21. Liu, M., Fang, S., Dong, H., Xu, C.: Review of digital twin about concepts, technologies, and industrial applications. J. Manufact. Syst. **58**, 346–361 (2021). https://doi.org/10.1016/B978-0-12-823657-4.00010-5
22. Shade, S., Madhani, P.H.: Android GNSS measurements - inside the BCM47755. https://doi.org/10.33012/2018.16001
23. https://shapesensing.com/. Accessed 25 Mar 2002. Visited 01 June 2022

Developing a Technology Readiness Level Template for Model-Based Design Methods and Tools in a Collaborative Environment

Roman F. Bastidas Santacruz[1], Brendan P. Sullivan[1], Sergio Terzi[1], and Claudio Sassanelli[2](✉)

[1] Department of Management, Economics and Industrial Engineering, Politecnico di Milano, Piazza Leonardo da Vinci 32, 20133 Milan, Italy
{romanfelipe.bastidas,brendan.sullivan,sergio.terzi}@polimi.it
[2] Department of Mechanics, Mathematics and Management, Politecnico di Bari, Via Orabona 4, 70125 Bari, Italy
claudio.sassanelli@poliba.it

Abstract. Currently, companies in the manufacturing field are experiencing the need to go digital, compelled by rising competitivity and efficiency requirements. Digitalization implies the development and implementation of complex systems in manufacturing plants as well as in the delivery of product-service systems and solutions, asking both for the adoption of Model Based Design (MBD) tools and methods. In this context, the assessment of suitability of MBD tools is vital for the companies that try to digitalise their operations. Due to the high relevance that this characteristic has for users and providers, a vital part of the implementation process is assesing the level of development or maturity of the tools. This paper presents and proposes a Technology Readiness Level (TRL) template developed in the HUBCAP project. This template aims to support MBD tools providers (guiding them in the description of the tool added on the platform), the platform management (easing governance tasks) and its users (clarifying the tool description for them) along the upload, update and control processes of the MBD tools in the collaborative platform.

Keywords: Technology readiness level · Model-based design · Cyber-physical system · Collaboration platform · Digital innovation hub

1 Introduction

Manufacturing companies produce goods in large scale to meet customer requirements and to manage to stay competitive in the market. In this context, they must constantly adopt new computation technologies and develop innovative solutions in their production lines [1]. This constant evolving context increases both plants and product/service's complexity offered by the companies [2]. Commonly, asset lifecycle is composed by different phases, starting from conception and planning, going through design and engineering,

Published by Springer Nature Switzerland AG 2023
F. Noël et al. (Eds.): PLM 2022, IFIP AICT 667, pp. 237–249, 2023.
https://doi.org/10.1007/978-3-031-25182-5_24

and then construction, validation, verification, and commissioning [3]. Through each one of these phases, it is highly relevant for the vendor and the asset user to clearly identify the development stage of the same. This is even more important while introducing in their solutions technologies belonging to the Industry 4.0 domain (as Cyber-Physical Systems (CPS)), vital for the constant development of the manufacturers production systems [4]. The development phase can be empowered by the adoption of Model-Based Design (MBD) approaches to which each company can have access to, depending on their level of expertise, flanked by related tailored tools [5]. Small and Medium Enterprises (SMEs) face several barriers when adopting digitalization [6]. Some of the areas where these barriers are most challenging are in the adoption of model-based driven models and tools (MBD assets) [7]. SMEs usually lack of expertise in the implementation of MBD assets due to the high investment costs that they imply (e.g., licensing, development, training, etc.). To encourage the technological transformation, ease the utilization of MBD and address cross-cutting challenges along European SMEs, the HUBCAP (Digital Innovation HUBs and CollAborative Platform for cyber-physical systems) project was funded by the European Commission (EC) under the Smart Anything Everywhere initiative. The objective of HUBCAP is to provide a one-stop-shop for European SMEs that intend to adopt CPS through MBD assets (through different techniques assimulation, model checking, contract-based analysis, model-based safety assessment). The project wants to create a growing and sustainable network where SMEs can undertake experiments, seek investment, access expertise and training, and network with other companies and institutions with support of specialized DIHs. In this perspective, the awareness on the level of development of the tools deployed in the platform becomes highly important for both users and asset providers. For the users, it represents more complete and accurate information to better understand the suitability of the tool to their specific purposes and applications. On the other hand, the asset providers have to be aware on the next steps of development they should follow to offer a complete product in the platform. To this aim, the well-known and consolidated Technology Readiness Level (TRL) models coming from NASA [8], and its modified EC [9] version, were considered. Due to the need to adapt these standards to the main assets of the HUBCAP platform, i.e. MBD tools, the aim of this paper is to present and propose the TRL template developed in the HUBCAP project to support the uploading, updating, and control of the methods and tools added to the collaborative platform by the solution providers. The template can play a key role in supporting the lifecycle management of manufacturers' assets, fostering the exploitation of the related data and knowledge since the development phase of the CPS technologies employed. The paper is structured as follows. Section 2 presents the TRL standards. Section 3 shows the research methodology needed to develop such an adaptation of TRL standards to the MBD domain (shown in Sect. 4). Section 5 provides concluding remarks.

2 The Research Context: Technology Readiness Level

The metric of TRL was defined as a relevant parameter to be included in HUBCAP with the intention to give its users and stakeholders a better understanding of the level of development of the MBD assets offered on the platform. More specifically, it is intended to be implemented in the tools offered as are the digital assets for which its level of development can be valutated.In the context of NASA [8], the model was conceived in 1974 with 7 levels intended to assess the level of maturity of technologies to be utilized in a specific field, space missions. In 1990, the model gained recognition and was extended to 9 levels, beings broadly utilized in NASA [10], including relevant parameters for NASA technologies such as (i) test and demonstration, and (ii) proven operation. In 1995, [8] offered a more detailed definition to each level and from this year until 2006, the model gained recognition internationally. Based on the white paper published by NASA to define the TRL metric, the model can be adapted to different environments as long as five main steps of the development process of the application are always considered [8]: 1. "Basic" Research in the technology or concept, 2. Focused technology development addressing specific technologies or concepts for an application, 3. Technological development of the application, 4. System development, 5. System "launch". With time, TRL gained recognition in other fields of different countries and industries being utilized in some cases as-is, while in others it has been adapted to meet specific requirements [11]. The European Space Agency (ESA) was one of the firsts in implementing the TRL model as-is but adding some additional levels with additional standards [12]. Also the Department of Energy (DoE) implemented the nine level model with small variations in the initial and in the last levels [13]. The last level of the DoE's TRL specifies that the technology must be proven in a "full range" of spected conditions. In other cases, the TRL was modified to assess readiness of not only technology, but the process of incremental innovation. In [14], the Innovation Readiness Level is defined. In the manufacturing industry, this metric gained high levels of attention, until being specifically addressed by the EC who proposed a new version in the context of EU Horizon programs with some slight differences from the original one from NASA [9]. Table 1 shows the two models and their main differences in each of the 9 steps.

Table 1. TRL models proposed by NASA and European Commission

TRL	NASA	EU
1	Basic principles observed and reported	Basic principles observed
2	Technology concept and/or application formulated	Technology concept formulated
3	Analytical and experimental critical function and/or characteristic proof-of-concept	Experimental proof of concept

(continued)

Table 1. (*continued*)

TRL	NASA	EU
4	Component and breadboard validation in laboratory environment	Technology validated in laboratory
5	Component and/or breadboard validation in relevant environment	Technology validated in relevant environment (industrially relevant environment in the case of key enabling technologies)
6	System/subsystem model or prototype demonstration in a relevant environment (ground or space)	Technology demonstrated in relevant environment (industrially relevant environment in the case of key enabling technologies)
7	System prototype demonstration in a space environment	System prototype demonstration in operational environment
8	Actual system completed and "flight qualified" through test and demonstration (ground or space)	System complete and qualified
9	Actual system "flight proven" through successful mission operations	Actual system proven in operational environment (competitive manufacturing in the case of key enabling technologies or in space)

In the EU Horizon programs context [9], the TRLs were defined with the intention to narrow down the topics of the H2020. Additionally, the definition of this metric, gives an idea on the next steps that the project must follow and an initial iteration on the distance of the technology from the market. In the context of HUBCAP, the definition of the TRLs became vital since it can support stakeholders to define the current development step of the tool in discussion, identify its possible benefits and limitations, and give a broad idea on the further steps that it will have.

3 Research Method

In this section, the process of development of a TRL template adapted for HUBCAP's MBD assets is described, based on the initial NASA [8] and EC [9] models. The research process is composed by 5 main steps (Fig. 1):

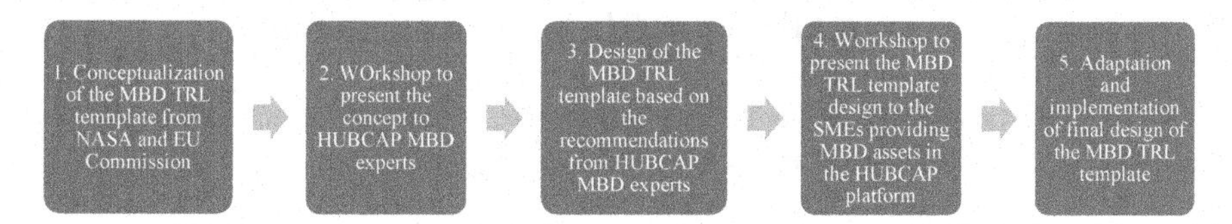

Fig. 1. TRL adaptation process for HUBCAP context

1. the conceptualization of the HUBCAP TRL template was done starting from the NASA and EC versions, adding three parameters to each level of the template (a. considerations in the stage and some examples, b. what question should the tool provider answer, c. expected output documentation that justifies the selected TRL).
2. To perform additional activities to verify the suitability of the TRL with the MBD assets provided in HUBCAP, a workshop to receive feedback about the template concept was conducted involving MBD experts belonging to the HUBCAP project.
3. The feedback gathered from the workshop with the experts involved can be wrapped up in Table 2 reporting the relevant points of improvement:

Table 2. Feedback gathered from meetings and countermeasures for HUBCAP TRL

N	Feedback gathered during meetings	Countermeasure
1	Inclusion of output for each level of the TRL was requested by the partners. It is highly relevance as the platform counts with	The output for each TRL level was defined as an additional parameter to the TRL model. Nevertheless, as it can have high variations from one provider to another, it was defined as a general output
2	Corrections and clarifications of the definitions of each level. More specifically	The column "Development actions taken" were expanded to clarify them and make it clearer each TRL level to the tool providers
3	The level 9 could lead to misunderstandings as it was not clear what is the last level of development that a technology can have	The level 9 was better defined and the context of the HUBCAP tools better explained. It was specified that the last level of the HUBCAP TRL model is reached when a tool is for first time successfully deployed in a real application for a customer

4. A further workshop was done to explain the MBD TRL template to the SMEs providing MBD assets on the HUBCAP platform (the users of the TRL template). Some additional corrections (description and levels of specifications) were gathered.
5. The new feedback were implemented first in the template itself and then in the platform on which it had to be embedded.

4 Results: TRL Template for MBD Methods and Tools

The MBD TRL proposed is structured in 9 levels, classified in three main phases, as the traditional one. Phase 1 (MBD asset concept definition (levels from 1 to 4) focuses on the definition of the tool concept and its validation.Phase 2 (MBD asset concept validation and test (levels from 5 to 8)) focuses on piloting activities and demonstration. Phase 3 (MBD asset deployment (level 9)) is reached after the last "bug fixing" aspects. Table 3 specifies the name of the level, the development action taken (i.e. the steps that the MBD tool provider have taken in order to reach the TRL), what to consider in this stage and examples (in this column is mentioned the main thing to consider when evaluating the tool under the TRL including also an example to guide the provider), question for tool provider (i.e. a question for the provider to assess TRL), output (possible outputs that the provider should have when the TRL is reached).

5 Discussion

The adoption of the MBD TRL template defined for HUBCAP brought successful results as each provider was able to better define their current level of development, provide a short description of the asset state of development, and additionally, generate further documentation that can be accessed and exploited by their stakeholders. However, since the TRL template is a new platform feature, not all the MBD asset providers have been able to offer a complete documentation of their current TRL yet. From the experience that each asset provider had in the definition of the asset TRL, the time required to perform the assessment depends on the availability of information that they actually had internally. In most of the cases the tool providers were able to identify their TRL but were not able to offer the right documentation that justifies their selection. For this reason, additional time were required by them to develop the new documentation required. For the asset providers that had available the documentation, the time required for the assessment varied depending on the completeness of the information that each MBD asset provider has available.

The providers that had complete information that justifies the TRL of their tool required approximately 1 h to identify and justify it in the platform. The providers that did not have available documentation in regard to their TRL required additional time, this due to the fact that they needed to prepare from scratch the documentation to justify it.

Table 3. MBD TRL template

Phase	TRL	Development stages	Development action taken	What to consider in this stage and examples	What question should the tool provider answer?	Output
Phase 1: Tool concept research and validation	1	Basic principles observed and reported	Scientific research translated into applied research and development. Most relevant topics related with the tool end goal were investigated and applied to the tool methodology. This includes physical principles or basic topics over which the intended tool will be built on	The basic principles of the tool are researched. Ex: The analysis of the different types of the battery control methodologies	[Is the tool basic principle already researched?]	• Documents about basic physical principles or basic topics over which the intended tool will be built on • Further information about how scientific research (with related references) has been translated into applied research and development
	2	Tool concept and/or tool formulation	The tool is discussed and defined around the basic principles defined in the previous level. The previous topics converge in the definition of the tool objective, being merely speculative: without experimental proof or detailed analysis that supports the conjecture	Practical applications of the basic principles are invented or identified. Ex: the potential application of a certain control methodology is explored, like its utilization for electric trucks	[The applicability of the basic principles has been proposed?]	• Document reporting the tool objective connected with the physical principles and basic topics found in the previous stage

(*continued*)

Table 3. (*continued*)

Phase	TRL	Development stages	Development action taken	What to consider in this stage and examples	What question should the tool provider answer?	Output
	3	Analytical and experimental critical function and/or characteristic proof of concept	Active R&D starts. Starting from the basic principles investigated and the definition of the tool, an appropriate context of the tools is defined, and studies are performed to validate that the analytical predictions are correct. This is translated into a "Proof of concept" of the tool concept	A proof of the concept over which the tool is build had been developed. Ex: If the tool is based on the implementation of a battery management system (BMS) for electric trucks, then this stage is achieved when it has been theoretically proven or modelled that the theoretical model behind the tool is feasible for the application. Ex.: the theory behind the battery control has been proved	[Is the theory behind the tool and the application already theoretically proven for this type of applications?]	• Proof of concept of the tool • Document reporting the validation of the concept of the tool (feasibility analysis of tool/model) • Document reporting requirements of potential real applications of the tool
	4	Tool concept validation in laboratory digital environment	A validation in a digital environment must support the concept formulated in the previous step being consistent with the requirements of potential real applications	This validation can be done in a digital environment without considering all variables as in real application cases. Ex.: Testing the tool with a virtual battery for a vehicle or truck	[Is the Tool already tested in a digital environment proving that it is feasible to be applied?]	• Report of the validation conducted in a digital environment

(*continued*)

Table 3. *(continued)*

Phase	TRL	Development stages	Development action taken	What to consider in this stage and examples	What question should the tool provider answer?	Output
Phase 2: Tool pilot line and demonstration projects	5	Tool concept validation in relevant environment	The basic tool elements must be integrated with reasonably realistic environments. The demonstration might represent an actual system application, or it might be similar to the planned application, but using the same technologies	The validation in this stage must be done by implementing the tool in a similar environment than the application case. Ex.: Testing the model for a truck battery control with a real truck or a similar vehicle	[Is the tool already tested and its results were taken to a similar environment to the real environment?]	• Tool prototype (Tool concept validated) • Report of the validation of the tool concept in a relevant environment, showing conformity with the requirements previously defined
	6	Tool prototype demonstration in a relevant environment	A model or prototype system would be tested in a relevant environment. At this level, considering the NASA definition, if the only "relevant environment" is the space, then the model/prototype must be demonstrated in space. The demonstration might represent an actual system application, or it might only be similar to the planned application but using the same technologies. At this level, several-to-many new technologies might be integrated into the demonstration	In this stage all the variables of the application are considered. Ex: In this stage, not only the model behind the tool must be tested, but also including additional variables such as movement, time, etc. which could affect the model of the tool	[The previous test had been performed including external environment variables?]	• Report of the demonstration of the tool prototype in a relevant environment considering additional variables of the application • Validation of integrability with new technologies

(continued)

Table 3. (*continued*)

Phase	TRL	Development stages	Development action taken	What to consider in this stage and examples	What question should the tool provider answer?	Output
	7	Tool system prototype demonstration in an operation environment	T his TRL is not always implemented. In this case, the prototype should be near or at the scale of the planned operational system and the demonstration must take place in the specific conditions of the real operational environment of the tool. The driving purposes for achieving this level of maturity are to assure system engineering and development management confidence (more than for purposes of technology R&D). Therefore, the demonstration must be of a prototype of that application. Not all technologies in all systems will go to this level. TRL 7 would normally only be performed in cases where the tool and/or subsystem application is mission critical and relatively high risk	This stage is important in case that the tool is related with safety or critical risks for the customer company. Ex.: If the battery control system includes the modelling of the support of security systems in the truck, the experiment must be performed including all the variables that can influence this	[In case the tool has critical risk or security concerns, it has been tested in a real environment with the variables that affect these risks?]	• Tool system prototype (Proven in risk environments) • Document reporting the demonstration of the tool/system prototype in an operational environment granting that critical risks and security concerns had been considered

(*continued*)

Table 3. (*continued*)

Phase	TRL	Development stages	Development action taken	What to consider in this stage and examples	What question should the tool provider answer?	Output
	8	Actual tool system completed with a test made in an environment similar to potential customer applications	In this stage, the tool had been already tested in applications alike to the customer applications. In other words, the tool has been tested in a situation or application similar to the one that the potential customers would have	The tool has to be tested in a real environment. Ex.: Tested in a set of truck batteries	[Is the tool already tested in a real environment as the ones that customers would use it on?]	• Functional tool • Document reporting the test made in a similar environment to the one that the potential customers would have
Phase 3: Tool (initial) deployment	9	Actual tool system proven to have successful application for customer needs	All the tools that are being implemented by companies are in TRL 9. This stage is reached after the end of the last "bug fixing" aspects of the tool development	The tool had been successfully implemented in real environments. Ex: The tool for battery control simulation had been tested in the application of an electric truck of a company with successful results in meeting customer needs	[Is the tool already proven with customers and showed that it successfully meets their requirements?]	• Report of tool deployment in a customer application environment

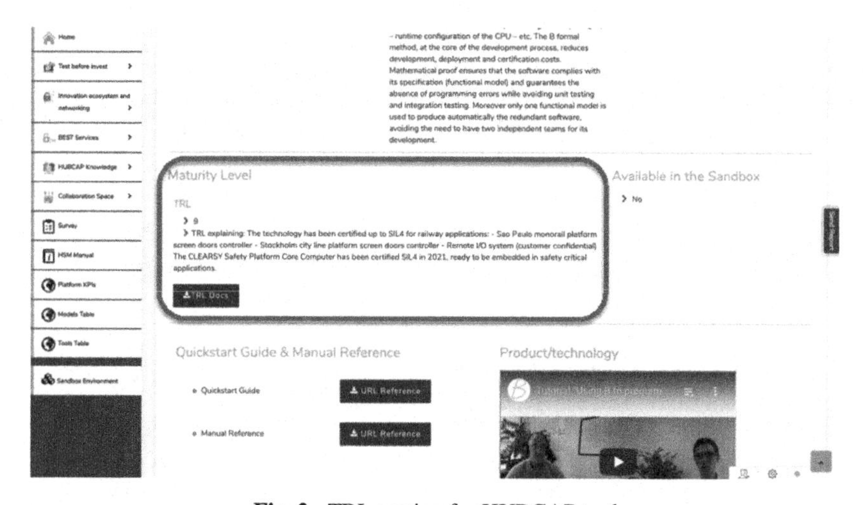

Fig. 2. TRL section for HUBCAP tool

To better support the adoption and implementation of the TRL template in the HUBCAP platform, additional input boxes for each one of the assets available were added. In this way, the TRL defined by the asset provider can be found by the platform users while they access the asset webpage in the catalog on the platform (Fig. 2). As the process is still in an early stage of adoption in the platform, most of it must be done manually by the MBD asset provider, implying possible human errors and further revisions on the quality of the TRL assessments.

6 Conclusions

This paper presented a new TRL template for updating and controlling MBD methods and tools of a collaborative platform. This new template can be considered an improvement and adaptation to the MBD domain of the EC and NASA TRL standards. After the implementation of the tool, new feedback from the tool providers were gathered.

1. inclusion of additional details for the documents to be uploaded into the platform: at the moment, the format is not specified as it can highly vary from company to company (it is needed a list of types of verification documents that can be utilized in the various stages of the TRL or a standardized document template that asset providers will feed with the information related to their tools and methods).
2. inclusion of the verification of TRL definition in the Quality Assurance process, currently being designed for the HUBCAP platform (due to the fact that even when specifications regarding the TRL proving documents were given in the template, some users did not comply with them, triggering a manual verification).
3. exploration of the feasibility of the template extension to additional TRLs to consider additional phases of the MBD assets' life cycle.

In addition to the previous improvements to be done in the current implementation of the TRL template, some weaknesses are identified in the paper. Currently, the verification of the correct implementation has been done in some of the assets currently offered in the platform (13 of 49). The model must still be applied to all the tools added on the HUBCAP platform to further verify its applicability. Furthermore, improvements are intended to be implemented to the platform with the intention to make easier the assessment and verification of the assets' TRL, intended to be executed by the HUBCAP platform providers. Nevertheless, additional effort will be done to address the sustainability of the TRL definition process and the verification procedure.

References

1. Stasiak-Betlejewska, E.R., Parv, A.L., Gliń, E.W.: The influence of Industry 4.0 on the enterprise competitiveness. Sciendo **1**(1), 641–648 (2018). https://doi.org/10.2478/mape-2018-0081
2. Dalenogare, L.S., Benitez, G.B., Ayala, N.F., Frank, A.G.: The expected contribution of industry 4.0 technologies for industrial performance. Int. J. Prod. Econ. **204**, 383–394 (2018). https://doi.org/10.1016/J.IJPE.2018.08.019

3. Stark, J.: Product Lifecycle Management, 2nd edn. Springer, London (2011). https://doi.org/10.1007/978-0-85729-546-0_16
4. Rüßmann, M., et al.: Industry 4.0: the future of productivity and growth in manufacturing industries (2015). https://doi.org/10.1007/s12599-014-0334-4
5. Jensen, J.C., Chang, D.H. Lee, E.A.: A model-based design methodology for cyber-physical systems. In: IWCMC 2011 - 7th International Wireless Communications and Mobile Computing Conference, pp. 1666–1671 (2011). https://doi.org/10.1109/IWCMC.2011.5982785
6. Sassanelli, C., Panetto, H., Guedria, W., Terzi, S., Doumeingts, G.: Towards a reference model for configuring services portfolio of digital innovation hubs: the ETBSD model. In: Camarinha-Matos, L.M., Afsarmanesh, H., Ortiz, A. (eds.) PRO-VE 2020. IAICT, vol. 598, pp. 597–607. Springer, Cham (2020). https://doi.org/10.1007/978-3-030-62412-5_49
7. Lachiewicz, S., Matejun, M., Pietras, P., Szczepańczyk, M.: Servitization as a concept for managing the development of small and medium-sized enterprises. Management **22**(2), 80–94 (2018). https://doi.org/10.2478/manment-2018-0024
8. Mankins, J. C.: Technology readiness levels: a white paper, January 1995 (1995). http://www.hq.nasa.gov/office/codeq/trl/trl
9. Héder, M.: From NASA to EU: the evolution of the TRL scale in public sector innovation. Innov. J. **22**(2), 1–23 (2017)
10. Banke, J.: NASA, technology readiness levels demystified (2010)
11. Mankins, J.C.: Technology readiness assessments: a retrospective. Acta Astronaut. **65**(9–10), 1216–1223 (2009). https://doi.org/10.1016/J.ACTAASTRO.2009.03.058
12. Tec-shs, *technology readiness levels handbook for space applications.* European space agency (ESA) (2008)
13. U.S. Department of energy, technology readiness assessment guide (2009)
14. Lee, M.-C., Chang, T., Chien, W.-T.C.: An approach for developing concept of innovation readiness levels. Int. J. Manag. Inf. Technol. **3**(2), 18–37 (2011). https://doi.org/10.5121/ijmit.2011.3203

Measuring Static Complexity in Mechatronic Products

Ruben Dario Solarte Bolaños[1(✉)], Sanderson César Macêdo Barbalho[2(✉)], Antonio Carlos Valdiero[1(✉)], Joao Carlos Espindola Ferreira[1(✉)], and Alan Mavignier[1(✉)]

[1] Federal University of Santa Catarina, Florianopolis, SC 88040900, Brazil
rubendariosolarte@gmail.com, antoniocvaldiero@gmail.com, j.c.ferreira@ufsc.br, alanmavignier@gmail.com

[2] University of Brasilia, Brasilia, DF 70910900, Brazil
sandersoncesar@unb.br

Abstract. The complexity present in products does not only affect development time, it also has impacts on production, for example: production costs, manufacturing lead times, quality and customer satisfaction. The complexity of the product will have a profound impact on the manufacturing organization and the product management style. A complex product generally consists of a large number of components, elements or agents, which interact with one another and with the environment. A system or product would be more complex, if there are more parts or components, and more connections between them. The main objective of this article is to propose a methodology to measure the complexity in a mechatronic product. In the course of proposing this methodology, several methodologies used by different authors to measure this variable are studied. The proposed methodology is applied to measure the complexity of four products manufactured and marketed by a Brazilian company. The proposed methodology uses tools such as the DSM (Design Structure Matrix) to support the calculation of the complexity between the interconnections of the subsystems of the products.

Keywords: Product complexity · Mechatronic product · Production systems

1 Introduction

The current economic environment is characterized by increasing customer requirements on product performance, quality, and price, leading to a highly dynamic product design environment with shortened development and product life-cycles [1]. Currently NPD (New Product Development) teams are in a constant struggle to reduce development times, because the product life cycles are becoming shorter. Shorter life cycles lead to greater complexity in the areas of product and process design, factory implementations and production operations [2].

Complexity is an inseparable aspect of complex products [3], particularly in mechatronic systems with a large number of components and interconnections, interactions,

F. Noël et al. (Eds.): PLM 2022, IFIP AICT 667, pp. 250–261, 2023.
https://doi.org/10.1007/978-3-031-25182-5_25

and interfaces [4, 5]. Modelling the complexity is increasingly challenging in these systems [6], being related to the number of product components/parts and their interaction [7–11], the number of different disciplines necessary for designing the product [12], the degree of innovation and the technology novelty [7, 13, 14], and the complexities involved on customer interfaces [4, 15]. A high product complexity can impact manufacturing, inventory and distribution areas and consequently production costs [16]. Given this scenario, this article provides a method to measure the complexity in a mechatronic product, as a first step to propose a methodology to manage the complexity in a mechatronic product and thus mitigate the impacts of these variables in the manufacturing or product development process.

The complexity of a product is usually determined by the number of components, elements or agents that interact with each other and with the environment [17]. This paper addresses elements to deepen the understating of the product static complexity. Next section presents a review of literature listing works that strive to find solutions for managing product complexity. In Sect. 3, the method for complexity determination is exposed. Section 4 presents the features of products studied here. Section 5 shows the application of method in four products, which are produced by the same company. Finally, Sect. 6 presents the conclusions and future works to give continuity to this research topic.

2 Theoretical Revision

To identify indicators for measuring complexity, the IEEE, SCOPUS and SCIENCE-DIRECT databases researched. The keywords of this research were Mechatronic Product and Complexity. These works are presented below.

According to Barbalho, Chapman, Novak and Danilovic in [8, 14, 18, 19, 20, 21], a mechatronic product complexity can be calculated on the basis of the number of product components, the extent of interaction, the degree of innovation embedded in the product, and the number of different kinds of technology employed.

In [22], Pugh uses math to calculate product complexity with the variables: number of parts, number of types of parts, number of interconnections and interfaces, and number of functions that the product must perform. With these variables, the authors calculate the product complexity for forecasting the production costs. Hobday in [23], presents an example of assessing the complexity of an air traffic control system, and a flight simulator, as a way to wonder about product costs. Hobday uses 16 indicators to evaluate the product complexity associated with the development cost.

McCarthy et al. [24] defined a range from one to five for a set of product complexity elements with the intention of numerically measuring the total complexity of a company. Likewise, Meysam et al. [25] presents a table to measure the complexity of a mechatronic product, explaining the concepts that are used to evaluate it. Moulianitis et al. [12] present the characteristics of a mechatronic product in terms of an indicator vector that contains variables of intelligence, flexibility and complexity. To assess complexity, Moulianitis uses the number of components, number of internal interconnections, number of design alternatives, number of feedback loops, number of knowledge base, degree of customization, and degree of intelligence. In [14], Danilovic represents a product by implementing the DSM (Design Structure Matrix) to manage the dependencies and relationships of the product components and the DMM (Domain Mapping Matrix) to compare two different products or projects. It is possible to observe that DSM and DMM matrices allow viewing the whole product with all components and interconnections, offering great support for estimating and visualizing the complexity of a mechatronic product.

For Zhang [26], the complexity of the product is determined by the analysis of its technology and size to optimize product design and evaluate design alternatives for different production systems. Hehenberger in [27], defines an equation that allows to calculate the probability of technical success by quantifying the complexity of the project. Ahmadinejad [28] used the DMM and DSM matrices to obtain the structural model of an intelligent gas meter. Tastekin [29] uses the Gray measurement methodology to measure the complexity of five software products. Medina [30] presents a study to measure design complexity (DC) in medical devices by using the device functions and components with their interactions and variations. Park an Kremer in [31] define metrics to correctly calculate the static complexity for both design and manufacture processes. These proposed complexity metrics are based on the concept of similarity of products and processes. Diagne in [32], proposes a methodology based on DSM matrix principles to evaluate the performance of a complex product.

According to Elmaraghy et al. [33, 34] the type of complexity may be classified as: (i) static or (ii) dynamic. Static complexity is time-independent complexity due to the product and systems structure and dynamic complexity is time-dependent and deals with the operational behavior of the system [35].

Park and Kremer [31] argue that since static complexity is based on a functional analysis of product design, it is useful to clearly identify the direct impact of complexity on manufacturing performance. Static complexity is made up of factors (a) structural, (b) computational and algorithmic, (c) size, volume, and quantity, and (d) network interaction [34]. Park focuses on static complexity, occurring from the structural configuration of a system, from the perspectives of both product design and manufacturing, and scrutinizes the impact of complexity on manufacturing performance.

Regarding the visualization of the internal interconnections between the subsystems of a product, a tendency to use the DSM tool and its derivatives can be observed. Therefore, it is proposed to use this tool to measure the complexity of the internal structure of the mechatronic product.

For the full product complexity determination, it is possible to observe the tendency to use tables, where a value is simply given to an indicator. These evaluations of the indicators are directed to count the number of components, knowledge bases and others. So, the challenge of this research becomes to unify a method that can express the complexity derived from the internal structure of the product (exchange of in-formation (software)-energy-material-space), the complexity derived from the physi-cal components (number of components, degree of customization and others) and the complexity derived from the interdisciplinary nature of the product (areas of knowledge and others involved).

In accordance with the main objective of this research, a methodology to measure product complexity should be used. Starting then from the literature review of the most used indicators to measure the complexity of a technological product, from the definition of static complexity and the similar to [31], in this proposition a time independent analysis is made (calculation of static complexity).The works presented in this section (fourteen papers) study the complexity from different points of view. Table 1 presents the mentioned indicators and the frequency with which they are used by the referenced authors.

Table 1 shows that the most frequent indicators to measure the complexity in a product, according to the bibliographic review of this section are the indicators 1, 2 and 9 that correspond to: (a) the quantity of the components, (b) the degree of complexity of the interconnections between components and (c) The degree of customization of the final system.

Table 1. Indicators of complexity in technological products.

Indicators	Referenced works	Percentage (%)
1. Quantity of components	13/14	92.86
2. Interconnections Complexity	9/14	64.29
3. Financial scale of the project	1/14	7.14
4. Product volume	2/14	14.29
5. Number of knowledge bases for product design	6/14	42.86
6. Degree of technological maturity of the process	5/14	35.71
7. Software complexity	1/14	7.14
8. Degree of variety of components	3/14	21.43
9. Degree of customization of the final system	7/14	50
10. Feedback cycles of later phases	1/14	7.14
11. Intensity of the client's participation in the design	1/14	7.14
12. Uncertainty/alteration in user requirements	1/14	7.14
13. Intensity of participation of suppliers	1/14	7.14
14. Regulatory standards	3/14	21.43
15. Quality of the staff	3/14	21.43
16. Number of departments to develop the product	1/14	7.14
17. Competition in the market	1/14	7.14
18. Transformation of information	1/14	7.14
19. Resource allocation	1/14	7.14
20. Number of suppliers and customers	1/14	7.14
21. Number of Product Functions	3/14	21.43
22. Complexity of manufacturing	1/14	7.14

To obtain better results in the evaluation of the complexity in a mechatronic product, the indicators found in the review should be taken into account, however this work was limited to choosing the indicators whose information was provided by the company. Nevertheless, the insight presented in this chapter provides the reader with indicators, tools and bibliography for determining the complexity of a product.

3 Proposal to Measure the Static Complexity in Products

Based on Table 1, Table 2 is proposed to measure the static complexity. The indicators listed on Table 2 were chosen according to the applicability (information provided by the company). The indicators were identified in the company documentation, and analyzed by an experienced engineer. In column 5 of Table 2 it is explained how the authors evaluated each indicator.

The indicators 2 and 6 (Table 2) represent the complexity provided by the mechatronic nature. The mechatronic nature refers to the interaction and integration of different disciplines that involve a mechatronic product. Indicator 2 lists the number of areas involved in the design of the mechatronic product and indicator 6 evaluates the degree of interaction of these areas within the mechatronic product. These areas are found in the structure of the product represented by subsystems, modules or others. The degree of integration is directly proportional to the degree of interaction of the areas and the number of areas.

Table 2. Complexity determination table

Indicators (I)	Evaluation (E)	Maximum limit (ML)	Complexity (C)	How to calculate
I.1. Number of Fuctions				Counting the number of primaries functions for which the prototype is designed
I.2. Number of knowledge bases				Listing the number of areas of knowledge that are necessary for the implementation of the main functions of the prototype
I.3. Components number				Counting all the components of the prototype

(*continued*)

Table 2. (*continued*)

Indicators (I)	Evaluation (E)	Maximum limit (ML)	Complexity (C)	How to calculate
I.4. Degree of variety of components				Counting the total number of components without repetition in the prototype
I.5. Degree of customization of the final system				Counting the number of components, the company had to design and manufacture (unique components)
I.6. Degree of complexity of the interconnections between components				Dividing a prototype into its main subsystems and calculating the strength of the interconnection from 1 to 5 in space dependencies, information exchange, material ex- change and energy exchange, based on [36]. In order to analyze the ex- change of information, it was necessary to study the algorithms of each product, thus addressing the complexity of the software
Total complexity (TC)				Adding all values in column 4

For the measurement of complexity, the maximum values (column 3) were calculated as the *maximum (Indicator1.Product1, Indicator1.Product2, Indicator1.Product3, Indicator1.Product4)* and so on for each indicator: (I.1) 1, (I.2) 4, (I.3) 2474, (I.4) 927,

(I.5) 475 and (I.6) 374. These maximum values correspond to the highest data in the evaluation of the indicators in the products studied.

In the bibliographic review carried out in this section, it was observed that the works [14, 32, 36, 37], used the DSM (Design Structure Matrix) to visualize and understand the operation of a complex product. In the evaluation of the complexity proposed in this work, DSM is used to evaluate the indicator "Degree of complexity of interconnections between subsystems". Figure 1 presents an example of the DSM of a product having 4 subsystems (a, b, c and d). Each evaluation of the complexity of an interconnection is evaluated considering the interaction guidelines defined in [36] as: *Spatial,* associations of physical space and alignment between the components or subsystems, *Energy, t*he interaction of the Energy type identifies needs of a physical phenomenon between two elements, *Information,* the interaction of the type of information identifies the need to exchange information or signal between two subsystems or components and *Material, t*he interaction of the type of material identifies needs of interfaces or components between two elements.

The interconnections represented by each position of the matrix receive a rating from 1 to 5 depending on their complexity (0 represents a null value of relations between subsystems and 5 the maximum value of relation between subsystems). At the end, the values of each position of the matrix are added and the result of this sum is the value of complexity of the interconnections of the product. To fill the DSM, specifically the Information Exchange (Information) item, it is necessary to study the algorithms of each product, which considers the software complexity as an influential indicator in the complexity determination.

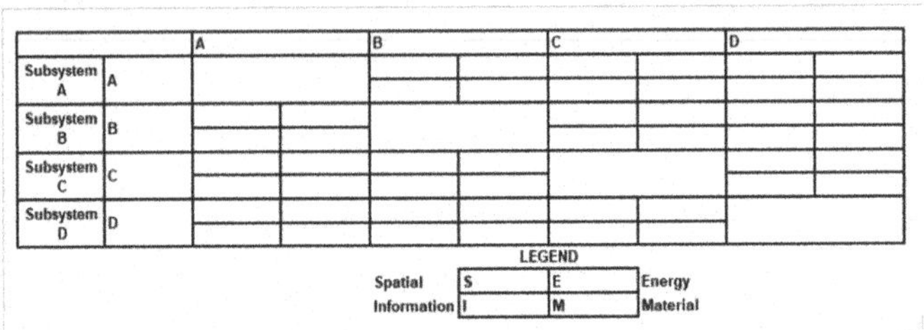

Fig. 1. DSM for the evaluation of the complexity of interconnections between product components.

4 Products Features

The company studied is a Brazilian company in the field of optics, which operates in medical, industrial, optical components, aerospace and defense. The structural and functional characteristics of the products were extracted through the analysis of the documentation provided by the company. Table 3 was filled with the characteristics of the products and shows the values of the indicators of each product, necessary for the calculation of the static complexity.

Table 3. Indicators information for the complexity evaluation

Indicators	Product 1	Product 2	Product 3	Product 4
I.1. Product functions number	1	1	1	1
I.2. Number of knowledge bases for product design	4	4	2	4
I.3. Number of components	2474	1515	46	2340
I.4. Degree of variety of components	927	404	30	750
I.5. Degree of customization of final system	475	150	25	400
I.6. Interconnections Complexity	324	374	72	345

5 Methodology Application

This section presents the application of the methodology for the measurement of static complexity, specifically for product 1 to exemplify the step-by-step process. By analyzing the internal structure and identify the product 1 subsystems, the DSM was designed and filled to determination of (I.6). This matrix is shown in Fig. 2. According to Fig. 2, the complexity of the interconnections for product 1 is 324, based in the methodology proposed to evaluate this indicator. Then, total complexity of product 1 is then calculated according to the indicators used in this article. Table 4 shows the values obtained to calculate the product 1 complexity.

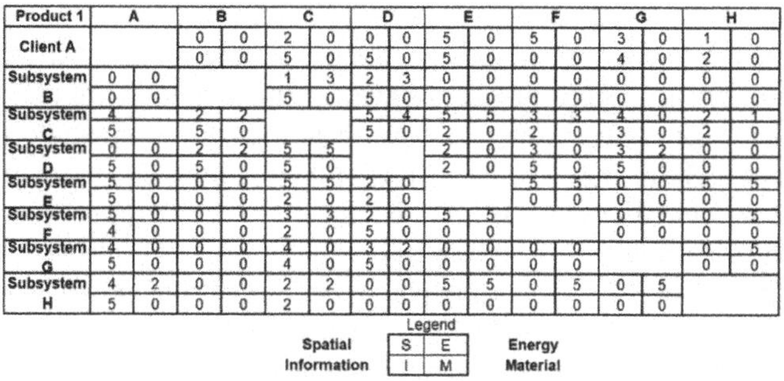

Product 1	A		B		C		D		E		F		G		H	
Client A			0	0	2	0	0	0	5	0	5	0	3	0	1	0
			0	0	5	0	5	0	5	0	0	0	4	0	2	0
Subsystem B	0	0			1	3	2	3	0	0	0	0	0	0	0	0
	0	0			5	0	5	0	0	0	0	0	0	0	0	0
Subsystem C	4		2	2			5	4	5	5	3	3	4	0	2	1
	5		5	0			5	0	2	0	2	0	3	0	2	0
Subsystem D	0	0	2	2	5	5			2	0	3	0	3	2	0	0
	5	0	5	0	5	0			2	0	5	0	5	0	0	0
Subsystem E	5	0	0	0	5	5	2	0			5	5	0	0	5	5
	5	0	0	0	2	0	2	0			0	0	0	0	0	0
Subsystem F	5	0	0	0	3	3	2	0	5	5			0	0	0	5
	4	0	0	0	2	0	5	0	0	0			0	0	0	0
Subsystem G	4	0	0	0	4	0	3	2	0	0	0	0			0	5
	5	0	0	0	4	0	5	0	0	0	0	0			0	0
Subsystem H	4	2	0	0	2	2	0	0	5	5	0	5	0	5		
	5	0	0	0	2	0	0	0	0	0	0	0	0	0		

Legend

Spatial	S	E	Energy
Information	I	M	Material

Fig. 2. DSM to evaluate the complexity of the internal interconnections for product 1.

For the Number of Product Functions indicator (I.1), the main function of product 1 is to allow the physician to quickly and clearly visualize the background of the eye, allowing accurate examination of the retina. For the indicators Number of Knowledge bases (I.2), Number of components (I.3), Degree of variety of components (I.4), and Degree of customization of final system (I.5), it is only necessary to count the components of the product. As can be seen in Table 4, the static complexity of the product 1 is 587. Tables 5, 6 and 7 show the complexity tables for Product 2, Product 3 and Product 4, respectively.

Table 4. Complexity of product 1

I	E	M	C
I.1	1	1	100
I.2	4	4	100
I.3	2474	2474	100
I.4	927	927	100
I.5	475	475	100
I.6	324	374	87
TC	$\sum C$		587

In this investigation, each Indicator is evaluated from 0 to 100, with respect to the highest values assigned to the indicator. Therefore, if a more complex product is added to the study, the limit values of the table will be modified. For further studies, tools will be implemented to visualize the weight in greater detail of each indicator in the design of a mechatronic product. In this investigation, the complexity of 4 products produced in the same company was measured, with the future objective of evaluating the effect of complexity on factors such as manufacturing time, development time, quality and costs, to propose predictive models that support the *Decision Making*.

6 Conclusions

This article proposes a methodology to measure complexity in mechatronic products, but it can be used in other types of products. The methodology addresses a structural and functional study of the designs of each product. Indicators found in the bibliographic review were not taken into account because there was no information the indicators from the company. The methodology to measure the complexity in a mechatronic product presented in this article contributes to investigating the effects of the products complexity in their respective manufacturing processes. For future work, it is planned to analyze the dynamic complexity in a manufacturing process of mechatronic products starting from the determination of the static complexity presented here. It is also important to consider adding to the methodology the component of complexity that provides connectivity and the insertion of enabling technologies of industry 4.0.

Table 5. Complexity of product 2.

I	E	ML	C
I.1	1	1	100
I.2	4	4	100
I.3	1515	2474	61
I.4	404	927	44
I.5	150	475	32
I.6	374	374	100
TC	$\sum$C		***437***

Table 6. Complexity of product 3.

I	E	ML	C
I.1	1	1	100
I.2	2	4	50
I.3	46	2474	2
I.4	30	927	3
I.5	25	475	5
I.6	72	374	20
TC	$\sum$C		***180***

Table 7. Complexity of product 4.

I	E	ML	C
I.1	1	1	100
I.2	4	4	100
I.3	2340	2474	95
I.4	750	927	80
I.5	400	475	84
I.6	345	374	93
TC	$\sum$C		***552***

References

1. de Carvalho, R.A., da Hora, H., Fernandes, R.: A process for designing innovative mechatronic products. Int. J. Prod. Econ. **231**, 107887 (2021). https://doi.org/10.1016/j.ijpe.2020.107887
2. Ferreira, F., Faria, J., Azevedo, A., Luisa, A.: International journal of information management product lifecycle management in knowledge intensive collaborative environments : an application to automotive industry. Int. J. Inf. Manage. **37**(1), 1474–1487 (2017). https://doi.org/10.1016/j.ijinfomgt.2016.05.006
3. Bakhshi, J., Ireland, V., Gorod, A.: Clarifying the project complexity construct: Past, present and future. Int. J. Proj. Manag. **34**, 1199–1213 (2016). https://doi.org/10.1016/j.ijproman.2016.06.002
4. Bolaños, R.D.S., Barbalho, S.C.M.: Exploring product complexity and prototype lead-times to predict new product development cycle-times. Int. J. Prod. Econ. **235**, 108077 (2021). https://doi.org/10.1016/j.ijpe.2021.108077
5. Bolaños, R.D.S., Valdiero, A.C., Rasia, L.A., Ferreira, J.C.E.: Identifying the trend of research on mechatronic projects. In: Canciglieri Junior, O., Noël, F., Rivest, L., Bouras, A. (eds.) Product Lifecycle Management. Green and Blue Technologies to Support Smart and Sustainable Organizations. PLM 2021. IFIP Advances in Information and Communication Technology, vol 640. Springer, Cham (2022). https://doi.org/10.1007/978-3-030-94399-8_3

6. Kellner, A., Hehenberger, P., Weingartner, L., Friedl, M.: Design and use of system models in mechatronic system design (2015). https://doi.org/10.1109/SysEng.2015.7302747
7. Shenhar, A.J., Dvir, D.: Reinventing project management: the diamond approach to successful growth and innovation by aaron shenhar and dov dvir. J. Prod. Innov .Manag. - J PROD Innov. Manag. **25**, 635–637 (2008). https://doi.org/10.1111/j.1540-5885.2008.00327_2.x
8. Novak, S., Eppinger, S.: Sourcing by design: product complexity and the supply chain. Manage. Sci. **47**, 189–204 (2001). https://doi.org/10.1287/mnsc.47.1.189.10662
9. Kim, J., Wilemon, D.: Sources and assessment of complexity in NPD project. R&D Manag. **33**, 15–30 (2003). https://doi.org/10.1111/1467-9310.00278
10. Ahmadinejad,A., Afshar, A.: Complexity management in mechatronic product development based on structural criteria. In: Mechatronics (ICM) , pp. 7–12. IEEE (2011). http://ieeexplore.ieee.org/xpls/abs_all.jsp?arnumber=5971266
11. Pugh, S.: Total Design Integrated Methods for Successful Product Engineering. Addison-Wesley Publishing Company, Boston (1991)
12. Moulianitis, V.C., Aspragathos, N.A., Dentsoras, A.J.: A model for concept evaluation in design - An application to mechatronics design of robot grippers. Mechatronics **14**(6), 599–622 (2004). https://doi.org/10.1016/j.mechatronics.2003.09.001
13. Meyer, M., Utterback, J.: Product development cycle time and commercial success. Eng. Manag. IEEE Trans. **42**, 297–304 (1995). https://doi.org/10.1109/17.482080
14. Danilovic, M., Browning, T.R.: Managing complex product development projects with design structure matrices and domain mapping matrices. Int. J. Proj. Manag. **25**(3), 300–314 (2007). https://doi.org/10.1016/j.ijproman.2006.11.003
15. Clark, K., Fujimoto, T.: Product development performance: strategy. Organ. Manag. World Auto Ind. **15**(2). Boston (1991)
16. Jacobs, M., Swink, M.: Product portfolio architectural complexity and operational performance: Incorporating the roles of learning and fixed assets. J. Oper. Manag. - J OPER Manag **29**, 677–691 (2011). https://doi.org/10.1016/j.jom.2011.03.002
17. Braun, S.C., Lindemann, U.: The influence of structural complexity on product costs. In: 2008 IEEE International Conference on Industrial Engineering and Engineering Management IEEM 2008, pp. 273–277. IEEE (2008). https://doi.org/10.1109/IEEM.2008.4737873
18. Bolaños, R.D.S., Barbalho, S.C.M.: Analisis de la literatura sobre la complejidad de productos mecatrónicos e impacto en factores críticos de suceso en proyectos de desarrollo de productos. *Congeso. Iberoamecano Ing. Proy.* (2016)
19. Barbalho, S.C.M., de Carvalho, M.M., Tavares, P.M., Llanos, C.H., Leite, G.A.: Exploring the relation among product complexity, team seniority, and project performance as a path for planning new product development projects: a predictive model applying the system dynamics theory. *IEEE Trans. Eng. Manag.*, 1–14 (2019). doi: https://doi.org/10.1109/TEM.2019.2936502
20. Barbalho, S.C.M., Rozenfeld, H.: Mechatronic reference model (MRM) for new product development: Validation and results [Modelo de referência para o processo de desenvolvimento de produtos mecatrônicos (MRM): Validação e resultados de uso]. Gest. e Prod. **20**(1), 162–179 (2013). https://doi.org/10.1590/S0104-530X2013000100012
21. Chapman, R., Hyland, P.: Complexity and learning behaviors in product innovation. Technovation **24**(7), 553–561 (2004). https://doi.org/10.1016/S0166-4972(02)00121-9
22. Pugh, S.: Total Design: Integrated Methods for Successful Product Engineering. Addison-Wesley Publishing Company, Boston (1991)
23. Hobday, M.: Product complexity innovation and industrial organisation. Res. Policy **26**(6), 689–710 (1998). https://doi.org/10.1016/S0048-7333(97)00044-9
24. McCarthy, I.P., Tsinopoulos, C., Allen, P., Rose-Anderssen, C.: New product development as a complex adaptive system of decisions. J. Prod. Innov. Manag. **23**(5), 437–456 (2006). https://doi.org/10.1111/j.1540-5885.2006.00215.x

25. Mousavi, S.M., Tavakkoli-Moghaddam, R., Vahdani, B., Hashemi, H., Sanjari, M.J.: A new support vector model-based imperialist competitive algorithm for time estimation in new product development projects. Robot. Comput. Integr. Manuf. **29**(1), 157–168 (2013). https://doi.org/10.1016/j.rcim.2012.04.006
26. Zhang,Z., Luo, Q.: A grey measurement of product complexity. In: Conference *Proceedings* - International Conference on Systems, Man and Cybernetics, pp. 2176–2180 (2007). https://doi.org/10.1109/ICSMC.2007.4413624
27. Hehenberger, P., Poltschak, F., Zeman, K., Amrhein, W.: Hierarchical design models in the mechatronic product development process of synchronous machines. Mechatronics **20**(8), 864–875 (2010). https://doi.org/10.1016/j.mechatronics.2010.04.003
28. Ahmadinejad, A., Afshar, A.: Complexity management in mechatronic product development based on structural criteria. In: 2011 IEEE International Conference on Mechatronics, *ICM 2011 – Proceedings,* April. 2011. https://doi.org/10.1109/ICMECH.2011.5971266
29. Tastekin, S.Y., Erten, Y.M., Bilgen, S.: Software product complexity estimation using grey measurement. In: *Proceedings - 39th Euromicro Conference on Software Engineering and Advanced Applications SEAA 2013*, pp. 308–312 (2013). https://doi.org/10.1109/SEAA.2013.42
30. Medina, L.A., Collet, M., Cruz, G., Pacheco, T.N., Kremer, G.E.O.: Developing design complexity metrics for medical device development, pp. 2562–2571 (2013)
31. Park, K., Kremer, G.E.O.: Assessment of static complexity in design and manufacturing of a product family and its impact on manufacturing performance. Int. J. Prod. Econ. **169**, 215–232 (2015). https://doi.org/10.1016/j.ijpe.2015.07.036
32. Diagne, S., Coulibaly, A., De Beuvron, F.D.B.: Complex product modeling based on a Multi-solution eXtended Conceptual Design Semantic Matrix for behavioral performance assessment. Comput. Ind. **75**, 101–115 (2016). https://doi.org/10.1016/j.compind.2015.06.003
33. Badrous, S., Elmaraghy, H.: A model for measuring complexity of automated and hybrid assembly systems. Int. J. Adv. Manuf. Technol. **62**, 813–833 (2012). https://doi.org/10.1007/s00170-011-3844-y
34. Elmaraghy, W., Elmaraghy, H., Tomiyama, T., Monostori, L.: Complexity in engineering design and manufacturing. CIRP Ann. - Manuf. Technol. **61**(2), 793–814 (2012). https://doi.org/10.1016/j.cirp.2012.05.001
35. Frizelle, G., Woodcock, E.: Measuring complexity as an aid to developing operational strategy. J. Oper. & Prod. Manag. (1995)
36. Sharman, D.M., Yassine, A.A.: Characterizing complex product architectures. Syst. Eng. **7**(1), 35–60 (2004). https://doi.org/10.1002/sys.10056
37. Schlick, C.M., Beutner, E., Duckwitz, S., Licht, T.: A complexity measure for new product development projects. In: 2007 IEEE International Engineering Management Conference, pp. 143–150 (2007). https://doi.org/10.1109/IEMC.2007.5235079

Towards a Requirements Co-engineering Improvement Framework: Supporting Digital Delivery Methods in Complex Infrastructure Projects

Yu Chen(✉) and Julie R. Jupp

Faculty of Engineering and IT, University of Technology Sydney, Sydney, Australia
yu.chen-4@student.uts.edu.au

Abstract. To support the delivery of cyber-physical systems of complex infrastructure assets, different requirements (e.g., physical system requirements, asset information requirements) must be developed and managed properly during the lifecycle of the assets. However, there is a lack of integrated and continuous approach to support the co-development and co-management of physical systems requirements and asset information requirements. Adopting a design science research methodology, this paper develops the structure of Requirements Co-engineering Improvement Framework for complex infrastructure projects. This framework defines five maturity levels for requirements relevant process, protocol and supporting software tools. Further validation will be conducted using the Delphi Method in future research.

Keywords: Requirements co-engineering · Capability maturity · Process · Protocol · Technology

1 Introduction

As modern transport infrastructure projects have grown in size and complexity, there has been an increasing need to move away from traditional spreadsheets and find more efficient and reliable methods of managing large volume of requirements. For example, as a part of a linear network, rail transport assets can each be defined as a complex cyber-physical system. Requirements engineering (RE) in the context of the delivery of these projects is increasingly complex. This is particularly true when it comes to 'mega projects', where requirements can number in the hundreds of thousands. All of which must be managed and traced across multiple stakeholders, work packages and interfaces in a complex, high-pressure environment where errors, changes and delays can cost millions of dollars.

Different types of requirements about the cyber systems and physical systems of complex infrastructure assets must be developed and managed throughout the planning and delivery phases. Requirement types include, amongst others; high-level capability

F. Noël et al. (Eds.): PLM 2022, IFIP AICT 667, pp. 262–273, 2023.
https://doi.org/10.1007/978-3-031-25182-5_26

requirements defining system architecture capabilities; current and future operational requirements; system-, sub-system-, and unit- level requirements spanning functional, physical and performance-based needs; and business case requirements. To support more strategic approaches to digital asset management, during the planning and acquisition of complex infrastructure projects, asset information requirements describing physical systems, their virtual replicas, and real-time behaviours must also be developed and managed.

Yet, often due to issues related to the scale, complexity, and emergent properties of the cyber-physical systems being developed, the different (and evolving) requirement types of cyber systems and physical systems and corresponding asset information requirements are increasingly difficult for current engineering practices to handle. The digital delivery of complex infrastructure projects increases the need to implement more integrated and continuous approaches to RE that recognised the importance of asset information requirements review processes and verification traceability methods.

Based on the Software Engineering Institute's (SEI's) Capability Maturity Model for Integration (CMMI), this paper develops a Requirements Co-engineering Improvement Framework for complex infrastructure projects. The proposed framework presents an approach for the organisation or project team to understand the current maturity level of requirements management capabilities and a potential pathway for improvement. The paper proceeds in Sects. 2 with an overview of current requirement engineering maturity models. Section 3 describes the design science research methodology, and Sect. 4 presents the development of the Improvement Framework. Section 5 concludes the paper with future research plan to further development and verification of the framework and the limitation of the research.

2 Requirements Engineering Maturity Models

The effectiveness level of an organisation to develop quality products or services is directly related to the maturity of their processes [1]. In the context of this research, measuring the maturity level of RE processes offers a solution to organisations who are seeking for RE process improvement. This section investigates the capability of existing capability maturity model, especially the relatively new and up to date RE process maturity models and discusses their applicability in supporting capability measurement of RE in rail infrastructure.

Several RE capability models have been proposed based on Software Process Improvement (SPI) standards and framework such as Software Engineering Institute's (SEI's) old Capability Maturity Model (CMM) and relatively new Capability Maturity Model for integration (CMMI). Existing models supporting RE process maturity assessment include Requirements Capability Maturity Model (RCMM) [2, 3], Market-Driven Requirements Engineering Process Model [4], Requirements Engineering Good Practice Guide [5] and Requirements Engineering Process Maturity Model [6]. However, all these four models arose in software industry and were built based on the retired and unsupported Software-CMM or CMM [7]. Some of them are not completely developed and validated while others are difficult to implement [8, 9].

In this research, Requirements Engineering Process Assessment and Improvement Model (REPAIM) and Capability Maturity Model improved (R-CMMi) are selected for further review as they are both cited as two of the most popular maturity models [8, 9]. However, REPAIM uses the continuous representation model to describe capability levels of processes, while R-CMMi adopts the staged representation model to define maturity levels, which characterise the organisation's behaviour. These two types of representation models are both defined in the CMMI standard (see Table 1) [10].

Table 1. REPAIM and R-CMMi levels in CMMi

	REPAIM	R-CMMi
Level	Capability levels Continuous/Process	Maturity levels Staged/Organisation
Level 0	Incomplete	–
Level 1	Performed	Initial
Level 2	Managed	Managed
Level 3	Defined	Defined
Level 4	–	Qualitatively managed
Level 5	–	Optimised

In REPAIM, four levels of maturity are defined as incomplete, performed, managed and defined. The REPAIM shows its capability to access RE processes and prioritise their improvement, adapt and complement existing maturity standards and assessment approaches, and adapt to the demands of different organisations [7]. However, there are two identified drawbacks of REPAIM. One of the main drawback is that training is still required by the practitioner in order to understand the model [7]. Another drawback is that it appears to need further examples, templates, and instructions to inform an effective implementation by potential users [7].

Five maturity levels are defined as initial, managed, defined, qualitatively managed and optimised in R-CMMi which shows a high consistency with CMMI standard [9]. The drawback of R-CMMi is similar with REPAIM in terms of validation, implementation issue, training, future work of instructions, examples and templates.

Furthermore, there are important interrelationships between maturity levels and capability levels. These organisational maturity levels are dependent on the capability levels of their processes. To research a certain maturity level, the organisation have to successfully achieve the objective of the targeted process areas to that level [1]. The maturity levels reflect the current status of RE in an organisation or a project, while capability levels provide with an improvement pathway to the higher maturity level. Thus, a combination of these two types of representation model would potentially resolve these issues and is adopted in this research.

3 Design Science Research Methodology

A design science research methodology (DSRM) was adopted as this whole research project seeks to extend the boundaries of human and organisational capabilities by creating new and innovative artefacts. Figure 1 adapts the design science research framework of Information System proposed by Hevner et al. [11] and overlays three inherent research cycles – relevance cycle, rigor cycle, and design cycle [12].

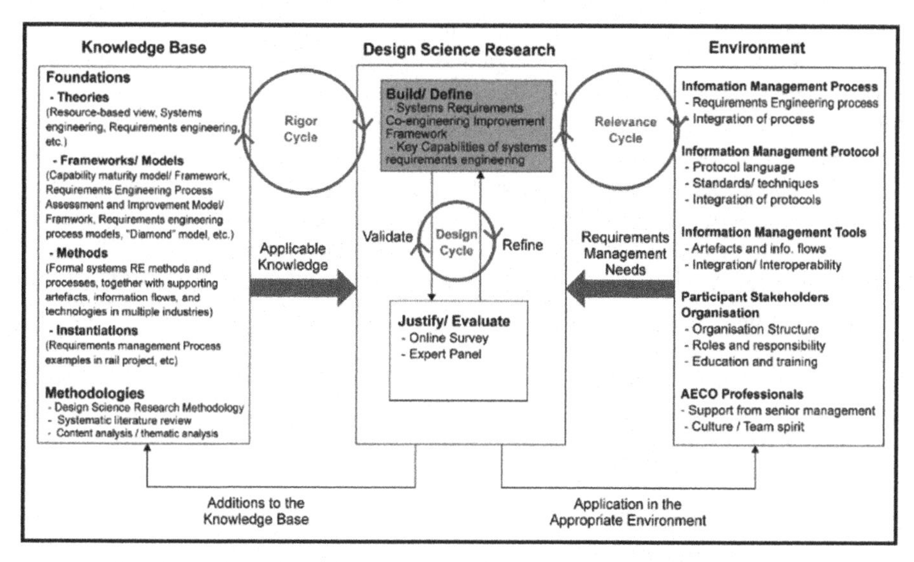

Fig. 1. Design science research conceptual framework [Adapted from 11]

Figure 1 reflects three main areas of the whole research project including the general knowledge base of the research area, the environment of research problem we are focusing on, and the core design science research activity. The content in each area has been adapted based on the research scope, nature of research problem, industrial context, as well as the availability of resource. The content of this paper focus on the Build and Define activity as highlighted in green in Fig. 1.

The central Design Cycle iterates between core activities of developing the Improvement Framework, its evaluation, and subsequent feedback to refine the framework [12]. Findings drawn from the systematic literature review form 'Knowledge Base' and semi-structured interviews form 'Environment' foundation of the design. Section 4 of this paper focuses on presenting the development of the structure of Requirements Co-engineering Improvement Framework. In the next stage of this research, survey among a group of experts will be implemented as evaluation methods to test and validate key elements of the framework and their significance with regards to supporting higher levels of integration and model-based approaches to requirements management in complex infrastructure.

4 Requirements Co-engineering Improvement Framework

4.1 System Requirements and Information Requirements

Different types of requirements about the cyber systems and physical systems in the built environment must be developed and managed throughout the Planning and Acquisition Phases of complex infrastructure projects. Requirement types include, high-level capability requirements defining system architecture capabilities; current and future operational requirements; system-, sub-system-, and unit- level requirements spanning functional, physical and performance-based needs; and business case requirements. To support more strategic approaches to digital asset management, during the Acquisition Phase, asset information requirements describing physical systems, their virtual replicas, and real-time behaviours must also be developed and managed.

Asset information requirements (AIR) is the precise description of the information required to operate and maintain a specific built asset through its lifecycle. The information required in AIR focuses on the as-built state. It defines not only what information is required (content) but also how it should be delivered (form and accepted formats of deliverables). The AIR is a subset of the overall project brief. The processes of delivering the assets and the associated data and information are parallel and connected (see the Fig. 2 below).

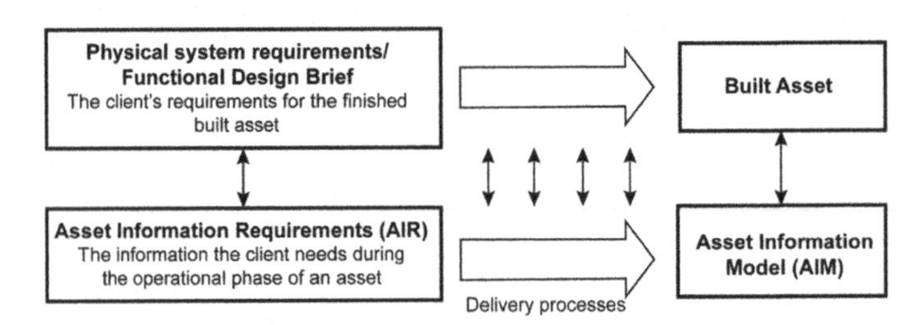

Fig. 2. Parallel delivery of built asset and asset data [28]

In complex infrastructure projects, especially in rail sector, the development and management of physical system requirements usually follow the process of the traditional systems engineering approach which also be described in the traditional "V" model. In order to understand the co-development and co-management of physical system requirements and AIR, a "Diamond" model (see Fig. 3) has been developed based on Boeing [29]and TfNSW [30]. The lower V reflects the classic systems engineering process of the physical system, while the mirror reflection of the V above represents the digital twins modelling and simulation [29, 30]. The inverted V represents the design and realisation of the behavioural simulations [29].

4.2 Contemporary Approaches to Requirements Engineering

Previous work of this research includes a systematic literature review which explored contemporary and state-of-the-art RE approaches to supporting the creation of complex software dependent systems (e.g., digital twins and cyber-physical systems) and semi-structured interviews with industry experts from transport infrastructure. Main capabilities supporting RE are identified and categorised into three main aspects including process and protocol, technology, organisation and people (as shown in Table 2). Based on this, the structure of Requirements Co-engineering Improvement Framework is developed and described in the following section.

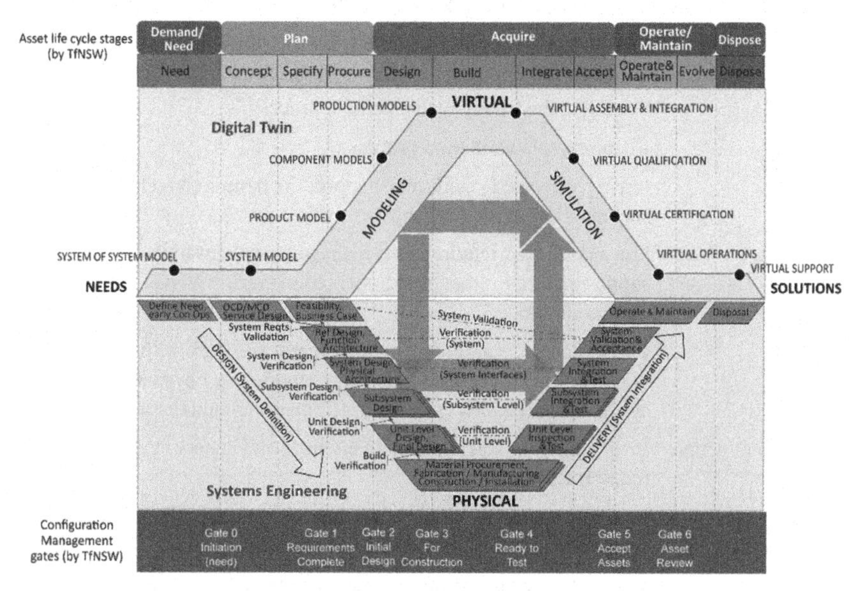

Fig. 3. "Diamond" model reflecting interaction of physical system requirements and AIR

4.3 Structure of Improvement Framework

Main capabilities identified in Table 2 have been developed into 14 requirements co-engineering sub-capabilities. Figure 4 presents the matrix displaying the score system for these sub-capabilities in different maturity levels. Once the maturity levels of capabilities are evaluated, a final sore of three categories (i.e., process and protocol, technology, organisation and people) will reflect the status of requirements co-engineering practices in a project or an organisation. 'Requirements' in Fig. 4 refer to physical systems requirements and asset information requirements.

Activities contained in each capability will be further developed and verified in the next stage of research which will then support identifying improvement pathway of requirement co-engineering capability in an organisation or a project. Moreover, the derivation of a sorting or rating system for all capabilities supports a modular and flexible approach assessment. A final weighting system is yet to be identified according to capabilities domains.

Table 2. Main capabilities supporting requirements engineering

Category	Main capabilities	Sources
Process & protocol	Requirements elicitation process	[13–15]
	Requirements analysis and prioritisation process	[14, 16, 17]
	Requirements allocation and verification process	[14, 16]
	Negotiation of conflicting requirements amongst stakeholders	[14, 15]
	Requirements change management process	[18–20]
	Requirements validation process	[18, 21]
	Use a recognised standard to support the definition and specification of requirements	[15]
Technology	Use a dedicated requirements management software supporting requirements documentation, verification, and validation	[16, 22, 23]
	Integration of requirements management software with 3D modelling software support the handling of Physical system requirements and Asset information requirements	[19, 24, 25]
Organisation & people	Involvement of stakeholders in eliciting and analysing requirements	[21, 26, 27]
	Formally defined roles and responsibilities for handling requirements in multiple phases of a project	[19, 21]
	Training in requirements software in support of requirements handling in multiple phases of a project	[18]

4.4 Development of Requirements Co-engineering Maturity Levels

In this research, a combination of REPAIM and R-CMMi maturity levels are adopted for the definition of processes and protocols maturity levels. While the definition of supporting technology maturity levels is concluded based on findings from semi-structured interviews with industry experts. Table 3 presents the description of maturity levels of process and protocol, as well as technology related capabilities supporting requirements co-engineering. Organisation and people related capabilities are relative intangible compared with process and technology. This requires a different way to describe its maturity. For example, the frequency of formally defined roles and responsivities, and the frequency of trainings. Because of the page limitation, organisation and people related maturity levels will not be included in this paper.

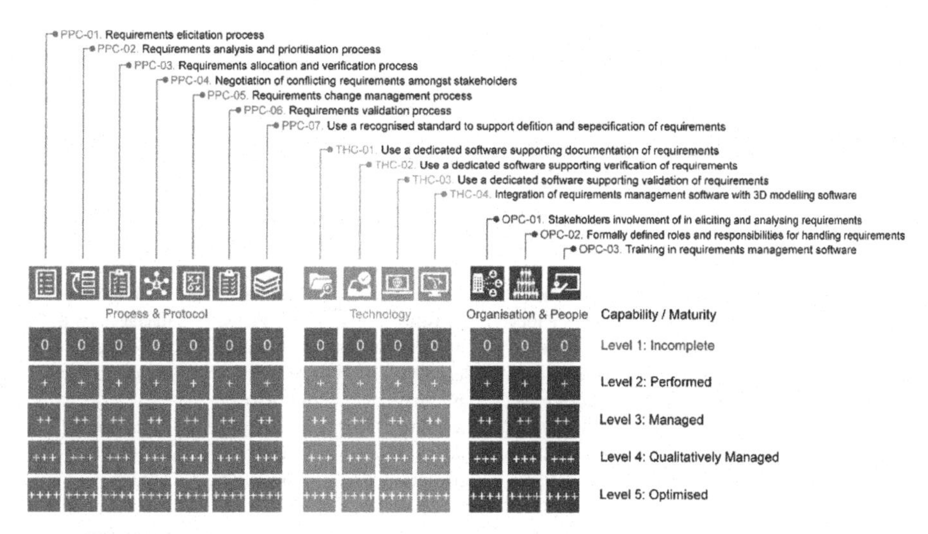

Fig. 4. Matrix displaying the scores for 14 capabilities in different maturity levels

Table 3. Description of requirements co-engineering maturity levels

Main capabilities	Maturity levels				
	Level 1: Incomplete	Level 2: Performed	Level 3: Managed	Level 4: Qualitatively managed	Level 5: Optimised
Process & protocol					
Process	There is no formal process defined or implemented	An ad hoc process is implemented during project delivery	An organisational standard that describes a generic process exists	Level 3 + Process is monitored, and performance is assessed	Level 4 + Continuous process improvement enabled by performance feedback loop
Protocol	There is no standard used to define and specify requirements	Standards supporting definition and specification of requirements based on individual delivery-side stakeholder approach	An industry sector-specific standard is used to define and specify requirements	A standard specified by the Government Agency/Client is used to define and specify requirements)	International standards are utilised (E.g., ISO Standard used to define and specify requirements)

(continued)

Table 3. (*continued*)

Main capabilities	Maturity levels				
	Level 1: Incomplete	Level 2: Performed	Level 3: Managed	Level 4: Qualitatively managed	Level 5: Optimised
Technology					
Software tools	There is no dedicated requirements software tool utilised by project delivery team organisations	Separate requirements software tools are utilised across the various project delivery team organisations	An integrated requirements software tool/platform is **used by a minority** of relevant project delivery team organisations	An integrated requirements software tool/ platform is **used by a majority** of relevant project delivery team organisations	Level 4 + Requirements managed by a dedicated project role, e.g., requirements engineer, systems engineer, digital engineer, BIM manager
Integration of requirements management software and 3D modelling software	Neither requirements management nor 3D modelling software is used	Only requirements management software is used but 3D modelling software is not	Separate and distinct requirements management and 3D modelling software is used, however there are no digital links between them	There is basic integration enabled between the requirements management and 3D modelling software utilised, e.g., providing spatially enabled requirements mapping, linking requirements with 3D objects, and automating a basic level of spatial requirements verification	There is a high level of integration enabled between the requirements management and 3D modelling software utilised, supporting the use of configuration management to establish and maintain consistency of system performance, functional, and physical attributes with its requirements, design, and operational information

5 Discussion and Future Research

Adopting a design science research methodology, a preliminary requirements co-engineering capability improvement framework is developed. This framework includes two parts: i) capabilities supporting requirements co-engineering, and ii) maturity levels of capabilities. The main capabilities are developed based on findings from a systematic literature review and semi-structure interview survey with industry experts. They are then categorised into process and protocol, technology, organisation and people. Maturity levels of process and protocols in this research follow the structure of REPAIM and R-CMMi models while technology related maturity levels conclude from semi-structured interview survey.

During our semi-structured interview survey, we identified that there exists a great dependency on a co-engineering approach during the creation of complex and adaptive systems, where "co" requires the project team to work towards the virtual deliverables (e.g., digital twin, or cyber-physical systems) as a common goal [31]. Co-engineering therefore addresses both collaborative and concurrent engineering concepts. The impacts of the implementation of systems and co-engineering approaches can be identified at two levels; the organisation and project relative to the "mind-set" and sharing of the digital twin system objectives and vision. Thus, adopting systems and co-engineering approaches is identified as a key criterion for complex and adaptive systems when the lifetime of the asset extends over several decades [31]. For complex infrastructure projects the co-engineering of information requirements is key to support the delivery of both physical and virtual assets with decades long lifespans.

Thus, the development of this improvement framework is aimed at identifying main capabilities supporting requirements co-engineering and presenting an approach for the organisation or project team to understand the current maturity level of requirements management capabilities and a potential pathway for improvement. Further development and validation of this improvement framework will be conducted using the Delphi Method through an expert panel consisting of experienced practitioners and researchers in the area of complex infrastructure, systems engineering and requirements engineering.

Acknowledgement. This research is supported by an Australian Government Research Training Program scholarship.

References

1. Ramos, V.S.S.A.: A CMMI-compliant requirements management and development process. Ph.D. thesis (2014)
2. Beecham, S., Hall, T., Rainer, A.: Building a requirements process improvement model. 20 (2003)
3. Beecham, S., Hall, T., Rainer, A.: Defining a requirements process improvement model. Softw. Qual. J. **13**, 247–279 (2005). https://doi.org/10.1007/s11219-005-1752-9
4. Pettersson, F., Ivarsson, M., Gorschek, T., Öhman, P.: A practitioner's guide to light weight software process assessment and improvement planning. J. Syst. Softw. **81**, 972–995 (2008)
5. Sommerville, I., Sawyer, P.: Requirements Engineering: A Good Practice Guide. Wiley, Hoboken (1997)
6. Gorschek, T., Tejle, K.: A method for assessing requirements engineering process maturity in software projects (2002)
7. Solemon, B., Sahibuddin, S., Ghani, A.A.A.: A new maturity model for requirements engineering process: an overview (2012)

8. Rana, M.E., Dauren, J., Kumaran, S.: An improved requirements engineering framework for cloud based application development. In: 2015 IEEE Student Conference on Research and Development (SCOReD), pp. 702–709 (2015)
9. Solemon, B., Shahibuddin, S., Ghani, A.A.A.: Re-defining the requirements engineering process improvement model. In: 2009 16th Asia-Pacific Software Engineering Conference, pp. 87–92. IEEE, Batu Ferringhi (2009)
10. Software Engineering Institute (2010) CMMI for Development, Version 1. 3
11. Hevner, A.R., March, S.T., Park, J., Ram, S.: Design science in information systems research. MIS Q. **28**, 75–105 (2004)
12. Hevner, A.R.: A three cycle view of design science research. Scand. J. Inf. Syst. **19**, 87–92 (2007)
13. Johnson, A., Heaton, J., Yule, S., et al.: Informing the information requirements of a digital twin: a rail industry case study. Proc. Inst. Civ. Eng. - Smart Infrastru. Constr. **174**, 1–13 (2021)
14. Heaton, J., Parlikad, A.K., Schooling, J.: A building information modelling approach to the alignment of organisational objectives to asset information requirements. Autom. Constr. **104**, 14–26 (2019)
15. Cavka, H.B., Staub-French, S., Poirier, E.A.: Developing owner information requirements for BIM-enabled project delivery and asset management. Autom. Constr. **83**, 169–183 (2017)
16. Baldauf, J.P., Formoso, C.T., Tzortzopoulos, P., et al.: Using building information modelling to manage client requirements in social housing projects. Sustainability **12**, 2804 (2020)
17. Nekvi, M.R.I., Madhavji, N.H.: Impediments to regulatory compliance of requirements in contractual systems engineering projects: a case study. ACM Trans. Manage. Inf. Syst. **5**, 15:1–15:35 (2014)
18. Ramos, C.: Requirements management-how 110,000 requirements are managed on northwest rapid transit. In: CORE 2018: Conference on Railway Excellence. Railway Technical Society of Australasia (RTSA) (2018)
19. Jallow, A.K., Demian, P., Anumba, C.J., Baldwin, A.N.: An enterprise architecture framework for electronic requirements information management. Int. J. Inf. Manage. **37**, 455–472 (2017)
20. Koltun, G.D., Feldmann, S., Schütz, D., Vogel-Heuser, B.: Model-document coupling in aPS engineering: challenges and requirements engineering use case. In: 2017 IEEE International Conference on Industrial Technology (ICIT), pp. 1177–1182. IEEE (2017)
21. Arnaut, B.M., Ferrari, D.B., Souza M.L.D.O.: A requirements engineering and management process in concept phase of complex systems. In: 2016 IEEE International Symposium on Systems Engineering (ISSE), pp. 1–6 (2016)
22. Gebru, H.M., Staub-French, S.: Leveraging data to visulize and assess space planning compliance (2019)
23. Junior, J.S., Baldauf. J.P., Tzortzopoulos, P., et al.: The role of building information modelling on assessing healthcare design. In: CIB World Building Congress 2019: Constructing Smart Cities. International Council for Research and Innovation in Building and Construction, pp. 1659–1672 (2019)
24. Arayici, Y., Fernando, T., Munoz, V., Bassanino, M.: Interoperability specification development for integrated BIM use in performance based design. Autom. Constr. **85**, 167–181 (2018)
25. Fucci, D., Palomares, C., Franch, X., et al.: Needs and challenges for a platform to support large-scale requirements engineering: a multiple-case study. In: Proceedings of the 12th ACM/IEEE International Symposium on Empirical Software Engineering and Measurement, pp. 1–10. ACM, Oulu (2018)
26. Jupp, J., Awad, R.: BIM-FM and information requirements management: missing links in the AEC and FM interface. In: Ríos, J., Bernard, A., Bouras, A., Foufou, S. (eds.) PLM 2017.

IAICT, vol. 517, pp. 311–323. Springer, Cham (2017). https://doi.org/10.1007/978-3-319-72905-3_28

27. Navendren, D., Mahdjoubi, L., Shelbourn, M., Mason, J.: An examination of clients and project teams developing information requirements for the asset information model (AIM). Build. Inf. Modell. (BIM) Design Constr. Oper. **149**, 169 (2015)
28. NATSPEC BIM - ABAB AIR Guide. https://bim.natspec.org/documents/abab-air-guide. Accessed 19 Mar 2022
29. Hatakeyama, J., Seal, D., Farr, D., Haase, S.: An alternate view of the systems engineering "V" in a model-based engineering environment. American Institute of Aeronautics and Astronautics, Orlando, FL (2018)
30. TfNSW: T MU AM 06006 ST Systems Engineering (2017)
31. Chen, Y., Jupp, J.R.: Challenges to asset information requirements development supporting digital twin creation. In: Canciglieri Junior, O., Noël, F., Rivest, L., Bouras, A. (eds.) PLM 2021. FIP Advances in Information and Communication Technology, vol. 639, pp. 474–491. Springer, Cham (2022). https://doi.org/10.1007/978-3-030-94335-6_34

Reliability of Design Data Through Provenance Management

Tim G. Giese(✉) and Reiner Anderl

Technische Universität Darmstadt, 64287 Darmstadt, Germany
{giese,anderl}@dik.tu-darmstadt.de

Abstract. In today's virtual product development, huge amounts of data are shared and exchanged between a large number of experts in collaborative and highly complex design tasks. While in these processes designers are oftentimes working with data, which were not created by themselves, they are missing knowledge and transparency about the origin and reliability of the data source. Approaching this problem, we identified the necessity of a responsible and transparent generation and usage of design data. Therefore, we developed a concept, which tracks and stores the historical record of a data item and its modifications in order to identify and evaluate the source of the data item. The concept proposes a novel provenance model, which consists of a provenance graph, design criteria and evaluation criteria. To validate the concept, a prototypical implementation was conducted and evaluated. We came to the conclusion, that the presented concept can be used effectively to model and evaluate the historical record of a data set in the virtual product development in order to create a transparent and reliable use and generation of design data.

Keywords: Data provenance · Data literacy · Virtual product development

1 Introduction

In the modern product lifecycle management (PLM), data is seen as one of the most valuable assets. Massive amounts of data are collected, stored and distributed since the majority of the workforce in PLM are interacting with data on a daily level [1, 2]. However, research indicates that a vast number of workers are lacking essential skills to properly interact with data and use its full potential. This aspect is addressed by the research area Data Literacy [2]. Moreover, with the exchange of bigger amounts of data, processes in the design of virtual products are becoming more complex and untransparent. In global and collaborative projects, designers are often working with data, which were not created by them. Furthermore, there oftentimes exists missing knowledge about the origin and reliability of data they are working with [3]. This lack of transparency leads to a missing trust in the data and hence the potential of data is not fully used. Consequently, research proposes a transformation of the PLM towards a responsible generation and usage of design data to create a reliable and transparent

Published by Springer Nature Switzerland AG 2023
F. Noël et al. (Eds.): PLM 2022, IFIP AICT 667, pp. 274–283, 2023.
https://doi.org/10.1007/978-3-031-25182-5_27

product lifecycle. In order to do so, all participants of the product lifecycle have the responsibility to deliver reliable and high quality data [4]. However, suitable solutions in order to introduce Data Literacy to product development and provide designers with more competence in terms of gaining knowledge about a data source, are still under development [2].

To approach this issue and help designers not having to blindly rely on the quality of data, we developed a concept which provides a designer with reliable data in terms of identifying and evaluating the source of a data item. This concept consists of the development of a completely novel provenance model for virtual product development and focuses on tracking and storing the historical record of the generation and modification of data sets.

2 State of the Art

In the following, current research on Data Literacy and Data Provenance is described as well as the relevant previous work conducted by us.

2.1 Data Literacy

Although the research topic "Data Literacy" is tremendously gaining in significance, there does not exist a generally accepted definition of the term. However, what definitions have in common is the fact that the term is used to describe the ability to collect, critically assess and consciously apply data in a given context [5]. Nevertheless, this competence is not focused on particular scientific fields only, but being data literate rather describes an interdisciplinary data expertise [6]. The key competencies, which a data literate individual is considered to be equipped with are: exploration, prediction and inference. Using these competencies useful conclusions from large and diverse data sets can be drawn [6]. Despite the importance of these competencies a lack of these skills can be identified in the engineering workforce. Recent studies show that companies are increasingly looking for employees with an expertise in the interaction with data [7]. Furthermore, the skillset of a data literate individual is not limited to academia or the workforce only, but additionally Data Literacy focuses on skills in order to solve "real-world" problems as well. This is emphasized by the fact, that the varying definitions of Data Literacy additionally describe a skillset regarding the interaction of data, which can be used for everyday thinking and reasoning [5]. Since daily interactions with data are nowadays inevitable and common place throughout all age groups and all areas of life, Data Literacy is tremendously gaining in significance. Due to the ongoing digitalization, researchers highlight the necessity of knowledge about a proper interaction with data and even compare the importance of being data literate with the ability of how to read [5].

2.2 Data Provenance

Due to the increasing digitalization, the amount of data which is exchanged and shared between participants of the supply chain is significantly increasing. Furthermore, the growing amount of participants often cause unclear situations about the origin or owner of a data item and its reliability [8]. The scientific area which is focused on the detailed description about the origin and authenticity of data is called "Data Provenance" or "Data Lineage". The provenance of a data item can be seen as the historical record of its derivation [9]. This record contains information about the source, processes and the current representation of the data item. Especially in collaborative, multidisciplinary projects (like the product lifecycle), in which designers often work with data which were not created by themselves, knowledge about the origin, transformations, which were applied to a certain data item and it's connection to other data sets are of great interest [8]. With knowledge about changed parameters, conducted simulations and individuals involved throughout the design process, the quality and reliability of a data item can be determined [3]. In literature models, which are able to save, depict and manage all decisions, procedures and results, which lead to the current state of a data set, are called provenance models [8].

There exist a great variety of different provenance models, which are able to receive the historical record of a data item. The majority of provenance models are graph-based since this creates the possibility to link data sets with defined relationships. Commonly well-known graph-based provenance models are the PROV W3C recommendation [10] and the Open Provenance Model [11]. The commonality of these models is, that they describe a meta provenance graph to describe the transformations a data item undergoes. These graphs generally consist of two categories: relationships (also called edges) and nodes (also called vertices) [10]. Relationships define the connection between two nodes. The second category – nodes – are further divided into 3 subcategories – entities, agents and activities [10]. Entities represent a physical, digital or conceptual unchanging state of a unit. Examples are documents, graphics or data sets. Activities describe an action which has been performed on or caused by an entity and are creating new entities while using already existing ones. An example is an action or a process. An agent can be understood to be taking on a role in an activity, which means, that an agent is responsible for an activity. Examples for an activity are a person, organization or software [10].

In comparison to provenance graphs, modern product data management systems (PDMS) are able to track and display the historical record of a data item in great detail as well. However, for tracking provenance information PDMS require that every contributor works on the same PDMS [12]. Once a data set leaves that PDMS the historical record of the data item cannot be tracked anymore [13]. Consequently, provenance models are needed next to PDMS since in the modern supply chain data is exchanged across company borders involving a great variety of different systems [12].

2.3 Previous Work

In the modern PLM an increasing amount of data is exchanged and shared among an increasing amount of participants [1]. This is oftentimes causing ambiguity of where a data item originates from and whether that source is reliable [3]. Even though in virtual product development the majority of the workforce are interacting with data on a daily basis, research indicates that, designers are missing skills and essential knowledge about the proper interaction with data originating from an external source in order to use that data to its fullest potential [3, 4, 7].

To approach this issue in a previous research paper [14], the authors developed a concept for the introduction and application of Data Literacy to virtual product development in order to create transparency and awareness for a responsible generation and use of design data [14]. This concept is called "Design Data Literacy" and the architecture is depicted in Fig. 1.

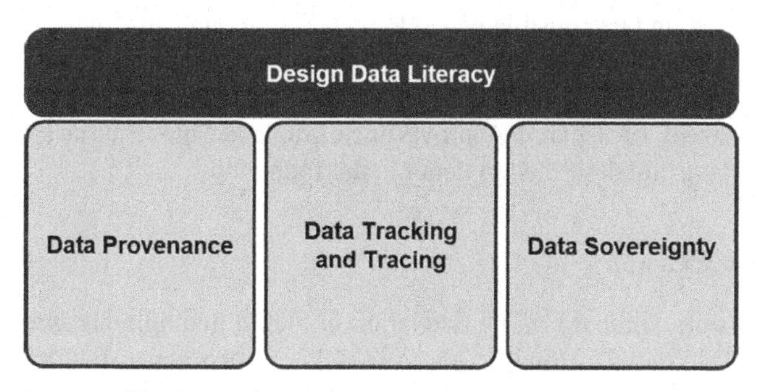

Fig. 1. Architecture of design data literacy [14]

In order to develop a comprehensive approach for the introduction of Data Literacy to virtual product development the concept is based on 3 major components – *Data Provenance*, *Data Tracking and Tracing*, as well as *Data Sovereignty*. Each component represents a different aspect of transparency and awareness and combined they convey the key competencies of Data Literacy adjusted to product development. The component Data Provenance is focused on identifying the origin and reliability of a data item. After a virtual product has been designed and further distributed, the component Data Tracking and Tracing is creating an overview about the receiver of the virtual product and the modifications which have been conducted. The third component – Data sovereignty – regulates the ownership and rights when the virtual product is further distributed [14].

In order to display and store relevant provenance information, a tree structure was identified to be best suited [14]. Since the development of the overall concept of Design Data Literacy, the component Data Provenance has been worked out in detail in terms of a development of a comprehensive data provenance model, which is described in the following.

3 Conceptual Approach

In order for designers in virtual product development to use design data to their fullest potential, they need to be provided knowledge about the origin and the quality of a data item. As identified by the authors in a previous research paper, more information about the historical record of a data set, which is exchanged across companies and several systems, can be achieved by a provenance model (see Fig. 1). The more knowledge a designer of a virtual product has about the origin and modifications of a data item, the more responsible and target-driven data can be used. Information about the origin and conditions under which a data set was produced generates high-quality and reliable design data without having to blindly rely on the creator.

Therefore, the goal of this concept is to create a comprehensive provenance model which generates a transparent and reliable use of design data within the product lifecycle. In order to achieve this goal three requirements are defined.

- R1: Provision of reliable and high-quality data
- R2: Identification and evaluation of the source of data
- R3: Transparent and responsible generation and usage of design data

To develop a comprehensive provenance model which addresses all of the requirements, the concept consists of 3 parts: A provenance graph, design criteria and evaluation criteria. The 3 parts are described in detail in the following.

3.1 Provenance Graph

To track the historical record of the derivation of a data item graph-based provenance models are well suited [3]. However, these models can be seen as meta models, which suggest a structure and are not yet adjusted to a certain scientific area or problem. In order to develop a provenance graph for virtual product development in which large numbers of experts from a great variety of disciplines are working collaboratively together, our model is mainly based on the PROV W3C recommendation [10], but has been extended by several additional structures. First, all relevant criteria to model and track the provenance of a data item in virtual product development were identified and subsequently integrated into a graph structure by linking the nodes with relationships (Fig. 2).

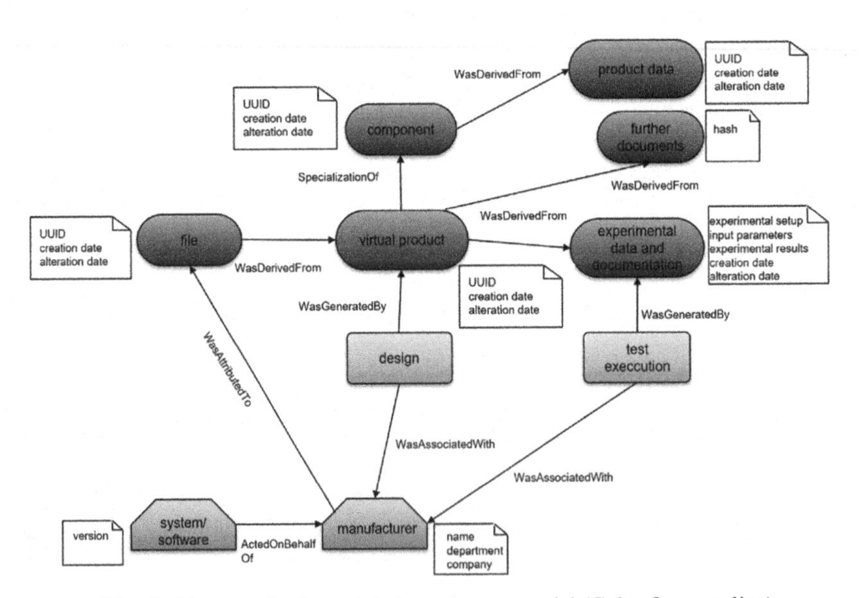

Fig. 2. Meta graph of presented provenance model (Color figure online)

For the model depicted above, the general layout of the PROV W3C recommendation has been adopted, consisting of entities, agents, activities and the associated relationships. The graph illustrates all relevant agents (orange), activities (blue) and entities (yellow), which are necessary in order to track provenance information in virtual product development. Additionally, attributes of a node are displayed in the white boxes. An agent represents a manufacturer or system involved in the design process. Moreover, the entity nodes are divided into 3 categories (file level, component level and product data level). The virtual product represents a 3D-data model of a product. The component node represents a part or an assembly of a virtual product. Moreover, product data describe the characteristics of a data item (e.g. material, density, weight, elastic module, surface, volume). Furthermore, entities which are necessary for a comprehensive description of a historical record of a data item are experimental data and documentation (documentation of the conducted simulations and their results) as well as further documents like measuring reports for example. The relevant actions involved in the design process of a virtual product are represented by the design and test execution nodes. Additionally, the relationships describe how 2 nodes in the design process are linked to each other. At last, for a clear identifiability of a data item and its current state the entities have a universally unique identifier (UUID) as well as creation and alteration dates, which are assigned to the nodes in terms of attributes.

The presented graph represents the structure of all relevant design data and their relationships to each other which are necessary to evaluate the quality and reliability of design data. To determine the origin and creator of a specific data item as well as transformations it underwent across its life cycle this graph still needs to be implemented and filled with specific information.

3.2 Design Criteria

Although with the presented provenance graph it is possible to track the historical record of a part or assembly, it can only display information which are stored in the file of the respective component. This means, if a previous designer did not specify from which standard a screw was taken for example (no entry in the file history), it is not possible to track this information. Consequently, this leads to the conclusion that a concept for a comprehensive, transparent and responsible use of design data cannot be guaranteed by usage of a provenance graph only. This highlights the necessity of further criteria, which need to be adhered to in order to guarantee a responsible design process. An extract of additional design criteria is depicted in Table 1.

For a responsible use of design data, every participating designer has the responsibility to create and promote trust. This means that conducted tests or simulations have to be documented for a subsequent designer, including time stamps, input parameters and test results. This creates the ability for a subsequent designer to re-run a test and therefore create trust in the data. Furthermore, inserting a PMI (Product Manufacturing Information) in a data model (consisting of a time stamp and information about the used standard) guarantees the ability to double check if a model was designed according to the correct standard and if that standard was still valid at that time. Moreover, in the product lifecycle documents like measuring reports are oftentimes exchanged among several

Table 1. Extract of design criteria

Design criterion 1	Verifiably documented tests and simulations
Design criterion 2	Test documentation must be available for subsequent designer
Design criterion 3	Standard-compliant design must be verifiable
Design criterion 4	External suppliers have to document provenance data
Design criterion 5	Further documents have to be securely encrypted

system participants. Securely encrypting these documents with hash values creates the advantage, that a recipient can verify, if the document was modified without permission.

The presented design criteria are included in the provenance graph as attributes of the respective entities (see Fig. 2).

3.3 Evaluation Criteria

In order to develop a data provenance model which automatically determines and evaluates the reliability of a source of a data item, evaluation criteria need to be identified. Therefore, the term reliability was defined first: "Reliability is the traceability of the history (place, author and time), conditions of origin and context under which a datum was created or modified in the product development process, so that the creation of the datum can be traced and reconstructed, and thus the quality of the data can be assessed, creating trust in the data". Subsequently, evaluation criteria were defined based on which algorithms search the provenance graph for a specific component. Following that, the reliability of the origin can be evaluated based on the selected criterion.

3.4 Architectural Approach

In this paragraph, the overall architecture of the provenance model is presented which is derived from the previous presented partial concepts (depicted in Fig. 3).

For the development of a comprehensive provenance model for virtual product development, it is crucial to consider the various phases of data processing: collection, storage, usage and transfer [4]. For the model presented in this paper, the phase of transfer is out of scope and the phase of usage is further divided into access and analysis. In order for a designer to determine whether the received data is reliable, the source of the data needs to be identified and evaluated. To achieve this, in the phase of collection a designer receives a data model, which is usually a virtual 3D-model of a product. With the usage of an Application Programming Interface (API) all relevant provenance data is exported in a machine-readable format and stored in a data base (in the presented concept a graph-data base is used and more specifically the provenance graph (see Sect. 3.1)). Additionally, further documents (like measuring reports) are encrypted with a hash value and the respective value is stored in the data base as well (phase storage). For a defined access to the data base a graphical user interface (GUI) was developed (phase access). In the GUI, a user can select a predefined evaluation criterion which results in an algorithm searching the graph for a specific part. Once the respective data item is found and the reliability

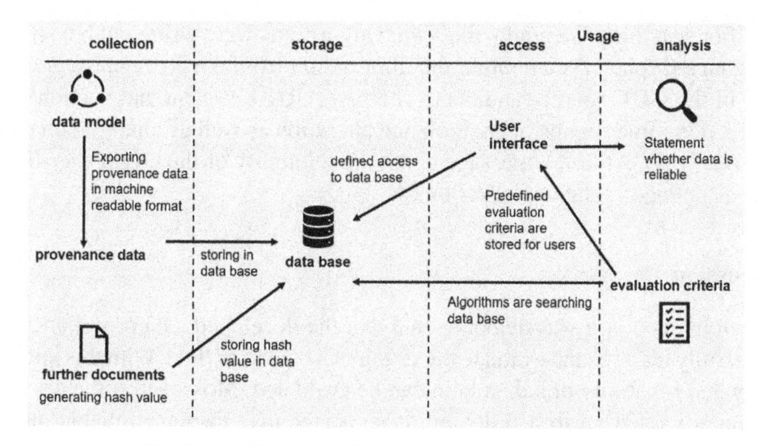

Fig. 3. Architecture of the overall provenance model

of its origin is evaluated information whether the source is reliable will be displayed in the GUI (phase analysis). As a specific application to the PLM domain, this provenance model can be used when a designer receives design data from a previous participant of the supply chain to determine where the data originates from, who was involved in creating the data and to rerun conducted experiments (like finite element, kinematic or dynamic analysis for example). The design data which the concept evaluates can be a file, virtual product or component of a product. Being able to determine the creator and conditions under which a data item was created provides designers with the possibility to evaluate the reliability and quality of data.

4 Validation

To validate the presented concept of a provenance model for virtual product development, the concept was prototypically implemented and the results are discussed.

4.1 Prototypical Implementation

For the implementation a sample virtual product was modelled in Siemens NX 12, consisting of parts modelled by several different designers and as a sample experiment a finite-element analysis was run on the product with predefined input parameters (the FEA is used as a sample experiment to retrieve its provenance information later on). After we received the 3D CAD model of the sample part, the provenance information had to be retrieved and exported from the data model. To achieve this, the API NX Open was used and the relevant provenance information exported into a CSV file. The application for the export was written in C#. Subsequently, the CSV file was imported into the graph data base using Neo4j (filling the provenance graph with specific information). Once the provenance information was stored in the data base an evaluation criterion could be selected from a drop-down menu in the developed GUI (which was written in Python 3). By selecting a criterion, an algorithm searched the graph data base for the respective data item. For searching the graph, the search algorithms were written in Neo4j's graph query language Cypher. As an output the information from the provenance graph were displayed in the GUI (source (author) of the part, UUID, system and version the part was modelled in, time stamps of creation and alteration as well as input parameters and results of the FEA). At last, a message about the reliability of the origin was displayed based on the defined evaluation criterion.

4.2 Discussion

In the previous section it was demonstrated that the developed data provenance model can successfully identify and evaluate the origin of a data item (R1). With this knowledge the quality and reliability of a data item can be evaluated. Moreover, if the provenance information is passed on to a following designer or manufacturer reliable and high-quality design data is guaranteed (R2). Furthermore, keeping track of and displaying the historical record of every component used in a design creates a transparent and responsible generation and usage of design data (R3).

Even though we successfully developed a provenance model for the transparent and reliable use of design data within the product lifecycle, there are further aspects one has to take into account. Securely encrypting a document with a hash value only guarantees security to a certain degree. Even when assigned a proper digital signature, there always remains a risk that the hash value is forged and modified by a man in the middle [15]. Moreover, the goal of a transparent and reliable use of design data along the product lifecycle can only be guaranteed if all participants contribute to that [4]. When the design criteria proposed in Sect. 3.2 are adhered to this is ensured, however, the possibility remains that some individuals do not obey to these rules. In that case a transparent and reliable use of design data cannot be guaranteed anymore.

5 Conclusion

We successfully developed and implemented a novel concept for a transparent and reliable use and generation of design data along the product lifecycle by tracking and storing the historical record of any data item used in the design process of a virtual product. In the product lifecycle many experts from a great variety of disciplines are working collaboratively. However, the origin of a data item is often unclear and whether the data is reliable. The more complex and untransparent a design process gets, the more a designer of a virtual product has to blindly rely on the reliability of the data. However, if the quality of data is not known the potential cannot be fully used. To improve this issue, designers need to be provided with reliable and high-quality data. To approach this problem and provide this information our concept consists of a provenance graph, which tracks and stores the historical record of design data. With the provenance graph, it is possible to identify the origin and to understand which transformations a dataset underwent and who was responsible for the modifications. Moreover, to evaluate the quality and reliability of the data, evaluation criteria were developed. Using these criteria, the provenance graph can be searched for a specific data set and the reliability of the source and quality of the data is automatically evaluated. Additionally, to guarantee a responsible generation and usage of design data along the whole product lifecycle, further design criteria were developed as a guideline of how to generate and exchange data with one another to achieve trust in data and a responsible and transparent product lifecycle.

Having considered all these aspects, we developed a comprehensive data provenance model for a responsible generation and use of design data in order to create transparency and reliability when exchanging data in virtual product development.

References

1. Collins, V., Lanz, J.: Managing data as an asset. CPA J. **89**, 22–27 (2019)
2. Pothier, W.G., Condon, P.B.: Towards data literacy competencies: business students, workforce needs, and the role of the librarian. J. Bus. Financ. Librariansh. **25**(3–4), 123–146 (2020). https://doi.org/10.1080/08963568.2019.1680189
3. Reitenbach, S., Vieweg, M., Hollmann, C., Becker, R.-G.: Usage of data provenance models in collaborative multi-disciplinary aero-engine design. In: Turbomachinery Technical Conference and Exposition (2020)
4. Die Bundesregierung – Bundeskanzleramt Deutschland: Datenstrategie der Bundesregierung (2021)
5. Frank, M., Walker, J., Attard, J., Tygel, A.: Data literacy - what is it and how can we make it happen? J. Commun. Inform. 4–8 (2016)
6. Adhikari, A., DeNero, J.: Computational and inferential thinking: the foundations of data science. University of California, Berkeley (2019)
7. Tech Partnership, Employer Insights: Skill Survey. https://www.techskills.org/globalassets/pdfs/research-2015/tec_employer_skill_survey_web.pdf
8. Glavic, B., Dittrich, K.R.: Data provenance: a categorization of existing approaches (2007)
9. Schreiber, A.: Provenance für workflows und prozesse, 4 March 2011
10. Groth, P., Moreau, L.: PROV-Overview: an overview of the PROV Family of documents. https://www.w3.org/TR/2013/NOTE-prov-overview-20130430/. Accessed 17 Aug 2021
11. Moreau, L., Freire, J., Futurelle, J., McGrath, R.E., Myers, J., Paulson, P.: The open provenance model: an overview. In: Freire, J., Koop, D., Moreau, L. (eds.) Provenance and Annotation of Data and Processes. IPAW 2008. Lecture Notes in Computer Science, vol. 5272, pp. 323–326. Springer, Heidelberg (2008). https://doi.org/10.1007/978-3-540-89965-5_31
12. Mesihovic, S., Malmqvist, J., Pikosz, P.: Product data management system-based support for engineering project management. J. Eng. Design **15**, 389–403 (2004)
13. Sebes, E.J., Stamp, M.: Solvable problems in enterprise digital rights management. Inf. Manag. Comput. Secur. **15**(1), 33–45 (2007). https://doi.org/10.1108/09685220710738769
14. Giese, T.G., Anderl, R.: Design data literacy – impact of data literacy in virtual product development. In: 2021 IEEE Asia-Pacific Conference on Computer Science and Data Engineering (CSDE), Brisbane, Australia, pp. 1–8 (2021)
15. Hasan, H.R., et al.: A blockchain-based approach for the creation of digital twins (2020)

Developing a Roadmap Towards the Digital Transformation of Small & Medium Companies: A Case Study Analysis in the Aerospace & Defence Sector

Federica Acerbi(✉), Marco Spaltini, Anna De Carolis, and Marco Taisch

Department of Management, Economics and Industrial Engineering, Politecnico di Milano, Milan, Italy
{federica.acerbi,marco.spaltini,anna.decarolis, marco.taisch}@polimi.it

Abstract. The contemporary era is pushing companies worldwide in undertaking a digital transformation path to keep high their competitive advantages acquired throughout the years thanks to their engineering competencies. Companies, especially Small & Medium ones, are getting forced to set up clear roadmaps towards an enhanced digital maturity level to address their strategic objectives. Roadmaps must support them in being both efficient and effective to keep stable their competitive advantage. Therefore, the goal of this contribution is to clarify the key elements representing the basement towards an improvement digital path and apply them to a case study. The key elements emerged to be: i) the clarification of firms' strategic objectives, ii) the awareness about firms' current internal digital maturity level to benchmark themselves in respect to competitors as well as their expected "desired TO-BE scenario", and iii) the investigation of the causes and related effects that may harm the reaching of the strategic objectives. Hence, in this contribution, these three steps are deeply investigated to design a structured and tailored roadmap leading a company to reach an increased level of digital maturity facilitating the achievement of the strategic objectives. The roadmap supports companies in evaluating the most appropriate technologies to overcome the internal inefficiencies identified hindering the achievement of the corporate results. The roadmap development process was studied from the extant literature and it has been applied in this contribution in a case study, specifically in an Italian company operating in the Aerospace & Defense (A&D) sector.

Keywords: Digital transformation · Aerospace & Defence case study · Roadmap

1 Introduction

Nowadays, companies are asked to keep high their competitive advantage optimizing their resource consumption, and digital and Industry 4.0 (I4.0) technologies may help them to move towards this direction [1]. Indeed, all the sectors are forced to improve

Published by Springer Nature Switzerland AG 2023
F. Noël et al. (Eds.): PLM 2022, IFIP AICT 667, pp. 284–293, 2023.
https://doi.org/10.1007/978-3-031-25182-5_28

their personalization performances to cover all their customers' needs and they are highly facilitated by the introduction of specific I4.0 technologies [2]. Nevertheless, they all need to find the proper balance of investments in technologies according to their strategies and the available financial resources. Different tools have been proposed in the extant literature with the goal to facilitate the Digital Transformation (DT) of manufacturing companies. For instance, [3] proposed a qualitative roadmapping tool to support Small & Medium Enterprises (SMEs) by evaluating the actions over five dimensions: business and strategy, product, customers and suppliers, production processes, factory and infrastructure. [4] proposed a Maturity Model (MM) assessing the adoption of I4.0 technologies and, [5] proposed a MM, named DREAMY4.0, aiming at evaluating the readiness of manufacturers in undertaking a DT. Although these models have been already applied in different contexts, they still lack a detailed and objective investigation over the main criticalities charactering the company to support the choice of a determined DT path. Among all, this need emerged especially in the A&D sector which is characterized, on one side by strict regulations due to the great complexity, on the other side the process digitalization lags behind the product digitalization due to the extensive product development life cycle and the limited production volumes [6]. For this reason, the research objective of this contribution is to create a robust, complete, and objective model to analyze and cover the criticalities emerged in manufacturing companies in an objective way after a maturity assessment, to provide them precise roadmaps for DT to justify the huge investments. Therefore, starting from DREAMY4.0, which emerged to be the most consolidated operative model (i.e., several citations and industrial adoptions) in the extant literature about MM towards DT, an extension has been developed in this contribution to perform an objective investigation of the major criticalities to build a structured roadmap. Moreover, considering the needs emerged in the A&D sector, the developed extended model has been applied in a manufacturing company operating in the A&D sector. The paper is structured as follows: Sect. 2 describes the theoretical background focusing on the description of the DREAMY4.0 model to highlight the key characteristics and the rooms for improvement, Sect. 3 provides the methodology employed, Sect. 4 elucidates the extended version of the model, Sect. 5 analysis the A&D case study in which the extended model has been applied and Sect. 6 concludes the papers highlighting the key contributions and the main limitations opening the way for further improvements.

2 Theoretical Background: DREAMY4.0

As just mentioned, the starting point of this contribution relies in the DREAMY4.0 digital readiness maturity model [5]. This assessment tool was developed by the Manufacturing Group of the School of Management of Politecnico di Milano and it has been already validated in multiple Italian and international-based companies (both SMEs and Large Enterprises) [5]. DREAMY4.0 aims to investigate the digital readiness of manufacturing companies by assessing six processes (i.e. design & engineering, production, quality, maintenance, logistics and supply chain) evaluating them based on four analysis dimensions namely: (i) *Execution* (i.e., how the processes are carried out within the company), (ii) *Organization* (i.e., information regarding the organizational aspects of the processes), *Control* (i.e. how a processes are monitored and controlled), *Technology* (i.e. information regarding the ICT systems, hardware and/or software, used in support of the process). The maturity is assessed along five levels of maturity [5]: from 1 (minimum level of digital readiness) to 5 (maximum level of digital readiness) in conformance to the CMMI framework [7]. DREAMY 4.0 enables to give a big picture about the current digital readiness of a manufacturing company by delineating a qualitative description of the potential rooms for improvement towards higher levels of maturity. Nevertheless, it lacks an objective quantification of the main weaknesses leading to a limited analysis of potential detailed solutions to be suggested for improvements.

3 Methodology

The present contribution aims at developing an extended version of the DREAMY4.0 especially focused on complex sectors such as the A&D. To achieve this objective, it has been relied on a workshop-based approach backed by a literature review to create the PCIM (Priority Criticality Index Mapping). Indeed, according to [8], the definition of a DT journey requires mainly a workshop-based approach. Such kind of involvement of company's employees described in the case study allows to maximize the level of commitment of all the relevant stakeholders that will implement or benefit from DT approach. Additionally, it facilitates the interviewers (i.e. the authors in this case) in the identification of the critical elements of the processes analyzed and their related solutions. Indeed, relying on an action research approach to conduct the analysis, allows to be more concrete and actively involve the case study company through constant feedbacks [9]. Figure 1 shows the 6 main steps followed to conduct the analysis to create an objective roadmap.

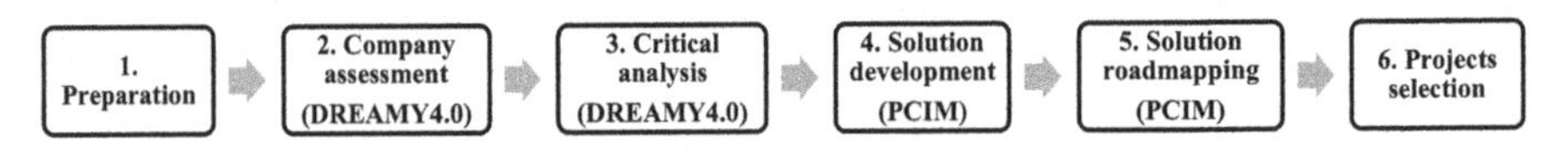

Fig. 1. Extended methodology to create an objective roadmap

4 Extended Model Development

In the first step depicted in Fig. 1, named "Preparation", there is a limited engagement of the firm, and thus, additionally to the external interviewers (here the authors) only the project champion/promoter and the CEO need to be involved. In particular, in this phase it is requested to collect all the relevant information regarding: (i) the industry in which the firm operates (e.g. trends, competitors, market approach etc.), and (ii) the records of the information about the company under analysis (e.g. balance sheets, newspapers, story of the firm, etc.). Within step 1, also the definition of the boundaries of the analysis, the signature of the NDAs and the definition of the people to be involved was carried out. Indeed, the assessment deployed is structured to cover the six processes reflecting those of the DREAMY4.0, thus the managers of the following functions were involved:

System Quality and Continuous improvement, Program, Operations, Production and Planning, Supply Chain, Quality, Logistics.

The second step, "Company assessment", requires a greater involvement of company's employees with particular regards of the functions' representatives. In this phase, the assessment method employed was the DREAMY4.0 (described in detail in the "theoretical background" section). Therefore, the authors carried out a set of interviews, both qualitative and quantitative, with the aim of collecting valuable information to understand the context in which the firm operates, to clarify its objectives, to assess the current level of digitalization by identifying its main criticalities and strengths.

The third step of the methodology, "Critical analysis", consists in the extrapolation of the key elements describing processes and technology adopted for each process, collected during the step 2, to identify the elements that might represent real barriers for the achievement of company's objectives.

Given the complexity and the needs of the sector, and more in general, the capacity and budget constraints of companies in undertaking a roadmap toward DT [10], a tool specifically designed to identify the most relevant criticalities for the case has been developed entering the fourth step "Solution Development" (see Fig. 1). Here, the extended part of the DREAMY 4.0 has been named PCIM and it aims at prioritizing criticalities and consequences in a quantitative form. Therefore, taking as input the weaknesses collected thanks to the DREAMY4.0, the PCIM methodology starts with a further sorting into: root weakness (the criticality) or weakness generated by the firsts (the effect). Afterwards, such criticalities and effects are linked through arrows which specify whether a block (criticality or effect), positioned at the tail of the arrow, has an impact on another block (effect), positioned at the head of the arrow. According to [11], graphical representation of problems facilitate the process of depicting alternatives (and scenarios) more easily and effectively. The map is then converted into a matrix which represents the occurrence of a criticality, row, in impacting a given effect, column. The translation of the map into a matrix is useful to numerically detect which criticalities generates most of the problems within the set of the six processes analyzed and consequently to propose a set of solutions able to address the most critical areas. Based on that, the fifth step of Fig. 1, "Solution roadmapping", is reflected into the prioritization of the solutions to be proposed based on the key criticalities identified and based on a further quantitative analysis of inefficiencies (e.g., waiting times, time dedicated to not value adding activities etc. extracted from database exports). More specifically, the authors identified 3 areas of intervention: 1) *Process*: including those actions of improvement which require a redesign of the processes and/or a change in the approach adopted to carry them out; 2) *Technology*: including those actions of improvement which involve the selection, implementation and integration of Information Systems and/or I4.0 technologies, 3) *Training*: including all those actions of improvement aimed at enlarging the competences of workers, at every hierarchical level, both in technological and procedural terms and/or at setting up a digital culture. This categorization was developed not only to clarify the drivers of the solutions but also to stress the importance of a multi-layered approach to DT. Indeed, the technological evolution of the system must be anticipated by reviewing processes to be digitalized to minimize the risk of digitalizing the inefficiencies [12]. On the other hand, the change in the technological configuration of the system as well as the re-engineering of the processes must be supported by a proper education of the individuals operating in the system [13]. The last step, "Projects selection", required the active involvement of both the authors and the firm's stakeholders. Indeed, it consisted in the prioritization of the solutions to be developed, the identification of the expected efforts and logical constraints and the chronological distribution of the solutions themselves. As suggested by [8], the activity needs to be conducted collectively to ensure the maximum commitment of all the managers involved and alignment with corporate objectives. In this phase, the objective was to define the actual action plan and ensure its robustness and feasibility. In particular, in terms of robustness, the authors have focused on the logical and technological priorities that might link different projects [14]. Hence, the requisite of robustness was mainly functional to the definition of the macro-phases of the roadmap developed and the positioning of solutions within each of them. In terms of feasibility, greater attention was devoted of the level of expected effort for each process and consequently on levelling them according to the actual resources available. It is worth specifying that the evaluation of effort, although shared with the managers of the case study firm, was defined in a qualitative way, and considered the following drivers [15, 16]: (i) Expected Full Time Equivalent (FTE) needed; (ii) Capital invested; (iii) Cross-functional coordination needed; (iv) Numerousness of the functions involved; (v) Numerousness of the processes modified; (vi) Coordination with external stakeholders needed.

5 Results from the Extended Model Application

In this section, the results of the application of the extended model is presented. The company subjected to the analysis is an Italian family-owned SME operating in the A&D sector. It is specialized in the design, development, production, maintenance and logistical support of defense equipment, structural components and ground support equipment of fixed-wing and rotary-wing aircraft and employs around 200 workers. Given the peculiarity of the industry, where the requisites are often fixed and rarely negotiable, the company adopts an Engineer to Order (ETO) approach. However, an element of stability in design and engineering effort is provided by the long lasting of the programs (i.e. the products) contracted. The CEO wanted to perform this assessment since within the organization some criticalities (unknown yet) were perceived as affecting processes efficiency. Hence, first it was posed the attention in formalizing and making evident those problems. For this reason, the scope of the analysis consisted in the definition and prioritization of the solutions addressing the criticalities collected to cope with the long-term corporate objectives: i) Increase in control over the processes; ii) Increase in process management efficiency; iii) Preserving the high product's quality standards; v) Supporting the expected increase in volumes. Due to the peculiarity of the industry, some information collected and analyzed, even if crucial for the prioritization of the intervention, will be omitted. The case study was conducted in four steps reflecting the structure of the next sections: **Task 1:** AS-IS critical analysis of processes; **Task 2:** Identification and prioritization of existing critical issues; **Task 3**: Identification and prioritization of improvement projects; **Task 4**: Roadmap development.

5.1 Digital Readiness and Criticalities Identification

The analysis of the AS-IS situation shows an overall maturity index equal to 3 as depicted in Fig. 2. First, it emerged that the "engineering" and the "supply chain" processes were those more ready towards a DT. This was especially driven by the commitment of the managers of these functions. On average the different processes were instead characterized by a poor control over the processes and clear difficulties in terms of coordination and collaboration among departments.

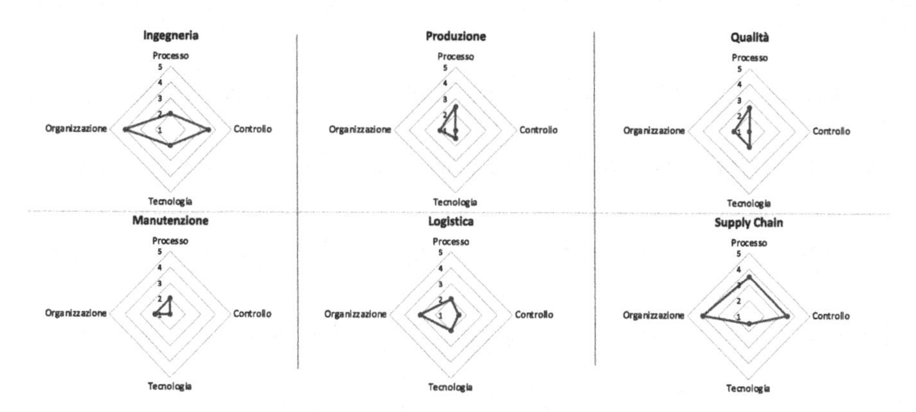

Fig. 2. Process maturity index

No processes were considered advanced from a technological perspective thus proving a pervasive inadequateness of ICT systems. Such gaps, in parallel to the low reliability on data quality, resulted in a general adoption of locally-stored files (mostly Excel documents) developed and used by individuals to manage their processes [17]. In addition, it emerged that the spread of home-made files turned out to reinforce the misgiving towards the IT systems. Overall, 38 main criticalities were detected and based on an iterative process these were clustered according to six macro categories: 1) Misalignment between purchasing and production Lead Times (LTs) and LT deemed acceptable by the market: 2 criticalities; 2) Process and product data not sufficiently disclosed among the functions and generated and managed outside the IT systems: 6 criticalities; 3) Process and product data not sufficiently reliable, manually input in the IT system and not unique: 8 criticalities; 4) Unsuitability of IT systems: 13 criticalities; 5) Lack of physical and logical Track & Trace of products and components throughout the procurement, production and delivery cycle: 3 criticalities; 6) Lack of codified and standardized procedures and methodologies: 6 criticalities. The identification of criticalities was backed by a quantitative approach which consisted in the analysis of 25 locally-stored files exploited by the departments during their activities, the ERP system and more than 40 documents ranging from procedures to follow up to performance reports. This analysis was beneficial for two main reasons:, it allowed the company to understand the real impact of the processes inefficiencies and, it supported the authors in estimating the expected benefits of the solutions proposed. To report an example, the time spent for not value adding activities (e.g., photocopying, manual delivery of documents etc.) proved to be to most critical. Design & Engineering areas, namely Research & Technology and Engineering, proved to be acutely affected by such waste which accounted, for Engineering, for up to 71% of the time spent by an FTE.

5.2 Prioritization of Criticalities

Once the criticalities were identified, the analysis of the links among criticalities and the related effects has been performed by using the PCIM model. Figure 3 shows the map in which all the linkages have been identified. All the criticalities are represented through a colored square based on the 6 macro-categories of criticalities defined while the effects are represented by pink hexagons. Overall, 32 effects coming from 50 criticalities have been collected. The links in the maps, once jointly validated with firm's managers, have been elaborated, into a matrix, through the PCIM model to quantify the occurrences of each criticality impacting on each effect.

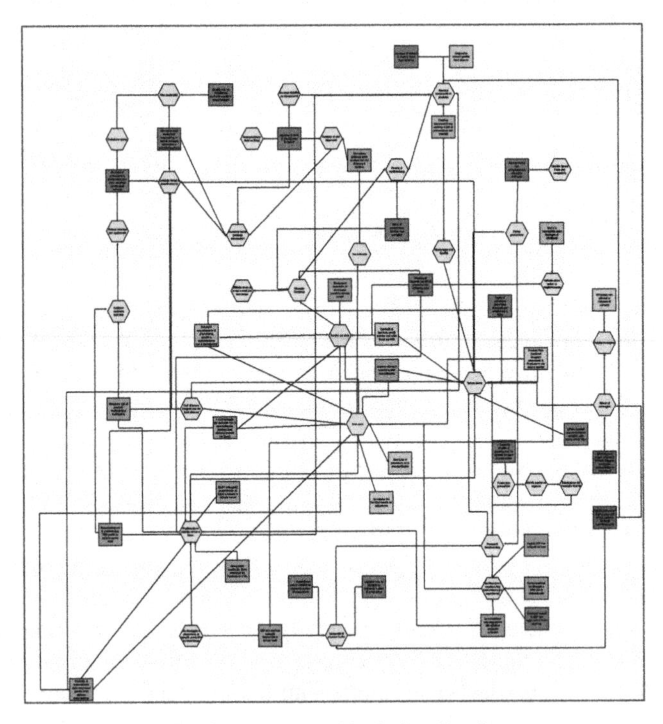

Fig. 3. PCIM graphical visualization

More specifically, each criticality has been mapped across the effects as well as each effect has been mapped across all the effects emerged to quantify the secondary effects weight. This enabled to quantitative evaluate how many times a criticality creates an effect looking also to the secondary effects. Last, to tailor the possible solutions to be proposed in accordance with the main criticalities observed, the matrix developed, linking the criticalities and the effects, has been divided into four quadrants based on the

Pareto law. With regards to the most relevant critical issues in terms of relationships with the effects found, it was noted that 50% of them account for 80% of the concatenations detected. Shifting the analysis on the most impacted effects, there is a high preponderance of those concerning the increase in times spent in not value adding activities. More specifically the main criticalities emerged are the following: Misalignment between time to market and supplier's delivery LT (C1), Data manually managed and not visible to functions (C2), Unreliable and not unique data (C3), ICT systems' inadequacy (C4), Poor materials and products track & trace (C5), Lack of standardized procedures and methodologies (C6).

5.3 Solution Development

Based on the PCIM, 50 different detailed solutions have been proposed. Afterwards, they were clustered according to 6 main areas of interventions described below:

1. PLM (**S1**): Introduction of a PLM system capable of supporting the efficiency of the process of data sharing from the technical department towards the others and of developing different Bills of Materials (BoM), like E-BoM, M-BoM etc.. This system needs also to be designed for the documentation management of the product currently managed by the various functions (e.g. Quality); (covering C2, C4)
2. ERP (**S2**): Update or replacement of ERP in use to meet the business requirements found (e.g. robust robustness, reliability of data, ability to manage different BoMs, purchasing procedures, planning, and warehouse); (covering C2, C4, C5)
3. ERP-MES (**S3**): Introduction of a system for visualizing and monitoring the production processes and assets which guarantees the product tracking. The system, identified as the MES, needs to be integrated with the ERP to enable consistent decisions based on the actual assets' performances; (covering C2, C4)
4. PROCESS REDESIGN (**S4**): Review of part of the procedures and process logics in use aiming to optimize and align them with the needs of the market. This project was designed also considering to the TO-BE information systems; (Covering C1, C6)
5. I4.0 (**S5**): Introduction of systems based on business intelligence and Machine Learning in order to make internal business processes more effective and efficient; (Covering C3, C5)
6. TRAINING (**S6**): Definition of training programme for staff at all hierarchical levels to raise awareness about the functionalities and benefits of digital technologies. (Covering C1, C2, C3, C4, C5, C6)

These proposed solutions addressing the key criticalities were also mapped along a timeline to highlight the prioritization in terms of both urgency and links. Hence, such blocks were organized as depicted in Fig. 4. The roadmap generated was composed by three transversal projects: cybersecurity, ICT systems integration and training which supported the projects on the organization, the update of the already existing systems and the implementation of the new ones. It is worth clarifying that the definition of such roadmap does not represent a standard of clustering of solution but rather was jointly defined with the management to support them to tackle the DT project into manageable and self-sustained sub-projects, defined based on an analytical and objective model.

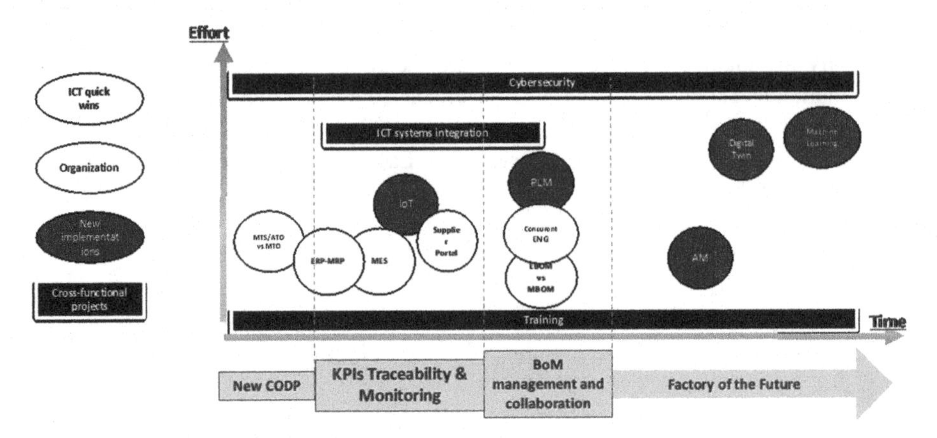

Fig. 4. Final roadmap

6 Discussion and Conclusions

This paper aimed to support manufacturers, operating in complex contexts like A&D sector, in undertaking a properly design DT. Therefore, based on literature review and workshop-based research, a new method was developed to quantify the criticalities emerged from the assessment to design a roadmap. This method, the PCIM, objectively clarifies the key criticalities to address through specific solutions. The proposed extended model, DREAMY4.0-PCIM, relies on six main phases and was applied in one A&D manufacturing company. The manuscript evidenced a successful integration of the tools which resulted into a comprehensive, cross-functional, and multi-objective roadmap for manufacturers operating in complex contexts. It enabled to first evaluate strengths and weaknesses of the different processes and quantify the criticalities. The following research presented some limitations related to the adoption of the single-case study approach and the limited focus on quantitative benefits achievable from the solutions proposed. Hence, the authors suggest to consider the following activities to expand the research: to implement a complete review of the organizational structure that may consider also the departments supporting operations with the aim of maximizing the effectiveness of the solutions, to deepen the micro-processes of the areas investigated to eliminate the macro-issues not tackled in this study, to support the firm in redesigning processes and selecting and implementing the adequate IT and I4.0-related tools suggested.

References

1. Spaltini, M., Poletti, A., Acerbi, F., Taisch, M.: A quantitative framework for industry 4.0 enabled circular economy. Procedia CIRP **98**, 115–120 (2021). https://doi.org/10.1016/j.procir.2021.01.015

2. Wang, Y., Ma, H.-S., Yang, J.-H., Wang, K.-S.: Industry 4.0: a way from mass customization to mass personalization production. Adv. Manuf. **5**(4), 311–320 (2017). https://doi.org/10.1007/s40436-017-0204-7
3. Verhulst, E., Brenden, S.F.: Strategic roadmapping towards industry 4.0 for manufacturing SMEs. In: Dolgui, A., Bernard, A., Lemoine, D., von Cieminski, G., Romero, D. (eds.) APMS 2021. IAICT, vol. 631, pp. 3–12. Springer, Cham (2021). https://doi.org/10.1007/978-3-030-85902-2_1
4. Wagire, A.A., Joshi, R., Rathore, A.P.S., Jain, R.: Development of maturity model for assessing the implementation of Industry 4.0: learning from theory and practice. Prod. Plan. Control (2020). https://doi.org/10.1080/09537287.2020.1744763
5. De Carolis, A., Macchi, M., Negri, E., Terzi, S.: A Maturity Model for Assessing the Digital Readiness of Manufacturing Companies. In: Lödding, H., Riedel, R., Thoben, K.-D., von Cieminski, G., Kiritsis, D. (eds.) APMS 2017. IAICT, vol. 513, pp. 13–20. Springer, Cham (2017). https://doi.org/10.1007/978-3-319-66923-6_2
6. Carou, D.: Aerospace and Digitalization: A Transformation Through Key Industry 4.0 Technologies. Springer, Heidelberg (2020)
7. Lacerda, T.C., von Wangenheim, C.G.: Systematic literature review of usability capability/maturity models. Comput. Stand. Interfaces **55**, 1339–1351 (2018). https://doi.org/10.1016/j.csi.2017.06.001
8. Kerr, C., Phaal, R., Probert, D.: Addressing the cognitive and social influence inhibitors during the ideation stages of technology roadmapping workshops. Technol. Roadmapping, 655–686 (2018). https://doi.org/10.1142/9789813235342
9. Andrews, K., et al.: This is Service Design Thinking: Basics — Tools — Cases. BIS Publishers, Amsterdam (2011)
10. Cotrino, A., Sebastián, M.A., González-Gaya, C.: Industry 4.0 roadmap: implementation for small and medium-sized enterprises. Appl. Sci. **10**(23), 1–17 (2020). https://doi.org/10.3390/app10238566
11. Kerr, C., Phaal, R., Probert, D.: Cogitate, articulate, communicate: the psychosocial reality of technology roadmapping and roadmaps. R D Manag. **42**(1), 1–13 (2012). https://doi.org/10.1111/j.1467-9310.2011.00658.x
12. Valamede, L.S., Akkari, A.C.S.: Lean 4.0: a new holistic approach for the integration of lean manufacturing tools and digital technologies. Int. J. Math. Eng. Manag. Sci. **5**(5), 854–868 (2020). https://doi.org/10.33889/IJMEMS.2020.5.5.066
13. Ameri, F., Stecke, K.E., von Cieminski, G., Kiritsis, D. (eds.): APMS 2019. IAICT, vol. 567. Springer, Cham (2019). https://doi.org/10.1007/978-3-030-29996-5
14. Phaal, R., Farrukh, C.J.P., Mills, J.F., Probert, D.R.: Customizing the technology roadmapping approach. In: Portland International Conference on Management of Engineering and Technology, pp. 361–369 (2003). https://doi.org/10.1109/picmet.2003.1222814
15. Azzone, G., Bertelè, U.: L'impresa. Sistemi di Governo, Valutazione e Controllo, 5th edn. (2011)
16. Spina, G.: La Gestione Dell'impresa. Organizzazione, Processi Decisionali, Marketing, Acquisti E Supply Chain, 3rd edn. (2012)
17. Spaltini, M., Sassanelli, C., Rossi, M., Terzi, S., Taisch, M.: Using wastes as driver to integrate digital and engineering practices maturity in the product development process: an application case. In: 24th Summer School "Francesco Turco" – Industrial Systems Engineering (2021)

"Metro Map" Illustrating the Digitalisation in Industry

Daniel Schmid[1(✉)] and Felix Nyffenegger[2]

[1] Zurich University of Applied Sciences (ZHAW), Winterthur, Switzerland
daniel.schmid@zhaw.ch

[2] Eastern Switzerland University of Applied Sciences (OST), Rapperswil, Switzerland
felix.nyffenegger@ost.ch

Abstract. This paper presents an illustrated model to explain digitalisation in industry in a holistic, non-technological way. It reflects the dimensions of the product lifecycle and the supply chain. More specifically, it discusses the interaction between customer, product, production, and suppliers along a product's life and how digitalisation supports these interactions. Particular attention is given to the transition from type to its product instances. Especially in the case of complex products, mastering this transition plays a key role in understanding digitalisation requirements. Reclining the concept of type and instance, different element of the digital twin, as digital master and the different shadows, are classified. Developed initially to structure a course in an engineering master's program, the model has evolved to be used in different courses and currently supports consulting and applied research projects to orchestrate digitalisation initiatives in the industry.

Keywords: Education · Digitalisation · Industry 4.0 · Holistic view · Metro map

1 Introduction

Digitalisation in industry is moving forward at a fast and continuous pace. Underlying technologies of connectivity, interoperability, Industrial Internet of Things (IIoT), could computing, artificial intelligence (AI), machine learning (ML), or mixed and virtual reality (AR/VR) are advancing rapidly. At the same time, digitalisation has also changed the way we look at our processes and organisations. To give a few examples, agility and cross-disciplinary have become important concepts in product development [1], the collaboration between customer and producer was brought to a new level [2], and the approach of the smart factory is rapidly changing our perception of processes and data [3].

The educational sector needs to follow these developments and is challenged with a high degree of complexity [4]. On top of the traditional fundamentals of engineering, today's students need to grasp the essence of digitalisation and learn to create value proposition in real-world cases. To do so, not every technology must be understood in detail in the first place. Rather, a concept of the different approaches, interfaces und and

Published by Springer Nature Switzerland AG 2023
F. Noël et al. (Eds.): PLM 2022, IFIP AICT 667, pp. 294–303, 2023.
https://doi.org/10.1007/978-3-031-25182-5_29

use cases in the context of the product lifecycle and along the product's supply chain is required. Still, it remains a huge challenge to squeeze all these aspects into a single course. Thus, we need simplification without losing context.

In analogy to a metro map, perfection is not the aim, but orientation, connection, timing, and transfer. The goal of the presented work is to illustrate the simplified complexity of digitalisation. Yet ensuring that the students understand relations, dependencies, mechanisms, and the different dimensions of business processes and applied technologies. Whatever type of engineer they become, from mechanical to software engineers, they will contribute to the digitalisation with their holistic views and considering the other specialist fields. This is the inevitable base in the present and the future of interdisciplinary engineering co-operations.

Today, industrial digitalisation plays a major role in most engineering degree programs. Interesting enough, an overreaching model to explain the different approaches of digitalisation along the products lifecycle in its context could not be identified by the authors. Instead, academic work on specific concepts, such as "Smart Factories" or didactic approaches such as gamification is well described in the context of engineering education. In addition, previous experience and work from the authors in the context of PLM education, closed-loop PLM, and smart factory testbeds incorporate the presented model [3–5].

2 Educational Model "Digitalisation in Industry"

A good didactical design of a module/course addresses the content in different ways to the participants. In principle, knowledge is divided into factual and structural knowledge [6]. In general, factual knowledge is on a deep taxonomy level of 1 or 2, while structural knowledge leads to higher levels. The latter is divided into declarative (what), procedural (how) and conditional (when) knowledge [7]. The metro map supports teaching accordingly by allowing locating the theory elements in the big picture and thus addressing the semantic memory (concepts, network of concepts) of course participants. The associated exercises, in turn, support knowledge acquisition via the procedural memory system. To complete the course, guest lectures from the industry are also included in the lessons, which are explaining real-life use cases and thus address episodic memory. The metro map always forms the basis and thus becomes a "table of contents" for the whole module/course and eases the recall of knowledge accordingly.

2.1 Introducing the Model

The model of digitalisation in industry has been illustrated in similarity to the metro map as e.g. known from the "standard tube map" from the London Underground [8], the map from the Tunnelbana from Stockholm [9] or the S- and U-Bahn map from Berlin [10]. Not only the concept indicating different metro lines has been adapted, but also the interchange stations, the internal interchanges, and the different zones.

Metro Lines. Four different metro lines are incorporated plus an additional direct connection.

- The line "Digitalised Product" (DP) represents the classic development methodology [11].
- The line "Digitalised Manufacturing" (DM) marks the manufacturing of the product.
- The line "Operating Product" (OP) represents the customer journey from initial requirement to maintenance.
- The line "Manufacturing Equipment" (ME) represents supplier involvement, e.g., of manufacturing equipment.
- The line "Requirement Engineering" (RE) is illustrated by a feedback loop form operation data (line OP) to the requirement engineering in the digitalised product (DP).

Zones. The digital twin model (DT) is composed of the physical product in real space, the virtual product in virtual space, and the connections of data and information exchange [12, 13]. In the model of digitalisation in industry the DT has been matched to the product lifecycle covering the digital master and the digital shadow of the product in production and operation [14, 15]. In other terms, this represents the product's stages "as designed" (instance), "as built", and "as maintained". Properly applied, the digital twin's value proposition leads in a wide range of areas to related benefits [16].

Supply Chain Management (SCM). The supply chain is indicated on the right side of the illustrated model. It consists of the supplier, the manufacturer, and the customer. As it has recently been differentiated between supplier relationship management (SRM) and customer relationship management (CRM) the present approach is combining both in the holistic approach of supply chain management (SCM) [17]. Eventually, both alternatives can be brought into connection with the model.

Internal Interchange Stations. Master data and configurations are two internal interchange stations. They are shared by the two lines digitalised product (DP) and digitalised manufacturing (DM) and do not cross borders of companies (customer, manufacturer, and supplier). It is undisputed that product design and production must rely and operate on identical data having one single point of truth.

Interchange Stations. In contrast to the internal interchange station, the interchange stations are bridging the border between manufacturer, customer, and/or supplier. This requires related interfaces and in the case of an applied electronic data interchange (EDI), managed file transfer (MFT), and/or enterprise filesharing services (EFS) management [17] of access rights and data format. A typical enabler for these kinds of communication between partners within a supply chain are the enterprise resource planning (ERP) systems. Following internal interchange stations have been implemented into the model of digitalisation in industry.

2.2 Illustration of Use Cases

Smart Factory. The Smart Factory is a concept to support the envisioned goal of digitalisation in manufacturing [18]. But for achieving a fully connected manufacturing system, mainly operating without human force by generating, transferring, receiving, and processing necessary data to conduct all required tasks for producing all kinds of goods, the related products must be designed accordingly.

Mechatronic products are getting predominant as they offer the opportunity for related data allocation in production and operation and being part of the (industrial) internet of things ((I)IoT). In consequence, the manufacturing does not end by delivering a physical entity but continues in collecting data, creating insights, and offering services. Therefore, taking the model of digitalisation in industry the students are instructed, to focus on the metro lined "Digitalised Product" (DP) and "Digitalised Manufacturing" (DM).

Digital Twin's Shadow – Data-Driven Services. Services of the future will be based on data analytics offering e.g. condition-based maintenance. The aim is, to illustrate that data-driven services are offered throughout the whole product development and manufacturing process. Depending on the point of view it is vertical or in a kind tilted. Furthermore, the illustration gives the opportunity to discuss the issue that operation data of production means are required to enable services of the product in operation. An example could be traceability preventing consequential damage to the manufacturer's customer. If e.g. an insufficient surface quality can be linked to an abnormality of a milling machine during a certain period, this data needs to be matched accordingly: Which parts have been produced while the abnormality occurred? In which products are they? And which applications require the specific surface quality (related applications)?

Product Lifecycle: While many authors agree on three principle phases in the product lifecycle (beginning of life, mid of life, end of life) [19], we choose to focus on the two stages of a product's life accruing to RMAI: the lifecycle of product type and the lifecycle of the product instances [5, 20]. The underlying concept of type and instance can well be explained by the discussion of their representation as digital master or digital shadows receptively. So, it becomes obvious that we need a properly released type as the baseline to created instances without the high price of manual work. Even more in the case of product variance, the systematic creation of instances based on the customer's configuration becomes essential to build data-driven services that relies on the product instance. While the result of the configuration remains a filtered view on the digital master, it serves as a template to create an instance. At this very moment, the lifecycle of the instance in its physical and digital appearance starts to live. The purple metro line illustrates how we can learn from product instances to improve our product types. This concept is also known as closed-loop PLM [21] (Fig. 1).

Requirement Engineering Based on Load Profiles. Using load profiles of applied products for a next-generation or just an upgrade is essential for defending and probably extending existing market shares. The present technical specification can be taken and optimise the product and the manufacturing accordingly. The efficiency gets increased ("Doing the things right.") and leads to a better margin or a reduced price aiming for a higher quantity in sales and therefore scaling effects (economy of scale). The model of digitalisation in industry and especially this metro line "Requirement Engineering" illustrates to the students the value of fundamental data analysis avoiding over or under-engineering a product.

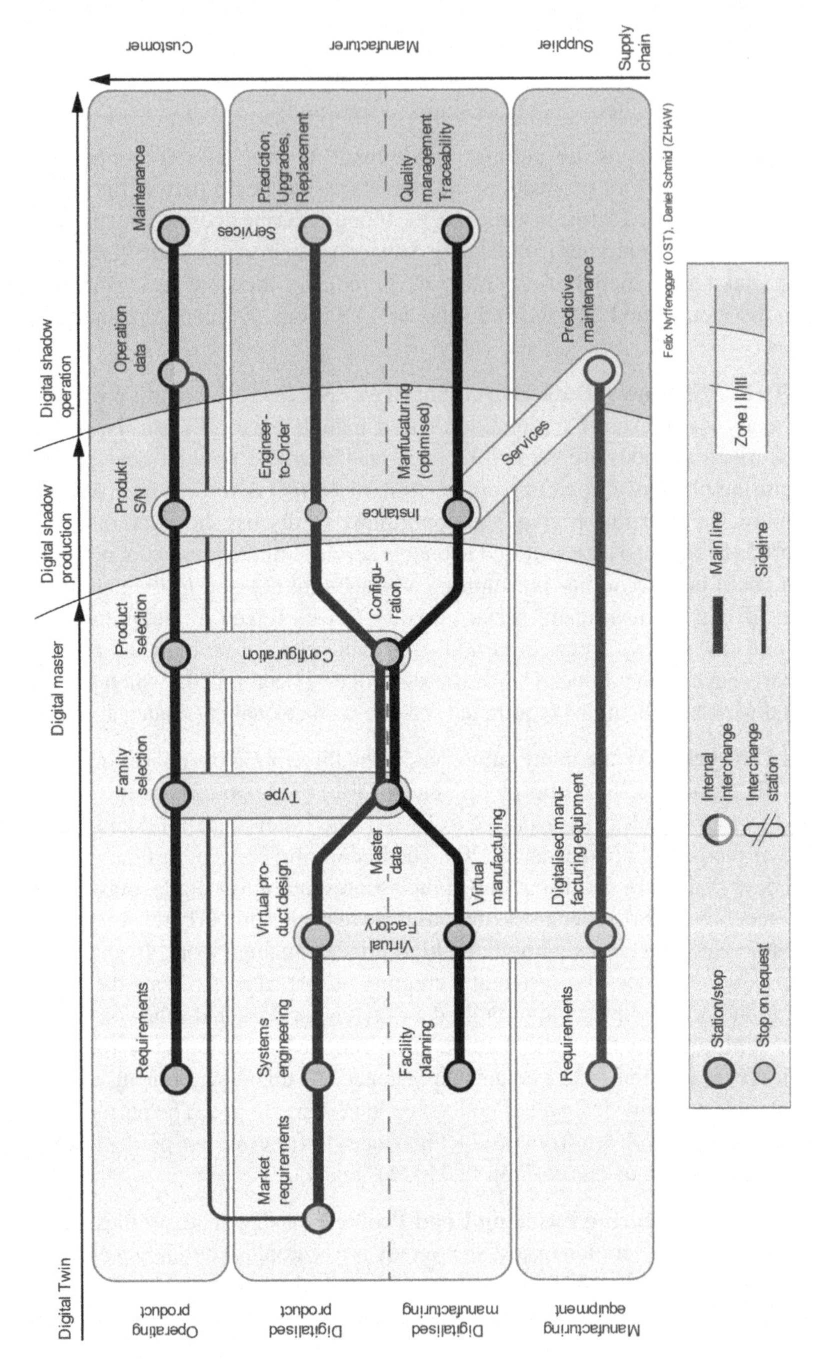

Fig. 1. Illustration of the digitalisation in industry in the style of metro map

3 Application of the Model in Education

The presented model was co-created while elaborating a didactical concept for a new module "Digitalisation in Industry" for the master's program of Swiss Universities of Applied Sciences (Master of Science in Engineering, MSE). In this section, some insights are given into applications of the model as a framework for the module.

3.1 "Digitalisation in Industry" (Master of Science in Engineering)

As mentioned, the module (course) is part of a master's degree program that is open to different types of engineering. The students might have different backgrounds, from mechanical to software to food engineering. The education goal of the course can be summarised as follows:

1. Enabling the students to contribute to digital transformation in the industry. They learn about digital transformation's fundamental concepts and technical and organisational requirements. They will be able to ask the right questions in a conceptual discussion.
2. The students obtain an overview of the processes, data structures and information flows based on different product strategies inside a company.
3. They are qualified to evaluate different approaches to a company's product strategy, architecture, production processes, and the deployed IT solutions. Relying on this, they can identify and apply optimization strategies.
4. They are familiar with state-of-the-art concepts of digitization to classify efficiency and transparency in production processes (industry 4.0).
5. They are familiar with the basic concepts of digitized products (Internet of Things) and how these are linked to the processes and data streams of the original company to increase the range of product-related services or business models.
6. They can rationally decide between "digital" and "non-digital" solution concepts.

Given this aim and target audience, it becomes obvious that not the technical details but a holistic understanding of all elements are required: technical, organisational, and business-related. We expect students to meet the taxonomy level 3 to 4 according to the classification of Bloom [22]. The following learning units were chosen to complete the required holistic view:

- End-to-end process from market requirements to recycling
- Digital twin concepts: Digital master (type) and digital shadows (instance)
- Dealing with product complexity and variants
- Challenges related to developing mechatronic products and IoT connectivity
- Sales to delivery and smart sales approaches
- Efficiency in production applying "Industry 4.0"
- Data-driven services for predictive and condition-based maintenance
- Adaption to business models (e.g., pay per use and subscriptions)
- IT tools along the process chain (CAD, PLM, ERP, MES, IoT Backbones)d

Figure 2 illustrates how these topics are reflected on the metro map. Each learning unit is structured into theoretical and practical parts (exercises). While the topics change pretty rapidly over the course, the exercises always rely on the same two elaboration examples:

An industrial pressure sensor is used as a first example showing the complexity of integrated and cross-disciplinary IoT products. It can be used in two representations. The first one links the complexity of the real product's issues in detail. The other is a highly abstracted conceptual model of the sensor based on 5 Lego bricks to discuss the link between mechatronic disciplines and customer-driven variants. The second example is an invented scenario of a company producing sorting machines and its customer that runs a local package delivery service. This example gives the full context of a supply chain between component suppliers, OEM, and industrial customers. A Lego-based machine represents the simplified product yet gives an accurate level of product structure complexity.

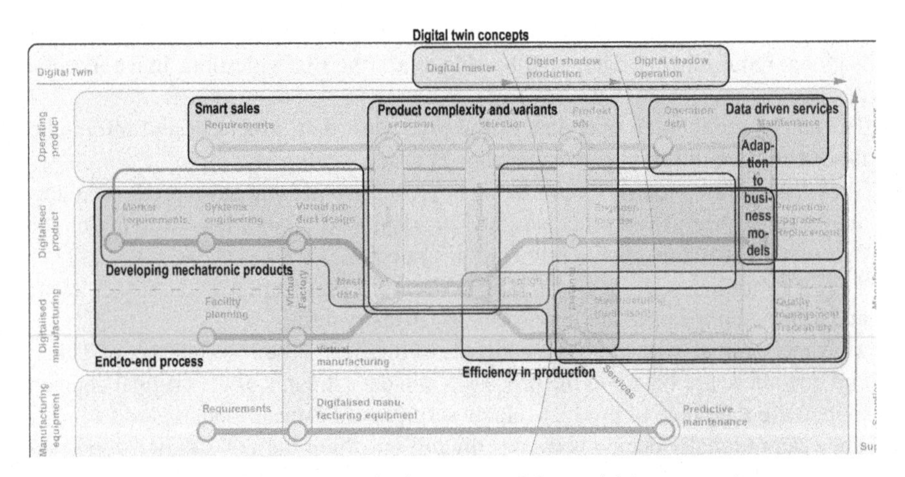

Fig. 2. Learning units in the context of the model (metro map)

Having enough understanding of these topics, elaborated by the examples, the students will be able to ask the right questions in a conceptual discussion. This last step in the didactical concept is consolidated by a final exercise where the students must elaborate and implement a concept for an IoT driven service. The given problem is a real-world problem of a local airport. The implementation, however, will be realised by the usage of prepared low-code tools. It will convey a feeling of implementation complexity without losing too much time. Of course, the groups must defend their concepts to improve their reasoning.

3.2 An Overview of Further Application

Besides the example of the application discussed above, the metro map was used in various courses to support orientation (Table 1).

Table 1. Further application of the metro map

Degree	*Title*: Educational target	Main topics	Tax. level
BSc	***Smart Factory***: The students are getting an overview of product development, manufacturing technologies, and Industry 4.0. It shall lead to the understanding that development and production need to co-operate accordingly, utilising the full potential of the smart factory	- Development process - Manufacturing technologies from turning and milling to additive manufacturing - Modern development tools and how creating the master data - Concept of the digital twin - Cost management	2–3
BSc	***Smart Products and Production 1&2***: The students get a sound understanding of complex developments, product configuration, and state of the art tools as such CAx. Furthermore, they can select and apply suitable manufacturing technology and use data from development, manufacturing, and product operation to create data-driven product functions and services	- Variant and cost management - Non-linear kinematic systems - Manufacturing technologies - Quality management and usage of related data - Basics about sensor technologies and applications - Digital twin and introduction into "Industry 4.0"	2–4
BSc	***Product Data Engineering***: This course aims at the hands-on experience of mechatronic product development, including IoT aspects. The students must develop a remote-controlled IoT toy that shows interaction with its remote friends. On the way of development, the students work in groups	- Mechatronic development tool chain - Systems Engineering/SysML - IoT Stack/Communication - Understanding of the views on a product from different disciplines - Instantiation of products	4–5
BSc	***Applied Digitalisation in Industry***: Embedded into a smart factory setup for unihockey ball production (injection modelling, configuration, assembly) the students learn about the value and implementation of different use cases in industry 4.0	- Use Cases of Industry 4.0 - Discrete event simulation for production planning - Introduction into IoT communication in an industrial environment - Introduction to AI/ML for prediction of factory events	3

(*continued*)

Table 1. *(continued)*

Degree	*Title*: Educational target	Main topics	Tax. level
CAS	***Production Management***: This certificate of advanced studies addresses the management level of the industry, particularly in the manufacturing sector. They will learn about the latest movements, concepts, and technologies in production management	- Digitalisation in industry - Lean manufacturing principles - Integrated data management in an industrial context - Complexity management - Use cases of industry 4.0	4

4 Conclusion

As stated in the introduction, the presented work is not the result of actual research, but rather the outcome of a creative process. However, it proved to be helpful in organising and planning courses in the field of digitalisation in industry. Particularly, it helped to create a common understanding among the different involved lecturers and experts from the industry. Feedback from the annual student's survey 21/22 of the module mentioned in 3.1 particularly shows high appreciation for the structure of the course and the alignment among the lecturers.

Based on project reports and exams by the students, we can state that our illustration provides good support in the argumentation of technical concepts. However, our students struggle to create clear value propositions for their suggested approaches and use cases regarding digitalisation in industry. It might be a general lack in the education of engineering programs. Yet, the holistic context helps to discover and discuss such gaps. In addition, we are investigating enriching the our model with an extra layer to explain the value proposition of different use cases.

Due to positive feedback, we also started to use this model for strategic planning of digitalisation initiatives in consulting and applied research projects. Hence, it helps to argue links between different digitalisation initiatives and supports a proper road map setup. Eventually, the positive feedback from the management level moved us to publish our approach.

References

1. Riesener, M., Doelle, C., Perau, S., Lossie, P., Schuh, G.: Methodology for iterative system modeling in agile product development. Procedia CIRP **100**, 439–444 (2021)
2. Boldosova, V.: Telling stories that sell: the role of storytelling and big data analytics in smart service sales. Ind. Mark. Manag. **86**, 122–134 (2020)
3. Hänggi, R., Nyffenegger, F., Ehrig, F., Jaeschke, P., Bernhardsgrütter, R.: Smart learning factory – network approach for learning and transfer in a digital & physical set up. In: Nyffenegger, Felix, Ríos, José, Rivest, Louis, Bouras, Abdelaziz (eds.) PLM 2020. IAICT, vol. 594, pp. 15–25. Springer, Cham (2020). https://doi.org/10.1007/978-3-030-62807-9_2

4. Fradl, B., Sohrweide, A., Nyffenegger, F.: PLM in education – the escape from boredom. In: Ríos, J., Bernard, A., Bouras, A., Foufou, S. (eds.) PLM 2017. IAICT, vol. 517, pp. 297–307. Springer, Cham (2017). https://doi.org/10.1007/978-3-319-72905-3_27
5. Nyffenegger, F., Hänggi, R., Reisch, A.: A reference model for PLM in the area of digitization. In: Chiabert, P., Bouras, A., Noël, F., Ríos, J. (eds.) PLM 2018. IAICT, vol. 540, pp. 358–366. Springer, Cham (2018). https://doi.org/10.1007/978-3-030-01614-2_33
6. Jonassen, D.H., Beissner, K., Yacci, M.: Structural Knowledge: Techniques for Representing, Conveying, and Acquiring Structural Knowledge. Routledge, New York (1993)
7. Dubs, R.: Lehrerverhalten: Ein Beitrag zur Interaktion von Lehrenden und Lernenden im Unterricht, 2. Auflage. Zürich (2009)
8. (London and TF), "Standard Tube Amp" (2022). https://content.tfl.gov.uk/standard-tube-map.pdf. Accessed 25 Mar 2022
9. (AB Storstockholmsk) Lokaltrafi, "Stockholm Spårtrafikkartor" (2022). https://sl.se/global assets/sl-spartrafik.pdf. Accessed 25 Mar 2022
10. (BVG), "Interaktives Liniennetz für S-Bahn, U-Bahn, Regio" (2021). https://sbahn.berlin/lin iennetz/. Accessed 25 Mar 2022
11. Ingenieure, V.D.: Design of technical products and systems - model of product design. Düsseldorf, Germany: VDI Verein Deutscher Ingenieure e.V. (2019)
12. Grieves, M.: Digital twin: manufacturing excellence through virtual factory replication. White Pap. **1**, 1–7 (2014)
13. Grieves, M., Vickers, J.: Digital twin: mitigating unpredictable, undesirable emergent behavior in complex systems. In: Kahlen, F.-J., Flumerfelt, S., Alves, A. (eds.) Transdisciplinary Perspectives on Complex Systems, pp. 85–113. Springer, Cham (2017). https://doi.org/10.1007/978-3-319-38756-7_4
14. Bauernhansl, T., Hartleif, S., Felix, T.: The digital shadow of production - a concept for the effective and efficient information supply in dynamic industrial environments. Procedia CIRP **72**, 69–74 (2018)
15. Schuh, G., et al.: Effizientere Produktion mit Digitalen Schatten. ZWF Zeitschrift fuer Wirtschaftlichen Fabrikbetr. **115**, 105–107 (2020)
16. Rasheed, A., San, O., Kvamsdal, T.: Digital twin: values, challenges and enablers from a modeling perspective. IEEE Access **8**, 21980–22012 (2020)
17. Thiel, C., Piai, G., Ihlenburg, D., Meierhofer, J.: Nutzerbasierter digitalisierungsnavigator - wie KMU ihre digitalisierungsstrategie selbst entwickeln können. St. Gallen (2019)
18. Osterrieder, P., Budde, L., Friedli, T.: The smart factory as a key construct of industry 4.0: a systematic literature review. Int. J. Prod. Econ. **221**, 107476 (2020)
19. Terzi, S., Bouras, A., Dutta, D., Garetti, M., Kiritsis, D.: Product lifecycle management – from its history to its new role. Int. J. Prod. Lifecycle Manag. **4**(4), 360 (2010)
20. VID/VDE. Reference architecture model industrie 4.0 (RAMI4.0). Igarss 2014, (1), 28 (2015)
21. Kiritsis, D.: Closed-loop PLM for intelligent products in the era of the Internet of things. Comput. Des. **43**(5), 479–501 (2011)
22. Bloom, B.S., Krathwohl, D.R.: Taxonomy of educational objectives: the classification of educational goals. In: Handbook I: Cognitive Domain (1956)

Integrating User Experience and Business Process Management for a Better Human Interaction and Software Design

Mariangela Lazoi, Angela Luperto(✉), Aurora Rimini, and Marco Lucio Sarcinella

Department of Innovation Engineering, University of Salento, 73100 Lecce, Italy
angela.luperto@unisalento.it

Abstract. The digital transformation is leading any field of social and working life. New technologies are more and more designed considering the relevant role of the users for providing efficient experience in the application of software solutions. A current research stream is analyzing the potentialities to integrate user experience techniques and business process modelling for reaching better results in software design. In this paper, an application of the integration between user experience and business process management is described for providing more insights in the state of art and suggest future directions also in the design of Product Lifecycle Management systems.

Keywords: User experience · Business process management · Human interaction · Software design

1 Introduction

Product Lifecycle Management needs new insights and feedback for leading future research and providing software solutions closer to the user needs. In the field of software design, several techniques and methods are used to support the IT architect, software designer and user experience expert to define better and better software solution. Among these, a current research trend integrates the user experience design with the business process management approach. The aim is to provide in the user experience new ways for understanding the users' needs and workflow of activities and providing a graphical and immediate overview of roles, responsibilities and tasks to be performed that can lead the software design.

The research question of this paper is: *in the current scenario characterized by an intense need of digital technologies in any field, how can the software design be improved for leading a better human interaction?*

The context of study for addressing this research question is the project SAFE (Safe Approach For Event management) that aimed to provide innovative solution for supporting Performing Arts organizations. In the Performing Arts (PA) sector the pandemic deprives spectators of the cultural experience and puts artists and cultural operators in

Published by Springer Nature Switzerland AG 2023
F. Noël et al. (Eds.): PLM 2022, IFIP AICT 667, pp. 304–313, 2023.
https://doi.org/10.1007/978-3-031-25182-5_30

difficulty, interrupting the demand growth. Forecasts range from the ending of previous social habits to a slow and gradual return to normality. It is therefore necessary to seize the opportunities for innovation offered by technology and research to accelerate the recovery. By exploiting the potential of VR, the paper is based on the design of a PaaS (Product-as-a-Service) for users to increase accessibility to the cultural event and support the economic activities of cultural operators. The solution involves the creation of an application from which the user has the possibility to select an event and purchase a virtual ticket for the show.

This paper focused on the definition of actors of a VR platform to be developed and their behaviors and goals in using it. BPM modelling and User Experience tools went hand in hand with software design, improving the consistency of the overall results. The outcome of this study consists in a set of BPMN processes for each target actor of the platform, describing the provisional overall user experience.

The paper is structured as follows. The next section introduces the current state of art of study merging user experience and Business Process Management (BPM) for improving the technological human interaction and software design. Section 3 introduces the main steps of the study and Sect. 4 describes the main results. A final section ends the paper.

2 Theoretical Background

In order to understand the level of innovativeness of the approach in the solution development, the scientific literature have been analyzed on the combined use of User Experience and BPM to support the design of an IT solution.

2.1 Designing New Technologies: User Experience and BPM

New technologies require more complex systems that are difficult to use. A good interaction between the user and the system, which is provided by the User Interface (UI), is needed to manage this complexity. The development of a UI is not easy because it is difficult and important to understand both the characteristics of the users and the tasks they perform with the system [1]. This requires User Experience (UX) which is defined as "a person's perception and response resulting from the anticipated use and/or use of a product, system or service" [2]. It is determined whether it is easy or difficult to interact with the elements of the user interface. UX designers are responsible for determining how the user interface works, which means that they have to determine the structure, the functionalities, the usability (i.e., quality attribute that evaluates how easy user interfaces are to use), the organisation and how all parties interact with each other (see Fig. 1). If the user interaction is complicated or non-intuitive, the user experience may be very bad [1].

On the other hand, BPM is defined as a set of concepts, methods, tools and technologies used to design, implement, analyse and control operational business processes, including people, systems, functions, businesses, customers, suppliers and partners [3, 4]. Thus, BPM is a process-centric approach to improve performance that combines information technologies with governance processes and methodologies. It involves collaboration between people and information technologies to promote effective, agile and transparent business processes [3] and the creation of a platform where processes are managed and monitored [4].

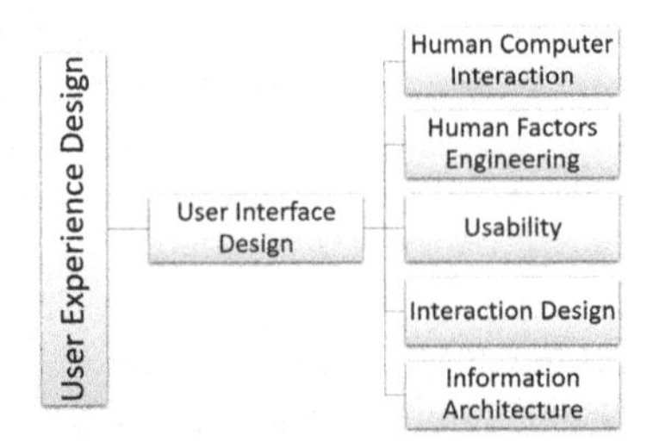

Fig. 1. User experience interrelations [3].

In this context, Díaz et al. [5, 6] proposed the generation of user interfaces using BPMN (Business Processes Modelling Notation) models in order to cover the gap between what is described in the BPMN models and what it is really implemented in the interface. The first paper [5] proposed a method to generate user interfaces using the BPMN model for extracting information about the navigation between interfaces and their behaviour, and the class diagram for identifying what information is required in each interface. In detail, the proposed method is based on the identification of several rules and uses stereotypes to extend the BPMN. Indeed, these stereotypes are used to generate interfaces based on business process models. However, the approach has some limitations: there is a strong dependency between BPMN and class diagrams and not all BPMN models have a class diagram thus reducing the applicability of the approach. Other limitation is that the approach has not been validated in a real generation of interfaces.

The second paper [6], discussed the problem of the design of graphical user interfaces (GUIs) to support the business process. The analyst usually defines the interfaces according to the BPMN models. Since he/she does not design the BPMN models, he/she has to make an effort to define the interfaces according to these models. For this reason, Díaz et al. [6] performed a group of experiments to establish a method that can automatically generate Graphical User Interfaces (GUIs) from extended BPMN models with stereotypes. Furthermore, they compared this method to the one in which GUIs were generated manually by considering the experiment's response variables of effort, accuracy and satisfaction (in terms of perceived ease of use (PEOU), perceived usefulness (PU) and intention to use (ITU)). This study has resulted in GUI widgets from more accurate BPMN models but with higher effort in the creation of interfaces, automation of the GUI generation process and less time in understanding the BPMN models.

Traetteberg H. [7] dealt with an approach for modelling business applications using BPMN and Diamodl (i.e., dialog modelling language for user interfaces), where BPMN is used for both process and task modelling and Diamodl for the structure and behaviour of user interfaces (UI). In particular, he focused on how the BPMN can be extended to include tasks and expanded with extra information regarding the object life cycle in order to improve the connection between the approach dealt with, the process and the dialogue modelling.

From this work it emerged that for the design of model-based user interfaces an adaptation of the existing process-oriented notation, that is BPMN, to task modelling makes more sense than introducing other notations that would add further costs.

Pavlickova et al. [8] carried out a usability study combining BPMN notation and the process modelling tool DEMO. They used a collaborative approach based on experience of a series of studies and the interaction of the participants in order to test how easy the BPMN business process modelling tools are to use. To verify this, five usability indicators were used: (1) Learnability - How friendly it is for users to perform tasks they encounter on the design for the first time; (2) Efficiency - Velocity of executing tasks after learning the design; (3) Memorability - How easy it is to re-establish competency after a period of non-use; (4) Errors - Number and severity of errors occurred and how easy it is to resolve errors; (5) Satisfaction - how comfortable it is to use the design. From this study it was identified that the usability testing method can be used to measure, evaluate and design metrics for process model quality and that therefore user experience (UX) approaches are valid for evaluating process model quality (mainly usability) [8].

As Pavlickova et al. [8] also Quiñones et al. [9] used BPMN notation to evaluate aspects different from usability that are related to user experience (UX). In several case studies, they applied a formal methodology through BPMN notation to develop usability/user experience heuristics. The new methodology presents tables summarising the inputs, outputs and activities to be performed on BPMN diagrams, modeling each step. They concluded that if a methodology is not used to develop heuristics, the heuristics created may be (1) difficult to understand, (2) difficult to use, and/or (3) not successful in usability/UX evaluation of a specific application. It emerged that this methodology is applicable to develop heuristics for UX-related aspects such as playability, communicability and learnability. It also has great potential in developing tools for further quality attributes such as security and adaptability.

Greunen et al. [3] stated that when user interface design integrates the way users think and learn, accepts their needs, and meets their expectations for comfort and convenience, the human-machine interaction becomes more productive. That is why they focused on factors that influence the UX and the user interface (UI) when using BPM tools to complete tasks in support of business processes. In order to detect these factors and to have a better understanding of the interactions and dependencies between people and technology, a focus group was conducted. It emerged that it is fundamental to have an interface that supports the information needs of organisations in order to meet their critical success factors and achieve organisational goals. Moreover, the identified user experience (UX) factors that have an impact on BPM are divided into 3 categories: (1) People factors: visual design, user needs and interface design; (2) Business factors: tasks supported and business processes; (3) Technology factors: interaction and information design. Finally, the study showed a lack of frameworks and tools to assess and determine UX and a lack of skills in the use of BPM tools [3]. In the following Table 1 the advantages and disadvantages of each technique found in the literature are summarised.

Table 1. Summary of the domain of study in the literature.

Techniques	Results	Disadvantage	Reference
Graphical User interface and BPMN model with Class Diagram	- Information on the interface behaviour - Identity information for each interface - Interface generation on BPMN models	- Strong dependency of the BPMN and class diagrams - Not validated in a real interface generation	[5]
Graphical User Interfaces and BPMN model	- Generating more accurate GUIs - Automating the GUI generation process - Less time to understand BPMN models	- The study is based on few experiments - Higher effort for GUI implementation	[6]
User Interface design and BPMN with Diamodl	- Design UI makes sense with BPMN - Low costs for modelling	- Strong dependency of the BPMN and Diamodl	[7]
Usability study and BPMN modelling tools	- Identify usability of BPMN tools - UX methods valid to assess BPMN quality	- Determines only usability in the process model	[8]
Usability/UX and BPMN	- Identification of UX-related aspects of usability	- Determines only some usability aspects	[9]
User Experience and BPM tools	- Identify UX factors that impact on BPM - Better understanding of human-technology interactions	- Lack of frameworks and tools to assess and determine UX - Lack of skills in BPM tools	[3]

3 Research Method

Based on the analyzed literature, in this paper, user experiences techniques have been combined with Business Process Management to enrich the obtained result. Mock-up is a very wide term that can be used to indicate an explanatory representation of a system. A Mock-Up for a VR reality solution to allow the participation in streaming to a live events has several issues to explore related to the activities to be performed by the events' organizers and by the spectators, the technological choices to do (e.g. the type of visors) and the final layout that a such solution should have. To obtain this results, the research team has faced several issues and the main research phases are:

1. Defining Processes for the User Experience. BPM modelling and User Experience tools went hand in hand with software design, improving the consistency of the overall results. The outcome of this phase consists in a set of BPMN processes for

each target actor of the platform, describing the provisional overall user experience. In particular, in BPM method, it's usually to attribute al level 0 a high level detail (Mega), at level 1 a more detailed process and so on.

2. Interviews and Think Aloud with SAFE consultants. The third phase of the methodology started with the design of a Think Aloud section with each SAFE's consultants. During the section the Blended Experience Timeline was presented and feedback was collected to validate the overall user experience emerging by the previous phase. In this phase, it has been possible to identify a needs assessment on desired functionalities on platform and to the final user needs and type.
3. User Interfaces Definition. It aims to design the SAFE VR platform's user interfaces, based on the refined user experience model from the previous phase. User Interface (UI) design is commonly known as the design of the user interface of any IT system that interact with a user through a screen. The BPM notation was adapted to define the views, data and functional elements of the software to be developed, essential elements useful to mockup the related user interfaces.

4 Results

The SAFE platform is defined as an IT architecture providing a Virtual Reality live streaming video service as a PaaS (Product-as-a-Service) product that allows cultural operators in the PA sector to provide live cultural events, which can be attended remotely. The Blended Experience Timeline (i.e. Blended Experiences are the embodiment of the interactions between tools, space and practices. Blended Experiences, as illustrated encompass the dynamic and diverse ways in which the people interact with digital tools and spaces to evoke a perception of social facilitation) (see Fig. 2) The purpose of this conceptualization is to describe the Blended Experience provided by SAFE platform for all its kind of users, described as follow under "roles" tag.

Furthermore, the user experience and the user interface of the SAFE VR platform was carried on and formalized with BPMN. Three roles were defined:

- Administrator: He/she is the SAFE Platform orchestrator. Through the use of SAFE Platform he/she can:

  - Manage the events' organizers: he/she can visualize and manage the Events' organizers users on platform;
  - Manage programming schedules: he/she can visualize and manage the programming schedules created on the platform;
  - Manage events: he/she can visualize and manage the events created on the platform by event's organizers;
  - Manage booking: he/she can visualize and manage the booking made from Spectator on the platform;
  - Manage VR Headset: he/she can manages the reservations for events and can manage the VR headset reservation made by Spectators.

- Event's organizers: he/she focuses on the administrative relationships with stakeholders inside and outside the organization, such as public administration and suppliers. Event's organizer, a user can access to the following functionalities:
  - Manage the Organizer profile: Event's manager can manage his profile and can give more information about the organization;
  - Mange the Programming schedules: Event's manager can manage the programming schedules about events provided;
  - Manage events: Event's manager can create, edit and delete events on his profile;
  - Manage a booking: Event's manager can manage all bookings and purchases for virtual participation in an event that are been made on his profile by a spectator.
  - Mange a campaign: Event's manage can create, edit and obtain reports about advertising and discount campaigns.
- User "Spectator": he/she is the final user and represent the spectator of live event. The target user can access to the following functionalities:
  - Search for an event: Spectators can search on the platform all the events available on SAFE. He/she can make personalized research based on Event's Manager, based on Location and so on.
  - Make a reservation: Spectators can make a reservation to watch an event through the SAFE VR Platform (Some screenshots are available here: shared folder). He also can choose if buy or rent a VR Headset to have a more immersive experience;
  - Browse an event: Spectator can enjoy the event trough a different mode, such as VR Live Streaming (by using VR device) or Live Streaming (by using a device such as smartphone or smart tv), he can interact with Participants trough game or can interactive functionalities;
  - Manage a reservation: A spectator can visualize and manage all his reservation about events.
  - Visualize the menu: A spectator can go to the home and visualize all functionalities.

For each role, processes have been modelled and validated. Three groups of processes arise for each role: Administrator's flow, Event's organizer flow and User's flow. Every workflow has been analysed in depth into 3 level.

- **Level 0** shows a bird's-eye view of each actor interacting with the IT platform and has been modeled for Administrator's flow, Event's organizer flow and User's flow.
- **Level 1** For each subprocess defined in the level 0, a second detailed level (Level 1) has been modeled. In this scenario, the aim is to define in a better way the interaction between user and the platform's functionalities trough the task definition.
- **Level 2** At this detail level are modeled the interaction between users and the platform. Thanks to the process model its' possible to have a support for the wireframe design.

As follow an example are provided.

This level 0 (see Fig. 3) is composed by seven subprocesses, and every one of them describe the interaction between user "Administrator" with a platform's functionality.

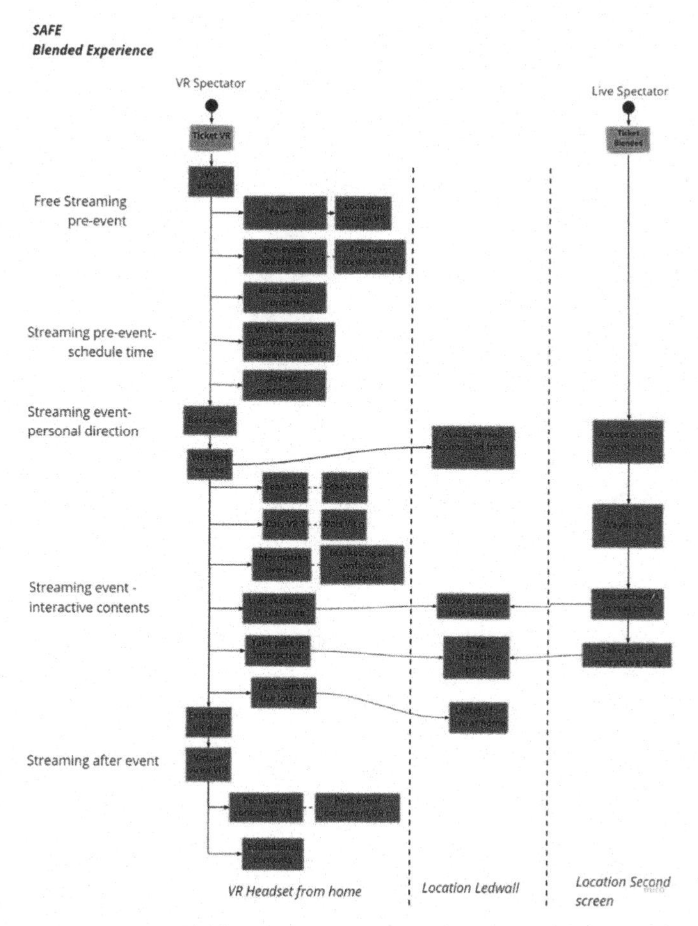

Fig. 2. Blended experience timeline.

Event's organizer's flow. This level 0 is composed by seven subprocesses, and every one of them describe the interaction as the user "Event's organizer" with a platform's functionality. At level 1, the subprocess "Manage the events' organizers" (see Fig. 4), describe the interaction between user and SAFE Platform. At Level 2, is better detailed the activity made by Administrator when "Browse the event's organizers list" (see Fig. 5).

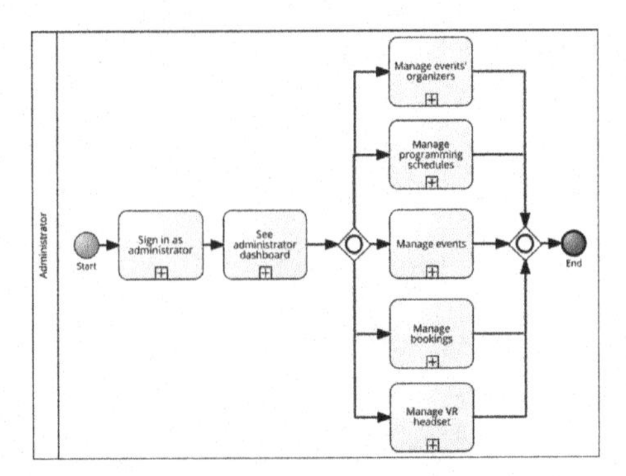

Fig. 3. Administrator's flow

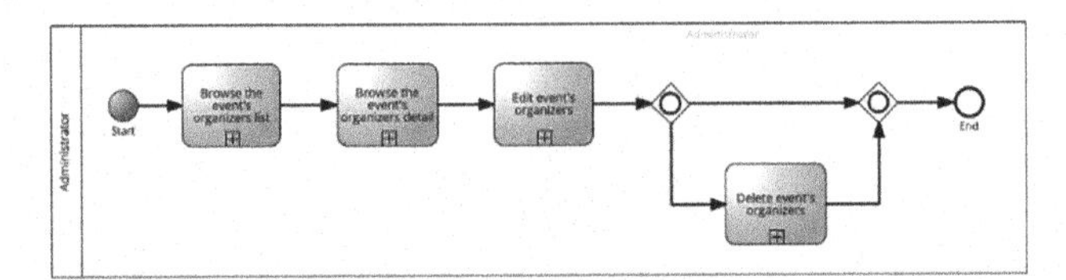

Fig. 4. Manage event's organizers

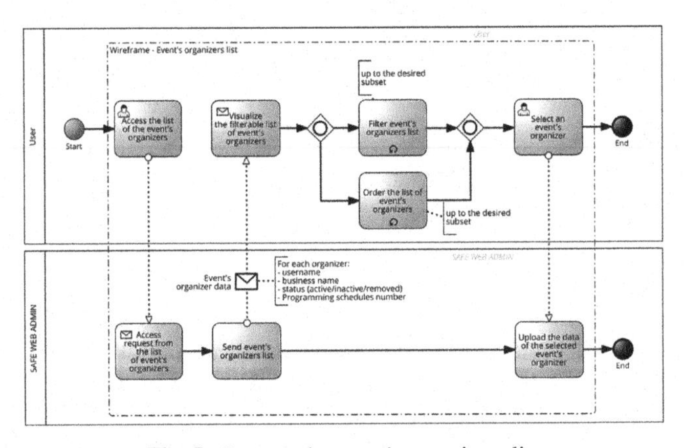

Fig. 5. Browse the event's organizers list

5 Conclusion

The results of the research phases are the method used to design of the SAFE platform's user interfaces based on the refined user experience model from the previous phase. The BPM notation was used to identify the views, data and functional elements of the software to be developed, essential elements useful to mockup the related user interfaces. Thus, each BPMN diagram includes a subset of elements, which coincides with the wireframe of the SAFE platform views. The method used to design the platform is innovative. Indeed, an experiment was conducted to design the functionalities: starting from a process analysis, a possible interaction process between user and platform has been drowned using BPMN notation, and thanks to the BPMN is possible to understand how to have interaction between VR technology and the platform conceptualized. Processes process of the user interface have been designed for the main actors of the platform (administrator, event's organizers and user). Thus, processes have been used as wireframes of the platform mockup. Due to the ease of BPMN and its versatility, it is easy understandable to everyone, and at the same time it is very rigour. However, other notations for process modelling (e.g., EPC) can be considered and applied with limits and strengths of each specific notation. The proposed method has a positive impact on PLM because it contributes to the reduction of product design time, in this case the software. In particular, the codesign activities conducted with the possible end user thanks to BPMN for requirements design not only shorten design time, but also optimise the subsequent development and testing phases, since it is the user preferences modeled in BPMN that guide the development activities. The described result extends the current state of art about the combined use of User Experience and BPM and provide further evidence for future research in this field.

References

1. Pratama, M., Cahyadi, A.: Effect of user interface and user experience on application sales. In: IOP Conference Series: Materials Science and Engineering, Bandung, Indonesia (2020)
2. Rusu, C., Rusu, V., Roncagliolo, S., González, C.: Usability and user experience: what should we care about? Int. J. Inf. Technol. Syst. **8**, 1–12 (2015)
3. Greunen, D., Merwe, A., Kotzé, P.: Factors Influencing BPM tools: the influence on user experience and user interfaces. Int. J. Comput. ICT Res. **4**(1), 47–57 (2010)
4. Rangiha, M.E., Karakostas, B.: Towards a meta-model for goal-based social BPM. In: Lohmann, N., Song, M., Wohed, P. (eds.) BPM 2013. LNBIP, vol. 171, pp. 104–112. Springer, Cham (2014). https://doi.org/10.1007/978-3-319-06257-0_9
5. Diaz, E., Panach, J.I., Rueda, S., Pastor, O.: Towards a Method to Generate GUI Prototypes from BPMN. In: 12th International Conference on Research Challenges in Information Science, RCIS 2018, Nantes, France (2018)
6. Diaz, E., Panach, J., Rueda, S., Distante, D.: A family of experiments to generate graphical user interfaces from BPMN models with stereotypes. J. Syst. Soft. **173**, 110883 (2021)
7. Trætteberg, H.: UI design without a task modeling language – using bpmn and diamodl for task modeling and dialog design. In: Forbrig, P., Paternò, F. (eds.) HCSE/TAMODIA -2008. LNCS, vol. 5247, pp. 110–117. Springer, Heidelberg (2008). https://doi.org/10.1007/978-3-540-85992-5_9
8. Pavlickova, P., Pavlicek, J.: Business process models (BPMN and DEMO notation) - usability study. In: Pergl, R., Babkin, E., Lock, R., Malyzhenkov, P., Merunka, V. (eds.) EOMAS 2019. LNBIP, vol. 366, pp. 167–174. Springer, Cham (2019). https://doi.org/10.1007/978-3-030-35646-0_13
9. Quiñonesa, D., Rusua, C., Rusu, V.: A methodology to develop usability/user experience heuristics. Comput. Stan. Interfaces **59**, 109–129 (2018)

Transversal Tools: Artificial Intelligence, Advanced Visualization and Interaction, Machine Learning

Sustainable Artificial Intelligence: In Search of Technological Resilience

Norbert Jastroch(✉)

MET Communications, 61352 Bad Homburg, Germany
norbert.jastroch@metcommunications.de

Abstract. Great demands are placed on the role of digital technology and artificial intelligence for sustainable development. External shocks, like the current pandemic, as well as creeping degradation, like the effects of carbon economy on the climate, need convincing concepts to control lasting negative effects on society, environment, and economy. This paper is intended to contribute to the search for ways of enabling resilience through technology.

High expectations are being put on data driven artificial intelligence in this respect. We consider that artificial intelligence tends to fall short of scientific rigor regarding cause-and-effect relations and discuss the inherent limitations of so-called formal systems that are at the bottom of artificial intelligence systems. We take into view what data analysis and reasoning can deliver regarding the discovery of empirical phenomena, arguing that targeted, reflective data reasoning can well help discover correlations worth further theoretical investigation. We suggest combining established methods of epistemic knowledge generation with data driven artificial intelligence, i.e. human intelligence with machine-based algorithmic intelligence, in support of advanced human-systems integration. For this concept of hybrid intelligence, we provide a procedural framework.

This methodological approach gets exemplified by the description of recently published cases of a technical application resp. Scientific practice, illustrating the potential of hybrid intelligence for the scientific as well as technical solution of problems. Concluding remarks finally draw the line to future work on sustainable artificial intelligence as a pathway to resilience delivered by technological means.

Keywords: Artificial intelligence · Formal systems · Data reasoning · Hybrid intelligence · Human-systems integration

1 Introduction

External shocks are often perceived as being sudden surprises. However, this only holds true under superficial observation or deficient reflection. So have the causes and effects of climate change been named since decades, but it took long to establish public insight, based upon scientific rigor. Likewise, the emergence of the SARS-CoV-2 virus and the subsequent global pandemic were only surprising as to the specific virus type – there have been numerous indications and warnings regarding potential virus pandemics since

Published by Springer Nature Switzerland AG 2023
F. Noël et al. (Eds.): PLM 2022, IFIP AICT 667, pp. 317–326, 2023.
https://doi.org/10.1007/978-3-031-25182-5_31

many years. And geopolitical tensions were virulent for long, including the risk of severe political and economic confrontation. While their extension into military action has been well observable, it was hardly believed until the day it turned into reality, leading to a major assault on a sovereign country which was a surprise to many in Europe. Although not coming out of the blue, these cases generated shock waves across the globe. They originated from technological (carbon economy), natural (virus evolution), or societal (geopolitical claims of imperial power) grounds.

And they are what the International Risk Governance Council (IRGC) calls emerging risks, resulting from the complexity of the concerned dynamic, non-linear systems they are part of. With focus on systemic risks in socio-economic systems, Helbing in [7] elaborated on the effects of complexity, in particular their cascading spread involved with network interactions. In their report [8] the IRGC built on his observations specifying several contributing factors that make up the fertile ground for risk emergence. Among these is the issue of technological advances. Changes in technology may become a source of risk if their impacts are not scientifically investigated in advance or surveyed after deployment - even more so if there is insufficient regulatory framework in place[1]. Thus, the IRGC are strongly arguing for ex ante as well as ex post risk assessment. The aim can be taken as a kind of sustainability evaluation, in the broad sense of securing sustaining, desirable implications whilst avoiding undesired side effects.

Interestingly, new technologies appear likewise to support the ability to adapt to future shocks. Brunnermeier [19], with a general societal perspective, points out that dealing with risk can either mean trying to avoid it, or to accept it in accordance with a framework of institutions, rules, and processes that are bound to enable recovery from external shocks. The first option remains constrained though, because total robustness, which covers any conceivable emergency, can hardly be realized as it would normally involve unacceptable high cost. The latter approach in fact finds increasing interest these days in the concept of resilience. A most relevant question then is to what extent can technology contribute to resilience, respectively become a driving force to it.

The currently most prominent technological field in this respect is digital transformation along with the resumed concept of artificial intelligence. The availability of massive data via digitization enables novel ways of empirical investigation. Techniques for their analysis do not only offer new approaches for applications in domains as diverse as health, mobility, manufacturing, agriculture, finance, energy, public administration etc. They also drive the development of what may be called Artificial Intelligence for Resilience, or Artificial Intelligence for Risk Governance.

This paper is intended to contribute to the search for technological ways of enabling resilience in that sense. The availability of massive data from digitization, along with powerful algorithms for analysis and reasoning, are the means to pro-actively assess risk scenarios and prepare for adequate responses of choice. We argue for the combination of human intelligence with algorithmic intelligence for the purpose of expanding theoretical knowledge from a theoretical as well as practical perspective. Our starting point

[1] For the technological field of Artificial Intelligence, and more general the digital transformation on its way these days, we have presented an investigation of risk mitigation matters in [10], comprising functional, societal, and cybersecurity risks. And their relation to regulative frameworks in the EU.

are reflections on formal systems, which are the methodological foundation of every scientific domain, and their limitation that must be accounted for in any application of artificial intelligence. A formal system is a set of axioms and rules for inference, and the resulting space of propositions, which altogether govern a domain of knowledge. It cannot be complete, free of contradiction, and closed at the same time. We then take into view what data reasoning, in particular machine learning can deliver for the discovery of empirical phenomena. As algorithmic machine-driven systems can become sources of novel security issues, need arises for precautionary searching of potential flaws by use of theoretical models. Respective procedural interrelations get illustrated in the framework for hybrid intelligence we suggest in the subsequent section, showing the combination of formal deductive methodology and probabilistic approaches. We continue by recurring to recently published cases of technical resp. Scientific examples showing the practical benefits, and close with concluding remarks as to future work on sustainable artificial intelligence.

2 Formal Systems are Limited

The scientific way of generating knowledge rests upon well-established processes that comprise the formulation of concepts, the generation of assumptions and their testing, the validation of findings and their alignment within a theory, and their incorporation into a coherent set of theorems and propositions which make up the body of domain knowledge [cf. 12]. Science is, in terms of methodology, extensively while not completely determined work, according to Kuhn [12]. Scientific progress is incremental and sometimes even disruptive when legacy paradigms get shifted – where the rationale of such a shift remains unclear in Kuhn's work [12]. Rovelli [16], on the contrary, argues that science and its progress work through continuity, not discontinuity. He identifies two origins of conceptual shifts in science: new data exerting decisive rationale for change, e. g. in the case of Kepler getting to his ellipses by mathematical analysis of empirical data of planet's courses, and informed investigation of contradictions within an existing theory, e. g. heliocentrism of our solar system. Following his observation of philosophy having contributed essentially to scientific development especially in the case of physics, Laplane et al. [13] more generally localize this impact of philosophy in the conceptual clarification and the critical assessment of assumptions or methods in a scientific discipline. We consider this a little deeper.

As mentioned, a scientific theory consists of a coherent set of theorems and propositions that are derived from a set of axioms with the help of rules of inference. The question of the validity of a theory in the sense of 'being true' has been subject to a wealth of debates in science theory. While the question 'What is truth?' could not be finally answered by philosophers throughout millennia up to now, Penrose [15] provided a comprehensive elaboration on the simplified question 'What is mathematical truth?'. The interest of our paper here are formal systems as part of artificial intelligence systems, i.e. abstract sets of axioms and rules for the purpose of inferring propositions to build a knowledge base to a certain domain. Mathematics is considered the perhaps most basic manifestation of such an axiomatic system, so it is safe to refer to Penrose's work [15].

Penrose [15] draws on the finding by Goedel formulated in his incompleteness theorem which applies to any formal system, consequently to any attempt of founding an artificial intelligence system on such a formalism. We content ourselves to refer the line of argument in brief without extensive detail: To ensure only valid propositions - in the sense of being mathematically proven - be derived, mathematical reasoning must be free of contradiction. Goedel showed that any such mathematical system, of whatever type, which is free of contradiction must include statements that are neither provable nor disprovable by the means allowed within the system. So full truth cannot be achieved within an axiomatic system by methods of proof. Penrose [15] pointed out that there is a way to get to the validity of a proposition he calls the reflection principle. By repeatedly reflecting upon the meaning, we can *see* it is true although we cannot derive it from the axioms. This *seeing* requires a mathematical *insight* that is not the result of deductive proof, or purely algorithmic operations which could be coded into some mathematical formal system. Admittedly, the status of this *insight,* as a mental procedure, remains unclear except its non-algorithmic nature. Its applicability appears as and insofar it leads to a coherent mathematical theory. At the very end, the consequence most interesting in the context of artificial intelligence - taken as a machine based algorithmic inference engine – may be: "...the decision as to the validity of an algorithm is not itself an algorithmic process" [15, p. 536]. Hence the inherent theoretical limitation of formal systems, which is depicted in Fig. 1: the object of human intelligence along with epistemics are domains of the real world, which include the physicalist part ('knowledgeable' in Fig. 1) and the non-physical qualities ('qualia')[2]. A formal system builds a proper part of a knowledgeable domain.

The algorithmic nature of purely machine-based Artificial Intelligence systems allows for formal procedures that cannot fully cover the related knowledge domain, not even the knowledgeable subset. Autonomous AI applications therefore will always have a blind spot area of propositions. The engineering of such AI systems needs to take care respectively, e.g. by ruling out their application in situations which might bear the risk of touching this area, or by calling in human guidance in such a situation. These issues get addressed under the concept of operational design domains, ODD, where functional constraints are introduced to avoid system states of that nature (cf. The case of autonomous vehicles engineering). Or by utilization of digital twins, which enable the investigation of such states for mitigation purposes (cf. The case of robotics).

[2] The difference between real domains and what is physically explainable is known as the ontological or epistemic gap – 'knowledgeable' then is a proper subset of reality. This philosophical distinction is not addressed in our context here.

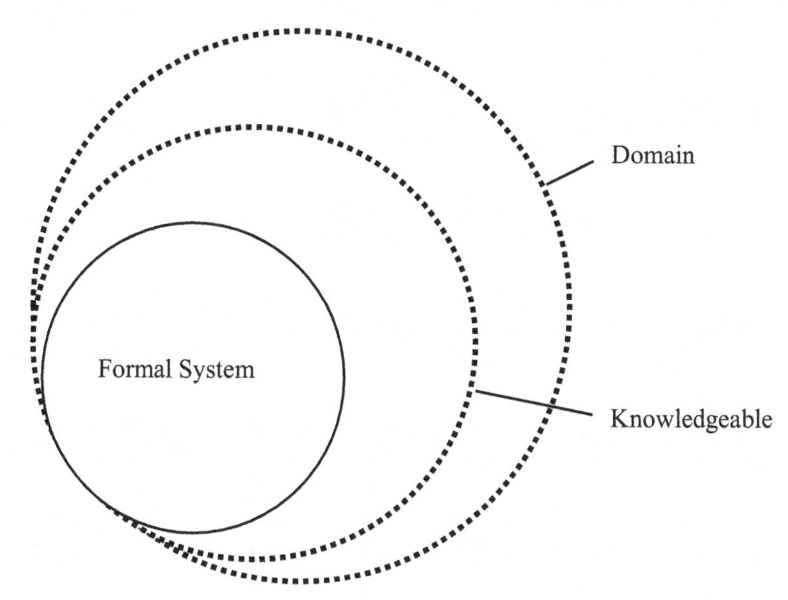

Fig. 1. The set of propositions of a formal system is a subset of the knowledgeable content of a domain which is a subset of the domain itself

3 Data Reasoning and Discovery

With the rise of digitization came data analytics as a source of innovation in science, industry, administration, and commercial marketplaces. Increasing utilization of massive data for automated decision making, machine learning and autonomous systems lead to several issues to be taken care of, among them data privacy (including data sovereignty, protection, safety) and data quality [cf. 9], and the validity of processes for their analysis and reasoning [cf. 10]. Uncontested, though, is the valuable potential of data driven artificial intelligence. The expectations are high: novel applications supporting societal and or economic progress; enhanced decision making based on algorithms; advancement of computational methods in sciences. In the context of this paper the latter both are of interest.

Our understanding of problems with our making decisions has been fundamentally enriched by the work of Kahneman et al. [11]. Deficient judgements appear to have two kinds of sources that are not related to the quality of information available: bias, the systematic deviation from neutral assessment, triggered by personal preferences; and noise, the statistical variance of judgements resulting from personal disposition[3]. They raise the question if and how bias and noise can be overcome with the help of metric criteria and parameter, and whether machine-based algorithms are per se more suited. Their answer is: they can be, it depends on the availability of parameters, their correct measurement, and the selection of parameters and measurement process being

[3] Eren and Mocan [6] provided an impressive empirical investigation of the correlation of unexpected football match losses with the length of sentences of judges, showing the impact of emotions in one domain on human decisions made in a completely unrelated domain.

free from bias and noise. However, Ludwig and Mullainathan [14] presented evidence of the fallibility of algorithms – be they controlled by humans or machines – depending on the algorithm as such, i.e. its fragility resulting from deficient construction. They resume that it is possible to reduce bias by well-built algorithms. If done right, artificial intelligence has the potential to undo human fallibility. They suggest human plus machine combination to be the best choice, though. A similar conclusion is drawn by Athey et al. [3]. Discussing when and how humans and machine-based algorithms should collaborate and who would be best to have formal decision authority, they argue for the AI system under optimal conditions as to data and algorithms employed, but for the combination of AI and human agent knowledge if that cannot be guaranteed.

The second option of harvesting the potential benefits of data reasoning lies in the generation of scientific novelty forced by data. It is not restricted to Physics if Rovelli [16] highlights sophisticated use of induction based upon accumulated empirical and theoretical knowledge as the most promising way forward in science. Regardless of whether massive data in a domain becomes available through digitization as such or by targeted experimental collection, it can be used to discover patterns and detect correlations that suggest new cause and effect relations (or propositions within the formal theory) to be tested for their validity. This is nothing new (recall the Kepler case mentioned in Sect. 2), but modern machine learning operations open a wealth of opportunities of that kind as they can find out conspicuous patterns in seconds instead of years of calculation. The basic process description is visualized in Fig. 2, as compared to Fig. 1 in Sect. 2. For exemplification, we present two recently published cases, one from pharmaceutical and another from historical research, in Sect. 5.

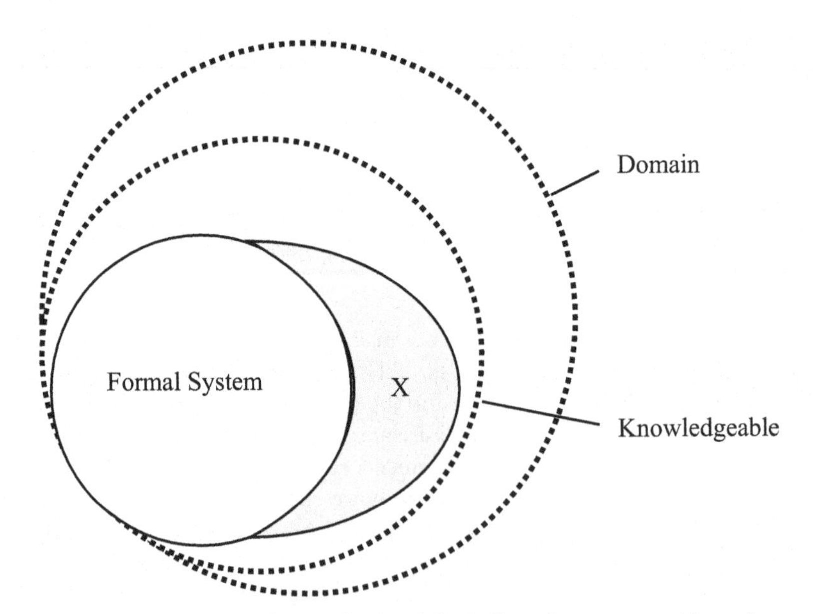

Fig. 2. A new empirical observation X stimulated the finding of a new proposition, what expands the formal system

4 Hybrid Intelligence and Its Potential

The limitation of Artificial Intelligence is a matter of principle, not of practicability [cf. 15]. Formal systems cannot completely cope with reality due to missing instantiations of e.g. 'understanding' and 'meaning', mental capabilities that are fundamentally human. But Artificial Intelligence enables the detection of correlations by data analysis techniques a human would not be capable of in terms of quantity of data to process. These techniques are at the heart of machine learning operations, with algorithmic procedures as their major building blocks. Algorithmic processes are the very kernel of Artificial Intelligence. Human intelligence, however, comprises these and also what Penrose called 'reflection' and 'insight'. Both human and artificial intelligence are subject to bias, while artificial intelligence appears to be free of noise, other than humans. This is the rationale of combining human and artificial intelligence in the concept of hybrid intelligence (Fig. 3).

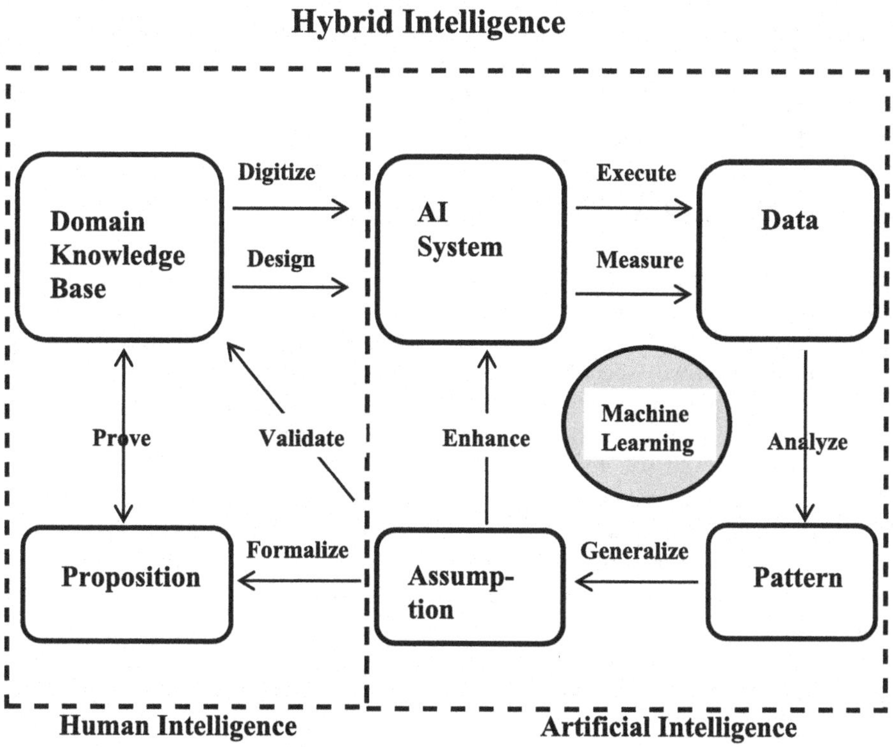

Fig. 3. Combining human and artificial intelligence

The Artificial Intelligence side of this concept is mainly building on data collection and analysis, and on pattern recognition and related reasoning. Altogether they make up the machine learning operation of the AI system. If such a system operates completely autonomous, the developing experience base can become instable and get out of order, as examples with text generation applications let show. Rudin [17] therefore strongly suggests machine learning to be tailored specifically to the domain of interest, thus enabling interpretability, as opposed to pure black box machine learning. This approach becomes focus in the concept of federated learning. Likewise, Athey [2] argues for big data applications in the policy field. She contests the usefulness of machine learning 'off-the-shelf', which is without understanding of underlying assumptions that require domain expertise to verify. Thus, she makes a strong point for supervised machine learning as it would be inherent to hybrid intelligence. And for computational social sciences Watts, Beck et al. [18] claim that data reasoning for predictive purposes must rest upon consistent estimation of causal relation, which requires the involvement of human domain knowledge.

Human supervision is represented in the left part in Fig. 3. It has controlling impact on the machine-based operations via digitization and design. Reversely, data reasoning on the AI side influences the advancement of the domain knowledge base by formalization and validation of assumptions drawn from patterns that were detected. It is worth noting here that in certain contexts of application human supervision can be important as it enables the purposeful introduction of bias which is explicitly desired. Cirillo et al. [5] discuss this for the field of precision medicine, where there is the need to consider sex related differences with physiological parameters or digital biomarkers when empirical patient data is collected for analysis.

5 Exemplifying Cases and Concluding Remarks

It is worthwhile to note that hybrid intelligence, as illustrated above, is not only meant to apply to scientific fields. It is appropriate in the industrial or public administration sphere, too, and also in commercial applications. In terms of innovation through knowledge generation however we expect the most valuable impact on science. Two cases of this kind that were subject to recent publications shall be described briefly here.

The first one is related to the COVID19 pandemic. When the SARS-CoV-2 virus emerged in the year 2019 and spread worldwide since 2020, a most unique innovation in pharmaceutics was accomplished in record time: the development of mRNA vaccines. They turned out the most powerful preventive measure against the COVID19 disease. However, numerous virus variants started to develop quickly and steadily. The mutant viruses showed significantly different levels of infective potential and severity of symptomatic illness. Soon questions arose about the effectiveness of vaccines and eventual need for adjustment. In the past the evaluation of new mutant risk rested on ex post observation of manifest infections, which required significant lead time for empirical investigation. Recently, though, as a novel approach a hybrid intelligence concept was put into action. The preprint of first results became available in December 2021. Beguir, Sahin et al. [4] describe their approach to the early computational detection of high-risk virus variants. It builds upon the combination of human domain expertise, as to the relation of specific virus structures and their potential implications, and the so called *in silico* assessment of their risk level with the help of supervised machine learning technique. Results of the machine learning application were reviewed by human domain experts as to their validity, and to enhance the model employed. The researchers report significant improvements regarding the time needed to detect dangerous variants, on average two months ahead of WHO sourced respective warnings. Furthermore, they demonstrated the applicability of their approach to real-time risk monitoring of mutations by an Early Warning System.

The second case is about historical research into the restoration of ancient text inscriptions that are damaged or remain preserved only partially, as most recently published [1]. The established scientific methods of epigraphy are constrained to the use of repositories of textual and contextual parallels and bring along high levels of generalization but low certainty of results. Assael et al. [1] applied a deep learning software based on neural networks that, with the help of human domain expertise, was carefully tailored to the epigraphic tasks to accomplish. They conducted experimental evaluation by applying their approach to a couple of ancient Greek inscriptions, using a specific metric to evaluate the performance achieved. They found substantial improvements of accuracy and speed of the restoration tasks under use of the combined human-machine intelligence concept. In fact, they report that the combination of human and artificial intelligence in an iterative process achieves significantly better results than human or artificial intelligence only.

These examples point to how technological resilience can be achieved with the help of hybrid intelligence approaches, be it pro-active risk governance, e. g. in health, or life-cycle advancement of ODD's, e.g. with autonomous vehicles, or enhancement of simulation models using digital twins, e. g. in smart manufacturing.

References

1. Assael, Y., et al.: Restoring and attributing ancient texts using deep neural networks. Nature **603**(7900), 280–283 (2022). https://doi.org/10.1038/s41586-022-04448-z
2. Athey, S.: Beyond prediction: using big data for policy problems. Science **355**, 483–485 (2017). https://doi.org/10.1126/science.aal4321
3. Athey, S., Bryan, K., Gans, J.: The allocation of decision authority to human and artificial intelligence. NBER working paper no. 26673 (2020). http://www.nber.org/papers/w26673
4. Beguir, K., Sahin, U., et al.: Early computational detection of potential high-risk SARS-CoV-2 Variants. bioRxiv preprint, 27 December 2021. https://doi.org/10.1101/2021.12.24.474095
5. Cirillo, D., et al.: Sex and gender differences and biases in artificial intelligence for biomedicine and healthcare. NPJ Digital Med. **3**(1) (2020). https://doi.org/10.1038/s41746-020-0288-5
6. Eren, O., Mocan, N.: Emotional judges and unlucky juveniles. Am. Econ. J. Appl. Econ. **10**(3), 171–205 (2018). https://doi.org/10.1257/app.20160390
7. Helbing, D.: Systemic risks in society and economics. International risk governance council workshop on emerging risks. Geneva, December 2009. (2010). https://irgc.org/wp-content/uploads/2018/09/Systemic_Risks_Helbing2.pdf
8. International Risk Governance Council: The Emergence of Risks: Contributing Factors. Geneva (2010)
9. Jastroch, N.: Trusted artificial intelligence: on the use of private data. In: Nyffenegger, F., Ríos, J., Rivest, L., Bouras, A. (eds.) PLM 2020. IAICT, vol. 594, pp. 659–670. Springer, Cham (2020). https://doi.org/10.1007/978-3-030-62807-9_52
10. Jastroch, N.: Applied artificial intelligence: risk mitigation matters. In: Junior, O.C., Noël, F., Rivest, L., Bouras, A. (eds.) Product Lifecycle Management. Green and Blue Technologies to Support Smart and Sustainable Organizations: 18th IFIP WG 5.1 International Conference,

PLM 2021, Curitiba, Brazil, 11–14 July 2021, Revised Selected Papers, Part I, pp. 279–292. Springer International Publishing, Cham (2022). https://doi.org/10.1007/978-3-030-94335-6_20
11. Kahneman, D., Sibony, O., Sunstein, C.: Noise. Siedler Verlag, München (2021)
12. Kuhn, T.S.: Die struktur wissenschaftlicher revolutionen. 23rd impression. Suhrkamp Verlag, Frankfurt a. M (2012)
13. Laplane, L., et al.: Why science needs philosophy. In: PNAS 5 March 2019, vol. 116, no. 10, pp. 3948–3952 (2019). https://doi.org/10.1073/pnas.1900357116
14. Ludwig, J., Mullainathan, S.: Fragile algorithms and fallible decision-makers: lessons from the justice system. J. Econ. Perspect. **35**(4), 71–96 (2021). https://doi.org/10.1257/jep.35.4.71
15. Penrose, R.: The Emperor's New Mind. Revised impression 9, Oxford Landmark Science. Oxford University Press (2016)
16. Rovelli, C.: Physics needs philosophy. philosophy needs physics. Found. Phys. **48**(5), 481–491 (2018). https://doi.org/10.1007/s10701-018-0167-y
17. Rudin, C.: Stop explaining black box machine learning models for high stakes decisions and use interpretable models instead. Preprint. arXiv: 1811.10154v3. 22 Sep 2019 (2019)
18. Watts D.J., Beck, E., et al.: Explanation, Prediction, and Causality: Three Sides of the Same Coin? OSF Preprint. 31 Oct 2018. https://doi.org/10.31219/osf.io/u6vz5
19. Brunnermeier, M.K.: Die resiliente Gesellschaft. Aufbau Verlage, Berlin (2021)

NMR Spectroscopy and Learning-Based Classification of Acerola Samples

Louis Combis[1], Asma Bourafai-Aziez[2], Baudouin Dafflon[3(✉)], Maxime Gueriau[4], Philippe Bonal[2], Emmanuel Cassin[2], and Yacine Ouzrout[1]

[1] Univ Lyon, Univ Lyon 2, INSA Lyon, Université Claude, Villeurbanne, France
[2] Evear-Extraction, 49320 Couture, France
[3] Univ Lyon, Université Claude Bernard Lyon 1, INSA Lyon, Univ Lyon 2, DISP-UR4570, 69621 Villeurbanne, France
baudouin.dafflon@univ-lyon1.fr
[4] INSA-Rouen, Saint-Étienne-du-Rouvray, France

Abstract. Acerola (Malpighia emarginata DC) is an exotic fruit that has a high agro-industrial potential. It is known to be rich in ascorbic acid, phenolic compounds, and carotenoid pigments. These nutrients make acerola one of the best sources of natural antioxidants, helping to prevent many conditions and delay aging. Acerola fruit is transformed into concentrate juice then powder to be incorporated into nutritional supplements. The natural ascorbic acid content of juice powders must be between 16 and 17%. Unfortunately, the origin of ascorbic acid in acerola-based products is not always natural. That is to say, some food manufacturers add synthetic ascorbic acid to reach the recommended values (16 to 17%), which can be considered as a falsification of the product. Since a decade, the control of the life cycle and the quality of foodstuffs is an increasingly important concern. In this context, EVEAR Extraction (French company) establishes a high level of traceability of its extracts by combining sourcing, extraction processes and laboratory controls throughout the production process. The determination of the composition of raw material and final products can be determined by spectrometric analysis and more precisely by Nuclear Magnetic Resonance (NMR) spectroscopy. However, spectral analysis remains a tedious and time-consuming task requiring an expert.

In this study, the feasibility of discriminating acerola-based product was investigated using 1H NMR spectroscopy in combination with a supervised classification procedure consisting of several steps: principal component analysis (PCA), a fast Fourier transform (FFT) and a neuronal network classification. A total of 6 classes (Colored Acerola powder, Acerola concentrate, Acerola powder, Ascorbic Acid, Acerola with added ascorbic acid, Other extract) were examined. Following the classical approaches, we opted for a convergent network using hidden layers and a divergent output. The results demonstrate that 1H NMR spectroscopy combined with ANN analysis is an effective tool for verifying the nature of Acerola samples.

Keywords: Neuronal network · NMR spectroscopy · Tracability

Published by Springer Nature Switzerland AG 2023
F. Noël et al. (Eds.): PLM 2022, IFIP AICT 667, pp. 327–336, 2023.
https://doi.org/10.1007/978-3-031-25182-5_32

1 Introduction

Acerola (Malpighia glabra L.) is a small tree that grows in dry deciduous forests. It is native to central and northern South America and has been cultivated in large areas of Brazil [4]. Its red fruit, which resembles the European cherry, contains about 80% juice and a large amount of ascorbic acid (vitamin C), but it is also rich in other nutrients such as carotenes, thiamin, riboflavin, niacin, proteins, and mineral salts, mainly iron, calcium and phosphorus [2,3]. Acerola's high vitamin C content makes it one of the best sources of natural antioxidants, helping to prevent many diseases and delay aging [5]. Vitamin C is involved in several biological functions, such as enhancing collagen formation [8] and is considered one of the major vitamins required by the human body because of its antioxidant properties [15]. Indeed, increased antioxidant intake has been associated with a lower risk of cardiovascular disease [7]. As a result, acerola concentrate is used in the manufacture of many dietary supplements and their quality depends on the quantity of key active components and the absence of undesirable materials such as adulterants and residual solvents. Claims of benefit depend on the presence of specific molecules in the extracts, which must therefore be identified and quantified with great precision. Recently, NMR spectroscopy has been widely used as a qualitative and a quantitative tool to characterize plant extracts. NMR spectroscopy-based metabolome analyses can be highly effective in identifying and quantifying novel and known metabolites [6,14,18,20]. However, the spectral analysis remains a tedious and time-consuming task requiring an expert. Proton nuclear magnetic resonance (1H NMR) allows to obtain a metabolomic profile of the analyzed sample but does not allow to detect the addition of synthetic ascorbic acid in acerola products. Moreover, the addition of 1 to 2% ascorbic acid changes the metabolomic profile slightly but it is not possible to see this modification with the human eye. Hence the interest of using artificial intelligence. The idea is to train the model with real acerola concentrates, concentrates transformed into powder (without addition of ascorbic acid) and concentrates transformed into powder with addition of ascorbic acid, then query it to classify spectra of unknown products. To date and to the best of our knowledge, No NMR method coupled with artificial intelligence has been implemented for the classification of acerola products according to their composition in order to detect the addition of synthetic ascorbic acid.

The determination of the nature of the extracts can be summarized as a classification problem. Data classification is the process of analyzing structured or unstructured data and organizing them into categories based on the type and content of the signals. There are several types of classification: unsupervised and supervised (Logistic Regression, SVM, etc.) techniques [12]. Among the methods in the literature, the classification proposed in this work is based on Convolutional Neural Networks (CNNs) [1]. Indeed, CNNs have shown their efficiency in the creation of feature maps. These maps are a strong point for NMR spectrum analysis since they are invariant to the small transformation introduced by the measurement.

To detail this approach, the paper is decomposed as follows: first, a state of the art on spectral analysis methods and on neural network classification is proposed. In Sect. 3, we will explain how the data are collected and how our model is made. Section 4, that details the first results of our approach, is followed by a discussion in the conclusion.

2 State of the Art

2.1 Overview

To ensure the authenticity of herbal extracts or herbal products, the process of standardization is a lengthy one, requiring proper sample preparation, time-consuming analytical method development for the resolution of an analyte peak from the complex natural extract, and more importantly, a pure authentic natural product. As a solution to these issues, the creation of a reliable and simple method is needed as an alternative to standard analyses.

Metabolomic profiling is a discipline that focuses on the detailed description of the metabolite composition of herbal extract. Metabolomics thus focuses on the analysis of metabolites that represent the final phenotype of the extracted herb. These metabolites are low molecular weight molecules (molecular weight $<$ 1500 Da) and can be sugars, amino acids or fatty acids, their levels reflect changes in the genome, transcriptome and proteome. Proton nuclear magnetic resonance (1H NMR) is used in this kind of study because it is a highly reproducible technology that offers information about all metabolites in a herbal extract sample that are over the limit of detection. While artificial intelligence has been widely used in the pre-processing of NMR data, peak identification, peak integration, its use in metabolomics is not as developed as it is in other omics domains like as genomics [16]. In this study, artificial intelligence techniques such as artificial neural networks, genetic algorithms, and genetic programming will be applied to metabolomic data.

2.2 Nuclear Magnetic Resonance (NMR)

Nuclear Magnetic Resonance (NMR) spectroscopy has evolved into a powerful tool for metabolomic analysis of plant extract [6]. it's non destructive, fast, accurate, quantitative and information-rich analytical method. It is a highly repeatable and reproducible method when compared to mass spectrometry. It is possible to compare, distinguish, or classify samples using NMR spectra. However, because of the high level of signal overlap, especially in one-dimensional NMR spectra, this approach has been limited in its application. Indeed, NMR appears to be a good fit for artificial intelligence techniques because of this. The most typical process in NMR data handling is data pre-processing, which involves converting the free induction decay (FID) to a matrix of chemical shift and intensity, baseline correction, normalization and peak alignment [16]. On an NMR spectrum, each metabolite has characteristic peaks whose position is well defined and whose intensity correlates with the amount of this metabolite.

In this study, NMR spectra of acerola samples under different formulations and of other extracts are recorded. The spectra are processed and transformed into a matrix with the chemical shifts on the x-axis and the intensity on the y-axis. The classification of the spectra of large series is generally done using unsupervised (PCA) or supervised statistical (PLS) methods.

2.3 Classification

In machine learning applications, the goal is to train a model (or enable it to learn) from the data it is given, and thus improve its output (results). Machine learning techniques are generally split into two families: supervised and unsupervised learning methods. Unsupervised models learn from unlabelled data: the trained model works on raw data and looks for patterns in the given information. They are very convenient because the input data does not need to be tagged or labelled. However these techniques can be limited when the task is to distinguishing between many complex classes. Supervised models require labelled data and they usually need more computing capacity (and more data). Given the exact information of the expected class for each data (label), the model is able to discern the specificity of two closely related classes since it knows they must be different. However, when using supervised techniques, one must be careful about over-fitting. When the training of the model is not well established (for instance, when the input data is too specific), it can identify false relations that bias the reasoning, resulting in wrong classifications.

The NMR spectra of some molecules are very close and are quite difficult to differentiate for humans. As a result, the use of unsupervised models seems inconsistent with the high similarity of the data. In our case, supervised model seem the most appropriate. The distinction between two close NMR spectrum seems too complicated to be done by the model itself.

In this context, many algorithms for spectral matching have been developed since the 1970s [9,11]. Most of these algorithms were developed for mass spectrometry. Their applications quickly extended to vibrational spectroscopy. These conventional spectral matching approaches are iterative techniques. They are based on the identification of the largest similarities between the unknown spectrum and the reference spectrum. These approaches document the tools and techniques that are used to automate this classification through AI. The first results of 1D CNN for spectral data analysis [21] revealed the potential of using machine learning methods for spectral analysis, from classification of a substance to identification of components in a mixture in various scientific fields [13]. The majority of recent publications use 1D CNNs for various spectral applications.

The first results of 1D CNN for spectral data analysis [21] revealed the potential of using machine learning methods for spectral analysis, from classification of a substance to identification of components in a mixture in various scientific fields [13]. The majority of recent publications use 1D CNNs for various spectral applications. However, a few studies have highlighted the need for further research in order to address the problem of not having enough samples compared to the number of features.

3 Material and Methods

3.1 Global Overview

This section provides a description of the dataset used, containing NMR signal recordings of the different samples to be analyzed, as well as the extracted features used to generate the classification model. Then, the methodology used all along the experimentation is described. The design and implementation of the ANN presented in this work relies on Python programming language and on Keras and Tensorflow libraries, that are among the most used for deep learning applications. The key steps in the proposed classification process for sample tractability are: data preparation, network construction, training and evaluation.

3.2 Data

1D 1H NMR spectrum was acquired for each sample. Spectra were acquired on a Bruker Advance-400MHz spectrometer using a 5mm broad-band probe tuned to detect 1H resonances at 400.15 MHz. Data were collected without sample rotation at 300K, as à 64K complex points using a noesygppr pulse sequence with 90° pulse length and pre-saturation to remove the residual water signal. The number of scan was set at 16. The receiver gain was set to 90.5 and the spectral width was fixed to 20ppm. Obtaind FID were converted to spectra using Topspin 3.5 software. Spectra were processed (phase correction, baseline correction) and the signal of TSP set at 0 ppm was used as an internal reference for chemical shift measurement. Finally, spectra were converted to csv files using Mnova software. The csv files contain the chemical shifts on the x-axis and the corresponding intensities on the y-axis.

The data are distributed in the following way (Table 1).

Table 1. Dataset distribution

Class	Number of spectrum
Colored acerola powder	1390
Acerola concentrate	1990
Acerola powder	1040
Acerola powder with added ascorbic acid	1290
Ascorbic acid	2720
Other extracts	520

The following figure (Fig. 1) show two spectra that are visually close but belong to two different classes.

3.3 Model

When exploring existing work in Sect. 2, we identified two techniques that could be relevant for our classification problem:

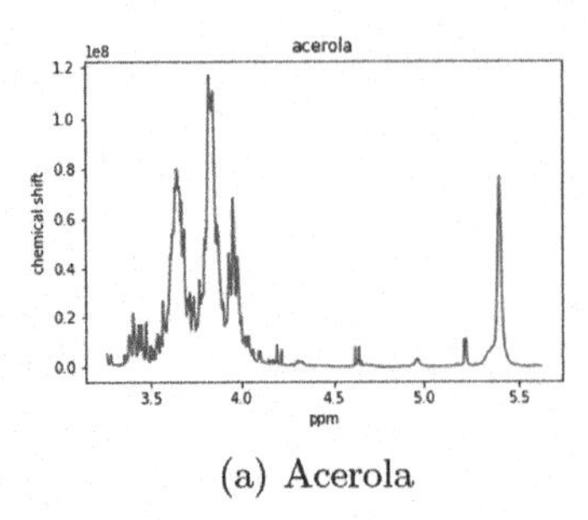

(a) Acerola

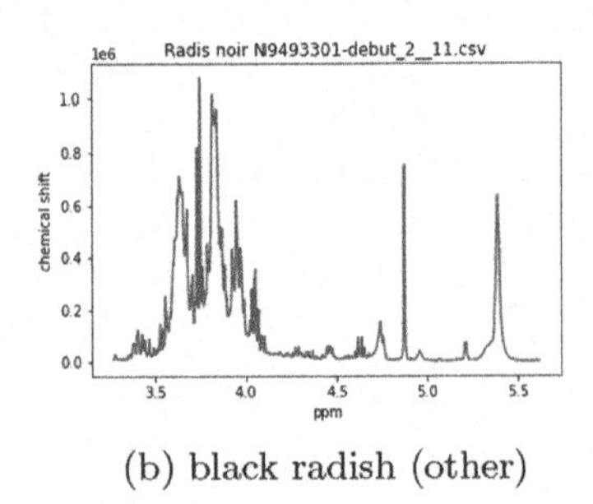

(b) black radish (other)

Fig. 1. Examples of spectra from two different classes.

Deep Neural Network (DNNs) are fully connected networks, all neurons in each neuron in each layer are connected to all neurons in the following layer. The input of a neuron is therefore equal to the number of neurons in the previous layer. The classic DNN structure consists of an input layer, several hidden layers and an output layer. When data enters the input layer, the output values are calculated layer by layer in the network. In each hidden layer, after receiving a vector consisting of the output values of each neuron in the previous layer, it is multiplied by the weights associated with each neuron in the current layer to obtain the weighted sum. The activation function of a neuron is specific to its layer. The functions are chosen according to the type of problem. Sigmoid functions are non-linear as opposed to Rectified Linear Unit (ReLu) for example.

Convolutional Neural Networks (CNNs) are a particular architecture of deep networks [10]. They are designed to process data from several sources: 1D for sequences, 2D for images and 3D for videos [17]. They are very suitable for shape or pattern recognition while being insensitive to scale factor, rotation, etc. In opposition to DNNs, convolution networks are not strongly connected. The hidden layers are separated by layers acting like filter. These filters represent weights and biases. In general, the basic structure of CNNs consists of convolution layers, nonlinear layers and pooling layers. To avoid combinatorial explosion, all neurons in a convolution layer share the same filter, i.e. the same weights and biases, in order to reduce the number of training parameters. As with DNNs, the outputs of these filters are then passed through nonlinear layers that typically use the ReLU function. The role of pooling layers is to aggregate semantically similar features to identify complex features by creating maximal or average subsamples in the feature maps. Sometimes pooling layers are also used to avoid network overfitting and improve model generalization. Given the excellent ability of CNNs to analyze spatial information, they can be applied to NMR spectra reconstruction, denoising, and chemical shift prediction. A convolution network is more appropriate in our case of study because it will allow us to build a feature map corresponding to the different parts of the NMR spectrum.

When using Keras library to implement an ANN in Python, it is necessary to specify the type of model to be created. There are two ways to define Keras models: Sequential and Functional. A sequential model refers to the fact that the output of each layer is taken as the input to the next layer, and this is the type of model developed in this work. The objective is to build a feature map representing the different metabolites characteristic of a class of plant extract (Fig. 2).

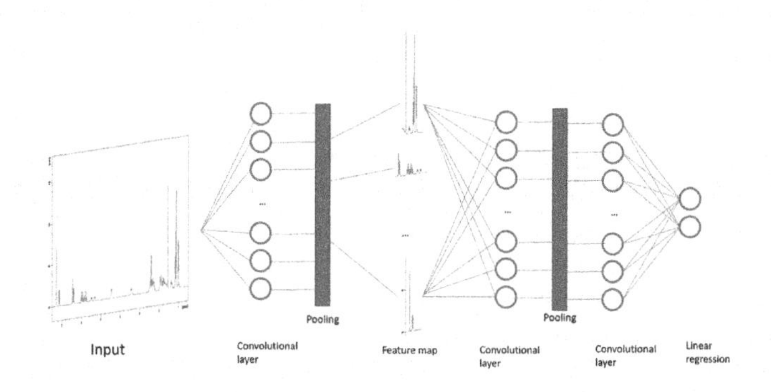

Fig. 2. Feature extraction

The built network has 7 layers. The input layer is composed of as many neurons as there are points on the spectrum. It has been reduced to 1300 datas thanks to Evear's expertise. The following layers reduce little by little the vector to propose an output vector of 6 neurons, as summarized in Table 2.

Table 2. Network topology

Layer (type)	Output shape	Param #
Input (Dense)	(None, 1300)	75401300
Layer2 (Dense)	(None, 800)	1040800
Dropout (Dropout)	(None, 800)	O
Layer4 (Dense)	(None, 500)	400500
Layer 5 (Dense)	(None, 300)	150300
Layer 6 (Dense)	(None, 100)	30100
Outupt (Dense)	(None, 6)	606
Total params: 77,023,606		

A network is associated with several metrics:

- Optimization algorithm: The ANN will use an optimization algorithm to calculate the weight of each neuron. There are several ways to do this. The most common one is to minimize (or maximize) an objective function $E(X)$ which is a mathematical function depending on the internal training parameters of the model where X are features. The result of $E(X)$ is used to compute the objective values Y of the set of training parameters where Y ars labes. The most commonly used optimization algorithms in ANNs are gradient descent. In our case, we use a classical backpropagation proposed by Tensorflow.

- Loss function: The loss function, also known as the cost function, is a function that measures the quality of the network response. A high result indicates that the ANN is performing poorly and a low result indicates that the ANN is performing positively. This is the function that is optimized or minimized when backpropagation is performed. There are several mathematical functions that can be used, the choice of one of them depends on the problem to be solved. The most suitable function for classification is the cross-entropy function. The cross-entropy loss, or log loss, measures the performance of a classification model whose output, Y, is a probability value P, between 0 and 1, and is calculated using the following equation (Eq. 1). The cross-entropy loss increases as the predicted probability deviates from the actual label. This function is used for classification problems.

$$-(ylog(p) + (1-y)log(1-p)) \tag{1}$$

4 Results

Our dataset contains a set of 8960 spectra distributed as describe in Table 1. Following a similar procedure as in traditional approaches [19], the dataset was randomized and then separated into two samples: Training 80% and Validation 20%. The learning (training) phase is a backpropagation repeated thirty times.

The indicators of the training phase are presented in this section. Reviewing learning curves of models during training can help to diagnose problems with learning, such as an underfit or overfit model, and if the training and validation datasets are suitably representative.

We first observe the accuracy of the model (how well it is able to guess the expected class for a given input spectra). This learning curve is calculated from a hold-out validation dataset that gives an idea of how well the model is generalizing.

As illustrated in Fig. 3(a), the accuracy increases with the repetition of the training batch. At the end of this step, we can see that our network reaches an accuracy level of 93% on the validation dataset. On samples that are not part of the original dataset, we observed similar results. The loss curve, depicted in Fig. 1(b), shows that our network is learning well and is converging to an optimum. There would not be much benefit in extending the learning phase.

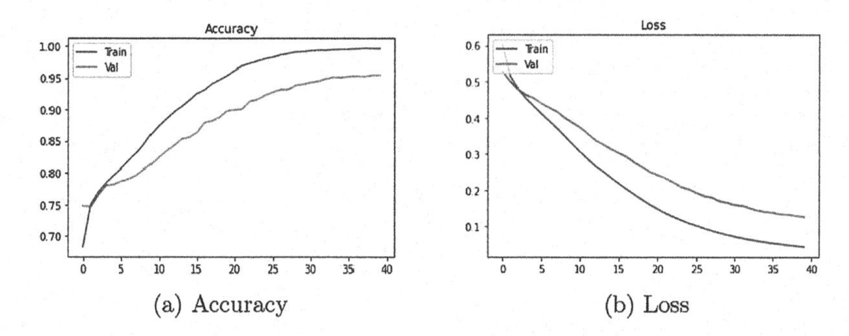

(a) Accuracy (b) Loss

Fig. 3. Learning curves

5 Conclusion and Future Work Direction

The aim of this work was to create a classification model of acerola samples using an NMR spectrum as a data source. Once the data was made usable by the classical techniques of chemistry, we could use it to ensure a better traceability. The advantage of this model is to save time and precision to the Evear-extraction team. Through the introduction of a CNNs, we were able to define and train a model capable of meeting our expectations. The training game was discriminating enough to properly train the 7 layers of our CNN. However, from an industrial point of view, it will be necessary to increase the capabilities of the POC to take into account more complex plant extracts. However, from an industrial point of view, it will be necessary to increase the capabilities of the POC to take into account more complex plant extracts.

However, this work is only a proof of concept. The proposed classification method of plant extracts could be significantly improved by exploring the following directions:

- using different (and more) classes for the network. Indeed, the acerola is only one example of the plants to be traced. For this, it will be necessary to increase the precision of the network and rework its topology.
- increasing the reliability of the traceability by being able to define the metabolites and their quantity present in a spectrum.
- strengthening our knowledge of the spectrum by analyzing the parameters described in the architecture of the neural network or using feature selection techniques for the dataset.

References

1. Aghdam, H.H., Heravi, E.J.: Guide to convolutional neural networks. New York NY: Springer **10**(978–973), 51 (2017)
2. Albertino, A., Barge, A., Cravotto, G., Genzini, L., Gobetto, R., Vincenti, M.: Natural origin of ascorbic acid: validation by 13C NMR and IRMS. Food Chem. **112**(3), 715–720 (2009)

3. Alves Filho, E., Silva, L.M., Canuto, K.: Metabolomic profiling of acerola clones according to the ripening stage. Food Mesure **15**, 416–424 (2021)
4. Anand, P., Revathy, B.: Acerola, an untapped functional superfruit: a review on latest frontiers. J. Food Sci. Technol. **55**, 3373–3384 (2018)
5. Belwal, T., et al.: Phytopharmacology of acerola (Malpighia spp.) and its potential as functional food. Trends Food Sci. Technol. **74**, 99–106 (2018)
6. Deborde, C., Moing, A., Roch, L., Jacob, D., Rolin, D., Giraudeau, P.: Plant metabolism as studied by NMR spectroscopy. Prog. Nucl. Magn. Reson. Spectrosc. **102–103**, 61–97 (2017)
7. Ellingsen, I., Seljeflot, I., Arnesen, H., Tonstad, S.: Vitamin c consumption is associated with less progression in carotid intima media thickness in elderly men: a 3-year intervention study. Nutr. Metab. Cardiovasc. Dis. **19**, 8–14 (2009)
8. Findik, R., Ilkaya, F., Guresci, S., Guzel, H., Karabulut, S., Karakaya, J.: Effect of vitamin C on collagen structure of cardinal and uterosacral ligaments during pregnancy. Eur. J. Obstet. Gynecol. Reproductive Biol. **201**, 31–35 (2016)
9. Grotch, S.L.: Matching of mass spectra when peak height is encoded to one bit. Anal. Chem. **42**(11), 1214–1222 (1970)
10. Gu, J., et al.: Recent advances in convolutional neural networks. Pattern Recogn. **77**, 354–377 (2018)
11. Knock, B., Smith, I., Wright, D., Ridley, R., Kelly, W.: Compound identification by computer matching of low resolution mass spectra. Anal. Chem. **42**(13), 1516–1520 (1970)
12. Lorena, A.C., Garcia, L.P., Lehmann, J., Souto, M.C., Ho, T.K.: How complex is your classification problem? A survey on measuring classification complexity. ACM Comput. Surv. (CSUR) **52**(5), 1–34 (2019)
13. Lussier, F., Thibault, V., Charron, B., Wallace, G.Q., Masson, J.F.: Deep learning and artificial intelligence methods for Raman and surface-enhanced Raman scattering. TrAC Trends Anal. Chem. **124**, 115796 (2020)
14. Pauli, G., Jaki, B., Lankin, D.: Quantitative 1 h NMR: development and potential of a method for natural products analysis. J. Nat. Prod. **68**, 133–149 (2005)
15. Podmore, I., Griffiths, H., Herbert, K., Mistry, N., Mistry, P., Lunec, J.: Vitamin c exhibits pro-oxidant properties. Nature **392**, 559 (1998)
16. Pomyen, Y., Wanichthanarak, K., Poungsombat, P., Fahrmann, J., Grapov, D., Khoomrung, S.: Deep metabolome: applications of deep learning in metabolomics. Comput. Struct. Biotechnol. J. **18**, 2818–2825 (2020)
17. Shamsaldin, A.S., Fattah, P., Rashid, T.A., Al-Salihi, N.K.: A study of the convolutional neural networks applications. UKH J. Sci. Eng. **3**(2), 31–40 (2019)
18. Smolinska, A., Blanchet, L., Buydens, L., Wijmenga, S.: NMR and pattern recognition methods in metabolomics: from data acquisition to biomarker discovery: a review. Anal. Chim. Acta **750**, 82–97 (2012)
19. Tan, J., Yang, J., Wu, S., Chen, G., Zhao, J.: A critical look at the current train/test split in machine learning. arXiv preprint arXiv:2106.04525 (2021)
20. Ward, J., Baker, J., Beale, M.: Recent applications of NMR spectroscopy in plant metabolomics. FEBS J. **274**, 1126–1131 (2007)
21. Yang, J., Xu, J., Zhang, X., Wu, C., Lin, T., Ying, Y.: Deep learning for vibrational spectral analysis: recent progress and a practical guide. Anal. Chim. Acta **1081**, 6–17 (2019)

State-of-Art and Maturity Overview of the Nuclear Industry on Predictive Maintenance

Amaratou Mahamadou Saley[1,2], Jérémie Marchand[1], Aicha Sekhari[2], Vincent Cheutet[2](✉), and Jean-Baptiste Danielou[1]

[1] INEO Nucléaire, Lyon, France
[2] University Lyon, INSA Lyon, Univ. Lyon 2, Université, Claude Bernard Lyon 1 DISP UR4570, Villeurbanne, France
vincent.cheutet@insa-lyon.fr

Abstract. In order to ensure as continuous as possible nuclear energy production, it is necessary to guarantee the availability of all equipment involved in the production chain, in all its complexity. The safety of all these equipments is based on a good command of the maintenance policy with very specific conditions related to the demanding regulations of the nuclear field. It is about having facilities that are safe, available with a good quality and a limitation to radiological risk.

Today, with the advancement of digital technology, substantial improvements have occurred in the tools that can be applied in the maintenance and monitoring of Structures, Systems and Components (SSCs), enabling an understanding of equipment performance far beyond that available only a few decades ago. Therefore, predictive maintenance becomes a subject of prior interest for the nuclear industry.

In this paper, we will emphasize the incentives and obstacles of predictive maintenance deployment in a nuclear context. The objectives here are to draw a clear picture of what can be the practices of predictive maintenance in the nuclear industry context and to identify the requirements and the needs to implement a predictive maintenance model on the lifecycle management of a nuclear facility.

Keywords: Lifecycle management · Predictive maintenance · Nuclear system

1 Introduction

Nuclear energy is one main source of electricity worldwide and the safe operation of nuclear industry is generally acknowledged as contributing to society's success and promoting economic performance. However, managing nuclear energy production is very complex. In order to maintain a stable energy, it is necessary to guarantee the availability of all the equipment involved in this production chain.

F. Noël et al. (Eds.): PLM 2022, IFIP AICT 667, pp. 337–346, 2023.
https://doi.org/10.1007/978-3-031-25182-5_33

IAEA (International Atomic Energy Agency) provides a PLiM (Plant Life Management) program for long-term operation, which aims at identifying all of the factors and requirements for the overall life cycle of the plant for all [16]. One of the important factor of this program is the maintenance prioritization. Maintenance in nuclear industry remains a major challenge for operators: a majority of nuclear power plant (NPP) systems are aged. The inherent risk of the activity imposes strict regulations and compliance. This consumes money time and hard efforts. We need to put in front of it reliable means, methods and tools to respond to it. The regulatory activities on maintenance are defined by various authorities and institutions working for nuclear safety. Among the standards referred to the maintenance, we have NF EN 13306 which defines the types of maintenance and organization of maintenance, the IAEA safety standards especially for the operation, maintenance and safety of NPP [14,16].

The research interest regarding maintenance strategy to optimizing availability/reliability, minimizing risk and cost has been growing up time by time. According to Web of Science, since 1975, 18857 journals and articles discussing maintenance in nuclear context are found using the keywords "Maintenance" and "Nuclear".

However, there is a wide gap concerning the implementation of maintenance in research and practice. Although many models, methods, and strategies have been proposed, most of the nuclear industries in the world are still using traditional approaches [22]. These traditional approaches help in minimizing the risk of failure. Nevertheless, they do not guarantee a full availability of all the critical equipments. At the same time, we are now facing the development of industry 4.0 strategy linked with the emergence and maturity of digital. Known as the fourth Industrial Revolution, it is characterized by the widespread use of digital and physical environments called Cyber Physical Systems. This convergence achievement of information technologies, Internet Of Things (IoT), big data, cloud, communication systems (SCADA), Cybersecurity, data analytic, with additional accelerators, such as Artificial Intelligence and cognitive science makes it possible to **improve the traditional maintenance strategies with the Predictive Maintenance (PdM)** [12,23]. Regarding NF EN 13306 standard, the PdM is a "Condition-based maintenance carried out following a forecast derived from repeated analysis or known characteristics and evaluation of the significant parameters of the degradation of the item". The condition based maintenance is a "preventive maintenance which include assessment of physical conditions, analysis and the possible ensuing maintenance actions". Its implementation requires a large number of concepts and tools such as Industrial IoT [5], data mining, cloud, big data, and artificial intelligence [23].

Nevertheless, deploying such technologies in nuclear industry is not an easy task due to the regulatory constraints. Also, the maturity of PdM in nuclear facilities is not the same compared to other industries. The nuclear industry can be seen as a conservative one that is usually strictly regulated [15]. Also, the equipment involved in the production chain are very specifics, with very long lifespans. The current fleet of NPP is old, with a majority of plants built and commissioned during the 70 s and 80 s and an initial lifespan of around 40 years. In addition, of this ageing effect, the equipments that make up the facility are diverse, varied and do not have the same level of criticality. Some systems are very critical, with highly regulated owing to the risks associated with plant accidents and radiation exposure to the public. The objective of this paper is to **establish an overview of maintenance practices in the nuclear industry to evaluate its maturity and to identify opportunities and threats for PdM deployment.**

The paper is structured as follow: Sect. 2 is dedicated to the related works of maintenance in nuclear facilities. Section 3 introduces the PdM as a whole. Section 4 focuses on PdM in a nuclear context by identifying the current evolution. Section 5 makes a critical analysis of the maturity of PdM in the nuclear industry and introduces further research directions.

2 Maintenance in Nuclear Facilities

As seen in the introduction, 18857 articles were found in the Web of Science database considering "Maintenance" and "Nuclear" keywords. The process for relevant papers collections and selections consisted of three main phases. First, we have filtered 18857 articles by categories, considering only papers that are closed to our research area. Then, we keep papers written in English that are journals and proceedings and removed duplicated articles. At least, a total of 8882 research papers were short-listed for classification. In order to deepen our investigation of the short-listed papers, VOSviewer software [21] was used to carry out an analysis of the co-occurrence of authors keywords (Fig. 1a).

Firstly, we can highlight a cover of keywords around maintenance, CBM, PdM, fault diagnosis in the NPP. This cluster shows the scientific orientation on the NPP over the last ten years. Then a second group of keywords around ITER remote maintenance, remote handing is observed. ITER (International Thermonuclear Experimental Reactor) is an international nuclear fusion research and engineering megaproject aimed at replicating the fusion processes of the sun to create energy on the Earth. A final keyword overlay around the maintenance is visible. This cluster is linked to all the other ones and is very representative. It can be understood that the focus on the maintenance in nuclear industry is not only on nuclear fission (NPP) but also on nuclear fusion, and the topics are the same and centered on the PdM, optimization, remote maintenance, equipments availability, fault diagnosis.

From this state of the art, we observe that whatever the industry, the challenges remain the same on the maintenance strategies and methods. IAEA highlight the importance of the Reliability Centered Maintenance (RCM) in a nuclear facility optimization [14] as demonstrated by EDF in France [8]. The RCM was introduced in the late 1960 by the airline industry [14]. The core of the RCM philosophy is that maintenance will be performed only after evaluating the consequences of failures at component level. It deals with optimization of preventive maintenance activities considering failure consequences. The RCM approach is

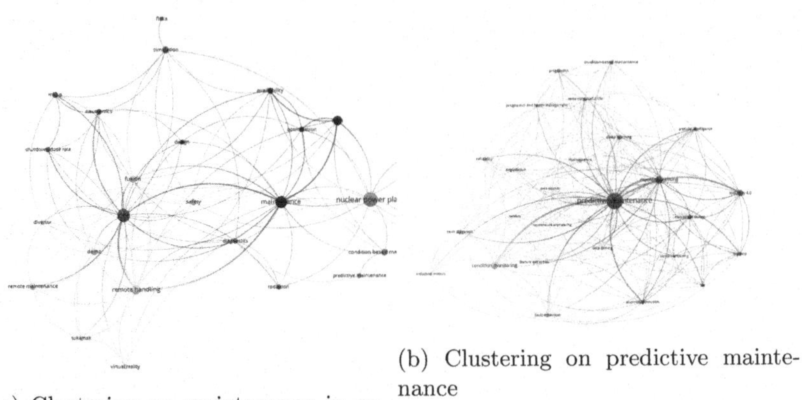

(a) Clustering on maintenance in nuclear context

(b) Clustering on predictive maintenance

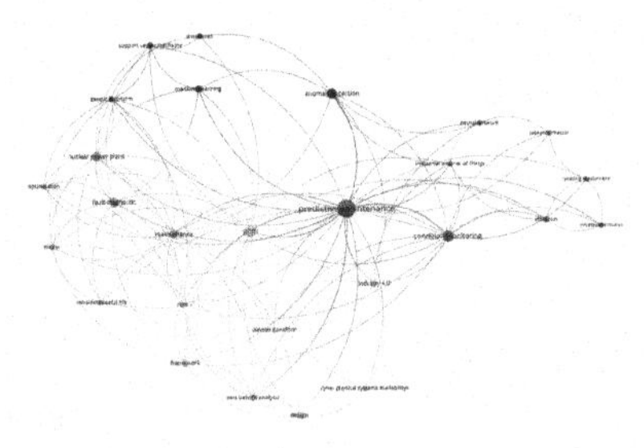

(c) Clustering on predictive maintenance in nuclear context

Fig. 1. Clustering of co-occurrence of authors keywords. source: VOSviewer Clustering

system-oriented and may be implemented free of a company's organizational culture. It helps in establishing system boundaries and developing a critical equipment list. An equipment is said to be critical if it performs a function or if its failure can affect functioning of equipment that impacts nuclear safety, prevention of release of radioactivity to the environment, personnel safety and continued power production. Thus the RCM facilitates the decision making on the type of maintenance to be implemented for each equipment.

The maintenance in nuclear industry are how to perform to the optimum level of the capability of equipments and plants. The literature describe several aspects of maintenance. They may be grouped into the technical aspect, commercial one and management one. This implies that maintenance decisions need to be made in a framework that takes into account these issues from an overall business perspective.

The overall maintenance strategies can be categorize in two types. **Corrective maintenance** is carried out after failure detection and aims to restore an asset to a condition in which it can perform its intended function [17]. In the literature, authors identify two types of corrective maintenance: the planned and the unplanned corrective maintenance [17]. **Preventive maintenance** aims at doing the maintenance just before a component fails so the system downtime and the emergency will not happen, by scheduling machine repairs or rebuilds [23].

Two types of preventive maintenance are discussed in the literature [16]: the **time-based maintenance (TBM)** aiming at performing periodic maintenance on the equipment and the **condition-based maintenance (CBM)** carried out in response to a significant deterioration in a machine as indicated by a change in a monitored parameter of the machine condition.

The procedure for performing corrective or preventive maintenance in a nuclear environment is complex. Indeed, it requires a lot of preparation up, given the regulatory safety aspects. First of all, the maintainers must assess the radiological environment, the chemical risks of the process and facility concerned by the maintenance. Then, it is necessary to examine the operating procedure. This allows to optimize the intervention time since this time is very limited and strongly correlated to the radiation exposure. Often, given the complexity of some processes, the failure is difficult to detect.

Maintenance is thus becoming more complex and detecting failures is becoming harder and more time-consuming. The labor cost to carry out maintenance has also been increasing. As a result, maintenance will continue to evolve and the two main drivers for this are (i) technology and (ii) management.

3 Predictive Maintenance as a Whole

The literature highlight many research areas related to PdM. As in the previous case, web of science is used for relevant articles collection. 3750 articles were found using the key word "Predictive maintenance". We filtered the most relevant ones using the same process as in the maintenance section. At least, a total of 2453 research papers were short-listed for classification.

Figure 1b draws up three main clusters around PdM literature. The first one is on the maintenance management. It includes all the research studies related to the reliability, fault detection, fault diagnosis, maintenance engineering for maintenance optimization. The second cluster covers research topics on the PdM techniques and tools. This cluster is the most representative one, as it is linked to all the others clusters. It can be understood that PdM in the literature is an association of three main research field: maintenance management, IoT and artificial intelligence. The last cluster relies on the prognostics, health management and the remaining useful life (RUL). Very present in the literature, it is a part of PdM which aims at equipment heath monitoring.

PdM has the potential to make the production more reliable and improve service provisioning. However, the maturity for a proper implementation of PdM in research and in practice is not the same. That is why we conducted a multi-case study to underline the conditions and technological aspects for implementing a PdM system in practice. We found that PdM initiatives are triggered by severe impacts of failures on revenue and profit. Furthermore, successful PdM initiatives require that some pre-conditions are fulfilled: Data must be available and accessible. Also, the support by the management is very important. We identified four important factors for PdM implementation: Data integration (highly facilitated by Cloud-based mechanisms), failure detection (enabled by advanced analytics), PdM operations execution (supported by data-driven process automation) and the visualization [18].

It is important to note that the failure modelling approaches for PdM are a bit common to the maintenance and fault computing. PdM also relies on two distinct approaches:

- **The first approach** is the probabilistic approach. It is based on a statistical and reliability analysis. The input data is retrieved from the Computerized Maintenance Management Information System (CMMIS) software. The CMMIS centralizes maintenance information and facilitates maintenance operations processes. Moreover, the CMMIS optimizes the use and equipment availability.
- **The second approach** is the deterministic approach. It is characterized by the use of some sensors data combined with data mining techniques and artificial intelligence algorithms to predict a machine failure.

The second approach was of great interest to the researchers because it made it possible to optimize maintenance costs better than the first approach. When CBM was introduced in 2002, many implementation methods were developed. These methods involved at measuring certain process parameters from sensor data. The most used methods at the beginning of CBM are: Vibration analysis, Acoustics analysis, Lubrication oils analysis, Corrosive analysis, and Thermal analysis or Infrared thermography [4,19].

As methods have evolved since 2010, new PdM techniques have emerged. A new method called Prognostic and Health Management (PHM) was introduced [2,23]. The PHM is an engineering process where algorithms are used to detect anomalies, diagnose faults and predict RUL. Moreover, the financial benefits such as operational and maintenance cost reductions and extended lifetime are achieved with this method [20].

To implement a PHM, many steps are required. Author of [2] addresses the Open System Architecture for Condition-Based Maintenance (OSA-CBM), and draws up six steps for performing a PHM based on ISO 13374: Data acquisition (Access to digitized sensors), Data manipulation (perform single and/or multichannel signal transformations), State detection or Fault Detection, Diagnosis step, Prognostics assessment and Advisory generation [2,4,6,23]. The architecture system for the PHM includes a database which stores data at each step and a Human Machine Interface for visualization.

CBM and PHM are sometimes confused: CBM includes data acquisition, data processing, fault detection and isolation, and failure diagnostics. Whereas PHM is more a prognosis, an estimation of the RUL and a decision-making process for the implementation of PdM and associated logistics [5,23].

The predictive modelling techniques used in diagnostic and prognostic functionalities of PHM can be categorized into three: physical modelling (data-based models), artificial intelligence methods (machine learning and deep learning) and statistical-based methods [1,5,7,23].

4 Predictive Maintenance in a Nuclear Context

The state of the art on maintenance has shown that sometimes, a maintenance intervention can be complex especially when a failure is not anticipated. That is why nuclear industry is today moving towards PdM to overcome this problem. The explosion of digital technologies has enabled the collection of sensor data to better model the failure of an equipment. However, is this industry enough mature in deploying such maintenance strategy?

In this sense, we made a literature review based on Web Of Science. 90 articles have been found considering the combination of the following keywords: "predictive maintenance" and "nuclear". By using the same process as seen previously for relevant papers collection, we finally get a total of 44 research papers. VOSviewer software makes it possible to analyze the co-occurrence of authors keywords.

Overall, Fig. 1c draws up all the topics covered in the literature around PdM in the nuclear industry. We identify four main clusters. First, we observe a grouping of keywords around the PHM, characterizing thus the relevant topics. Those keywords (RCM, RUL, wavelet transform, condition monitoring, maintenance, availability, machine learning) show the research orientation of the PHM in nuclear industry. We can also see a significant link between the PHM and the RCM standard. It can be understood that PHM highly impact on the RCM. The second cluster covers various topics around PdM. The topics are: condition based monitoring, industrial IoTs and so one. It is the most representative cluster because it is highly correlated to all the other clusters. The third cluster gathers different topics around the NPP: optimization using fault diagnosis, remaining useful file. Finally, the last cluster includes all the machine learning techniques for PdM implementation. These are: genetic algorithm, support vector machine, neural network and anomaly detection.

The specificity of PdM in nuclear industry lies in its implementation. First of all, equipment instrumentation is governed by regulations that are highly correlated to the radiation exposure and various risks associated with the facility. The location of the sensors on the equipment depends mainly on the risk associated with the process. This is why a risk analysis on the overall process must first be carried out before any intervention. Second, data is often subject to military regulations requiring authorization before any accessibility. Also, predictive models must be in line with the expectations of the process experts and their deployment should be in accordance with the confidentiality rules as defined by the operator and the various safety authorities.

It is important to note that before the introduction of PdM, the nuclear industry, as in all production industries, used traditional techniques such as

Infrared Thermography, Ultra sound analysis, Motor current signature analysis, Oil analysis, and Partial discharge to maintain all the facilities. Also, the CMMIS data is mainly used to compute the risk associated with operating the facility, expressed in terms of various statistic metrics related to the different levels of failure to the facility (e.g. core failure frequency), or its environment (e.g. societal or individual risk) [13,16]. However these techniques were not enough efficient in predicting a failure. This gave rise to the PdM methods proposed by [11].

PdM has Positively Impacted the RCM Standard in Nuclear Industry. This positive impact is justified by a good improvement in plant productivity and a reduction in maintenance costs. In [12], authors state that a correct implementation of RCM analysis based on PdM in NPP could bring great benefits in many aspects. Moreover, it is considered to be the best maintenance strategy [12]. Three main lines of research stand out when referring to PdM in the nuclear industry.

The first line is based on the **appropriate methods and techniques used.** The concepts are almost the same as what we saw in the previous section. Many researchers have made a state-of-art of PdM techniques in nuclear industry. Among them, Hashemian was more interesting. He describes three PdM techniques in identifying the onset of equipment failure in NPP: sensor-based technique, test-sensor-based technique and test-signal-based technique [4].

The second line focuses on the **development of Machine learning algorithms for PdM**. The topics were centered on the continuous machine learning for abnormality identification to aid condition-based maintenance [3,3] and prediction of the RUL of a component [3]. The models work as PHM models and the anomaly detection method is the most used in the reality.

The last line focuses on **data acquisition through the Wireless sensor architecture and applications**. This topic is of great interest to many researchers especially Hashemian [10,11]. He reported a research and development program to implement wireless sensors for equipment condition monitoring and other application. He proposed an architecture system which integrate a signal in the existing wireless sensors and news wireless sensors for a holistic view of the equipment health and process. He also proposed some wireless sensors for PdM of rotating equipment [10,11].

In the same way, authors of [9] developed a PdM architecture for nuclear infrastructure using machine learning. They developed the architecture in two parts: a data acquisition part (Offline modelling) and a representation learning part (Online evaluation) with the integration of machine learning algorithms.

5 Critical Analysis and Conclusions

The research literature conducted in this paper enhance various research aspect of maintenance in nuclear. To come with clear and constructive outcomes, we moved from exploring the maintenance in the nuclear industry with special focus on PdM as a whole and PdM in nuclear industry.

The first finding of this literature study is the **gap in the use of PdM technologies between industries**. In fact, the maturity is not the same. All the papers that introduced PdM using AI in the nuclear industry are very recent in comparison with the airline industry. In order to maintain and optimize their facilities, nuclear operators use traditional methods based on failure probability analysis, failure modelling, most often computed using data retrieved from CMMIS. These analyses contributed in computing some classic performance criteria such as MTBF, MTTR. All these methods help in minimizing the risk of failure. Nevertheless, they do not guarantee a full equipment availability. **RCM has become a mandatory standard** that has been implemented by the entire NPP in the world. It allowed them to have a solid methodology to optimize the efficiency of their facilities. PdM technologies are mainly used at NPP; not at the whole fuel cycle.

On the modelling part, most of researchers used some simulated data to implement some machine learning algorithms for PdM. The problem here is that we are not sure to get the same performance on real data. Also many ad-hoc algorithms haven been proposed. **The algorithms are very specific to each equipment** of a nuclear plant; thus do not include a generality.

The Information System for PdM has also Taken an Important Part in this Scientific Literature. The proposed architecture do not integrate nuclear requirements. The architecture for deploying PdM in nuclear context is not well explained: there is no information about how or where the sensors are placed in the equipment, how they communicate with Supervisory control and data acquisition system, and how a PdM solution could be deployed in a manufacturing execution system. Furthermore there is no distinction between cloud or on premise PdM software solution. Some old wireless technologies for vibration analysis, acoustics analysis are mentioned in many studies. Not enough studies have been done on the new wireless technologies such as IIoT or 5G.

This study reveals also **a methodological gap for the deployment and implementation of PdM in nuclear context**. This methodology shall include all the nuclear stakeholders intended to create, manage and exploit PdM technologies. This methodology included in the lifecycle management will increase the speed of deployment and reduce the cost of this technology.

References

1. Ayo-Imoru, R.M., Cilliers, A.C.: A survey of the state of condition-based maintenance (CBM) in the nuclear power industry. Ann. Nucl. Energy **112**, 177–188 (2018)
2. Cachada, A., et al.: Maintenance 4.0: Intelligent and predictive maintenance system architecture. In: 2018 IEEE 23rd International Conference on Emerging Technologies and Factory Automation (ETFA), vol. 1, pp. 139–146. IEEE (2018)
3. Çınar, Z.M., Abdussalam Nuhu, A., Zeeshan, Q., Korhan, O., Asmael, M., Safaei, B.: Machine learning in predictive maintenance towards sustainable smart manufacturing in industry 4.0. Sustainability **12**(19), 8211 (2020)

4. Coandă, P., Avram, M., Constantin, V.: A state of the art of predictive maintenance techniques. IOP Conf. Ser.: Mater. Sci. Eng. **997**(1), 012039 (2020)
5. Compare, M., Baraldi, P., Zio, E.: Challenges to IoT-enabled predictive maintenance for industry 4.0. IEEE Internet Things J. **7**(5), 4585–4597 (2020)
6. Das, S., Hall, R., Herzog, S., Harrison, G., Bodkin, M., Martin, L.: Essential steps in prognostic health management. In: 2011 IEEE International Conference on Prognostics and Health Management, PHM 2011 - Conference Proceedings (2011)
7. Davari, N., Veloso, B., Costa, G.D.A., Pereira, P.M., Ribeiro, R.P., Gama, J.: A survey on data-driven predictive maintenance for the railway industry. Sensors **21**(17), 5739 (2021)
8. Despujols, A.: Optimisation de la maintenance par la fiabilité (OMF). Ed. Techniques Ingénieur (2004)
9. Gohel, H.A., Upadhyay, H., Lagos, L., Cooper, K., Sanzetenea, A.: Predictive maintenance architecture development for nuclear infrastructure using machine learning. Nucl. Eng. Technol. **52**(7), 1436–1442 (2020)
10. Hashemian, H.M.: Wireless sensors for predictive maintenance of rotating equipment in research reactors. Ann. Nucl. Energy **38**(2–3), 665–680 (2011)
11. Hashemian, H.M., Bean, W.C.: State-of-the-art predictive maintenance techniques. IEEE Trans. Instrum. Meas. **60**(10), 3480–3492 (2011)
12. Huang, L., Chen, Y., Chen, S., Jiang, H.: Application of RCM analysis based predictive maintenance in nuclear power plants. In: 2012 International Conference on Quality, Reliability, Risk, Maintenance, and Safety Engineering, pp. 1015–1021. IEEE (2012)
13. IAEA: Applications of probabilistic safety assessment (PSA) for nuclear power plants (2001)
14. IAEA: Application of Reliability Centered Maintenance to Optimize Operation and Maintenance in Nuclear Power Plants (2007)
15. IAEA: Safety of Nuclear Power Plants : Commissioning and Operation (2011)
16. IAEA: Implementation Strategies and Tools for Condition Based Maintenance at Nuclear Power Plants (2012)
17. Mobley, R.K.: An Introduction to Predictive Maintenance. Elsevier, Amsterdam (2002)
18. Möhring, M., Schmidt, R., Keller, B., Sandkuhl, K., Zimmermann, A.: Predictive maintenance information systems: the underlying conditions and technological aspects. Int. J. Enterp. Inf. Syst. (IJEIS) **16**(2), 22–37 (2020)
19. Selcuk, S.: Predictive maintenance, its implementation and latest trends. Proc. Inst. Mech. Eng., Part B: J. Eng. Manuf. **231**(9), 1670–1679 (2017)
20. Sutharssan, T., Stoyanov, S., Bailey, C., Yin, C.: Prognostic and health management for engineering systems: a review of the data-driven approach and algorithms. J. Eng. **2015**(7), 215–222 (2015)
21. Van Eck, N., Waltman, L.: Software survey: vosviewer, a computer program for bibliometric mapping. Scientometrics **84**(2), 523–538 (2010)
22. Wayan Ngarayana, I., Do, T.M.D., Murakami, K., Suzuki, M.: Nuclear power plant maintenance optimisation: models, methods & strategies. J. Phys.: Conf. Ser. **1198**(2), 022005 (2019)
23. Zwingelstein, G.: La maintenance préventive - Méthodes et technologies. Techniques de l'Ingénieur **33**(MT9571 v1) (2019)

Smart Contracts Auto-generation for Supply Chain Contexts

Bajeela Aejas[1(✉)], Abdelhak Belhi[2], and Abdelaziz Bouras[1]

[1] Computer Science and Engineering, College of Engineering, Qatar University, Doha, Qatar
{ba1901053,abdelaziz.bouras}@qu.edu.qa

[2] Joaan Bin Jassim Academy for Defence Studies, Al Khor, Qatar

Abstract. The introduction of blockchain technology into Supply Chain management has opened the possibility of faster and more secure transactions of commodities and services. As for every blockchain, Smart Contracts are the tool for controlling the transactions in blockchain-based supply chains. In this paper, we introduce a method for automating the implementation of natural language contracts into Smart Contracts in the Supply Chain context. The basic idea here is to extract information from a natural language contract using two Natural Language Processing (NLP) techniques, the Named Entity Recognition (NER) and Relation Extraction (RE), and then use this extracted information to automatically create a corresponding Smart Contract. This is an ongoing project, and we implemented the first phase of NLP, i.e., NER. The main issue we are facing here is the limited availability of annotated contract datasets. To tackle this challenge, we created an annotated legal contract dataset dedicated to the NER task. The dataset is analyzed with the deep learning method (BiLSTM) and transformer-based method (BERT). As per the generation of smart contracts, our approach consists of identifying meaningful entities and the relations between them and then representing them as business logic that can be directly incorporated into computer code as blockchain smart contracts.

Keywords: NLP · NER · RE · Dataset · Deep learning · Legal domain

1 Introduction

Irrespectively of its size, any business is supported by the establishment of written contracts. These contracts are created and accepted between the parties involved in the business or any kind of transaction. As the number of involved parties and the size of the business increase, contract management becomes more and more complicated and prone to error. Supply Chain management is one of the largest domains that need to handle numerous contracts every day. As it contains a lot of different types of transactions and actors, the manual management of contracts is very complicated and may lead to conflicts and loss of money and time.

After the rise of the AI era, law firms and other companies are trying to automate the analysis and management of legal documents such as contracts and court order

Published by Springer Nature Switzerland AG 2023
F. Noël et al. (Eds.): PLM 2022, IFIP AICT 667, pp. 347–357, 2023.
https://doi.org/10.1007/978-3-031-25182-5_34

documents by making use of various deep learning methods on contract datasets. The automation speeds up the procedures and reduces the risk of conflicts and loss of money that may arise in manual document analysis. Thus, if we can apply the concept of contract management automation in the Supply Chain, it can save a lot of effort, money, and time in the process.

Another emerging technology that shakes different aspects of business and transactions is the blockchain which came to the light in 2009 along with Bitcoin, the first cryptocurrency [1]. Blockchain ensures security and transparency in transactions by providing the concept of immutable and distributed ledgers that are shared between every participant and cannot be manipulated. The transactions and the states are controlled by software scripts called Smart Contracts. All participant nodes must validate the preconditions of the Smart Contract, before its execution. The transactions cannot be changed after the deployment of the Smart Contract.

This paper presents a novel idea for introducing Supply Chain related contract management to the blockchain. The initial step is the automation of contract management by using NLP techniques, Named Entity Recognition (NER) and Relation Extraction (RE). The next phase involves the automatic generation of the Smart Contract based on the extracted results from the initial step. This automatic conversion of natural language contracts to Smart Contracts helps to avoid all the shortcomings of manual contract management in the Supply Chain and provides safer contract management.

The remainder of this paper is organized as follows. In Sect. 2, we explain the Smart Contract lifecycle. Section 3 explains the proposed methodology, Sect. 4 gives a detailed description of the dataset creation. In Sect. 5, we provide some technical details about our methodology to generate Smart Contracts. In Sect. 7 we draw our conclusions and provide some perspectives for future work.

2 Smart Contract Life Cycle

This section explains the various stages that the Smart Contract development process goes through. The life cycle of a Smart Contract consists of 4 steps [2, 3], namely creation, deployment, execution and finalization. Figure 1 shows various stages involved in the lifecycle of a Smart Contract.

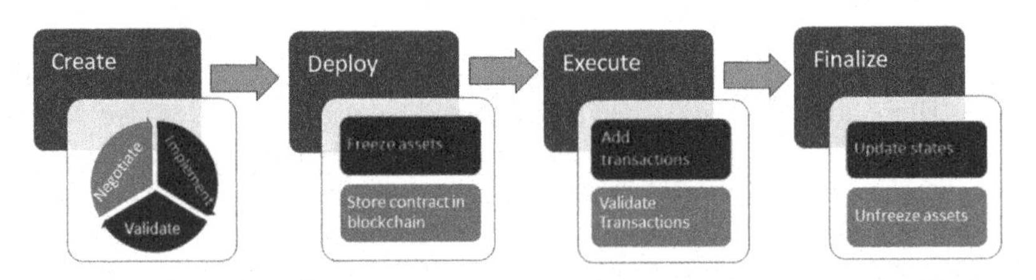

Fig. 1. Accuracy and Loss graph

- **Creation**. In this initial phase, all the parties are involved in the contract negotiated for the obligations and terms. After all the parties reach an agreement, the software engineers implement the Smart Contract using languages such as JavaScript and Go. After the implementation phase, the parties must validate and agree on the coded version of the Smart Contract.

- **Deployment.** Now, as all the nodes agree, the Smart Contract will be included in the distributed ledger, and thus it gets published on the blockchain. At this phase, the Smart Contract is available to all parties. Once, the Smart Contract got published, no one can modify it. The digital assets of the involved parties are frozen.
- **Execution.** After the deployment, the clauses will be evaluated, to check if the pre-conditions are met. The Smart Contract will be automatically executed once it triggers a condition. Thus, transactions will be added and validated through a consensus protocol.
- **Finalization.** After the execution, the new states of involved parties will be updated. The transactions and states will be stored in the blockchain. The frozen digital assets of the parties will be unblocked, and digital asset transfer will be carried out between the parties.

3 Proposed Methodology

The proposed architecture is shown in Fig. 2. Here we are using two NLP techniques for automatically creating Smart Contracts from natural language contracts, namely:

- Named Entity Recognition (NER) and
- Relation Extraction (RE)

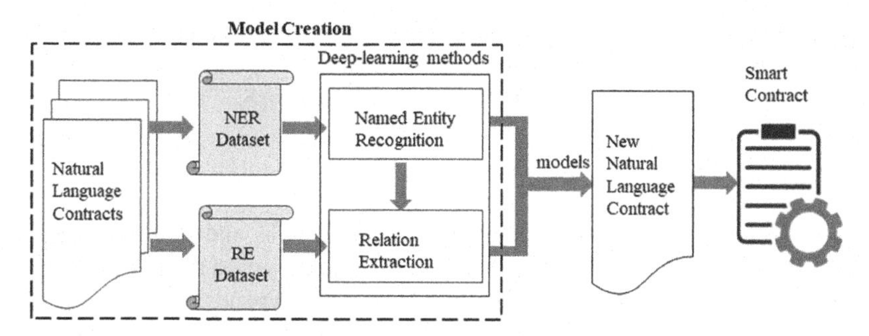

Fig. 2. Proposed architecture

NER is a method for extracting specific entities from texts such as person, organization, place, date etc. In this case of contracts, we are extracting the legal entities such as parties, effective date, governing law etc. Relation extraction identifies and extracts the relation between the entities. Models created from these methods are used to generate a Smart Contract for the corresponding Supply Chain contract. We are currently working on the NER part that is explained in detail in the upcoming session.

3.1 NER and Extracting Entities from Contracts

NER is one of the fundamental tasks of NLP for identifying and extracting generic named entities (NE) such as the name of a person, organization, dates, locations, etc. from the natural language text. The term "Named Entity" was first introduced by R.Grishman and Sundheim [4] in 1996 at the Sixth Message Understanding Conference (MUC-6)[1]. Many NER software tools are available for identifying generic NEs. Some of the most used NER tools available are Stanford NER[2], OpenNLP[3], and SpaCy NER[4]. There are many publicly available datasets for generic NER such as MUC-6 [4], CONLL 2002 [5], OntoNotes [6], and WikiNER [7].

NER is also used to recognize domain-specific entities from various domains. For example, identifying entities from historical documents [8], finding protein names from biomedical articles [9], identifying tags like parties, effective date, termination date and governing law [10] from legal documents such as court decisions and contracts. That means the dataset and the entities to be identified completely depend on the specific domain. Thus, the annotation process and entity identification are more difficult in domain-specific NER compared to generic NER.

4 A Legal Contract Dataset

The main issue with the NER research in the legal domain is the lack of annotated domain-specific datasets. This negatively impacts the overall research progress in this domain. The main reason for dataset scarcity is the confidentiality nature of the documents [11]. When we take the case of legal contracts, most researchers cannot openly publish datasets due to security and privacy reasons. Realizing the necessity of the task, this paper attempts to manually create an annotated contract dataset for NER and makes it publicly available to help researchers in the field. The natural language contracts for annotation are collected from the public source, the Electronic Data Gathering, Analysis, and Retrieval System (EDGAR)[5] of the U.S Security and Exchange Commission (SEC).

4.1 Structure of the Contract

A contract is a legal document that defines the obligations and rights of the parties involved in the agreement. Contracts usually follow a pattern even though we cannot say it is a perfect template. The size of the contracts varies and depends on the need of parties involved in the agreement. Some contracts are only a few pages long, while others are hundreds of pages long. The first section of the contract is the preamble, which contains the title, the effective date of the contract and the parties involved in the contract with their address. Some contracts may also have a cover page before the preamble with this information along with the table of contents. Recitals or background section gives some

[1] https://www-nlpir.nist.gov/related_projects/muc/.

[2] http://nlp.stanford.edu/software/CRF-NER.shtml.

[3] https://opennlp.apache.org/.

[4] http://spacy.io.

[5] https://www.sec.gov/edgar/searchedgar/ accessing-edgar-data.htm.

background details of the preamble contents. After that, the contract contains articles or sections which mention the clauses. Usually, the first article is "definition" which defines each term clearly. Each article heading mirrors its contents. For example, the termination date will be under the "Termination" clause. To get which country's law the contracts abide by, we need to check the "Governing Law" section. We can also see different heading terms on different contracts. For example, some contracts may use "Applicable Law" instead of "Governing Law". Thus, the legal entities can easily be recognized under the corresponding headings [11].

4.2 Dataset Description

Our source of data is a set of natural language contracts. Since we are creating a domain-specific dataset, the corresponding entities are different from normal NEs. To create a dataset, we need to perform various operations that are explained below in detail:

Collection of Contracts. We collected different types of publicly available contracts from EDGAR of the U.S. Security and Exchange Commission (SEC). From this collection, 200 contracts have been selected for annotation. The contracts are in pdf format.

Cleaning and Segmentation. Before annotating the contracts, we pre-processed the contracts as follows:

1. The pdf contracts are converted to.txt format.
2. The contract files are cleaned Cleaning of text files for minor structural corrections.
3. Documents segmentation to make one sentence per line: the sentence segmentation is performed using SpaCy. Figure 3 shows the number of tokens per sentence in the dataset.

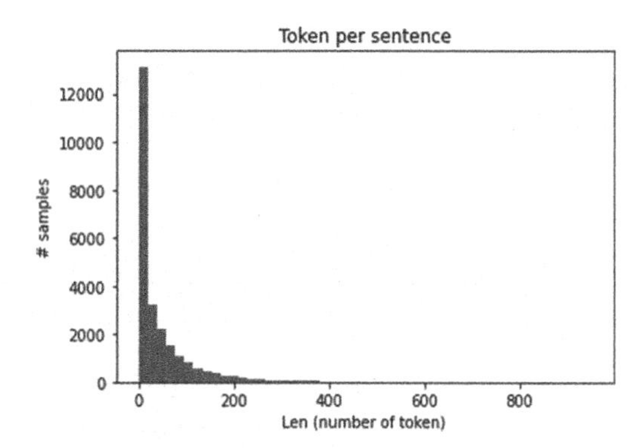

Fig. 3. Token length per sentence

Table 1. Labels and their descriptions.

No	Label	Description
1	Title	Title of the contract usually specifies the type
2	Parties	Parties involved in the agreement
3	AgreementDate	Date on which the contract is signed
4	EffectiveDate	Date from which the contract is effective
5	TerminationDate	Date of termination of the agreement
6	TermLength	Duration in which the contract will be effective
7	NotificationPeriod	Period of notification for termination of the contract
8	GoverningLaw	A country whose laws apply
9	Renewal	Renewal period of the contract

Selection of Labels. We selected 9 contract-related labels or tags for annotation. All these labels are very important for contract analysis. Each contract is thoroughly analyzed and annotated with these labels. The tags we selected are shown in Table 1.

Manual Annotation. Now the documents are ready for annotation. We used the annotation tool called Datasaur[6] for annotating the text documents. In this tool, we need to select the NEs and assign the associated NE from the list of NEs. Figure 4 shows the annotation process of a contract from the dataset.

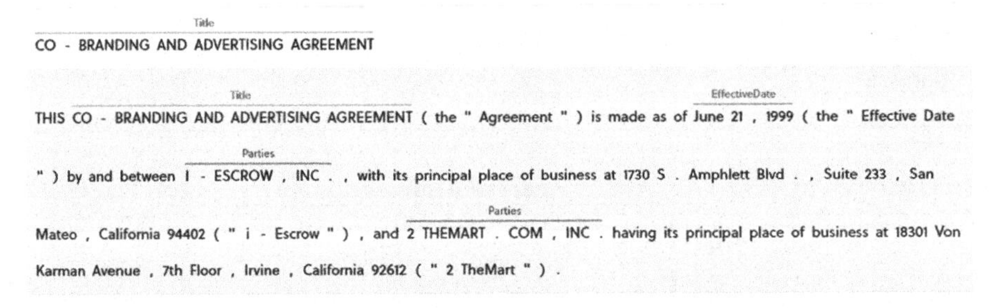

Fig. 4. Annotated contract from the dataset

On finishing the annotation, the file is ready to be exported to our system. There are different tagging schemes available for annotation. This dataset follows Inside- Outside-Beginning (IOB) [12] format. If an NE contains multiple tokens, each token is annotated with the tag along with prefixes B or I. The B-prefix of a tag indicates that the tag is the beginning of an NE, and the I-prefix indicates that the tag is inside an NE. All the tokens that are not an NE are indicated as "O".

6 https://datasaur.ai/.

4.3 Experiments and Results

We are still working on the annotation of contracts to increase the size of the dataset. The initial dataset is evaluated and compared using the deep learning method, BiLSTM as well as the hugging face transformer model BERT. We opted these two methods as both methods proved their efficiency in providing state of the art result in many NLP datasets. In both experiments, we divided the dataset into 80% training set and 20% test set. The accuracy and loss graphs for training for BiLSTM are shown in Fig. 5 below. Even though accuracy is high, the precision, recall and F1 values for each tag are not that promising in the initial results because of multiple reasons:

- Since the majority of the token in contracts comes under "O" tags, the accuracy can be very high
- Compared to the "O" tags, the remaining labels are very few in number, and the dataset is highly imbalanced. This affects the recall and precision negatively.

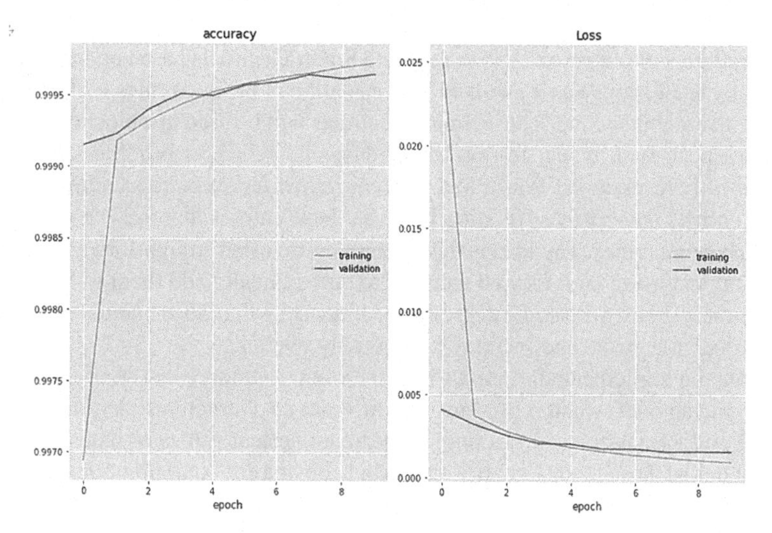

Fig. 5. Accuracy and loss graph

Table 2 shows the labels that received comparatively higher precision, recall and F1 scores.

4.4 Relation Extraction

Like NER, RE is a subtask in NLP that identifies the relationship between different entities in the text. The RE part is very crucial in this process as it is necessary to identify the relation between the entities for creating the conditions in the Smart Contract. As for NER for contracts, RE for contracts also needs a dedicated dataset annotated with required relations. As there are no public datasets available for RE in contracts, we need

Table 2. Accuracy, Loss, Precision, Recall and F1 score

Method	Accuracy	Loss	Labels	Precision	Recall	F1 score
BERT	0.994	0.023	GoverningLaw	0.70	0.86	0.77
			Parties	0.80	0.90	0.85
			Title	0.70	0.87	0.78
BiLSTM	0.996	0.0016	GoverningLaw	0.73	0.70	0.72
			Parties	0.79	0.78	0.79
			Title	0.73	0.80	0.76

to create one from the scratch and this is going to be our next future task in this ongoing research.

5 Generation of Smart Contracts

The overall goal of our research is to generate Smart Contracts based on legal contracts. The task by itself is huge as it involves the translation of business rules written in natural language to computer code. These Smart Contracts will be used in a blockchain network to ensure aspects such as peer-to-peer trust, transparency, tamper-proofness, and security.

We mostly leverage NLP as a tool to analyze raw legal contracts aiming at finding relevant entities and business relation between these entities that may be automatically added as business rules. The idea is that such rules are often straightforward and easy to understand and using an AI-based solution to automatically add them to Supply Chain information systems, with some degree of confidence, will reduce manual modifications and will decrease errors and increase productivity overall.

As per the implementation, at a high level, our solution consists of establishing a common metamodel which represents the different data structures identified following the NER and RE tasks. As an example, all the numerical values or dates, etc. and the rules that define these fields integrity check and constraints. At a subsequent stage, these rules can even be mapped to predefined functions such as payments and penalties which will be directly translated from natural language to Smart Contract code.

Since we intend to use blockchain technology as a platform for both Supply Chain data and business logic, Smart Contracts offer the ability to represent both. One can write object-oriented classes similarly to any language with both data structures and methods or establish relations between these classes.

As per the representation of this information, we rely on a business rules representation which consists of specifying rule conditions and actions in a standardized way. The task consists mostly in translating data field names and types from a JSON representation into real-world actions. In our approach, we rely on a business rules representation framework written initially for Python to make this representation. The task consists of identifying conditions and actions and then populating the generated JSON or XML rule to relevant data models. Figure 6 represents a rule written in JSON and XML that affects product prices based on inventory and expiration dates.

```json
{
  "conditions": {
    "all": [
      {
        "name": "expiration_days",
        "operator": "less_than",
        "value": 5
      },
      {
        "name": "current_inventory",
        "operator": "greater_than",
        "value": 20
      }
    ]
  },
  "actions": [
    {
      "name": "put_on_sale",
      "params": {
        "sale_percentage": 0.25
      }
    }
  ]
}
```

```xml
<?xml version="1.0" encoding="UTF-8" ?>
<root>
    <conditions>
        <all>
            <name>expiration_days</name>
            <operator>less_than</operator>
            <value>5</value>
        </all>
        <all>
            <name>current_inventory</name>
            <operator>greater_than</operator>
            <value>20</value>
        </all>
    </conditions>
    <actions>
        <name>put_on_sale</name>
        <params>
            <sale_percentage>0.25</sale_percentage>
        </params>
    </actions>
</root>
```

Fig. 6. An example of a rule definition using our approach (JSON format on the left and XML format on the right)

6 Results and Discussion

Following its representation in JSON or XML, the rule is propagated to the established smart contract. The related data model enforced checks related to the rule for records present in the blockchain. For our proof of concept, Fig. 7 and 8 represent a sample of the meta-model file for the rule's definition. This example rule is an `OnChange` trigger that changes a value to 100 if another value below 5 is given as input.

```
rules = [
    {"conditions": {"all": [
        {"name": "current_value",
         "operator": "less_than",
         "value": 5,
         },
    ]},
        "actions": [
            {"name": "more100",
             "params": {},
             },
        ],
    }]
```

Fig. 7. Sample of a meta business rule

```javascript
async changecustomerValue(ctx, customerNumber, newValue) {
    console.info('============= START : changecustomerValue ===========');
    const customerAsBytes = await ctx.stub.getState(customerNumber);
    if (!customerAsBytes || customerAsBytes.length === 0) {
        throw new Error(`${customerNumber} does not exist`);
    }
    const customer = JSON.parse(customerAsBytes.toString());
    customer.value = newValue;
    if(parseInt(customer.value, 10)< 5 )
    {

        console.info('Value less than 5');
        console.info('Customer not created 5');
    }
    else{
        await ctx.stub.putState(customerNumber, Buffer.from(JSON.stringify(customer)));
        console.info('============= END : changecustomerValue===========');
    }

}
```

Fig. 8. Translation to a Smart Contract

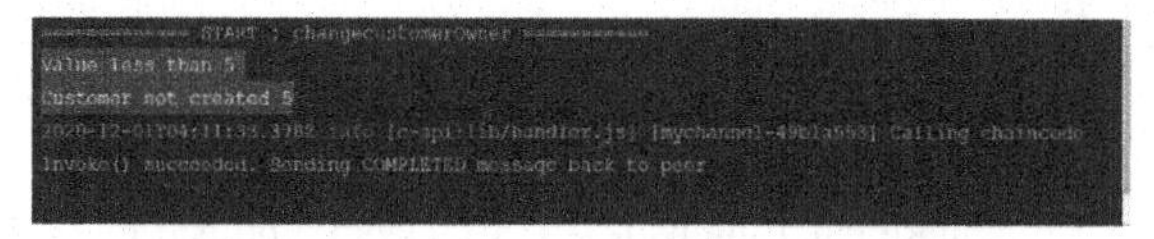

Fig. 9. Blockchain peers log following the invocation of the Smart Contract changeValue method. The reply was that the value was less than 5

Figure 9 shows the Log of one of the blockchain peers following the invocation of the Smart Contract `changeValue` method with a value less than 5 as an argument (Method depicted in Fig. 8).

At the current stage, our approach consists of predefining a set of actions and then mapping NER and RE results, namely data values, condition and action as input to the smart contract generation process. The process is accurate but further tests need to be carried on real example as the data we used for our test was synthetically generated due to the lack of RE results at this stage. We also aim at experimenting with undefined relations and action in future iterations of this ongoing project.

7 Conclusion

A Supply Chain involves the management of a large number of contracts throughout its lifecycle. Manual contract analysis is very complicated, time-consuming and at the same time may cause conflicts and disputes between the partners. This paper proposes a method to automate and implement the contract management related to Supply Chain in blockchain converting the natural language contracts into Smart Contracts. This extracted information is then used for generating corresponding Smart Contracts. The challenging situation here is the unavailability of annotated contract datasets for both NER and RE tasks. As a first phase of the research, we have manually created a NER dataset using the public contract database, EDGAR. The dataset is being evaluated using BiLSTM and BERT. This is an ongoing project, and the dataset will be widened by annotating more contracts from EDGAR. The next phase is the creation of the RE contract dataset. After the relation extraction model is created, both NER and RE models will be used to create the Smart Contract corresponding to a natural language contract for the Supply Chain.

Acknowledgement. This publication was made possible by NPRP grant NPRP11S-1227-170135 from the Qatar National Research Fund (a member of Qatar Foundation). The statements made herein are solely the responsibility of the authors (www.supplyledger.qa).

References

1. Nakamoto, S.: Bitcoin P2P e-cash paper. Mail Arch. 1 (2008)
2. Sillaber, C., Waltl, B.: Life cycle of smart contracts in blockchain ecosystems. Datenschutz und Datensicherheit - DuD **41**(8), 497–500 (2017). https://doi.org/10.1007/s11623-017-0819-7
3. Aejas, B., Bouras, A., Belhi, A., Gasmi, H.: Smart Contracts Implementation Based on Bidirectional Encoder Representations from Transformers BT - Product Lifecycle Management. Green and Blue Technologies to Support Smart and Sustainable Organizations. Presented at the (2022). https://doi.org/10.1007/978-3-030-94335-6_21
4. Grishman, R., Sundheim, B.: Message understanding conference-6: a brief history. In: Proceedings of the 16th Conference on Computational Linguistics, vol. 1, pp. 466–471. Association for Computational Linguistics, USA (1996). https://doi.org/10.3115/992628.992709
5. Tjong Kim Sang, E.F.: Introduction to the coNLL-2002 shared task: language-independent named entity recognition. In: {COLING}-02: The 6th Conference on Natural Language Learning 2002 (CoNLL}-2002) (2002)
6. Weischedel, R., et al.: Ontonotes release 5.0 ldc2013t19. Linguist. Data Consortium, Philadelphia 23 (2013)
7. Nothman, J., Ringland, N., Radford, W., Murphy, T., Curran, J.R.: Learning multilingual named entity recognition from wikipedia. Artif. Intell. **194**, 151–175 (2013). https://doi.org/10.1016/j.artint.2012.03.006

8. Aejas, B., Bouras, A., Belhi, A., Gasmi, H.: Named entity recognition for cultural heritage preservation. In: Belhi, A., Bouras, A., Al-Ali, A.K., Sadka, A.H. (eds.) Data Analytics for Cultural Heritage, pp. 249–270. Springer, Cham (2021). https://doi.org/10.1007/978-3-030-66777-1_11
9. Ramachandran, R., Arutchelvan, K.: Named entity recognition on bio-medical literature documents using hybrid based approach. J. Ambient Intell. Humaniz. Comput. (2021)https://doi.org/10.1007/s12652-021-03078-z
10. Aejas, B., Bouras, A., Belhi, A., Gasmi, H.: A review of contract entity extraction. In: Yang, X.-S., Sherratt, S., Dey, N., Joshi, A. (eds.) Proceedings of Sixth International Congress on Information and Communication Technology. LNNS, vol. 217, pp. 763–771. Springer, Singapore (2022). https://doi.org/10.1007/978-981-16-2102-4_69
11. Chalkidis, I., Androutsopoulos, I., Michos, A.: Extracting contract elements. In: Proceedings of the 16th Edition of the International Conference on Articial Intelligence and Law, pp. 19–28. Association for Computing Machinery, New York, NY, USA (2017). https://doi.org/10.1145/3086512.3086515
12. Ramshaw, L., Marcus, M.: Text Chunking using Transformation-Based Learning. In: Third Workshop on Very Large Corpora (1995)

Augmented Reality-Assisted Quality Control Based on Asset Administration Shells for Concrete Elements

Mario Wolf(✉), Jan Luca Siewert, Oliver Vogt, and Detlef Gerhard

Digital Engineering Chair, Ruhr University Bochum, 44801 Bochum, Germany
mario.wolf@ruhr-uni-bochum.de

Abstract. The implementation of the Digital Twin concept in the form of Asset Administration Shell (AAS) enables smart flow production even in fields that currently do not incorporate such methods, such as in the fabrication of precast concrete elements. The authors present an approach to implement architecture, engineering and construction-specific AAS submodels for flow production using Building Information Modeling (BIM) data and further enhance these with a new stepwise, task-based quality control process utilizing smart devices as a human machine interface to incorporate Augmented Reality visualization. The concept briefly describes the general production process for concrete wall elements, in which the detailed process of creating quality control tasks is embedded, as well as the meta-models for those components communicating using AAS. Utilizing both product type-specific and product instance-specific data, knowledge about the process, the product's state in the process, and the physical setup enable a reliable and integrated quality control system with clear data flows, automatic documentation, and improved traceability. Finally, a prototypical implementation shows the integration between BIM-Software, the created AAS templates and submodels, and a prototypical quality control setup with smart device support in the proposed process.

Keywords: Augmented reality · Asset Administration Shell · BIM integration

1 Introduction

To enable a fully digitized flow production of precast concrete elements, data from the production process and the current state of the precast element being produced must be continuously recorded, compiled, and made available. The data and communication structure for these functionalities were developed in previous work in the form of the widely used representation as so-called Asset Administration Shells (AAS). Based on the continuously recorded data, the current production status of the precast concrete element can be monitored and tracked with ease.

The underlying AAS approach generally enables machine-to-machine communication and envisaged using the data and communication options in the event of production-related deviations to check further use elsewhere in the structure or make automatic corrections in the production process.

Published by Springer Nature Switzerland AG 2023
F. Noël et al. (Eds.): PLM 2022, IFIP AICT 667, pp. 358–367, 2023.
https://doi.org/10.1007/978-3-031-25182-5_35

Ultimately, agile control of the planning and production of the precast concrete element is made possible by the context-sensitive management shells. For the people in the production environment, the application of the AAS concept results in the possibility of retrieving information on processes, products, machines, and efficiencies ad hoc via suitable end devices. Based on the information collected from product planning and design in Building Information Modeling (BIM) software and the digitally supported production process, individual sub-process can be coordinated more precisely, and people can be supported in their work, while results can be fed back in a process-safe manner.

This paper focuses on the support potential through Augmented Reality (AR) applications using multiple visualization methods and data supply through the developed data representations and interaction mechanisms based on the AAS concept. In particular, the purely smart device-based AR, in which image acquisition, registration, overlay, and display take place entirely on a mobile device, is to be compared with the use of stationary imaging devices, external calculation of the overlay, and a subsequent visualization via a smart device or as a projection in stationary installations.

The problems in the development of AR solutions are manifold and must therefore be considered from different dimensions, which concern both the implementation of AR itself (image acquisition, registration, overlay, display), as well as the implementation alternatives in each individual use case. Likewise, environmental specifics, device protection classes, privacy, and similar concerns in real production environments must be considered.

After an overview of related works, the contribution describes in detail the concept and the specified AAS models used throughout the process as well as the integration into existing BIM workflows. The prototypical implementation is then used to validate the described data models by describing an AAS export from Autodesk Revit, as well as a simplified AR application based on ARKit and a custom registration setup. The work closes with an outlook on the next steps and further work.

2 Related Work

2.1 Quality Control and Digital Twin Approaches in Civil Engineering

Several approaches have been presented to combine BIM and quality control measures, such as the QC approach to testing the compliance of the final products with quality standards [11]. To support workers in the construction phase, Puri and Turkan suggest additional, textual information to help to interpret the information provided in tolerance specifications [12].

An additional current problem with BIM's concept and implementation in the industry is the focus on the design and construction phases, while lacking guidelines for later life cycle phases and the efficient use of existing data [8], especially in the context of off-site construction. Lee et al. show an approach to combine factory-side quality control with on-site control after transportation [7].

A digital twin is a virtual representation of components or structures that goes beyond a BIM model. According to Boje, Guerriero et al. [4], a digital twin consists of three main components: a physical component, a virtual representation, and a data link. In their literature review, Shahzad et al. [16] investigate concrete requirements for digital twins in civil engineering. A key finding is the dynamic enrichment and possibility of communication between digital twins. Akanmu et al. [1] explore the use of digital twins in cyber-physical systems and present use cases for improving productivity, health, and safety. They outline key technologies and barriers to implementation. Currently, many ideas for building digital twins and specific solutions exist in the construction industry. However, there is a lack of flexible data models and interfaces, as they already exist in the stationary industry.

This is where the digital twin in form of the AAS concept offers the flexibility to connect to the current state of the art in BIM while enabling flow production, product instance-specific data storage, and quality control, as shown by the authors in [6, 19].

2.2 Augmented Reality in Civil Engineering and Quality Control

Today, AR is widely used for various tasks in engineering, including quality control [14]. In such an AR quality control process, the technology is used to give workers visual guidance throughout the task. This is regularly attributed to lowering the time required for such tasks [2] and ensuring the quality of the required process and lowering the number of mistakes made by workers [15]. Complex tasks can be broken down into smaller, step-by-step instructions that closely guide a worker with visual aid. For such tasks, manual tools to create and manage the associated data exist, making content creation easy-to-use and straightforward [17]. For mass production applications, however, it is necessary to integrate the AR systems into the existing IT infrastructure and existing data models [9, 12].

The fundamental idea behind AR is to overlay virtual information over the real world [3]. Milgram et al. [10] differentiate between optical-see-through systems, where virtual content is shown on a half-transparent display between the users' eyes and the real world, and video-see-through systems, where both the virtual information and an image of the real world are observed on a screen, and systems where the virtual information is directly projected onto the real world. Additionally, the information can be observed through head-mounted displays, on mobile devices like smartphones and tablets, or through stationary displays, where a static camera observes the environment.

For this approach, the authors additionally differentiate between mobile AR, which is shown on a device the operator moves around, and stationary AR, where a fixed monitor displays an overlay on top of the image of a static camera [18]. Such fixed setups are often used in quality control settings, where workstands are often static, easing tracking and other AR challenges, as shown in [2].

AR can be used during the design, construction, and operation phases. Common areas include design assessment, progress monitoring, defect detection, quality management, assembly instructions, and maintenance [5]. [13] show an example of quality control based on BIM models. Here, the model is displayed on a smartphone, allowing the user to move around the building, view information on specific parts of the building, reviewing schedules, and adding notes to BIM objects. However, no registration between the building and the model takes place.

3 Concept

At the end of the design phase, a state-of-the-art BIM model is rich in information about the overall building structure, and each individually planned concrete element. To reach the aim of a fully digitized production of precast concrete elements, each element, and machine in the factory should have its digital twin, describing all relevant aspects of the physical asset. Each individual digital twin is able to reference BIM data and production systems and exchange necessary process information.

The AAS in this scenario is the functional representation of a digital twin and can be divided into different submodels, which in turn describe a certain aspect of the concrete element with a variety of properties. The concept developed in this work is based on the structure of the AAS already designed by the authors in previous publications, as stated earlier.

Figure 1 shows the interconnections between BIM, AAS, and the new process for creating quality control support information, which can be visualized using augmented reality. Based on the BIM data, information about each element type, element instance, and the associations between instances can be transferred to individual AAS with their specialized submodels. For an integrated quality control process, geometric features of elements (edges or faces) in the BIM model can be marked as relevant for future quality control steps. These custom properties yield the basis for the association between individual feature points of elements and the later task definitions, that when fully filled out, make up the AR-supported quality control workplan.

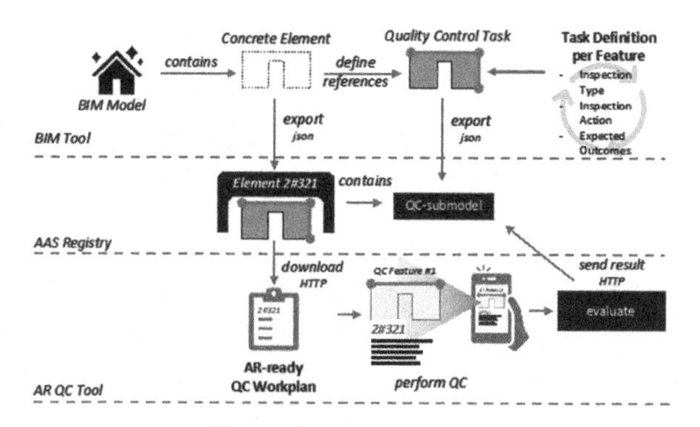

Fig. 1. Concept overview

After being exported from the BIM tool in an open exchange format (like JSON or AASX), the administration shells and the defined QC submodel are made available through the AAS registry. From there they are accessed by the AR QC tool. When a submodel as specified below is found, it is interpreted as a QC workplan and is shown to the operator in AR. After the worksplan has been completed, the results are evaluated and sent back to the registry, where they are stored in the submodel for later reference.

3.1 Meta-model for Quality Control Sub-models with AR Support

As described earlier, the AAS of a concrete element consists of multiple submodels each describing a specific aspect. In the following, the submodels *DetailedDesign* and *QualityControl* are described in further detail.

In the *DetailedDesign* submodel, the geometry representation of the concrete element is described in the *GeometryElement*. For visualization and other related tasks, like object detection, a simplified, tessellated version of the geometry is saved in a file and referenced in this submodel element. For its simplicity and widespread adoption, the Wavefront OBJ format was chosen for this.

To precisely reference and associate the faces and edges in the quality control steps, these are also saved as explicit submodel element collections. The *GeometryElement* holds a collection of vertices, which are described as cartesian coordinates with an x, y, and z coordinate. These vertices are then in turn referenced by the individual *GeometryFaces*, which describe a polygon in 3D space, and *GeometryEdges* which are described by a series of vertices. As these are described as *SubmodelElementCollections*, they are referable across the whole administration shell. Both *GeometryEdges* and *GeometryFaces* do not store the actual vertices used to describe the face and edge, but only store a reference to the local identifier of vertices in the specific *GeometryElement* (Fig. 2).

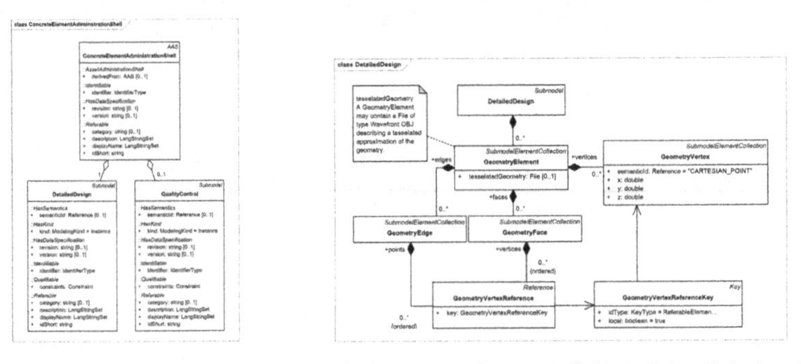

Fig. 2. UML class diagram of the AAS for concrete elements (left) and detail design (right)

The *QualityControl* submodel closely follows the metamodel of quality control for concrete elements outlined in. It consists of a series of inspection tasks that are described as co-managed entities. A task is further broken down into individual actions, which either are actions to measure the flatness of a certain face of the geometry, the length in a certain dimension, or an angle between two edges. The faces and edges are stored as local references to the edges and faces stored in the aforementioned *DetailedGeometry* submodel, on which the *QualityControl* submodel depends. After an action is completed, an *InspectionResult* is created. There, a datetime and a reference to the working operator are stored together with the corresponding measurement result. Those can be described by either a length for dimensions and flatness checking or the measured angle for angle checking actions. An associated *EvaluationTask* then compares the given result either to default evaluation criteria defined for all tasks in the *QualityControl* submodel or specific evaluation criteria that may be attached to a specific *GeometryElement*. Each checking action results in a passed, rework-required, or failed evaluation based on the measurement and the evaluation criteria. The final evaluation result of the complete inspection task is determined by the worst evaluation of the individual checking actions.

The quality control process is limited to checking dimensions, flatness, and angles between edges. However, the concept can be expanded to other dimensioning and tolerancing criteria as long as they can be described by a relationship between the individual vertices, edges, and faces of the concrete module.

The approach breaks down the quality control activity into individual tasks, which are then further described by a series of actions. This stepwise approach is well suited to be supported by an AR application. An AR interface can show a suitable, guiding overlay and input mechanism by referencing the individual edges and faces in each *InspectionAction*. Because the edges and faces reference local vertex coordinates, this information can be attached to the physical concrete element as long as a suitable tracking approach maps world coordinates captured by the AR system to the local coordinates for the model. For that to work, both model-tracking based on the tessellated OBJ geometry and edge model tracking based on the vertex information can be used (Fig. 3).

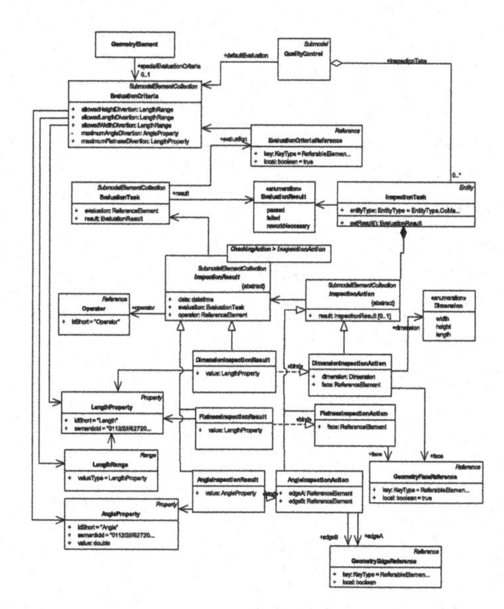

Fig. 3. UML class diagram of the AAS for quality control

3.2 BIM Integration

As shown in Fig. 1, the integration of the concept into BIM requires some additional information about each quality control relevant feature. Each element in BIM has its own set of properties, which are divided into type properties and instance properties. For example, an instance of a wall of a certain type always has certain defined properties, which are supplemented by instance properties that only apply to this specific wall. This is the basis for the creation of type- and instance-specific AAS.

The proposed quality control process requires the quality control features of an element to be uniquely identifiable and offer geometric information with their respective measuring points and criteria. For this reason, the integration should provide user-defined properties to add a "marker" to geometric features in BIM, so that further down the toolchain and within the AAS submodels, every feature can be referenced correctly.

In addition to the new data points, the integration should also provide a tessellated 3D model of the elements for visualization purposes.

An inspection task can be created from these measuring points and instructions, which, in addition to the then defined measuring points and instructions, also contain the target dimensions and permissible deviations already stored in BIM and the automatically created AAS.

4 Prototypical Implementation

4.1 Autodesk Revit Plugin for BIM Integration

Autodesk Revit 2022 was chosen to be the BIM solution that the BIM integration is conceptionally based on. The data model behind each BIM system is highly specific, so the concept might need to be adjusted to fit other BIM systems. For demonstration purposes, a BIM model of a single-family house was created in Revit 2022, with the accompanying prototype plugin being based on the Revit API C# template.

The plugin offers the user the possibility to choose the elements to be exported in the 3D model of the building in Revit. After the selection has been confirmed, the element IDs of the selection are passed to the plugin. The instance properties are queried via the unique ID of the element. One of these properties is the ID of the element type. Since many properties are stored in the element type in Revit, these are also queried and stored as type properties for the element. To export material properties of an element, all materials of the element must be queried first, since a wall can consist of masonry and insulation material, for example. Thus, the query of the materials in turn provides a list of IDs of all materials. Those are used to retrieve the material-specific general, thermal and physical properties. Other properties are retrieved identically to populate the AAS for the elements and their types. For each property, its name, its unique Revit ID, its data format, and its value are stored. This information is relevant for the creation of a property in the AAS.

It must be considered that the values of the properties are exported in the internal Revit standard units. A conversion to the required unit is therefore necessary and is performed by the plugin. In addition to the properties, the 3D geometry of the element is relevant for quality control. For this reason, the Edges and Faces and a tessellated 3D model of the element are required. For each element, the faces and edges can be retrieved via the API, where Revit offers the option to tesselate each face. The tessellation provides the point information and the normal, which lets the plugin generate an OBJ file format compatible export. At last, user-specific properties are created to identify elements and geometric features to be included in creating an AR-supported inspection task. Those properties are exported to the corresponding submodel of each element's AAS.

4.2 Laboratory Experiment and Validation

To validate the designed data model, a simple tablet-based AR application has been implemented using an Apple iPad and the ARKit framework. The application loads the AAS and allows the user to browse through the various submodels and properties. In particular, the application recognizes the defined semantic identifier of the special quality control and detailed design submodel elements and shows a specialized view for each of them. For the quality control assistance, the application launches an AR view shown in Fig. 4.

A sidebar displays the various inspection actions that need to be performed by the user. Upon selecting a specific action, the edge gets highlighted on the model of the concrete element. Additionally, this model can be placed in AR and overlayed over the concrete element. For this visualization, the referenced vertices, edges, and faces are extracted from the detailed design submodel.

Results and measurements can be input by the worker and saved in the result submodel element. The result is then automatically evaluated based on the defined evaluation criteria.

Fig. 4. The prototype of the AR assistance system showing the edge that needs to be measured and allows the operator to input measured values.

The system was implemented using the ARKit 5 SDK on a 6th generation iPad Mini. ARKit only supports model tracking based on captured point clouds. To work around this limitation, the authors plan to use a separate, static object tracking setup that is responsible for calculating the transformation from world space to model space. Registration between the tablet and the world can then happen through a simple image marker that is attached to the camera setup. For the validation, only a scaled, 3D-printed model of the concrete element was used. Nevertheless, this demonstrator shows that the data model is suitable both for automatic exports from BIM systems, to create overlay information in an AR guidance system, and to save and document performed quality control steps.

In the next phase, the system is evaluated in an industrial setting. Additionally, an outside-in model-tracking system will be deployed for more reliable and flexible AR overlay information. There, a static camera system will capture images of the production line and calculate the specific pose of the concrete element automatically based on the provided tessellated geometry.

5 Conclusion

The goal of the paper at hand was to create a foundation to set up a quality control workplan with AR content, which is associated with the digital features of a BIM model and anchored to the equivalent physical features later in the production and quality assurance process of precast concrete elements. The flexible nature of the AAS concept enables the enhancement of the previously discussed AAS for precast elements with new submodels for specialized content such as stepwise AR instructions and associations between digital and physical features.

While the AR implementation is currently a laboratory-only demonstrator, the general feasibility of the IT process in its entirety has been proven.

In future work, the authors plan to validate the approach on both regular precast concrete elements, as well as more specialized high-performance concrete elements, in real manufacturing environments.

Funding. This research is funded under Grant No. 423963709 by the German Research Foundation as part of the priority project "SPP2187 - Adaptive modularized constructions made in flux".

References

1. Akanmu, A.A., Anumba, C.J., Ogunseiju, O.O.: Towards next generation cyber-physical systems and digital twins for construction. J. Inf. Technol. Constr. **26**, 505–525 (2021). https://doi.org/10.36680/j.itcon.2021.027
2. Alves, J.B., Marques, B., Dias, P., Santos, B.S.: Using augmented reality for industrial quality assurance: a shop floor user study. Int. J. Adv. Manuf. Technol. **115**(1–2), 105–116 (2021). https://doi.org/10.1007/s00170-021-07049-8
3. Azuma, R.T.: A survey of augmented reality. Presence: Teleoperators Virtual Environ. **6**(4), 355–385 (1997)
4. Boje, C., Guerriero, A., Kubicki, S., Rezgui, Y.: Towards a semantic construction digital twin: directions for future research. Autom. Constr. **114**, 103179 (2020). https://doi.org/10.1016/j.autcon.2020.103179
5. Chen, K., Xue, F.: The renaissance of augmented reality in construction: history, present status and future directions. Smart Sustain. Built Environ. (2020). https://doi.org/10.1108/SASBE-08-2020-0124
6. Kosse, S., Vogt, O., Wolf, M., König, M., Gerhard, D.: Digital twin framework for enabling serial construction. Front. Built Environ. **8**, 864722 (2022). https://doi.org/10.3389/fbuil.2022.864722
7. Lee, S., Kwon, S., Jeong, M., Hasan, S., Kim, A.: Automated On-Site quality inspection and reporting technology for Off-Site Construction(OSC)-based precast concrete members. In: Proceedings of the 37th International Symposium on Automation and Robotics in Construction (ISARC). Proceedings of the International Symposium on Automation and Robotics in Construction (IAARC). International Association for Automation and Robotics in Construction (IAARC) (2020). https://doi.org/10.22260/ISARC2020/0158

8. Leygonie, R., Motamedi, A.: quality management processes and automated tools for FM-BIM delivery. In: Proceedings of the 38th International Symposium on Automation and Robotics in Construction (ISARC). Proceedings of the International Symposium on Automation and Robotics in Construction (IAARC). International Association for Automation and Robotics in Construction (IAARC) (2021). https://doi.org/10.22260/ISARC2021/0120
9. Masood, T., Egger, J.: Adopting augmented reality in the age of industrial digitalisation. Comput. Ind. **115**, 103112 (2020). https://doi.org/10.1016/j.compind.2019.07.002
10. Milgram, P., Takemura, H., Utsumi, A., Kishino, F.: Augmented reality: a class of displays on the reality-virtuality continuum. In: SPIE Proceedings, Telemanipulator and Telepresence Technologies, pp. 282–292 (1994)
11. Motamedi, A., Iordanova, I., Forgues, D.: FM-BIM preparation method and quality assessment measures. In: 17th International Conference on Computing in Civil and Building Engineering (ICCCBE), Tampere, Finland (2018)
12. Porcelli, I., Rapaccini, M., Espíndola, D.B., Pereira, C.E.: Technical and organizational issues about the introduction of augmented reality in maintenance and technical assistance services. IFAC Proc. **46**(7), 257–262 (2013). https://doi.org/10.3182/20130522-3-BR-4036.00024
13. Ratajczak, J., Schweigkofler, A., Riedl, M., Matt, D.T.: Augmented reality combined with location-based management system to improve the construction process, quality control and information flow. In: Mutis, I., Hartmann, T. (eds.) Advances in Informatics and Computing in Civil and Construction Engineering, pp. 289–296. Springer, Cham (2019). https://doi.org/10.1007/978-3-030-00220-6_35
14. Röltgen, D., Dumitrescu, R.: Classification of industrial augmented reality use cases. Procedia CIRP **91**, 93–100 (2020). https://doi.org/10.1016/j.procir.2020.01.137
15. Segovia, D., Mendoza, M., Mendoza, E., González, E.: Augmented reality as a tool for production and quality monitoring. Procedia Comput. Sci. **75**, 291–300 (2015). https://doi.org/10.1016/j.procs.2015.12.250
16. Shahzad, M., Shafiq, M.T., Douglas, D., Kassem, M.: Digital twins in built environments: an investigation of the characteristics, applications, and challenges. Buildings **12**(2), 120 (2022). https://doi.org/10.3390/buildings12020120
17. Siewert, J.L., Wolf, M., Böhm, B., Thienhaus, S.: Usability study for an augmented reality content management system. In: Auer, M.E., May, D. (eds.) REV 2020. AISC, vol. 1231, pp. 274–287. Springer, Cham (2021). https://doi.org/10.1007/978-3-030-52575-0_22
18. Wang, X., Kim, M.J., Love, P.E., Kang, S.-C.: Augmented reality in built environment: classification and implications for future research. Autom. Constr. **32**, 1–13 (2013). https://doi.org/10.1016/j.autcon.2012.11.021
19. Wolf, M., Vogt, O., Huxoll, J., Gerhard, D., Kosse, S., König, M.: Lifecycle oriented digital twin approach for prefabricated concrete modules. In: Semenov, V.U., Scherer, R.J. (Hrsg.): ECPPM 2021 – eWork and eBusiness in Architecture, Engineering and Construction, pp. 305–312. CRC Press, London (2021). https://doi.org/10.1201/9781003191476-42

An Application of a Wearable Device with Motion-Capture and Haptic-Feedback for Human–Robot Collaboration

Jongpil Yun, Goo-Young Kim, Mahdi Sajadieh, Jinho Yang, Donghun Kim, and San Do Noh(✉)

Sungkyunkwan University, Suwon 16419, South Korea
sdnoh@skku.edu

Abstract. Research on human–machine collaboration in Industry 5.0 has attracted significant attention in the manufacturing sector. Although human–robot collaboration can improve work efficiency and productivity, the design of its process is time consuming and cost intensive. The digitalization of machines facilitates automation and improves the intelligence of mechanical tasks. However, the digitalization of humans to improve the intelligence of operation functions for workers is difficult. To solve the difficulty of human digitalization, this study designed a human–robot interaction application based on motion capture using a digital human and virtual robot. The proposed framework supports process managers and shop-floor workers. Process managers can design the optimal collaborative process by interacting with robots and identifying their movements in the virtual world. Shop-floor workers can avoid collision accidents with robots by checking the future movements of the robot in the virtual world, on the basis of which they become proficient in the collaborative task to be actually performed in the physical world. An experiment was conducted on a virtual shop-floor that was modeled based on a physical shop-floor. The experimental results showed that a worker can avoid collision with the help of the proposed framework. Thus, the proposed framework can prevent collisions and accidents during the human–robot collaboration process in the real world.

Keywords: Human–robot interaction · Virtual human · Digital human · Immersive analytics · Haptic feedback · Force feedback

1 Introduction

Manufacturers believe that the 4M (method, machine, materials, and man) are the most important aspect in assembly tasks [1]. Humans are essential resources for manufacturing systems, but they are affected by uncertain factors such as labor intensity, proficiency, or physical conditions, which can increase the risks involved in the manufacturing field [2]. Industry 4.0 has facilitated the automation of factories and realized intelligent factories. However, this has led to major changes in human workload, as complex assembly lines still require manual labor and thus involve human errors. In

Published by Springer Nature Switzerland AG 2023
F. Noël et al. (Eds.): PLM 2022, IFIP AICT 667, pp. 368–377, 2023.
https://doi.org/10.1007/978-3-031-25182-5_36

addition to human issues, human–machine interaction errors result in operational problems. Human–machine interaction refers to communication and collaboration between humans and machines through a user interface [3]. Four elements are critical for a reliable and faultless human–machine interaction: machine behavior, operational goals (or task specifications), a model that describes machine behavior for the user (called the user model), and the user interface [4, 5]. As Industry 5.0 approaches, collaborative robots play an important role in the manufacturing industry, and manufacturers have been considering the use of collaborative robots to increase the flexibility and responsiveness of production processes in facilities [6]. Accordingly, research on human–machine collaboration has attracted attention. In the actual field, human operators are still considered as separators because they lack the necessary collaboration skills. Robots have high physical strength and are sophisticated in terms of handling tasks, whereas humans are intelligent and have problem-solving skills [7]. Thus, humans and machines can enhance each other's operations and improve overall performance. However, there are some problems with human–machine interaction, which are discussed as follows:

- In the past, robots were typically surrounded by a protective fence and separated from operators located in a different zone. However, in this scenario, much of the fenced space is not utilized, leading to an increase in cost [8]. The robot and operator might not be able to interact because of the separation, which also leads to efficiency issues.
- The real-time monitoring capability of depth sensors can be utilized to measure and track the distance between human operators and robots [7]. However, the presence of obstacles may lead to the view of the camera being obstructed; this leads to a difficulty in creating a virtual model for workers and robots, which may result in an accident.

To solve these problems above, this paper presents a framework involving a wearable device with motion-capture and haptic-feedback. The proposed framework enables seamless interaction and collaboration between humans and robots with several functionalities, such as the calculation of real-time distance or detection of the number of collisions using a haptic feedback module. Thus, shop-floor workers can avoid collision accidents by maintaining an appropriate safe distance from a robot surrounded by a virtual fence. The application can increase human safety and accessibility if a digital human is implemented in the virtual world.

2 Literature Review

2.1 Operator 5.0 in Smart Manufacturing

Recently, the term "digital twin" has been used to refer to digital representations of humans and objects that can be replicated, merged, and exchanged as well as saved and recorded, representing the advantages of digitalization [9]. Industry 4.0 focuses primarily on automation, whereas Industry 5.0 aims to combine the advantages of humans and robots through collaboration. Robotic co-workers will enable humans to work harmoniously with robots without fear; furthermore, the knowledge that robots understand them well and can collaborate effectively with them will help improve work efficiency

[10]. Romero [11] stated that Operator 4.0 included the social sustainability and human-centricity necessities of Industry 5.0, while Operator 5.0 completes the Industry 5.0 requirements by adding resilience. This has two main dimensions: self-resilience and system resilience. The former focuses on the natural, physical, cognitive, and mental wellbeing and safety as well as efficiency of each operator; the latter focuses on alternative methods for human–machine systems to continue functioning by sharing and trading control between humans and machines to ensure performance and system stability [11].

2.2 Motion Capture and Haptic Feedback

In the motion capture (MOCAP) technology, the posture and movement of a human are measured on the basis of their position and orientation in a 3D space and recording the information in a form that can be used in a computer-configured digital human model (DHM) [12]. The MOCAP technology is used to improve the working environment in terms of worker safety management, worker process design and operation, improvement of manual task productivity, and training for unskilled workers through the collection of worker motion data and using the DHM of workers in the manufacturing field [13]. Bortolini et al. digitalized human body movements during assembly manufacturing and analyzed the control volume for operator performance evaluation [14]. Nam et al. digitalized the motions of workers in the physical environment to the virtual environment and measured the working time and difficulty involved in the assembly process [15]. Geiselhart et al. utilized MOCAP systems to calculate the production performance in the actual process and compared it with the performance predicted by simulation [16]. Jun et al. developed automating human modeling technology using Kinects. This technology can reduce the costs and time required when using the DHM and engineering simulation [17]. The MOCAP systems used in existing research are advantageous because they collect the position and rotation values of each joint from skeleton data. However, they have a limitation in that they cannot provide direct feedback to the worker wearing the device. To overcome this limitation, the Teslasuit device is used for motion detection and haptic feedback, which enables the digital human to interact with other objects across the physical and virtual worlds [18].

3 Proposed Method and Application

3.1 Modules and Proposed Framework

The proposed framework is illustrated in Fig. 1. The framework is divided into several modules. An operator wearing a motion-capture haptic-feedback suit is connected to a virtual human through a wireless network. The operator's actual movement is reflected to the digital human in the virtual world, and collision detection is tested with the robot movements according to the defined assembly sequences. The safety distance between the human and the virtual fence of the robot is measured by the minimum Euclidean distance between the human body and the robot's end effector. When a collision event occurs in the virtual world, the operator can feel a haptic feedback response to their actions through electrostimulation. The movement of the robot or actual human is first corrected to avoid collision, and the movement is performed if it is determined to be safe. During the validation, the real-time distance and a warning message are displayed on the dashboard.

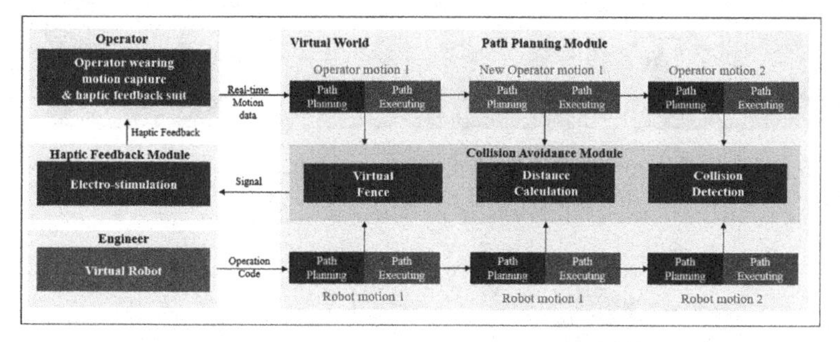

Fig. 1. Framework for application of human–robot interaction.

3.2 Virtual Fence of the Robot

To solve the existing problems mentioned above, this study proposed a solution that can protect the operator. A virtual fence is installed in the robot joint and end effector. The collision of the operator detected during path execution refers to a contact between the digital human and virtual fence rather than with the robot. Polyhedral shapes are usually used for collision detection; however, in this study, a spherical shape was used to create a virtual fence because the minimum distance can be easily calculated using a sphere. The calculation involving the polyhedral model becomes complicated because many cases, such as vertex–vertex, vertex–edge, edge–edge, and vertex–surface, need to be considered for the calculation of short distances [19] The virtual fence follows the formula of the *volume* of a sphere. The *radius* is measured from the center of the robot's end effector to the robot's maximum reachable distance. Since the robot's end effector is mostly located close to the human, the virtual fence radius was measured from the center of the end effector.same as that of a real human

$$Volume = \frac{4}{3}\pi r^3 \tag{1}$$

Considering a scenario in which the robot moves unexpectedly with a speed of 30 cm/s or moves differently opposed to the controller's intention, the robot's virtual fence should cover the area up to which the robot arm can reach the farthest. Figure 2 shows the representations when (a) the robot is in the initial state and (b) the robot arm reaches the farthest. Figure 2(c) shows the scenario in which a virtual fence is installed at the end effector (Fig. 3).

The radius of the virtual fence is calculated as follows. The *distance* from Joint 2 to the top of the robot along the z axis is considered the radius of the virtual fence shield, which is 240 mm [20, 21].

$$Radius\ of\ the\ Virtual\ Fence = safety\ distance \tag{2}$$

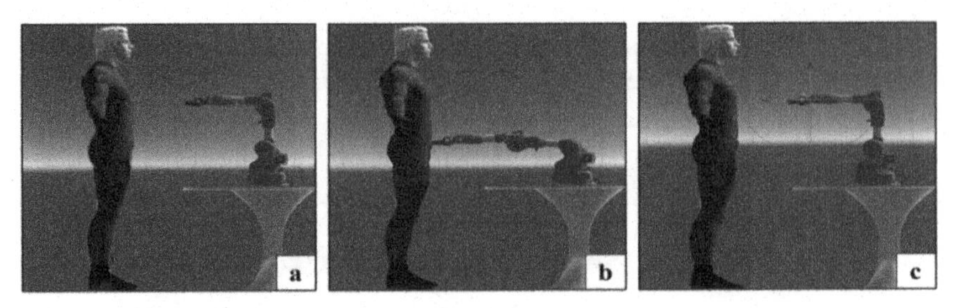

Fig. 2. (a) Robot's initial state, (b) robot's abnormal state, (c) robot with virtual fence.

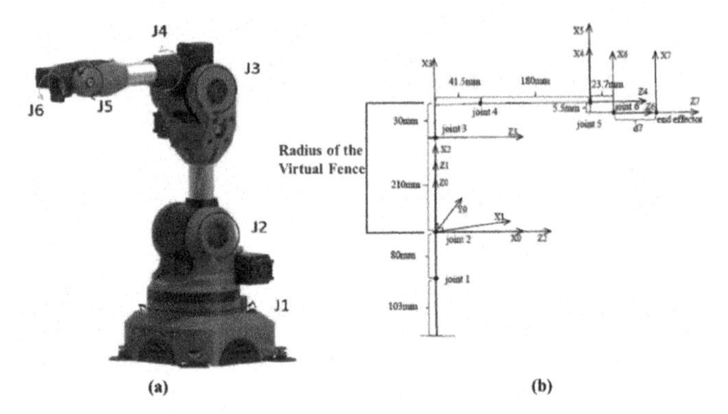

Fig. 3. (a) Directions of joint rotation. (b) Coordinate and mechanical dimensions of robot arm.

3.3 Calculation of Distance Between Human and Robot

The size of a virtual human is the same as that of a real human. For a safe and efficient collaboration, the human and robot should maintain a safe distance for seamless interaction. The minimum Euclidean distance of two given convex sets H_i and R_j is used, where $i = 1, \ldots, p$ and $j = 1, \ldots, \sigma$. Here, i indicates the upper-body joints of the virtual human shown in Fig. 4(a), and j indicates the sphere of the robot in Fig. 4(b) [21]. Since there are many cases in which the operator is likely to collide, nine body parts are designated for measurement with the robot hand to prevent potential accident. The nine points for the operator are on the head, neck, shoulders, elbows, and hands. The robot has one point corresponding to the end effector. Two vectors measure the values at these points (one from the operator and one from the robot) to calculate the distance [22].

The minimum distance is calculated using the following equation:

$$d_{min}\left(H_i, R_j\right) = \min\left\{\|h_i - r_j\| : \forall h_i \in H_i, \forall r_i \in R_i\right\}. \tag{3}$$

For example, the shortest distance is selected as follows. Since Eq. (3) yields the shortest distance, it is selected as the criteria before a warning message appears.

$$\sqrt{|h_H - R_h|^2} = \sqrt{\{(x, y, z) - (a, b, c)\}^2} = 310\,\text{mm},$$

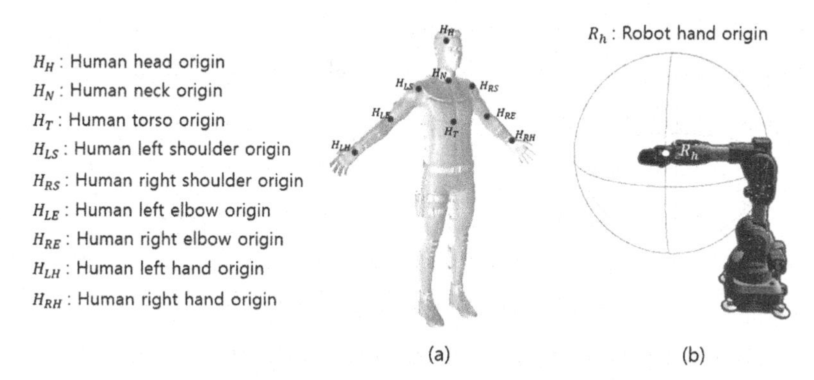

Fig. 4. (a) Upper body joints of humans, (b) joint in a robot.

$$\sqrt{|h_T - R_h|^2} = \sqrt{\{(x, y, z) - (a, b, c)\}^2} = 230\,\text{mm},$$

$$\sqrt{|h_{LH} - R_h|^2} = \sqrt{\{(x, y, z) - (a, b, c)\}^2} = 140\,\text{mm}.$$

3.4 User Interface

The dashboard shown in Fig. 5 combines general information that is useful to a supervisor, such as the production line number, operator's name, current job order, and real-time operation data during human–robot interaction. The real-time distance is calculated from the designated points of the human to the robot's end effector. The interface shows the closest distance between the human and robot in Fig. 5(a). A warning message in Fig. 5(a) with a graphic shows the position where the collision may occur, indicated in red in Fig. 5(b).

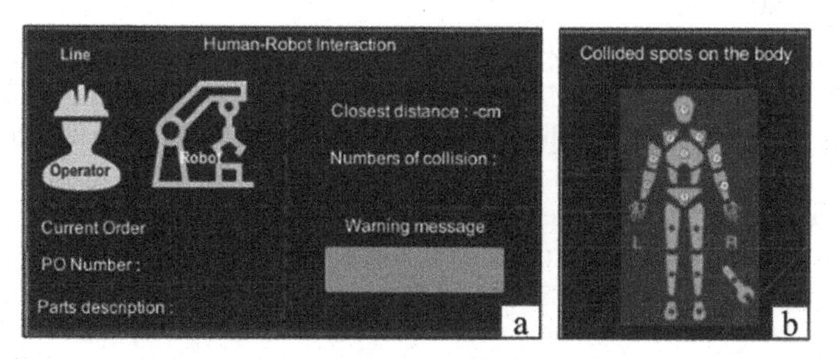

Fig. 5. Dashboard shows (a) Useful information include distance between the human and robot and numbers of collision and (b) Collied spots on the body.

This study uses a Teslasuit as a wearable device with motion capture and haptic feedback. The Teslasuit's haptic feedback system is integrated within the suit and can be activated during actions, on demand, or in response to MOCAP comparison. Teslasuit's sensitization system works by sending tiny pulses of different amplitudes, frequencies, and voltages to the electrodes. The two electrostimulation systems, TENS, and EMS, of Teslasuit stimulate nerve endings; these stimuli are experienced on the skin surface. In the event that a virtual part of the body collides with a virtual object, the commutation unit delivers pulses from a pulse generator to electrodes in the same vicinity on the wearer's real body [23]. There are 68 haptic points that realize sensation on certain areas of the body. Haptic mapping is designed to interact with Teslasuit using a coordinate system that is independent of the suit configuration. In the API, this process is called target mapping. Hit events work with the default source mapping. When a target map is loaded, hit coordinates are processed in a target mapping coordinate system to be transformed into the source mapping coordinates. In the virtual world, when either a robot or an operator approaches within a certain distance, a warning message is generated on the dashboard. Haptic feedback is retrieved to the actual area simultaneously in the physical world. The operator can identify an abnormal situation while interacting with the robot. Figure 6 shows the scenario in which the robot responds when an operator collides with the virtual fence.

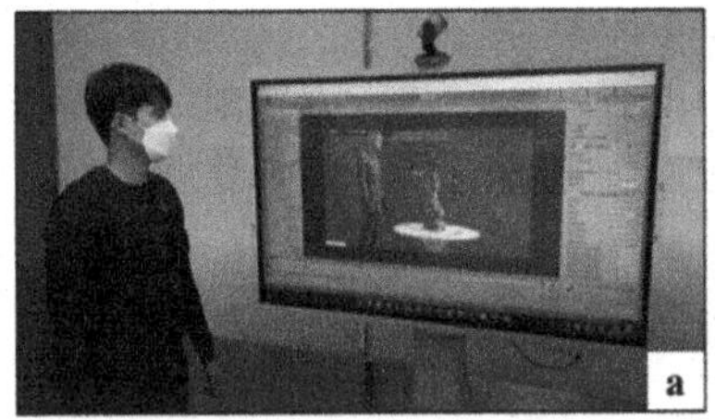

Fig. 6. Scenarios in which (a) there is no contact, and (b) the robot avoids the human.

4 Implementation and Case Study

4.1 Experiment

The proposed framework is implemented in a virtual testbed with an assembly process producing fan filter units. Figure 7 shows that a worker located near the manufacturing work-center is connected through a conveyor line. This work-center depicts a typical example of an assembly process, which picks equipment and performs tasks using the upper body at a fixed location. The bolting and screwing processes for the product in this virtual testbed are basic assembly processes. To validate and evaluate the proposed application, we selected the testbed in a virtual factory production line where produce fans and human–robot collaboration processes exist. This virtual factory is a real floor-shop-like environment. The experimental setup aims to validate the following functionalities when an operator collides with robot during production.

4.2 Results and Analysis

Virtual fence and haptic feedback are utilized in the application, and the supervisor can monitor real-time data, such as distance calculation and the number of collisions shown in the dashboard in Fig. 7. In addition, the current order information is shown while the human and robot are handling the job. During the production hours, the operator and robot should execute a defined motion after they are tested in the validation process. When an operator does not maintain the safety distance, collision is detected, and the area of collision is marked as red and displayed in the skeleton figure.

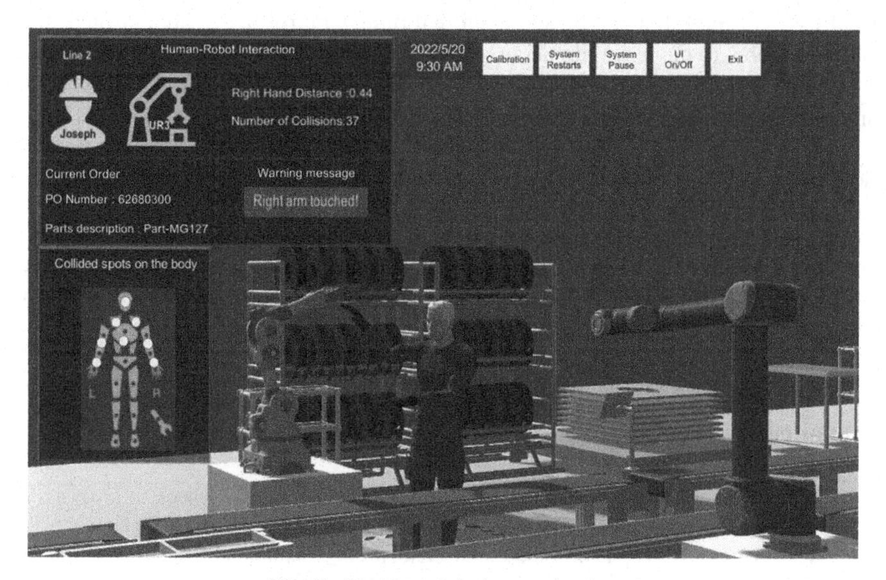

Fig. 7. Dashboard during production

5 Conclusions

The paper presents a framework of human–robot interaction using a wearable device with motion-capture and haptic-feedback that tracks human body interaction in the virtual world. The first aspect of the framework is to indicate a collided body part in the dashboard while the human is wearing a haptic suit. The second aspect is a method of using a virtual shield and calculating distance in real-time to protect human during collaboration. This framework enables humans and robots to share the same workspace and protects operators from severe accidents while combining their complementary strengths to improve work efficiency. Through the proposed framework, human safety in proximate human–robot collaboration is guaranteed through real-time human body motion capture, robot's virtual fence installation, and human-robot virtual-fence distance measurement. A dashboard and haptic sensors are used to provide feedback to the user. The results show that this solution is feasible for improving safety in real applications. This study developed the human–robot interaction system by supplementing the existing limitations through the proposed framework. However, posture correction must be performed through calibration whenever the sensors in the Teslasuit can be heavily affected by ambient electromagnetic waves. Moreover, a battery equipped should not be exposed to heat or water vapor. Currently, after a collision occurs between a machine and a person, the person receives feedback and manually designs the next path; however, efficiency might be improved if an optimized path is automatically created. In the future, route optimization will be applied as a reinforcement learning method and modified to an optimal route. For a better visualization using an MR device, HoloLens will be suitable to enable an operator to work in entirely new ways; for example, the virtual fence of the robot can be shown in real time. Controlling a robot through ROS enables the robot to move simultaneously in the virtual and real worlds. Detouring paths are currently designed manually, but training data for optimization using reinforcement learning algorithms will help to achieve automatic robot movements.

Acknowledgement. This work was supported by project for Smart Manufacturing Innovation R&D funded Korea Ministry of SMEs and Startups (Project No. RS-2022–00140261) and supported by Institute of Information & communications Technology Planning & Evaluation (IITP) grant funded by the Korea government (MSIT) (No.2022–0-00866, Development of cyber-physical manufacturing base technology that supports high-fidelity and distributed simulation for large-scalability).

References

1. Favi, C., Germani, M., Marconi, M.: A 4M approach for a comprehensive analysis and improvement of manual assembly lines. Procedia Manuf. **11**, 1510–1518 (2017)
2. Kampa, A., Gołda, G., Paprocka, I.: Discrete event simulation method as a tool for improvement of manufacturing systems. Computers **6**(1), 10 (2017)
3. Ke, Q., Liu, J., Bennamoun, M., An, S., Sohel, F., Boussaid, F.: Computer vision for human–machine interaction. In: Computer Vision for Assistive Healthcare, pp. 127–145. Academic Press (2018)
4. Degani, A., Heymann, M.: Formal verification of human-automation interaction. Hum. Factors **44**(1), 28–43 (2002)
5. O'malley, M.K.: Principles of human-machine interfaces and interactions. Life Sci. Autom.: Fundam. Appl., 101–125 (2007)
6. Wang, X.V., Keméy, Z., Váncza, J., Wang, L.: Human–robot collaborative assembly in cyber-physical production: classification framework and implementation. CIRP Ann. **66**(1), 5–8 (2017)
7. Liu, H., Wang, L.: Collision-free human-robot collaboration based on context awareness. Robot. Comput.-Integr. Manuf. **67**, 101977 (2021)
8. Fryman, J., Matthias, B.: Safety of industrial robots: from conventional to collaborative applications. In: ROBOTIK 2012: 7th German Conference on Robotics, pp. 1–5. VDE (2012)
9. Kawamura, R.: A digital world of humans and society—digital twin computing. NTT Tech. Rev. **18**(3), 11–17 (2019)
10. Mourtzis, D., Angelopoulos, J., Panopoulos, N.: Operator 5.0: a survey on enabling technologies and a framework for digital manufacturing based on extended reality. J. Mach. Eng. **22**, 43–69 (2022)
11. Romero, D., Stahre, J.: Towards the resilient operator 5.0: the future of work in smart resilient manufacturing systems. Procedia CIRP **104**, 1089–1094 (2021)
12. Menache, A.: Understanding Motion Capture for Computer Animation and Video Games. Morgan Kaufmann, Burlington (2000)

13. Menolotto, M., Komaris, D.S., Tedesco, S., O'Flynn, B., Walsh, M.: Motion capture technology in industrial applications: a systematic review. Sensors **20**(19), 5687 (2020)
14. Bortolini, M., Faccio, M., Gamberi, M., Pilati, F.: Motion analysis system (MAS) for production and ergonomics assessment in the manufacturing processes. Comput. Ind. Eng. **139**, 105485 (2020)
15. Nam, Y.W., Lee, S.H., Lee, D.G., Im, S.J., Noh, S.D.: Digital twin-based application for design of human-machine collaborative assembly production lines. J. Korean Inst. Ind. Eng. **46**(1), 42–54 (2020)
16. Geiselhart, F., Otto, M., Rukzio, E.: On the use of multi-depth-camera based motion tracking systems in production planning environments. Procedia CIRP **41**, 759–764 (2016)
17. Jun, C., Lee, J.Y., Noh, S.D.: A study on modeling automation of human engineering simulation using multi kinect depth cameras. Korean J. Comput. Des. Eng. **21**(1), 9–19 (2016)
18. Teslasuit. https://teslasuit.io. Accessed 23 Mar 2022
19. Balan, L., Bone, G. M.: Real-time 3D collision avoidance method for safe human and robot coexistence. In: 2006 IEEE/RSJ International Conference on Intelligent Robots and Systems, pp. 276–282 IEEE (2006)
20. Niryo One Mechanica Specifications. https://niryo.com. Accessed 23 Mar 2022
21. Li, F., Huang, Z., Xu, L.: Path planning of 6-DOF venipuncture robot arm based on improved a-star and collision detection algorithms. In: 2019 IEEE International Conference on Robotics and Biomimetics (ROBIO), pp.2971–2976. IEEE (2019)
22. Secil, S., Ozkan, M.: Minimum distance calculation using skeletal tracking for safe human-robot interaction. Robot. Comput.-Integr. Manuf. **73**, 102253 (2022)
23. https://developer.teslasuit.io. Accessed 23 Mar 2022
24. Unity-Robotics. https://github.com/Unity-Technologies/Unity-Robotics-Hub. Accessed 23 Mar 2022

Design and Development of a G-Code Generator for CNC Machine Tools Based on Augmented Reality (AR)

Dimitris Mourtzis(✉), Panagiotis Kaimasidis, John Angelopoulos, and Nikos Panopoulos

Laboratory for Manufacturing Systems and Automation, Department of Mechanical Engineering and Aeronautics, University of Patras, 26504 Rio Patras, Greece
mourtzis@lms.mech.upatras.gr

Abstract. Modern manufacturing systems rely on the utilization of Computer Numerical Control (CNC) for the manufacturing of products and components. By extension, engineers must be capable of programming machine tools (through G-code) in order to achieve the desired results. However, this process is time consuming and requires highly skilled and trained personnel. On the other hand, under the Industry 4.0 framework, several cutting-edge digital technologies have been developed in an attempt to improve user perception, such as Extended Reality (XR), which encapsulates Augmented Reality (AR), Mixed Reality (MR), and Virtual Reality (VR). Furthermore, the above-mentioned technologies enable engineers to extend Human-Machine Interaction (HMI) into new dimensions. Therefore, in this research work the design and development of a framework for facilitating engineers during the generation G-code script for machining operations based on the utilization of AR is presented. The MR functionalities will facilitate develop a better perception of the machining processes, as well as support the generation of virtual instructions for shop-floor operators. Further to that, advanced communication functionalities are provisioned based on Cloud technologies, in order to automate the transmission of G-code scripts/instructions to the machine tools.

Keywords: Mixed Reality (MR) · Computer Numerical Control (CNC) · G-code · Human-Machine-Interaction (HMI)

1 Introduction

Towards the Mass Personalization era manufacturing companies aim to reduce the length and cost of the product development cycle [1]. A paradigm shift in manufacturing from 'real' to 'virtual' production has resulted in the field since the early 1990s. However, with the pillar technologies of Industry 4.0, it is now possible to simulate some of the activities of a physical manufacturing system. The main goal is to understand and simulate the behavior of a specific manufacturing system prior to physical production, reducing the amount of testing and experiments on the shop floor [2]. Trial and error is frequently

Published by Springer Nature Switzerland AG 2023
F. Noël et al. (Eds.): PLM 2022, IFIP AICT 667, pp. 378–387, 2023.
https://doi.org/10.1007/978-3-031-25182-5_37

used in the planning of manufacturing processes. Engineers use physical experimentation to verify alternative plans in the case of complex processes. This approach frequently prevents engineers from achieving the desired cycle time and cost reductions [3]. Additionally, less material is wasted and interruptions to the actual machine on the shopfloor can be avoided by using a virtual system (Owen JV, 1997). Moreover, there is less concern about safety issues and cloud-based collaboration can be achieved for education and training purposes [4]. Focusing on Computer Numerical Control Systems (CNC) and more recently to Cyber Physical Machine Tools [5], part programming can be accomplished either manually or using automated methods such as Computer Aided Manufacturing (CAM) [6]. In Smart Manufacturing landscape, there is a critical need to have appropriate tools of human-machine interaction (HMI) to support customization and personalization needs [7]. Therefore, the following research question arises: "what should the HMI system look like that individual can access and interact within Machine Tools to design and manufacture customized products? "Thus, the contribution of this research work is focused on the design and development of a method for the visualization of G- and M- code scripts, based on the utilization of Cloud technologies and XR. The MR functionalities will facilitate develop a better perception of the machining processes, as well as support the generation of virtual instructions for shop-floor operators. Finally, the proposed framework has been validated in an industrial CNC milling machine.

The rest of the paper is structured as follows. In Sect. 2 the most relevant and recent related research works are investigated. Then, in Sect. 3 the architecture of the proposed method is presented and discussed. Then, In Sect. 4 the implementation of the method is presented. Finally, in Sect. 5 conclusions are drawn, and points for further research are discussed by the authors.

2 State of the Art

Voxels can be realized as the equivalent of pixels in the 3D space, forming perfect cubes [17]. During the last decades various applications in the field of Extended Reality (XR) as well as in 3D rendering, voxel-based architectures have been implemented. A classic example is the HP (Hewlett Packard) 3D printer, which uses voxels sized at 25-micron building blocks (for reader's reference the size of the voxels is equivalent of 1/4 the thickness of a single human hair). Voxel-based architectures can be found in the literature in two main categories, either as regular voxel grids, or as octree structures. A variety of visualization methods and techniques, in particular i) raytracing, ii) topological analysis, iii) volume estimation, iv) collision detection, are based on voxels in an attempt to efficiently process more complex 3D geometries [8]. The importance of voxel-based architectures is underlined by the fact that they are simple representations of the 3D volumes, instead they can carry additional information about the volumes, mimicking the structure of real-world objects (e.g. inclusion of material properties). Therefore, voxels can be realized as data structures capable of carrying semantic labels, facilitating engineers to model 3D systems in a more natural/efficient way [9]. In the available literature there are several examples of voxel engines. Among them VoxelPrint, which is a Grasshopper plug-in is focused on the simulation of 3D concrete printing constructions as presented in [10]. Nie et al. in [11] presented a voxel-based method is proposed for enabling the calculation of accuracy regarding Material Removal Rate (MRR), based on graphics processing units (GPGPU) computing. Furthermore, similar approaches have been proposed by the authors [10, 11] to improve the simulation of five-axis milling processes by calculating and visualizing the material deformation induced by the cutting force exerted on the workpiece. As such, voxel-based architectures can be utilized to improve the simulation, visualization and modelling of systems, in order to represent more closely their natural behavior. By extension, simulation techniques [12] and voxel techniques need to be further examined as a coupled system.

On the other hand, Computer Aided Manufacturing (CAM) which is based on the visualization of the G- and M- code, can be further improved by new digital technologies, such as AR, in order to facilitate engineers visualize in a more intuitive manner the various material removal processes and whenever possible to optimize the process parameters. To that end, virtual instructions were overlayed in the user's AR Head Mounted Display (HMD) field of view, while virtual CAD models were displayed around physical objects [13]. Furthermore, AR has been investigated for interacting with robots for path planning [14], spatial programming [15] and trajectory planning [16]. Similarly, AR has been used to simulate manufacturing processes. Various manufacturing processes, such as metal casting [17] and CNC machining [18], have been simulated in an AR environment for process plan verification, material cost reduction, and novice mechanist training [19]. However, the significant contribution of this research work compared to the abovementioned is that the MR functionalities will develop a better perception of the ma-chinning processes, as well as support the generation of virtual instructions for shop-floor operators. Further to that, advanced communication functionalities are provisioned based on Cloud technologies, in order to automate the transmission of G-code scripts/instructions to the machine tools.

Lin and Fu in [20] developed the Virtual Machine Tool Structural Modeller (VMSM) system, which can be used to model machine tools in a virtual reality environment. The architecture of the system is made up of several modules, including component modules, module shape libraries, combination rule libraries, and structure libraries. The system allows a user to choose the structure of the machine tool, modify it, control it, and simulate the machining process. Moreover, Training is a crucial part of the current challenging production environment. Kao et al. [21] developed the VRCNC system to support virtual training. The VR CNC structure and the VR CNC controller were built as two separate modules. The machine table, servomotor, driving system, and feedback system are all part of the VR CNC structure. Next, in the development of a VR-based training system, both immersion and desktop VR can be used. One such system in literature is the Products Virtual Analysis System (PVAS) [22]. There were four main functions used. They are as follows: (a) virtual reality with stereo glasses, (b) product kinematics analysis and optimization, (c) dynamics, and (d) a virtual interactive environment for virtual machining and assembly processes.

3 Architecture of the Proposed Method

In this Section of the manuscript the general system architecture will be presented and discussed. Consequently, it is necessary to break-down the architecture into key modules, and by extension to analyze the information flow. In Fig. 1 the general system architecture is illustrated.

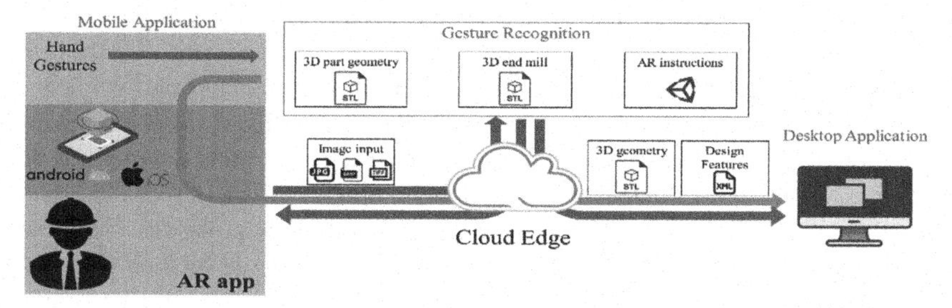

Fig. 1. General system architecture

3.1 Cloud Platform and Database

The Cloud platform can be realized as a virtual workspace for hosting the services required for the operation of the proposed method. More specifically, in the Cloud platform, there has been provisioned a space for the handling of users, registration and login operations. Secondly, on the Cloud platform there have been developed several libraries, containing cutting tools (end-mills) along with the 3D CAD of each cutting tool, so that they can be retrieved during the generation of the G- and M- codes. This service is entitled "Tool Handling Service - THS". The THS also supports the addition of new tools through the corresponding user interface. At this point, it is stressed out that the 3D geometries of the cutting tools have been designed fully parametrically, in order to further minimize the space requirements on the Cloud platform. In the following Fig. 2 the modelling of the end mills implemented in the architecture is presented.

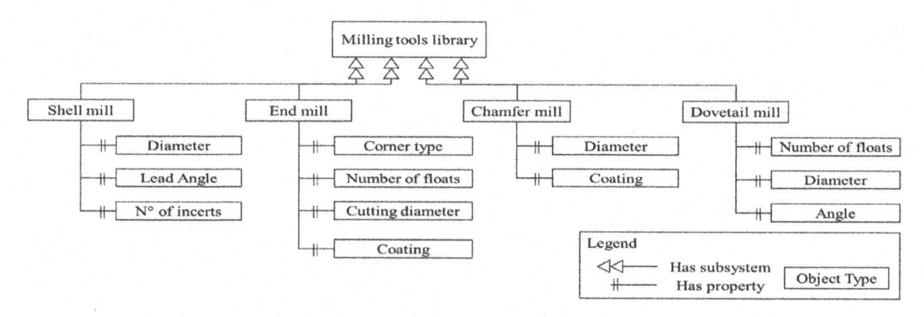

Fig. 2. UML diagram for end mills

3.2 Voxel Generator

As a recap of the previous paragraphs, voxels can be realized as 3D pixels. The application was created using Unity by Unity Technologies. Unity is a game development tool that also provides us with the ability to create 3D applications in both AR &VR. In our case, Unity AR Foundation is utilized. Other systems could have been used, but Unity provides with a much friendlier and more efficient user interface, making development easier and highly modifiable. The "AR Tracked Image" in Unity corresponds to the real-world position of the preloaded image to that of the "parent" cube. A 3D mesh of voxels is generated using C#. Voxels provide a digital simulation of the real-world analog behavior of the workpiece. The realism of the simulation is in direct correlation with the computational capabilities of each device, as a more realistic simulation requires more and smaller voxels, which comes at the cost of computational resources. The size of the voxels in turn defines the scale of the cutting tool along with the real-world scale of the simulation. In the current implementation of the proposed method the applied voxel size has a volume of 1.0*E−9 m^3. Each voxel has a cube collider (triggered) attached to it allowing Unity Engine to recognize collisions between the cutting tool (which has "Rigidbody" property) and the 3D geometry of the material. Consequently, whenever collision is detected, the colliding voxels are deactivated. Even though using colliders is not the most efficient way of recognizing collisions between the cutting tool and each voxel, they are necessary as every tool has a deferent geometry which cannot be calculated by a generic algorithm for every type of tool. Additionally, the utilization of voxels provides the user with access to easily interchangeable materials to simulate different workpieces, in the following examples Aluminum is used. The material can be altered by changing the Unity material properties of the "parent" cube. Regarding the memory allocation, each the voxel requires a minimum of 1 byte, for a workpiece of 1.0 m*0.5 m*0.5 m (25*E−2 m^3), constantly 0.232831 Gb of RAM memory is the minimum requirement for the voxel storage. Consequently, in an attempt to avoid the possibility of memory drain (taking into consideration the use of mobile devices), voxels are grouped in chunks, with a chunk size of 16×16×16 voxels which requires 4096 bytes (4 MB) of RAM memory. Furthermore, the access to a specific voxel can be facilitated by the storage of the voxel chunks in a serial array, requiring only knowledge of the chunk ID and the voxel ID. Finally latency is an issue that should be addressed, and we have so far adjusted the framerate to minimize the problem, until further development.

3.3 Augmented Reality Module

The Augmented Reality (XR) module is one of the key modules of the proposed method since it constitutes the Human-Machine Interface (HMI). Through the AR module the users are capable of interacting with the virtual components and setup the CAM instructions, based on the utilization of intuitive Graphical User Interface (GUI) elements. When the user begins a new session, the first selection is the type of material to be used. Then the initial dimensions as well as the form of the billet/casting are inserted. Upon selection of the basic parameters, the Image recognition module is activated allowing the XR visualization to appear on the user's Field of View (FOV). The material as well as the Cutting tool is then displayed on the table of the machine tool (or chuck, wherever applicable). When the user taps on "GCode Input" a panel is activated that allows the serial input of G-commands. Upon pressing continue the panel is deactivated and the simulation is realized. The material removal prosses is simulated by disabling the voxels when they collide with the cutting tool. Unity 3D physics engine is responsible for detecting the collisions using "Rigidbody" properties. In the flowchart presented in Fig. 3, the main modules of the architecture are presented, highlighting the interactions and the information flow.

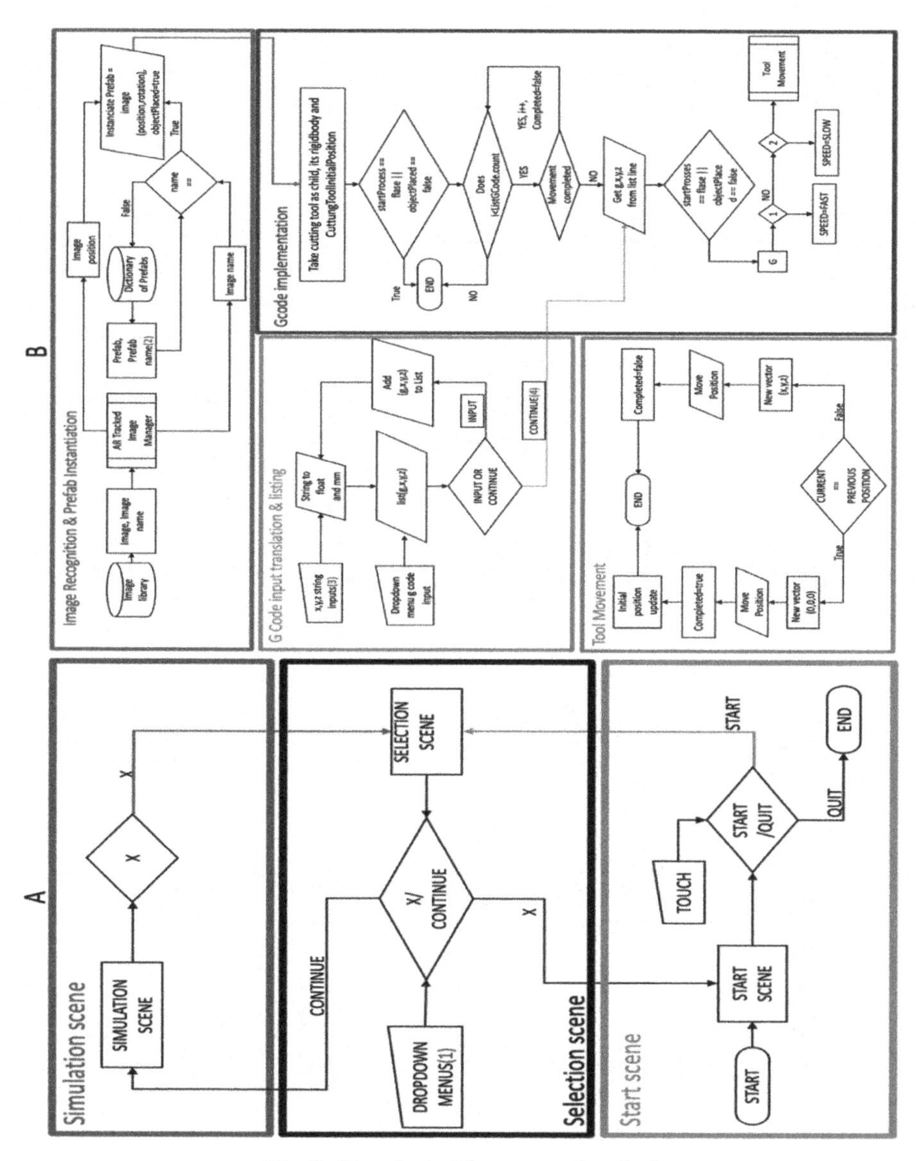

Fig. 3. Flowchart of the proposed method

Regarding the content of Fig. 3, (A) represents the scene generation, and (B) represents the G-/M-Code simulation in AR. Concretely in (1) the users inputs are gathered as presented in Fig. 5 (1, 2). Then in (2) the cutting tool prefab and name are predetermined by the user in selection scene, as shown in Fig. 5 (2). The G-code commands are input by the user in simulation scene, as shown in Fig. 5 (3). Finally the G-/M-Code is input as a list and the tool simulation is visualized (refer to Fig. 5 (3)).

4 Implementation

The first part of the implementation process requires the connection of the AR tool with the Cloud platform in order to gain access to the initial CAD files of the part to be produced, and the cutting tools/inserts available for the CNC milling machine. The initial billet is loaded as a geometry in a suitable file format (Collada file). Afterwards, a list of available G- and M-Operations is loaded from the Cloud database. Further to that, in an attempt to assist the users to setup the machine correctly, a set of AR steps for the correct installation of the cutting tools. It is stressed out that the tool, automatically detects the tool based on its name, thus the corresponding steps and specifications are loaded and displayed on the user's FOV.

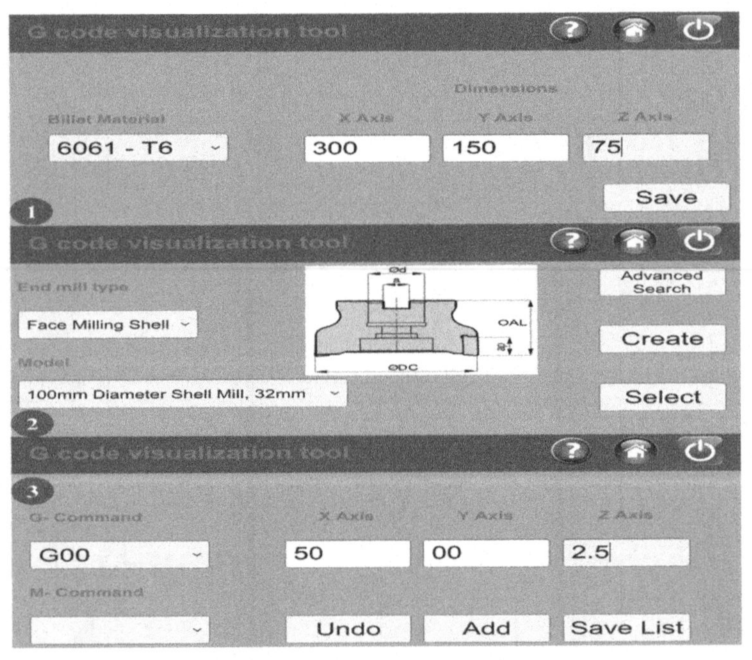

Fig. 4. 1) Material selection GUI; 2) End mill selection GUI; 3) G-/M- code command insert GUI

For the development of the AR module the Vuforia Engine API (Application Programming Interface), which provides the necessary algorithms for the environmental understanding and the registration of the 3D information in the user's FOV. In order to provide a better/more intuitive user experience tool animations have also been implemented, following the path as it is generated by each G- command. The development of the necessary GUIs (Graphical User Interfaces) has been accomplished via the utilization of the Unity 3D game engine (See Fig. 4). The code scripts have been developed in C# with the use of Microsoft Visual Studio IDE (Integrated Development Environment). The current implementation of the proposed method is compatible with smart mobile devices based either on Android or iOS operating systems.

5 Experimental Case Study

The applicability of the developed method is tested on a use case of a CNC milling machine. Currently, the engineers use standard CAM tools in order to generate the code required for the CNC milling machine. Although the current approach is functional, expert, well-educated/experienced engineers are required in order to setup processes for the CNC milling machine. However, the manufacturing company has identified that less-experienced engineers should be trained in order to be capable of producing such code script, as well as to facilitate more experienced users to visualize the process and eliminate any errors (such as use of proper tooling, proper clamping/securing the part on the machining table etc.) which could be facilitated via the proposed approach.

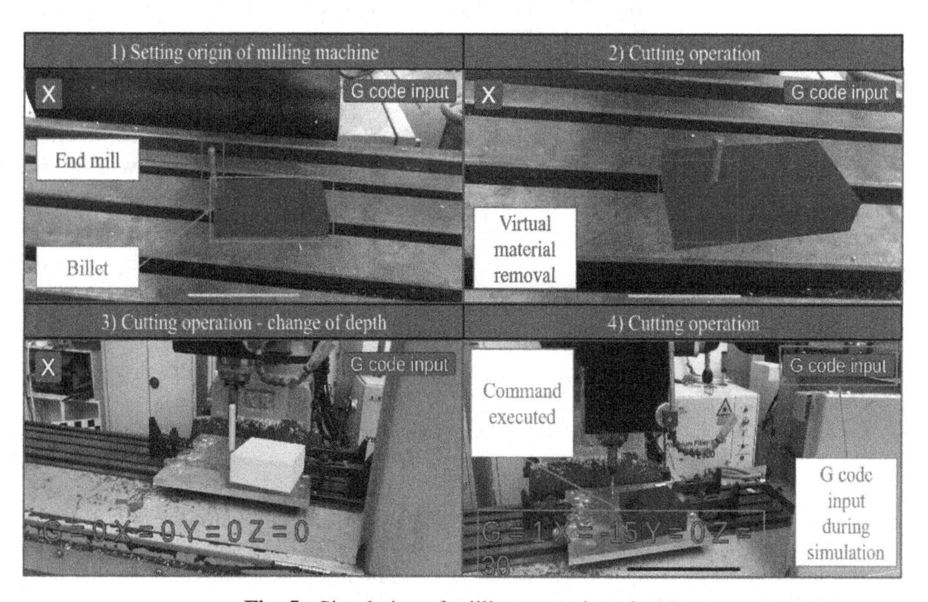

Fig. 5. Simulation of milling operation via AR

6 Concluding Remarks and Future Work

The Virtual Machine Shop Application for G- and M- codes import and visualization in AR environment was created to aid in the assessment of various decision variables in a virtual manufacturing environment. For example, the user could stabilize the virtual workpiece using real world work holding devices, as shown in Fig. 5. It enables a design, verification, and redesign loop to be carried out in a manner similar to that used in the real world. In addition, the intuitive interaction between the human and the process allows for the reliable verification of human-related parameters. The proposed framework can also be used to aid in the early stages of machine tool operator training. This simulation environment provides realistic and safe training conditions for immersive and interactive hands-on practice with the process, in contrast to physical training, which can be dangerous and expensive. Furthermore, AR allows for far greater adaptability, as the trainee practices on the real-world machine, making the educational experience for effective. The user, before the utilization of the application must follow the same safety routine that he would follow if he was using the real machine. That empowers the educational purpose of this AR application and builds important safety habits to the user. Finally, the smartphone version is a better option than a desktop one as they are cheaper, more widespread and portable [23].

To summarize, in this research work the development of a method for the composition of G- and M- codes based on AR visualizations. The key outcome of the presented research work is the development of a mobile application for the support of generation of machine code to aid in manufacturing processes and improve education of new engineers in CAM. Through the AR visualizations, engineers are capable of visualizing each step of the process, including the motion of the cutting tool as well as the removal of material. Along these lines, engineers be actively involved in full automation of the production segment of PLM.

Future research work will be focused on the implementation of physics in order to enable the accurate visualization of chip formation. An additional step will be the support of Head Mounted Displays (HMD).

References

1. Mourtzis, D.: The mass personalization of global networks. In: Design and Operation of Production Networks for Mass Personalization in the Era of Cloud Technology, pp. 79–116 (2022). https://doi.org/10.1016/B978-0-12-823657-4.00006-3
2. Kadir, A.A., Xu, X., Hämmerle, E.: Virtual machine tools and virtual machining—a technological review. Robot. Comput. Integr. Manuf. **27**(3), 494–508 (2011). https://doi.org/10.1016/j.rcim.2010.10.003
3. Chryssolouris, G., Mavrikios, D., Fragos, D., Karabatsou, V., Pistiolis, K.: A novel virtual experimentation approach to planning and training for manufacturing processes–the virtual machine shop. Int. J. Comput. Integr. Manuf. **15**(3), 214–221 (2002). https://doi.org/10.1080/09511920110034978
4. Owen, J.V.: Virtual manufacturing. Manuf. Eng. **119**, 84–90 (1997)
5. Liu, C., Xu, X.W.: Cyber-physical machine tool – the era of machine tool 4.0. Procedia CIRP **63**, 70–75 (2017). https://doi.org/10.1016/j.procir.2017.03.078
6. Mourtzis D., Makris S., Chryssolouris G.: Computer-aided manufacturing. In: Chatti S., Laperrière L., Reinhart G., Tolio T. (eds) CIRP Encyclopedia of Production Engineering. Springer, Berlin, Heidelberg (2018). https://doi.org/10.1007/978-3-642-35950-7_6550-4
7. Why Multi-Material Additive Manufacturing Will Change Our Approach To Part Design. https://aerosint.com/why-multi-material-additive-manufacturing-will-change-our-approach-to-part-design/. Accessed 14 Mar 2022

8. Gebhardt, S., Payzer, E., Salemann, L., Fettinger, A., Rotenberg, E., Seher, C.: Polygons, point-clouds and voxels: a comparison of high-fidelity terrain representations. In Simulation Interoperability Workshop and Special Workshop on Reuse of Environmental Data for Simulation—Processes, Standards, and Lessons Learned, p. 9 (2009)
9. Regenbrecht, H., Park, J.W., Ott, C., Mills, S., Cook, M., Langlotz, T.: Preaching voxels: an alternative approach to mixed reality. Frontiers ICT **6**(7), 7 (2019). https://doi.org/10.3389/fict.2019.00007
10. Cherabier, I., Häne, C., Oswald, MR., Pollefeys, M.: Multi-label semantic 3D reconstruction using voxel blocks. In: 2016 Fourth International Conference on 3D Vision (3DV), Stanford, CA, USA, pp. 601–610 (2016). https://doi.org/10.1109/3DV.2016.68
11. Nie, Z., Lynn, R., Tucker, T., Kurfess, T.: Voxel-based analysis and modeling of MRR computational accuracy in milling process. CIRP J. Manuf. Sci. Technol. **27**, 78–92 (2019). https://doi.org/10.1016/j.cirpj.2019.07.003
12. Mourtzis, D.: Simulation in the design and operation of manufacturing systems: state of the art and new trends. Int. J. Prod. Res. **58**(7), 1927–1949 (2020). https://doi.org/10.1080/00207543.2019.1636321
13. Makris, S., Pintzos, G., Rentzos, L., Chryssolouris, G.: Assembly support using ar technology based on automatic sequence generation. CIRP Ann. Manuf. Technol. **62**(1), 9–12 (2013). https://doi.org/10.1016/j.cirp.2013.03.095
14. Fang, H.C., Ong, S.K., Nee, A.Y.C.: A novel augmented reality-based interface for robot path planning. Int. J. Interact. Des. Manuf. (IJIDeM) **8**(1), 33–42 (2013). https://doi.org/10.1007/s12008-013-0191-2
15. Lambrecht, L., Kleinsorge, M., Rosenstrauch, M., Kruger, J.: Spatial programming for industrial robots through task demonstration. Int. J. Adv. Robot. Syst. **10**(5), 254 (2013). https://doi.org/10.5772/55640
16. Fang, H., Ong, S., Nee, A.: Interactive robot trajectory planning and simulation using augmented reality. Robot. Comput. Integr. Manuf. **28**(2), 227–237 (2012). https://doi.org/10.1016/j.rcim.2011.09.003
17. Watanuki, K., Hou, L.: Augmented reality-based training system for metal casting. J. Mech. Sci. Technol. **24**(1), 237–240 (2010). https://doi.org/10.1007/s12206-009-1175-9
18. Zhang, J., Ong, S.-K., Nee, A.Y.: Development of an AR system achieving in situ machining simulation on a 3-axis CNC machine. Comput. Anim. Virtual Worlds **21**(2), 103–115 (2010). https://doi.org/10.1002/cav.327
19. Mourtzis, D., Zogopoulos, V., Katagis, I., Lagios, P. Augmented reality based visualization of CAM instructions towards industry 4.0 paradigm: a CNC bending machine case study. In: 28th CIRP Design Conference 2018, Procedia CIRP, pp. 368–373 (2018). https://doi.org/10.1016/j.procir.2018.02.045
20. Marinov, V.P., Seetharamu, S.: Virtual machining operation: a concept and an example. In: Intelligent Systems in Design and Manufacturing, vol. 5605, pp. 206–213 (2004).https://doi.org/10.1117/12.571333
21. Kao, Y.C., Chen, H.Y., Chen, Y.C. Development of a virtual controller integrating virtual and physical CNC. In: Materials Science Forum, vol. 505, pp. 631–636 (2006).https://doi.org/10.4028/www.scientific.net/MSF.505-507.631
22. Wang, J., Chen, E., Yang, J., Wang, W.: Applications of virtual reality in turn-milling centre. In: 2008 IEEE International Conference on Automation and Logistics, pp. 2302–2305 (2008). https://doi.org/10.1109/ICAL.2008.4636550
23. Doolani, S., et al.: A review of extended reality (XR) technologies for manufacturing training. Technologies. **8**(4), 77 (2020). https://doi.org/10.3390/technologies8040077

A Telexistence Interface for Remote Control of a Physical Industrial Robot via Data Distribution Service

Damien Mazeas[1(✉)], John Ahmet Erkoyuncu[1], and Frédéric Noël[2]

[1] Centre for Digital Engineering and Manufacturing, Cranfield University, Cranfield, UK
Damien.j.mazeas@cranfield.ac.uk
[2] G-SCOP, Grenoble University, Grenoble, France

Abstract. The increasing adoption of remote working practices and internet-based systems highlights the need for a new mode of interaction and communication interfaces between humans and machines. Current human-machine interfaces (HMI) are based on complex controllers that are not intuitive for a novice or remote operators. The increase in skills required in today's manufacturing environment allows the use of telexistence capabilities to facilitate human-robot collaboration. We propose a digital-twin based framework with a telexistence interface as a means to reduce the complexity of operating and programming an industrial robot. To remotely control the industrial robot, sensors data are shared via a data distribution service (DDS). The immersive virtual reality (VR) interface is deployed for effective control and monitoring of the industrial robot. Telexistence capabilities allow intuitive manipulation and are combined with real-time data visualization of the robot through the digital twin. The interface is implemented with the Unity 3D engine and connected to a console application that collects sensor data and shares them via a DDS connectivity framework. A simple experiment with a physical FANUC M20-IA industrial robot and Azure Kinect RGBD cameras shows the reliable performance when a robot path request was sent by a remote operator through DDS and the immersive VR interface. We then discuss future work and use cases for this telexistence platform to support maintenance activities in manufacturing contexts.

Keywords: Telexistence · Virtual reality · Digital twin · Immersive interface

1 Introduction

Working environments can be potentially hazardous to humans in many sectors such as nuclear engineering, manufacturing, or defence. Operators are exposed to heavy equipment, fast-moving machinery, or harmful materials. To maintain a safe working environment and provide flexible solutions while improving efficiency throughout the product life cycle, companies are responding with a smart factory vision. Although the main objective of robots in automation is to provide the highest level of autonomy to perform complex repetitive tasks such as disassembly or sorting, when it comes to unplanned

F. Noël et al. (Eds.): PLM 2022, IFIP AICT 667, pp. 388–398, 2023.
https://doi.org/10.1007/978-3-031-25182-5_38

situations, automation is limited. Providing industrial robots with the ability to deal with unexpected events is essential for the effectiveness and productivity of the factory. Humans can exploit experiences in their fields to provide innovative solutions and use maintenance robots to do so. To avoid wasting time and money, it is important to keep the Human in the loop by enabling safe remote collaboration between the human and the robot. We hypothesize that the introduction of telexistence capability into remote work on tasks such as maintenance can enhance the capabilities and way of working for remote maintainers since it could lead to a more intuitive user interface (UI) and better spatial awareness, due to advantages such as sense of presence and more advanced visualization of the remote asset to maintain with the use of the immersive digital twin. A prototype that reflects the real-time operating conditions of a physical asset is a twin [1], Grieves [2] has described the concept of digital twin as a real and virtual spaces with a link for data flow from the real to the virtual space and a link for information flow from the virtual space to the real space. Lack of appropriate visualization of interaction forms is one of the main gaps in digital twins' literature reviews and immersive technologies are often cited to overcome these issues [3]. Conventional methods such as video feed-based teleoperation do not provide sufficient environment awareness and an efficient UI for the various data involved. A 2D control screen can be confusing for an operator of a remote robot [4]. Extended reality (XR), which refers to immersive 3D technologies such as virtual reality (VR) has raised the interest to support remote maintenance workers [5, 6]. Human-robot interaction in the work environment is a widely researched topic in robotics. XR has been identified as a useful tool to provide the operator with immersive real-time feedback to facilitate remote robot control [7–9]. Combining haptic devices and sensors allows for better operability of the system and a sense of presence for the operator in the simulated remote environment, whilst enhancing telexistence capability. This concept was proposed by Professor Susumu Tachi in 1980 [10], it is described as a fusion of VR and robotics which enables a sense of existence in another place.

Therefore, the overall goal of the present study is to develop a reliable framework via data distribution service (DDS) and implement telexistence capability for an industrial robot FANUC. Industrial robots are fully automated, pre-programmed for their tasks without human interaction, while collaborative robots (cobots) work with humans. As part of the industry 4.0, we want to bring human and industrial robot in a same collaborative and single workplace thanks to a telexistence interface. Our framework aims to offer usability and functionality to make it suitable specifically for operating and monitoring a physical industrial robot through its immersive digital twin in real-time. Through the design of this framework, we want to investigate whether telexistence-based remote monitoring and control interface can be applied to an industrial robotic arm using a modern data centric protocol and the immersive visualization of its digital twin.

2 Telexistence DDS-Based Framework

2.1 Interface Architecture

The following interface architecture has been developed to remotely enable the communication between the FANUC industrial robot and the human operator as well as to visualize the spatial context. This spatial context includes the static CAD model of the robot equipped with a gripper, dynamic data from the robot actuators and Kinect video/3D sensors and the operator orientation and position through the use of a VR Head Mounted Display (HMD).

In the beginning, a search and comparison was conducted on communication standards/protocols widely used in the industry. Thus, OPC unified architecture (OPC UA), message queuing telemetry transport (MQTT), advanced message queuing protocol (AMQP) and data distribution service (DDS) were identified as relevant for our framework among others. These protocols are used to design industrial Internet of Things (IoT) applications to efficiently share data. The majority of these protocols are designed for simplicity and can support a very limited set of use cases except for DDS. Our choice to use DDS is justified by its wide field of application in real-time and scalable systems. The protocol DDS provide features in term of data-centric approach, predictability, real-time quality of service settings and respect to security [11]. The ROS2 robotic interface is also based on top of DDS. In addition, it facilitates development efforts and allows data to be efficiently delivered to systems using a publish-subscribe pattern and is an Object Management Group (OMG) machine-to-machine standard.

Figure 1 shows the architecture of our DDS-based framework. To send and receive data and commands, a publisher node is creating a topic and a subscriber node is receiving data from a specific topic. Moreover, DDS integrate a concept of Quality of Service (QoS) to configure the parameters of participants inside the system.

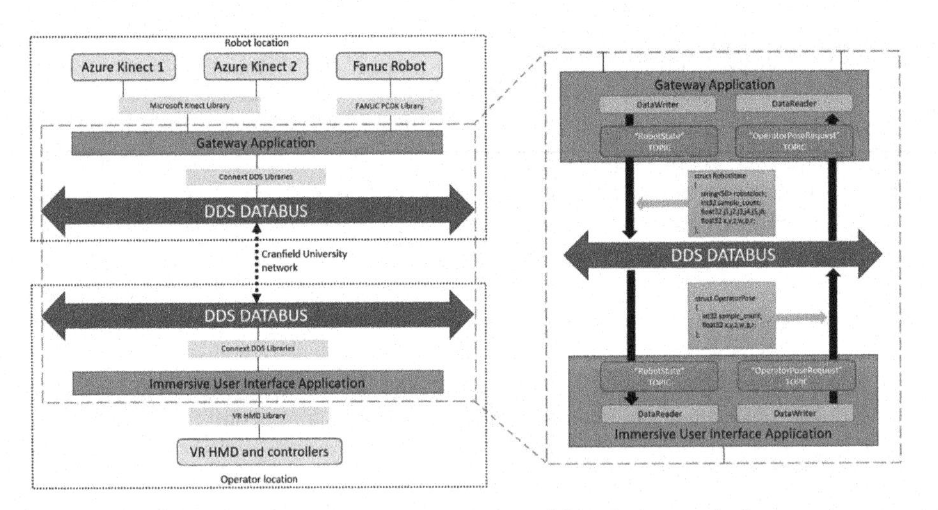

Fig. 1. Architecture and components overview of the telexistence rig framework

Subsequently, DDS was set up by creating data types sent in samples. A DDS sample is a combination of a Topic (distinguished by a Topic name) and the actual data type defined by the user. The ROS libraries, the most used operating system for robotics provides common interface messages organised in modules such as sensor devices or geometric primitives. It also offers the possibility to create custom messages. For our framework, since ROS is not implemented and for simplifying the development, we have created our own data structures. However, it could make sense in the future to match

ROS data types for interoperating with ROS applications. We created ours to provide a means to manage data efficiently for our framework especially for large amount of data such as the point cloud. Only the necessary data is transmitted to achieve the desired function and reusable through different functions (Table 1).

Table 1. Data structures description of samples transmitted

	Data structure name	Data structure members
From the FANUC robot to remote operator	Robot_State	- Clock (string - robot internal timestamp) - Sample (int - samples count) - J1, J2, J3, J4, J5, J6 (floats - joints angles in degree) - X, Y, Z (floats - robot world position) - W, P, R (floats - robot world rotation)
	Robot_Point_Cloud	- Clock (string - robot internal timestamp) - Sample (int - samples count) - Sequence Memory (floats - sequence containing points position and Colour data)
	Robot_Image	- Clock (string - robot internal timestamp) - Sample (int - samples count) Memory (bytes - sequence containing MPEG image data)
	Robot_Alarm	- Clock (string - robot internal timestamp) - Sample (int - samples count) - Message (string of alarm message)
	Robot_Reachability_State	- Clock (string - robot internal timestamp) - Sample (int - samples count) - IsReachable (bool - state regarding reachability for a position by the robot)

(*continued*)

Table 1. (*continued*)

	Data structure name	Data structure members
From the remote operator to the FANUC robot	Operator_Request	- Clock (string - operator computer timestamp) - Sample (int - samples count) - Buttons (enumeration of buttons, RESET, ABORT, HOME, PATH)
	Operator_Teleop_Target	- Clock (string - operator computer timestamp) - Sample (int - samples count) - X, Y, Z (floats - operator target position) - W, P, R (floats - operator target rotation)
	Operator_Path_Point	- Clock (string - operator computer timestamp) - Sample (int - samples count) - ID (int - added/modified pose ID) - IsUpdating (boolean - sample is for updating point values) - IsDelete (boolean - sample is for deleting pose) - X, Y, Z (floats – pose position values) - W, P, R (floats - pose rotation values)

2.2 Immersive Operator Station

Operating an industrial robotic arm is not possible without experience, keeping that in mind we started developing our telexistence rig keeping functions as simple as possible for an untrained user. For this purpose, the following functionality are expected:

- A functionality to interact with the immersive UI.
- A functionality to interact with the robot digital twin.
- A functionality to jog[1] the robot.
- A functionality to register a user pose.
- A functionality to switch between robots' programs.
- A functionality to abort the current robot motion.

[1] Jogging the industrial robot is the term used to describe the act of manually moving the robot via the user interface.

The interaction in the 3D environment is done as shown in Fig. 2, a controller button is dedicated to the teleport function (left) as well as another one for a pointer (middle) and one for direct manipulation of the 3D cursor (right). The main way of programming a robot without an offline simulation software is to jog the robot using the teach pendant device, the 3D cursor will aim to replace the jogging keys on the regular FANUC HMI. The emergency stop is a controller button which is faster to trigger than the "ABORT" 3D button in the environment UI in case the user wants to stop the robot motion.

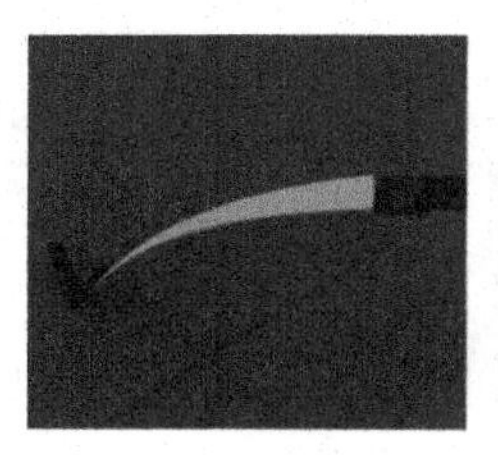

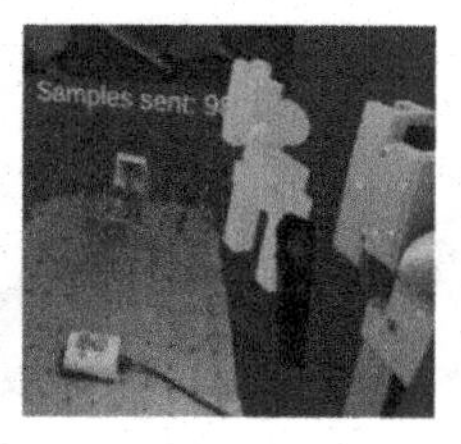

Fig. 2. Interaction methods of the framework

Figure 3 shows the elements of the UI, each element can be described as follows:

1. Four 3D buttons for the following features and the robot clock:

   - The possibility to call the home program thanks to the "HOME" button to bring the robot into its default pose.
   - The possibility to teleoperate the robot and create a path thanks to the "PATH" button. The teleoperation of the robot is done by moving the 3D cursor as seen in Fig. 2 (right). When the 3D cursor is green the robot can reach the new pose, when the 3D cursor is red the robot cannot reach the new pose and the pose is not updated in the robot program to prevent it to go into fault mode. This same "PATH" button also allows sending poses that will be used to create a path program to repeat a specific trajectory chosen by the user. This program is created on the robot gateway application by creating a Karel[2] file with a pose sent by the user and registered in the robot controller. This Karel file is then sent to the controller which compiles and executes it when the user wants to play the trajectory created.
   - The "RESET" button to reset the current robot controller fault.
   - The "ABORT" button to abort the current controller robot task.

2. A 2D video feed viewer of the first Kinect with the remote robot current ping in milliseconds. On the bottom right of the viewer, there is the number of samples received from the robot.
3. A 3D visualization of the 3D data received from the second Kinect. To do so, we have used additional Unity packages, High-Definition renders pipeline and Visual Effect (VFX) Graph, a node-based visual logic graph creator. The VFX graph created for this project takes a mesh generated from the received Kinect data by a unity C# script

[2] Lower-level language similar to Pascal and used to write FANUC robots programs.

as an input, using vertices of this mesh to output particles in cube shape. Position and Color are updated 26 times per second. The Kinect is currently calibrated manually in the 3D environment.

4. Angles value of the 6 joints of the robot in degree, world pose in millimeters and degree of the robot as well as the last alarm message are also displayed to the user.

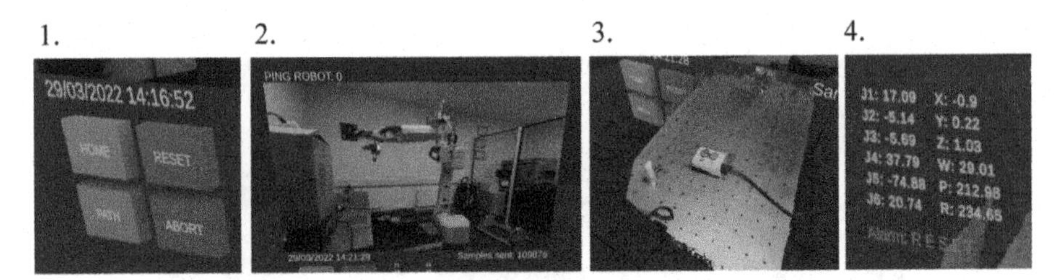

Fig. 3. UI of the framework, 3D buttons (1), 2D video feed viewer (2), point cloud (3) and robot pose values (4)

2.3 Coordinate Systems

The industrial FANUC robot's pose consists of a position XYZ in millimeters and three angles WPR in degrees with W rotation around the x-axis, P rotation around the y-axis and R rotation around the z-axis. The coordinate systems of Unity and FANUC as shown in Fig. 4 are different. The transformation from FANUC XYZ WPR to Unity 3D can be done by converting angles in degrees and then to a quaternion. We also have inverted x-axis and P, R angles to match coordinate systems.

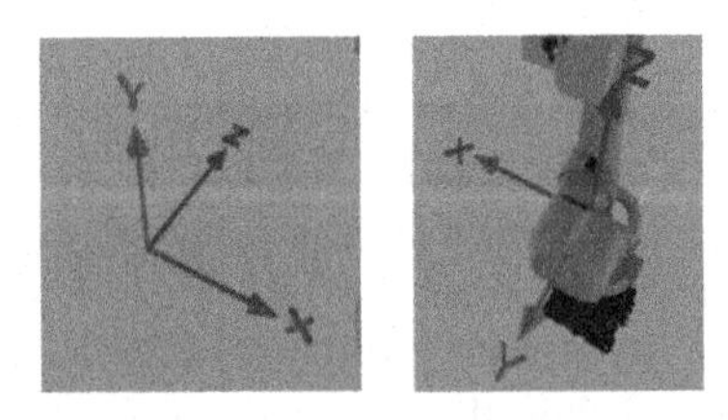

Fig. 4. Coordinate systems of Unity 3D (left) and Fanuc (right)

3 Case Study

To illustrate the proposed framework, we used a FANUC M-20iA (Fig. 5) of the Centre for Digital Engineering and Manufacturing (CDEM) at Cranfield University. It is controlled by an R-30iB mate cabinet and equipped with a pneumatic gripper. The objective

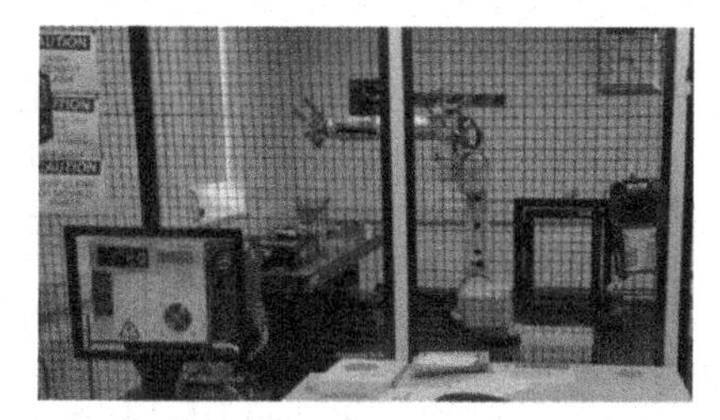

Specification		Motion Range	
Axes	6	J1	±370°
Payload	20 kg	J2	±260°
H-Reach	1811 mm	J3	±458°
Repeatability	±0.08 mm	J4	±400°
Mass	250 kg	J5	±380°
Structure	Articulated	J6	±900°
Mounting	Floor		

Fig. 5. Cranfield University FANUC robot setup (left) and the robot specifications (right)

is to test the DDS protocol over the university networks. Safety limits have been set to prevent the robot from colliding with the environment.

In the first instance, the robot is connected via Ethernet to a computer running the gateway application (Fig. 1), a static local IP address has been defined on the robot and on the computer to allow a ping command between the two. The computer is also connected to the university networks using Wi-Fi.

In a second step, at a remote place of Cranfield University a computer connected also to the university network is running the immersive UI application (Fig. 1) and is equipped with a VR HMD Varjo XR-3.

The two computers remotely located but on the same network have an XML file that describes the QoS profile of our DDS system including the 'initial peers list' (hostname of the two machines) as part of the participant phase of the discovery process. Other properties of the QoS profile are set, for instance the name participants ("Operator" and "Fanuc") and the buffers size limits. Our QoS profile is based on BEST_EFFORT_RELIABILITY_QOS which is good for periodic updates of sensor values, fast and efficiently with the least resource-intensive method. It gets the newest value for a topic from DataWriters to DataReaders (Fig. 1) without taking care of previous samples sent but it is not guaranteed that the data sent will be received. It may be lost due to connectivity issues or loss by the physical transport layer.

Finally, two Azure Kinect cameras are positioned around the robot, one targeting the table in front of the robot for the point cloud and the other targeting the robot itself for a wider view of the operating area, the two-camera run different configurations and are connected to the computer on the robot side.

The overall UI application can be started on any computer equipped with a VR HMD. Regarding the deployment of our framework on different robots, further work will be need. It is possible to keep the framework as it is for all FANUC robots only by importing and setting up the new robot CAD files in unity 3D. However, for different branded robot, the gateway application must be re-developed with the relevant libraries. Also, FANUC SDK is using a specific robot server program which must be installed to access robot controller and enable communication with the robot using TCP/IP and to the gateway application through an object (COM) interface.

4 Discussion

We developed a framework that offers simple teleoperation and trajectory creation of a FANUC industrial robot using telexistence capability and DDS-based. We focused on the interface using DDS which is an industrial protocol standard for scalable and real-time distributed systems. Our contributions include the development of the telexistence framework composed on the robot side a gateway application in C#.NET and on the remote operator side an immersive UI application in C# developed on Unity 3D game engine. This telexistence provide a safe immersive working environment because the robot and the human operator are interacting in a virtual environment, avoiding potentially harmful operations. Furthermore, the complexity of operating the robot in immersive environment can be evaluate and compare with traditional HMI or other 2D monitor based interface with our framework. Manipulating the 3D cursor in the immersive environment required less training and time than using the regular Fanuc HMI. This could be particularly advantageous when a maintenance operation must be carried out under tight delay. A further contribution is the development of the immersive digital twin for the robot set up in Unity 3D and in real-time with DDS. This is usually achieved through offline simulation software like ROBOGUIDE[3]. Lastly, we proposed a framework without the use of a Robot Operating System (ROS) which is most of the time used in a project involving robotics, especially for research, however, ROS2 has introduced the DDS protocol. We employed RTI Connext [12] which is a DDS distributor because they provide academics free of cost license and a C# SDK that is perfect for integration with Unity 3D as well as use with FANUC SDK.

When developing an immersive digital twin interface, difficulties remain with interoperability of commercial and open-source tools. The development is not straight forward, the FANUC SDK is for instance operating system dependent. The company does not provide official ROS packages; however unofficial packages exist using Karel for communication [13]. Garg et al. [14] have also recently developed a digital twin application of a FANUC robot using Karel for communication. Most of the research involving teleoperation of industrial robotics like FANUC is limited to simulation, we have implemented our framework in a real physical robot. Videos demonstrations are available[4].

Our DDS-based telexistence framework will be easy to update for other experiments, as all selected data structures are dynamically created at runtime in our applications have been created by ourselves. It has also shown reliable performance during our test using the university Networks. Further tests should be carried out on a wider network using a VPN in the future.

Limitations can also be highlighted already regarding the lack of accuracy in the VR environment. Compared to traditional interface in which user can enter precisely the value of the 3D position, immersive environment does not offer as high level of accuracy which could potentially influence operator ability to reach a certain position. To overcome these issues, we can hypothesize that different perspectives combine with the use of 3D data can improves the accuracy for complex remote tasks. Additional evaluations with participants including more complex tasks are required.

As a next step, the functionality of our framework will be enhanced and applied to an inspection maintenance scenario, developed in association with Dstl and with utility for deployed maintenance and repair operations. The 3D UI will be adapted to this use

[3] Offline motion and command simulation for FANUC branded robots.

[4] https://www.youtube.com/channel/UCyuSQ1JzesH9KpYrImPu4Fw.

case. Task performance assessment of telexistence for remote maintenance will also be performed.

5 Conclusion

The integration of telexistence capability for activities such as maintenance has the potential to offer a range of advantages such as enhanced visualizations, an enhanced sense of presence for a user operating in a remote location, as well as an intuitive interaction. It speeds up the handling of the robot without the need for extensive training with the regular FANUC HMI, we tested simple trajectory creation. The present developed framework shows the possibility of using a physical industrial robot with telexistence thanks to its immersive digital twin. The result is a hybrid immersive framework concept that allows an operator to interact with a robot normally surrounded by a fence and to program it with direct manipulation as if he were inside the fence near the robot, and this, in complete security and at a distance. Our framework is based on the Unity 3D game engine, a tool that is being more and more used in the industrial environment and for creating interactive 3D environments. Further research will be conducted to apply this framework in maintenance context tasks with user evaluation of the proposed UI, including time to complete the task, a functionality questionnaire, and a usability test.

Acknowledgment. This research is funded by the joint Anglo-French PhD scheme, organized by the respective nations' Ministry of Defence, with Dstl acting as the Technical Partner.

References

1. Rosen, R., von Wichert, G., Lo, G., Bettenhausen, K.D.: About the importance of autonomy and digital twins for the future of manufacturing. IFAC-PapersOnLine **48**, 567–572 (2015). https://doi.org/10.1016/j.ifacol.2015.06.141
2. Grieves, M., Vickers, J.: Digital twin: mitigating unpredictable, undesirable emergent behavior in complex systems. In: Kahlen, F.-J., Flumerfelt, S., Alves, A. (eds.) Transdisciplinary Perspectives on Complex Systems, pp. 85–113. Springer, Cham (2017). https://doi.org/10.1007/978-3-319-38756-7_4
3. El Saddik, A.: Digital twins: the convergence of multimedia technologies. IEEE Multimed. **25**(2), 87–92 (2018). https://doi.org/10.1109/MMUL.2018.023121167
4. Fung, W.K., Lo, W.T., Liu, Y.H., Xi, N.: A case study of 3D stereoscopic vs. 2D monoscopic tele-reality in real-time dexterous teleoperation. In: 2005 IEEE/RSJ International Conference on Intelligent Robots and Systems, IROS, pp. 181–186 (2005). https://doi.org/10.1109/IROS.2005.1545299
5. Linn, C., Bender, S., Prosser, J., Schmitt, K., Werth, D.: Virtual remote inspection - a new concept for virtual reality enhanced real-time maintenance. In: Proceedings of the 2017 23rd International Conference on Virtual Systems and Multimedia, VSMM, pp. 1–6 (2018). https://doi.org/10.1109/VSMM.2017.8346304
6. Masoni, R., et al.: Supporting remote maintenance in industry 4.0 through augmented reality. Procedia Manuf. **11**, 1296–1302 (2017). https://doi.org/10.1016/J.PROMFG.2017.07.257
7. Burdea, G.C.: Invited review: the synergy between virtual reality and robotics. IEEE Trans. Robot. Autom. **15**, 400–410 (1999). https://doi.org/10.1109/70.768174

8. Stotko, P., et al.: A VR system for immersive teleoperation and live exploration with a mobile robot. In: IEEE International Conference on Intelligent Robots and Systems, pp. 3630–3637 (2019).https://doi.org/10.1109/IROS40897.2019.8968598
9. Chen, J., Glover, M., Li, C., Yang, C.: Development of a user experience enhanced teleoperation approach. In: ICARM 2016 International Conference on Advanced Robotics and Mechatronics, pp. 171–177 (2016). https://doi.org/10.1109/ICARM.2016.7606914
10. Tachi, S.: Telexistence. In: Brunnett, G., Coquillart, S., van Liere, R., Welch, G., Váša, L. (eds.) Virtual Realities. LNCS, vol. 8844, pp. 229–259. Springer, Cham (2015). https://doi.org/10.1007/978-3-319-17043-5_13
11. Talaminos-Barroso, A., Estudillo-Valderrama, M.A., Roa, L.M., Reina-Tosina, J., Ortega-Ruiz, F.: A machine-to-machine protocol benchmark for eHealth applications – use case: respiratory rehabilitation. Comput. Methods Programs Biomed. **129**, 1–11 (2016). https://doi.org/10.1016/J.CMPB.2016.03.004
12. Connext Product Suite for Intelligent Distributed Systems. https://www.rti.com/products
13. Fanuc - ROS Wiki. http://wiki.ros.org/fanuc
14. Garg, G., Kuts, V., Anbarjafari, G.: Digital twin for Fanuc robots: industrial robot programming and simulation using virtual reality. Sustainability **13**, 10336 (2021). https://doi.org/10.3390/SU131810336

Predicting Car Sale Time with Data Analytics and Machine Learning

Hamid Ahaggach[1(✉)], Lylia Abrouk[1(✉)], Sebti Foufou[1(✉)], and Eric Lebon[2(✉)]

[1] LIB Laboratory, University of Burgundy, Dijon, France
{Hamid.ahaggach,lylia.abrouk,sfoufou}@u-bourgogne.fr
[2] Syartec, Aix en Provence, France
elebon@syartec.com

Abstract. There is no doubt that marketing is an important step in Product Lifecycle Management (PLM) and obviously decreasing time-to-market is crucial to reduce storage costs and increase profit. This paper aims to improve marketing strategies in the automotive field for car dealers and car selling supply chain. Due to the cost of new cars and the high risk of car value depreciation it becomes necessary for car dealers to know which type of cars can be sold faster than others, this will allow dealers to adapt their marketing strategies and satisfy the need of their customers. We propose to use data analysis and machine learning algorithms to address this problem and create models to help these companies in their decision-making processes. In our experiments, we used sale data from two big dealers of multi-maker cars. The dataset contains the sale history of around 73200 cars over a period of 8 years. We compared the different machine-learning algorithms and got promising results classifying cars into different predicted sale time ranges.

Keywords: Product Lifecycle Management · Data analysis · Machine learning

1 Introduction

Product Lifecycle Management (PLM) is an important strategy that helps companies in the representation, storage, and processing of product information throughout its lifecycle phases [1]. At every stage of the lifecycle, PLM solutions ensure integrated processing of all relevant information. Although the term PLM is much more used in the manufacturing sector than any other sector, PLM is expanding to cover more and more broader applications such as software industry and marketing strategies.

PLM organizes and integrates data to provide a detailed view of each product manufactured and how it is received in the marketplace, maximizing efficiency in the following areas [2]:

- Design and manufacturing integration: A company's production process can use a range of software applications for design and manufacturing. With PLM,

Published by Springer Nature Switzerland AG 2023
F. Noël et al. (Eds.): PLM 2022, IFIP AICT 667, pp. 399–409, 2023.
https://doi.org/10.1007/978-3-031-25182-5_39

the entire production process can be optimized in real time. Without PLM, valuable data may not be shared across all these systems and people.
- Virtual environments that support global operations: PLM provides a central repository for product data management (PDM) that integrates the entire global process from concept to customer.
- Accessible data: PLM makes all product information available to each department, improving production efficiency and allowing for a closed-loop for all teams involved.
- Product commercialization: PLM assures products are ready for global rollout, with effective, reliable data, document management and process governance. Unified data and collaborative workflows across the organization keep things running smoothly and allow teams to respond quickly when challenges arise.

In this paper, we consider the car commercialization aspect of PLM to help solving the inventory problems of car dealers. Indeed, car dealerships buy cars from the manufacturers and sell them to their customers to make profit. Obviously, dealers cannot just send the unsold cars back to the manufacturer. Keeping unsold cars in the parking lots for more than six months is extremely costly and may even threaten the financial prosperity of these companies. Therefore, they will have to find a way to get rid of these cars and get prepared to receive newer models, but this cannot be done without losing a lot of money. This paper proposes to use data analytics and machine learning to predict the time required to sell car models. Car properties and sales history are taken into consideration to compute ML models capable of making the correct prediction in most of the cases. The contributions of this paper can be summarized as follows: (i) using data analytics to find car properties with the highest influence on car sales, (ii) build several ML algorithms to predict car selling time, (iii) conduct experiments to test these algorithms, compare and discuss their results.

The rest of the paper is organized as follows. Section 2 presents the state of the art of using Machine-learning in PLM. Section 3 describes our method to improve marketing strategies in the automotive field for car dealers. Section 4 presents and discusses the experimental results obtained with the proposed Machine Learning based algorithms. Section 5 concludes the paper and gives few perspectives to extend and improve this work.

2 PLM and Machine Learning

As Big Data becomes more prevalent, Machine Learning (ML) is opening up new opportunities for data processing to support decision making in all areas of manufacturing, from customer engagement to design and supply chain to product lifecycle management. The combination of Big Data and advanced, low-cost computing systems has made machine learning viable in many real world work applications, e.g. cultural heritage [23], leading to positive changes in the product development lifecycle [3]. Today, companies are entering a new era of digital transformation in product lifecycle management (PLM) where ML is one of the enablers of forth Industrial Revolution (usually called *Industry 4.0*), which is almost twice as likely to be used as any other tool [3]. Currently, ML models used to enhance the PLM power in offering additional insights into the product lifecycle [3].

2.1 Machine Learning in Product Design

ML algorithms are used to support product design at least in two major aspects: analyze market trends and assist designers to achieve a fast and personalized design processes. The conceptual design stage is increasingly seen as an essential step in product development and customization. Effective conceptual design is inseparable from proper market investigation, as it has a critical impact on market prospects, customer acceptance and product lifecycle, among the works realized in this context we mention [4,5]. Market analysis aims to identify target customers, recognize their requirements, and translate these requirements into product features. Compared to manual market analysis, market research based on data mining can discover the implicit associations of market data by integrating ML and analytic algorithms such as Support vector machine [6], Apriori [7], ARIMA [8].

2.2 Machine Learning in Product Manufacturing

ML improves the product manufacturing framework, including material procurement, resource configuration, production scheduling, machining, assembly, quality control, warehousing, logistics, etc. ML contributes to improving the manufacturing phase of products in two ways: optimizing the information flow within the manufacturing processes, and monitoring smart devices to execute repetitive tasks. In addition, the integration of ML and production systems enables significant advances in human-machine collaboration, defect prediction, intelligent decision-making, and other aspects, Among the works realized in this context we mention intelligent supplier selection decision [9,10], Human-robot collaborative manufacturing control [11,12].

2.3 Machine Learning in Product Services

The combined application of Deep Learning and recommendation algorithms (e.g., Deep Neural Networks (DNN) and collaborative filtering) can improve the personalized decision level of recommendations by analyzing user and product group information associated with demand [14]. Moreover, the time series network can extract preference features based on the user's historical purchase behavior and make recommendations by modeling the content of the product sale network and user profile [15]. Moreover, product services need technical guidance for product maintenance according to consumer requirements. Traditionally, it is difficult for customers to perceive gradual changes in the condition, quality, and performance of products. Therefore, unnoticed product deterioration affects the user experience and reduces the quality of customer service. That is why product status monitoring is an important part of product lifecycle information processing as it is used to evaluate equipment performance and prevent product failures [16,17].

3 Methodology

This section presents the proposed solution to help car dealers in their effort to car inventory problems. As stated in the introduction section dealerships cannot afford keeping unsold cars in their parking lots for an extended period, so effective solutions must be offered to find out which cars could be sold within a short period of time, and which cars could require longer sale time. This classification will allow dealerships to put better marketing strategies for example reducing the number of car models that are not easy to sell.

Time series could be used as a tool to predict which cars will be sold in the next months, but this method is not practical in our case because the sales information, in the datasets we are using, do not follow a precise pattern. For example, if we take car C that was sold once since the company was established, where $\{c_1, c_2, \ldots, c_t\}$ are the monthly sale values, then most of the values are equal to zero. Given the sale values $\{c_s, \ldots, c_e\}$ over a period $[s, e]$, where c_s and c_e are respectively the sale value of the start and the end of the period, if we train a network based on sequential models like *(LSTM, RNN ...)*, or based on a statistical analysis model like *ARIMA* to predict $Y(t+1) = F(c_s, \ldots, c_e)$, the sale value for the next period $t+1$, then we will get wrong results. Therefore, we propose to use car characteristics to predict selling time. So we have dataset of pairs $D = \{(x_1, y_1), \ldots, (x_n, y_n)\}$ where $X = \{x_1, x_2, \ldots, x_n\}$ contains the characteristics of cars such as the color, the price, the power of the engine... and $Y = \{y_1, y_2, \ldots, y_t\}$ is the time taken to sell the vehicles (see Table 1). We need to find function $f(x_i) = y_i$ that will be able to predict the time needed to sell a car, this function can be any machine-learning algorithm. As shown in Fig. 1 the proposed system operates in three steps: (i) data processing, (ii) dimensionality reduction, and (iii) model training. In the next sections, we will describe the dataset, the machine learning models and highlight their similarities and their differences.

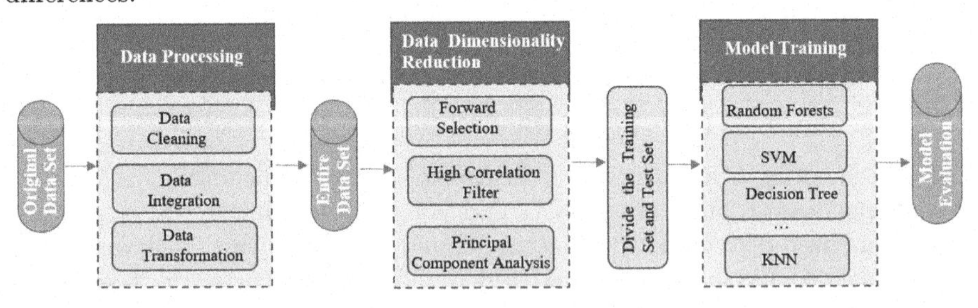

Fig. 1. The system architecture of the proposed solution

3.1 Dataset

The dataset used in this work was provided by two major multi-brand car dealerships in the European Union. The dataset covers the dealerships activities for a period of 8 years from Oct. 2013 to Nov. 2021. It contains more than 73200 data samples and 33 attributes for each data sample. A brief description of these attributes is given in Table 1.

Table 1. Attributes of car sales dataset.

Attributes	Description	Example
Label	Label of vehicle: 0: if selling the vehicle takes less than 3 months 1: if selling the vehicle takes between 3–6 months 2: if selling the vehicle takes between 6–8 months 3: if selling the vehicle takes between 8–12 months 4: if selling the vehicle takes more than 12 months	0
Entry_date	The vehicle entry date	2013-10-10
company_name	Name of company	Com1
Sale_date	The vehicle sale date	2013-12-02
Carrosserie	Vehicle bodywork	COUPE
C02	C02 emissions in g per km	185
Couleur_Vehic	Color of vehicle	GRIS F
Cylindree	The volume of gas that can be burned in the cylinders	1995
Date_1ere_Cir	Date of 1st entry into service	20040629
Depollution	Depollution device	OUI
Empat	Wheelbase of vehicle	273
Energie	Fuel type of vehicle	GAZOLE
Genre_V	Type of vehicle	VP
Immat	Registration of vehicle	BX368SQ
Marque	Marker of vehicle	BMW
Modele	Model of vehicle	SERIE 3
Nb_Cylind	Number of cylinder	4
Nb_Pl_Ass	Number of seats or payload	5
Nb_Portes	Number of doors	2
Nb_Soupapes	Number of valves	4
Nb_Vitesses	Number of speeds for manual gearboxes	5
Propulsion	Propulsion	ARRIERE
Puis_Ch	Real power in steam horsepower	150
Puis_Fisc	Fiscal power in fiscal horsepower	9
Tp_Boite_Vit	Gearbox type	B.V.A.
Turbo_Compr	Presence of turbo	TURBO
Version	Version of vehicle	320 CD
Code_Moteur	Motor code	204D4/M47TUD20
Cons_Urb	Urban consumption	9,7
Cons_Exurb	Extra-urban consumption	5,4
Cons_Mixte	Mixed consumption	6,9
Prix_Vehic	The selling price of the vehicle	28 800 €
C_code	If the vehicle is used or new	NV

3.2 Data Processing

This step is very important and sensitive because it directly affects the results of the model; in our case, we noticed missing data for several attributes. For example attributes such as *Couleur_Vehic* and *Cons_Exurb* are lucking about 5000 data, but luckily the missing data is not so important for all the attributes, e.g. There are less than 10 missing data for attributes *Tp_Boite_Vit*, *Energie* and *Code_Moteur*. Depending on the influence of different attributes on car sales and the amount of missing data, we treat the problem in different ways, for attributes with a high number of missing data and little impact on car sales, for example, *Color_Vehic*, we first group the data by *Label*, then fill each group's attribute by the mode value. For attributes with some missing data but having a significant impact on car sales, we design an accurate filling method. For anomalous data and missing data that cannot be filled, we opt for zero filling to minimize their impact on prediction accuracy. In addition, some data elements are anomalous due to possible recording errors and must be filtered out. We also noticed the presence of data of the same type but written in different formats for example *Sale_date* and *Entry_date* written respectively in *YYYY-MM-DD* and *MM/DD/YY* formats. A data integration step is therefore required to address these kinds of discrepancies and bring attributes of the same type into a unique format. In addition, non-numerical data cannot be used directly for prediction as the same information is rarely expressed in a unique way e.g. passenger front door, passenger side door, front right door, right front door are all used to refer to the same information, therefore, it is essential to transform the non-numeric data into numeric data in a way that best preserves the information and facilitates feature extraction.

3.3 Dimensionality Reduction

With such a large amount of data, the variety of attributes and their formats, we must carefully select the intrinsic features to achieve a high prediction accuracy while reducing computation costs. We also need to apply a dimensionality reduction method to allow visualizing the data in a reduced space. We distinguish two classes of dimensionality reduction methods: Methods of the first class keep only the most important features in a dataset and eliminate the rest; in this case, the features are not transformed. Backward elimination, Forward selection and Random forests are examples of this method. Method of the second type find a new combination of features, in this case, features are transformed, and the new set of features contains different values instead of the original values. We can divide this class further into linear and non-linear methods. The Principal Component Analysis (PCA) and Linear Discriminant Analysis (LDA) are examples of linear dimensionality reduction methods. t-distributed Stochastic Neighbor Embedding (t-SNE) is an example of non-linear dimensionality reduction methods, In Fig. 2, we display the data of new vehicles of the first company in 2D space with PCA.

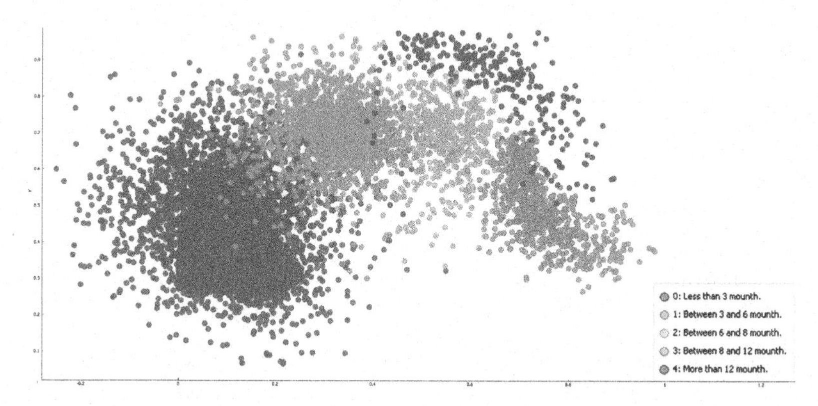

Fig. 2. Scatter plot of data of the new vehicles for the first company in 2D space after PCA dimensionality reduction.

3.4 Machine Learning Algorithms

To predict whether a car would, or would not, be sold in a predetermined period of time we use classification algorithms. To this end, the different classes of cars are identified, and a label is assigned to each car in the dataset to mark to which class it belongs. This type of learning is named "supervised" because we give our algorithm the data with their labels to learn, and once the model is learned, it will be able to predict the label of a car never seen before. Among few others, we have tried the below classification algorithms.

Support Vector Machines (SVM) is the best-known form of kernel methods inspired by Vladimir Vapnik's statistical theory of learning. SVM is a method of classification by supervised learning introduced by Vapnik in 1995 [18]. This method searches for the hyperplane that separates the positive data samples from the negative ones, ensuring that the margin between the closest positive and negative is maximal. This ensures a generalization of the principle because new examples may not be too similar to those used to find the hyperplane but may be located on one side or the other of the border. The interest in this method is the selection of support vectors, which represent the discriminant vectors by which the hyperplane is determined. The examples used during the search for the hyperplane are then no longer useful and only these support vectors are used to classify a new data sample, which can be considered as an advantage for this method.

Decision trees (DT) is a supervised learning technique that can be used for classification and regression problems, it is called a decision tree because, similar to a tree, it starts with the root node, which grows on other branches and builds a tree structure. In a decision tree, there are two type of nodes, which are the decision nodes and the leaf nodes. The decision nodes are used to make any decision and have several branches, while the leaf nodes are the output of these decisions and do not contain other branches. To build a tree and find out the attribute that should be selected for its initialization one can use the CART algorithm [19], which is based on the Gini index, or the ID3 or C4.5 algorithms, which are both based on the notion of entropy [20,21].

Random forests (RF) are a combination of tree predictors such that each tree depends on the values of a random vector sampled independently and with the same distribution for all trees in the forest [22]. The generalization error for forests converges to a limit as the number of trees in the forest becomes large. The generalization error of a forest of tree classifiers depends on the strength of the individual trees in the forest and the correlation between them.

K-Nearest Neighbors (KNN) is one of the simplest machine learning algorithms, based on the supervised learning technique. KNN stores all available data and classifies a new data point based on similarity. This means that when a new data point appears, it can be easily classified into a well-fitting category using the KNN algorithm.

4 Experimental Results

To test the above algorithms and compare their prediction results, we use a large-scale dataset provided by two car dealership companies covering their car sale activities for the period between Oct. 2013 and Nov. 2021. The dataset has 33 attributes and more than 73200 entries. The dataset of the first company contains 40700 among these cars there are 18800 new cars and 21900 used cars, and for the second company there are in total 32500 cars among which there are 18700 new cars and 14000 used cars. Our goal is to predict the time margin that a car will stay in stock before being sold, we will build two models for each company, one model for used cars and the other for new cars. The dataset is randomly split into two parts: training set (80% of the dataset), is used to train and test set (20% of the dataset) to evaluate our model, during the train we validate our training process using 10-Folds cross-validation. The accuracy and training-time are considered as a comparison criterion between algorithms on the test set. The accuracy in our case is defined as follows:

$$Accuracy = The\ number\ of\ well\text{-}classified\ cars / The\ total\ number\ of\ cars \quad (1)$$

The results of our experiments are reported in Table 2. To train our models we used a Dell OptiPlex-7070 computer with Intel(R) Core(TM) i7-9700 @ 3.00GHz 8-Core CPU and 16GB DDR4 RAM on Windows 10 Pro 64-bit.

Table 2. Results of the prediction on the datasets of the two companies.

Company	Vehicle type	Metrics	Models			
			KNN	*SVM*	*DT*	*RF*
1	VN	Accuracy	0.971	0.951	**0.990**	**0.990**
		Training time	**0.096 s**	8.431 s	0.996 s	0.140 s
	VO	Accuracy	0.849	0.854	0.814	**0.863**
		Training time	0.058 s	13.52 s	**0.057 s**	0.647 s
2	VN	Accuracy	0.967	0.944	0.987389	**0.994**
		Training time	**0.079 s**	20.76 s	0.140 s	1.145 s
	VO	Accuracy	0.845	0.862	0.802	**0.870**
		Training time	**0.066 s**	47.15 s	0.187 s	1.484 s

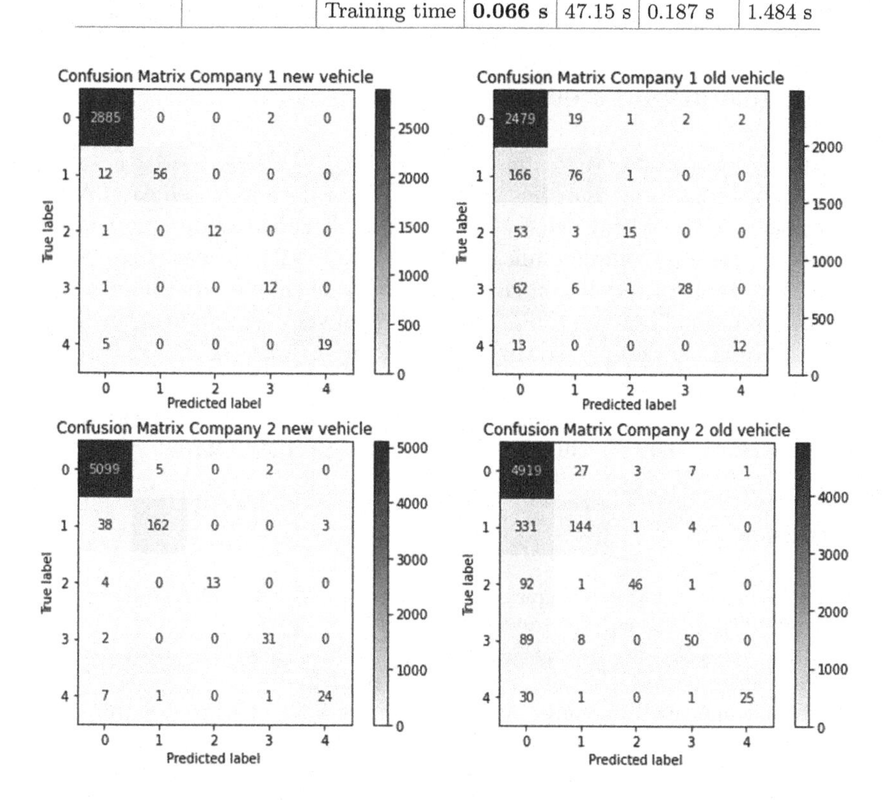

Fig. 3. Confusion matrix of the prediction on the two datasets

Table 2 shows that the random forest gives much better results in comparison with other models, and this is because RF is composed of several decision trees that collaborate with each other. In the case of the first company, both the DT and RF give the same accuracy score on a new vehicle because the data in this case are easy to be discriminated by the decision tree. We also note that KNN generally gives good results because it is based on data. SVM takes a lot of time to learn in comparison with other models because SVM tries to find a hyperplane that separates the margin between the nearest samples of each class, generally maximization problems take more time and depend on the performances of the machine used to train the model. In Fig. 3, the confusion matrix for each type of cars for the two companies with the model that we found the best score. One can see that the models are able to classify the cars in the correct category. There is a little confusion between class 0 and 1. That is because we can find two cars of the same type, one sold in three months and the other in three months and a day, for the model, they are two cars of two different classes. This kind of confusion are solvable with regression models: instead of predicting the sales time intervals we will predict the exact number of days to sell a car, but generally companies are not interested in predicting the exact day but the time interval of sales.

5 Conclusion and Perspectives

In this paper, we proposed to implement SVM, DT, KNN, and RF machine learning algorithms to predict the time required for dealers to sell cars. A large-scale car sales dataset provided by two multi-maker dealership companies has been pre-processed to complete missing data and identify the car characteristics that have the greatest impact on car sales. This sale time prediction gives companies better ideas about the commercialization of vehicles and hence help them putting the right marketing strategy to avoid buying cars that are not easy to sell. In future work, we intend to extend this work towards customer behavior analysis to build a recommendation system based on association rules, to target customers who can buy specific cars based on the profile of former customers.

References

1. What is product lifecycle management. https://www.sap.com/insights/what-is-product-lifecycle-management.html. Accessed 9 Dec 2021
2. Product lifecycle management. https://www.propelplm.com/articles/what-is-product-lifecycle-management. Accessed 2 Mar 2022
3. PLM and machine learning meet. https://www.aberdeen.com/featured/blog-when-plm-machine-learning-meet/. Accessed 29 Feb 2022
4. Hu, X., Hu, J., Peng, Y., Cao, Z.: Constrained functional knowledge modelling and clustering to support conceptual design. Proc. Inst. Mech. Eng. C J. Mech. Eng. Sci. **226**(5), 1326–1337 (2012)
5. Liu, X., Liu, H., Duan, H.: Particle swarm optimization based on dynamic niche technology with applications to conceptual design. Adv. Eng. Softw. **38**(10), 668–676 (2007)
6. Pal, R., Kupka, K., Aneja, A.P., Militky, J.: Business health characterization: a hybrid regression and support vector machine analysis. Expert Syst. Appl. **49**, 48–59 (2016)
7. Kumar, V.S., Renganathan, R., VijayaBanu, C., Ramya, I.: Consumer buying pattern analysis using apriori association rule. Int. J. Pure Appl. Math. **119**(7), 2341–2349 (2018)

8. Gurnani, M., Korke, Y., Shah, P., Udmale, S., Sambhe, V., Bhirud, S.: Forecasting of sales by using fusion of machine learning techniques. In: 2017 International Conference on Data Management, Analytics and Innovation (ICDMAI), pp. 93–101. IEEE (2017)
9. Keskin, G.A., İlhan, S., Özkan, C.: The Fuzzy ART algorithm: a categorization method for supplier evaluation and selection. Expert Syst. Appl. **37**(2), 1235–1240 (2010)
10. Parkouhi, S.V., Ghadikolaei, A.S.: A resilience approach for supplier selection: using fuzzy analytic network process and grey VIKOR techniques. J. Clean. Prod. **161**, 431–451 (2017)
11. Neto, P., Simão, M., Mendes, N., Safeea, M.: Gesture-based human-robot interaction for human assistance in manufacturing. Int. J. Adv. Manuf. Technol. **101**(1), 119–135 (2018). https://doi.org/10.1007/s00170-018-2788-x
12. Cwikla, G., Sekala, A., Wozniak, M.: The expert system supporting design of the manufacturing information acquisition system (MIAS) for production management. In: Advanced Materials Research, vol. 1036, pp. 852–857. Trans Tech Publications Ltd. (2014)
13. Zhang, J., Yang, Y., Zhuo, L., Tian, Q., Liang, X.: Personalized recommendation of social images by constructing a user interest tree with deep features and tag trees. IEEE Trans. Multimedia **21**(11), 2762–2775 (2019)
14. Lecouteux, B., Vacher, M., Portet, F.: Distant speech recognition in a smart home: comparison of several multisource ASRs in realistic conditions. In: Interspeech 2011 Florence, pp. 2273–2276 (2011)
15. Kulkarni, C.S., Bhavsar, A.U., Pingale, S.R., Kumbhar, S.S.: BANK CHAT BOT–an intelligent assistant system using NLP and machine learning. Int. Res. J. Eng. Technol. **4**(5), 2374–2377 (2017)
16. Shen, J., Wan, J., Lim, S.J., Yu, L.: Random-forest-based failure prediction for hard disk drives. Int. J. Distrib. Sens. Netw. **14**(11), 1550147718806480 (2018)
17. Kalsoom, A., Maqsood, M., Ghazanfar, M.A., Aadil, F., Rho, S.: A dimensionality reduction-based efficient software fault prediction using Fisher linear discriminant analysis (FLDA). J. Supercomput. **74**(9), 4568–4602 (2018). https://doi.org/10.1007/s11227-018-2326-5
18. Cortes, C., Vapnik, V.: Support-vector networks. Mach. Learn. **20**(3), 273–297 (1995). https://doi.org/10.1007/BF00994018
19. Breiman, L., Friedman, J., Olshen, R., Stone, C.: Cart. Classification and Regression Trees (1984)
20. Quinlan, J.R.: Induction of decision trees. Mach. Learn. **1**(1), 81–106 (1986). https://doi.org/10.1007/BF00116251
21. Salzberg, S.L.: C4. 5: programs for machine learning by J. Ross Quinlan. Morgan Kaufmann publishers, Inc., 1993 (1994)
22. Breiman, L.: Random forests. Mach. Learn. **45**(1), 5–32 (2001). https://doi.org/10.1023/A:1010933404324
23. Belhi, A., Bouras, A., Foufou, S.: Leveraging known data for missing label prediction in cultural heritage context. Appl. Sci. **8**(10), 1768 (2018). https://doi.org/10.3390/app8101768

Automatic Transformation of HVAC Diagrams into Machine-Readable Format

Noah Mertens(✉), Tommy Wohlfahrt, Nick Hartmann, and Chethan Babu Venkata Reddy

Fraunhofer IIS EAS, Dresden, Germany
noah.mertens@eas.iis.fraunhofer.de
https://www.eas.iis.fraunhofer.de/

Abstract. Accounting for around 40% of total energy consumption, buildings in industrialized countries are a particular focus for saving energy. Especially digital twins and building information modeling (BIM) can serve as a mean to improve operational processes, detect faulty ventilation and air conditioning (HVAC) components which lead to an increased energy consumption and evaluate future renovation possibilities during a buildings life cycle. To enable these advantages a transition from existing HVAC diagrams to a machine-readable representation including spatial and functional interrelationship of all components as well as their semantic relationship is necessary. Especially in retrofit cases the manual transition is a time-intensive and an error prone process. The present work aims to develop a procedure for the (partially) automated recognition of HVAC diagrams and extraction of information about their intercorrelation. In order to achieve the desired results, we use multiple approaches of computer vision, both conventional as well as more recent ones using artificial intelligence such as "Faster R-CNN". The developed pipeline consists of three basic steps: Find symbols of components, find connecting lines, combine the extracted information into a machine-readable representation. Due to privacy issues, only very few real diagrams are available. To evaluate our approach we developed a data generator to train and test our pipeline. The obtained results show that to a large extend it is already possible to transfer relevant information from technical diagrams into a machine-readable format in order to reduce the effort of creating and validating digital twins and BIM for retrofits.

Keywords: Ontology · HVAC diagram recognition · Machine learning · Computer vision · Building information modelling

1 Introduction and Problem

In industrialized countries the building sector is accountable for around 40% of the total energy consumption. Therefore it becomes a particular focus for saving energy and reducing its environmental impacts. In order to attain a more

F. Noël et al. (Eds.): PLM 2022, IFIP AICT 667, pp. 410–419, 2023.
https://doi.org/10.1007/978-3-031-25182-5_40

efficient use of resources [12], cost and time reduction [11] during a buildings lifecycle, multiple stakeholders like architecture, engineering, construction and facility management (FM) depend on the exchange of information across organizational boundaries. Especially digital twins and building information modeling (BIM) can serve as a mean to improve operational processes, detect faulty HVAC components which lead to an increased energy consumption and evaluate future renovation possibilities during a buildings life cycle. To enable these advantages, a transition from existing HVAC diagrams to a machine-readable representation including spatial and functional interrelationship of all components as well as their semantic relationship is necessary. Creating digital twins and implementing BIM into the daily workflow is a difficult challenge, especially in retro fit cases where documentation exists mostly in a non-machine-readable format.

The present work aims to develop a procedure for the (partially) automated recognition of HVAC diagrams and extraction of information about their intercorrelation for real world applications. In previous research on this topic a multistep approach was developed [8]. It contains the recognition of only four types of symbols (damper, heat exchanger, pump, valve), as well as the recognition of connecting lines and a basic identification of line intersections. Although it is a promising approach, it lacks important characteristics to apply for real world use-cases. Especially the number of symbols, but also the capability and robustness of line and intersection detection are the main limiting factors. Also commonly seen problems with noise such as manually added text, folding lines or dirt on scanned images are not taken into consideration. Even though, it created a starting point for the development of our pipeline, which mainly focuses on the improvement of robustness of the overall process and applicability for real use-cases.

Our developed pipeline consists of three basic steps, which are discussed more detailed in Sect. 4:

1. Find symbols of components
2. Find connecting lines (including intersections)
3. Combine the extracted information into a machine-readable representation

Our approach could reduce the effort for creation and validation of digital twins and BIM significantly by providing a promising starting point for real world applications as well as further research towards a fully automated process.

At the time of this research we could not identify a suitable public data set containing HVAC diagrams. In order to develop and evaluate our procedure we created a comprehensive data generator, which is described in Sect. 3. It includes multiple kind of symbols, combinations of components and line intersections as well as different kind noise. In Sect. 4 our approach is described in detail and the achieved results are discussed in Sect. 5.

2 Machine-Readable Representation

A digital representation of the HVAC diagram contains all relevant technical components, as well as their connections. Even though most existing semantic ontologies in the building sector put their focus on different goals [5], they are all based on the idea of graph structures. They consist of three concepts: Classes, relationship and attributes. To allow a flexible deployment of the extracted information for streamline applications we decided to use a basic graph approach. A conversion to more domain-specific ontology or API-friendly formats for further usage like creating a digital twin or implementing BIM is easily possible. Furthermore, by remaining on a high-level graph structure we avoid limitations of more specific ontologies such as limited types of components or attributes.

Graphs capture interactions (edges) between units (nodes) which allows storing and accessing relational knowledge about interacting entities [2]. The graph represents all relevant components as nodes and their connecting lines as edges. The developed process for line detection provides a robust way to describe the structure unequivocally. As of now the graph is undirected and therefore contains no information about the direction of flow inside the system.

3 Data Generator

3.1 Data Structure

Due to privacy aspects, the number of publicly available HVAC diagrams is limited. For the purpose of evaluation and testing of our approach a data generator was created. The most common components of HVAC systems are grouped to a pool which can be extended with new symbols at any time. Out of this pool, the generator chooses randomly symbols and connection line colors to generate the diagram. The aim of the data generation is the optimization and evaluation of the overall approach, hence the technical usefulness of the generated diagrams are not always given. Therefore it not only creates the diagram, but also the corresponding ground truth information. The generation can be divided into the following steps.

Basic Shape. A NxM grid structure forms the base graph of every diagram, where a node represents a component and the edges represent the connection lines between them (cf. Fig. 1a). The size and shape of the graph are freely selectable.

Removing Edges. A parameter determines the number of randomly selected edges (connecting lines) to be removed in the graph (cf. Fig. 1b). Therefore not all nodes contain exactly 4 connections and thus represent a more realistic diagram structure. If there are several completely separated graphs after this step, only the graph with the most nodes is used for the further procedure.

Create Connection Nodes. In order to realize L-, T- and 4-way intersections in the diagram, a parameter determines how many nodes represent only a intersection instead of a component. In the case of a 4-way intersection, the decision for a crossing without a connection or a connection of all four edges is taken randomly. In Fig. 1b the connection nodes are marked with a C.

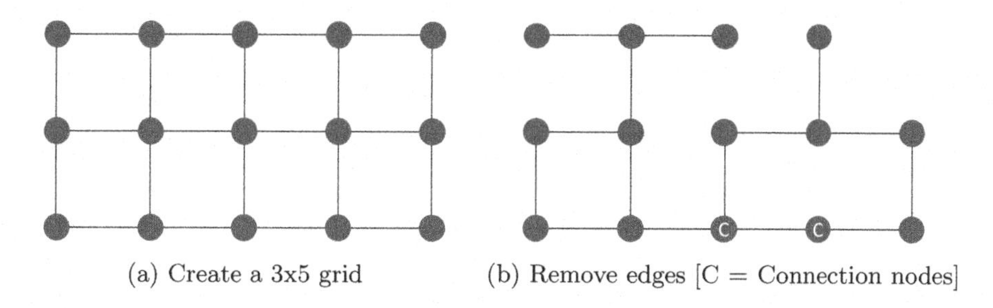

(a) Create a 3x5 grid (b) Remove edges [C = Connection nodes]

Fig. 1. Data generation

Add Components. Each node of the graph (which does not represent a intersection node) is assigned to a component. The random selection of components can be restricted by the number of edges of a node (e.g. heat exchanger are usually supplied with 4 connections).

Derive Diagram from Graph. The diagram is generated from the created graph. The selected symbols are drawn on a white background and the connection lines between the symbols are added, whereas each symbol size is randomly scaled (cf. Fig. 2).

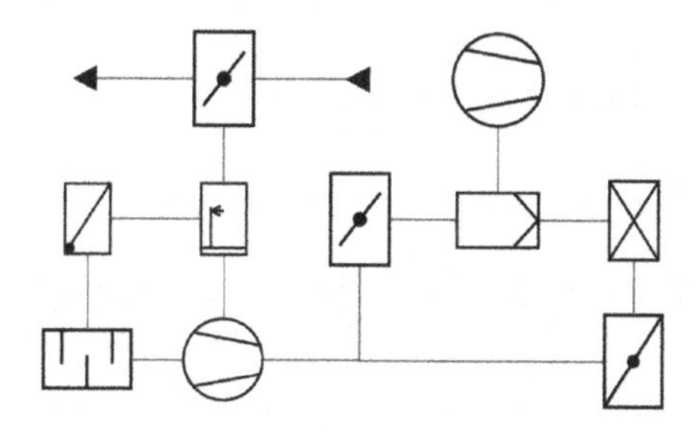

Fig. 2. A generated diagram based on a 3×5 grid without noise

3.2 Noise

Discussions with domain experts as well as the review of provided real HVAC diagrams show that the quality varies greatly and various interfering factors can occur. Digitized building plans can be available as scans, photographs or even as hand drawings. In addition to reduced resolution, scanned and photographed images often contain folded edges, slight noise or rotation, distortion or information added by hand. In order to simulate these issues in our synthetically generated diagrams and to maximize the robustness of the application, several disturbance factors were implemented as filter and can be added to the diagrams.

A random line filter, for example, represents the folded edges and inserted rectangles the grouping of strongly interacting system components. Slight rotations of the input diagram, colored backgrounds and connection lines and general noise like Gaussian or Jpeg compression are only a few more possible filters. An example of a diagram with noise is shown in Fig. 3.

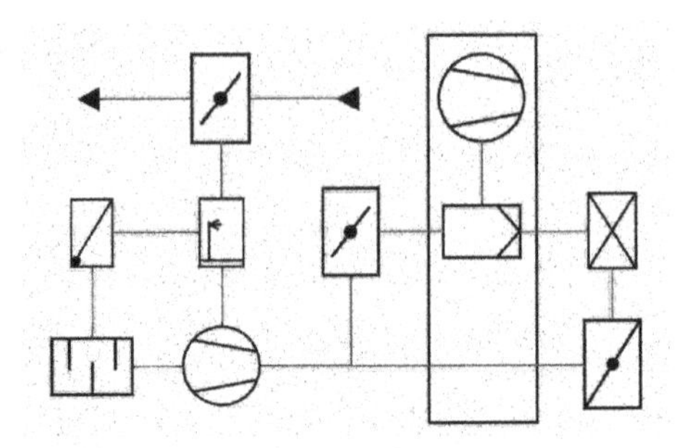

Fig. 3. A generated diagram based on a 3×5 grid with noise

Besides the changeable diagram size and multiple intersection types, especially the variation of the symbol size within a diagram and the addable noise should increase the robustness of the object detection algorithm.

4 Developed Approach

4.1 Symbol Detection

Template Matching. A conventional and still common approach for detecting objects in images or videos is Template Matching (TM) [9,10]. A given template image is compared with a section of the same size as the template in a larger input image and the areas most similar to the template are identified. TM is characterized by a simple algorithm and requires little computational power for simple applications. Initial tests and results can be implemented and evaluated within a few lines of code using Python libraries like OpenCV [3]. With simple geometric shapes (e.g. squares, circles), the TM provides good results but difficulties arise even with even small differences (e.g. aspect ratio) between the object and the template. The scaling or rotation of the object can also lead to a lack of correspondence between template and object and thus strongly influences the object recognition results.

A brief manual evaluation of the TM approach was conducted for the generated, as well for real world HVAC diagrams. In the case of the generated diagrams with several superimposed noise filters and varying component sizes, almost all of the components were recognized. In contrast, the TM approach for components of real world diagrams has not delivered satisfying results. The very few existing real diagrams demonstrate, that even for diagrams from the same manufacturer there are small variations between similar components regarding the aspect ratio. Even an optimization approach to rotate the templates and vary the aspect ratio of the templates slightly did not lead to better results. Driven by the poor performance in real world applications the focus was switched on a Deep Learning approach using *Faster R-CNN* as described in the following.

Faster R-CNN. Deep Learning approaches are state of the art for many complex computer vision problems such as image segmentation, object detection and classification. This is possible due to the level of abstraction achieved by those models, which not only improves the robustness of a system, but also allows a great transferability between different domains of application. In order to locate and classify components within the HVAC diagram, we have selected a commonly used architecture of deep neural networks called *Faster R-CNN* [7]. In our context it is used to fulfill two tasks: perform object detection (create bounding boxes) and classification of objects (e.g. pump, valve, etc.). On an abstract level the network architecture is composed of three neural networks [4,7]:

1. *Feature Network* to generate representing features from the image
2. *Region Proposal Network (RPN)* to generate bounding boxes for region of interest (ROI)
3. *Detection Network* to inference object class for each ROI

The model is trained with data from our data generator (cf. Sect. 3) and the characteristics in Table 1.

Table 1. Characteristics of generated dataset and model training

Number of images (train)	1000
Number of images (test)	500
Symbol pool size	18 unique components (e.g. valve, damper)
Grid size for symbol placement	Height: 3–5, width: 3–5
Image resolution [px]	Height: 600–900, width: 600–900
Batch size	4
Base learning rate	0.001
Learning rate decay	0.05
Epochs	2000

For inference a single HVAC diagram is fed into the network, which returns the results in the Common Objects in Context (COCO) data format [6]. It contains the information of detected components as well as the corresponding bounding boxes in form of a defined JSON-format.

4.2 Line Detection

To create a robust line detection we decided to follow a multi-steps approach as described in the following.

Preprocessing. Real world diagrams can have all kinds of noise and artifacts. To improve the robustness of the line detection the first step is to slightly blur the image. In a second step we apply object detection to get the bounding boxes of all symbols. Subsequently the bounding boxes are replaced by a white rectangle with a black border. By replacing the original components we reduce the complexity of the image in terms of number of lines and receive a "skeleton" of the diagram. In the next step the image gets gray-scaled and binarized.

Processing. From the binarized image a base graph is created using a skeleton network [1] which subsequently gets corrected and improved by multiple postprocessing steps as described in the following.

Postprocessing. The base graph can contain several incorrect connections and is cleaned based on different rules. Some edges have the same start and ending nodes, which has no meaning in our context and thus they are removed. Sometimes multiple nodes are found very close or within the region of a detected symbol (bounding box). Adding a margin in all directions of the bounding box, these nodes are united into a single node in the center of the bounding box. The new node is named based on the symbol type and position to match the naming schema of the data generator. Subsequently all nodes that do not represent components are checked if they are either 3- or 4-way connections and the node names are updated accordingly to match the data generator naming schema. In a final step all two-way connections are removed from the graph, as they don't offer important information.

5 Results

The evaluation of results is performed on the final graph, which describes the connection of all detected components within the processed diagram. As a metric we use the *jaccard similarity* in order to describe how similar the created graph and the ground truth graph (derived from data generator) are. Given two sets, A and B, the *jaccard similarity* is defined as the size of the intersection of set A and set B (i.e. the number of common elements) over the size of the union of set A and set B (i.e. the number of unique elements). Using graphs, either the edges or the nodes can considered to be the elements, whereas each edge represents a connection between two components and nodes represent symbols. Two edges are considered "equal" if they share the same starting and endpoint. The *jaccard similarity* returns a value between 0% and 100% and can thus be easily interpreted.

The evaluation is performed with 500 test diagrams produced by the data generator. Each diagram is processed individually and the *jaccard similarity* is calculated. As it can be seen in Table 2, a distinction is made between the evaluation of nodes (symbols) and edges (connections) and whether the whole diagram is correctly recognized or not. Out of 500 diagrams, almost 98.5% were recognized completely correct, meaning no manual effort is needed for correction. Even in the case of incorrect detected diagrams, on average most symbols (83%) and connections (79%) are recognized and only little manual editing would be required for correction.

Table 2. Evaluation for 500 generated test diagrams

	Correct detected	Incorrect detected
Total number	492 (98,4%)	8 (1,6%)
Jaccard similarity nodes [%]	100	82,75
Jaccard similarity edges [%]	100	79,38

For errors in the final diagram we identified two possible reasons for which a description and an example is given in the following. The data generator creates recursive connections (connecting a component to itself, cf. Fig. 4a), a scenario which we most probably won't face in real applications and therefore no further efforts were made to solve the problem. Another possible reason for errors can be found in the usage of the Faster R-CNN for symbol detection. All input images are resized to a defined resolution before the symbol detection is applied. For big images (high resolution) which contain small symbols (e.g. compressor), the

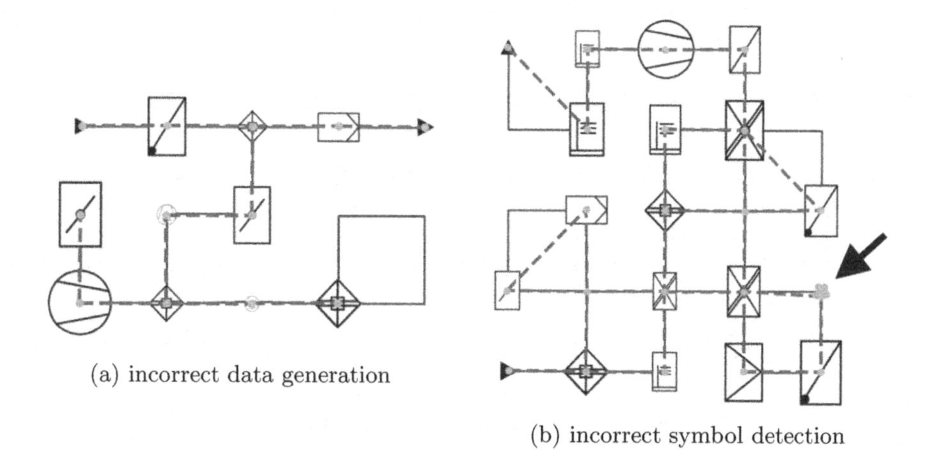

Fig. 4. Example of detected errors - - - - - - detected lines (graph edges) • detected symbols (graph nodes) ——— real connection lines (Color figure online)

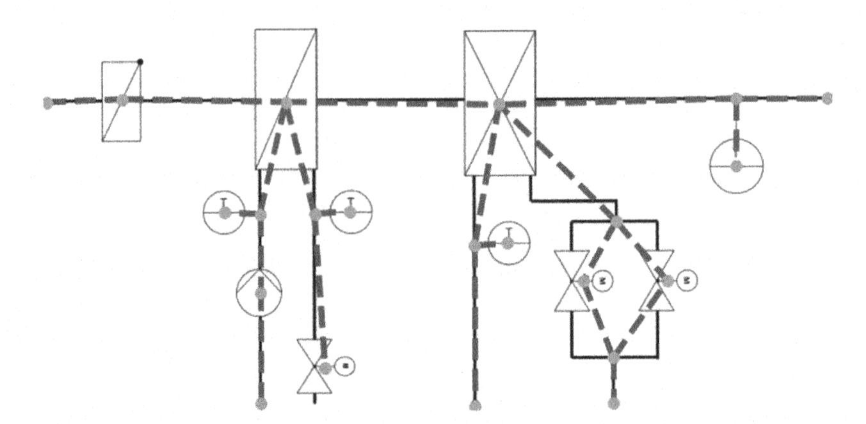

Fig. 5. Results with real diagram - - - - - - detected lines (graph edges) • detected symbols (graph nodes) ——— real connection lines (Color figure online)

resizing removes information to such an extend, that the model can't detect the symbol anymore. For the given example in Fig. 4b the original symbol for the compressor is hidden underneath multiple (incorrectly) detected graph nodes.

Additionally to the automated evaluation of generated data we manually inspected the results for a real diagram. Text elements were removed before processing, but no other manual steps were conducted. The final graph in Fig. 5 seems to represent the diagram almost completely correct. Reviewing the detected symbols we found only a single incorrect match: The heater was classified as a damper. Taking into consideration the probabilities, a clear trend can be observed: All correct symbols were detected with a probability over 95%, whereas the incorrectly classified damper reaches less than 75%.

6 Conclusion

The developed approach for symbol and line detection achieves very good results with diagrams created by the data generator and converts them into a machine-readable graph. Using graphs enables a flexible utilization for downstream applications. The achieved results with real world diagrams the developed approach are promising to significantly reduce the amount of work of implementing BIM and creating digital twins during a buildings lifecycle, especially in retrofit cases. Despite the promising results, the applicability for real diagrams has not yet been extensively tested. Especially in cases of real diagrams with a high resolution the symbol detection becomes a bottleneck in our developed approach. Although it is a first step towards an automated system, in future research additional aspects need to be put into consideration:

- Tackle detection problems of small symbols in large diagrams with a stepwise approach using a sliding window

- Improve the automation of the overall process by increasing the number of symbols and using data augmentation for the data generator
- Implement optical character recognition (OCR) to extract text information such as datapoint names (enables automatic datapoint mapping for building automation systems)

Acknowledgements. Supported by: BMWi (Federal Ministry for Economic Affairs and Energy), promotional reference 03ET1567A.

References

1. Skeleton network. https://github.com/Image-Py/sknw
2. Angles, R., Gutierrez, C.: Survey of graph database models. ACM Comput. Surv. **40**(1), 1–39 (2008). https://doi.org/10.1145/1322432.1322433
3. Bradski, G.: The openCV library. Dr. Dobb's J. Softw. Tools Prof. Program. **25**(11), 120–123 (2000)
4. Chen, X., Gupta, A.: An implementation of faster RCNN with study for region sampling. arXiv preprint arXiv:1702.02138 (2017)
5. Gilani, S., Quinn, C., McArthur, J.J.: A review of ontologies for smart and continuous commissioning. arXiv preprint arXiv:2205.07636 (2022)
6. Lin, T.-Y., et al.: Microsoft COCO: common objects in context. In: Fleet, David, Pajdla, Tomas, Schiele, Bernt, Tuytelaars, Tinne (eds.) ECCV 2014. LNCS, vol. 8693, pp. 740–755. Springer, Cham (2014). https://doi.org/10.1007/978-3-319-10602-1_48
7. Ren, S., He, K., Girshick, R., Sun, J.: Faster r-CNN: towards real-time object detection with region proposal networks. Adv. Neural Inf. Process. Syst. **28** (2015). https://doi.org/10.48550/ARXIV.1506.01497
8. Stinner, F., Wiecek, M., Baranski, M., Kümpel, A., Müller, D.: Automatic digital twin data model generation of building energy systems from piping and instrumentation diagrams. arXiv preprint arXiv:2108.13912 (2021)
9. Sun, Y., Mao, X., Hong, S., Xu, W., Gui, G.: Template matching-based method for intelligent invoice information identification. IEEE Access **7**, 28392–28401 (2019). https://doi.org/10.1109/ACCESS.2019.2901943
10. Thomas, L.S.V., Gehrig, J.: Multi-template matching: a versatile tool for object-localization in microscopy images. BMC Bioinform. **21**(1), 1–8 (2020). https://doi.org/10.1186/s12859-020-3363-7
11. Vanlande, R., Nicolle, C., Cruz, C.: IFC and building lifecycle management. Autom. Constr. **18**(1), 70–78 (2008). https://doi.org/10.1016/j.autcon.2008.05.001
12. Volk, R., Stengel, J., Schultmann, F.: Building information modeling (BIM) for existing buildings—literature review and future needs. Autom. Constr. **38**, 109–127 (2014). https://doi.org/10.1016/j.autcon.2013.10.023

Intent Detection for Virtual Reality Architectural Design

Romain Guillaume[1,2](✉), Jérôme Pailhès[1], Elise Gruhier[1], Xavier Laville[2], Yvan Baudin[2], and Ruding Lou[3]

[1] Institut de Mécanique et d'Ingénierie (I2M), Avenue d'Aquitaine, 33170 Gradignan, France
romain.guillaume.2015@gadz.org

[2] Airbus Operations SAS, 316 Route de Bayonne, 31027 Toulouse, France

[3] Arts et Métiers Institute of Technology LISPEN, HESAM University, UBFC, 71100 Châlon-Sur-Saône, France

Abstract. In the context of optimization and cycles reduction for product design in industry, digital collaborative tools have a major impact, allowing an early stage integration of multidisciplinary challenges and oftentimes the search of global optimum rather than domain specific improvements. This paper presents a methodology for improving participants' implication and performance during collaborative design sessions through virtual reality (VR) tools, thanks to intention detection through body language interpretation. A prototype of the methodology is being implemented based on an existing VR aided design tool called DragonFly developed by Airbus. In what follows we will first discuss the choice of the different biological inputs for our purpose, and how to merge these multimodal inputs a meaningful way. Thus, we obtain a rich representation of the body language expression, suitable to recognize the actions wanted by the user and their related parameters. We will then show that this solution has been designed for fast training thanks to a majority of unsupervised training and existing pre-trained models, and for fast evolution thanks to the modularity of the architecture.

Keywords: Virtual reality · Machine learning · Body language · Intent detection · Computer-aided design

1 Introduction

Products complexification and the need for shorter time to market through a reduction of design loops foster the development of methods and tools enhancing product visualization and cross-disciplinary collaboration. Airbus has thus developed DragonFly – an internal virtual reality (VR) tool – to take advantage of an immersive environment at scale. This tool is primarily suitable for such architectural design tasks like space allocation and design reviews. Although efforts have been made to improve the interface, many experts refuse to learn to use a new design tool often associated to hard-to-master interfaces. We aim at fostering the adoption of the software by decreasing related learning phases. We plan to infer the actions intended by the users thanks to the analysis

F. Noël et al. (Eds.): PLM 2022, IFIP AICT 667, pp. 420–430, 2023.
https://doi.org/10.1007/978-3-031-25182-5_41

of their body language while minimizing VR equipment intrusiveness. More precisely, due to the availability of different motion capture systems and due to the low equipment intrusiveness constraint, we focus in what follow on the language and posture analysis.

In this paper, we present a conceptual framework for a natural body language understanding specialized in the activities realized in DragonFly. Literature provides methods for inferences based on natural language and natural gesture separately [1], but also propose methods to synchronize and find relations between time series of different natures [2] – e.g. audio and video. Literature finally propose empirical descriptions of high level statistical analysis of body language (gesture and voice) [3]. Our proposition fills the gap between language and posture analysis, thus presenting a rich representation of the body language of a person over time and taking into account relationships between both inputs. In a first approach, we make the hypothesis that the action intended by a user can be inferred independently from any environmental variables and so that the information retrieved from the user body language is sufficient for this task.

2 Related Work

In this section, we discuss the different existing methods for body language understanding and the biological variables chosen in literature. More precisely, we focus on the body language empirical analysis, on its links with design actions in a virtual reality environment and we will show that a simple statistical analysis based on high-level features is not sufficient. However, we will see how the addition of a temporal component could help at eliciting simple user state of action in a context of pure gesture analysis. On the other hand, present how to handle speech through widely used "Natural Language Understanding" techniques but also how to identify entities – parameters of an action – in a sequence. Then, we tackle the multimodality of the input analyzing existing methods bringing together inputs of different natures. Finally we highlight existing solutions dealing with limited labelled datasets.

2.1 Body Language Empiric Analysis

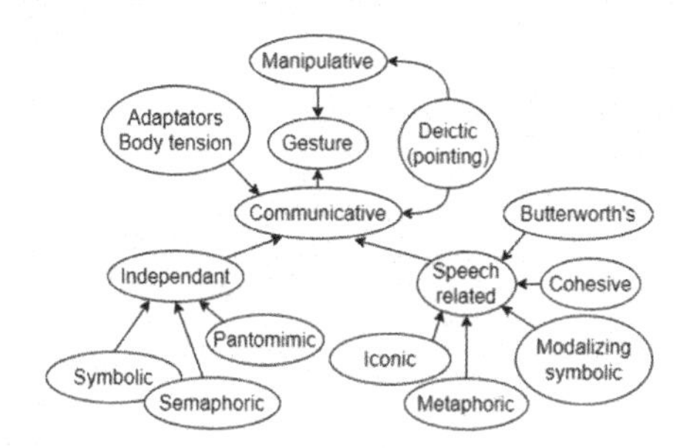

Fig. 1. Gesture classification (Vuletic T. et al. 2019)

Gestures are aimed at two purposes: manipulate, and communicate [4] (see Fig. 1). The first category includes the movements performed when interacting directly with a physical, virtual or imaginary object generally with our hands whereas the second one relates to information sharing. More precisely, in this classification, the speech-related gestures are linked to communication; but the "modeling symbolic" gestures can carry meaningful element from a manipulative point of view when the speech itself carry such a manipulative message such as "draw a ball this big" spacing hands to a precise distance from each other thus suggesting the size of the ball. Moreover, a Computer Aided Design (CAD) task can be naturally performed with gesture different ways (see Fig. 2) [3] and some gesture patterns can correspond to multiple actions. This puts emphasis on the need for additional information to identify the action. Nevertheless, we can observe that metrics about the symmetricity of the arm's activity (not clearly defined in literature) and also the hand posture often give characteristic distributions (see Fig. 2 for arm's activity) for the different gesture patterns according to each action to be performed. This tends to show that these metrics are good discriminators for our purpose, even though they are not sufficient. Finally, the most relevant descriptors for hand posture are geometrical (articulation angle) and topological (pinch) [5].

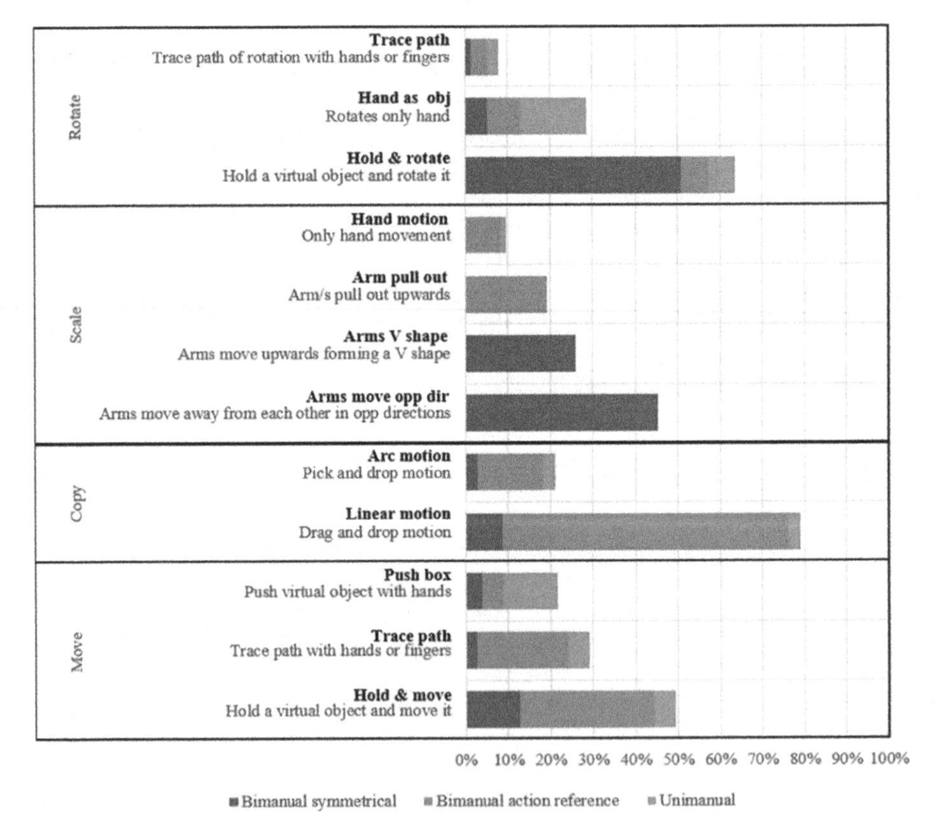

Fig. 2. Themes in manipulation groups (Khan S. and Tunçer B., 2019)

Besides, the distinction between simple user states (for instance waiting for a drink or in our case manipulating an object) can be deduced from other parameters: head orientation, body trunk inclination and the relative position of the user from others [1]. More precisely, the trunk inclination appears to be correlated to the implication of the user and to his will to interact. Finally, the addition of voice to the gesture ease the understanding of the actions to be performed especially for non-tech users [3].

2.2 Sequence Analysis

As seen in sub Sect. (2.1), the static analysis of video and audio inputs seems to be insufficient for a good action inference. Literature proposes methods to add a temporal component to this analysis. Let's first consider sequence labelling, consisting of properties discrimination for each element of a sequence (for instance function identification for each word in a sentence), and then of sequence understanding which yields to a global property of the sequence (like the prediction of the evolution of a share).

Sequence Labelling. A commonly used solution is a hidden Markov model (HMM) [1, 6] which picks up in real time the most probable sequence of states in time according to previously observed variables. This artificial intelligence (AI) model relies on a transition table showing the most likely transitions from one state to another, and an emission table depicting the likelihood of a specific input for a given state. The model then learns from a tagged dataset these probabilities by counting the different transitions and emissions. The issue with this model is that the function optimized during training is not the likelihood of a given sequence of labels given a particular sequence of inputs, but rather the likelihood to get both at the same time. A solution that outperforms HMM in most cases for discriminative tasks is called conditional random fields (CRF) [7]. The idea there is to assume a log-linear probability distribution for the likelihood of a given sequence of labels given a particular sequence of inputs with respect to observations. The model is then trained to maximize this likelihood by optimizing parameters of this model (weights and biases) instead of frequencies in the HMM. CRF are thus harder to train but give better predictions. Even though both technologies have been primarily made for discrete observations, it is possible to convert inputs into Gaussian probabilities and thus to deal with continuous observations [1, 8].

Sequence Understanding. The idea here is to process the first element of the sequence and reuse this output as a part of the input for the next element [9], and so on until the end of the sequence. This design has been further improved with long-short term memory RNN (LSTM) [10] and gated recursive units (GRU) [11] thus keeping a longer track of the context given by the previous elements. Despite these improvements, these models suffer two flaws. Firstly, it remains hard to detect long term dependencies between elements [12]; secondly, sequence of inputs treatment cannot be parallelized due to the nested outputs for each new element computation. The current state of the art architecture for sequence understanding is based on the transformer architecture [13] initially developed for language tasks. This is a self-attention based auto-encoder [14] architecture that is to say that AI looks specifically for relations between each pairs of tokens. By construction, each interaction related to one token can be computed independently from the others, thus enabling parallel computing. In fact, the position of the token has to be encoded because the model is permutation independent: the input is treated as an unordered set of inputs, thus allowing the detection of long-term dependencies. Thanks to the diversity of relation, transformers are used for high level image processing and transfer learning (pre-trained transformers [15]). The main limitation of this architecture is that it supports, by construction, a fixed maximum input sequence size and this size increase causes a quadratic growth of the computations.

2.3 Multimodality

A shared representation mixing several modalities (e.g. image and text) has been first done by concatenating all the data of the different modalities into one large input vector, but has been shown to be inefficient and not suitable for capturing dependencies and relations between the different modalities [16]. This is why research then focused on various techniques for a more reliable unification of different modalities. The most recent studies on cross-modality representation usually imply the projection of each modality input into a same semantic (or latent) space [16, 17, 18]. This projection can be done after a preprocessing of the raw inputs [18] and it seems more suitable to project low and medium-level features of the input in order to have a better similarity recognition between the inputs. However, the literature presents usually methods for generative tasks or for discriminative pairing tasks (e.g. associating text to a given image) that have not really been used for time constraint classification tasks. A model proposes to gather each modality in a same input vector where the nature of the modality is identified by a keyword before each new sequence [16].

2.4 Limited Labelled Dataset

Dealing with limited datasets in machine learning is more and more made possible by pre-trained auto-encoders (see Fig. 3) [19], popularized by the transformer architecture [13]. Auto-encoders are pairs of models – an encoder and a decoder – that are usually trained to reproduce the input after condensing the data in a latent space. This unsupervised approach is feature based: the models learn the principal key features of the data. The encoder learns the discriminant features contained in the data whereas the decoder learns the common features found in the dataset. Once trained, either the encoder is kept for discriminative tasks or the decoder is kept for generative tasks. The main advantage of this technique is that it does not require labelled data – the expected output being the input.

Once a model is pre-trained, additional layers are added on top of it to be trained for the initial task – this phase is called "fine-tuning". This special case of transfer learning [19] speeds up speeds this second phase up because the learning effort is focused on the final task, and not on the learning of contextual data. For instance, in natural language processing, the pre-training learns the grammar and the semantics of the word whereas the fine tuning learns to classify a sentence.

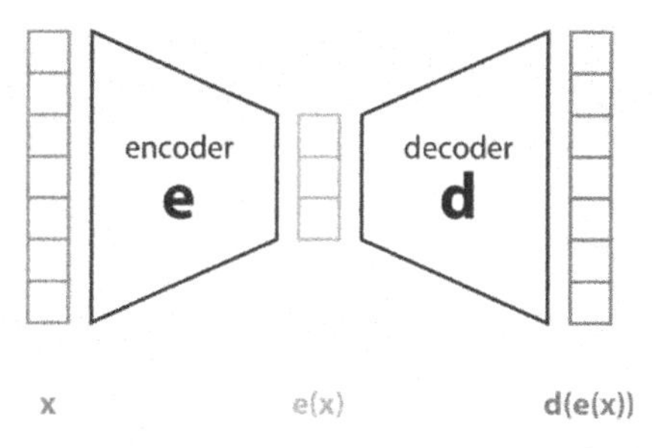

Fig. 3. Principle of a simple auto-encoder (Joseph Rocca, 2019)

3 Proposed Approach

In this section, we describe our new approach tackling multimodal design intent detection. We first describe the conceptual framework, the general architecture and then we describe more precisely each component. In this paper we don't tackle the translation of inferred actions into actual DragonFly commands.

3.1 Conceptual Framework

As a first approach, we propose the set of inputs described in Table 1. The proposed inputs contain all the information for the measures discussed in the literature review with additional information position-wise, so as to enable action parameters (e.g. a pointing direction) extraction. It allows also a clearer definition of such proposed measurements like the symmetricity of the activity of the arms. This inputs selection has two objectives. Firstly we reduce the dimensionality of the problem by exposing only relevant features to the model: doing so, it does not have to learn how to extract these lowest level features. Secondly, it helps understanding the model by making possible to see which input has been more active in a particular context, fostering the maintenance and further improvements for the model.

Table 1. Input selection proposition

Input type	Device	Measures
Hand posture	Leapmotion	For each hand: Flat, Flat hold, Fist, Thumb up, Index, L, C, Pinch or Undefined
Upper body posture	Kinect Azure	Relative positions in 3D to the body referential* of neck, shoulders, elbows, hands and pelvis
Head orientation	HTC Vive Pro	Relative Euler angles of the head with respect to the body referential*
Speech	HTC Vive Pro	Sequence of words obtained through a speech recognizer

*: the sagittal plane is defined as the median plan between shoulders and the frontal plane contains the segment from left shoulder to right shoulder

The outputs of our models correspond to the most used elementary capabilities of DragonFly. Several interviews with DragonFly users and experts, the observation of work sessions on the tool have drawn the following set of actions as our target classes:

- Change position (pan and rotate)
- Hide/unhide
- Grab
- Measure distance
- Make a section
- Select several objects
- Select parent object of selected object
- Take a picture
- Draw a primitive shape (box or cylinder)

3.2 General Architecture

We propose a similar architecture to what we have seen in literature [18] but introducing specific features dealing with the asynchronous nature of the different inputs – the different devices don't deliver new data neither at the same time neither nor at the same frequencies – but also adapt the method to classification tasks and not to similarity cross-modal pairing or to generative tasks. The general approach is described in Fig. 4.

First of all, we define two modalities for our model: gesture and speech. We extract from these some low level features in the form of time series of one dimensional vector[1] and embed each token of each time series into a same dimensional space before being concatenated as seen in literature [10, 16]. This concatenation is than transformed into a rich and high level representation of these tokens with the full body language context. This representation is directly used for the action classification and each of the new rich features are concatenated to their corresponding low level features in order to identify the most relevant features through modality specific conditional random fields (CRF). Finally, a solver gathers the extracted actions and parameters to make a decision.

In addition, we take into account the non-random chaining of actions during a session as testified during the sessions observations and the interviews: we propose an additional input for the unified modality transformer encoding the previous action performed by the user. This idea is also motivated by the results obtained in literature [1] using Hidden Markov Models (HMM). In fact, this implementation tends to mimic the behavior of a transition diagram by adding information about the previous state – in our case the states are the different possible actions (see Sect. 3.1) – to find the most probable following action.

3.3 Component Details

Gesture Low Level Features. The low level representation of a user's gesture is built by taking regular snapshots of a one dimensional posture vector to create a sequence

[1] The frequencies of gestures' features and speeches' ones are uncorrelated and thus may be completely different (see **Shared Representation.**).

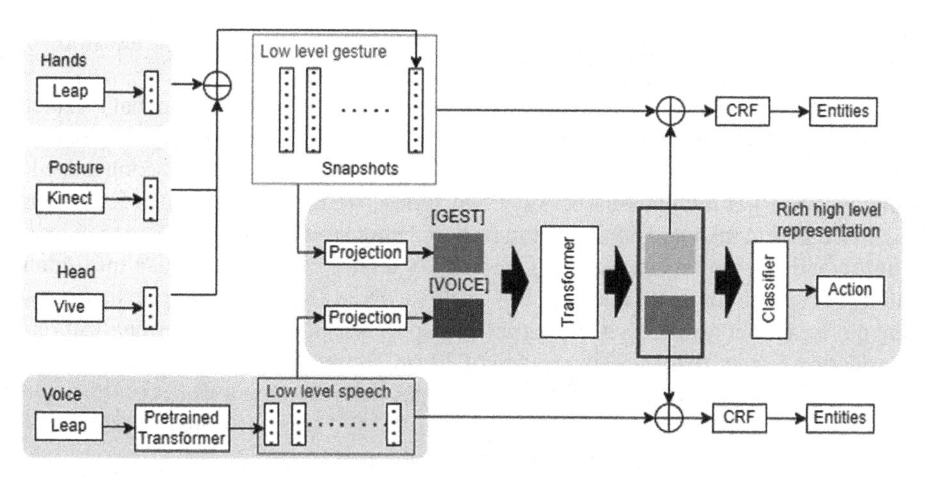

Fig. 4. General architecture of the proposed solution (the last detected action and the solver described below are not represented)

of postures. The frequency of these snapshots may be fine-tuned but we can deduce its order of magnitude considering two criteria. First of all, the interval between snapshots must be sufficient to have a real change of the vector between two consecutive shots. We discuss below the update of this vector but it is directly dependent on the frequencies of involved devices. The slowest device is the Kinect Azure with 30 Hz. Moreover, we observe empirically that an action takes around 5 s to be performed by a regular user – rounded to 10 s to ensure a better context understanding; the size allocated in the shared representation transformer is in the order of 100 tokens. We can thus predict a 10 Hz frequency for this second estimation, giving us a pretty good estimation of the possible range of the snapping frequency: from 10 to 30 Hz.

The posture vector is obtained by the asynchronous concatenation of the different inputs described in Table 1. Each input is obtained as follows: the hands postures are presented as a one hot encoding vector – a vector filled with zeros with a one at the position of the corresponding class – which is obtained from Boolean operations on the geometric and topologic characteristics of each hand (the detailed operations are not presented in this paper) [5]; the relative positions of the upper body articulation are concatenated in a fixed order and all the dimensions are normalized by the pelvis-neck distance; the head Euler angles in radians are also normalized by a factor $\frac{\pi}{4}$. Each part of the vector updates the global posture vector at its device frequency.

Speech Low Level Features. Once the sound has been captured by the HTC Vive Pro microphone, it goes through a voice recognizer (we use the built-in C# library System.Speech.Recognition) to obtain a list of strings. We then use a BERT pre-trained transformer [19] accessible online to get our low-level features for the voice input.

Shared Representation. This part of the process runs in a separate thread. It creates a rich body language representation of the different low level features. First, the low-level features are loaded and projected on a same dimensionality plane by an embedding layer (the dimensionality is primarily set to the original dimension proposed in the transformer method [13] or 512×1 for each individual low-level feature).

Then, as proposed by the literature, the different modality sequences are put together using keyword separators – for our purpose we define [GEST] and [SPCH] at the beginning of each modality sequence, [PREV] at the beginning of the previous action embedding and [SEP] at the junction between two series. We define a minimum length for each modality. A standard adult communication being around 150 words per minute in English and a common action taking empirically around 5 s, we propose a maximum number of speech-allocated tokens of at least 50 (or about 15 s of normal speech), to capture more easily context dependencies and fast speeches. Similarly for gesture features, considering a snap frequency (see above) of 20 Hz, we propose to dedicate at least 300 tokens to gesture features. Finally, we proposed the following input word considering a 512 long input sequence for the transformer: "GEST[432 gesture tokens]SEP|SPCH[74 speech tokens]SEP|PREV[previous action token]".

The sequence described above is then transformed by a unified modal transformer [16] to reveal a rich cross-modality representation of each of the input features. More precisely, each input feature is projected on a same semantic plane. This transformer can be pre-trained an unsupervised manner using its auto-encoder form with data gathered during real sessions. Indeed, the previous layer of the solution is already trained so we can gather words during various sessions before using them as a training dataset. The operation is repeated as soon as the following classification and parameter extraction are done, independently from the low level features update frequencies.

Classification and Parameters Extraction. The previous representation is directly used by a classifier to infer the action. We propose to use a simple sparse linear shallow neural network trained on labelled data for fast supervised training.

For the parameters extraction, we use residual connections as proposed in literature [18] by concatenating the low level features with their corresponding rich representation computed above. These augmented features go through conditional random fields trained with labelled data. The labels of the dataset correspond to the different natures of arguments of the different actions. The construction of this dataset needs discussions with the users, so as to identify which part of their body language expression at which moment determines the parameter – for instance a pointing finger can be the direction of an object: this posture is then inferred as an object parameter.

Once the action and the parameters have been extracted, a solver uses this data alongside the rich representation, so as to decide Dragonfly action is to be performed. It first decides if the user has finished to communicate his intent or not and then if there are missing parameters or conflicts. It finally pilots DragonFly accordingly.

4 Conclusion and Perspectives

This paper presents a method for architectural design intent recognition in VR yet to be implemented. This model does not take into account the virtual environment context and is not able to infer several intents at the same time. Nevertheless, it deals with the asynchrony of the input, and with the multimodality used for augmented representation while having a modular architecture for future improvements.

A first prototype is currently under development and will be built incrementally alongside with the methodology presented in this paper. This prototype will have a limited vocabulary to test the relevance of our model and retrieve the missing order of magnitude and hyper-parameters of our current approach. We will also define the precise architecture of the solver. Finally, we would like to define relevant environmental variables to be added to our model to increase even further the F1-score of the model.

We are also preparing a statistical analysis of user's body language when performing design actions in DragonFly. The objectives are to build a reference to assess the recognition performance of our solution, to validate the assumptions made on the biological variables we consider for our model and identifying confusing factors not highlighted by the literature. Moreover, it builds a first dataset for the training of our prototypes. Preliminary results show the necessity of having an assistant-like bot mainly because of the difficulty to speak naturally to a screen without expecting an answer.

References

1. Gaschler, A., Jentzsch, S., Giuliani, M., Huth, K., de Ruiter, J., Knoll, A.: Social behavior recognition using body posture and head pose for human-robot interaction. In: 2012 IEEE/RSJ International Conference on Intelligent Robots and Systems, pp. 2128–2133. IEEE, Vilamoura-Algarve, Portugal (2012)
2. Ngiam, J., Khosla, A., Kim, M., Nam, J., Lee, H., Ng, A.Y.: Multimodal deep learning. In: 28th International Conference on Machine Learning (ICML), pp. 689–696 (2011)
3. Khan, S., Tunçer, B.: Gesture and speech elicitation for 3D CAD modeling in conceptual design. Autom. Constr. **106**, 102847 (2019)
4. Vuletic, T., Duffy, A., Hay, L., McTeague, C., Campbell, G., Grealy, M.: Systematic literature review of hand gestures used in human computer interaction interfaces. Int. J. Hum. Comput. Stud. **129**, 74–94 (2019)
5. Laviola, J.J.R., Zeleznik, R.C.: Flex and pinch: a case study of whole hand input design for virtual environment interaction. In: Second IASTED International Conference on Computer Graphics and Imaging, Innsbruck, Austria, pp. 221–225 (1999)
6. Ghojogh, B., Karray, F., Crowley, M.: Hidden markov model: tutorial. engrXiv (2019)
7. Lafferty, J., McCallum, A., Pereira, F.C.N.: Conditional random fields: probabilistic models for segmenting and labeling sequence data. In: 18th International Conference on Machine Learning (ICML), pp. 282–289 (2001)
8. Bevilacqua, F., Zamborlin, B., Sypniewski, A., Schnell, N., Guédy, F., Rasamimanana, N.: Continuous real-time gesture following and recognition. In: Kopp, S., Wachsmuth, I. (eds.) Gesture in Embodied Communication and Human-Computer Interaction, pp. 73–84. Springer, Berlin Heidelberg, Berlin, Heidelberg (2010)
9. Elman, J.L.: Finding structure in time. Cogn. Sci. **14**, 179–211 (1990)
10. Hochreiter, S., Schmidhuber, J.: Long short-term memory. Neural Comput. **9**(8), 1735–1780 (1997)

11. Cho, K., van Merrienboer, B., Bahdanau, D., Bengio, Y.: On the properties of neural machine translation: encoder-decoder approaches. arXiv:1409.1259. (2014)
12. Gradient flow in recurrent nets: the difficulty of learning long term dependencies. In: A Field Guide to Dynamical Recurrent Networks. IEEE (2001)
13. Vaswani, A., et al.: Attention is all you need. In: Advances in Neural Information Processing Systems, pp. 5998–6008 (2017)
14. Kramer, M.A.: Nonlinear principal component analysis using auto associative neural networks. AIChE J. **37**(2), 233–243 (1991)
15. Esser, P., Rombach, R., Ommer, B.: Taming transformers for high-resolution image synthesis. In: 2021 IEEE/CVF Conference on Computer Vision and Pattern Recognition (CVPR), pp. 12868–12878. IEEE, Nashville, TN, USA (2021)
16. Li, W., et al.: UNIMO: towards unified-modal understanding and generation via cross-modal contrastive learning. In: Joint Conference of the 59th Annual Meeting of the Association for Computational Linguistics and the 11th International Joint Conference on Natural Language Processing (2021)
17. Radford, A., et al.: Learning transferable visual models from natural language supervision. arXiv:2103.00020. (2021)
18. Liu, A.H., Jin, S., Lai, C.I.J., Rouditchenko, A., Oliva, A., Glass, J.: Cross-modal discrete representation learning. arXiv:2106.05438. (2021)
19. Devlin, J., Chang, M.W., Lee, K., Toutanova, K.: BERT: pre-training of deep bidirectional transformers for language understanding. arXiv:1810.04805. (2019)

A Framework to Optimize Laser Welding Process by Machine Learning in a SME Environment

Jean-Rémi Piat[1,2(✉)], Baudouin Dafflon[1], Mohand Lounes Bentaha[1], Yannick Gerphagnon[2], and Néjib Moalla[1]

[1] Université de Lyon, Université Lumière Lyon 2, Université Claude Bernard Lyon 1, INSA Lyon, DISP, E4570, 69676 Bron, France
jean-remi.piat@univ-lyon2.fr
[2] FAYOLLE SAS, Group PHEA, ZI La Guide, 43200 Yssingeaux, France

Abstract. The Metallurgical industry, and in particular the make-to-order industry, are extremely dependent on the composition of the raw material. As a subcontractor, the company is limited by customer constraints and must be able to adapt the process as fast as possible. Therefore, machine programs need to be adapted to control material reactions (i.e., deformations, vibrations, burrs, porosity, etc.) following the laser cutting and welding. Material reactions depend on the gap between theoretical and intrinsic mechanical characteristics of the materials, on the geometry and on the tools used to maintain the parts. The variability of this gap leads to scrap, machine delay and productivity loss. To improve our knowledge, Industry 4.0 provides several technologies enabling the automation of complex tasks. For instance, Artificial Intelligence (IA), supervised or unsupervised, provides opportunities for optimizing and customizing machine programs considering material reactions. The case study presented in this paper proposes to use machine learning (ML) to define a decision support system. Two ways are explored: quality and productivity optimization. The objective is to free up resources (operators, machines) time for other strategic projects for the company.

Keywords: Machine learning · Parameter optimization · Laser welding · Productivity

1 Introduction

Laser welding has been widely used in a variety of industrial applications due to the advantages of deep penetration, high speed and small heat affected zone [1–3]. These advantages enable high seam quality with fine and deep weld seams and strong mechanical properties, under the condition that parameters are optimized for the order and the raw materials [3–6]. This optimization is particularly challenging because the relationship between process parameters and the output seam characteristics are complexes and non-linear which makes prescribed parameters not optimal [7, 8]. This optimization also

Published by Springer Nature Switzerland AG 2023
F. Noël et al. (Eds.): PLM 2022, IFIP AICT 667, pp. 431–439, 2023.
https://doi.org/10.1007/978-3-031-25182-5_42

depends on raw materials that are subject to variability in terms of element composition and historic of transformations and transportations. The optimization and setting of process parameters represent a puzzling problem for the operators who can only select the welding process parameters according to experience, charts, and handbooks in the practical production. Tests on actual machine tool are generally needed to complete verification and to improve performance [6, 9, 10]. However, trial-and-error method often leads to sub-optimal solution, especially for new welding process, and requires time, materials, and competent human resources [1, 2, 6–9, 11].

To improve industrial capacities, Industry 4.0 provides several technologies enabling a shift in production process. For instance, Artificial Intelligence (AI) perceives and learns, sometimes more effectively than humans, by means of a variety of Machine Learning (ML) technologies. ML technologies provide opportunities to predict, detect, control, optimize, etc. complex processes like laser welding. ML technologies regroup unsupervised approaches to find cluster and discover hidden structures in the data (e.g., k-means methods), and supervised approaches to learn from a training dataset and predict the outcome (e.g., decision trees or neural networks). Supervised ML technologies can for example be applied for Computers-Aided-control (CNC) process parameter optimization with the prediction of the welding quality and the optimization of these parameters. The resulting models enable the construction of a decision support system which can help, during trial-and-error method, to reduce the time needed to find optimal parameters, to reduce the materials needed to validate the optimal parameters and therefore to reduce the effort and time provided by competent human resources.

This paper is organized as follows. Firstly, a state-of-the-art is presented about the approaches used to optimize laser welding parameters and process. Secondly, a framework is proposed to integrate these solutions in a production environment. Subsequently, a case study is proposed to implement the framework and the technologies used are explained. After a discussion on the expectation of the case study, a conclusion is drawn, and a perspective work is discussed.

2 State of the Art

In a context of make-to-order laser welding industry, raw material characteristics constitute constraints impacting the production process and the process plans are challenging to optimize due to variabilities in the orders. These industries need to optimize their process plans to save time, materials, and labor efforts [12]. Different optimization objectives can be distinguished in laser welding, like parameter optimization, feature prediction, seam tracking, defect classification, simulation validation or adaptive control [13]. Optimization of laser welding process is multifactorial. Considering the diversity of possible actions to perform such work, we have chosen to focus on the optimization of the welding machine parameters. The other aspects will be discussed in the perspectives section. The prescribed parameters for welding process are the result of the provider tests and statistical optimization [14] but they generally require tests on an actual machine tool for complete verification and to improve performance [6, 9, 10]. Statistical optimizations are challenging to apply to various practical situations because the relationship between the process parameters and the seam characteristics are non-linear and are usually dependent on the specific experimental results [3, 7]. Therefore, machine learning methods are gradually popular to model physical correlations and obtain optimal responses.

The typical workflow to optimize parameters of industrial processes contains the following four steps [12]:

1. Generating a database with few experiments or run simulations with Design of Experiments (DoE) methods.
2. Modeling the physical correlations between the process parameters and the quality criteria with statistical or machine learning methods.
3. Optimization of the process parameters using the created process model.
4. Adjusting the process parameters manually or automatically.

The Taguchi DoE has been widely used in the field of laser welding because of its distinctive design of orthogonal arrays used to study the entire process parameters with a acceptable number of experiments [7].

A promising strategy in the second step is to develop metamodels, which are also called approximation models as they provide "model of model" fitting the relationship between the input process parameters and the output performance for the purpose of welding design optimization [2]. Comparative studies of the accuracy of metamodels have demonstrated that different metamodels perform well in different cases [15]. Arbitrarily selecting a metamodel to fit the relationship between the input parameters and output responses may increase the risk of adopting an inappropriate metamodel because of the nature of the relationship and the current training data [15]. Different metamodels have been proposed in the literature, like Kriging [2, 15, 16], artificial neural networks (ANN) [2, 5, 7, 8, 16, 17], Radial basis function (RBF) [2, 15] or support vector regression (SVR) [2, 15].

The metamodels only learn about the correlation to predict the output with different input and can be used for root cause analysis, the early prediction of manufacturing outcomes, and diagnostic systems to optimize product quality or process efficiency [12]. Otherwise, the models need to be combined with optimization algorithms to find the optimal production parameters for a specified objective. Optimization algorithms are generally particle swarm optimization (PSO) or genetic algorithms (GA) [5, 12, 13].

The main challenges for parameters optimization are [2, 18]:

- The representativeness and reliability of the data which need to contain problem-relevant variables and to be in high quantity to be able to describe the relationship between the process parameters and output performance.
- The variation in manufacturing orders which make it challenging to assure the accuracy of the constructed metamodel.

The issue of representativeness and reliability are addressed by DoE methods and in the choice of monitoring variables. For example, many researchers choose to realize their experiments on a specific raw material and obtain a particular model that do not consider the impact of different elements on the welding process.

As for the variation in production, parameters optimization can be combined with monitoring solution to control the real output and the accuracy for new processes. In the field of laser welding, high weld depth is particularly important for high welding seam quality but the welding depth is currently only known through destructive quality tests [19]. New monitoring solutions have been studied to collect the different signals during laser welding, such as acoustic signals, optical signal, or thermal signals [13]. The acoustic signals are one of the most monitored signals for the weld penetration depth due to the low cost, high responsible speed, and convenient features [19]. The acoustic signal analysis can be divided in time domain analysis and frequency domain analysis to extract the acoustic signature of the welding process and link it to the depth of the seam [20]. The acoustic signatures analysis can result into a classification of penetration like full penetration, overheat penetration and half penetration [19, 21], but quantification has also been studied through machine learning to find he relationship between acoustic signals and penetration depth [19, 20]. For example, [20] and [22] use ANN to characterize the weld penetration with the acquired acoustic signatures and find that it worked well under different laser welding parameters. Noise reduction methods are also needed to improve the quality of the acoustic signal and overcome the limit of noisy and hostile environment [19–21].

The state-of-the-art shows that the selection of the proper process parameters for high quality welding seam quality is challenging and that the optimization of this process can lead to major time, materials and competent human resources saving. A workflow has been presented to help in the process of parameter optimization and some techniques have shown remarkable results. For instance, Taguchi DoE and a combination of meta-models are efficient to describe the relationship between input welding parameters and seam characteristics. Likewise, approaches using GA exhibit great capability with the optimization of many configurations of parameters. However, these studies are often on specific cases and materials and are not adapted for production environment. This paper will therefore present a framework integrating these approaches to a decision support system for parameter optimization in the production context of an enterprise (a SME). This framework will also incorporate a monitoring system to extend the capacities of the decision support system considering the variabilities of customer orders.

3 Proposition

Two main challenges have been identified to integrate ML algorithms in a production context. Firstly, ML technics, with meta-models and optimization algorithms, need to consider problem-relevant data about the laser welding process, such as laser power, welding speed, gas flow rate, focal position, nozzle height, etc., but also raw materials composition, like carbon rate, thickness, etc. The raw materials composition are uncontrollable variables but influences the relationship between the controllable process parameters and the output. Secondly, the customer order variabilities reduced the usefulness of the optimization model in production environment, so a monitoring system is needed to help operators during the optimization process and overcome the lack of accuracy. These tools are integrated in a framework representing the industrial environment, see Fig. 1.

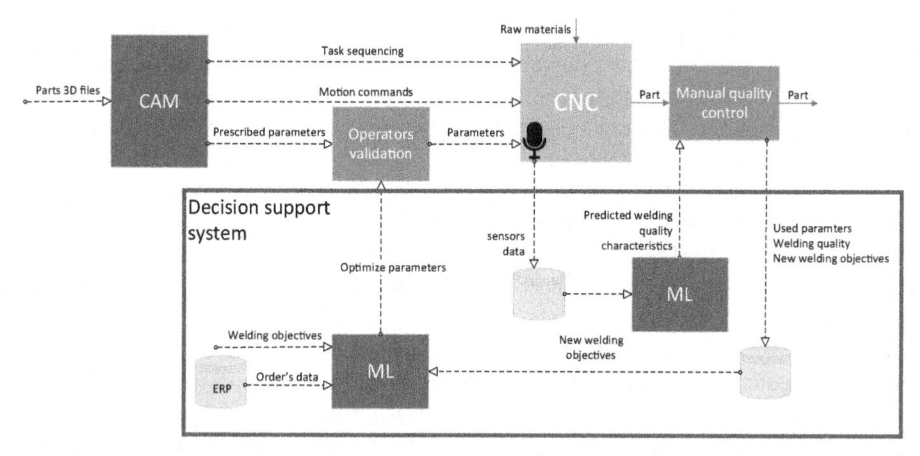

Fig. 1. Integration of the sensors and ML tools to the industrial environment

This framework represents a value chain of the laser welding process, from the customer order, with the parts 3D files, order's data, and the raw materials, to the produced parts. The first step in the process is to transform the customer files with Computers-Aided-Manufacturing (CAM), which is a design-expert system, to generate the process plan with prescribed motion commands, prescribed process parameters and the sequencing of each task. As seen in the state-of-the-art, ML tools can be used in a decision support system to optimize the process parameters and reduce the trial-and-error method which is a time-, materials- and competent human resources consuming task. The first ML tool of the decision support system generate optimized parameters from the order's data and welding objectives. The prescribed and optimized parameters are compared by an operator to choose one. The main objective is to avoid significant gap between the two set of parameters and control accuracy issues of the model. Sensors are also integrated in the CNC environment to collect the in-process signals and store them in a database. A second ML tool is used to predict welding quality characteristics. Then the systematic manual quality control and the predicted quality control are used by the operators to evaluate the fitness of the parameters with the customer order. Depending on the quality control, either the parameters are confirmed to be optimal for the order or new objectives are set up in the first ML tool.

4 Case Study

The framework is being adopted in a B2B metallurgical SME in France to support laser welding. The CNC used is Trumpf laser welder and cutter with a disk laser Yb:YAG with a wide power range (2–6 kW). This SME is a make-to-order company realizing trial-and-error for all their new orders. The trials are realized either on customers prototyping raw materials, on company's materials or on additional production raw materials. In either case, the SME needs to provide the best performance with the minimal iterations to reduce operator time and the materials used.

The first tool is a ML algorithm for parameters optimization. Following the workflow described in the state of the art, Taguchi DoE is used to generate the training points as it can provide good uniform and space-filling sample points in the design space. The input data controllable process parameters, such as laser power, welding speed, gas flow rate, nozzle height and uncontrollable raw material characteristics, such as carbon rate and the material thickness. These inputs are selected by an expert to contain the problem-relevant variables of most of the production, i.e., many different compositions of steel sheet for butt welding. The process parameters are decimal numbers and are available in the process program saved on the network. The material characteristics can be fetched from the company ERP database, which is, in our case study a HFSQL database in the company server. The output data are the welding depth and the seam width because they represent the quality characteristics generally required by the customers. For each input variable, five levels with an interval are considered to cover the design space. The current problem is a six-factors and five-levels design problem, hence an L32 orthogonal array could be generated. The four process parameters can be selected in the CNC while the two raw material characteristics represent the materials used for the experiment. For the output, the seam width can be calculated manually with a caliper, but destructive tests are needed to collect the weld depth. Based on the experiment data, ensemble of two metamodels including SVR and ANN are constructed to estimate the seam geometry. Their accuracy is evaluated with relative maximum absolute error and root mean square error (RMSE) and if their accuracy is checked then a non-dominated sorting genetic algorithm (NSGA-II) are implemented to obtain the optimum process parameters. During the NSGA-II optimization, the fitness values assigned to the populations are generated according to the predicted value using metamodels and the objectives correspond to the welding depth and the seam width required by the customer. This combination of metamodels and NSGA-II have already proven their capabilities in [2]. The termination criteria for the optimization are (1) the relative distance between the optimal process parameters of two successive iterations is below $1e{-3}$; (2) the pre-specified maximum number of iterations is reached.

The model can then be used in production environment where the optimum process parameters are compared with prescribed ones by the operators because of the possible lack of accuracy of the model, due to the variation between the order and the training data. An in-process monitoring system is also used to check the welding depth and assure the quality for the new process. This second tool takes as input the in-process acoustic signals to predict the welding depth. The training data for the welding depth are generated during the L32 Taguchi DoE and the corresponding acoustic signals are collected from a microphone such as WeldMIC which are welding acoustic sensors designed by Xiris Automation Inc. The microphone delivers an analogic signal that is first processed to a numerical signal to be able to analyze the temporal domain and the frequency domain. These two domains are then used in a back-programming neural network (BPNN).

5 Discussion

To improve the welding process, a framework has been proposed regrouping a first ML tool to generate optimized parameters and a second ML tool to link the in-process acoustic signals to one of the main welding quality characteristics, the welding depth. These ML tools are integrated in a decision support system to help the operator in the widest range of orders. Nowadays, the trial-and-error represents a little less than 10% of the welding time. This time is used by a competent human resource to optimize parameters and consume customers or company raw materials. The expectation for the support decision system is to reduce this time to less than 5% which will improve the productivity of the welding machine, reduce the use of raw materials and free up time of the competent human resources for other strategic projects.

The first limit of this system is the variabilities of the knowledge on the raw materials composition because some customers do not track the composition of the raw materials. One solution can be to take standard composition for the type and grade of the materials and to use the result of the welding to adjust the welding objectives by decreasing or increasing them. In addition, the carbon rate does not characterize the whole raw materials. For example, the historic of transformation of the raw materials are not considered but can be very impactful for the welding. The last limit is the need of additional sensors to monitor the welding depth. The monitoring of this characteristic is not integrated in the provided machine and need destructive tests, which are a waste in the production process. Besides, the case study works with a 3D welding machine, so the environment and the working space define significant constraints. These constraints have impact in the choice of monitoring the sound pressure to avoid space constraints, but the environment noise remains a key challenge for the accuracy of the models.

6 Conclusion

In a make-to-order company, trial-and-error methods are generally used to set CNC parameters, but the results are sub-optimal, and the process represent a waste of time, materials, and competent human resources. The challenge to optimize the parameters with the orders come from the variabilities in the raw materials and the complexes relationships between the different parameters and the welding quality characteristics. Machine learning particularly have been studied in different contexts to improve industrial process, such as parameter optimization for CNC production. Workflow, DoE, and optimization algorithms have successfully been used to predict and optimize the welding process, but these studies are often applied to cases and materials and are not adapted for production environment. This paper proposed a framework to integrated ML models in a production environment and consider the raw materials as an input to improve the scalabilities of the models. A monitoring system has also been introduced to overcome the lack of accuracy of the models for particular processes. The implementation of the framework can reduce the time to set up the optimal parameters considering the requirements of the client. One of the perspectives is to take advantage of the industrial context to regularly evaluate the accuracy of the models and define new design space to extend the models. New design spaces could consider wider range of variables, like the type of joint, others raw materials element (like aluminum or chromium) or other parameters (like focal position, gas type, head orientation, root gap, beam width, etc.). The models can also be extended for new processes in the same CNC machine such as the cutting or wire-feeding laser welding.

References

1. Zhao, Y., et al.: Optimization of laser welding thin-gage galvanized steel via response surface methodology. Opt. Lasers Eng. **50**(9), 1267–1273 (2012)
2. Zhou, Q., et al.: Optimization of process parameters of hybrid laser-arc welding onto 316L using ensemble of metamodels. Metall. Mater. Trans. B **47**(4), 2182–2196 (2016)
3. Ai, Y., et al.: The prediction of the whole weld in fiber laser keyhole welding based on numerical simulation. Appl. Therm. Eng. **113**, 980–993 (2017)
4. Tercan, H., Khawli, T.A., Eppelt, U., Büscher, C., Meisen, T., Jeschke, S.: Improving the laser cutting process design by machine learning techniques. Prod. Eng. Res. Devel. **11**(2), 195–203 (2017). https://doi.org/10.1007/s11740-017-0718-7
5. Nagesh, D.S., Datta, G.L.: Genetic algorithm for optimization of welding variables for height to width ratio and application of ANN for prediction of bead geometry for TIG welding process. Appl. Soft Comput. **10**(3), 897–907 (2010)
6. Altarazi, S., Hijazi, L., Kaiser, E.: Process parameters optimization for multiple-inputs-multiple-outputs pulsed green laser welding via response surface methodology. IEEE (2016)
7. Sathiya, P., Panneerselvam, K., Abdul Jaleel, M.Y.: Optimization of laser welding process parameters for super austenitic stainless steel using artificial neural networks and genetic algorithm. Mater. Design (1980–2015) **36**, 490–498 (2012)
8. Ai, Y., Wang, J., Jiang, P., Liu, Y., Liu, W.: Parameters optimization and objective trend analysis for fiber laser keyhole welding based on Taguchi-FEA. Int. J. Adv. Manuf. Technol. **90**(5–8), 1419–1432 (2016). https://doi.org/10.1007/s00170-016-9403-9
9. Lynn, R., et al.: The state of integrated computer-aided manufacturing/computer numerical control: prior development and the path toward a smarter computer numerical controller. Smart Sustain. Manuf. Syst. **4**(2), 20190046 (2020)
10. Trumpf: Collecte de données, Trulaser Cell Série 7000, vol. D737fr. Trumpf (2014)
11. Preez, A.D., Oosthuizen, G.A.: Machine learning in cutting processes as enabler for smart sustainable manufacturing. Procedia Manuf. **33**, 810–817 (2019)
12. Weichert, D., Link, P., Stoll, A., Rüping, S., Ihlenfeldt, S., Wrobel, S.: A review of machine learning for the optimization of production processes. Int. J. Adv. Manuf. Technol. **104**(5–8), 1889–1902 (2019). https://doi.org/10.1007/s00170-019-03988-5
13. Cai, W., et al.: Application of sensing techniques and artificial intelligence-based methods to laser welding real-time monitoring: a critical review of recent literature. J. Manuf. Syst. **57**, 1–18 (2020)
14. Ruggiero, A., et al.: Weld-bead profile and costs optimisation of the CO2 dissimilar laser welding process of low carbon steel and austenitic steel AISI316. Opt. Laser Technol. **43**(1), 82–90 (2011)
15. Yang, Y., et al.: Multi-objective process parameters optimization of hot-wire laser welding using ensemble of metamodels and NSGA-II. Robot. Comput.-Integr. Manuf. **53**, 141–152 (2018)
16. Jiang, P., Cao, L., Zhou, Q., Gao, Z., Rong, Y., Shao, X.: Optimization of welding process parameters by combining Kriging surrogate with particle swarm optimization algorithm. Int. J. Adv. Manuf. Technol. **86**(9–12), 2473–2483 (2016). https://doi.org/10.1007/s00170-016-8382-1
17. Lin, H.-L., Chou, C.-P.: Modeling and optimization of Nd:YAG laser micro-weld process using Taguchi Method and a neural network. Int. J. Adv. Manuf. Technol. **37**(5–6), 513–522 (2008)
18. Tercan, H., Guajardo, A., Meisen, T.: Industrial transfer learning: boosting machine learning in production. IEEE (2019)

19. Yusof, M.F.M., Ishak, M., Ghazali, M.F.: Feasibility of using acoustic method in monitoring the penetration status during the Pulse Mode Laser Welding process. IOP Conf. Ser. Mater. Sci. Eng. **238**, 012006 (2017)
20. Huang, W., Kovacevic, R.: A neural network and multiple regression method for the characterization of the depth of weld penetration in laser welding based on acoustic signatures. J. Intell. Manuf. **22**(2), 131–143 (2011)
21. Huang, W., Kovacevic, R.: Feasibility study of using acoustic signals for online monitoring of the depth of weld in the laser welding of high-strength steels. Proc. Inst. Mech. Eng. Part B J. Eng. Manuf. **223**(4), 343–361 (2009)
22. Lee, S., Ahn, S., Park, C.: Analysis of acoustic emission signals during laser spot welding of SS304 stainless steel. J. Mater. Eng. Perform. **23**(3), 700–707 (2013). https://doi.org/10.1007/s11665-013-0791-9

Product Development: Design Methods, Building Design, Smart Products, New Product Development

Positive Emotions for Responsible Consumption Through Product Design: A Review

Kassia Renata da Silva Zanão[1], Osíris Canciglieri Junior[1], and Guilherme Brittes Benitez[1,2(✉)]

[1] Industrial and Systems Engineering Graduate Program, Polytechnic School, Pontifical Catholic University of Parana (PUCPR), Curitiba, Brazil
guilherme.benitez@pucpr.br

[2] Organizational Engineering Group (Núcleo de Engenharia Organizacional – NEO), Department of Industrial Engineering, Universidade Federal do Rio Grande do Sul, Porto Alegre, Brazil

Abstract. Stimulating positive emotional experiences in the user that result in a change to responsible consumption behavior is essential to achieve a balance between emotional, social, economic, and environmental aspects. Our goal is through bibliographical research to understand the emotional behavior of the users in relation to their purchase decision making, exploring how positive emotions (vs. negative) can play an environmentally and socially responsible appeal to the consumer. To this end, we propose a conceptual framework that allows visualizing the interaction of positive emotions in the user experience for the promotion of responsible consumption. Future studies are suggested that may help identify the necessary elements for the design and development of products through positive emotions to pursue responsible consumption.

Keywords: Positive emotions · Responsible consumption · Emotional experience · User experience · Product design

1 Introduction

Emotions can have a great influence on decision making and consumer behavior [1–3], because through them we are prompted to behave by avoiding or approaching a decision. This is already a good reason to justify how important it has been to apply a deep understanding of user emotions to product consumption behavior. However, emotions directly influence usage behavior and the positive quality of user experiences, they foster brand loyalty, product attachment, and can improve users' physical and mental well-being.

Encouraging conscious consumption of products through positive emotions and an optimistic approach becomes an extremely profitable and beneficial opportunity for all ties involved in the consumption cycle. Through anticipated emotions it is possible for consumers to foresee the positive emotional consequences of the outcome of their decisions.

Published by Springer Nature Switzerland AG 2023
F. Noël et al. (Eds.): PLM 2022, IFIP AICT 667, pp. 443–452, 2023.
https://doi.org/10.1007/978-3-031-25182-5_43

As human beings, we are automatically motivated to make the decisions that bring us the most pleasure. In other words, we almost unconsciously make choices that minimize the possibility of negative emotions, and because of this we see a great several studies portraying how the influence of negative emotions towards a behavior for customers can be the trigger for product consumption [2]. This explains why this negative approach is so widely used both in advertising and marketing content, and as a source of study among researchers in the field of emotional and ethical consumer behavior. Arli et al. [4] specifically studied guilt and shame, and reported that these anticipated negative emotions influence consumers' perceptions of ethical and unethical behavior. Negative emotions, such as disappointment, upset, or guilty, deter consumers from unethical behavior.

Usually when we refer to the Triple Bottom Line (TBL) in product design the main focus happens through negative emotions due to social pressures. Our study looks through another lens, the positivist one, where companies already start from the positive emotional aspects of their customers. Shiota et al. [5], points out in his study that positive emotions can have adaptive behaviors and thus be functional, in the sense of being helpful or having a desirable effect, in several different ways. They increase feelings of well-being, predict positive life outcomes, and may even promote physical health. In contrast, negative emotions can signal danger or loss, promote caution, and mobilize the energy needed to escape or ward off a serious threat.

Therefore, we understand the importance of these approaches in building more environmentally, socially, and emotionally responsible consumers. Conversely, we suggest the best path for product design is rather than preventing unethical behavior is to encourage the ethical behavior. Thus, we are questioning if enterprises which use the studies of positive emotions in relation to the consumption of products, addressing how the need for complete well-being (emotional, social, environmental and economic) affects the user's emotions in order to stimulate a responsible consumption. So, our research question is: it is possible to stimulate the conscious consumption of products by using positive emotional effects as triggers in product design? The following objective was established for this study: to understand the user's emotional behavior when choosing products, exploring how positive emotions can play an environmental and socially responsible role in appealing to the consumer for a more conscious consumption of products.

2 Emotional User Experience

We are highly influenced by emotion, both in our interactions with people and with the products we consume. For this reason, the role of positive and negative emotions in human-product interaction and the benefits of consuming products that evoke positive emotions have been widely discussed in the literature. Several studies talk about the ability of products to provoke a diversity of positive emotions; however, despite these emotions being pleasant, each one conveys a different meaning and feeling, which influences people's consumption behavior in different ways [6, 7].

Therefore, a conceptual framework which illustrates the interplay of user emotions in relation to responsible consumption was created. (see Fig. 1). The framework shows the three areas analyzed in this study, the first section addresses the area of user emotions that are both motivators and outcomes of consumption behavior [8], and this is the main motivator of the study. This section analyzes the user as a consumer and among the various human aspects (body, sensory senses, psyche…), the main focus will be on emotions within the purchase decision making scenario, and these emotions of the consumer at the time of decision making can be positively or negatively driven.

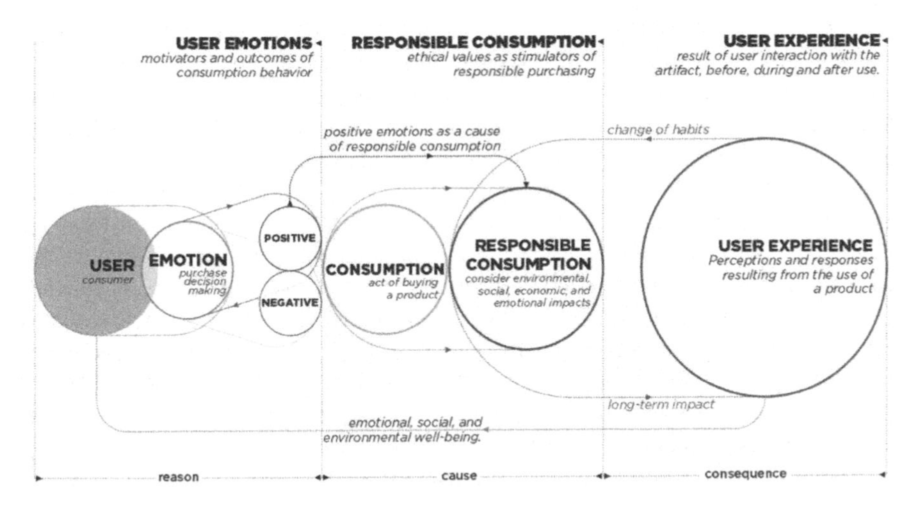

Fig. 1. Initial framework of the study concept.

In the second section of the framework, consumption is illustrated, the act of consuming a product, but the cause that this study addresses within the consumption scenario is ethical values as enablers of conscious and responsible shopping and talks about the act of buying a product considering its environmental, social, economic, and personal/emotional impacts [8, 9]. The link of the first section (user emotions) with the second section (responsible consumption) happens through the proposition of this study, that positive (vs. negative) emotions can act as better causers of responsible consumption.

This proposal about positive emotions being better stimulators, demonstrates the consequence that can be achieved through user experience, represented in the third section of the framework. This experience is the result of the user's interaction with the artifact, before, during, and after use. Experiences are people's perceptions and responses resulting from the use of a product, considering emotions, beliefs, preferences, physical and psychological responses, behaviors, and achievements [10]. When the user's experience is affected through a positivist communication and stimulates in the user a positive feeling in relation to the product and the company's brand, the bond becomes something more intense, and a change of habits for the responsible consumption of other products happens, and this can generate long-term impacts. This whole scenario of change for the responsible consumption behavior of products through positive emotions can bring the user emotional, social, economic, and environmental well-being.

2.1 Emotions as Motivators of Decision Making

Studies from affective neuroscience and psychology have reported that human affect and emotional experience play a significant and useful role in human decision making for product success [11]. Knutson et al. [12] reveal that it is possible to predict which was the motivation for consumer decision making by observing which area of the brain was most active when considering a purchase, through brain images which show a choice happens between two opposite feelings, the pleasure of owning and the pain of spending the money. This is because in relation to consumer behavior, positive feelings (such as pleasure and happiness) are linked to the propensity to approach, and negative feelings (such as pain and guilt) are linked to the propensity to avoid.

Mogilner et al. [13] bring the definition of happiness and how this feeling can affect consumer choice. Individuals tend to choose more interesting options when focusing on the future, and more calming options when focusing on the present. Given this, the question raised in their study is whether happiness actually drives purchasing, or whether this depends on what happiness means to each individual. The authors conclude that it seems to be the temporal focus, rather than age itself, that changes individuals' experiences of happiness. In addition, it is important to incorporate the distinction between exciting and calming activities, along with the temporal focus, to get a more assertive view of happiness. And the results of the experiments reveal that the specific meaning of happiness that individuals adopt determines the choices they make. The work by McFerran et al. [14] brings pride as the protagonist and analyzes why consumers buy luxury brands and proposes two distinct faces of this feeling in consumption: authentic pride (achievement and confidence) which leads to a desire for luxury brands and arrogant pride (arrogance and pretension) is the result of these purchases and is the form of pride signaled to observers by these purchases. An interesting finding, they make is that even if the purchase is made with authentic pride motivation it can give rise to another, presumably less desirable form, the arrogant pride.

Chethana et al. [15] say that each specific emotion is associated with a set of cognitive evaluations that drive the influence of emotion on decision making through differentiated psychological mechanisms. In other words, when relating consumer decision making to emotions, emotional influences can be divided between those that are triggered by marketing appeals intended to influence the decision (integral emotions) and those that are already part of a particular decision (incidental emotions). Emotion plays a significant role in linking consumers to products; a purchase motivated by emotional satisfaction is more likely to result in the possession of durable goods for longer periods [16].

Studies point out that it is not only experienced emotions that affect consumer decision-making, but anticipated emotions can also have great effects on cognitive processing [2]. Establishing how emotions are directly linked to the products that are consumed is key to explaining how consumer behavior can be detrimental not only to the user's emotional well-being, but also to social, environmental, and economic well- being. Thus, understanding how emotions affect the user's decision making helps us diagnose a much more intimate consumption behavior, and more focused on the main element of the relationship, the human being. Then, the next step is to use this resource in a way that brings benefits to companies and the economy, but mainly, that brings benefits to the user's emotional health and to achieve sustainability during the process.

3 Interaction of Emotions with Responsible Consumption

Excessive consumption is one of the greatest causes of the problems of contemporary society. In the last year, Brazilian people consumed 74% more natural resources than the planet can regenerate [17]. This scenario reinforces the importance of encouraging consumers to be more aware and responsible of what they are buying. They should consider that everything they consume is part of a context that involves production, purchase, use, and disposal, and that besides bringing consequences (positive or negative) to themselves, it also affects the economy, society, and especially the environment. Conscious (or ethical) consumption is a social movement based on the impact of purchasing decisions on the environment, health, and consumer well-being [18]. Long et al. [19] define ethical consumption as *"the act of buying products that have additional attributes (e.g., social, environmental, political, health, etc.) beyond their immediate use value, to signify commitment to their values and/or support changes in unfair market practices"*.

Many researchers have explored the relationship between emotion and sustainable consumption behavior, and the results clearly show that the former has a remarkable impact on the latter [20]. Escadas et al. [2] found in their study that more than feeling good, consumers do not want to feel bad, this makes ethically favorable decisions more practiced to avoid experiencing negative emotions in the future. A wide range of positive and negative emotions have been considered in relation to their influence on ethical decision making [21].

Emotions are both the generators and the outcomes of ethical and unethical behavior [8]. Ethical values are increasingly the key drivers of purchasing decisions [9]. Winterich et al. [22] talk about unethical judgments and the crucial role of emotions in forming moral judgments about ethical behaviors. Arli et al. [4] specifically studied guilt and shame, and reported these anticipated negative emotions influence consumers' perceptions of ethical and unethical behavior. Negative emotions, such as feeling disappointed, upset, or guilty, deter consumers from unethical behavior.

We understand the importance of these approaches through negative emotions in building more environmentally, socially, and emotionally responsible consumers, but suggest that the best path lies on a fine line, where more important than preventing unethical behavior is to encourage ethical behavior. Therefore, we understand a company/consumer relationship that starts with a positive feeling is the primary basis for building trust and loyalty and consequently help the firm to achieves better results. Thus, we propose that positive emotional approaches may have as much effectiveness in stimulating conscious consumption behavior as negative emotional approaches.

4 Methodology

The Systematic Literature Review was conducted with a focus on the emotional user experience and responsible consumption. The main intent is to understand the user's emotional behavior in relation to their purchase decision making, exploring how positive (vs. negative) emotions can play an environmental and socially responsible appeal on the consumer.

The literature review methodology applied was based on Unruh [23] and the PRISMA (Preferred Reporting Items for Systematic Reviews and Meta-Analyses) model, as described and illustrated in Fig. 2, in 10 steps. The first step was the definition of the research problem, which was to understand how design can stimulate the conscious consumption of products through the bias of emotional values, in order to generate a better emotional experience in the user. The second step was to objectify the research in order to understand the user's emotional behavior when choosing products, exploring how positive emotions can play an environmental and socially responsible role in appealing to the consumer for a more conscious consumption of products.

Fig. 2. Literature review process.

The third step was to define the macro keywords of the research: emotions, responsible consumption and user experience. The fourth step was the initial research, searches were conducted in the Scopus and Web of Science databases making the following crossings: 1. emotion* and conscious consumption; 2. emotion* and user experience; 3. emotion* and conscious consumption and user experience. The fifth step considered some correlated words of the research, such as: emotional experience, emotional design, positive emotions, negative emotions, conscious consumption, conscious behavior, ethical consumption, green consumption, product design, product experience, user centered design, experience design.

The sixth step was to conduct further research by cross-referencing the words, only articles in English, from the last 10 years, and peer-reviewed were considered. The seventh step was the definition of the inclusion criteria (address the theme of user emotions in the title; depth of the theme responsible consumption in its abstract). The eighth step is the definition of exclusion criteria (having the focus on consumption of food, services, or digital products; superficial or secondary approach to user emotions; not having access to the full text online).

The ninth step included articles from previous studies and results obtained from other sources (as suggested in the PRISMA model). At the end of this process, 25 articles were selected and then the research was conducted by reading and analyzing the selected articles. And finally, the tenth step was the synthesis of the articles, analysis and organization of the results obtained in data tables to define the main gaps in the studies that will serve as opportunities to advance knowledge in the area.

5 Results

The business market has already understood that one should not value only the economic aspect, because the customer, despite considering this aspect important, lately has highlighted the need for more care with the environmental and social issues, when related to their purchasing decision power, and this responsible positioning of the consumer directly impacts their emotional well-being. Therefore, when this responsible stimulus comes from designers, even if by pressure from stakeholders, the bond that is established becomes something much more comprehensive.

As previously mentioned, many studies not only from the areas of product design and production engineering, but also from marketing, psychology, neuroscience, and administration for example, have already shown a growing interest in the area of user emotions in relation to their consumption behavior. Donald Norman [24] lists three emotional levels that a well-designed product needs to achieve to be successful: beauty (visceral level), fun (behavioral level) and pleasure (reflective level). Visceral design is about appearances, behavioral design is about enjoyment and effectiveness in use, and reflective design considers the rationalization and intellectualization of a product. Each of the three levels of design plays its part in shaping our experience, and each level requires a different style of design. Tavares [25] proposes a methodology that serves to evaluate and translate the cognitive and affective experience of the product into consumer demands to better understand the cognitive and affective dimensions of consumers and the resulting purchase intention. The author's results show that product affectivity is important when what is being measured is its hedonism. Whereas the product's cognitiveness is important when its functionality is being measured. But few studies focus only on positive emotions.

Many studies in production engineering and product design also explore the development of new products aligned with sustainability and the stimulation of responsible consumption, as this is notably increasingly a major concern of the population. The survey conducted by Akatu [17] investigated the evolution of Brazilians' level of awareness in consumption behavior, quantitatively presenting the main challenges, motivations, and barriers to the practice of conscious consumption. The vast majority of respondents stated that they feel more motivated to join the practice of conscious consumption when motivated by emotional triggers, especially when related to other people who feel affection, than by more concrete and self-centered triggers. The study also pointed out that many people don't even know how to explain what sustainability is, and that the main barriers to a more responsible consumption of products are related to the high cost of more sustainable products, the distrust in the credibility of brands, and the need for effort to change habits.

However, there are very few methodologies in Product Design aiming to unite the concepts of emotional user experience and responsible consumption. The following framework presents the opportunity found for future studies within the area of Product Design, User Experience with a focus on the emotional user experience and responsible consumption of products (see Fig. 3).

The first section of the framework illustrates Product Development that considers the user experience in its process and aims to encourage sustainability for both production and consumption. This development process is based primarily on the user's positive emotions. Its development happens under three main perspectives: analysis of the main emotional barriers to responsible product consumption; analysis of the main emotional triggers; and survey of the benefits achieved.

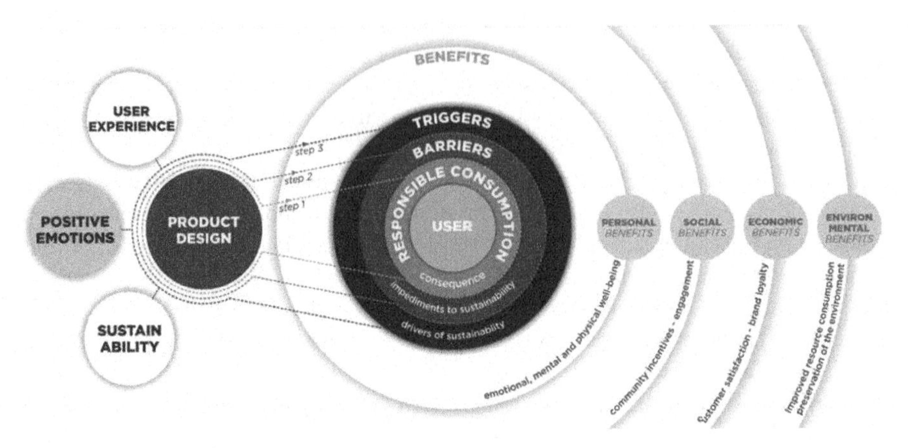

Fig. 3. Barriers, triggers and benefits of responsible consumption through positive emotions.

The second section brings the three steps to be considered in this product development, firstly one must adapt to the responsible consumption concepts, respecting its environmental, social, economic, and emotional attributes, the second step is to identify the main barriers. Several studies point out the barriers to responsible consumption of products [1, 4, 17], in general, these barriers that users often highlight are related to the need for cognitive, behavioral and financial effort (doubts about product quality, need to understand environmental issues, need for a daily change of habits, higher cost of sustainable products), distrust in companies in the government and in the products (the user concludes that it is no use just doing his part if for him the government and the companies are not totally transparent, the neighbors and close friends do not consider the theme relevant, the products do not seem to have the same quality as those the user is used to), the deprivation of daily pleasures (giving up products that offer small pleasures for environmental responsibility is a barrier that may affect the beginning of the acceptance of responsible consumption).

And the third step is to use the main triggers that influence conscious consumption, these are also explored in the literature [5, 14, 15]. The emotional ones include mainly aspects that concern the neighbor (contribute to a better future for my family, build a better world, opportunity to evolve as a human being, I give up small pleasures thinking of the good of all, a way of making a positive difference to the world), while the concrete aspects are more related to the benefits of responsible consumption itself (financial benefits, practicality and comfort in the routine, prevent reprimands from third parties, health benefits). In general, adherence to responsible consumption of products happens much more through emotional triggers, through positive emotions. This evidence serves as great support for the evolution of the current research, since it is the emotional triggers and benefits that most fuel responsible consumption, making it extremely possible that positive (vs. negative) emotions may in fact be more efficient as motivators for responsible product consumption.

The benefits reach four fundamental pillars for the balance of well-being: personal (emotional, mental, and physical well-being), social (community incentives, engagement), economic (customer satisfaction, brand loyalty) and environmental (improved resource consumption, preservation of the environment).

6 Conclusion

This article identified positive emotions as the main motivators of conscious and responsible product consumption. During the process a conceptual framework was built, which helps to visualize the fundamental aspects of product design through the user's emotional experience, providing a broader view of the consumption scenario and guidelines that relate to the application to product design.

From this analysis, some study opportunities were verified for the development of products that seek positive and optimistic emotional results. A first opportunity for study is to understand in more depth the barriers and triggers of responsible consumption behavior, since these are essential mechanisms for understanding purchase decision making. In addition, another opportunity is to deepen the understanding of positive emotions, approaching quantitatively which ones are more influential when faced with responsible consumption. It is still possible to highlight a final opportunity with the multidisciplinary of the proposed study, which encompasses the areas of design, engineering, marketing, sociology, and psychology, mainly because modeling the user's affect is a challenging task due to the complexity of understanding the resources of individuals and their subjectivity, involving interdisciplinary studies is essential for a complete approach and understanding.

References

1. Ladhari, R., Tchetgna, N.M.: Values, socially conscious behaviour and consumption emotions as predictors of Canadians' intent to buy fair trade products. Int. J. Consum. Stud. **41**(6), 696–705 (2017)
2. Escadas, M., Jalali, M.S., Farhangmehr, M.: What goes around comes around: the integrated role of emotions on consumer ethical decision-making. J. Consum. Behav. **19**(5), 409–422 (2020)
3. Muhammad, S., Ahmadi, H., Mortimer, G., Sekhon, H., Kharouf, H., Jebarajakirthy, C.: The interplay of positive and negative emotions to quit unhealthy consumption behaviors: insights for social marketers. Australas. Mark. J. **28**(4), 349–360 (2020)
4. Arli, D., Leo, C., Tjiptono, F.: Investigating the impact of guilt and shame proneness on consumer ethics: a cross national study. Int. J. Consum. Stud. **40**(1), 2–13 (2016)
5. Shiota, M.N., Neufeld, S.L., Danvers, A.F., Osborne, E.A., Sng, O., Yee, C.I.: Positive emotion differentiation: a functional approach. Soc. Pers. Psychol. Compass **8**, 104–117 (2014)
6. Desmet, P.M.A.: Faces of product pleasure: 25 positive emotions in human-product interactions. Int. J. Des. **6**(2), 1–29 (2012)
7. Desmet, P.M.A., Schifferstein, H.N.J.: Sources of positive and negative emotions in food experience. Appetite **50**(2–3), 290–301 (2008)

8. Gregory-Smith, D., Smith, A., Winklhofer, H.: Emotions and dissonance in 'ethical' consumption choices. J. Mark. Manag. **29**(11–12), 1201–1223 (2013)
9. Sudbury-Riley, L., Kohlbacher, F.: Ethically minded consumer behavior: scale review, development, and validation. J. Bus. Res. **69**(8), 2697–2710 (2016)
10. ISO 9241-210:2019 Homepage. https://www.iso.org/standard/77520.html. Accessed 05 Apr 2000
11. Ahn, H.: Modeling and analysis of affective influences on human experience. Prediction, Decision Making, and Behavior. Ph.D. thesis, MIT, Boston, MA (2010)
12. Knutson, B., Rick, S., Wimmer, G.E., Prelec, D., Loewenstein, G.: Neural predictors of purchases. Neuron **53**(1), 147–156 (2007)
13. Mogilner, C., Aaker, J., Kamvar, S.D.: How happiness affects choice. J. Consum. Res. **39**(2), 429–443 (2012)
14. McFerran, B., Aquino, K., Tracy, J.L.: Evidence for two facets of pride in consumption: findings from luxury brands. J. Consum. Psychol. **24**(4), 455–471 (2014)
15. Achar, C., So, J., Agrawal, N., Duhachek, A.: What we feel and why we buy: the influence of emotions on consumer decision-making. Curr. Opin. Psychol. **10**, 166–170 (2016)
16. Shigemoto, Y.: Designing emotional product design: when design management combines engineering and marketing. In: Fukuda, S. (ed.) AHFE 2019. AISC, vol. 952, pp. 28–39. Springer, Cham (2020). https://doi.org/10.1007/978-3-030-20441-9_4
17. Akatu Homepage. https://akatu.org.br/por-que-consumo-consciente. Accessed 05 Apr 2022
18. Ministério do Meio Ambiente Brasil Homepage. https://antigo.mma.gov.br/component/k2/item/7591. Accessed 25 Jan 2021
19. Long, M.A., Murray, D.L.: Ethical consumption, values convergence/divergence and community development. J. Agric. Environ. Ethics **26**(2), 351–375 (2013)
20. Wang, J., Wu, L.: The impact of emotions on the intention of sustainable consumption choices: evidence from a big city in an emerging country. J. Clean. Prod. **126**, 325–336 (2016)
21. Schwartz, M.S.: Ethical decision-making theory: an integrated approach. J. Bus. Ethics **139**(4), 755–776 (2016)
22. Winterich, K.P., Morales, A.C., Mittal, V.: Disgusted or happy, it is not so bad: emotional mini-max in unethical judgments. J. Bus. Ethics **130**(2), 343–360 (2015)
23. Unruh, G.U.: A proposed model for analysing and assessing human needs in the product development process. Hum. Factors Des. **9**(18), 52–77 (2020)
24. Norman, D.: Emotional Design: Why We Love (or Hate) Everyday Things. Basic Books, New York (2005)
25. Tavares, D.R.: Methodology for evaluating and translating cognitive and affective product experience into consumer requirements. Thesis (doctorate) – Pontifícia Universidade Católica do Paraná, Curitiba, Bibliography, pp. 142–154 (2021)

The Importance of Preparing Customized TRIZ Matrix to Accelerate the Innovation for Design Buildings

Hamidreza Hassanijajini and Mickael Gardoni(✉)

ÉTS - École de Technologie Supérieure, Université du Québec, Montréal, QC, Canada
Hamidreza.Hassanijajini.1@ens.etsmtl.ca,
Mickael.gardoni@etsmtl.ca

Abstract. This paper shows that the Lack of TRIZ implementation in the Construction industry is significant. Establishing specialized contradiction matrices extracted from TRIZ could help the designer solve further innovative proto-types than standard TRIZ. In this trend, some of the principles are more serviceable and more meaningful than the other principles that persuade us to recommend and highlight this research gap. Numerous papers have been written about TRIZ, but a restricted number of these essays are about the construction industry; especially in architecture and building manufacturing. The suggestion is that presenting a detailed methodology for invention in the construction industry's architectural field is remarkable. This paper focuses on several papers' literature reviews that helped us find our research gaps and research problem. Utilizing TRIZ can escalate the construction process's technical innovations. The utility of TRIZ directly in the building design process is not easy. In some fields, researchers extracted Eca-Triz from the original TRIZ, but Eca-Triz is not practical in the building design process [1]. So, establishing a customized contradiction matrix that has been extracted from the original TRIZ for building design is essential and valuable as a future work. Construction experts do not utilize formal or systematic design approaches in most cases. This circumstance results in several drawbacks (for a typical case, it is time consuming to discover an innovative and suitable solution). A systematic innovation approach could suggest to avoid such minuses that came out of TRIZ.

Keywords: TRIZ · Customized matrix · Design building · Innovation · Architecture

1 Introduction

TRIZ is the Theory of Inventive Problem Solving, a creative and innovative approach to problem-solving, which means problems can be solved by applying new ideas based on data and logic. [2]. Briefly, when an innovator wants to resolve a problem, he must procure an innovative approach to improve an element that will worsen another parameter. In other words, the inventor's innovatory solution has to endeavor to diminish or eliminate

Published by Springer Nature Switzerland AG 2023
F. Noël et al. (Eds.): PLM 2022, IFIP AICT 667, pp. 453–462, 2023.
https://doi.org/10.1007/978-3-031-25182-5_44

the worsening trend in another factor while improving one factor. Altshuller had several books, seminars, and articles, but he is famous because of his TRIZ contradiction matrix [3]. Since 2000, There has been much research in different fields utilizing TRIZ to find inventive solutions. Based on SCOPUS, the percentage of papers in the construction field that includes the word TRIZ is about two percent out of all TRIZ papers in all fields. Therefore, the lack of innovation in the construction industry is obvious. Various articles have been investigated with the keyword TRIZ, but the primary challenge is that when the search has limited to the words: construction engineering, architectural engineering, and building among TRIZ articles, the quantity of articles dramatically decreases. Furthermore, this limited collection of articles has either done joint work between QFD and TRIZ or presented examples of the usefulness of TRIZ. However, none of them has done straightforwardly preparing a special matrix for architectural engineering. Number 4 article is the only paper that worked on a multi-piece wall panel [4], which also did not work on specialized matrix. Rather, he has studied the exportable building that contains a segmented wall panel with the combined QFD and TRIZ methods, which also has differences from our proposed research. When numerous keywords have been examined, many articles have been founded, which have been supposed similar to current research in a glance, but the contrasts have been evidenced when the article's text has been read. Intact, there are a few papers like papers [5, 6], but they have not done a customized TRIZ contradiction matrix. The building and construction industry falls behind some other sectors (such as computers, IT, software, electronics, mechanical engineering, automotive industry, etc.) [2, 7–11]. Innovation can be explained as "the successful exploitation of new ideas" [2, 9]. The modern building industry is eager to use inventive design rather than traditional approaches to be more flexible and competitive in the novel construction market. The significance of building industry organizations' innovations is tremendous. Construction innovations can be appointed as a fourth dimension (4D) in the time ahead, parallel with the standard dimensions of time, cost, and quality. So, these companies could profit from market economy changes [12]. Observing the research mentioned above proved the significance of current research.

2 Literature Review

2.1 Lack of TRIZ Implementation in the Construction Industry

TRIZ has not been used vastly in developing new techniques or production in the industry. In other words, the lack of persistent and factual utility of TRIZ, an innovative approach to producing new products and designs, compared to its innate potential is significant [13–16]. Since 2000, There has been much research in different fields utilizing TRIZ to find inventive solutions.

Superior building artifacts must be innovative to be appointed as formidable on the market in performance, time, and cost-effectiveness. Furthermore, in a survey, 100% of respondents judged that innovation is pivotal for construction [17]. There are innumerable surveys, research, and literature regarding innovation in building or structure, and approximately all of them announce that innovations are essential in the construction field. Nevertheless, the question is how someone can become innovative. Utilizing TRIZ can escalate the construction process's technical innovations [18].

Construction experts do not utilize formal or systematic design approaches in most cases. This circumstance results in several drawbacks (for a typical case, it is time consuming to discover an innovative and suitable solution). A systematic innovation approach has suggested avoiding such minuses extracted from TRIZ. The procedure has 5 principles entitled "pillars": contradiction, resources, function, interfaces, and ideality [19].

2.2 Some Case Studies that Utilized TRIZ in the Construction Industry

The need for preparing specialized hypothesis for organizing innovation enhancement in the civil engineering field is obvious. Innovation is an essential side of the enrichment of construction techniques, but most procedures are based on the trial-and-error method. The Utility of TRIZ in discovering innovative solutions in Construction has been done in some subjects like tunnel construction [20, 21]. There are assorted articles in various construction fields, such as TRIZ in Evolution of Construction Techniques and Technologies [22–24], TRIZ in the Design of Construction Materials and New Structures [25–27], TRIZ in Value Engineering, and Construction Project Management [28–30]. However, a minimal number of them are relevant to our prospective research that I will particularly focus on them in the following chapters. Likewise, the innovation platform has been formulated by considering construction patents, and it gave more opportunities for obtaining inventive solutions in some sectors of this field [31].

Several design theories such as C-K theory, Coupled Design Process, Axiomatic Design, General Design Theory, Infused Design, and TRIZ are available.

Most of the theories are limited, which leads to inflexibility in the design process except for TRIZ and CK theory. TRIZ has principles for problem-solving and has special utility for our project. So, I choose TRIZ for our project. Ck theory is also an innovative design method, but it is needed knowledge and concept, and its knowledge is expanded during the design process. However, because CK theory could not solve our problem, we use TRIZ principles to solve our problem step by step.

None of them except TRIZ has compared 40,000 inventions to produce a matrix such as a TRIZ contradiction Matrix.

TRIZ has forty inventive principles extracted from 40000 inventions and should lead to solutions, especially in our case studies.

2.2.1 Case Study 1: The Implementation of the Theory of Inventive Problem-Solving in Architecture

In this article, TRIZ has been utilized for constructing a building more accessible for individuals with disabilities. A tool developed by Kishinev School of Moldova, the Innovation Situation Questionnaire (IQS), has been used to subdivide the problem into subproblems. The Problem Formulation Process (PFP) and inventive principles (MIP) tools have also been used with the contradiction matrix to find possible inventive solutions. Ultimately, the novel TRIZ contradiction Matrix was utilized to obtain solutions [32].

Table 1. Summarized Contradiction Matrix [32].

study					
FEATURE TO IMPROVE		UNDESIRED RESULT			
		...	PR-13 STABILITY OF THE COMPOSITION	...	PR-39
...	...				
PR-36	COMPLEXITY OF OBJECT		**IP-2, IP-22, IP-17, IP-19**		
...	...				
PR-39	...				

As it has been illustrated in Table 1, the principles IP-22 (convert harm into benefit), IP-2 (extraction), IP-19 (periodic action), and IP-17 (moving to a new dimension) have been extracted from the Contradiction matrix TRIZ. From the aforementioned principles, two of them (IP-22, IP-17) guide to results.

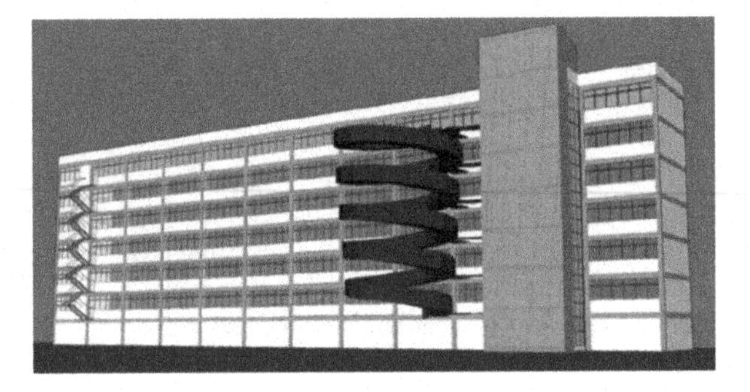

Fig. 1. TRIZ, IP – 17 application in the footbridges [32].

In Fig. 1, by inspiration from Principle IP-17, Spiral forms has been suggested for footbridge ramp [32].

2.2.2 Case Study 2: Development of an Exportable Modular Building System by Integrating Quality Function Deployment and TRIZ Method

The purpose of this article is to examine Quality function deployment and the use of the TRIZ technique in the manufacturing business. These solutions are exceptionally efficient in reducing costs and enhancing quality. In contrast to the regular manufacturing process, manufacturing and exportable modular construction systems include several concurrent subprocesses. Therefore, there is a limit to the efficiency attained if one of these approaches is used straight to product development. To solve this issue, the authors propose a novel technique that combines TRIZ with the deployment of quality functions. According to the findings of a case study, it is feasible to lower the volume of an exportable modular construction system compatible with ISO container shipping by 48% and the weight of structural steel by 30 percent by using the new approach [4].

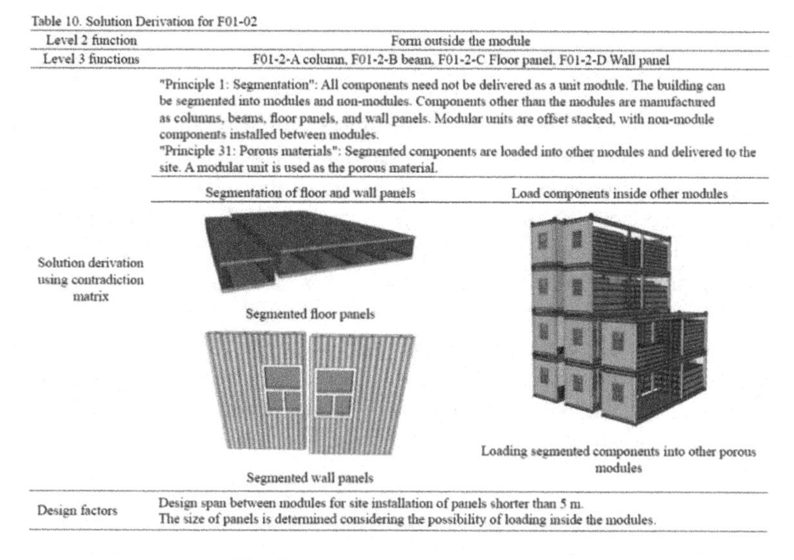

Table 10. Solution Derivation for F01-02

Level 2 function	Form outside the module	
Level 3 functions	F01-2-A column, F01-2-B beam, F01-2-C Floor panel, F01-2-D Wall panel	
Solution derivation using contradiction matrix	"Principle 1: Segmentation": All components need not be delivered as a unit module. The building can be segmented into modules and non-modules. Components other than the modules are manufactured as columns, beams, floor panels, and wall panels. Modular units are offset stacked, with non-module components installed between modules. "Principle 31: Porous materials": Segmented components are loaded into other modules and delivered to the site. A modular unit is used as the porous material.	
	Segmentation of floor and wall panels	Load components inside other modules
	Segmented floor panels Segmented wall panels	Loading segmented components into other porous modules
Design factors	Design span between modules for site installation of panels shorter than 5 m. The size of panels is determined considering the possibility of loading inside the modules.	

Fig. 2. Segmented wall panels [4].

As it has been illustrated in Fig. 2, On page 540 (JAABE vol. 16 no. 3 September 2017) in [4], the author (Seri Oh et al.) have used segmented wall panels for the Exportable Modular Building System, but it is just for transportable building design, and it is not what exactly we are going to do.

In Kiatake, M. and J.R.D. Petreche article, these following future works is mentioned:

"1 - Case studies of MIP applications to completely new design projects;
2 - Case studies of other TRIZ tools with experimentation in the Architectural design field;
3 - Studies on the interface of the TRIZ methodology and multi-criteria decision-aid methods;
4 - Development of architectural knowledge databases and design of computational support tools;
5 - Creation of a structure of the TRIZ theory concepts for application in architectural design education" [32].

These are the research gaps in TRIZ theory, precisely in the building and architecture industry, which shows our research problem and research gap. It is a worthwhile idea to adapt TRIZ to the conservative field of study like Construction and do some investigations to verify its performance.

In this context, TRIZ would be applicable more straightforwardly in the construction industry. To support this proposition, we can declare some comparable case studies have been done in papers [33–37] in other fields such as Redesign Service, Process Engineering, Quality Improvement, Related Context, and Electric Energy Storage Systems.

2.2.3 Case Study 3, 4

In a thesis, the author gives a number of examples demonstrating the effectiveness of TRIZ in underground constructions. The majority of case studies have been derived from real-world circumstances, and it has been shown that TRIZ methodologies aid in achieving creative conceptual outcomes. On the basis of TRIZ, a design framework for the technological innovation platform has been suggested by using patent knowledge in constructing projects. Some of the TRIZ concepts have not been used at all, whilst others have been used repeatedly [20].

After 3 years, the same author justified the advantage of TRIZ and its serviceableness in the Construction tunnel industry. Moreover, he did not consider the most suggestive principles in a customized new contradiction matrix to accelerate tunnel construction innovation by this methodology [20, 21] (Table 2).

Table 2. Contradiction matrix sample [21].

Parameter to be improved		Parameter that is getting worse				
		27	28	29	30	31
Engineering Parameters		Reliability	Accuracy of Measurement	Accuracy of Manufacturing	Harmful factors acting on object from outside	Harmful factors developed by an object
23	Loss of Substance			35, 10, 24, 31		
24	Loss of Information					
25	Waste of Time	10, 30, 4	24, 34, 28, 32	24, 26, 28, 18	35, 18, 34	35, 22, 18, 39
26	Amount of Substance			33, 30		
27	Reliability	Physical Contradiction		11, 32, 1		
28	Accuracy of Measurement		Physical Contradiction	.		

Resolution Principles	23) Feedback 24) Intermediary 25) Self-service	26) Copying 27) Inexpensive short-lived objects 28) Mechanics substitution

2.2.4 Case Study 5

Eca TRIZ has been suggested as a methodology to resolve some contradictions in eco-design. This methodology aims to aid small and medium-sized enterprises (SMEs) in developing products that will enable them to reach their eco-innovative goal. A qualitative matrix will enable the prioritization of all environmental impacts. Implementing the creative TRIZ principles on an individual basis will aid the researcher in selecting eco-innovative solutions. Based on an original contradiction matrix, a unique technique called

Ecatriz (ecology-friendly approach TRIZ) has been developed. It has been studied in various situations, including the "24 h of Innovation" competition and eco-innovative patents [1] (Table 3).

Table 3. The approach to obtaining the Ecatriz matrix [1].

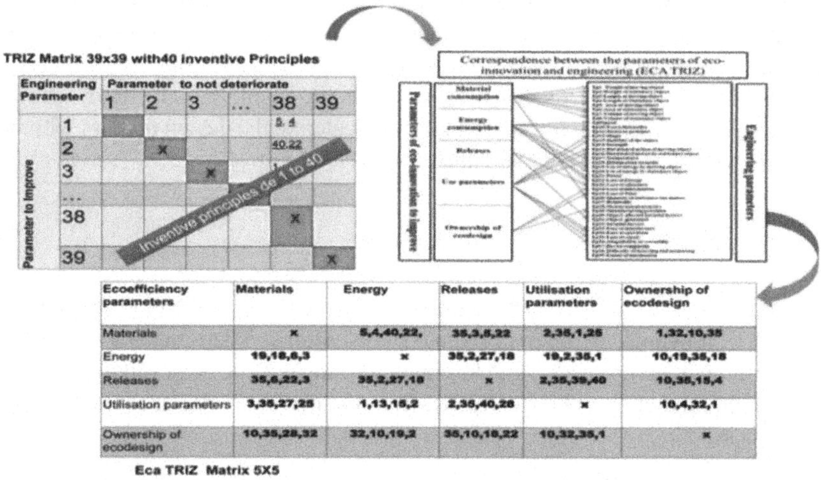

TRIZ Matrix 39x39 with40 inventive Principles

Engineering Parameter	Parameter to not deteriorate					
Parameter to Improve	1	2	3	...	38	39
1					5, 4	
2		x			40,22	
3			x			
...						
38					x	
39						x

Inventive principles de 1 to 40

Correspondence between the parameters of eco-innovation and engineering (ECA TRIZ)

Parameters of eco-innovation to improve

Material consumption

Energy consumption

Releases

Use parameters

Ownership of ecodesign

Engineering parameters

Ecoefficiency parameters	Materials	Energy	Releases	Utilisation parameters	Ownership of ecodesign
Materials	x	5,4,40,22,	35,3,5,22	2,35,1,25	1,32,10,35
Energy	19,18,6,3	x	35,2,27,18	19,2,35,1	10,19,35,18
Releases	35,6,22,3	35,2,27,18	x	2,35,39,40	10,35,15,4
Utilisation parameters	3,35,27,25	1,13,15,2	2,35,40,28	x	10,4,32,1
Ownership of ecodesign	10,35,28,32	32,10,19,2	35,10,18,22	10,32,35,1	x

Eca TRIZ Matrix 5X5

Based on these researches, we found that producing the customized contradiction matrix as the methodology in the architectural building design process with moveable walls would be a fascinating research gap that we can fill and add value to this field in the future.

3 Result

Previous surveys have concentrated on the efficacy of TRIZ in design various construction industries slightly and not on How to operate TRIZ to consider better the different customer needs base on considering customer elections by utilizing various architectural plans in the same apartment. TRIZ is one of the essential tools in the innovative design process. Especially for decreasing the cycles of design that is required to finalizing the architectural plan design process.

In other words, there are so many varied architectural plans that can be drawn for a unique apartment. How could we suggest an approach or tool that can provide different plans faster during the life cycle without wasting too much budget and time in renovating the apartment based on the new tenant's preferences? Consequently, this will diminish the price of the nonessential renovation of a building for interchanging its plan. The specialize matrix that could extracted from TRIZ could help the designers to accelerate the process of innovation in their design process.

4 Conclusion

Using TRIZ matrix in the building design process, the first challenge is that the plans produced right now by computer programs like Revit cannot consider all the needs of costumers perfectly. Moreover, when the property owner wants to rent out one apartment to a tenant. There are several choices for a tenant. Some tenants prefer a one-bedroom apartment rather than two bedrooms. Some others prefer to rent out two bedrooms or a studio without a bedroom, thinking about the problems and their possible solutions through the contradiction matrix Altshuler is significant. The main objective is to prepare and fulfill table by its 7 steps and prepare a specific contradiction Matrix extracted from the original TRIZ contradiction Matrix to accelerate the design process with more inventive solutions in the upcoming paper.

5 Future Works

As a future work, we will fulfill Customized Matrix and choose the most repetitive and relevant TRIZ principles to put in every part of this Matrix. Therefore, we hope that the work will become more specific, and fascinating results will be achieved by completing the specific contradiction matrix extracted from the original TRIZ matrix for building design process.

References

1. Cherifi, A., et al.: Eco-innovation and knowledge management: issues and organizational challenges to small and medium enterprises. AI EDAM **33**(2), 129–137 (2019)
2. Renev, I.A., Chechurin, L.S.: Application of TRIZ in building industry: study of current situation. Procedia CIRP **39**, 209–215 (2016)
3. TRIZ-CHANCE, W.o.t.O.G.S.A.w.d.b.O.S.a.K.S. Genrikh Saulovich Altshuller's Biography. https://www.altshuller.ru/biography/
4. Oh, S., Cho, B., Kim, D.-J.: Development of an exportable modular building system by integrating quality function deployment and TRIZ method. J. Asian Archit. Build. Eng. **16**(3), 535–542 (2017)
5. Potter, L., et al.: Developing a TRIZ-based design for flexibility tool for manufacturing facilities. In: Proceedings of IIE Annual Conference. Institute of Industrial and Systems Engineers (IISE) (2019)
6. Mann, D., Catháin, C.Ó.: Using TRIZ in architecture: first steps. TRIZ J. (2005)
7. Kulatunga, U., Amaratunga, D., Haigh, R.: Construction innovation: a literature review on current research (2006)
8. Cavallucci, D.: World wide status of TRIZ perceptions and uses a survey of results. Report at TRIZ Future (2009)
9. Dale, J.: Innovation in construction: ideas are the currency of the future. CIOB Survey (2007)
10. Blayse, A.M., Manley, K.: Key influences on construction innovation. Construction innovation (2004)
11. Ozorhon, B., et al.: Innovation in construction: a project life cycle approach. Salford Centre for Research and Innovation in the Built Environment (SCRI) Research Report, vol. 4, pp. 903–1012 (2010)

12. Asad, S., Khalfan, M., McDermott, P.: Promoting innovative thinking within construction. Salford (United Kingdom) In-House Publishing, pp. 62–71 (2006)
13. Goldense, B.: TRIZ is now practiced in 50 countries. Machine Design (2016)
14. Abramov, O., Sobolev, S.: Current Stage of TRIZ Evolution and its Popularity. In: Advances in Systematic Creativity, pp. 3–15. Springer (2019)
15. Chechurin, L.: TRIZ in science. Reviewing indexed publications. Procedia CIRP **39**, 156–165 (2016)
16. Chechurin, L., Elfvengren, K., Lohtander, M.: TRIZ integration into a product design roadmap. In: Proceedings of the 25th International Conference on Flexible Automation and Intelligent Manufacturing (FAIM), Wolverhampton, UK (2015)
17. Toole, T.M.: Technological trajectories of construction innovation. J. Archit. Eng. **7**(4), 107–114 (2001)
18. Ding, Z., Ma, J.: An exploration study of construction innovation principles: comparative analysis of construction scaffold and template patents. In: Wang, J., Ding, Z., Zou, L., Zuo, J. (eds.) Proceedings of the 17th International Symposium on Advancement of Construction Management and Real Estate, pp. 843–850. Springer, Heidelberg (2014). https://doi.org/10.1007/978-3-642-35548-6_86
19. Catháin, D.C.Ó., Mann, D.: Construction innovation using TRIZ. In: Global Innovation in Construction Conference (2009)
20. Mohamed, Y.: A framework for systematic improvement of construction systems. Ph. D. thesis, University of Alberta, Edmonton, AB, Canada (2002)
21. Mohamed, Y., AbouRizk, S.: Application of the theory of inventive problem solving in tunnel construction. J. Constr. Eng. Manag. **131**(10), 1099–1108 (2005)
22. Lin, P., Lee, Y.: Applying TRIZ to the construction industry. In: 10th International Conference on Civil, Structural and Environmental Engineering Computing, Civil-Comp (2005)
23. Coskun, K., Altun, C.: A conceptual model for improving in-situ construction techniques using triz and six sigma approaches. In: Vimonsatit, V., Singh, A., Yazdani, S. (Chair) Research Publishing. Symposium conducted at the meeting of the Research Development and Practice in Structural Engineering and Construction (ASEA-ISEC-1), Curtin University, Perth, Australia (2012)
24. Coşkun, K., Altun, C.: Applicability of TRIZ to in-situ construction techniques. In: 2nd International Conference on Construction and Project Management IPEDR (2011)
25. Yihong, L., Yunfei, S., Ting, C.: Study of new wall materials design based on TRIZ integrated innovation method. Manage. Sci. Eng. **6**(4), 15–29 (2012)
26. Lee, D., Shin, S.: Advanced high strength steel tube diagrid using TRIZ and nonlinear pushover analysis. J. Constr. Steel Res. **96**, 151–158 (2014)
27. Lee, D., Shin, S.: High tensile UL700 frame module with adjustable control of length and angle. J. Constr. Steel Res. **106**, 246–257 (2015)
28. Cui, H.: Exploring conflict resolution methods for construction projects: a TRIZ theory perspective. In: ICCREM 2014: Smart Construction and Management in the Context of New Technology, pp. 791–796 (2014)
29. Yang, J., Baeg, H., Moon, S.: Utilization of contradiction for creating design alternatives in construction value engineering. KSCE J. Civ. Eng. **18**(2), 355–364 (2014). https://doi.org/10.1007/s12205-014-0284-x
30. Moon, S., Choi, E., Hong, S.: Creation of robust design alternatives for temporary construction in value engineering. J. Constr. Eng. Manag. **142**(3), 04015086 (2016)
31. Ding, Z., Jiang, S., Wu, J.: Research on construction technology innovation platform based on TRIZ. In: Wen, Z., Li, T. (eds.) Knowledge Engineering and Management. AISC, vol. 278, pp. 211–223. Springer, Heidelberg (2014). https://doi.org/10.1007/978-3-642-54930-4_21
32. Kiatake, M., Petreche, J.R.D.: A case study on the application of the theory of inventive problem solving in architecture. Archit. Eng. Des. Manag. **8**(2), 90–102 (2012)

33. Pokhrel, C.: Adaptation of TRIZ method for problem solving in process engineering (2013)
34. Gazem, N., Rahman, A.A.: Improving TRIZ 40 inventive principles grouping in redesign service approaches. Asian Soc. Sci. **10**(17), 127–138 (2014)
35. Yang, C.-L., Huang, R.-H., Wei, W.-L.: Using modified TRIZ approach for quality improvement. Sci. J. Bus. Manag. **1**(1), 14–18 (2013)
36. Albers, A., et al.: Adaption of the TRIZ method to the development of electric energy storage systems. Procedia CIRP **21**, 509–514 (2014)
37. Gazem, N., Rahman, A.A.: Interpretation of TRIZ principles in a service related context. Asian Soc. Sci. **10**(13), 108 (2014)

Exploring a System Dynamics Approach to Develop Shared-Mobility Services Models: A Literature Review

Danilo Ribamar Sá Ribeiro[1](✉), Lúcio Galvão Mendes[1], Fernando Antônio Forcellini[1,2], and Mauricio Uriona-Maldonado[1]

[1] Graduate Program in Production Engineering, Federal University of Santa Catarina, Florianopolis, SC 88040-900, Brazil
danilo_saribeiro@hotmail.com

[2] Department of Mechanical Engineering, Federal University of Santa Catarina, Florianopolis, SC 88040-900, Brazil

Abstract. Shared Mobility Services (SHMS) are business models (BMs) that act as sharing economy platforms commonly based on mobile applications. They are conceived in a complex landscape characterized by the presence of information and the widespread presence of smart devices. In this sense, decision-makers, system developers, maintainers, and providers try to understand the interrelationships within this framework in terms of policy evaluation and/or impact on its successful adoption, analysis of all aspects of SHMS that affect the environment such as understanding patterns, user behavior, and improving the quality of this BM. To do this, models are built to capture the complex dependencies and interactions of these variables to guide decision-making. In this context, System Dynamics (SD) emerges as a useful approach to manage and capture the complexity/causality relationships of SHMS variables. Based on this, it developed a literature review to verify which research has applied SD to explore SHMS, analyzing existing models, their methods, variables, and impacts, as well as directions for future research. From a practical point of view, it is hoped that this research can be used as a guideline for urban planners and SHMS providers.

Keywords: Shared mobility services · System dynamics · Modeling

1 Introduction

Shared Mobility Services (SHMS) act as sharing economy platform commonly based on mobile applications, allow on-demand usage [16], and are conquering urban centers due it is beneficial for the Sustainable Development Goals (SDG 11) by promoting collective consumption behavior [37]. SHMS is part of the transportation system and also has some mutual influences, but it is difficult to estimate its impact on this system [17, 37]. Conceived in a complex scenario characterized by the presence of information and the wide presence of intelligent devices, understanding the interrelationships within the

Published by Springer Nature Switzerland AG 2023
F. Noël et al. (Eds.): PLM 2022, IFIP AICT 667, pp. 463–473, 2023.
https://doi.org/10.1007/978-3-031-25182-5_45

structure of this system provides an aggregate strategic view on the analysis of all aspects of SHMS that affect the total environment, which is crucial for the respective decision-makers, system developers, maintainers and providers [5, 34]. The use of simulation and modeling techniques is therefore considered necessary to conduct a consistent analysis and evaluation of different SHMS business models (BMs) in urban environments [9]. These can guide city planners to develop responsive policy interventions and service providers to develop SHMS [34]. In this context, System Dynamics (SD) emerges as a useful approach to manage and capture the complexity/causality relationships of variables [5, 6] that characterize many SHMS-related problems [34].

The interest in SD in the context of mobility is due to the growing challenges of intelligent configuration to understand and analyze the interactions between a set of dynamic factors that shape the patterns, behaviors, and impacts of these BMs [38]. [7] states that SD modeling aids in the construction of causal feedback relationships and consequent conceptual models, which can elucidate the behavior of various complex problems of a system or subsystems. This approach has been used in various applications and studies related to SHMS, as it is a potential tool for explaining the complex relationships of these systems [15, 29]. To understand the complexity of SHMS, new holistic and dynamic models of transportation demand must be developed [24].

In order to fill this gap, this paper aims to verify how SD has been used as an analysis tool for SHMS. More specifically, this paper answers the following questions: 1) Which studies have proposed dynamic models to understand the complexity of SHMS? 2) What are the main casual relationships and impacts studied by the models? 3) What are the main research directions on the application of SD in SHMS?

2 Research Background

2.1 Shared Mobility Services as a System

SHMS is a representation of a sharing economy, characterizing a BM that is linked to an emerging transportation strategy, smart urban infrastructure, and sustainable from social, economic, and environmental perspectives [5, 28]. It allows users to have short-term access to a transport mode as needed [37]. This new paradigm of a technology-mediated exchange system that promotes the sharing of underutilized assets, with no direct purchase and maintenance costs involved, and reduced fuel and parking expenses, is thus considered an economical alternative to car ownership [1].

SHMS are classified as a usage-driven product-service system (PSS), in which service providers sell the accessibility and use of specific products, retain ownership, and are responsible for the maintenance and performance of the products [14]. PSS is understood as a BM that combines products and services into a system that provides functionality for consumers [32]. Today, PSS become challenging in the context of Industry 4.0, leading to the integration of these BMs with digital technologies promoting digitalization and servitization [23]. Followed by the rapid progress of digitalization, SHMS has become an important and multifaceted topic [28, 37]. PSS and smart connected products are, from the companies' side, driving a market transition from the sale of products towards the sale of the use of solutions, and, from the customers' side, reshaping the concept of value [26]. Given that SHMS has a wide range of benefits, such as: reducing private vehicle use and congestion, providing cost savings through economies of scale, reducing emissions through the deployment of clean technology and fuels, facilitating more efficient land use (e.g. by reducing the number of required parking spaces), and increasing mobility options and connectivity between other modes of transportation [3, 12].

SHMS includes sharing of a vehicle, a passenger ride, or a delivery ride [27]. Carsharing, scooter sharing, and bike-sharing are services that enable the sharing of a vehicle. Ridesharing, on-demand ride services (including taxi e-Hail and ride-sourcing or transportation network companies - TNCs), and micro transit facilitate the sharing of a passenger ride. Finally, CNS enables the sharing of a delivery ride (i.e., a ride for cargo) [27, 37]. In addition, it can include sequential sharing, simultaneous sharing, and essential services [28]. The sharing modes are classified into Round-trip, One-way (one-trip, with an undetermined end), Peer-to-peer (users interact with each other), Business-to-peer, Station-based (the user can pick up and return the vehicle at fixed stations), and Free-floating [27].

2.2 System Dynamics

SD was developed by Jay Forrester in the 1960s at the Massachusetts Institute of Technology (MIT). It is defined as a set of conceptual tools used to study and understand the structure and dynamics of highly complex systems based on the foundation of feedback control theory [7, 29]. This approach deals with internal feedback loops and attractions that affect the behaviour of the system as a whole. Through various types of diagrams (causal; stock and flow), it is possible to graphically express a system (a well-delimited piece of an event) making it possible to see more clearly the dynamic complexity (over time) and the relationships between stakeholders [7].

Thus, SD seeks to simulate the behavior of the system over time, representing the relationship between identified key variables and is used for improvement-oriented decision-making or a better understanding of complex systems [29]. As a methodology, SD relies on both quality (e.g., survey and interview methods) and quantitative techniques (e.g., computer programming and simulation), emphasizes stakeholder involvement (to define mental models within the system), and encourages the researchers themselves to adopt a non-linear mental model (to seek and describe the feedback processes of a dynamic problem) [33].

[29] proposes a procedure that comprises five steps: (i) Identification and definition of the problem, representing the historical pattern to be described with the mathematical equations, what are its boundaries, the time horizon of the analysis, and the expected behavior of the system over time; (ii) Construction of the causal loop diagram (CLD), allow to map the mental models when analyzing systems, generate hypotheses about the causes of the dynamics of the system and communicate feedbacks considered responsible for the problem under analysis; (iii) construction of the simulation model to test the dynamic hypothesis, the stock and flow diagram (SFD) is drawn up and validation tests are performed; (iv) analysis of the experiments; (v) policy formulation and evaluation, used to better understand the roles and relative importance of model parameters in generating historical trends by creating entirely new strategies, structures, and decision rules.

3 Methodology

The method used in this research was conceptual-theoretical, based on a systematic literature review. This methodology promotes the collection and evaluation of knowledge through the theoretical synthesis of existing literature from a systematic sequence of steps and reproducible [31]. It used the Preferred Reporting Items for Systematic Reviews and Meta-Analyses (PRISMA) method that consists of the following steps: a) formulation of research questions; b) selection of databases and identification of keywords; c) evaluation and selection of studies; and d) analysis and synthesis [20]. The search term used to map the articles was: (*"smart mobilit*" OR "Shared mobilit*" OR "carsharing" OR "scooter sharing" OR "bike-sharing" OR "on-demand ride services" OR "ride-sourcing" OR "ridesharing") AND ("System Dynamics" OR "Behavio?r*" OR "dynamic model*" OR "Stock-flow diagram" OR "causal loop diagram"*). The choice of words was based on [27, 29] and defined from the research questions. This SLR is based on articles collected from the following databases: Scopus™ (Elsevier), Web of Science Core Collection™ (WoS), and Compendex - Engineering Village. Search engines were delimited to verify works related to the scope of this research. The inclusion and exclusion criteria used in the databases are shown in Table 1.

Figure 1 describes the process of inclusion and exclusion of papers during the evaluation and selection of studies. A total of 18 articles met the selection criteria and represent the bibliographic portfolio of this research. The synthesis of the results is a thematic analysis, a common technique for qualitative data, which comprises the organization of the studies into guiding axes for research [21].

Table 1. Inclusion and exclusion criteria.

Criteria		Explanation
Inclusion		The research efforts are related to studies that focused on SD in SHMS Document type: English peer-reviewed articles and conferences Time: Until February 2022
Exclusion	CE1	Papers that are not published in peer-reviewed journals or conferences
	CE2	Papers are written in a language other than English
	CE3	Duplicate papers
	CE4	Documents that are not available online
	CE5	It is used only as an example fact/ a part of research direction and future perspective/ a cited expression/ in keywords and/or references
	CE6	Non-adequacy to the scope of this study (papers applied other types of modeling/ simulation in SHMS)

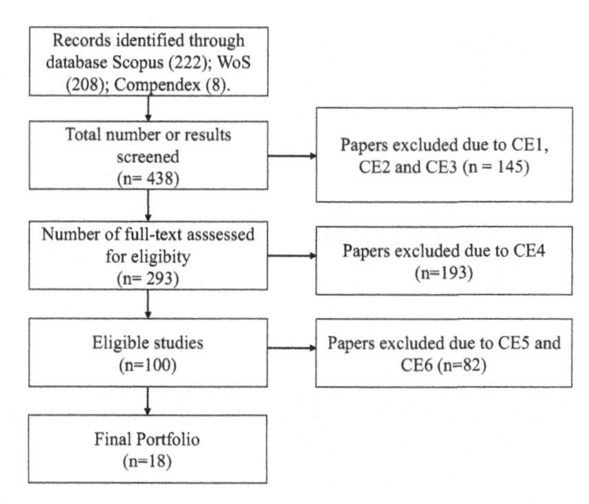

Fig. 1. PRISMA flowchart characterizing the search process.

4 Results and Discussion

Based on the methodology adopted, included 18 studies in the analysis to answer the research questions.

RQ1. Which Studies Have Proposed Dynamic Models to Understand the Complexity of SHMS? In this section, it will present the results according to the SHMS taxonomy provided by [27]. Most of the papers present models focusing on vehicle sharing, such as cars and bicycles as shown below. A few of the papers have focused on passenger sharing. Others discussed the general context of the theme.

Sharing of a Vehicle: [5] have built a CLD that shows the interconnections between carsharing, the car manufacturing industry, the environment, and key environmental and transportation regulations. [8] constructed a CLD that includes attributes that influenced the success of carsharing in Thailand and possible policy interventions through a group modeling building approach (GMB). [15] built an SD model to evaluate the long-term effects of e-carsharing policies on CO_2 emissions, the electric vehicle market in Fortaleza, and the conventional vehicle fleet. [39] presented an SD model to simulate the effect of introducing time-sharing electric vehicles in changing the user quantities in transportation tools, including public and private sectors, under different levels of government subsidies. [2] developed an SD model to analyze the adoption of self-consistent Connected and automated vehicles (CAV). [10] developed a conceptual SD model to analyze cybersecurity in the complex and uncertain deployment of CAVs. [36] developed an SFD that presents the variables used to model a dockless bike-sharing program (DBSP) and the causal relationships or linkages between them. The analysis focused on the economic profits of DBSPs in an environment of competition and government regulation. [35] developed an evolutionary game theory model combined with SD to demonstrate the interactive conflict between bike-sharing companies and government regulation. [30] developed an SD model based on the product life extension BM structure and business practices of Free-Floating bike-sharing in Beijing. [18] developed an SD model of commuter bicycling, through interviews and workshops with stakeholders, to simulate policy scenarios over the next 40 years in Auckland, New Zealand.

Shared of a Passenger Ride: [19] developed an SD model based on diffusion models to understand the influences that facilitate the growth of ridesharing by drivers and passengers. Using a sociotechnical systems-based SD model, [11] analyzed and predicted the market share between SHMS and cab services under three possible regulation scenarios to explore the relationship between innovation and information and communication technology (ICT) regulation. [22] modeled through an SFD the relationship between the dynamics of vehicle-hours-traveled (VHT), the vehicle-hours-dispatched (VHD), and the vehicle-hours-parked (VHP) using a shared mobility-on-demand Chicago cab system.

Generic Approaches: [25] explored the role of SHMS in activating SD of feedback loops in the context of sustainability transition. [13] analyzed the motorization process in China and then an SD model is designed for scenario analysis of urban traffic conditions and CO_2 emissions. [34] designed um a CLD based on causal feedback relationships between the various factors related to people's different mobility needs from the perspective of big data. [6] presented a decision support model using the SD method to evaluate urban mobility in Brazilian cities with policy influence. [4] evaluated the performance of several factors and attributes that influence a smart city: economy, mobility, and governance system, and recorded the stakeholders' views, for this they developed conceptual models of applied SD.

RQ2. What are the Main Casual Relationships and Impacts Studied by the Models? The papers included in this review are categorized into different aspects of SD, as shown in Table 2. The main variables of the models and their causal relationships from the papers are mentioned. Causality relationships are part of the representation of the relationship between two variables [29]. In addition, it classified the impacts of the models from social, economic, and/or environmental perspectives. In brief, impacts included policy/regulatory evaluation, subsidies, cost savings and convenience, reduction of vehicle miles traveled (VMT) or vehicle kilometers traveled (VKT), travel patterns and behavior as well as reduction of personal vehicle ownership as well as reduction of pollutant emissions such as CO_2. Most studies focus on the national or regional wideness of the results' impact. The main tools/ techniques for the construction and calibration of the models were research project [19], case study [4, 6, 15, 22, 34, 36], focus group and questionnaires [8, 13, 18, 19], series of searches of published literature [2, 5, 18, 19], primary datasets [2], historical data from the authoritative (secondary datasets) [6, 13, 15, 18, 22, 34]. Most studies developed the SD model and performed scenario analysis in software. In complementary SD, [13] performed a Sensitivity analysis and [35] performed an Evolutionary game analysis.

RQ3. What are the Main Research Directions on the Application of SD in SHMS? The authors referred to limitations and challenges associated with using SD modeling within the context of the problem being addressed. From the analysis of the final portfolio, it was possible to identify the predominant future directions:

Table 2. The main casual relationships and impact models of the studies.

Authors	Main causal relationships		Impact	Tool	Wideness
[19]	Ridesharing adoption	Market (drivers and passengers)	Social, Economic	Vensim	Regional
[2]	Adoption of CAVs	Energy Implications	Social, Environmental	Stella	National
[18]	Increase bicycle commuting in a car-dominated city	Effects of policies	Social, Economic	Vensim	Regional
[13]	Urban traffic conditions	CO_2 emissions	Social, Economic, Environmental	-	National
[34]	Different needs of people's mobility (smart mobility)	Optimizing decision making	Social, Economic	Vensim	Regional
[6]	Urban transport system	Environmental, economic, and traffic variables	Social, Economic, Environmental	Stella, iThink, PowerSim and Vensim	Regional
[22]	Taxi cab system	VHP, VHT, and VHD defined as a cost	Social, Economic	-	Regional
[36]	DBSPs	Maximizing the revenue	Economic	Vensim	Regional
[4]	Development of smart mobility	Access through ICT	Social, Environmental	-	Regional
[5]	Carsharing Services (considering the lifecycle of a shared vehicle)	Their environmental effects	Social, Environmental	Vensim	Global
[15]	E-carsharing	Electric vehicle adoption and CO_2 emissions	Social, Environmental	Stella	Regional
[35]	The conflict-handling strategy of bike-sharing companies	Government regulation	Social, Economic	Vensim	National

(continued)

Table 2. (*continued*)

Authors	Main causal relationships		Impact	Tool	Wideness
[39]	Development of Time-Sharing Electric Vehicles	Subsidy implications	Social, Economic, Environmental	Vensim	Regional
[8]	Carsharing operation	Viewpoints and expectations of stakeholders	Social, Economic	-	Regional
[30]	The business practices of DBSP	The product Lifetime extension	Social, Environmental	Vensim	Regional
[10]	Deployment of CAVs	Cybersecurity assessment	Social, Economic	-	Global
[11]	SHMS and taxi services	Regulation and the social benefits of mobility users	Social, Economic	AnyLogic	National

- Discuss the role of government in sustainable mobility [15]. Also, to be conducted on the interconnections between various modes of transportation and their relevant environmental effects on SHMS [5];
- Design and implement responsive policies according to the implemented scenarios in models [34];
- Explore the role of upscaling dynamics and feedback loops for the implementation of sufficiency-oriented transport policy measures [25];
- Explore the effectiveness of the models of the final portfolio, towards further validation of the parameters, and the refinement of the models, including new variables representing other facets of the problem or additional feedback loops, could be added [2, 11, 15, 19, 34, 39];
- Use methodologies that involve stakeholders' alignment to build SHMS-related mental models [8];
- Apply questionnaires to obtain the joint probability distribution of cost–convenience for users [39];
- Need for more studies about the urban transport system using the SD approaches [6, 18]. According to this review, most SD models focus on sharing vehicles (only cars and bikes) or passengers. In this sense, there is an opportunity to develop SD models to understand scooter sharing (adoption and its impacts) and model various categories of these BMs in combination [5].
- Cover a variety of SHMS focusing on product lifetime and its five key activities (improved product design, access, maintenance, redistribution, and recovery) [30].

5 Conclusion

This paper provides a comprehensive review of research that has applied SD to explore SHMS. A carefully chosen set of keywords was used to discover relevant studies. Following PRISMA guidelines, it performed a content analysis based on three research questions: Which studies have proposed dynamic models to understand the complexity of SHMS, what are the main casual relationships and impacts studied by the models, and what are the main research directions on the application of SD in SHMS. To answer these questions, the included records have been reviewed, and the essential aspects of SD models and the future directions of this thematic were defined. Based on the previously mentioned perspectives, it concludes that SD presents to be an appropriate approach to capture the causal relationships among variables in SHMS that is part of such a complex system as urban transportation [5]. These studies have theoretical and practical implications by contributing to the exploration of the applicability of SD and to the identification of value and main controls for assertive decision-making to get more consistent BMs of SHMS and to analyze and assess its multiple influences on the urban environment [19]. Furthermore, most studies analyze policies focused on the adoption of positive externalities of SHMS, highlighting efforts in the social, economic, and environmental spheres. In sum, this research presents itself as a useful starting point for researchers, practitioners, and stakeholders when it comes to this topic.

Acknowledgments. The authors acknowledge the Coordination for the Improvement of Higher Education Personnel (Brazil) – CAPES | Finance Code 001.

References

1. Acheampong, R.A., Siiba, A.: Modelling the determinants of car-sharing adoption intentions among young adults: the role of attitude, perceived benefits, travel expectations and socio-demographic factors. Transportation **47**(5), 2557–2580 (2019). https://doi.org/10.1007/s11116-019-10029-3
2. Bush, B., Vimmerstedt, L., Gonder, J.: Potential energy implications of connected and automated vehicles: exploring key leverage points through scenario screening and analysis. Transp. Res. Rec. J. Transp. Res. Board **2673**, 84–94 (2019)
3. Charkhgard, H., Takalloo, M., Haider, Z.: Bi-objective autonomous vehicle repositioning problem with travel time uncertainty. 4OR **18**, 477–505 (2020). https://doi.org/10.1007/s10288-019-00429-7
4. Das, D.: Perspectives of smart cities in South Africa through applied systems analysis approach: a case of Bloemfontein. Constr. Econ. Build. **20**, 65–88 (2020)
5. Esfandabadi, Z.S., et al.: Conceptualizing environmental effects of carsharing services: a system thinking approach. Sci. Total Environ. **745**, 141169 (2020). https://doi.org/10.1016/j.scitotenv.2020.141169
6. Fontoura, W., Chaves, G.: Ribeiro, GM: the brazilian urban mobility policy: the impact in são paulo transport system using system dynamics. Transp Policy **73**, 51–61 (2019)
7. Forrester, J.: Industrial dynamics. J. Oper. Res. Soc. **48**, 1037–1041 (1997)
8. Jittrapirom, P., et al.: Aligning stakeholders' mental models on carsharing system using remote focus group method. Transp. Res. Part D Transp. Environ. **101**, 103122 (2021). https://doi.org/10.1016/j.trd.2021.103122

9. Kamargianni, M., et al.: Incorporating the mobility as a service concept into transport modelling and simulation frameworks. In: Transportation Research Board (2019)
10. Khan, S., Shiwakoti, N., Stasinopoulos, P.: A conceptual system dynamics model for cybersecurity assessment of connected and autonomous vehicles. Accid. Anal. Prev. **165**, 106515 (2022). https://doi.org/10.1016/j.aap.2021.106515
11. Lee, J., et al.: The relationship between shared mobility and regulation in South Korea: a system dynamics approach from the socio-technical transitions perspective. Technovation **109**, 102327 (2022). https://doi.org/10.1016/j.technovation.2021.102327
12. Librino, F., et al.: Home-work carpooling for social mixing. Transportation **47**(5), 2671–2701 (2019). https://doi.org/10.1007/s11116-019-10038-2
13. Liu, S., et al.: Analysis of transport policy effect on co 2 emissions based on system dynamics. Adv. Mech. Eng. **7**, 323819 (2015). https://doi.org/10.1155/2014/323819
14. Lu, D., Lai, I., Liu, Y.: The consumer acceptance of smart product-service systems in sharing economy: the effects of perceived interactivity and particularity. Sustainability **11**, 928 (2019). https://doi.org/10.3390/su11030928
15. Luna, T., et al.: The influence of e-carsharing schemes on electric vehicle adoption and carbon emissions: an emerging economy study. Transp. Res. Part D: Transp. Environ. **79**, 102226 (2020). https://doi.org/10.1016/j.trd.2020.102226
16. Ma, Y., et al.: Co-evolution between urban sustainability and business ecosystem innovation: Evidence from the sharing mobility sector in Shanghai. J. Clean. Prod. **188**, 942–953 (2018). https://doi.org/10.1016/j.jclepro.2018.03.323
17. Machado, C., et al.: An Overview of Shared Mobility. Sustainability **10**, 4342 (2018)
18. Macmillan, A., et al.: The societal costs and benefits of commuter bicycling: simulating the effects of specific policies using system dynamics modeling. Environ. Health. Perspect. **122**, 335–344 (2014). https://doi.org/10.1289/ehp.1307250
19. Marco, A., Giannantonio, R., Zenezini, G.: The diffusion mechanisms of dynamic ridesharing services. Prog. Ind. Ecol. An Int. J. **9**(4), 408 (2015). https://doi.org/10.1504/PIE.2015.076900
20. Moher, D., et al.: Preferred reporting items for systematic reviews and meta-analyses: the PRISMA statement. Bmj. Res. Methods Rep. (2009). https://doi.org/10.1136/bmj.b2535
21. Munaro, M., Tavares, S., Bragança, L.: Towards circular and more sustainable buildings: a systematic literature review on the circular economy in the built environment. J. Clean Prod. **260**, 121134 (2020). https://doi.org/10.1016/j.jclepro.2020.121134
22. Papanikolaou, D.: Computing and visualizing taxi cab dynamics as proxies for autonomous mobility on demand systems. In: Lee, J.-H. (ed.) CAAD Futures 2019. CCIS, vol. 1028, pp. 183–197. Springer, Singapore (2019). https://doi.org/10.1007/978-981-13-8410-3_14
23. Pirola, F., et al.: Digital technologies in PSS: a literature review and a research agenda. Comput. Ind. **123**, 103301 (2020). https://doi.org/10.1016/j.compind.2020.103301
24. Reul, J., Grube, T., Stolten, D.: Urban transportation at an inflection point: an analysis of potential influencing factors. Transp. Res. Part D: Transp. Environ. **92**, 102733 (2021)
25. Ruhrort, L.: Reassessing the role of shared mobility services in a transport transition: can they contribute the rise of an alternative socio-technical regime of mobility? Sustainability **12**, 8253 (2020). https://doi.org/10.3390/su12198253
26. Sassanelli, C., et al.: Defining lean product service systems features and research trends through a systematic literature review. Int. J. Prod. Lifecycle Manag. **12**, 37 (2019)
27. Shaheen, S., et al.: Shared mobility: definitions, industry developments, and early understanding. Transp. Sustain Res. Center, Innov. Mobil Res. (2015)
28. Shokouhyar, S., et al.: Shared mobility in post-COVID era: new challenges and opportunities. Sustain Cities Soc. **67**, 102714 (2021). https://doi.org/10.1016/j.scs.2021.102714
29. Sterman, J.: Business Dynamics: Systems Thinking and MODeling for a Complex World. The McGraw-Hill Higher Education, Boston (2000)

30. Sun, S.: How does the collaborative economy advance better product lifetimes? A case study of free-floating bike sharing. Sustainability **13**, 1434 (2021). https://doi.org/10.3390/su1303 1434
31. Tranfield, D., Denyer, D., Smart, P.: Towards a methodology for developing evidence-informed management knowledge by means of systematic review. Br. J. Manag. **14**, 207–222 (2003). https://doi.org/10.1111/1467-8551.00375
32. Tukker, A.: Product services for a resource-efficient and circular economy - a review. J. Clean. Prod. **97**, 76–91 (2015). https://doi.org/10.1016/j.jclepro.2013.11.049
33. Turner, B., et al.: system dynamics modeling for agricultural and natural resource management issues: review of some past cases and forecasting future roles. Resources **5**, 40 (2016). https://doi.org/10.3390/resources5040040
34. Del Vecchio, P., et al.: A system dynamic approach for the smart mobility of people: implications in the age of big data. Technol. Forecast. Soc. Change **149**, 119771 (2019). https://doi.org/10.1016/j.techfore.2019.119771
35. Yang, H., et al.: Conflicts between business and government in bike sharing system. Int. J. Confl. Manag. **31**, 463–487 (2020). https://doi.org/10.1108/IJCMA-10-2019-0191
36. Yang, T., Li, Y., Zhou, S.: System dynamics modeling of dockless bike-sharing program operations: a case study of mobike in Beijing, China. Sustainability **11**, 1601 (2019). https://doi.org/10.3390/su11061601
37. Yi, W., Yan, J.: Energy consumption and emission influences from shared mobility in China: a national level annual data analysis. Appl. Energy **277**, 115549 (2020). https://doi.org/10.1016/j.apenergy.2020.115549
38. Zhang, Y., et al.: Smart mobility control agent for enhanced oil recovery during CO2 flooding in ultra-low permeability reservoirs. Fuel **241**, 442–450 (2019)
39. Zhou, B., Hu, H., Dai, L.: Assessment of the development of time-sharing electric vehicles in shanghai and subsidy implications: a system dynamics approach. Sustainability **12**, 345 (2020). https://doi.org/10.3390/su12010345

To Develop Capabilities for Patient-Specific Devices Through the Configuration of Product Life-Cycle Management Strategy

Clara Isabel López Gualdrón(✉), Israel garnica, and Javier Mauricio Martínez Gómez

Universidad Industrial de Santander UIS, Cra 27 Calle 9, 680006 Bucaramanga, Colombia
clalogu@uis.edu.co

Abstract. Technological shifts have changed the conception of new medical devices, such as orthopedic surgical devices. From a traditional approach to a flexible manufacturing concept for developing new products like patient-specific implants (PSI), biomodels, and surgical guides. However, process development becomes increasingly complex compared to traditional standards if a personalized product is required. Consequently, the configuration of the product's life-cycle management (PLM) was analyzed as a business process strategy considering technological implementation tools to achieve goals such as quality, cost, and time. Through this research, an innovation model has been defined to develop a PLM strategy. Process areas, as well as actors, tasks, and practices, had been defined. Those had guided the building of a reference framework for medical device development by case studies. From our sample, some were concepts while others were evaluated in a relevant environment. These cases were developed iteratively to reach a certain technology readiness level (TRL). Thus, practices applied as well as performance was observed. The research addresses value creation in new product development and how building capabilities were consolidated by PLM implementation. As a result, team members generated artifacts with an intermediate-high degree of personalization. Also, it was possible to define a flexible process based on-demand analysis. Additionally, training for low-cost technological integration could provide tools to produce viable products in local markets. Therefore, improvement opportunities and forward applicability in real scenarios must be addressed.

Keywords: New product development · Flexible factories · Product life-cycle management · PLM · Technology development capabilities

1 Introduction

The technological shift has been integrated into the real world by mixing physical and biological systems. This situation allowed us to alter reality and therefore our environment. So current reality has been the result to put together new studies of materials and technologies. Those had been applied to subjects like personalized products or bioprinting, redefining how processes, products, and value creation must be conceived, as

Published by Springer Nature Switzerland AG 2023
F. Noël et al. (Eds.): PLM 2022, IFIP AICT 667, pp. 474–483, 2023.
https://doi.org/10.1007/978-3-031-25182-5_46

mentioned by the World Economic Forum (WEF) [1]. Among technological trends, there are enabling technologies [2], additive manufacturing, and tailored product development [3]. Those trends could be integrated into a business strategy by implementing practices from 4.0 industries [4].

However, there is no business unless a value is perceived. Value creation, in the context of a new product design (NPD), is based on intellectual capital, process modeling, and operation management [5]. To maximize efficiency, information technology (IT) solutions are implemented to reduce time and cost in development, then increased customer satisfaction. A positive impact of IT on NPD throughout development can be attained in different ways: former projects on databases, project management applications, CAD/CAE/CAM tools, and interconnectedness among multidisciplinary faraway actors [6].

Likewise, in NPD different kinds of information must be managed based on inputs and outputs. Because different actors generated all sorts of data, a strategy must be outlined to control how the process flows, which activities must be carried on, and when collaborative work takes place among roles. Faced with this gap, product data management (PDM) as part of the PLM strategy has been configured like a business process strategy [7]. As an open innovation capability, PLM has been implemented in technology-based organizations for NPD as a collaborative framework [8]. Through PLM, data flows through in-out in a safe, reliable, and orderly way, to manage the creation and exchange of info related to the development of the NPD process [9]. Thus, PLM is a business strategy that integrates people, processes, and information to handle product maturity [10].

PLM is aligned with the concept of direct digital manufacturing (DDM) [11], allowing a network from dispersed resources like additive manufacturing technologies [12], to expedite development for NPD aimed at customization. A positive effect reported for implementing PLM was intermediate-high tailored NPD, time, and cost-saving [13]. PLM helps to define a flexible manufacturing process based on demand by supporting: human capital renovation and training with specific skills, technological resources management[14], and manufacturing control [15].

Previous advances offer flexibility to develop NPD, which is promising in the healthcare sector. Delimiting its application, a current problem noticed by researchers is how to develop medical devices (MD) for surgical procedures [16] in a low-income context [17]. For surgery, each patient requires a specific medical device (SMD) according to his/her biological topology, such as patient-specific implants (PSI), surgical guides for cutting, positioning, or drilling; and biomodels [18].

Despite PLM advantages in healthcare, few studies related to it have been found. Those were associated with: management by commercial PLM software of heterogenic data from neuroimaging and biomedical data in a laboratory [19]; offering 3d visualization for surgical pre-planning on a web-based platform for surgeons [20]; designing a framework to theorize relationships among actors in PSI development [21]. Those reports show how technological change could impact the conception of new SMD, although few studies explain how PLM solutions during NPD could be implemented in organizations with scarce resources [22].

Based on the previous outlook, we started with manufacturing trends like capability, customization, and PLM, then how could be relevant in SMD for surgical assistance in orthopedic, maxillofacial, or plastic surgery.

Thus, the statement research question was: **How a PLM strategy could be configurated to cope with building capabilities for using enabled technologies in SMD development that support complex surgical assistance?**

The description of this scenery intervened to shape the innovation strategy model proposed. Depiction of its main components was presented in the following sections, and how practices guided the construction of an innovation framework as a reference for SMD development for surgical assistance in our context.

2 Methodology

To achieve the building of development capabilities in SMD some concepts were intercepted for a theoretical foundation. Those were: the value drivers model [23]; the innovation capability model in firms [24]; and the PLM visualization model for SMD [25]. Thus, it was possible to involve capabilities such as technological development, manufacturing, and management; with value creation while the NPD's development evolves through process areas. Value definition was oriented on three milestones: first, the value discovery, to define who create it; second, the value proposal, to define how value is created by building capabilities and who made each activity; and third, value configuration, to obtain value by device materialization in each study case.

The first milestone, the value discovery, was related to technological development capability. In the beginning, epidemiology of traumatic injuries on the skull and leg was observed, as well to see among health staff who is involved and what kind of difficulties faces to offer well treatment. In the Plastic Surgical Department at Hospital Universitario de Santander, it was made an epidemiological profile with a sample convenience. Among the etiology, the vast majority were trauma; other causes were related to cancer or congenital disorders. According to medical experience, each case was examined to diagnose whether or not an SMD was required.

In the second milestone, the value proposal, it was defined how actors involved in SMD exchange data and how they relate with each other. Thus, the PLM visualization strategy was employed to build research and development (R&D) capability, throughout process areas, each one responsible step by step in NPD decisions making. The previous process was non-linear and iterative. As more case studies were involved, the strategy reach maturity.

R&D and PLM strategies were oriented towards value configuration by layers. Each layer is consecutive and represents a capability; to develop technology, make it operational, and manage all participant processes. In the R&D stage, the state of art in SMD was oriented toward competitive surveillance from similar companies and its patents. To reach SMD, enabling technologies were selected by low-cost considerations to configure a system of systems (SoS) by incorporating: computer-aided tools (CAx) for design, reverse engineering (RE) for patient geometry rebuilding, and additive manufacturing (3DP) to obtain iterative SMD for medical preplanning or surgical support. For PLM strategy configuration, a free PDM was selected to centralize data (support, training, and each study case), control data versions, formalize the format process, support different roles (collaborator, administrator, viewer), and allowed non-designer users a 3d data visualization. As a theoretical framework, designers and physicians collaborate by cocreation through design thinking methodology, as well learning-by-doing LBD and self-training, skills that depend on the kind of SMD that each case required by when: before, during, or after surgery.

In the third milestone, value configuration was achieved by: joining efforts to develop study cases depending on etiology. With every case, learned lessons were clarified on requirements and parameters, by proposing as practice a requirement library to allow reuse from previous knowledge. Another practice was to conform multidisciplinary teams by case. Thus, students from designing worked together with health staff or students, like physicians specializing in plastic surgery or orthopedic. Those students were members of research groups, INTERFAZ and GRICES, at Industrial Design School and Medicine School respectively, both from Universidad Industrial de Santander.

Once a case was evaluated as pertinent by epidemiology profile, students from GRICES presented each case's necessities to development members. Then INTERFAZ centralized data and execute the development. First roles were assigned, subsequently, the case's scope was delimited by time, resources, explicit requirements, and the kind of SMD. As consulted by the literature, four kinds of products were developed to materialize technological development and manufacturing capabilities: virtual preplanning, biomodels, surgical guides, and PSI [18]. INTERFAZ was dealing with how to design and build SMD, meanwhile, GRICES was responsible for thinking about surgical steps and how those must be taken into consideration in the final SMD. Depending on the risk, each functional requirement was checked in iterative steps among participants before surgery. Ethical considerations were evaluated in each study case, allowed by the university ethics committee in medical specialization projects.

3 Results

In the workflow, the process flows step by step identifying roles, tasks, and technology. Some practices applied are described below.

Technological and Competitive Surveillance: To understand the state of art in each SMD was required, as well as the value creation. This surveillance helps in research and development practices, as well as to detect potential ideas through process innovation practices in NPD. Voice of the literature (VoL) was oriented to looking for reported SMD and its performance. While the voice of publication (VoP) was conducted to identify key information by patent review. For each study case, surveillance showed us key details, so concepts generated by development teams could maintain a level of inventiveness distinguish from known SMD to get innovative potential in students' proposals.

Enabling Technologies: A technology assessment was proposed to configure a CAx architecture, to guarantee technology development capabilities therefore operational capabilities. Thus, the value configuration related to DDM and 4.0 industries could be viable by technological appropriation. However, it demanded an SoS by selecting each resource for compatibility and low cost, both software, and hardware. So, enabling technologies such as RE/CAx/3DP configure an SoS that could make a project viable.

Different projects, each one a case study, were developed by R&D, in some of those, SoS alternatives were implemented. Only with LBD and accumulative experience in each iteration, was possible to select an SoS according to performance. Due to the previous amount of data, it was necessary to configure a collaborative strategy to govern multidisciplinary information through PDM.

Workflow and Process Steps: A collaborative framework was proposed to enable knowledge transfer and data flow associated with three scenarios: research, development, and service. Employing a collaborative data management platform PDM, it was possible to structure key activities in a custom SMD development with actors and process areas. Thus, decisions making from start to end in each activity was delimited to understand, step by step, a filtering process to achieve the requirement accomplishment for the final product. Also, PDM allows us to secure communication channels among actors, and to keep patients' data privately.

Design Thinking (DT): The methodology applied in each study case looked to solve complexity in SMD. DT was also crucial to establishing a common vocabulary among designers, and physicians, to help them to empathize with each other skills. It was implemented because of the value proposal, meaning expectations from the idea must be met

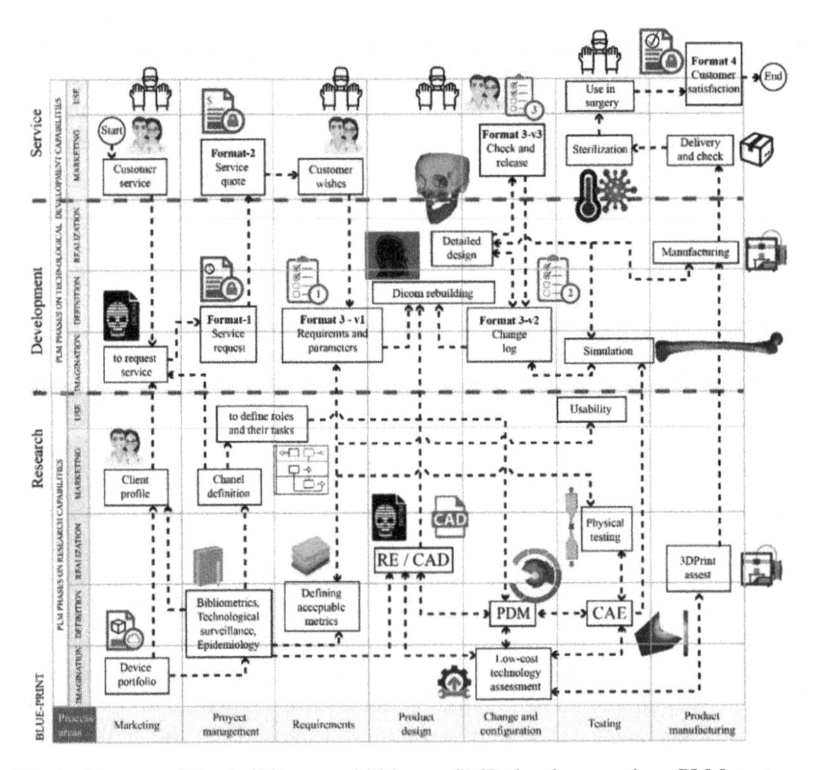

Fig. 1. Framework for building capabilities on SMD development by a PLM strategy.

in the final SMD. In the design process, medical knowledge was traduced on requirements and parameters, to help designers to cope with detailing aspects like tolerances, draft angles, and material performance, among others.

As a result of previous practices for SMD development, a framework in the initial phases of the product life cycle was proposed in Fig. 1. Relations between actors happened according to the case scope, meaning that a case study could be done only if at the research level capabilities were achieved. Thus, development could be accomplished, although decisions must be checked. To meet initial expectations, requirements were formalized in formats. Thus, iterative development was necessary for co-creation, to get a final SMD version, and finally to release for surgery.

Results derived from the previous framework showed us that deployment of innovations through the R&D process could finish on incremental innovation as well as efficiency and satisfaction. The value creation in the NPD process was enhanced through knowledge internalization in SMD creation, like how enabling technologies must be used, or when co-creation is needed. Collaboration between physicians and designers helped to achieve accurate SMD. Each process demanded empathy, trial-and-error, with LBD, and flexibility to iterate through the DT process.

Table 1. Cases and products description developed by this strategy.

Service request	Case	Surgical treatment	Virtual product		Tangible product		
			Biomodel	Pre-planning	PSI	Surgical guide	Biomodel
Diagnostic	Tibial plateau trauma	Fracture reduction	22				22
Surgical guides	Craniosynostosis	Cranial defect correction	1	1		1	
	Orbital-malar trauma	Fracture reduction	2	2	2		2
	Severe polytrauma pseudoarthrosis	Reconstructive surgery	1	1			1
	Maxillary retrognathia	Le Fort 1	3	3		3	
	Mandibular Fracture Sequelae	Segmental mandibulectomy	1	1	1	1	1
Patient-specific implant (PSI)	Cranioplasty	Bone defect reduction	3	2	3		3
	Type B hemipelvic fracture	Fracture reduction	1		1		
	Maxillary partial edentulism	Insertion of implants or prostheses	1		1		1
Total of SMD required			35	10	8	4	30

Thus, value configurations were materialized in different case studies depending on complexity, as summarized in Table 1. It also showed the service requested, the type of surgical treatment proposed, and the virtual and physical products obtained. That led to defining a flexible manufacturing process that adjusts to tailored production. SMD was produced to reduce uncertainty in surgical treatment, so the case studies had different customer requests. Those could be categorized as orthopedic, referring to leg trauma; or plastic surgery for skull and jaw injuries. The orthopedic cases were studied just for diagnosis and training by biomodel visualization. Instead, plastic surgery cases required not just diagnosis by biomodel, but also virtual surgical preplanning, cutting guides, and/or PSI. Thus, two kinds of products were observed: the first one was virtual intangible 3D biomodels (35 in total), and virtual surgical pre-planning (10 in total); and the second one was detailed design to obtain tangible SMD by 3Dp: biomodels (30 in total), PSI (8 in total), and cutting guides (4 in total).

In this sense, the value configuration through a workflow is adaptable to each case study according to its requirements. Thus, personalization criteria were how the value configuration was oriented to create value. The value chain was also incorporated to consolidate the first approach to value configuration. Thus, management capability was oriented by PLM strategy, to allow building capability by development stages and accumulative knowledge. As a result, different products, tangible, and intangible, were obtained. Those SMD were achieved on a natural scale (1:1) so that would be applicable in surgery. However, PSI was an exception, due to limitations in available technologies: our 3D printing hardware could not print on clinical material. So, PSI was obtained for geometrical comparison models and accuracy performance. On the other hand, biomodels and surgical guides, which required minor risk, were printed to verify geometrical accuracy and fitting on bone defect correction. So, some case studies could reach a certain maturity level of TRL because some artifacts could be used in relevant environments.

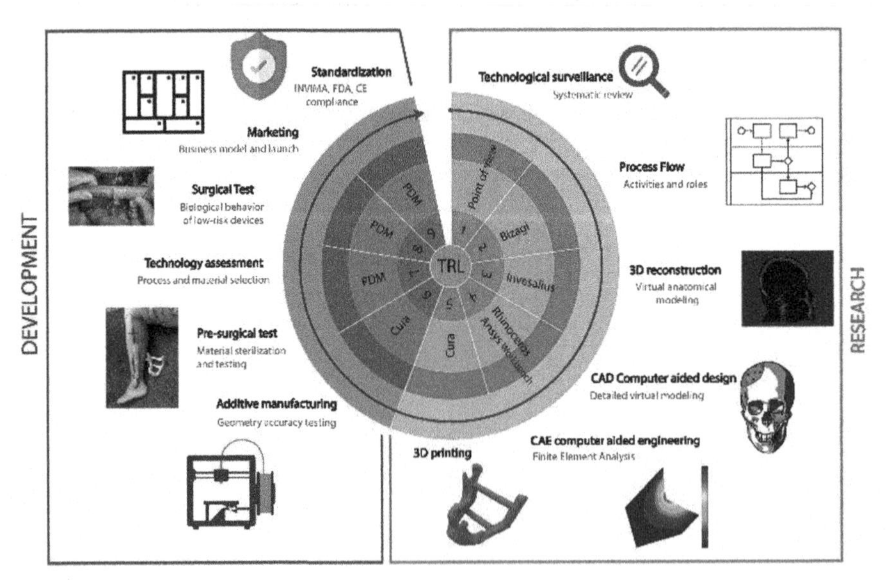

Fig. 2. A product development process model and its practices adjusted to TRL [14].

Figure 2 visualizes how a common process, even considering a diverse kind of SMD, could be oriented to make it repeatable.

Based on our experience, case studies were carried out consequently to physicians' expectations. TRL technology maturity reference framework [26, 27], ranging from level 1 to 9, was used to establish case study development. Thus, to reach a higher rank, practices were built to be followed up. Until TRL-5, practices were related to research; from TRL-6 onwards, practices were related to technology development.

In such a way, Fig. 2 shows the process in a particular case study where a female patient needed jaw reconstruction by using allograft. TRL-1 outlined technological surveillance to realize the state of the art. TRL-2 defined a plan and suitable SoS according to the case study. TRL-3 defined virtual reconstruction from a computer tomography CT into a 3D biomodel, by Invesalius®. TRL-4 runs through SMD design in CAD such as Rhinoceros®, and mechanical strength simulation in CAE like Ansys®. TRL-5 specified 3D procedures to obtain a model by 3DP by using a slider like Cura®. All data was centralized by using GrabCAD® PDM. TRL-6 specified a biocompatible material to obtain an SMD. TRL-7 involved laboratory tests on asset material behavior and pre-surgical verification. TRL-8 analyzed SMD in relevant environments like surgical settings. Finally, TRL-9, a business model could operate by manufacturer compliance.

4 Discussion and Conclusions

SMD such as biomodel, virtual surgical pre-planning, cutting guides, and PSI, were defined based on surveillance results. To build capability development for SMD several case studies were carried out. Literature reported that these devices can reduce surgical time by up to 22%, therefore, infection and anesthetic exposure; also, SMD help in decision-making by changing pre-surgical steps by up to 60% [28]. Biomodels could reduce time development and provide real-scale information, overcoming CT limitations [29]. Surgical guides reduce average errors from 6 to 7 mm, down to 1 or 2 mm, in cutting or drilling trajectory [30].

However, creating value in complex subjects like orthopedic and surgical assistance, it was required iteration, LBD, and collaboration. To achieve maturity, a strategy was outlined, to build capabilities and thus offer value for physicians and patients, as well as to build innovation capabilities in technology development for SMD. A workflow that compromised PLM process areas lead step by step an R&D process, where a low-cost SoS for enabled technologies was deployed to assist specialized activities: to get virtual reference biomodels with RE from CT; CAD models; 3DP files; and data management. All those activities, by using free or public software. Other software technologies could obtain suitable results but costs increase.

Management capability was built through the initial PLM stages (ideation, definition, realization) through its process areas. To achieve it, a workflow of processes for technology development was proposed as a framework from Figs. 1 and 2, as well as deliverables by steps and roles (collaborator, editor, viewer) among design and healthcare staff [25, 31]. Despite the viability of case studies and the use of a PLM strategy integrated with low-cost SoS, to provide further service in TRL 9, it is important to consider compliance costs. Those deliberations rely upon a higher risk for SMD. Additional work will require better technologies and infrastructure, Then, the integration of a new SoS for manufacturing should be evaluated.

Acknowledgements. The authors wish to express their gratitude to the work team from research groups at Universidad Industrial de Santander: INTERFAZ at Industrial Design School, GRICES at Medical School, specifically MD Genny Melendez, MD Jose Luis Osma, and MD Diana Navarro. In addition, to the cooperation between MINCIENCIAS (Ministery of Science, Technology and Innovation) and the Government of Santander, who provided great support.

References

1. Mahamood, R.M., Jen, T.C., Akinlabi, S.A., Hassan, S., Abdulrahman, K.O., Akinlabi, E.T.: Role of additive manufacturing in the era of Industry 4.0. In: Additive Manufacturing, pp. 107–126 (2021). Elsevier Ltd., Ed. Woodhead Publishing
2. Kermavnar, T., Shannon, A., O'Sullivan, K.J., McCarthy, C., Dunne, C.P., O'Sullivan, L.W.: Three-dimensional printing of medical devices used directly to treat patients: a systematic review. 3D Print. Addit. Manuf. **8**(6), 366–408 (2021)
3. Mourtzis, D., Angelopoulos, J., Panopoulos, N.: Collaborative manufacturing design: a mixed reality and cloud-based framework for part design. Procedia CIRP **100**, 97–102 (2021)
4. Ghobakhloo, M.: Industry 4.0, digitization, and opportunities for sustainability. J. Clean. Prod. **252**, 119869 (Apr.2020)
5. Duda, J.: Modelling of concurrent development of assembly process and system. In: Majewski, M., Kacalak, W. (eds.) IIRTS 2019. LNME, pp. 118–129. Springer, Cham (2020). https://doi.org/10.1007/978-3-030-37566-9_11
6. Corallo, A., Latino, M.E., Menegoli, M., Pontrandolfo, P.: A systematic literature review to exploretraceability and lifecycle relationship. Int. J. Prod. **58**(15), 4789–4807 (2020)
7. Bangalore, R., Rahul, G., Lundström, M.: The Impact of Cloud Plm On Smaller Manufacturing Companies. Chalmers University of Technology, Gothenburg, Sweden (2021)
8. Mousavi, A., Mohammadzadeh, M., Zare, H.: Developing a system dynamic model for product life cycle management of generic pharmaceutical products: its relation with open innovation. J. Open Innov. Technol. Mark. Complexity. **8**(1), 14 (2022)
9. Agostini, L., Galati, F., Gastaldi, L.: The digitalization of the innovation process: challenges and opportunities from a management perspective. Eur. J. Innov. Manag. **23**(1), 1–12 (2020)
10. Corallo, A., Del Vecchio, V., Lezzi, M., Luperto, A.: Model-based enterprise approach in the product lifecycle management: state-of-the-art and future research directions. Sustainability **14**(3), 1370 (2022)
11. Bi, Z.: Practical Guide to Digital Manufacturing. First-Time-Right for Designs of Products, Machines, Processes and System Integration (2021)
12. Chen, G., et al.: Three-dimensional printed implant for reconstruction of pelvic bone after removal of giant chondrosarcoma: a case report. J. Int. Med. Res. **48**(4), 1–10 (2020)
13. Ogunsakin, R.: Towards a Highly Flexible Manufacturing System for Mass Personalisation: Exploring Nature-Inspired Models. The University of Manchester, Manchester (2020)
14. Haleem, A., Javaid, M.: Different components, features of Industry 4.0 and their linkage to additive manufacturing. Int. J. Bus. Innov. Res. **26**(3), 366–389 (2021)
15. Narula, S., Prakash, S., Dwivedy, M., Talwar, V., Tiwari, S.P.: Industry 4.0 adoption key factors: an empirical study on manufacturing industry. J. Adv. Manag. Res. **17**(5), 697–725 (2020)

16. Thieringer, F.M., Sharma, N., Mootien, A., Schumacher, R., Honigmann, P.: Patient specific implants from a 3D printer – an innovative manufacturing process for custom PEEK implants in cranio-maxillofacial surgery. In: Meboldt, M., Klahn, C. (eds.) AMPA 2017, pp. 308–315. Springer, Cham (2018). https://doi.org/10.1007/978-3-319-66866-6_29
17. Corsini, L., Aranda-Jan, C.B., Moultrie, J.: Using digital fabrication tools to provide humanitarian and development aid in low-resource settings. Technol. Soc. **58**, 101117 (2019)
18. FDA, 3D Printing of Medical Devices (2020). https://www.fda.gov/medical-devices/products-and-medical-procedures/3d-printing-medical-devices. Accessed 29 May 2022]
19. Raboudi, A., et al.: Implementation of a product lifecycle management system for biomedical research. In: Canciglieri Junior, O., Noël, F., Rivest, L., Bouras, A. (eds.) Product Lifecycle Management. Green and Blue Technologies to Support Smart and Sustainable Organizations. IFIP Advances in Information and Communication Technology, vol. 640, pp. 185–199 (2022). https://doi.org/10.1007/978-3-030-94399-8_14
20. Popescu, D., Marinescu, R., Laptoiu, D., Deac, G.C., Cotet, C. E.: DICOM 3D viewers, virtual reality or 3D printing – a pilot usability study for assessing the preference of orthopedic surgeons. Proc. Inst. Mech. Eng. Part H J. Eng. Med. **235**(9), 1014–1024 2021
21. Ngo, T.N., Dang, P.V., Vo, N.T., Le, H.N., Do, L.H.T., Doan, L.A.: Towards an effective solution for medical treatment process based on product lifecycle management. In: ACM International Conference Proceeding Series, pp. 134–139 (2020)
22. Singh, S., Misra, S.C., Kumar, S.: Institutionalization of Product Lifecycle Management in Manufacturing Firms. IEEE Trans. Eng. Manag. 1–16 (2021)
23. Montemari, M., Taran, Y., Schaper, S., Nielsen, C., Thomsen, P., Sort, J.: Business model innovation or business model imitation-that is the question business model innovation or business model imitation-that is the question. Technol. Anal. Strateg. Manag. 1–15 (2022)
24. Pufal, N., Zawislak, P.: Innovation capabilities and the organization of the firm: evidence from Brazil. J. Manuf. Technol. Manag. **33**(2), 287–307 (2022)
25. Martínez Gómez, J. M., López Gualdrón, C. I., Murillo Bohórquez, A. P., Garnica Bohórquez, Israel: PLM strategy for developing specific medical devices and lower limb prosthesis at healthcare sector: case reports from the academia. In: Stark, John (ed.) Product Lifecycle Management (Volume 4): The Case Studies. DE, pp. 201–221. Springer, Cham (2019). https://doi.org/10.1007/978-3-030-16134-7_16
26. Helo, P., Hao, Y., Toshev, R., Boldosova, V.: Cloud manufacturing ecosystem analysis and design. Robot. Comput. Integr. Manuf. **67**, 102050 (2021)
27. Balestrucci, F.: Transition towards circular economy through a multi-readiness level model : an explorative study in the construction equipment industry. Mälardalen University (2020)
28. Thayaparan, G.K., Owbridge, M.G., Linden, M., Thompson, R.G., Lewis, P.M., D'Urso, P.S.: Measuring the performance of patient-specific solutions for minimally invasive transforaminal lumbar interbody fusion surgery. J. Clin. Neurosci. **71**, 43–50 (2020)
29. Wang, P., Kandemir, U., Zhang, B., Fei, C., Zhuang, Y., Zhang, K.: The effect of new preoperative preparation method compared to conventional method in complex acetabular fractures: minimum 2-year follow-up. Arch. Orthop. Trauma Surg. **141**(2), 215–222 (2020). https://doi.org/10.1007/s00402-020-03472-w
30. Hatz, C.R., Msallem, B., Aghlmandi, S., Brantner, P., Thieringer, F.M.: Can an entry-level 3D printer create high-quality anatomical models? Accuracy assessment of mandibular models printed by a desktop 3D printer and a professional device. Int. J. Oral Maxillofac. Surg. **49**(1), 143–148 (2020)
31. López, C., Bravo, I., Murillo, A., Garnica, I.: Present and future for technologies to develop patient-specific medical devices: a systematic review approach. Med. Devices (Auckl), **12**, 253 2019

Building Information Model and Safety Requirements for Spatial Program Validation

Aymen Ben Chaabane[1], Mahenina Remiel Feno[2](✉), Gilles Halin[3], Tommy Messaoudi[1], and Feriel Moalla[2]

[1] National School of Architecture of Nancy (ENSA), 2 rue Bastien Lepage, Nancy, France
[2] National Research and Safety Institute for the Prevention of Occupational Accidents and Diseases (INRS), 1 rue du Morvan, 54519 Vandœuvre-Lès-Nancy, France
remiel.feno@inrs.fr
[3] University of Lorraine, UMR MAP n°3045 CNRS-Culture, MAP-CRAI Team, 2 rue Bastien Lepage, Nancy, France

Abstract. The extensive use of digital mock-up and Building Information Models software packages is now within the reach of most practitioners in the Architecture, Engineering and Construction domain. The increasing use of BIM in construction projects may change the way risk prevention is approached. Potential risks can be identified in advance and the associated prevention measures can therefore be applied well before the work situations are set up. It is known that many safety risks are created in the early design stage of projects. Among the most effective means of handling a hazard is to eliminate it at source or provide protective means to mitigate the risks. A research project on workplace design for safety have investigated the missing link between design and safety in the context of a BIM based project. This paper discusses the contributions and limitations of BIM models and tools to consider and manage safety requirements from the architectural program to the design definition. An extension proposal of IfcSpace model is used to enrich space definition with the activities that will take place and the related topological requirements such as closeness, accessibility or minimum required area. A pre-formatted Excel ® document is used to fill in these requirements. The import of this document into the Autodesk Revit ® design tool allows to fill in the attributes associated with each space in a project structure. Several dynamo procedures have been developed to allow the designer to verify compliance with these requirements based on aisle width, door sizes, room area or minimum travel distance. These procedures have been applied to the design of a nursing home for elderly people. The main problem reported in these building is work equipment accessibility such as person lifter or handling system. We show that this approach can be helpful in sharing and validating safety related requirements in a BIM project.

Keywords: BIM · Spatial organization · Safety requirements

1 Introduction

Beyond the technical and economic aspects, the architectural design of a workplace must meet the needs of a functional organization. This consists in making the spatial

F. Noël et al. (Eds.): PLM 2022, IFIP AICT 667, pp. 484–494, 2023.
https://doi.org/10.1007/978-3-031-25182-5_47

organization and dimensions consistent with the activities that will take place. Spatial organization have a strong impact on working conditions and therefore the health and safety of future users [1].

A building design project is a complex process that involves a large amount of data from different domains: architecture, engineering, construction. Building Information Modelling (BIM) tools and processes makes it possible to design a building in a collaborative way by federating different actors around a single digital model [2]. BIM technology promotes a digital work process applied to a single digital model intended primarily for the building sector. It is a method of collaboration and information exchange between different actors throughout the life cycle of a building [3].

Previously reserved for complex projects, the use of digital mock-ups and BIM processes is now within the reach of most practitioners, thanks to the development of the IFC (Industry Foundation Classes) which is the standard data exchange format that facilitate the communication between different BIM software [4]. The IFC model contains a structure (a building) organized into components (walls, windows, doors, stairs…) each with its own characteristics (geometry, properties, links with another component…) and properties (materials, dimensions…).

The increasing use of BIM in construction projects [5] may change the way risk prevention is approached. Potential risks can be identified earlier. The associated prevention methods and risk analysis can be applied before constructions and use [6]. The implementation of this safe design principle can be facilitated by Knowledge Based Engineering techniques [7] that helps to identify, capture, structure, formalize safety requirements, and finally implement the associated knowledge to future projects. Research works on safety integration in engineering design have been conducted in the manufacturing domain thanks to the development of PLM (Product Lifecycle Management) systems [16]. Architecture domain presents new challenges to develop computer-aided tool for safe design within BIM platforms. The process of using BIM tools and IFC standard does not support occupational safety requirements of future workers in the building [8, 9]. How to integrate the requirements of occupational risk prevention into a BIM project?

In this context, the objective of this work is to confront the use of digital tools associated with BIM in architectural design with safety requirements for occupational risk prevention during the building use phase. More specifically, a design methodology is proposed to specify and validate safety related requirements that may affect the spatial organization. In this paper, we will focus on functional requirements such as access and circulation and technical requirements related to the physical environment such as lighting or acoustics.

The next section presents a synthesis of the works related to the requirements models representation for spatial organization. Section 2 presents a design methodology, which consists of three steps: requirements specification in a spatial program, the insertion of requirements in a design tool and the verification of these requirements. Section 3 presents an application of the design methodology to a nursing home for elderly people.

1.1 Spatial Program Validation

Since the late 1980s, several models of space representation have been created. In this section, we will cite some examples of conceptual models that have addressed the notion of architectural space and its requirements.

The first building model was mentioned around 1980 by Bjork [10]. This model described the representation of spaces but without considering the information about spatial requirements. Two Swedish researchers who were interested in architectural space and its geometric and quantitative requirements for the architectural design phase proposed the second model. These two researchers were the first to think about spatial requirements and properties related to the user's activity [11].

The third model [12] specifies two classes of spatial requirements (space-based requirements and user activities). This model mainly deals with the data of quantitative and not qualitative spatial requirements in the architectural programming phase.

More recent models consider spatial requirements, in both programming and design phases, such as IFC (IFC 2x4 RC3 (2011) and IFC4-Add2 (2016)). Currently, the IFC model is the most comprehensive as it encompasses the main components of a building for its entire life cycle. On the other hand, this model deals with spatial requirements of the quantitative order and not qualitative (the only qualitative spatial requirements present in IFCs are of the Boolean or label type).

In this paper we use a space model developed in a previous work that takes into consideration qualitative spatial requirements [11]. This model can be integrated with the existing IFC model to complement and make it more comprehensive for the programming and design phase.

1.2 Safety Issues Related to Workplace Design

Safety related requirements mainly concern the risks related to the operating phase of the building. The French National research institute in safety (INRS) provide a workplace design guidelines [13] that categorize safety requirements (Fig. 1).

- The physical environment requirements refer to the conditions necessary for an optimal comfort including light, ventilation, thermal and acoustics.
- The workplace safety requirements are essentials to ensure good accessibility and circulation inside and outside the buildings. These requirements are directly linked to the topological layout of spaces and workstations.
- The fire and explosion risk involve distance clearances, the smoke extraction facilities, the storage of flammable materials, means of protection against fire.
- The risks related to the structure of buildings consists to the maintenance and operation of structures such as the cleaning of roofs and facades.
- The technical installations requirements deal with the sanitary facilities, storage and electrical installations.

In the next sections, we will focus on work equipment accessibility to several spaces and relationship between workspace for spatial organization definition.

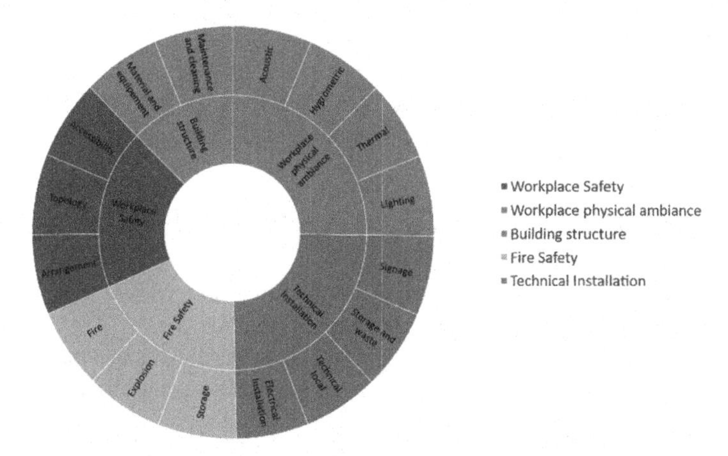

Fig. 1. Safety related requirement categories

1.3 BIM IfcSpace Model for Safety Requirements Definition

As mentioned above, the IFC model does not consider the qualitative spatial requirements, nor the requirements of risk prevention that we have identified. For this reason, we have chosen to use a IfcSpace model [14] to represent safety related requirements.

The model is based on the existing IFC model (Fig. 2). As Risk prevention requirements are indeed both qualitative and quantitative, new requirement entity <RequirementType> is added to consideration both types of requirements (qualitative and quantitative) via the enumerated type <QuantitativeOrQualitativeEnum>.

The new model is established according to the already predefined <RequirementType> category, which is structured into subclasses as shown in the figure below.

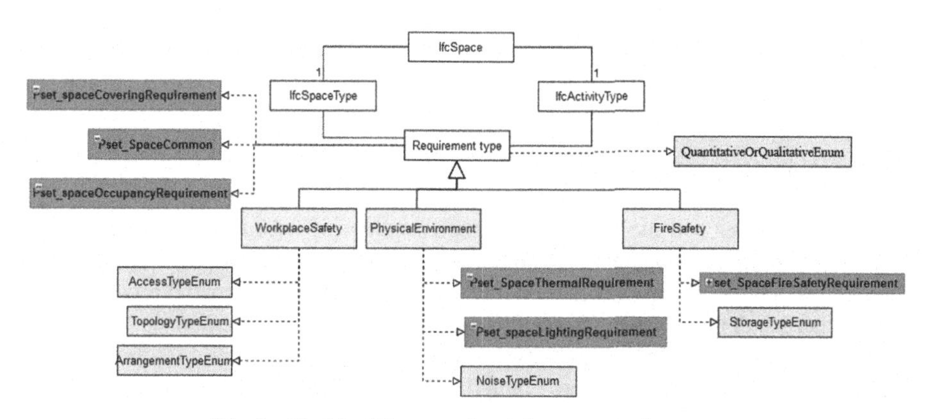

Fig. 2. Model of Space and activity type requirements

2 Design Process

To evaluate the applicability of the previously proposed space model, we followed a three steps design methodology that help to specify and validate the safety related requirements in BIM project context [11]:

1. Identify requirements according to the type of activity. Then, translate these requirements into a tabular database.
2. Import space list and related requirements in Revit architectural design software using "dynamo" scripts specific to each type of requirement.
3. Define a spatial organization with each space then verify if a specific requirement has been respected any time during the design process using the Dynamo scripts (Fig. 3).

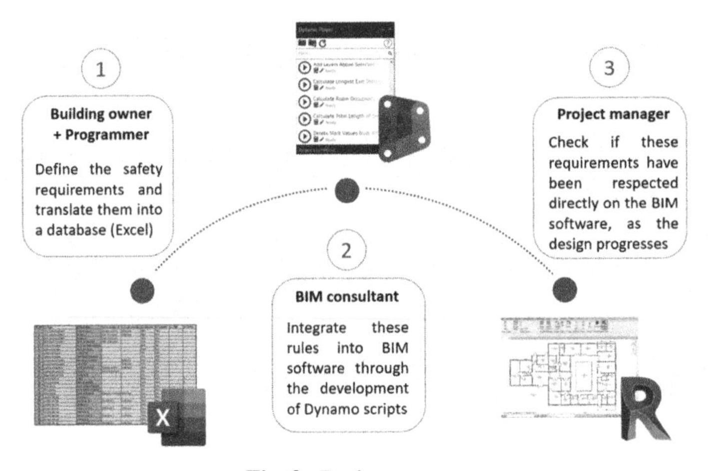

Fig. 3. Design process

2.1 Spatial Requirements Definition

The projects owner and the future users' representatives perform this task during the architectural programming phase. In collaboration, they develop the project program in a predefined tabular database, which includes on the one hand the spatial program, and on the other hand, the different categories of risk prevention requirements.

The first sheet of the table contains the functional "Program" that lists the different spaces (Type) and rooms (Sub-Type) (Table 1). The other sheets corresponds to a category of requirements. The sheets are composed of columns that represent the subsets of this category. Each subset corresponds to a rule of spatial requirements.

The project manager runs a predefined procedure to insert the different space and attributes of the program requirements into Revit software.

The project manager performs this task during the design phase by running dynamo procedures. The definition procedure ensures the setting up of the requirements. Three scripts allow the insertion of information from the Excel table to the Revit tool:

- A script to insert the list of spaces into Revit. The procedure executes the import of the Excel file and then the creation of a new list of rooms on Revit.
- A script to add space parameters to each space. The procedure extract the data from the spatial requirements, then adds it as a room parameters.
- A script to add values to the new room parameters. The procedure extracts the values from the requirements table, and then adds it for each room parameter.

As long as the database is shared and regularly updated, the project manager has access to all the program requirements directly in his CAD software. Safety related requirements are available as spatial object attribute.

2.2 Design Validation Procedures

The designer uses several scripts to verify compliance with safety related requirements throughout the design process. The project manager uses only the Revit design software to verify the adequacy of the project to the rules associated to each script.

These scripts are developed according to the types of requirements and the nature of the project. In this work, we have developed verification scripts related to aisle width, door size, room area, existing window for natural light.

3 Case Study

To demonstrate the applicability of this design methodology, this section presents a building project of Nursing home for elderly people [15]. We start from a functional program and then illustrate the steps for inserting requirements; the steps for importing requirements into a design tool and the verification of these requirements. The objective of this experimentation is to verify the perfect execution of the developed scripts during the study as well as their relevance with safety related requirement.

Workplace Safety Requirements

- Wheelchair accessibility: it is recommended to have minimum dimensions for the bathrooms to allow easy circulation of the person with reduced mobility as well as the accompanying worker for cleaning purpose.
- Accessibility to rooms: this requirement affects all rooms, to ensure the accessibility of people and equipment such as person lifter, handling systems, beds.
- Circulation: take into consideration the rules of horizontal and vertical circulation inside the building with the following criteria:

– Proximity: need of a direct relation between spaces.
– Space requirement: minimum required areas for each space depending on regular occupancy.

Physical Environment Requirements:

- Natural light: identify the spaces that requires natural lighting.
- Artificial light: identify the spaces that requires artificial lighting.
- Light level: this is the minimum recommended light level required for each space.

3.1 Spatial Program

The predefined tabular database is filled with the above requirements. Each sheet of the database is dedicated to a requirement category. All spaces of the program are translated into <IfcSpace> object and the attribute values corresponding to each requirement.

Table 1. Spatial program with accessibility requirements

	A	B	C	E	F	G	H	I	J
1	0.1 ID	0.1 Type	0.1 Sub-Type	0.1 Accessible from	0.1 Type of access	0.1 Secure	0.1 Controlled	0.1 For disabled person	0.1 Doors
2	1	Administrative pole	Lobby	Public area	Exterior	No	No		
3	2	Administrative pole	Office 1	Reception area	Indoor	No	No		
4	3	Administrative pole	Office 2	Lobby		No	No		
5	4	Administrative pole	Meeting room	Reception area		No	No		
6	5	Social premises	Catering room			No	Yes		
7	6	Social premises	Rest room			No	No		
8	7	Social premises	Locker rooms			Yes	No		
9	9	Technical rooms	Maintenance workshop		Outdoor	No	No		
10	10	Technical rooms	Air conditioning room		Outdoor	No	Yes		
11	11	Hygiene rooms	Garbage room	Public area	Outdoor	No	Yes		
12	12	Service area	Kitchen	Dining room		Yes	Yes		
13	13	Service area	Laundry room			Yes	Yes		
14	15	Collective life	Dining room			No	No		3
15	16	Care area	Cart return area			Yes	Yes		
16	17	Care center	Administrative area			Yes	Yes		
17	18	Care center	Technical area	Delivery area	Exterior	Yes	Yes		
18	19	Living units	Room 1			No	No		2
19	20	Living units	Sanitary R1	Room 1	Indoor	No	No		1
20	21	Life units	Room 2			No	No		2
21	22	Life units	Sanitary R2	Room 2	Indoor	No	No	Yes	1
22	23	Life units	Room 3			No	No		2
23	24	Life units	Sanitary R3	Room 3	Interior	No	No	Yes	1
24	25	Life units	Room 4			No	No		2
25	26	Life units	Sanitary R4	Room 4	Interior	No	No	Yes	1
26	27	Life units	Room 4			No	No		2
27	28	Life units	Sanitary R5	Room 5	Interior	No	No	Yes	1

Program | Accessibility | Topology | Lighting | Acoustic | Thermal

Prêt Accessibilité : consultez nos recommandations

3.2 Design Validation Procedure

These risk prevention requirement from the space model are translated into a graphical program so that every designer on the project is able to execute them at any time. The following rules have been translated in the validation program:

- The dimensions of the sanitary facilities in the rooms allow access for a wheelchair accompanied by a healthcare worker. Length ***a*** and width ***b*** of the spaces are verified as follows: "a > 2.4 m AND b > 2.7 m" OR "a > 2.7 m AND b > 2.4 m". The procedure selects two diagonally opposite points, calculate the width and length of the room through the two selected points, and then compares them to the predefined value.
- Verification of room accessibility: to ensure accessibility to equipment within the various rooms, the width of the entrance doors must be adapted to the size of the equipment and the nature of the activity planned in the room. To verify this condition, the procedure extract door sizes for all rooms, then compares them to the minimum widths recommended for each activity category.
- Verification of horizontal circulation: to allow a wheelchair and a bed to pass easily, it is recommended that the width of corridors be at least 2.20 m [15]. The procedure calculates the width of the room, and then compares it to the minimum width value.
- Verification of the proximity of rooms: two rooms are considered to be in proximity if the distance between the two access points does not exceed a certain value, the project manager according to the needs defines this value. The procedure verifies if the distance between the two doors is within the recommended minimum distance.
- Verification of the surface areas of the rooms: the surface areas of the rooms designed must not be less than the specified value spatial program. The procedure extracts the surfaces of the rooms modeled on Revit and compare them to the required surfaces in the program data.
- Verification of natural lighting: there must be at least one window or patio door that belongs to an exterior wall. The procedure selects all the windows of the rooms and check if the type of the host walls are of "exterior".

During the design phase, the project manager performs these scripts as the project progresses to ensure that safety requirements are met throughout the design process. At the end of the execution of each script, a "Pop Up" information message appears according to the results obtained, informing the user whether the rules have been respected or not.

The figure below shows an example of the execution of the script for checking the accessibility of parts showing the rooms that respect the accessibility requirement in green and the rooms that do not respect the requirement in red (Fig. 4).

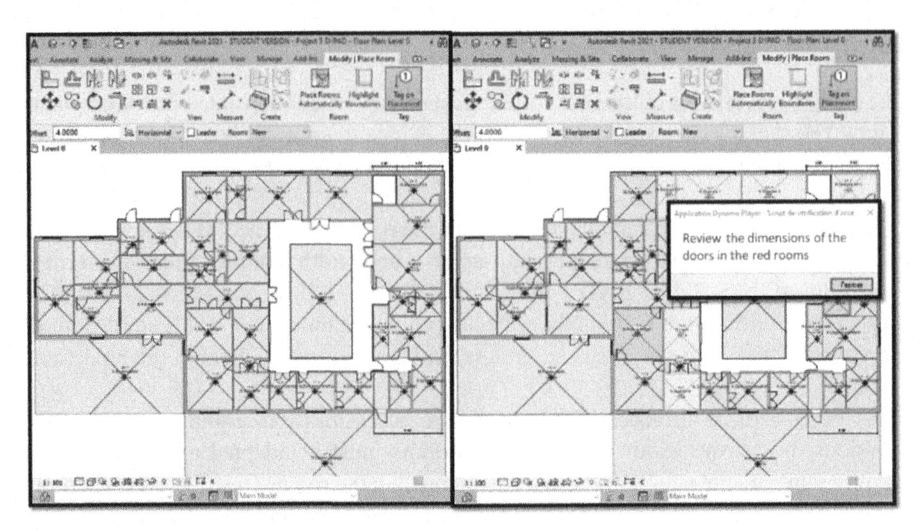

Fig. 4. Spatial design validation before and after the procedure execution (Color figure online)

4 Discussions

The proposed design method facilitates the definition and the validation of spatial program requirements. When the spatial program document is updated, the available data on the designer's workstation can be updated as well using the procedures.

The case study focused on a nursing home for elderly people building. Several type of requirements such as regulations or expert recommendations could be tested. The purpose was to demonstrate the feasibility of the design methodology in this context.

Some practical limitations about the use of the software package need to be clarified to improve the implementation of such design method. The experimentation was conducted under a single platform as a "Closed BIM". Reusing the developed procedures on another platform will require to translate the associated rules. The design method could be experimented in the future with several stakeholders on a larger scale in a so-called "Open BIM" context. IFC, being the reference standard, must fully meet the need for interoperability. There are two approaches in the literature: those aiming at extending the IFC (adding elements or attributes) and those aiming at semantic enrichment of the IFC (making it possible to specify and/or formalize the semantics of existing elements) [17].

5 Conclusions

This work focused on the use of BIM tools to take into account safety-related requirements involved in the spatial organization of a building during architectural design process. Using dynamo graphical programming tool associated with Autodesk Revit, we experimented the feasibility of integrating these requirements in a spatial layout design methodology. Among several ways to manage requirements related to IFC model object, an IfcSpace extension model is used to express these requirements.

The design method consists of expressing the future activities need and associated requirements in the IfcSpace model and verifying the compliance of a spatial layout with validation procedures. Through this approach, we succeeded in translating several rules related to occupational risk prevention.

As a conclusion, the use of BIM tools and processes for occupational risk prevention is gaining attention among academics. This work contributed to give a development path towards a better consideration and control of safety requirements in the design phase of a workplace building.

References

1. Heo, Y., Choudhary, R., Bafna, S., Hendrich, A., Chow, M.P.: A modeling approach for estimating the impact of spatial configuration on nurses' movement. In: 7th International Space Syntax Symposium, pp. 1–11 (2009)
2. Singh, V., Gu, N., Wang, X.: A theoretical framework of a BIM-based multi-disciplinary collaboration platform. Autom Constr. **20**(2), 134–144 (2011)
3. Eastman, C.M.: BIM Handbook: A Guide to Building Information Modeling for Owners, Managers, Designers, Engineers and Contractors. Wiley (2011)
4. ISO 16739-1: Industry Foundation Classes (IFC) for data sharing in the construction and facility management industries - Part 1: Data shcema. AFNOR, Paris, France (2018)
5. Hochscheid, E., Halin, G., Hochscheid, E., Generic, G.H.: Generic and SME-specific factors that influence the BIM adoption process: an overview that highlights gaps in the literature. Front. Eng. Manag. (2020)
6. Alomari, K., Gambatese, J., Anderson, J.: Opportunities for using building information modeling to improve worker safety performance. Safety **3**(1), 7 (2017)
7. Marroquin, R., Dubois, J., Nicolle, C.: Multiple ontology binding in a smart building environment. In: Proceedings Workshop on Linked Data in Architecture and Construction (2017)
8. Rodrigues, F., Antunes, F., Matos, R.: Safety plugins for risks prevention through design resourcing BIM. Constr. Innov. **21**(2), 244–258 (2020)
9. Kamardeen, I.: 8D BIM modelling tool for accident prevention through design. In: Proceedings of the 26th Annual ARCOM Conference, ARCOM 2010, pp. 281–289. Association of Researchers in Construction Management (2010)
10. Tolmer, C.-E.: Contribution à la définition d'un modèle d'ingénierie concourante pour la mise en oeuvre des projets d'infrastructures linéaires urbaines : prise en compte des interactions entre enjeux, acteurs, échelles et objets, Paris Est (2016)
11. Siala, A., Allani, N., Halin, G., Bouattour, M.: Toward space oriented BIM practices. In: Complexity & Simplicity - Proceedings of the 34th eCAADe Conference, vol. 2, pp. 653–662 (2016)
12. Cha, S.H., Steemers, K., Kim, T.W.: Modeling space preferences for accurate occupancy prediction during the design phase. Autom. Constr. **93**, 135–147 (2018)
13. INRS: Conception des lieux et des situations de travail - Santé et sécurité : démarche, méthode et connaissances techniques. ED 950, p. 193 (2021)
14. Siala, A.: Modelisation et representation des exigences spatiales qualitatives vers des pratiques bim orientees «espace». Thèse de doctorat de l'Ecole d'Architecture de Nancy (2019)
15. INRS: Conception et rénovation des EHPAD. Bonnes pratiques de prevention. ED6099 (2012)

16. Tedonchio, C.T., Nadeau, S., Boton, C., Rivest, L.: BIM and PLM-based management of occupational health and safety : a comparative literature review. In: CIB W78, pp. 892–901 (2021)
17. Steel, J., Drogemuller, R., Toth, B.: Model interoperability in building information modelling. Softw. Syst. Model. **11**(1), 99–109 (2010)

Automated Classification of Datapoint Types in Building Automation Systems Using Time Series

Noah Mertens(✉) and Andreas Wilde

Fraunhofer IIS EAS, Dresden, Germany
noah.mertens@eas.iis.fraunhofer.de
https://www.eas.iis.fraunhofer.de/

Abstract. To apply energy efficiency monitoring and optimizing systems for heating, ventilation and air conditioning (HVAC) systems at scale metadata of the system components must be extracted automatically and stored in a machine-readable way. The present work deals with the automated classification of datapoints using only historical time series data and methods from the field of artificial intelligence. The dataset used to conduct the research contains multiple time series from a total of 76 buildings and covers the months of January to November of 2018. Based on these data, relevant features of the time series are defined. Using these features, different classification models as well as the influence of seasonal effects are evaluated. Overall, our approach provides very good results and assigns on average about 92% of all datapoints to the correct class. The datapoints of the worst recognized class are still correctly classified to about 88% (validation data) and 90% (test data), respectively. Possible reasons for incorrect classified datapoints are discussed and a promising solution is proposed. The obtained results show that in practice these methods can reduce the effort of creating and validating digital twins and building information modeling (BIM) for retrofits.

Keywords: Datapoint · HVAC · Retrofit · Classification · Machine learning

1 Introduction

With the Paris Agreement's of a 1.5° target as a guideline efforts for reducing energy consumption must be increased. The building sector, with its 40% share of primary energy demand, is a particular focus. Since 90% of building floor space in non-residential buildings are not efficiently controlled [16], automation of heating, ventilation, and cooling is considered to have a realistic savings potential of 13% of building energy demand by 2035 [5].

Building automation systems (BAS) provide a lot of data about the condition of the building and the installed equipment. Often, errors in the design, programming or operation of the equipment lead to increased energy consumption in

F. Noël et al. (Eds.): PLM 2022, IFIP AICT 667, pp. 495–505, 2023.
https://doi.org/10.1007/978-3-031-25182-5_48

buildings. Using a digital twin with all relevant information on the components of the HVAC-system including their spatial and functional relationships, those errors can be detected and processes can be improved automatically. As of now the manual implementation of a digital twin is not only time consuming, but also prone to human errors [13,17]. One is the lack of norms and standards for assigning datapoint names. The various manufacturers of technical equipment follow different systems for naming the same component, which so far made it impossible to link the plant structure and software in a fully automated way.

In order to expand the use of digital twins and automation in the building sector in a scalable manner, the demand of automated linking of components and software arises. The systems setup with the spatial and functional relationships of the components and their semantic correlations should be recognized fully automatically at best, but at least minimizing human interaction.

This work focuses on datapoint types from ventilation systems. An important intermediate step is the recognition of datapoint classes, i.e. a classification of datapoint into e.g. temperature sensor, pressure sensor or damper position. The problem of automated metadata detection has been addressed in recent years [17]. This classification can be based on two different sources of information. On one hand, the time series of the datapoint can be used, on the other hand, the datapoint names are available. In [1], the ZODIAC system is presented that automatically extracts metadata using datapoint identifiers and features of the associated time series. Fütterer et al. [7] present a method that is, however, only tested on a single building. The sampling rate of the datapoints used in this process is with 1 min much higher then what usually is available in real cases. Bode et al. [2] present an approach for classification using unsupervised learning methods. However, the results obtained are significantly worse compared to methods based on supervised learning. Gao et al. [8] and Shi et al. [15] evaluate several approaches to classify datapoints based on the time series and compare the respective results for multiple buildings. In a more recent study, Chen et al. [3] use only features of time series to detect classes. The approaches presented achieve classification accuracies of 90% and 85%, respectively, but little attention is given to the practical aspects in real applications such as availability of data to train a model or seasonal effects.

The following use case is assumed for our investigations: Historical data is available from an unknown building but the amount of historical data can vary between a few days and multiple years. A classifier assigns a datapoint class and a classification error probability to each datapoint. For this purpose we developed a concept by selecting and combining existing techniques for a suitable classifier and realized its implementation. The influences of seasonal effects are taken into consideration to evaluate the portability of the developed solution towards different climate zones. Finally, the classifier will be tested with the data of two buildings and the result will be evaluated with regard to the practical applicability, especially the necessary amount of operational data.

2 Dataset and Classification Pipeline

Our approach for classifying datapoints is based on the properties and characteristics of the time series only and consists of the following steps:

1. All data of each datapoint is divided into sections of 24 h (daily time series, DTS).
2. Characterizing features for classification are extracted from the DTS.
3. Based on the extracted features, a classification model is trained.
4. Steps 2 and 3 are repeated with different combinations of features and classification algorithms. The best one is chosen (measured by weighted F1-score).
5. For the final classification of a datapoint each DTS is classified individually, then the class of a datapoint is determined according to the majority principle.

The two-step procedure for classifying the datapoint distinguishes the present study from previous approaches and drastically improves results and robustness. The underlying idea is that there are typical daily courses for each datapoint class, but atypical progressions occur on individual days. Using a day-by-day classification we reduce the influence of atypical courses on the overall results.

In the following Sect. 2.1 the dataset is described in more detail. Subsequently, Sect. 2.2 further explains the procedure for the classification of datapoints.

2.1 Description of the Dataset

The dataset used in the paper is from the Mortardata database (Modular Open Reproducible Testbed for Analysis & Research) [6], which was published in 2019. This database collects and unifies a variety of measurements and metadata from over 100 buildings. At the time of the study, the database only includes buildings from the area around San Francisco, California (USA). Possible influences due to the geographic location on the results are considered in Sect. 3. The data is predominantly available with a temporal resolution of 15 min (G. Fierro (UC Berkeley), personal communication, March 11, 2020). For this study all time series were sampled at this rate.

To account for seasonal effects when classifying datapoints, data from almost the entire year 2018 is used. In the building sector, typical patterns are often observed on a daily basis, so the length of each time series was set to 24 h (DTS). Thus in our data each DTS is represented by a vector with 96 elements. Thus, one datapoint provides up to 365 individual time series of measured values in one year. Time series which provide less than 96 measured values are discarded.

Figure 1 shows the classes and the corresponding number of datapoints and DTS. They come from a total of 76 buildings. The dataset is divided into training, validation and test data. The training data is used to train models and the validation data is necessary for feature and hyperparameter selection. Test data is used for a final performance evaluation only. The dataset is split based on buildings (holdout set) to represent a real application. The validation data comes from 12 buildings that contain all classes. The test data consists of 2 buildings, each one containing all classes.

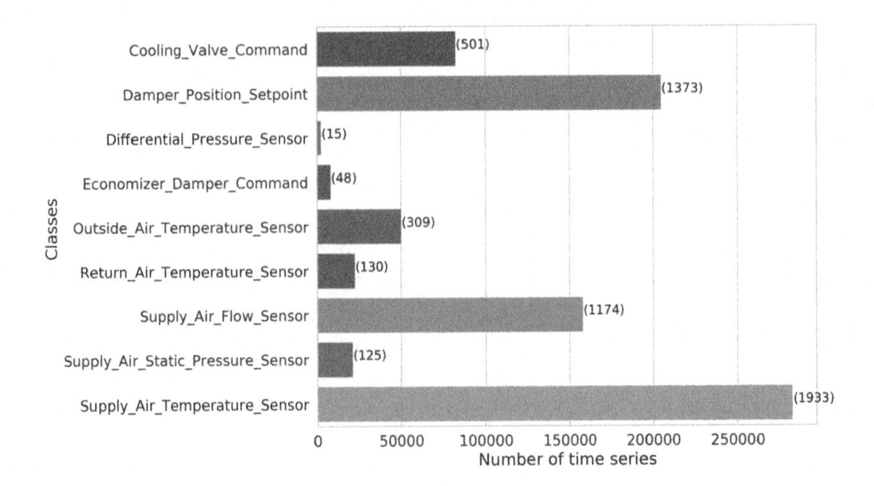

Fig. 1. Number of DTS per class (number of datapoints in brackets)

The dataset is highly imbalanced with respect to the number of DTS per class. To reduce negative effects on the models learning behavior, all classes in the training data are reduced to the size of the second smallest class (Economizer_Damper_Command) by undersampling (randomly removing samples) on a quarterly basis. The class sizes of both validation and test data remain unchanged.

2.2 Classification Procedure

To perform a classification features must be calculated from the DTS. In addition to standard features of descriptive statistics such as mean or extreme values, features based on patterns and shapes of the DTS are also taken into consideration. To find the best combination of time series features and classifications algorithms, we proceeded as shown in Fig. 2:

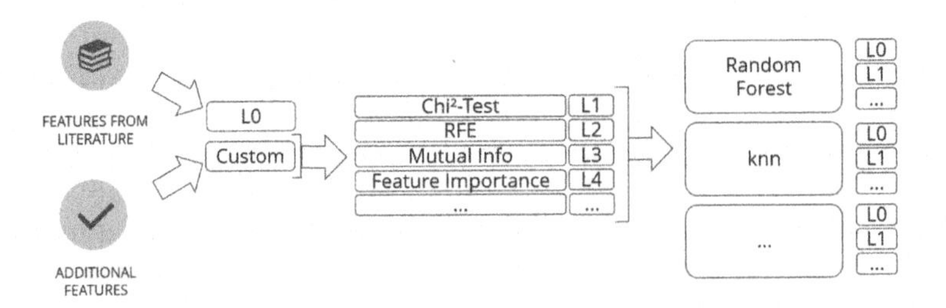

Fig. 2. Process of feature and model selection

An initial selection of features is made based on the literature (“L0”). Subsequently, the list “L0” is extended by multiple features from the library

TSFresh (0.16.0) [4] that we expected to bring high information gain, resulting in the list "Custom" with a total of 122 features. The additional features were chosen based on domain knowledge and are calculated using the Python library *TSFresh* with default settings.

We use different State of the Art approaches to determine feature relevance (Table 1). With each of these feature selection methods, all models from Table 2 are trained and evaluated respectively. For hyperparameter the default values from the Python libraries *Scikit-Learn*, *XGBoost* and *LightGBM* were used. The weighted F1-score is used as a measure of the performance.

Table 1. Feature selection approaches

Approach	#features	Encoding
Features from literature	60	L0
Extended list (L0 + additional features)	122	Custom
Chi2 Feature Importance	60	L1
RFE Default Random Forest	69	L2
Mutual Information	60	L3
Feature Importance Default Random Forest	58	L4
TSFresh Feature Selection	121	L5
Intersection of features of all approaches	23	L6
Features in at least 4 approaches selected	45	L7

Table 2. Used models

Model	Used in literature
Linear Discriminant Analysis (LDA)	[8,9]
k-Nearest Neighbor (kNN)	[7–9]
Random Forest	[7–12,14]
Gradient Boost	
ExtraTrees	
LightGBM	
XGBoost	

Comparing the results of all combinations of models and lists of features, the models *XGBoost* (F1 = 0.89) and *ExtraTrees* (F1 = 0.87) with the features of the list "L7" were selected based on the achieved results. *XGBoost* achieved the highest F1-score and is used for the final evaluation of our approach. *Extra Trees* has a significantly shorter training time and only slightly worse results and is therefore suitable for systematic tests concerning seasonal effects as described in the following section. Both models stem from the family of tree-based algorithms and therefore we expect a high transferability of the results.

3 Results

3.1 Influence of Seasonal Effects

Ideally, a classifier is trained with data from an entire year, thereby enabling classification of datapoints of unknown buildings independent of the recording date of the DTS. When considering the influence of seasonal effects, we investigate whether the performance of a model depends on the date of the time series. For this purpose, a model is trained with the training data of the whole year and evaluated for each month separately with validation data. In Fig. 3 the weighted F1 scores for each month are shown, where a red horizontal line represents the mean of all calculated F1 scores. Since neither training nor evaluation data are available for December, no result can be shown for this month. In the lower part of the figure, the number of DTS is plotted over the months.

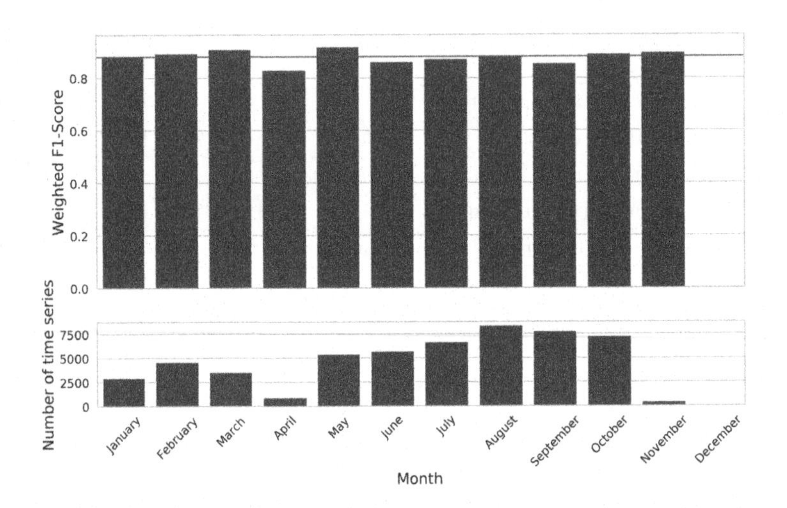

Fig. 3. Model trained with data of whole year, classification performed for each month individually (Extra Trees with "L7" Features)

It turns out that the performance of the classification is mostly independent of the date of the time series. Only in April there is a noticeable deviation from the mean value. One possible reason is the significantly small amount of training data for this month. On the other hand the training dataset contains very little data for November as well, but the performance in this month shows only a insignificant difference from the mean value. Therefore the number of training data does not seem to be the only influencing factor and seasonal effects could in fact play an important role.

The climate in San Francisco, California is subject to large fluctuations (as it can be seen in standard climate charts), especially in the months of spring and autumn. If only a few DTS are available during these months, information about possible fluctuations of e.g. sensor data could only be partially recorded. As a result, the classifier will receive less information from this period of the year. With an extended dataset of multiple years a more comprehensive understanding of the influence of seasonality could possibly be achieved.

In terms of flexibility for using the classifier for unknown buildings, the annual climate seasonality plays an important role. For the present dataset of buildings from the San Francisco area, the seasonal temperature pattern seem to play minor role(cf. Fig. 3) though. To apply the classifier in central european countries such as Germany, the influence of the seasons on the performance must be verified with an appropriate dataset.

3.2 Classification: From Time Series to Datapoints

This section examines the results of the classifier with test data which were previously completely unknown to the model and thus represents a real use case scenario. The test is conducted with the *XGBoost* model, which achieved the best results during model and feature selection. For this purpose, the model is trained with a combination of training and validation data to maximize the amount of training data. As described in Sect. 2.2, in a first step the time series of all datapoints are classified. Subsequently the datapoint class is determined by majority principle. Figure 4 (left) shows the weighted F1-scores for the classification of DTS and datapoints. The classification of the DTS already achieves a F1-score over 0.925. After the second step of the datapoint classification, the result improve significantly and reach a weighted F1-score of about 0.975.

For practical applications with limited availability of data the question arises how many time series per datapoint are necessary in order to make a reliable statement about the datapoint class. This question can be answered using the binomial distribution. Thus, the probability $P(X = k) = \binom{n}{k} p^k (1-p)^{n-k}$ that exactly k of n DTS are correctly classified is calculated, where p corresponds to the probability that a single time series is correctly classified. p varies for different datapoint classes. For a conservative estimation, the value of p is taken to be the worst datapoint class recognized on test and validation data ($p = 0.72$). For this approach, classification errors are assumed to occur randomly and the individual tests are assumed to be stochastically independent. The plot in Fig. 4 (right) shows that with 13 DTS per datapoint, the overall probability of determining the correct datapoint class is already greater than 95%.

When using the model with the data of unknown buildings, it is not only important to know the predicted class, but also the probability of the prediction being correct. This allows the class assignment to be fully automated for some classes and focusing expert knowledge on classes with low classification confidence. The prediction probability can be directly read from the main diagonal of the classification matrix in Fig. 5. They express the ratio of correctly assigned datapoints of a class to all datapoints of this class. More than half of the classes are recognized correctly in all cases. Only the three classes Damper_Position_Setpoint, Return_Air_Temperature and Supply_Air_Flow Sensor show classification errors at all.

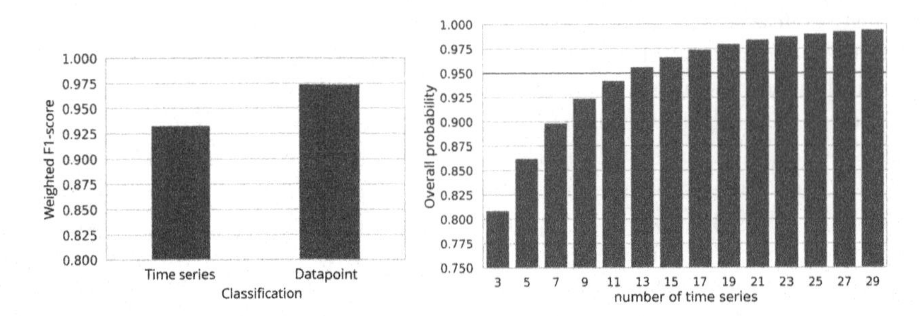

Fig. 4. Left: Comparison classification of individual time series and datapoints (XGBoost with "L7" features, test data), right: Overall probability over the number of time series per datapoint

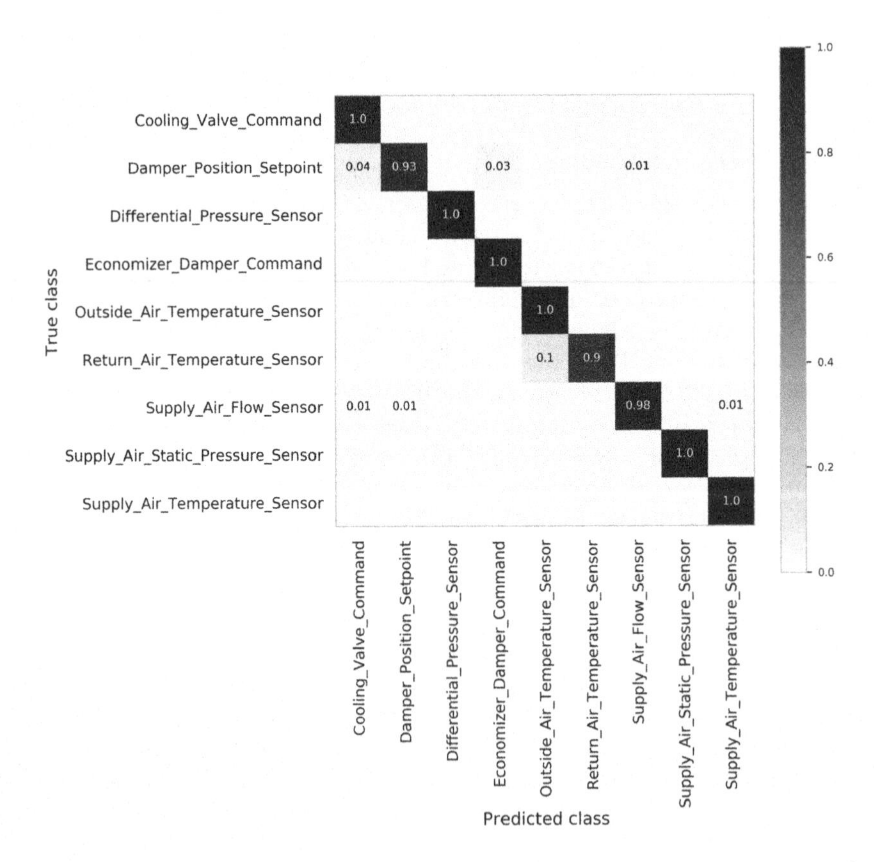

Fig. 5. Normalized classification matrix (XGBoost with "L7" features, Test data)

4 Discussion and Conclusions

For the data used in this research a very good recognition of datapoint types was achieved. The reliability increases with the number of available DTS, whereby on average a classification accuracy of 95% is exceeded if more than 12 DTS are available. In practice, this should lead to a drastic reduction of the expert effort in many cases. With respect to the season from which the DTS originate, the results are reliable for at least the investigated climate zone (San Francisco/USA). The transferability of this finding to other climate zones still has to be examined with an appropriate dataset. It is to be expected that problems can occur in climatic zones with more pronounced annual temperature variations: Heating systems in Germany, for example, do not operate for weeks in the summer, nor does cooling in the winter. In cases where the systems are not in operation, very little information can be extracted from corresponding time series. In theses cases, data of multiple seasons must be provided in order to achieve reliable results. As a rough guideline, however, it remains to say that about 2 weeks of operating data make the datapoint of the respective active systems well automatically identifiable (cf. Fig. 4).

It turned out that the classification results for individual classes such as "Damper_Position_Setpoint" (flap position) was comparatively poor. The reason for this could be that this datapoint class was imprecisely defined in the dataset. Dampers can be used at very different locations in a ventilation system and can have very different operating characteristics depending on their corresponding function. Thus, the comparatively poor detection performance indicates a fuzzy taxonomy of datapoint rather than problems with data availability, feature selection or the overall classification approach.

Also some datapoint of the Return_Air_Temperature class are not recognized correctly. More detailed investigations showed that the DTS of the incorrectly classified datapoint have a significantly larger range of values than the training data. One possible reason for this is the different usage profile of a single room in the existing dataset. It is expected that additional buildings with similar usage profiles in the training data may improve the classification results.

To create a useful digital twin of a building, all relevant technical components need to be included. Knowing the type of datapoints can enable (semi)automatic datapoint mapping or support evaluation of manually mapped datapoints to avoid errors (e.g. check if components are mapped in a unusual order). The results obtained in this work thus show that the classification of datapoints based on features of the time series could be a practical way to reduce the expert effort in creating and validation digital twins and BIM for retrofit applications in the building sector.

In this work, however, a number of other aspects such as the automatic detection of functional relationships and the affiliation to different parts of the plant are not considered. For these tasks, other sources of information such as datapoint identifiers or graphical representations of the plant structures will play a major role. However, since these sources of information are often incomplete and subject to errors, time series can offer valuable information for the plausibility check of information from other sources.

Acknowledgements. Supported by:

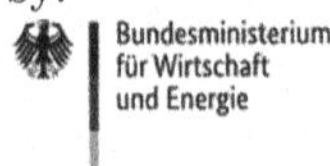

based on a resolution of the German Bundestag under the funding code *03ET1567A*.

References

1. Balaji, B., Verma, C., Narayanaswamy, B., Agarwal, Y.: Zodiac: organizing large deployment of sensors to create reusable applications for buildings. https://doi.org/10.1145/2821650.2821674
2. Bode, G., Schreiber, T., Baranski, M., Müller, D.: A time series clustering approach for building automation and control systems. https://doi.org/10.1016/j.apenergy.2019.01.196
3. Chen, L., Gunay, H.B., Shi, Z., Li, X.: A metadata inference method for building automation systems with limited semantic information. https://doi.org/10.1109/TASE.2020.2990566
4. Christ, M., Braun, N., Neuffer, J., Kempa-Liehr, A.W.: Time series Feature extraction on basis of scalable hypothesis tests (tsfresh - a python package). Neurocomputing **307**, 72–77 (2018)
5. Debusscher, D., Waide, P.: A timely opportunity to grasp the vast potential of energy savings of building automation and control technologies, p. 21
6. Fierro, G., et al.: Mortar: an open testbed for portable building analytics. https://doi.org/10.1145/3366375
7. Fütterer, J., Kochanski, M., Müller, D.: Application of selected supervised learning methods for time series classification in building automation and control systems. Energy Procedia **122**, 943–948 (2017)
8. Gao, J., Bergés, M.: A large-scale evaluation of automated metadata inference approaches on sensors from air handling units. https://doi.org/10.1016/j.aei.2018.04.010
9. Gao, J., Ploennigs, J., Berges, M.: A data-driven meta-data inference framework for building automation systems. https://doi.org/10.1145/2821650.2821670
10. Hong, D., Ortiz, J., Bhattacharya, A., Whitehouse, K.: Sensor-type classification in buildings. https://doi.org/10.48550/arXiv.1509.00498
11. Hong, D., Wang, H., Ortiz, J., Whitehouse, K.: The building adapter: towards quickly applying building analytics at scale. In: Proceedings of the 2nd ACM International Conference on Embedded Systems for Energy-Efficient Built Environments - BuildSys 2015 (2015). https://doi.org/10.1145/2821650.2821657
12. Koh, J., Balaji, B., Akhlaghi, V., Agarwal, Y., Gupta, R.: Quiver: using control perturbations to increase the observability of sensor data in smart buildings. https://doi.org/10.48550/arXiv.1601.07260
13. Koh, J., Balaji, B., Sengupta, D., McAuley, J., Gupta, R., Agarwal, Y.: Scrabble: transferrable semi-automated semantic metadata normalization using intermediate representation. https://doi.org/10.1145/3276774.3276795
14. Park, J.Y., Lasternas, B., Aziz, A.: Data-driven framework to find the physical association between AHU and VAV terminal unit - pilot study, p. 9

15. Shi, Z., Newsham, G.R., Chen, L., Gunay, H.B.: Evaluation of clustering and time series features for point type inference in smart building retrofit. https://doi.org/10.1145/3360322.3360839
16. Waide, P., Ure, J., Karagianni, N., Smith, G., Bordass, B.: The scope for energy and CO2 savings in the EU through the use of building automation technology
17. Wang, W., Brambley, M.R., Kim, W., Somasundaram, S., Stevens, A.J.: Automated point mapping for building control systems: recent advances and future research needs. Autom. Constr. **85**, 107–123 (2018)

How May Modular Construction Transform the Built Environment Ecosystem?

Vighneshkumar Rana(✉) and Vishal Singh(✉)

Indian Institute of Science Bengaluru, Bengaluru 560012, Karnataka, India
{vighneshkuma,singhv}@iisc.ac.in

Abstract. This paper presents a novel viewpoint on modular construction, seeking to create new opportunities to transform the built environment ecosystem by taking a product lifecycle and systems perspective. Based on a review of the challenges with modular construction and the recent social trends, it is proposed that the modular construction approach and modular spaces can offer a means to address one of the fundamental limitations of the existing built environment solutions– the coupling of products (built spaces) and their location. The propositions are based on a literature review and abductive reasoning. It is argued that in the current approach to modular construction, the modularity is typically lost after construction, that is, once the building is constructed. In contrast, novel modular construction approaches could seek lifecycle modularity goals as a part of the systemic built environment solution with the vision to achieve the decoupling of products (built spaces) from the location. It is argued that the decoupling of product and location will create new business and societal opportunities for modular construction, in keeping with trends such as sharing economy, resource effectiveness, adaptability, and flexibility of offerings for the end customers. This paper is based on theoretical arguments. Further research is ongoing to develop a proof of concept and test the technical feasibility, market viability, and social desirability of the proposed approach.

Keywords: Built environment · Technology · Modular construction · Global trends · Real estate innovation

1 Introduction

Offsite modular construction is when building components are constructed offsite in a factory with a controlled environment, transported to the construction site, and assembled. Modular construction has been shown to improve housing quality, reduce labor requirements, reduce construction time and uncertainty, and reduce environmental effects around the construction site. Consequently, in the last two decades, modular construction has attracted the attention of construction industry and the governments and regulatory agencies across multiple countries, especially developed countries [1, 2].

Despite the acknowledged benefits, support of the government agencies, and a range of enabling initiatives and efforts by different stakeholders to promote modular construction, the desired level of impact and adoption of modular construction in practice

Published by Springer Nature Switzerland AG 2023
F. Noël et al. (Eds.): PLM 2022, IFIP AICT 667, pp. 506–515, 2023.
https://doi.org/10.1007/978-3-031-25182-5_49

is lower than expected [1]. The private building sector still relies heavily on traditional construction methods [2]. The global modular construction market is still struggling to gain traction. Only 2–3% of the worldwide construction sector uses modular construction. The construction industry is by and large following the same methods and processes invented decades ago, with limited improvements in the quality of the buildings around us.

Some of the reported reasons for the low adoption of modular construction include lack of design guidelines, lack of training and investors, transportation difficulties, reliability of connection systems, and intensive capital requirements [2]. Despite exploration of approaches such as lean construction to address the identified issues, the gaps remain, especially in connecting technical advancements to industry practices and market reality [3]. Others continue to explore these challenges from the perspectives of construction technology and management. For instance, Liu et al. [4] propose a digital twin-based safety risk analysis of prefabricated buildings.

Hence, it is becoming apparent that there is something amiss in the current assessment of the reasons for the slower adoption of modular construction. For instance, Pasquale et al. [1, 5] suggest the need to change the way modularisation in construction is conceived and implemented. Similarly, this paper argues that a holistic, value-based, and market ecosystem-level understanding of the dependencies is required to identify other challenges and opportunities for the wider adoption of modular construction practices. Consequently, this research adopts an ecosystem perspective at the societal and built environment business ecosystem level and does not limit the analysis to construction technology and methodology issues. In particular, the research recognizes that the real estate factors, especially the dominance of the location of the property in the valuation of a building or housing, remain a critical factor. That is, the research builds on the argument that the eventual valuation of a typical housing is determined by its location to such an extent that it often does not matter whether a better-quality housing product (leaving the location factors aside) can be provided through one production method or the other, including modular construction.

Therefore, it is argued that the real estate market factors and the contemporary housing trends must be part of the modular construction strategy. We build on the premise that the technological effort toward modular construction is likely to see greater success if it can create new growth opportunities in the real estate market, not just by improving the production-centric technological methods. Hence, this paper revisits modular construction from the systemic and real estate business ecosystem perspective, focusing on reducing the coupling between the product (building) and its location. It is argued that the decoupling of the product and location will lead to an independent assessment of the product and the location, which otherwise tends to get entangled as one single offering where the value of the location tends to outweigh the value of the product.

2 Research Approach

This research is based on a literature review and theoretical arguments, building on abductive reasoning. Abductive reasoning seeks the best plausible explanations of observable patterns for scenarios where inductive reasoning or deductive conclusions are not adequate. Figure 1 shows the research steps and the flow of arguments leading to the research hypotheses and the planned next steps. Two main research hypotheses are proposed.

Hypothesis 1: The inclusion of real estate market factors in the conception of modular-construction-based solutions will generate greater value and market acceptance for modular construction.

Hypothesis 2: The decoupling of land and the built asset (product) will lead to an independent valuation of the built asset (product) and the location, resulting in increasing recognition of the quality of the built asset (product).

This paper provides preliminary theoretical evidence to justify the relevance of the hypotheses. The validation of the proposed hypotheses is outside the scope of this paper. The hypotheses have to be tested for technical feasibility, market viability, and social desirability as part of the ongoing research.

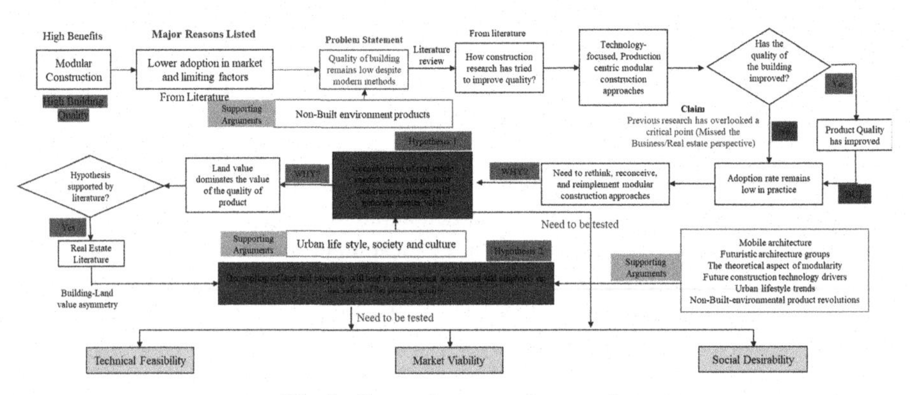

Fig. 1. Research steps and approach

3 Background and Review

This section briefly reviews and reflects on the literature and trends associated with modular construction and the built environment ecosystem.

3.1 Need to Adopt a Product and Product Lifecycle Point of View

There is increasing call for industrialized construction and the need to view buildings as products [6]. The current housing solutions do not demonstrate generational transformations that are observable in modern manufactured products such as cars and phones. Current housing solutions typically remain outdated, with incremental changes in construction technology and performance [5].

Considering housing as a product may offer opportunities to improve product quality. Nonetheless, product-centricity in housing and modular construction is currently production focused. The modular construction research is currently focused on manufacturing and assembly, improving the construction process, speeding the construction time, enhancing offsite manufacturing, and improving quality [1, 2]. The target value of such an approach is geared towards the construction process. It does not directly target the real-estate valuation, which is the most critical factor in choosing housing solutions. Others such as Pasquale et al. [5] have also argued the need to revisit modular construction as means to address changing societal trends. For instance, there is a demand for housing solutions that suit modern urban lifestyle. Thus, there is an apparent gap in the product-oriented approach to housing solutions. Pasquale et al. [5] suggest that modularity can help fill this gap.

In general, modularity is also desirable from the perspective of product lifecycle management. Ishi [7] assessed the impact of modularity on product lifecycle engineering and concluded that modular products offer high serviceability and recyclability, leading to high life cycle efficiency. Modularity provides means to respond to technological changes and flexibility requirements in products [7].

Different levels of modularity are possible in building housing [8], offering different benefits according to the stage in which modularity is introduced, as shown in Table 1. Panel-based approaches introduce early and more granular modules. However, the benefits are relatively low compared to Prefabricated Prefinished Volumetric Modules (PPVC), where modularity is more aggregated and late stage, but with higher benefits of cost, time, and quality. Nonetheless, in the current modular construction approaches, the PPVC modules typically lose their modularity once the building is constructed because the modules cannot be independently replaced unless the whole building is disassembled.

Table 1. Levels of modularity in housing solutions and delayed differentiation

Modular construction approach	Specification	Life cycle stage where modularity introduces	Benefits of postponement
Panelised modular system	Building wall panels manufactured in a factory and assembled on site	Early	Low
3D module manufacturing	3D structure (part modules such as modular bathrooms)	Middle	Moderate
PPVC	Prefabricated prefinished volumetric construction (fully completed apartment modules)	Last stage	High

The proposed solution approach seeks to leverage the benefits of PPVC modules, but alter the assembly process to seek lifecycle modularity. The objective is to not only introduce modularity in the last stage but to retain it across the lifecycle of the building.

This creates an opportunity to further the benefits of postponement because the PPVC modules could be replaced anytime during the lifecycle of the building. This argument forms the basis for the first hypothesis.

3.2 Need to Consider Real Estate Market Factors

Modular construction is argued to improve quality, and reduce labour requirements, construction time, and environmental effects [1, 2]. All these arguments are production-centric. Whereas, the real estate literature shows that the housing purchase behaviour is largely influenced by real estate factors, predominantly the location [9, 10]. Similarly, a review of popular housing portals shows that location is the most emphasized factor in housing deals. Hence, it is reasonable to argue that the modular construction approach needs to strategize around real estate factors for wider adoption.

Explaining the housing purchase behaviour, Wong et al.[9] state that real estate can be considered a hybrid product that combines the land and the building structure. The land value is largely transparent, whereas the building value is somewhat opaque to the buyer who has less information about the building than the seller. This asymmetry results in valuation largely determined by the location. Wong et al. [9] draw upon George Akerlof's lemon theory [11] to support their findings. They conclude that land drives the liquidity of the real estate market even if the buyers do not have uncertainty about the building.

Hence, it is argued that the coupling of the built asset and the land skews the valuation towards the land, reducing the importance of the product quality. Decoupling land and the built asset can be one of the ways to reduce this asymmetry and put independent emphasis on the product quality. This argument forms the basis for the second hypothesis. Hence, it is reasonable to conclude that the coupling of the built asset and the land skews the valuation towards the land, often leading to reduced importance of the product quality. Decoupling land and the built asset can be one of the ways to reduce this asymmetry and put independent emphasis on product quality. This argument forms the basis for the second hypothesis.

3.3 Need for Decoupling of Product and Location

Seeking solutions to decouple building products from the location is not a new endeavour. Several solutions that allow decoupling of building products from the location either exist or have been proposed earlier, albeit for reasons different from what is proposed in this article. For example, mobile homes, caravans, and portable units have existed for a long time. Nonetheless, the literature suggests that such mobile homes are not attractive from the real estate perspective as they are not considered building assets, which makes it difficult to procure loans or financing for such units [12]. The primary driver for such solutions, therefore, is affordability and lifestyle choices, and not really the drive to improve the product quality or change the real estate dynamics. Similarly, portable mobile shelters and units are often used as temporary shelters in emergency scenarios. Once again the objective is temporary housing and not the drive to improve product quality or change the real estate dynamics. Decoupled modular solutions have also been part of the visions of futuristic cities, typically driven by projected lifestyle and benefits of industrialization [13]. A few notable examples include Yono Friedman's concept of independent dwellings, and the Archigram and Metabolist concepts from the 1960s.

The Archigram concept proposed a 'Plug-in city' comprising an elevated city of personalised prefabricated capsule homes inserted into high-rise megastructures [13]. Archigram's plug-In city concept intended to give people the flexibility and choice to customise their capsule homes. Metabolist aimed to develop a living, self-generating system that could adapt over time [13]. Friedman's work provided the concept of mobile architecture, creating spaces that people could take with them [13]. These futuristic groups worked on concepts that sought decoupling of the product and the location. These concepts did not translate into reality, perhaps due to technological limitations at that time and a lack of social acceptance. However, the authors argue that such visions may have also failed because of the disconnections of the ideas from the practical realities of the real estate market and the housing purchase behaviour.

4 The Proposed Approach to Decoupling Product and Location via Modular Construction

Figure 2 shows a schematic of the proposed solution approach, which has similarities with Archigram's 'Plug-in City' [13] and Pasquale et al.'s [1] hybrid modular construction. The proposed solution comprises a building frame fixed on a location, but which can host independent, replaceable PPVC modules. The building frames have integrated hoisting mechanisms that allow PPVC modules to be plugged in or out anytime during the building lifecycle. The removable modules retain lifecycle modularity, decoupling them from the location.

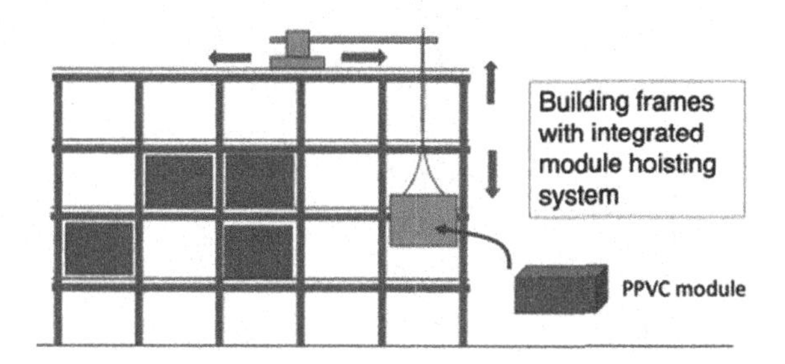

Fig. 2. Plug-in-plug-out system of modularity

The proposition to decouple the built asset from its location is driven by the following factors:

Improve the Housing Product Quality: Decoupling of the housing product and its location should lead to independent emphasis on product quality. As technologies and societal trends change, the built spaces will also require frequent updates to meet the desired performance and functions [1, 5]. With a decoupling approach, replacing the existing modules with advanced technical, societal, and functional units should be possible.

Enhance Variety and Choice: Decoupling will provide greater flexibility and choice for customers to select individual units that meet their changing needs, budget, and design requirements across the lifecycle.

Enhance Adaptability at The City Level: Decoupling the product from the location will create a system-level opportunity to adapt at the city.

Enhance Recourse Efficiency: The ability to share modules or locations will enable greater resource sharing and efficiency at the systems level, consistent with trends in sharing economy.

5 Discussion and Justifications

Figure 3 summarizes the observations and trends leading to the proposed objective of seeking greater value in the built environment through decoupling. The trends suggest that apart from the technological aspects, the real estate market, business, culture, and global trends are notable factors that influence the adoption of construction technologies in practice.

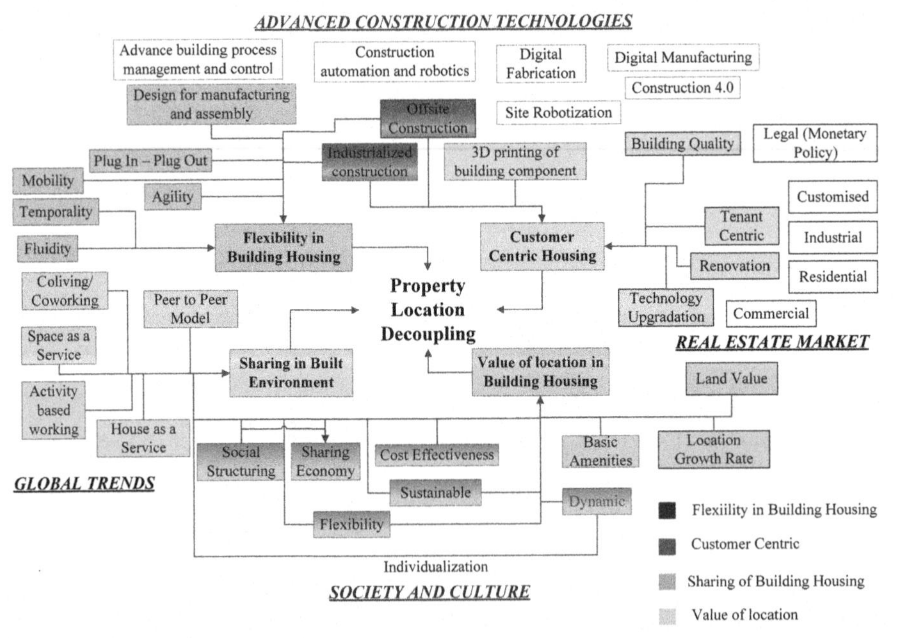

Fig. 3. Summary of trends leading to the need for greater value generation in the built environment

The following paragraphs explain how some of these trends and observations can be accounted for in the proposed solution approach that seeks to leverage modularity to decouple the building product from the location.

5.1 Advantages of Lifecycle Modularity and Delayed Differentiation

Delayed differentiation is postponing the products assembly late in the supply chain to have cost benefits, reduce inventory, tackle uncertainty in demands, and high product availability and variety [14, 15].

In traditional PPVC modular construction, all the PPVC modules in a single building are assembled at the construction point. Some level of delayed differentiation is available to the end customer until the point when the building is constructed. In contrast, the individual modular units are never permanently fixed to the building frame in the proposed plug-in-plug-out approach. The lifecycle modularity should allow greater delayed differentiation as the customers can choose their own customized units even after the building frame is constructed. The lifecycle modularity will enable reuse, recycling, and disposal [14] of PPVC units at any time, allowing a continued opportunity for differentiation.

5.2 Future Technology Drivers in Construction and the Global Social Trends

Singh [16] argued that the built environment research and practice needs a systemic approach along with a technology-based vision of the future of construction. Quoting a World Economic Forum report that identified several technological advancements and global trends that are likely to shape the future of construction, Singh [16] argues that the construction sector needs solutions that can combine technical advancements with global trends. Hence, it is desirable to seek approaches that can integrate technical advancements such as prefabrication and PPVC, digitalization, and wireless technologies with social trends such as sharing economy, resource efficiency, and lifecycle management.

Social trends are also reflected in the urban lifestyle. The younger generations want the spaces to be flexible, adaptive, resource-effective, and technology-oriented [1, 5]. Technology changes how we think, socialise, live, and work. Some noticeable global trends include temporality, mobility, agility, activity-based working, X-as-a-service, plug-in plug-out products, co-working, co-living, etc. [1]. Companies such as Airbnb and WeWork offer a flexible urban lifestyle. The real estate industry is going through a form factor change [17]. A similar desire for flexibility and resource efficiency is observable in other markets such as car-pooling because it adds value to the customer. Decoupling the housing units from the location can be one way to create such flexibility and adaptability in the real estate market.

The trends toward decoupling are also observable in areas such as mobile architecture, adaptable spaces, tiny houses, and modular pods[18, 19]. Each of these solutions can be clubbed in the category of volumetric modular solutions. Such flexible spaces are equally applicable in indoor and external environments. In parallel, outside the built environment discussion, modular volumetric solutions such as Mobile Health Units (MHUs) are increasingly used to improve public infrastructure in remote locations [20]. There is also a growing interest in the mobility domain of autonomous spaces, extending the scope of modular spaces. In addition to the physical spatial solutions, service-oriented spatial solutions such as space-as-a-service have emerged [21]. The ideas of shared economy and shared resources have emerged in built environment, mainly via innovations in real estate. Such service-oriented models also lead to greater use of modular spaces such as meeting pods.

The decoupling approach also seeks to bridge the silos of sub-disciplines within the construction industry, namely, architecture, construction, facilities management, and real estate and property management. For instance, Korkmaz [22] discusses the concept of kinetic architecture in which buildings are designed to allow parts of the structure to move. Dynamic design is an under-explored in construction. Whereas re-configurable spaces are increasingly desired in modern real estate. A built environment ecosystem that supports kinetic architecture at the city-level offers new growth opportunities in the construction industry.

5.3 Spatial Efficiency and Resource Efficiency

There is increasing housing demand in cities like Shanghai and Delhi with the growing urban population. Innovative solutions are needed for the effective use of new and existing spaces. Building occupancy rate is one way to assess the efficiency of building resources. The current occupancy rates across educational, residential and office buildings is less than 0.35. One crucial reason for such a low occupancy rate is that buildings are designed to serve a specific function. This results in building resources that remain unused for large amounts of time. In contrast, the proposed decoupling approach can enable multi-functional use of locations with adaptable and mobile spaces, raising the overall efficiency and occupancy rates.

6 Conclusions and Future Research

The coupling of building housing with location seems to be a fundamental limitation in construction. Decoupling the building housing from the location offers new growth opportunities, commensurate with trends such as sharing economy, resource efficiency, adaptability, and flexibility. Decoupling will build upon and offer lifecycle modularity in housing solutions. However, several challenges remain to be investigated and tested to take the proposed vision forward. First, the technical feasibility of the proposed approach needs to be tested. Second, the social desirability of such a solution needs investigation and validation. Third, the market viability of the proposed solution needs to be assessed to check whether a sound business case can be made for the proposed solution such that it creates and meets market demand.

References

1. Di Pasquale, J., Innella, F., Bai, Y.: Structural concept and solution for hybrid modular buildings with removable modules. J. Arch. Eng. **26**(3), 04020032 (2020)
2. Ferdous, W., Bai, Y., Tuan, D.N., Manalo, A., Mendis, P.: New advancement, challenges and opportunities of multi-storey modular buildings-a state-of-the-art review. J. Eng. Struct. **183**, 883–893 (2019)
3. Filomena, I., Arashpour, M., Bai, Y.: Lean methodologies and techniques for modular construction: chronological and critical review. J. Constr. Eng. Manag. **145**(12), 04019076-1 (2019)
4. Liu, Z., Meng, X., Xing, Z., Jiang, A.: Digital twin-based safety risk coupling of prefabricated building hoisting. Sensors **21**, 3583 (2021). https://doi.org/10.3390/s21113583

5. Di Pasquale, J.: PhD Thesis: Hybrid Modular Architecture: A Strategic Framework of Building Innovation or Emerging Housing Behaviours in Urban Contexts (2018). ABC-Department of Architecture, Built Environment and Construction Engineering
6. Thanoon, W.A., Peng, W.L., Abdul Kadir, M.R., Jaafar, M.S., Salit, M.N.: The essential characteristics of industrialised building system. In: International Conference on Industrialised Building Systems, KL, Malaysia (2003)
7. Ishii, K.: Life-cycle engineering design. ASME J. Mech. Des. **117**, 42–47 (1995)
8. Thai, T.H., Tuan, N., Brian, U.: A review on modular construction for high-rise buildings. Structures **28**, 1265–1290 (2020)
9. Wong, S.K., Yiu, C.Y., Chau, K.W.: Liquidity and information asymmetry in the real estate market. J. Real Estate Finan. Econ. **45**, 49–62 (2012)
10. Hassan, M.M., Ahmad, N., Hashim, A.H.: The conceptual framework of housing purchase decision-making process. Int. J. Acad. Res. Bus. Soc. Sci. **11** (2021). https://doi.org/10.6007/IJARBSS/v11-i11/11653
11. Akerlof, A.G.: The market for "lemons": quality uncertainty and the market mechanism. Q. J. Econ. **84**(3), 488–500 (1970)
12. Edward, C.L., Smith, D.L., Rhoades, M.: An analysis of default risk in mobile home credit. J. Bank Finan. **16**(2), 299–312 (1992)
13. Mallgrave, H.F., Goodman, D.J.: An Introduction to Architectural Theory: 1968 to the Present. Wiley, Hoboken (2011)
14. Ulrich, T.K., Eppinger, S.D., Yang, M.C.: Product Design and Development; Chapter 10: Product Architecture. McGraw Hill, New York (2020)
15. Newcomb, P.J., Bras, B., Rosen, D.W.: Implications of modularity on product design for the life cycle. J. Mech. Des. **120**, 483–490 (1998)
16. Singh, V.: Digitalization, BIM ecosystem and the future of built environment. Eng. Constr. Arch. Manag. (2018)
17. Basselier, R., Langenus, G., Walravens, L.: The rise of the sharing economy. Econ. Rev. issue iii, 57–78 (2018)
18. Bahamon, A. (ed.): PreFab – Adaptable, Modular, Dismountable, Light. Mobile Architecture. Harper Collins, New York (2002)
19. Kronenburg, R.: Houses in Motion: The Genesis, History and Development of the Portable Building, 2nd edn. Wiley, New York (2002)
20. Khanna, A.B., Narula, S.A.: Mobile health units: mobilizing healthcare to reach unreachable. Int. J. Healthc. Manag. **9**, 58–66 (2016)
21. Zervas, G., Proserpio, D., Byers, J.W.: The rise of the sharing economy: estimating the impact of airbnb on the hotel industry. J. Mark. Res. **54**(5), 687–705 (2017)
22. Korkmaz, K., Arkon C.: An analytical study of the design potentials in kinetic architecture. Izmir Institute of Technology, İzmir (2004)

An Overview of Smart Product Manufacturing Based on Classic Product Development Processes

Matheus Henrique Kupka Romano, Anderson Luis Szejka(✉), and Eduardo Rocha Loures

Industrial and Systems Engineering Graduate Program, Pontifical Catholic University of Parana, Curitiba, Brazil
matheus.kupka@pucpr.edu.br, {anderson.szejka, eduardo.loures}@pucpr.br

Abstract. The manufacturing landscape of the industry has been changing in recent years with the current disruptive technological advances. The industry 4.0 paradigm goes beyond the adoption of technology. It has accelerated the development of new concepts and methods that influence products and processes, gradually allowing the creation of intelligent products. Although these advances address the integration of Industry 4.0 with the intelligent product development process, the intelligent product concept does not consolidate yet. Likewise, the classical adjustment of product development processes to meet the prospects of developing and manufacturing smart products should be further addressed. This article contextualizes what has been studied about the concept of intelligent products, the intellectual process of product development, and the importance of enabling technologies of Industry 4.0 in the process. The analysis is carried out based on the study of relevant articles that focus on the themes, the concept, and the development of smart products. In addition, the main differences, and necessary adaptations in the classic product development processes about intelligent product development process frameworks are addressed. The study's main result is an overview of the existing gaps within the problematic issues. In addition, a discussion is carried out on the requirements and needs for future construction of the Intelligent Product Development Processes framework.

Keywords: Smart product · Product development process · Industry 4.0 · Problem analysis

1 Introduction

The industrial scenario has changed in recent decades due to successive technological advances; such advances have increasingly led to the development of products and services that include advanced computing platforms, increasingly characterizing them as intelligent or complex products [1, 2]. Smart products can be described as having autonomy, adaptability, multifunctionality, cooperation, and human interaction within

Published by Springer Nature Switzerland AG 2023
F. Noël et al. (Eds.): PLM 2022, IFIP AICT 667, pp. 516–525, 2023.
https://doi.org/10.1007/978-3-031-25182-5_50

the scope for which they were produced [3]. The challenge of promoting smart products, driven by global competition, increases the difficulty of determining customer requirements, being increasingly customizable and challenging to meet. This complexity increases effort and execution time in Product Development Process (PDP) [4, 5]. As a way of promoting agility to the PDP, using the enabling technologies of Industry 4.0, studies involving frameworks, methods, and approaches that use these digital tools to make the process smarter are on the rise [6, 7]. In addition, such frameworks aim to promote connectivity, communicability, and diagnostic capability to control throughout the Product Lifecycle Management (PLM) [8].

Although many studies address the applicability of industry 4.0 concepts in different domains, in the PDP, some ideas are considered nebulous and can bring some confusion [9, 10]. To corroborate this statement, the systematic literature review promoted by [3] about the smart product concept concludes that the high study of this concept by different domains has led to a dispersed and uneven body of knowledge. In [11], the authors highlight the need for framework adaptation between classical product development processes and intelligent product development process methods.

According to this context, this paper promotes an analysis and discussion about the concepts of smart products and the smart product development process based on a literature review. Additionally, it addresses the role of enabling technologies of Industry 4.0 around these themes to support the advancement of PDP.

This article is structured as follows: Sect. 2 addresses the problem statement regarding dispersion in the concept of smart products and the main differences between classical product development processes and innovative methodologies related to smart products. Section 3 presents an analysis of related works found after an initial survey of a theoretical framework. Section 4 discusses the main disadvantages and gaps in the associated results. Finally, Sect. 5 concludes and presents perspectives for further research.

2 Problem Statement

Through the study of the term smart products, it was possible to identify several descriptions of this term and generate diverse perspectives. In the survey developed by [12], the concept of a Smart Product is presented as one capable of retaining or storing data about itself. In [13], the authors point to intelligent products as those skilled in making decisions about themselves and establishing transport of information to other products or systems. The dispersion of the smart product concept impacts the agility of the development and production process of products with such characteristics. In addition, it retards the customer's defining the product requirements [13].

The advancement of emergent technologies, allied with industry 4.0, imposed an interaction between all stages of the classic development processes. With the increase in product complexity, the need to promote intelligence in product development became apparent the more it highlights the need for customized manufacturing requirements demanded by the customer [11, 13]. These contexts require agile methods and cheaper processes to launch innovative and technological products [14].

Therefore, this research intends to clarify the concept of smart products, pointing out the evolutions in the existed or conventional product development process. Additionally, this research aims to understand the new requirements to make the smart product process development process more agile and respect all customer's needs. Figure 1 summarizes all concepts involved in this research to active a Smart Product.

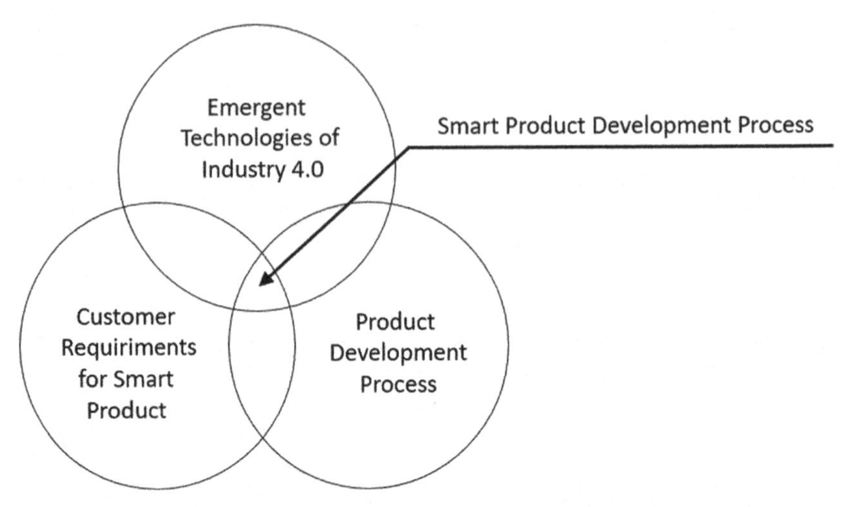

Fig. 1. Faces of the research problem.

Therefore, the starting point will be a previous survey of the theoretical framework considering three main dimensions:

i. The concepts and problems for the design of Smart Product.
ii. The main models of product development processes (Classic Models) and the models of intelligent product development processes.
iii. Industry 4.0 key concepts enable technologies to support smart product development.

This research considers these three dimensions with a discussion that identifies the main related works in this domain. It explores the differences and essential adaptations in the classical methods of product development focused on the agile process and the gaps that can be addressed in this context.

3 Related Works

The methodologies used to build the theoretical framework that supported the discussion about the topics presented were as follows:

i. A survey for articles and papers was carried out in the Scopus database using the following keywords (i) "Smart product development process" and (ii) "Smart products".

ii. The authors used multiple methods to rank the most recent articles and the most relevant research in the domain of this study.
iii. The selected articles were read and classified using four criteria to facilitate the analysis (the criteria will be detailed in Sect. 4 of this document).
iv. Finally, the authors analyzed articles to extract relevant information from the research. In addition, the study is divided into subtopics according to the context of the current research.

The following subsections summarize the extracted information and knowledge from the literature review that supports the discussion presented in Sect. 4. In addition, the subsections were structured according to the three research issues: (i) the Smart product concept issue; (ii) The Smart product development process adaptation issue; and (iii) Emergent Technologies Trends used in the Smart Product issue.

3.1 Smart Product Concept Issue

Although Smart Products are a widely used concept, there is still a need to clarify and centralize a single vision for a better understanding. A preliminary search made it possible to find many archetypes for these terms, and several examples used similar terms to describe different concepts. For example, [15] classify Philips Hue's bright lights, the Amazon Echo, and the self-driving car under the term "smart object," while [16] refers to in-car navigation systems, digital cameras, and cell phones as "smart products." In [17], the authors refer to "intelligent objects. According to [18], the authors consider the term "smart product" as "connected objects," and in [11], the authors introduce the term "internet-connected constituents." In [19], the authors use the term "Digitized artefact," while [18] use the term "smart thing." in a somewhat similar fashion, scholars have also used a variety of terms synonymously. This ambiguity is compounded by the fact that other concepts exhibit overlaps with the smart product but must be clearly distinguished in a conceptual sense.

Furthermore, there is no agreement on the definition and definition of "smart products" criteria. Existing reports are often based on packages of seemingly arbitrary features or features. It was also possible to note that recent research on smart products has focused on their created functions or service offerings and the larger ecosystems in which they operate [3, 19, 21, 22]. In this way, the preliminary study of intelligent products becomes relevant before analyzing and developing process methodologies and producing this new kind of product.

3.2 Smart Product Development Process Adaptation Issue

Within the literature, it was possible to find different perspectives of classic product development. In work developed by [20] are cited: [23] defines it as the process that converts customer needs and requirements into information to produce a product; and [24] defines it as a business process consisting of a set of activities that seek, from the needs of the market and the possibilities and technological constraints. In addition, the author considers companies' competitive strategies and strategies for the product to achieve product specifications.

In more recent research, approaches to intelligent product development were found, such as the study by [2] that compiles the primary studies for more competent production within industry 4.0. Within the research, a shortage of studies sought to adapt the traditional development processes to support intelligent products and services. Because of this, it becomes relevant to deepen the systematic review of the literature to prove this concern and delimit the requirements for such adaptation. Furthermore, in the study carried out by [8], it is already possible to notice an optimization approach for intelligent product service systems proposing a Digital Twin approach as an actuator in optimizing functions of intelligent systems and products.

3.3 Emergent Technologies Trends Used in Smart Product Issue

The industrial scenario has changed in recent decades due to successive technological advances. Industry 4.0, whose concept has been widely discussed by academics and organizations, is a new manufacturing paradigm combining intelligent factories, machines, systems, products, and processes in an integrated network that connects the physical and virtual worlds using Cyber-Physical Systems (CPS) technology. Industry 4.0 involves products and processes and is often related to CPS, the Internet of Things (IoT), and Smart Products, a new industrial approach encompassing future industry developments and technological advances that will increase productivity and business efficiency [24, 26].

Industry 4.0 is focused on creating smart products and processes by transforming conventional factories into smart factories. An intelligent factory environment is characterized by a networked intercommunication between human resources, machines, and objects, such as smart products. Smart products are integrated into the entire manufacturing process and actively support your manufacturing process by autonomously controlling individual production steps. Furthermore, like finished products, smart products are aware of the parameters in which they must be used, providing information on their status throughout their entire life cycle [25].

4 Discussion

This research is working to highlight the issues presented in Sect. 2 to ensure a better understanding of new product development processes and the concept of smart products in the face of the emergence of enabling technologies in industry 4.0. In this way, three dimensions of the problem are proposed to be discussed: the question of the concept of intelligent products, the adaptations in the classic processes of product development that the intelligent development frameworks have been promoting, and the importance of industry 4.0 technologies within this context.

A primary search was conducted where related works were found for each suggested problematic question. After a complete reading, each article was allocated based on the criteria in which it best fit. For a better understanding, subdivisions were made in each main question to guarantee a better perspective of the contribution of each article to this research. The criteria used in the subdivisions are described in Table 1, and the articles/papers classification according to the requirements can be seen in Table 2.

It is possible to notice in Table 1 that, within the selected articles, some themes have been little explored; for example, it is possible to note that few pieces addressed the differences between classic PDPs and intelligent product development processes or established a starting point for companies emerging companies that still use the classical process as a basis [20–22, 28]. In addition, it is possible to notice a high in the research of frameworks using enabling technologies in industry 4.0 aiming at the production of intelligent products but without the allocation of intelligence in the Development process, but with the intelligence focused on the product [17, 25, 27].

Table 1. Subdivisions of discussion questions

Issue	Subdivision	Criteria
Q1: Smart product concept	SP1	The article presents several definitions of smart products
	SP2	The article presents several definitions of smart products and offers as a conclusion a final definition
Q2: Classic PDP models vs. intelligent product development process frameworks	PD1	The article only covers classic product development processes
	PD2	The article only covers intelligent product development frameworks
	PD3	The article covers intelligent product development frameworks based on classic PDP models
Q3: Industry 4.0 enabling technologies	TH1	The article presents topics of enabling technologies of Industry 4.0
	TH2	The articles present topics about enabling technologies of Industry 4.0, linking them with intelligent product development

Another point that was possible to see is the extension of the smart product concept towards an earlier point in the product life cycle, thus leveraging the value-adding in-situ sensing capabilities of the products [28].

On the issue of the concept of smart products, it was possible to highlight the dispersion in the definitions of smart products because few studies centralize or conclude the purposes in a common one. A most detailed explanation was in the systematic literature review developed by [4], which presents four archetypes synonymously to define an intelligent product (digital, responsive, connected, and smart). Such conceptualization can describe the target product's level of complexity and the complexity necessary for its development.

These results represent a preliminary assessment of the issues raised, frameworks, and methodologies found within the literature. However, it is essential to consider that the requirements or conclusions established are not static; variations, advances, and

Table 2. Classification of articles based on the topics studied

Authors and publication year	Q1		Q2			Q3	
	SP1	SP2	PD1	PD2	PD3	TH1	TH2
R. Schmidt, M. Möhring, R.C. Härting et al. (2015) [1]	X					X	X
M.L. Nunes, A.C. Pereira, A.C. Alves (2017) [2]	X	X		X		X	X
S. Raff, D. Wentzel, N. Obwegeser (2020) [3]		X				X	
E. Rauch, P. Dallasega, D.T. Matt (2016) [4]	X				X		
X. Yang, R. Wang, C. Tang, L. Luo, X. Mo (2021) [5]	X			X			
C. Wu, T. Chen, Z. Li, W. Liu (2021) [7]	X				X		X
Z. Chen, X. Ming (2020) [8]	X			X			
M. Touzani, A.A. Charfi, P. Boistel (2018) [9]	X					X	
I.C.L. Ng, S.Y.L. Wakenshaw (2017) [10]	X			X		X	
D.J. Langley, J. van Doorn, I.C.L. Ng, S. Stieglitz, A. Lazovik, A. Boonstra (2021) [24]	X					X	
L. Bieler, T. Gruen, B. Akdeniz et al. (2018) [14]	X			X			
D.L. Hoffman, T.P. Novak (2016) [15]	X			X		X	
D. Beverungen, O. Müller, M. Matzner, J. Mendling, J. vom Brocke (2019) [19]		X				X	
Kärkkäinen et al. (2003) [25]	X			X			X
Kiritsis, D. (2011) [24]		X			X		X
Leitão et al. (2015) [27]		X		X		X	
Meyer et al. (2009) [17]		X		X		X	
Lenz et al. (2020) [28]		X			X		X

improvements in these issues may occur during the evolution of the studies. Within this context, it is necessary to explore more methodologies, research, and concepts to support the proposed questions. The authors consider this approach as the fourth issue to ensure consistency and coherence of system requirements.

5 Conclusions

This research presented a contextualization of adaptation of the traditional processes of product development aiming at the manufacture of intelligent products to guarantee the best use of enabling technologies of Industry 4.0. The issues and concepts addressed in this document are not static; that is, they may undergo changes, updates, or removals during the evolution of research.

To better understand the dimensions of the problem addressed, the authors proposed three key questions to be addressed: the question of the concept of intelligent products, the classic models of PDP versus the frameworks of intelligent processes of product development, and the importance of enabling technologies of Industry 4.0. in the evolution process of product development. With these questions, relevant articles were searched to determine which gaps were not explored and/or need further research. Within these gaps, the need for standardization or formalization of a concept of smart products was highlighted, in addition to the lack of research on smart product development methodologies that adapt the classic processes of product development and that are focused on promoting intelligence in the process of growth and not just the product.

The continuity of the research should, therefore, promote a systematic review of the literature to deepen and theoretically support the proposed questions to support the development of a framework for adapting the classic processes of product development aimed at the manufacture of intelligent products using enabling technologies of industry 4.0.

Acknowledgements. The authors would like to thank Robert Bosch CtP, the Pontifical Catholic University of Parana (PUCPR), and the National Council for Scientific and Technological Development (CNPq) for the financial support of this research.

References

1. Schmidt, R., Möhring, M., Härting, R.-C., Reichstein, C., Neumaier, P., Jozinović, P.: Industry 4.0 - potentials for creating smart products: empirical research results. In: Abramowicz, W. (ed.) BIS 2015. LNBIP, vol. 208, pp. 16–27. Springer, Cham (2015). https://doi.org/10.1007/978-3-319-19027-3_2
2. Nunes, M.L., Pereira, A.C., Alves, A.C.: Smart products development approaches for Industry 4.0. Procedia Manuf. **13**, 1215–1222 (2017). https://doi.org/10.1016/j.promfg.2017.09.035
3. Raff, S., Wentzel, D., Obwegeser, N.: Smart products: conceptual review, synthesis, and research directions. J. Prod. Innov. Manag. **37**, 379–404 (2020). https://doi.org/10.1111/jpim.12544
4. Rauch, E., Dallasega, P., Matt, D.T.: The way from lean product development (LPD) to smart product development (SPD). Procedia CIRP **50**, 26–31 (2016). https://doi.org/10.1016/j.procir.2016.05.081
5. Yang, X., Wang, R., Tang, C., Luo, L., Mo, X.: Emotional design for smart product-service system: a case study on smart beds. J. Clean. Prod. **298**, 126823 (2021). https://doi.org/10.1016/j.jclepro.2021.126823
6. Zhang, H., Liu, Q., Chen, X., Zhang, D., Leng, J.: A digital twin-based approach for designing and multi-objective optimization of hollow glass production line. IEEE Access. **5**, 26901–26911 (2017). https://doi.org/10.1109/ACCESS.2017.2766453
7. Wu, C., Chen, T., Li, Z., Liu, W.: A function-oriented optimising approach for smart product service systems at the conceptual design stage: a perspective from the digital twin framework. J. Clean. Prod. **297**, 126597 (2021). https://doi.org/10.1016/j.jclepro.2021.126597
8. Chen, Z., Ming, X.: A rough–fuzzy approach integrating best–worst method and data envelopment analysis to multi-criteria selection of smart product service module. Appl. Soft Comput. J. **94**, 106479 (2020). https://doi.org/10.1016/j.asoc.2020.106479
9. Touzani, M., Charfi, A.A., Boistel, P., Niort, M.C.: Connecto ergo sum! An exploratory study of the motivations behind the usage of connected objects. Inf. Manag. **55**, 472–481 (2018). https://doi.org/10.1016/j.im.2017.11.002

10. Ng, I.C.L., Wakenshaw, S.Y.L.: The Internet-of-Things: review and research directions. Int. J. Res. Mark. **34**, 3–21 (2017). https://doi.org/10.1016/j.ijresmar.2016.11.003
11. Huikkola, T., Kohtamäki, M., Rabetino, R., Makkonen, H., Holtkamp, P.: Unfolding the simple heuristics of smart solution development. J. Serv. Manag. **33**, 121–142 (2022). https://doi.org/10.1108/JOSM-11-2020-0422
12. Wong, C.Y., Mcfarlane, D., Zaharudin, A.A., Agarwal, V.: The intelligent product driven supply chain (2002). https://doi.org/10.1109/ICSMC.2002.1173319
13. Szalavetz, A.: The digitalisation of manufacturing and blurring industry boundaries. CIRP J. Manuf. Sci. Technol. **37**, 332–343 (2022). https://doi.org/10.1016/j.cirpj.2022.02.015
14. Bstieler, L., et al.: Emerging research themes in innovation and new product development: insights from the 2017 PDMA-UNH doctoral consortium. J. Prod. Innov. Manag. **35**(3), 300–307 (2018). https://doi.org/10.1111/jpim.12447
15. Hoffman, D.L., Novak, T.P.: Consumer and object experience in the Internet of Things: an assemblage theory approach. J. Consu. Res. **44**(6), 1178–1204 (2016). https://doi.org/10.1093/jcr/ucx105
16. Rijsdijk, S.A., Hultink, E.J.: How today's consumers perceive tomorrow's smart products. J. Prod. Innov. Manag. **26**(1), 24–42 (2009). https://doi.org/10.1111/j.1540-5885.2009.00332.x
17. Meyer, G.G., Främling, K., Holmström, J.: Intelligent products: a survey. Comput. Ind. **60**, 137–148 (2009). https://doi.org/10.1016/j.compind.2008.12.005
18. Pereira, R.M., Szejka, A.L., Canciglieri Junior, O.: Towards an information semantic interoperability in smart manufacturing systems: contributions, limitations and applications. Int. J. Comput. Integr. Manuf. **34**(4), 422–439 (2021). https://doi.org/10.1080/0951192X.2021.1891571
19. Beverungen, D., Müller, O., Matzner, M., Mendling, J., vom Brocke, J.: Conceptualizing smart service systems. Electron. Mark. **29**(1), 7–18 (2017). https://doi.org/10.1007/s12525-017-0270-5
20. Szejka, A.L., Junior, O.C.: The application of reference ontologies for semantic interoperability in an integrated product development process in smart factories. Procedia Manuf. **11**, 1375–1384 (2017). https://doi.org/10.1016/j.promfg.2017.07.267
21. Szejka, A.L., Canciglieri, O., Loures, E.R., Panetto, H., Aubry, A.: Requirements interoperability method to support integrated product development. In: Proceedings - CIE 45: 2015 International Conference on Computers and Industrial Engineering, pp. 1–10 (2015)
22. Szejka, A.L., Aubry, A., Panetto, H., Júnior, O.C., Loures, E.R.: Towards a conceptual framework for requirements interoperability in complex systems engineering. In: Meersman, R., et al. (eds.) OTM 2014. LNCS, vol. 8842, pp. 229–240. Springer, Heidelberg (2014). https://doi.org/10.1007/978-3-662-45550-0_24
23. Smith, D.W.: Proceedings of the 2002 Annual Midyear Meeting of the Engineering Design Graphics Division of the American Society for Engineering Education (2002)
24. Kiritsis, D.: Closed-loop PLM for intelligent products in the era of the Internet of Things. CAD Comput. Aided Des. **43**, 479–501 (2011). https://doi.org/10.1016/j.cad.2010.03.002
25. Kärkkäinen, M., Holmström, J., Främling, K., Artto, K.: Intelligent products - a step towards a more effective project delivery chain. Comput. Ind. **50**, 141–151 (2003). https://doi.org/10.1016/S0166-3615(02)00116-1
26. Langley, D.J., van Doorn, J., Ng, I.C.L., Stieglitz, S., Lazovik, A., Boonstra, A.: The internet of everything: smart things and their impact on business models. J. Bus. Res. **122**, 853–863 (2021). https://doi.org/10.1016/j.jbusres.2019.12.035

27. Leitão, P., Rodrigues, N., Barbosa, J., Turrin, C., Pagani, A.: Intelligent products: the grace experience. Control Eng. Pract. **42**, 95–105 (2015). https://doi.org/10.1016/j.conengprac.2015.05.001
28. Lenz, J., MacDonald, E., Harik, R., Wuest, T.: Optimizing smart manufacturing systems by extending the smart products paradigm to the beginning of life. J. Manuf. Syst. **57**, 274–286 (2020). https://doi.org/10.1016/j.jmsy.2020.10.001

Knowledge-Based Product-Service Ecosystems

Serdar Bulut(✉) and Reiner Anderl

Department of Computer Integrated Design, Technical University of Darmstadt, Otto-Berndt-Straße 2, 64289 Darmstadt, Germany
bulut@dik.tu-darmstadt.de

Abstract. Due to the ongoing digital transformation, businesses are exposed to work with a dynamic business environment. Traditional 1-to-1 relationships between manufacturers and end customers are transformed into a network of relationships between manufacturers, end customers, software developers, suppliers, and third parties. The network of partners evolves into an ecosystem. Using a smart product-service system as the central anchor of value creation for an ecosystem enables hybrid value creation for producers and consumers simultaneously. In this paper we develop a knowledge-based model of smart product-service systems in their corresponding ecosystems, that assists designers and developers in decision making during the product development process.

Keywords: Smart product-service systems · Product-service ecosystems · Knowledge-based engineering · Smart product engineering

1 Introduction

Product-Service Systems (PSS) offer a value proposition consisting of product and service components [1]. Traditionally, the value proposition is targeted towards end consumers forming a 1-to-1 relationship between manufacturer and end consumer. However, with emerging technology shaping the 4th industrial revolution, the rigid 1-to-1 relationship of manufacturers and consumers gets redefined. The potential of a network model arises, in which PSS take the role of an asset, that unites manufacturers, suppliers, consumers, and further third parties. All of these stakeholders may establish interdependent relations to one another based on the data that the PSS shares [2]. Suppliers may offer new spare parts to the customer based on condition monitoring, third party developers may offer new upgraded algorithms and functionalities as apps for the PSS, and manufacturers may analyze the behavior of the PSS in order to optimize the next generation of PSS. Ultimately, an ecosystem that shares PSS data is created; a product-service ecosystem (PSE) emerges. In order to achieve the data- and PSS-centric ecosystem, the PSS must become a Smart Product-Service System (SPSS). SPSS extend PSS by using smart services and offering a value proposition towards a variety of stakeholders via the Internet of Everything (IoE) [3]. Smart services are data-based services that allow a flexible and individual alignment with customer wishes and expectations [4]. Stakeholders are connected through an SPSS within a PSE. This paper introduces the definition

Published by Springer Nature Switzerland AG 2023
F. Noël et al. (Eds.): PLM 2022, IFIP AICT 667, pp. 526–535, 2023.
https://doi.org/10.1007/978-3-031-25182-5_51

and the knowledge-based representation of a PSE along with the support it provides to product developers and designers. The knowledge-based representation constitutes the foundation of a graphical assistance system. Section 2 summarizes the current state of the art regarding PSS, SPSS, PSE and modeling techniques for PSEs. Section 3 provides the selection of a modeling technique for PSE representation and introduces the core elements of the proposed PSE model with their corresponding potentials for developers and designers. Section 4 presents the use-case-centric implementation of the PSE model for a multi-purpose drone as the SPSS. Section 5 finishes with a discussion of the results and an outlook for future work.

2 State of the Art

2.1 PSS and SPSS

PSS have been widely discussed starting with the late 90s. The complementation principle of PSS is the foundation of PSS. The complementation principle states that a tangible product is extended by an intangible service to meet users' needs and vice versa [5–7]. The offered value proposition may be individualized to some extent, traditionally by adjusting the service component and maintaining the rigid product component. Industrie 4.0 allows for further individuality by using the concept of cyber-physical systems CPS. CPS extend conventional products by integrating embedded systems and an interface for communication and data exchange [8].

Therefore, further smart services such as remote condition monitoring and remote control are enabled transitioning PSS into SPSS [3]. Fundamentally, SPSS are PSS that are extended with data-based services and allow flexible and individual alignment of their linked, tangible product [4]. The communication of SPSS may function decentralized with direct communication, broadcast communication, mesh communication or may function centralized by using platform technology as the center of the infrastructure for information and communication technology.

2.2 Socioeconomic and Technical Ecosystems

PSEs are not directly referred in literature yet, as the term will first be introduced in this paper. However, in socioeconomics and especially in the era of Industrie 4.0, there are 3 key ecosystem types that have a significant impact on business models of manufacturers, software developers, suppliers, and commercial & private prosumers. Prosumers are stakeholders that may be producers and consumers simultaneously [9].

Platform-Based Ecosystems

Multi-sided platforms, on which multiple stakeholders and market participants operate, have recently gained significance, due to their business opportunities [10]. They form a platform-based-ecosystem: a network in which a platform provider offers business opportunities to third parties to develop innovative IT-solutions for the platform [11–14].

Knowledge Ecosystems and Business Ecosystems

Further ecosystem types, that are widely discussed in literature and applied by businesses and academies, are knowledge-ecosystems and traditional business ecosystems

[15]. Knowledge-ecosystems describe the union of businesses and institutions to collaborate for the sake of research & development: leveraging innovation is the driving intention [16]. A common use-case for knowledge-ecosystems is the collaboration for the standardization of emerging technologies such as OPC-UA. Ordinary business ecosystems describe any kind of business-related collaboration of multiple stakeholders to form common value creation & proposition [17].

2.3 Representation Techniques for PSS, SPSS, PSE

PSS consist of product and service components. SPSS extend PSS by incorporating shared data. PSE further incorporate a multitude of stakeholders who share the digital artifacts generated by, used by, and provided to the SPSS. Therefore, the representation of the entities PSS, SPSS, and PSE must incorporate interdisciplinarity in the fields of engineering, service design, data aggregation, and business development. UML, SysML, and IDEF Models are partially able to model these entities [18].

However, these technologies share a common trait. Their set of modeling rules, guidelines, statements, defined elements and their characteristics as well as relationships forcibly create restrictions, that limit the scope of representation capabilities for interdisciplinary entities [19–22]. Data Flow Diagrams may be used to represent the organization of PSS while IDEF Models are suitable for PSS manufacturing [23], UML 2.0 was used for PSS design and use case modeling [24–26], SysML further extends UML 2.0 by integrating enhanced traceability, rationalization and inter-team communication for co-designing PSS [27].

While these tools and technologies focus on certain aspects of PSS design, they lack a holistic approach on PSS engineering and organization regarding all functional components, life cycle phases, stakeholders, and value creation flow [28]. It is not possible to freely define types of system elements, types of semantic annotations, types of relationships and the characteristics of elements, annotations, and relationships. Hence, it is required to use a representation technique, that offers a higher degree of freedom. The interrelations between the system elements of a holistically analyzed PSS need further investigation [29]. These interrelations are responsible for the high degree of complexity and interdisciplinarity in PSS.

Ontologies, as conceptualizations of specific domains, offer more freedom due to their limitless possibilities to determine elements and relationships in their taxonomy and structure. Many ontologies have been developed for PSS design including service ontologies, engineering ontologies and product ontologies [30–33]. While these ontologies may represent PSS sufficiently, knowledge deduction and induction still remains insufficiently addressed, especially enabled through data-based communication of SPSS. Knowledge deduction and induction are crucial for interdisciplinary teams of analysts, engineers, designers and developers who aim to offer individual product and service solutions to a multitude of stakeholders within an ecosystem. Intuitive and knowledge-based assistance is required to support the development process of SPSS and ultimately PSE.

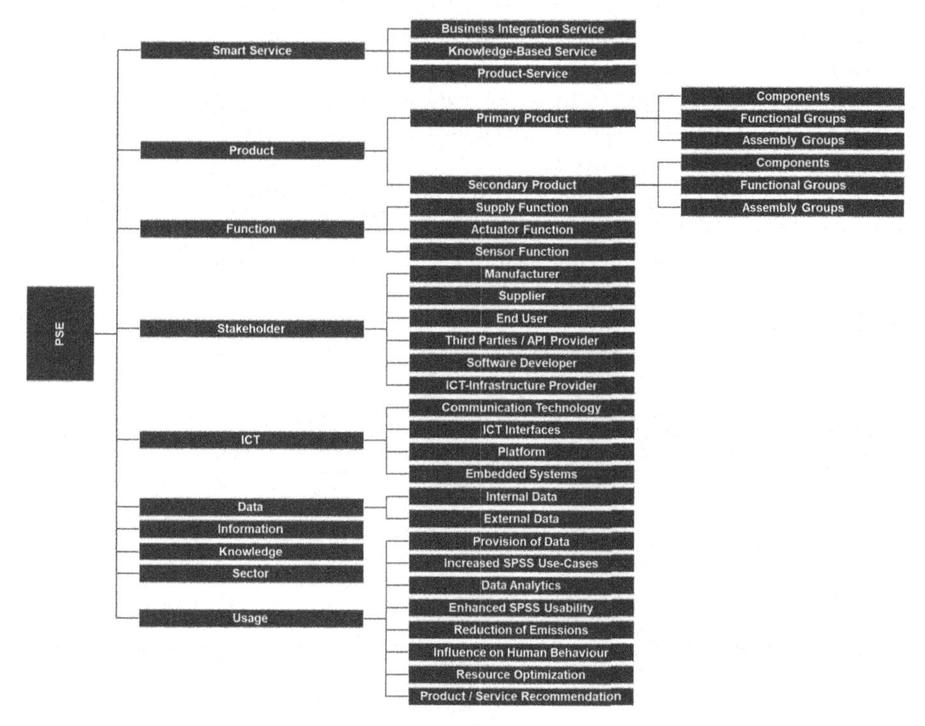

Fig. 1. Taxonomy of PSEs

3 Knowledge-Based Model of a Product-Service Ecosystem

Graphs are best suited to intuitively represent knowledge for humans. Graphical representations are possible with RDF-triples and labelled property graphs LPGs. Both concepts use ontologies to define their scope of representation. LPGs are superior in querying and storage. Especially human-initiated querying is essential as the goal of this paper is to assist human decision makers with a knowledge-based system in designing and developing a PSE. Hence, knowledge retrieval must be designed intuitively. Therefore, we choose LPGs as the foundation of the developed assistance system.

In order to determine the scope of the PSE domain, we first define PSEs. *A product-service ecosystem enables joint value proposition to multiple stakeholders based on offerings and needs of a central smart product-service system. The value proposition is embedded in tangible product as well as intangible service components. Value creation within the ecosystem is supported by information and communication technology.* A product-service ecosystem contains traits of knowledge ecosystems, platform ecosystems, and traditional business ecosystems. Hence, the PSE domain consists of the main elements: *product* and *smart service, functions, stakeholders, information and communication technology (ICT), data, information, knowledge,* generated *usage* and economic *sectors,* in which *stakeholders* operate. The economic sectors are important for the identification of further prosumers in different sectors that share similarities to the current prosumers for ecosystem expansion. Further analysis of the main elements leads to the taxonomy in Fig. 1.

After defining the elements that constitute the representation of PSEs, relationships are added. The relationships between the elements represent dependencies, value creation flows, hierarchical structures, specializations, compatibilities, necessities, and offerings. Therefore, the relationships are: *includes, is_a, aggregates_to, is_compatible, uses, offers, constitutes, requires, operates_in, enables, produces.* The elements with the corresponding relationships are seen in Fig. 2 as an LPG. Figure 2 represents the reference model for PSEs in Neo4j.

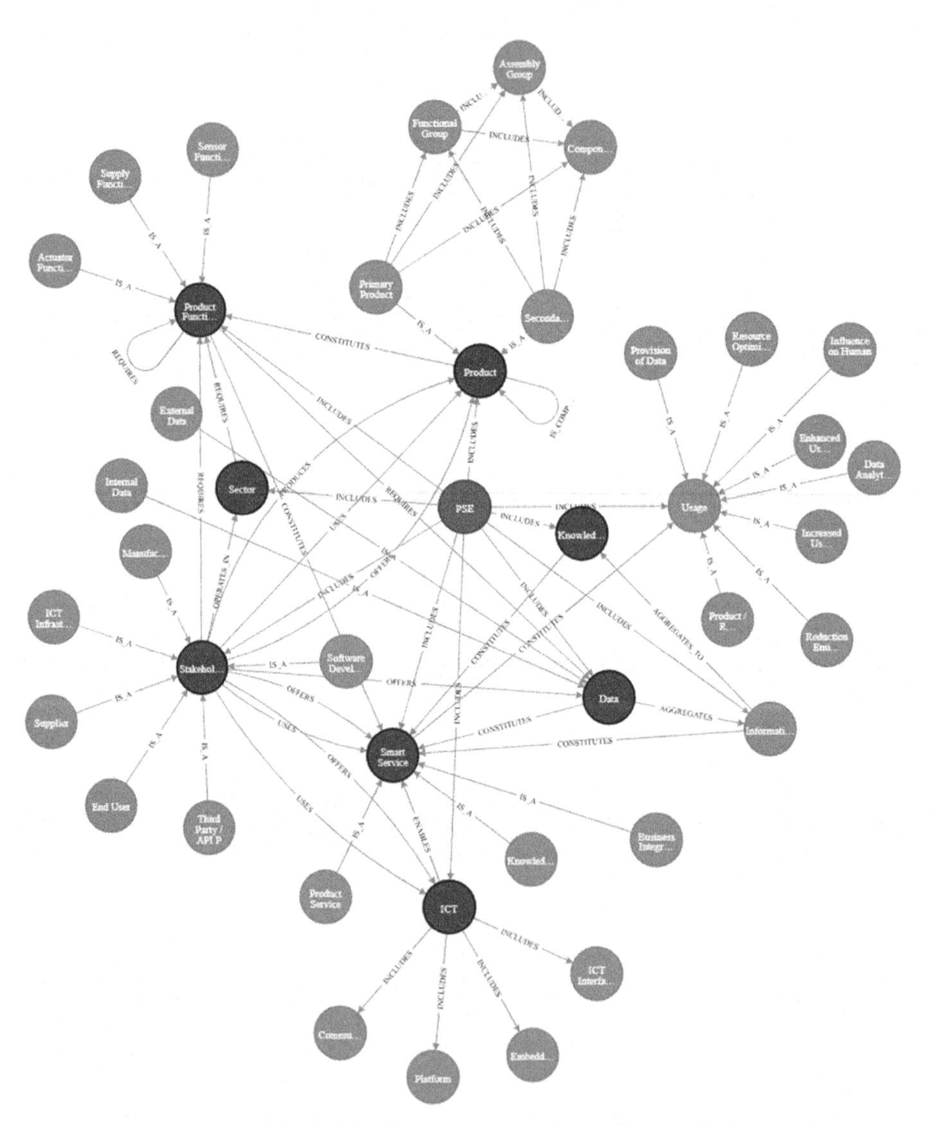

Fig. 2. Reference model of PSE with main elements in purple and sub-elements in gray

The proposed knowledge-based model of PSE consists of 2 sub-models: the reference model and the applied knowledge model. The reference model is seen in Fig. 2. Figure 2 contains the relationships and nodes of the LPG without depicting the attributes of the relationships and the attributes of the nodes. The attributes are out of scope of this paper. The applied knowledge model is presented in Sect. 4.

While the reference model is supposed to support decision makers in communication and on how to represent PSEs, the applied knowledge model contributes to the deduction and induction of knowledge. Both models combined unlock potentials to support the uses-cases: *production planning, compatibility check of modular products, service development, product development, similarity check of stakeholders and economic sectors for ecosystem expansion.*

4 Use Case for Applied Knowledge Model

Section 3 introduced the reference model for PSE and presented its potentials. This section presents how to apply the reference model to generate the applied knowledge model and how the applied knowledge model supports in compatibility check and ecosystem expansion. In order to use the applied knowledge model, we determine an SPSS and focus on a specific use-case. The SPSS is a multi-purpose drone, that is used for smart fertilizing of crop, that is difficult to access for terrestrial machines. Wine grape harvesting on cliffsides represents a sufficient use-case for multi-purpose drones. Figure 3 displays a segment of the applied knowledge model. The full model is out of scope for this paper. Starting with the Knowledge-Based Service Smart Farming the model tracks down the intelligence necessary to constitute the service. Knowledge for condition-based fertilizing is required which is constituted by various information segments such as pesticide effects, crop condition, fertilizer effects, geo localization, and also a data segment rain detection. The service is mostly used by farmers for resource optimization, reduction of emissions, data analytics, and provision of data. The service requires a supply function: the provision of fertilizer. The supply function is constituted by the tank.

A full tank significantly increases the mass of the drone, which has a direct effect on the drive. The drive consists of the rotors, motors, and the battery. Due to the structure of the physical components, there are compatibilities between selected components. The compatibilities narrow down the choices to realize a drive unit for the drone. Further requirements of the drive such as power and range are considered by querying the node attributes of the applied knowledge-model. Ultimately, the model leads to the only possible drive combination (red circles in Fig. 3): Hacker LiPo 6s, T-Motor Antigravity MN5208, and T-Motor NS17x5.8.

The insight deducted from the applied knowledge model directly supports decision makers whether or not a new supplier should be integrated into the ecosystem. Furthermore, additional use-cases for multi-purpose drones can be modeled to determine the economic impact of the tank & drive and the services they enable across sectors.

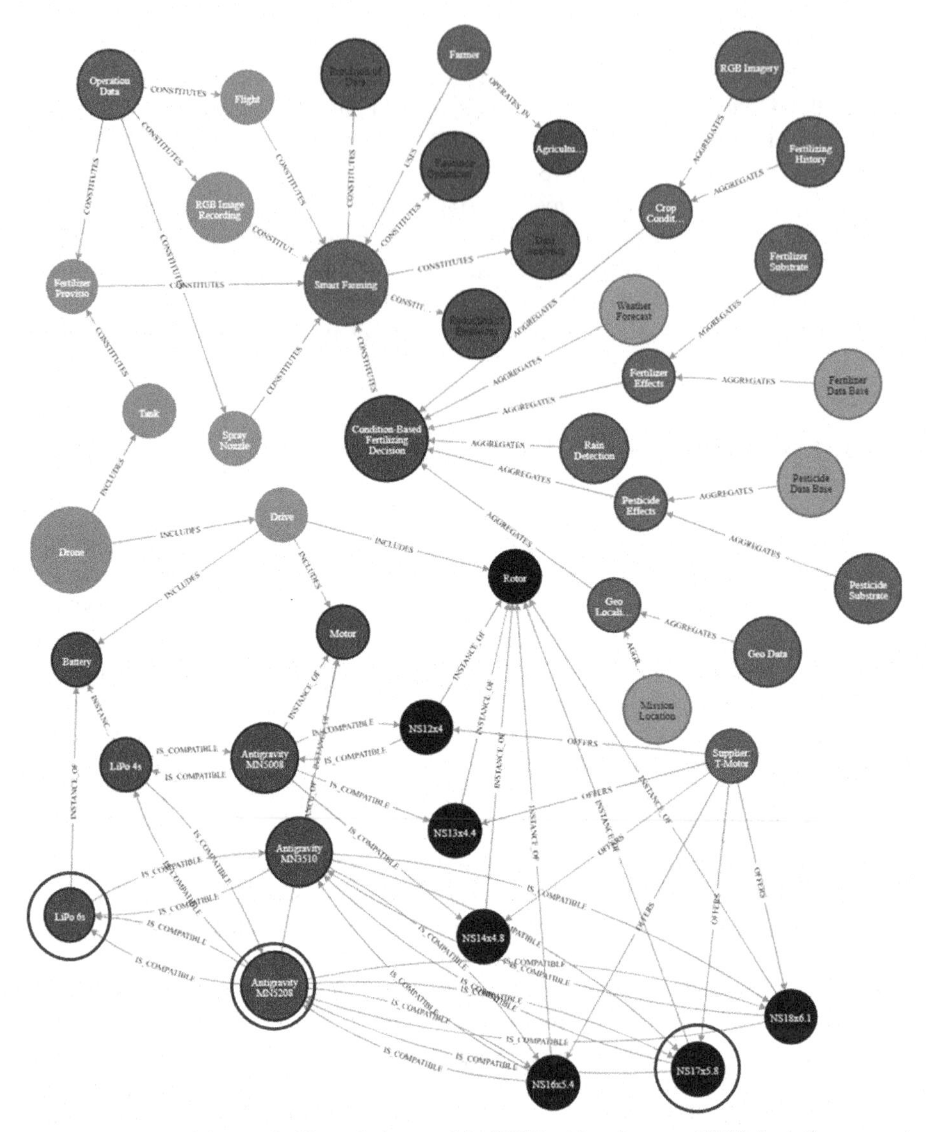

Fig. 3. Fraction of the applied knowledge model of PSE with a drone as SPSS depicting a smart farming service. Elements encircled in red are sufficient to constitute the functional group 'drive'.

5 Discussion of Results

In this paper we defined PSEs and used LPGs to model them. We developed a reference model for PSEs along with its instantiation as the applied knowledge model. The models are used for various purposes: representation of PSEs for common understanding, communication in interdisciplinary teams, inductive and deductive reasoning. Ultimately, the models assist in decision making in an interdisciplinary and dynamic environment.

The models support in production planning, compatibility check of modular products, service development, product development, similarity check for stakeholders & economic sectors, and ecosystem expansion. Section 4 showed the applied knowledge model and presented how compatibility check and ecosystem expansion are supported through the model. The models incorporate abstraction, instantiation, and generalization.

The parameters for tank size and tank weight are determined within the attributes of the tank. The attributes however are beyond the scope of this paper. This paper mainly focuses on the eligibility of LPGs to represent PSE with a knowledge base. The elements and relationships for the models are chosen to allow for sufficient, human-assisted reasoning through Cypher queries while minimizing the complexity of the model. The model represents human-readable interdisciplinary knowledge of the elements and relationships forming a PSE. With the proposed reference model, decision makers can model their own PSE and gain insight through querying the instantiated applied knowledge model. The complexity of the model increases with more distinct relationships and distinct nodes: balancing accuracy and granularity is key. Further research will focus on the diversity of potentials derived from querying the nodes and relationships. Furthermore, a guideline to determine the attributes of the nodes and relationships will be developed to enable sufficient querying and reasoning.

References

1. Tukker, A., Tischner, U. (eds.): New Business for Old Europe: Product-Service Development. Competitiveness and Sustainability. Taylor and Francis, London (2017)
2. Bulut, S., Anderl, R.: Framework approach for smart service development. Procedia CIRP **100**, 864–869 (2021). https://doi.org/10.1016/j.procir.2021.05.030
3. Anderl, R., Bulut, S.: Smart product-service systems engineering. In: International Seminar on High Technology, vol. 25, pp. 16–29 (2021)
4. Plattform Industrie 4.0: Engineering Smarter Produkte und Services: Plattform Industrie 4.0 STUDIE. Acatech – Deutsche Akademie der Technikwissenschaften (2018)
5. Goedkoop, M.J., van Halen, C.J.G., te Riele, H.R.M., Rommens, P.J.M.: Product-service systems – ecological and economic basis (1999)
6. Morlock, F.: Performance management der erbringung industrieller produkt-service systeme. Dissertation, Shaker Verlag
7. Schweitzer, E.: Lebenszyklusmanagement investiver produkt-service systeme. In: Aurich, J.C., Clement, M.H. (eds.) Produkt-Service Systeme, pp. 7–13. Springer, Heidelberg (2010). https://doi.org/10.1007/978-3-642-01407-9_2
8. Anderl, R.: Industrie 4.0-advanced engineering of smart products and smart production. In: International Seminar on High Technology, vol. 19 (2014)
9. Schwab, K.: The Fourth Industrial Revolution. Portfolio Penguin, London (2017)
10. Abdelkafi, N., Raasch, C., Roth, A., Srinivasan, R.: Multi-sided platforms. Electron. Mark. **29**(4), 553–559 (2019). https://doi.org/10.1007/s12525-019-00385-4
11. Gawer, A., Cusumano, M.A.: How companies become platform leaders. MIT Sloan Manag. Rev. **49**, 28–35 (2008)
12. Ceccagnoli, M., Forman, C., Huang, P., Wu, D.J.: Cocreation of value in a platform ecosystem! The case of enterprise software. MIS Q. **36**(1), 263 (2012). https://doi.org/10.2307/41410417
13. Cozzolino, A., Corbo, L., Aversa, P.: Digital platform-based ecosystems: the evolution of collaboration and competition between incumbent producers and entrant platforms. J. Bus. Res. **126**, 385–400 (2021). https://doi.org/10.1016/j.jbusres.2020.12.058

14. Parida, V., Burström, T., Visnjic, I., Wincent, J.: Orchestrating industrial ecosystem in circular economy: a two-stage transformation model for large manufacturing companies. J. Bus. Res. **101**, 715–725 (2019). https://doi.org/10.1016/j.jbusres.2019.01.006
15. Jacobides, M.G., Cennamo, C., Gawer, A.: Towards a theory of ecosystems. Strat. Manag. J. **39**(8), 2255–2276 (2018). https://doi.org/10.1002/smj.2904
16. Adner, R.: Match your innovation strategy to your innovation ecosystem. Harv. Bus. Rev. **84**(4), 98–107, 148 (2006)
17. Teece, D.J.: Explicating dynamic capabilities: the nature and microfoundations of (sustainable) enterprise performance. Strat. Manag. J. **28**(13), 1319–1350 (2007). https://doi.org/10.1002/smj.640
18. Vasantha, G.V.A., Roy, R., Lelah, A., Brissaud, D.: A review of product–service systems design methodologies. J. Eng. Des. **23**(9), 635–659 (2012). https://doi.org/10.1080/09544828.2011.639712
19. Morelli, N.: Developing new product service systems (PSS): methodologies and operational tools. J. Clean. Prod. **14**(17), 1495–1501 (2006). https://doi.org/10.1016/j.jclepro.2006.01.023
20. Chen, P.P.-S.: The entity-relationship model—toward a unified view of data. ACM Trans. Database Syst. **1**(1), 9–36 (1976). https://doi.org/10.1145/320434.320440
21. Alt, O.: SysML. In: Alt, O. (ed.) Modellbasierte Systementwicklung mit SysML, pp. 29–63. Carl Hanser Verlag GmbH & Co. KG, München (2012)
22. Alt, O.: Modellbasierte entwicklung. In: Alt, O. (ed.) Modellbasierte Systementwicklung mit SysML, pp. 19–28. Carl Hanser Verlag GmbH & Co. KG, München (2012)
23. Durugbo, C., Tiwari, A., Alcock, J.R.: A review of information flow diagrammatic models for product–service systems. Int. J. Adv. Manuf. Technol. **52**, 1193–1208 (2011). https://doi.org/10.1007/s00170-010-2765-5
24. Aurich, J.C., Fuchs, C., Wagenknecht, C.: Life cycle oriented design of technical Product-Service Systems. J. Clean. Prod. **14**(17), 1480–1494 (2006). https://doi.org/10.1016/j.jclepro.2006.01.019
25. Morelli, N.: Designing product/service systems: a methodological exploration. Des. Issues **18**(3), 3–17 (2002). https://doi.org/10.1162/074793602320223253
26. Aurich, J.C., Clement, M.H.: Produkt-Service Systeme. Springer, Heidelberg (2010). https://doi.org/10.1007/978-3-642-01407-9
27. Balmelli, L.: The systems modeling language for products and systems development. JOT **6**(6), 149 (2007). https://doi.org/10.5381/jot.2007.6.6.a5
28. Vasantha, G.V.A., Roy, R., Corney, J.: Advances in designing product-service systems. J. Indian Inst. Sci. **95**, 429–447 (2011)
29. Aurich, J.C., Fuchs, C., Wagenknecht, C.: Modular design of technical product-service systems. In: Brissaud, D. (ed.) Innovation in Life Cycle Engineering and Sustainable Development, pp. 303–320. Springer, Dordrecht (2006). https://doi.org/10.1007/1-4020-4617-0_21
30. Shen, J., Wang, L.: A new perspective and representation of service. In: 2007 International Conference on Wireless Communications, Networking and Mobile Computing, Shanghai, China, pp. 3171–3174 (2007)
31. Baxter, D., Roy, R., Doultsinou, A., Gao, J., Kalta, M.: A knowledge management framework to support product-service systems design. Int. J. Comput. Integr. Manuf. **22**(12), 1073–1088 (2009). https://doi.org/10.1080/09511920903207464

32. Li, Z., Raskin, V., Ramani, K.: Developing ontologies for engineering information retrieval. In: 27th Computers and Information in Engineering Conference, Parts A and B, Las Vegas, Nevada, USA, vol. 2, pp. 737–745 (2007)
33. Annamalai, G., Hussain, R., Cakkol, M., Roy, R., Evans, S., Tiwari, A.: An ontology for product-service systems. In: Hesselbach, J., Herrmann, C. (eds.) Functional Thinking for Value Creation, pp. 231–236. Springer, Heidelberg (2011). https://doi.org/10.1007/978-3-642-19689-8_41

Interconnecting Tailored Work Plans to Smart Products' Bills of Materials

Savarino Philipp(✉), Apostolov Christo, and Dickopf Thomas

CONTACT Software Ltd., Wiener Str. 1-3, 28359 Bremen, Germany
philipp.savarino@contact-software.com

Abstract. Holistic transition in information and communication technology led to a new generation of smart products. These products are cyber-physical systems augmented by internet-based services, so-called smart services, characterized by a high degree of interdisciplinary components. This results in a tremendous complexity in describing smart products using a tailored bill of materials (BOM) and the consecutive work plans that aim to provide, e.g., production or service instructions for the worker. This contribution focuses on smart product BOMs and their interconnection to work plans while considering new challenges derived from the smart products' characteristics, such as their high interdisciplinarity or real-time feedback data from the usage phase. Therefore, a holistic approach for the description of smart product BOMs and a generic technique for defining work plans and their breakdown structure to be coupled with the smart products' BOMs are proposed.

Keywords: Work planning · Bill of materials · Product lifecycle management · Enterprise IT

1 Introduction

Today's customer expectations of ever faster development of innovative, highly complex, and interdisciplinary smart products, combined with ever shorter production cycles on a global scale, lead to tremendous pressure on companies [1]. A cornerstone of this pressure is the increasing product individualization that changes product requirements between different product generations and within one product generation. Additional drivers are rapidly changing markets that require swift reactions of the companies regarding restructuring a product and thus the corresponding bill of materials (BOMs), which also implicates a restructuring of assembly or manufacturing lines. Further problems refer to the management of the actual smart product BOMs during the use phase, particularly for those parts that are serialized and thus can, and often must, be uniquely identified. The results are enormous challenges, particularly in globally distributed manufacturing sites with heterogenous IT systems in terms of consistent work or manufacturing planning, including the follow-up management during the smart products' usage [2, 3].

Published by Springer Nature Switzerland AG 2023
F. Noël et al. (Eds.): PLM 2022, IFIP AICT 667, pp. 536–546, 2023.
https://doi.org/10.1007/978-3-031-25182-5_52

Methodological concepts such as closed-loop engineering [4] as well as data-driven architectures in the sense of a digital thread, which link information from the entire product lifecycle, are becoming increasingly important as digital communication frameworks for streamlining design, manufacturing, and operating processes in order to plan, design, manufacture and maintain technical products more efficiently [5]. However, there is a lack of concrete implementation examples to bring these mostly abstractly described and generally held high-level concepts closer to small and medium-sized enterprises to use the potential of their data in a more targeted manner.

This paper will address these challenges by interconnecting smart product BOMs to work plans from a processual and systemic view. It will be located between different management systems commonly used in product design, resource planning, manufacturing, and usage environments: Product Lifecycle Management (PLM), Manufacturing Execution System (MES) or Manufacturing Operations Management (MOM), Internet of Things (IoT) or Customer Portals as well as Enterprise Resource Planning (ERP) (cf. Fig. 1).

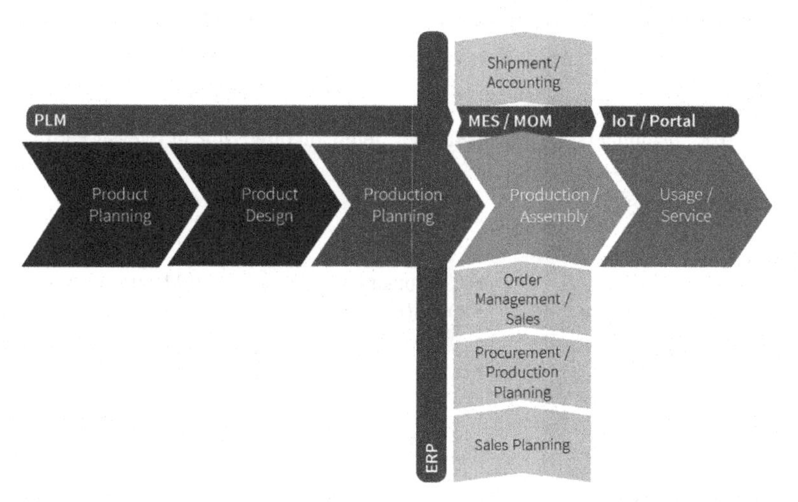

Fig. 1. Process and IT-system overview for the presented approach [6]

The structure of the paper at hand will give an initial overview regarding work planning in engineering and production as well as product structuring methodologies and shortcomings in terms of their interconnection. The following chapter will introduce a concept that considers especially the challenges regarding BOMs and work plans resulting from smart products' characteristics, such as their high complexity. Subsequently, it will be shown how the approach was prototypically implemented and validated. Finally, the last chapter summarizes the findings and provides an outlook on further research activities in the presented context.

2 Work Planning in Engineering and Production

Work planning involves all one-time planning activities for the manufacturing and assembling of products and product components. The goal is to define the sequence of steps for production processes and determine production times. As part of the work planning, it is defined how (i.e., by which method) and where (i.e., on which production equipment) a product or a component is to be manufactured or/and assembled. The work planning is done by creating process descriptions—i.e., work plans—for the production processes of a product from the perspective of the product itself. The plans include the working steps performed to produce an end item, the equipment and tools on which or with the help of which the tasks are to be performed, and times for preparation and performing the working steps [7].

The functions of work planning can be classified as short-term and long-term tasks. Short-term tasks include activities necessary for producing a specific product, such as creating working and routing plans, resource scheduling, bill of materials processing, and programming. The long-term tasks include material planning, investment planning, methods planning, and others. In addition, other activities such as cost planning and quality assurance can be classified as both short-term and long-term tasks [7] (Table 1).

Table 1. Work planning tasks, classified by the temporal dimension of relevance (after [7])

Tasks of work planning		
Short-term	Intermediate-term	Long-term
• Work plan creations • Production equipment planning • Production equipment planning • Programming • …	• Cost planning • Quality assurance • Work planning support • …	• Material planning • Investment planning • Methods planning • …

This contribution focuses on the short-term activities related to work planning and their methodological and tool-wise support; hence, these are described in more detail.

2.1 Short-Term Activities in Work Planning

Work and Routing Plan Creation

A core task of work planning is the creation of descriptions of the working activities and their executional sequence involved in the manufacturing or assembling of a product, an assembly component, or a single part. The basic information of a work plan includes the material to be used, the sequence of work activities, the workplaces at which the different activities are performed, the used tooling and operating resources, and the target times for conduction of the working steps. The definition of used materials in the working process relates to the BOM of an item to be assembled or manufactured. Preparing such bills is another main short-term task of work planning [7].

Bill of Materials Processing
The task of bill of materials (BOM) processing is preparing production-oriented BOMs out of the engineering ones (eBOM). While the eBOM is generally functionally structured and driven by the design structures in CAD, the production/manufacturing BOM (mBOM) considers and depicts product manufacturing stages and additional materials needed in the production process. The items of the mBOM are used to define the inputs and outputs of working activities.

Production Equipment Planning
Production equipment planning encompasses planning, developing, and producing or procuring equipment necessary to manufacture a component. The production equipment to be used in the different stages of the production process of an item is defined in the work plan.

2.2 Interdependence Between Work Plan and Bill of Materials

A work plan defines a process or a sequence of processes by which raw materials and single components are manufactured, processed, and assembled into finished products or parts. Conceptually, work planning can be supported by the Product-Process-Resource (PPR) model. PPR is an established concept for planning and modeling production systems commonly used in work planning methods and production systems modeling, adopted by many contemporary industrial software systems [8, 9]. The entity classes Product, Process, and Resource define the relevant aspect of a work plan, as depicted in the upper part of Fig. 2. A Product represents final and intermediate product types or additional residual materials from the production process. The Process is the central class that consumes intermediate products to produce a final one using resources to execute [10]. Finally, a Ressource is a hardware of software equipment involved in process execution [11].

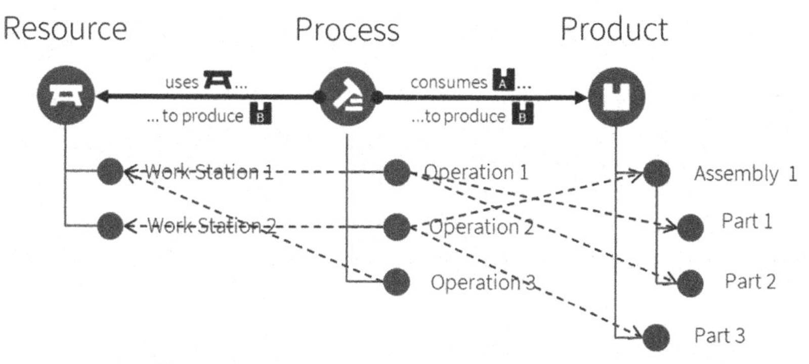

Fig. 2. Exemplary relations between elements of the work plan, materials of the product, and resources according to the PPR model

The entities of the three PPR entity classes are generally structurally decomposable. Further, a product can have different configurations, which vary in their composition.

Thus, there are different bills of materials contained by different product end-items. One such end-item is represented on the right-hand side of Fig. 2. A work plan is valid for a specific product configuration, i.e., it refers to a specific end-item. A work plan is built out of a sequence of operations. The operations refer to positions of the BOM (single parts and assemblies). The BOM positions are handled in the process and represent the material input and output from the process steps. In the simplified example above, Operation 1 assembles Parts 1 and 2 to Assembly 1, and then it is assembled in Operation 2 with Part 3 to the product end-item. Operation 3 is a working step without material input, such as testing or finishing actions.

In summary, a process consumes functions provided by a resource and thereby occupies it for a period of time. It also consumes intermediate products in order to produce an end product. The consumed products define the material transformation for value-creation, while consumed functions of resources define the production capacities. In combination, the two contain all necessary information for production, which is recorded in the form of work plans [10].

2.3 Concept for Integrated, Lifecycle-Spanning Work Planning and BOM Management

A concept for product lifecycle overreaching modeling and software support was developed based on the PPR model. The concept covers product modeling throughout the different phases of its life and considers its representation in a contemporary, integrated information management solution. The lifecycle phases covered by the concept are, as depicted in Fig. 3, the design and development phase, the process and work planning, production operations support, and the usage phase of the end product.

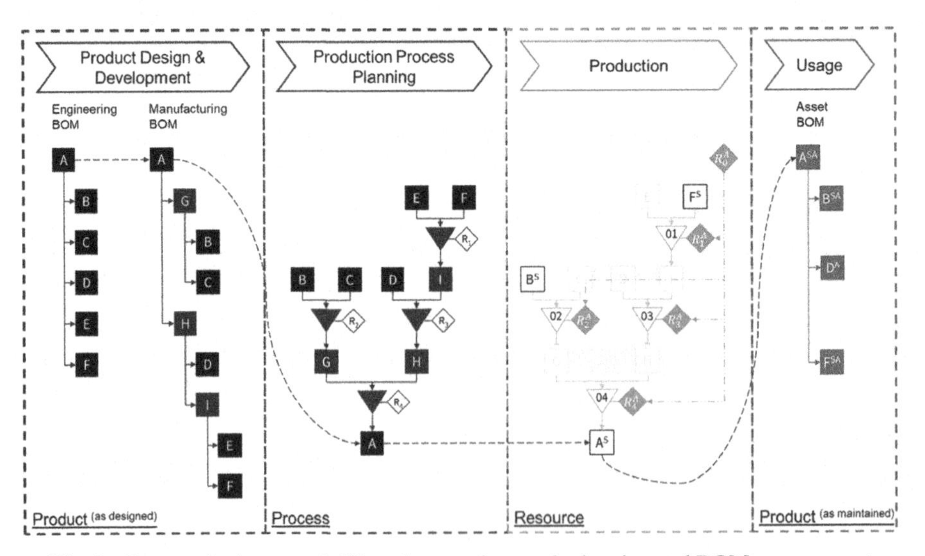

Fig. 3. Concept for integrated, lifecycle-spanning work planning and BOM management

Different product, process, and resource descriptions are necessary throughout these phases, each serving the specific goal of the respective phase. In the product design and development phase, the product is defined in terms of its composition using modular BOMs. An engineering BOM (eBOM) defines the product from a functional perspective, mostly driven by CAD. Next, a manufacturing BOM (mBOM) is derived, serving as a basis for the production process [12]. While it contains the materials of the eBOM, it includes additional intermediate assembly nodes following the assembly stages of the product. In the production process planning phase, the process of building the product is described. Thereby, the process steps consume the materials of the mBOM and transform them into wider assemblies while using the capabilities of the production resources. The process model, input and output material, and necessary resources amount to the work plan. The facility implementing the process is put in the foreground in the production phase. The resources – i.e., production equipment – of the production line model perform operations, which correspond to the process steps of the work plan. While the product and process descriptions of the former two phases have descriptive character, the production line model also serves the manufacturing execution and control. With the completion of the production phase, the unmounted/unprocessed material defined in the BOMs has been transformed into a finished end-product, which is ready to be put into service. The finished product requires a different model in the usage phase, which enables its monitoring, control, and maintenance record. The product in operation is represented by the asset BOM (aBOM), which reduces the structure of the eBOM fitted to the needs of the concrete use case.

2.4 Work Planning and Lifecycle-Spanning BOM-Management from an Enterprise IT-Landscape View

As seen from an engineering perspective, product design and development create several information objects that can be managed in product lifecycle management (PLM) IT systems (cf. Fig. 4). Objects are, for example, product or part-related specifications, describing single skills or functionalities to be satisfied by the product or part increments. Parts can describe generic products, assemblies, or single product-related items. Instantiated parts are represented by the serial object that allows individual identification of each product, assembly, or single product-related instance.

The objects of the work plan essentially describe the process of production process planning. The work plan is detailed by sequence objects, describing individual work processes and operations. Operations can be assigned to one or more sequences, use product parts as in-and outputs, and be allocated to a specific workplace that corresponds to a production resource. As part of the production process planning and the production itself, data from objects in enterprise resource planning IT systems can be relevant, such as from procurement or stocking. However, as the presented approach focuses on the engineering view, these objects will not be discussed further.

A production order is executed to initialize the production process, for example, as part of a machine execution system (MES) or manufacturing operations management system (MOM). This order can include batch size, timetables, or production sites. Each production step is referred to by the operation object, which gives detailed information about the conduction, such as the assigned machine, length, or description. The asset

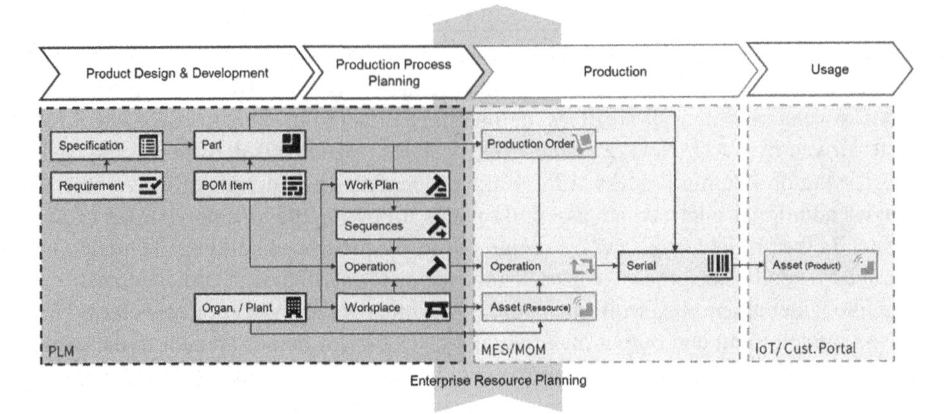

Fig. 4. Systemic view of product lifecycle IT objects from a work planning & production view

object describes the assigned machine (cf. Fig. 3 R^A_{0-n}) and thus the existing resources in the production process. It inherits information about the machine, such as properties, component structure, settings, or operating conditions. In addition, it can provide usage and environment data such as measurement data, production order, machine parameters, malfunctions and error codes, or user data.

Serialized objects (i.e., the Serial) refer to the produced product instance (cf. Fig. 3, A^S). These product instances can be tracked in their usage phases in an IoT platform or customer portal and thus managed as an asset (cf. Fig. 3, A^{SA}).

3 Implementation and Validation

The implementation and validation of the concept discussed in Sect. 3 were done by using data of an excavator in a prototypical system based on the CONTACT Elements technology. CONTACT Elements is a modular software suite for consistent product lifecycle information and process management combining management, planning, and control solutions from the early product design & development phases to the real-time data tracking in usage as part of an Internet of Things solution.

As part of the product design & development phase, the derivation of an mBOM from the eBOM (cf. ① and ② in Fig. 5) was realized for the excavator example product. The view is generated with the help of the so-called xBOM Manager application in the CONTACT Elements suite, in which a new BOM can be derived from an existing one, e.g., by restructuring, adding new parts and positions, and comparison of different BOM types. The mBOM is then used for the production process planning. In this example, a tracked undercarriage assembly of the excavator is shown. First, production process steps are created for the undercarriage, which reference the parts of the mBOM as consumed materials (cf. ④ in Fig. 5). Then, the workplan is linked to the undercarriage assembly product (cf. ③ in Fig. 5) assembled from the consumed parts. The created work plan consists of the main sequence, can include parallel and alternative ones, and can be visualized in the work planning application alongside further information such as the workplaces, as exemplified in section ⑤ of Fig. 5.

The execution of the production order and the management of the required production lines and their components are carried out in the Manufacturing Operations Management System, which incorporates IoT capabilities. For the production of a defined quantity of products (serials), the production order bundles the operations to be carried out (process), the production lines or machines required as assets (resources), as well as the parts produced or assembled (Material No.) in the operations (cf. ① in Fig. 6).

The production line can be managed via a superordinated asset (or an asset group), including information from the shopfloor, and is visualized via 3D models (cf. ② in Fig. 6). In addition, machines used as resources in the production process can be visualized in the 3D model, including information on their current state, operations, and condition, offering navigation shortcuts to a detailed view of the corresponding asset.

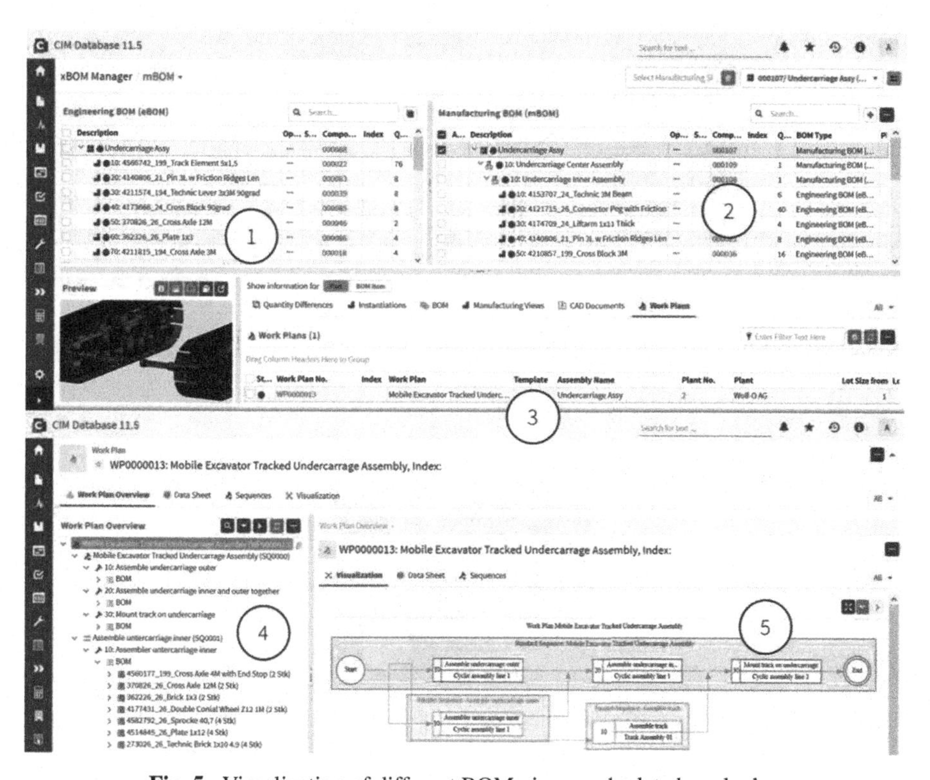

Fig. 5. Visualization of different BOM views and related work plan

The manufactured product can then be managed on a different IoT platform in a customer portal or asset management system to provide various stakeholders with information about the product's as-built and as-maintained states (cf. ③ in Fig. 6), its current state defined by field and analysis data (cf. ④ in Fig. 6), its location and driving route (cf. ⑤ in Fig. 6) or derived business applications such as generated service cases and their execution status. The result is a consistent management all relevant objects such as products, assemblies, or parts in joint BOMs and data structures from early engineering phases, through an integrated production line planning with resource scheduling and routing, to an possible, individual IoT-based asset management of smart product.

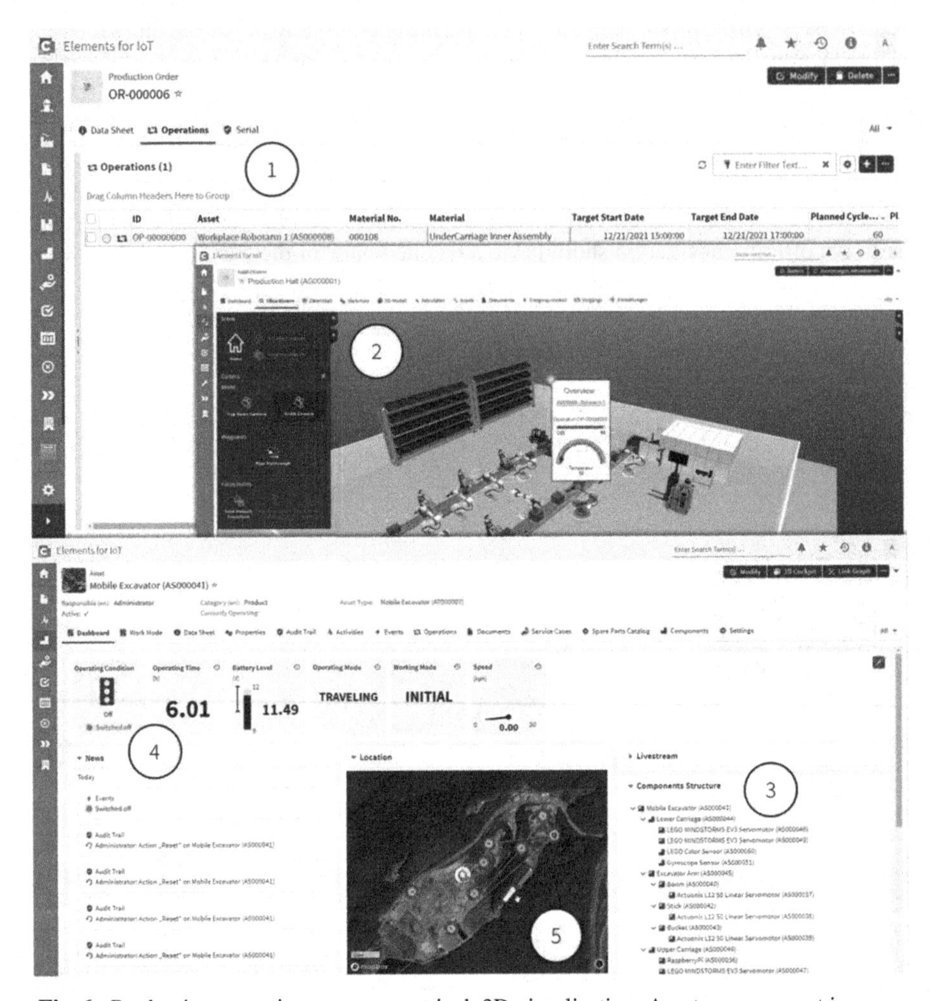

Fig. 6. Production execution management incl. 3D visualization; Asset management in usage

4 Conclusion and Outlook

The contribution presented an integrated, lifecycle-based approach to the information management of smart products focusing on work planning. First, the fundamentals of work planning and related bill of materials processing were elaborated from a methodical perspective, and a product lifecycle-spanning data integration approach was presented. Finally, the protypical implementation of the concept in a state-of-the-art, integrated software solution with a focus on data continuity was briefly illustrated.

Thereby, Sect. 3 dealt with the IT landscape of a manufacturing company. Smart products often consist of components from different manufacturers, and the product of one company can be a part of another product or can be a resource for the production of another company's product or a part of it. Such interconnections raise the question of how companies along a supply chain, especially in the age of Industry 4.0, can define their IT architecture and business models to enable the greatest possible value creation for themselves, their customers, and the end-user. This question will be dealt with in future publications.

Acknowledgments. This research and development project is funded by the German Federal Ministry of Education and Research (BMBF) within the "Innovations for Tomorrow's Production, Services, and Work" program and implemented by the Project Management Agency Karlsruhe (PTKA). The author is responsible for the content of this publication.

References

1. Abramovici, M.: Smart products. In: Lapierriére, L., Reinhart, G. (eds.) CIRP Encyclopedia of Production Engineering. Springer, Heidelberg (2015). https://doi.org/10.1007/978-3-642-35950-7_16785-1
2. Abramovici, M., Göbel, J.C., Savarino, P., Gebus, P.: Towards smart product lifecycle management with an integrated reconfiguration management. In: Ríos, J., Bernard, A., Bouras, A., Foufou, S. (eds.) PLM 2017. IAICT, vol. 517, pp. 489–498. Springer, Cham (2017). https://doi.org/10.1007/978-3-319-72905-3_43
3. acatech: Engineering smarter Produkte und Services. Plattform Industrie 4.0 Studie (2018). https://www.plattform-i40.de/IP/Redaktion/DE/Downloads/Publikation/hm-2018-fb-smart.html. Accessed 08 Apr 2022
4. Stübbe, R.: Product Lifecycle Management-Wo bleibt der Return-on-Invest? In: IT & Production Online (2019). https://www.it-production.com/produktentwicklung/wo-bleibt-der-return-on-invest/. Accessed 29 Mar 2022
5. Schmidt, J., Wieneke, F.: Produktionsmanagement, 4th edn. Europa Lehrmittel (2012). 978-3-8085-5314-5
6. Maropoulos, P.G., McKay, K.R., Bramall, D.G.: Resource-aware aggregate planning for the distributed manufacturing enterprise. CIRP Ann. Manuf. Technol. **51**(1), 363–366 (2002)
7. Cutting-Decelle, A.F., Young, R.I.M., Michel, J.J., Grangel, R., Le Cardinal, J., Bourey, J.P.: ISO 15531 MANDATE: a product-process-resource based approach for managing modularity in production management. Concurr. Eng. **15**(2), 217–235 (2007)
8. Schleipen, M., Drath, R.: Three-view-concept for modeling process or manufacturing plants with AutomationML. In: 2009 IEEE Conference on Emerging Technologies & Factory Automation. IEEE (2009)
9. Pfrommer, J., Schleipen, M., Beyerer, J.: PPRS: production skills and their relation to product, process, and resource. In: 2013 IEEE 18th Conference on Emerging Technologies & Factory Automation (ETFA), pp. 1–4. IEEE (2013)
10. Olhager, J., Wikner, J.: Production planning and control tools. Prod. Plann. Control **11**(3), 210–222 (2000)

11. Dickopf, T., Apostolov, C.: Closed-loop engineering approach for data-driven product planning. In: Proceedings of the 17th International Design Conference, 23–26 May 2022
12. Singh, V., Willcox, K.E.: Engineering design with digital thread. In: AIAA/ASCE/AHS/ASC Structures, Structural Dynamics, and Materials Conference, Kissimmee, Florida, 8–12 January 2018. American Institute of Aeronautics and Astronautics (2018). https://doi.org/10.2514/6.2018-0569

Strategic Planning in the Digital Engineering of Smart Product-Based System of Systems Reconfiguration

Sven Forte, Sebastian Weber(✉), and Jens C. Göbel(✉)

Institute of Virtual Product Engineering (VPE), TU Kaiserslautern, Gottlieb Daimler Str. 44, 67663 Kaiserslautern, Germany
{forte,weber,goebel}@mv.uni-kl.de

Abstract. Future product and service innovations are increasingly based on cross-system and cross-industry functionalities and components. As a result of the increasing connectivity and intelligence of products towards smart products, the system boundaries are also becoming increasingly fuzzy with the result that new smart product-based system of systems are emerging. In order to make these resilient and innovative in operation in the future, approaches for reconfiguring functionalities are playing an increasingly important role. For this purpose, the article presents a lifecycle-supporting method that enables the integration of dynamic external factors in the sense of strategic system of systems requirements into an existing system of systems with smart products as core components for the target-oriented reconfiguration. The focus here is on the identification and integration of dynamic external factors in terms of requirements from strategic management across the different granularity levels relevant to the system of systems. For this purpose, a multi-stage framework-model is presented which describes the strategic requirements integration at the system of systems level via a corresponding development of system of systems capabilities and their systematic break-down to functions (in sense of intended system behavior) and system logic up to the solution level in the sense of the smart product-instance. The framework supports the model-based reconfiguration based on the identified reconfiguration-need and derived reconfiguration potential from strategic management phases upstream the traditional product lifecycle management phases with the use of an associated integral System of Systems Engineering Lifecycle Management.

Keywords: Systems engineering · Reconfiguration · System of systems · Lifecycle management

1 Introduction

Products and services that were historically mostly monolithic are increasingly transforming into smart networked ecosystems with the help of steadily growing digitization [1]. Today, a large percentage of product innovation is already achieved through digitization and servitization [2]. However, increasing product intelligence and connectivity also

F. Noël et al. (Eds.): PLM 2022, IFIP AICT 667, pp. 547–556, 2023.
https://doi.org/10.1007/978-3-031-25182-5_53

means that system boundaries are becoming increasingly fuzzy. On the one hand, this enables the development of new potential business areas and market potentials; on the other hand, however, it also requires disruptive changes within the engineering of such smart products and the resulting smart product ecosystems, so-called smart product-based system of systems [3]. This not only affects the instantiated products themselves, but also their digital twins [4, 5] and the corresponding information technology management of the product models and components, as well as the context in which the products are embedded. This is a decisive factor in the provision of cross-industry and cross-product services and functions such as future mobility services including shared scooters, bicycles, trains, infrastructure and other services (e.g. payment) [6, 7] or smart home applications which can be integrated for example with vehicles, domestic objects, gardening equipment, home automation applications in the charging and energy sector, but also with other cross-sector services. This is where traditional engineering lifecycle approaches (like [8] for example), which are often optimized for specific systems, reach their limits. Contributing furthermore to preserve digital sovereignty by integrating engineering phases (such as innovation and roadmapping etc.) in advance of conventional product lifecycle management approaches, together with additional lifecycle integrations of participating products in the ecosystem context in order to facilitate the holistic development of smart product ecosystems in a comprehensive integrated model-based and information technology sense.

This also comes with new kinds of uncertainties engineering has to deal with that occur within the system of systems development phases, which for example could be addressed by certain reconfigurability in the utilization phase and therefore do not have been clarified deterministically in the design phase. With the help of system of systems reconfiguration [10–12] there is the possibility of increasing the resilience of the ecosystem and thus also the participating smart products. This also entails a certain conservation of resources, which addresses sustainability in the engineering and product-usage context. A central research question thereby is how smart product-based systems of systems in operation can react and implement dynamic external influences by identifying the need and reconfiguration potential supported by the integration of several aspects from strategic management together with an appropriate reconfiguration approach. In particular, the reconfiguration-supporting synergetic interaction between strategic management and system of systems engineering for identification of emerging reconfiguration potential will be discussed in more detail in the following article.

2 State of the Art

2.1 System of Systems Engineering Lifecycle

According to the International Council on Systems Engineering (INCOSE) a System of Systems (SoS) is "an System of Interest (SoI) whose elements are managerially and/or operationally independent systems. These interoperating and integrated collections of constituent systems usually produce results unachievable by the individual systems alone" [13]. To identify an SoS five properties according to [14] can also be applied. Those characteristics referring to the constituent systems are: independent operation (operational independence of the component); independent management of the subsystems (managerial independence of constituent systems); geographical distribution; emergent behavior; and evolutionary development processes. According to [14, 15] an SoS can be further classified as directed, collaborative, virtual and acknowledged SoS. A directed SoS, as focused on in this contribution due to the most common type in relation to smart product use cases nowadays, is characterized by the fact that the subsystems retain their independence but are controlled by a central SoS management and are subordinated to the objectives of the SoS.

During the lifecycle of systems (and system of systems), a huge amount of data is generated and processed. For example, in the design phase, (strategic) requirements development, configuration management, usage data (e.g. for monitoring purposes) and reconfiguration in the usage phase. In order to provide the highest possible innovations in the manner of cross-product related functions and services for the customer through the ecosystem of individual systems, it is essential to manage and provide relevant product-related data in an efficient, structured and demand-oriented manner throughout the different lifecycles of the participating systems. For this purpose, there are various forms of lifecycle models in the literature, e.g., generic ones such as ISO/IEC/IEEE 15288:2015 [16] with the phases of conception, development, production, use/support, and end of system. Although these can be adapted to the respective use case, they are primarily focused on a specific product or system and optimized for this purpose. For an SoS, it is necessary to consider all phases of the SoS as a system environment as well as those of the constituent systems with their own lifecycles in an integrated manner. A lifecycle approach that is specifically focused not only on SoS lifecycle management but also on the integration of smart products in the system environment is, for example, the System of Systems Engineering Lifecycle [9] with the phases SoS design (including innovation and strategic management), smart product engineering, SoS validation &

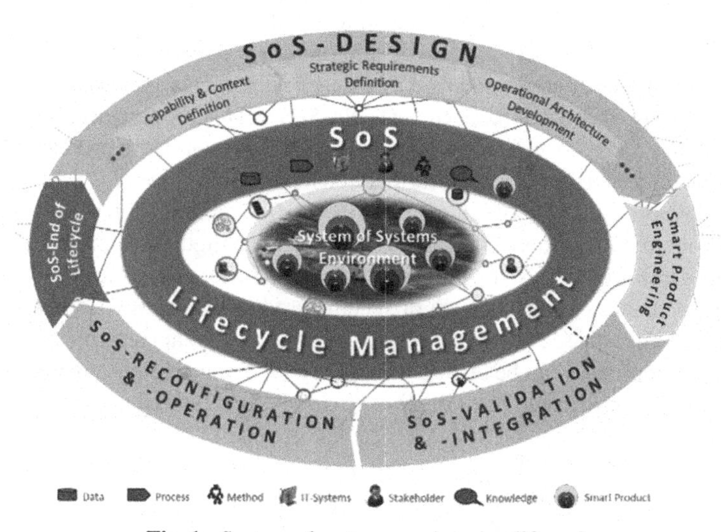

Fig. 1. System of systems engineering lifecycle

integration, SoS reconfiguration & operation, and SoS end of life, as shown in Fig. 1 below.

This approach is furthermore a bidirectional, multi-stage and supporting engineering-approach. The increasing importance of the integration of e.g. innovation and strategic management aspects with respect to the System of Interest (SoI) being developed is also shown by the V-model of VDI 2206 [17] in the revision from 2004, that originally defines the requirements definition as the starting point. The revision from 2021, the new V-model of VDI 2206 [18] however, is now defining the considerations of strategic planning and customer needs in form of business models and development-assignments, respectively, as the new entry point. And according to [3] the engineering-related information distribution associated with this is still one of the central challenges of smart lifecycle engineering in the context of smart product environments.

2.2 System of Systems Reconfiguration

SoS are confronted with dynamic (and disruptive) changes along their lifecycle which often cannot be considered or planned holistically at the early SoS design phases. A key challenge for SoS reconfiguration is to maintain the capabilities of the SoS, even in situations of disruptive change. The ability of a system to change is represented with reconfigurability (e.g. the extend of flexibility of a systems configuration) [19]. By the increasing number of connections between participating and associated systems within an SoS, resilience has become a relevant system property. Resilience can be described as "…the ability of the system to withstand a major disruption within acceptable degradation parameters and to recover…" [20]. Resiliency techniques are used to improve the SoS resilience to face these disruptive situations. Where dynamic reconfiguration is an adaptive resiliency technique, which focuses on responding to changes to maintain a minimum of operational capabilities [21].

The configuration refers to the actual integrated constituent systems of the SoS and their interconnections [11, 22]. This configuration is used to achieve a certain SoS capability. Changes in the environment or internal misalignments can cause a degradation of SoS capabilities. To react to these changes, the SoS can reconfigure itself to withstand these disruptions and recover its capabilities. For the reconfiguration process the actual configuration represents the basis for further considerations [11].

Reconfiguration can be furthermore described as the modification of existing products to meet new requirements [23], whereby dynamic reconfiguration refers to the ability of real time adaption of the system. Further, it can be characterized as the improvement of whole product classes or enhancement of specific product instances with new functionality during their use phase [10]. In the context of an SoS two main reasons to implement a reconfiguration process can be distinguished [11]. First, the configuration has to adapt to changes (e.g. a constituent leaves the SoS or reconfigure itself) due to the managerial independence of the constituent systems. Second, because of changing requirements to which the SoS must adapt. These two reasons can occur together and should not be considered strictly separated. A necessity for the reconfiguration process of the SoS is, that the actual and intended future states are identified and changes in the architecture are expressed. Changes in the architecture affecting its composition and furthermore the communication ways of the solution elements (e.g. constituent systems). The SoS engineering has to consider individual requirements of the constituent systems, due to their managerial and operational independence, and their therefore limited reconfiguration capability. The reconfiguration process is type-dependent in terms of behavioral independence of the constituent systems and their willingness reconfiguration operation (e.g. when to perform potential reconfiguration) [24]. To support these and other aspects within the SoS reconfiguration, strategic management methods provide a good basis for identifying and analyzing influencing factors that can support the reconfiguration engineering of the SoS and are described in more detail in the following.

3 Integrating Strategic SoS-Requirements by Using an SoS-Specific Reconfiguration Framework

SoS reconfiguration as part of SoS evolution often starts with the identification of external factors influencing the SoS in its target parameters. To ensure the future SoS operation in such case, the reconfiguration offers a good solution as described in Sect. 2.2. At the same time, the identification of such influencing factors as well as their engineering-specific structuring and their consistent transition in the sense of information technology into (model-based) requirements is especially in the given SoS context a major challenge. This is where strategic management has its strengths and can contribute decisive added value especially in conjunction with the engineering. For example, in the process of transferring strategic foresight analyses to downstream model-based requirements development phases in engineering. With the support of the strategic management, the SoS-influencing factors are identified and processed at the SoS top level (SoS Level 0) and converted finally into so-called strategic SoS requirements ready to be integrated in the systems models and further used by the various SoS engineering disciplines. In particular, the focus in the strategic planning is on the identification of the reconfiguration necessity. This contains four key relevant clusters: analysis, synthesis, portfolio management and potential assessment in the sense of strategic management. In the iterative and feedback-driven foresight process, external factors such as: technology drivers, political influencing factors, environmental factors, market analyses etc. are considered and transferred as input variables to the analysis phase, as shown in Fig. 2 on the left. Here the external factors are used to provide an as-is state of the actual situation in an environmental sense.

Information from the SoS itself and its environment are gathered. Among others, one of the crucial aspects for the analysis phase is to identify important challenges and opportunities for decision making in following lifecycle-related stages. Due to the importance of technology innovations for example in the usage phases, foresight techniques are applied to provide initial insights and (mega-)trends into possible future use cases (e.g. by applying delphi or scenario techniques) and thus offer added value for the development of new services and functions within the smart product ecosystem. Key aspect of the strategic analysis phase is to deliver a clear and common product-and portfolio-related understanding in terms of the SoS context across the involved disciplines as a base for following engineering phases.

The gathered and structured information in the analysis process is further used in the synthesis process phase, where business strategies are fostered. Based on the identified

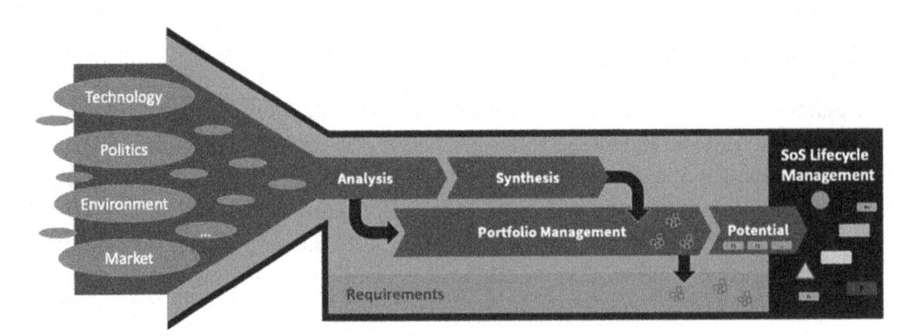

Fig. 2. System of systems reconfiguration-relevant strategic planning

as-is state from the analysis phase, a vision and goals for a to-be state are formulated. Furthermore, key attributes of the SoS and business strategies act as the foundation to formulate business models, which are getting planned and implemented (e.g. pay-by-use) and transferred accordingly with the accompanying portfolio management for integration into SoS as a strategic SoS requirement. Based on defined SoS-Goals and-Strategies, needed capabilities can then be derived on the SoS level.

These SoS-(L0-)Capabilities are hereby a stage gate to the formulated target-parameters (for example the reconfiguration-necessary to maintain the targets of the SoS in terms of resiliency). From the SoS-Analysis and-Planning phases the capabilities can be planned in a timely manner to introduce them when its needed delivered and managed in the SoS Lifecycle Management especially for the configuration management of the affected solution elements in the SoS context. This is furthermore also important due to the different SoS-related lifecycle phases of the constituent systems and the lifecycle of the SoS itself. Using the SoS Engineering Lifecycle Management, the results are then made available to the SoS top level (SoS Level 0) for further derivation of the corresponding SoS-relevant systems design and expression in the SoS-architecture. For the model-based development of the SoS within the design phases as well as the reconfiguration phases the Smart Product-based System of Systems Reconfiguration Architecture Framework (Fig. 3), a framework for the description of system of systems reconfiguration and architecture was developed based on the Concept of Kaiserslautern System Concretization Model (KSKM) [25]. The KSKM was originally designed and optimized for cyber-physical systems. In addition to the different levels within an SoS and the associated granularity and abstraction levels, the framework contains the according behavior and structure description as well as the respective derivation of the corresponding artifacts from the SoS top level (SoS Level 0) to the solution element level, so-called constituent systems (SoS Level 2). These are integrated and administered in SoS lifecycle management in accordance with SoS Lifecycle Engineering Concept. Thus, the Smart Product-based System of Systems Reconfiguration Architecture Framework describes the architecture of SoS Level 0 up to the integration at solution element level to the KSKM and thus forms the possibility of a continuous development from SoS Level 0 down to the Solution Elements (e.g. constituent systems) with corresponding administration of related information by SoS lifecycle management and the constituent system lifecycles integrated as part of it.

For the SoS reconfiguration phase and the corresponding integration of dynamic requirements for reconfiguration supported by strategic management, the focus is primarily on the description and implementation of capabilities and the corresponding derivation of requirements for the solution elements. Supported by the integration of the strategic management aspects as shown in Fig. 2, the reconfiguration of the SoS in operation is initiated after the assessment of the reconfiguration potential through the targeted application of strategic management methods. These are applied to the requirements space with the use of the SoS lifecycle management and form the initial basis for the reconfiguration process in the form of strategic SoS requirements. This takes place with the expression initially in the SoS architecture. Strategic SoS requirements are then derivable from the corresponding capabilities and used to describe the SoS capabilities that will be required in the intended future SoS state. In addition, capability gap analyses are carried out in parallel to address the first issues for the current reconfiguration. Following the modeling of the capabilities, these are included in the description of the required capabilities and semantically linked with the upstream strategic SoS requirements for the purpose of system configuration traceability.

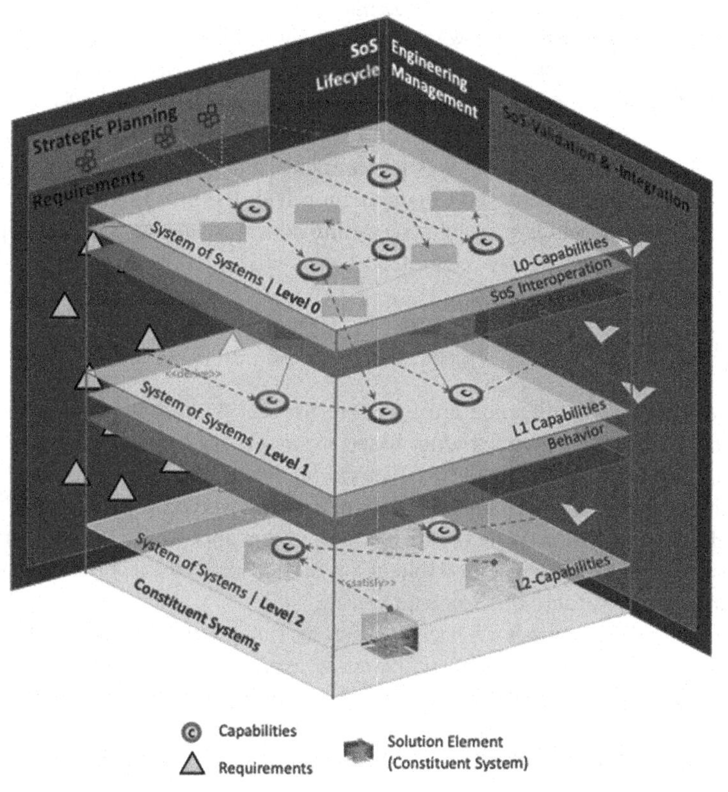

Fig. 3. Capability-specific view of the smart product-based system of systems reconfiguration architecture framework

Through the functional description especially for covering interoperational aspects of the future intended system state, initial simulation-ready architectural prototypes (in the sense of minimal viable products (MVP) [26]) can already be achieved. Furthermore, transferred and used with the available usage data from the digital twin context of the SoS, among other things, for initial validation and testing in the integration and validation space (Fig. 3 - right), so that finally the logical operational SoS architecture can be derived (structurally). Thus, the description required for the reconfiguration at SoS Level 0 has reached a level of maturity to address the subsystems and to serve as input for their capability development (L1 and L2). These are then mapped into the subsystem behavior as well as subsystem structure. However, this can differ considerably depending on the type and characteristics of the SoS. For instance, as the SoS moves further away from the directed type, the relationship between the subsystems can also vary in terms of their commitment and allocation. So, it can happen that on SoS Level 1 (subsystem level) only some subsystems find application and form direct relation to constituted systems. This means that not only subsystems communicate with subsystems in relationship manner. Individual subsystems also have possible direct relationships with other constituted systems outside their own subsystem. As a final step in the context of the reconfiguration-relevant architecture description, the required capabilities of the solution elements (L2-Capabilities) are described below based on the subsystem structures or (depending on the application scenario) based on the SoS Level 0 system artifacts. In this way, an SoS architecture for reconfiguration is established which also enables the semantically enriched modeling of the corresponding system levels and in this way a high level of consistency from SoS top level to solution element level (e.g. smart product instance) can be achieved in conjunction with the SoS Engineering Lifecycle Management and downstream systems modeling concepts such as KSKM (as part of a solution element) be seamlessly integrated.

4 Summary and Outlook

Communication and intelligence are two core aspects that allow products to evolve further and further into smart product-based system of systems. These also make it possible to bring new innovative cross-sectoral innovations to the consumers. In order to keep these ecosystems sustainably resilient and innovative in the sense of product functionalities in operation in the future, reconfiguration approaches in engineering play an increasingly important and central role. A key aspect here is the integration of strategic management processes into the SoS engineering.

The contribution describes a System of Systems Engineering Lifecycle Management driven method for integrating strategic planning aspects into the engineering of smart product-based SoS. In particular, the strategic management phases analysis, synthesis and portfolio management were considered and the integration into the downstream product-based engineering processes was discussed. Strategic management plays a major role in the identification and potential assessment of SoS-related reconfiguration processes. Since system of systems are usually long-term projects, which in most cases are undergoing significant and dynamic changes in their life cycle, it is therefore possible to operate them more resiliently and more sustainably in terms of their reconfiguration abilities with the integration of the strategic management. For this purpose, a multi-level framework is introduced, Smart Product-based System of Systems Reconfiguration Architecture Framework, which describes the integration of the SoS-relevant strategic requirements and their model-based integration at SoS level, e.g. in the form of capabilities (L0, L1 and L2) as well as the associated system architectures up to the handover and integration to the corresponding solution elements at product-instance level. This is supported by an integral System of Systems Engineering Lifecycle Management and, in this way, also encourages the digital sovereignty in the engineering context to be maintained throughout the entire system of systems reconfiguration process.

Further research is here needed for example in supporting validation-related aspects of the reconfiguration process e.g. through the targeted application of mixed reality methods for the early validation and virtual testing of the identified reconfiguration potential. This could also support and accelerate the reconfiguration for other SoS types in the future, in addition to the directed type as focused on in this paper.

References

1. Abramovici, M., Herzog, O.: Engineering im Umfeld von Industrie 4.0: Einschätzungen und Handlungsbedarf. Acatech Studie, Herbert Utz Verlag, München (2016)
2. Strahringer, S., Wiener, M.: Datengetriebene Geschäftsmodelle: Konzeptuelles Rahmenwerk, Praxisbeispiele und Forschungsausblick. HMD **58**, 457–476 (2021). https://doi.org/10.1365/s40702-021-00731-1
3. Tomiyama, T., Lutters, E., Stark, R., Abramovici, M.: Development capabilities for smart products. CIRP Ann. **68**(2), 727–750 (2019). https://doi.org/10.1016/j.cirp.2019.05.010
4. Göbel, J.C., Eickhoff, T.: Konzeption von Digitalen Zwillingen smarter Produkte. ZWF Zeitschrift für wirtschaftlichen Fabrikbetrieb **115**(special), 74–77 (2020). https://doi.org/10.3139/104.112301
5. Stark, R., Damerau, T.: Digital twin. In: Chatti, S., Tolio, T. (eds.) CIRP Encyclopedia of Production Engineering, pp. 1–8. Springer, Heidelberg (2019). https://doi.org/10.1007/978-3-642-35950-7_16870-1
6. Müller, C.: Metastudie: Die Mobilität von morgen – Eine Herausforderung für die Automobilindustrie. Whitepaper, Salesforce.com Germany GmbH, München (2020)
7. ACATECH: Transformation der Mobilität (acatech HORIZONTE), München (2021)
8. Eigner, M., Koch, W., Muggeo, C.: Modellbasierter Entwicklungsprozess cybertronischer Systeme – Der PLM-unterstützte Referenzentwicklungsprozess für Produkte und Produktionssysteme. Springer, Heidelberg (2017). https://doi.org/10.1007/978-3-662-55124-0
9. Forte, S., Göbel, J.C., Dickopf, T.: System of systems lifecycle engineering approach integrating smart product and service ecosystems. Proc. Des. Soc. **1**, 2911–2920 (2021). https://doi.org/10.1017/pds.2021.552
10. Savarino, P., Abramovici, M., Göbel, J.C., Gebus, P.: Design for reconfiguration as fundamental aspect of smart products. Procedia CIRP **70**, 374–379 (2018). https://doi.org/10.1016/j.procir2018.01.007
11. Petitdemange, F., Borne, I., Buisson, J.: Design process for system of systems reconfigurations. Syst. Eng. **24**(2), 69–82 (2021). https://doi.org/10.1002/sys.21567
12. Forte, S., Dickopf, T., Weber, S., Göbel, J.C.: Towards sustainable systems reconfiguration by an IoT-driven system of systems engineering lifecycle approach. In: 29th CIRP Conference on Life Cycle Engineering 2022 (LCE 2022), Leuven, Belgium, 04–06 April 2022 (2022). Procedia CIRP **105**, 654–659. https://doi.org/10.1016/j.procir.2022.02.109

13. Walden, D.D., Roedler, G.J., Forsberg, K., Hamelin, R.D., Shortell, T.M.: Systems Engineering Handbook. A Guide for System Life Cycle Processes and Activities. INCOSE-TP-2003-0
14. Maier, M.W.: Architecting principles for systems-of-systems. Syst. Eng. **1**(4), 267–284 (1998). https://doi.org/10.1002/(SICI)1520-6858(1998)1:4<267::AID-SYS3>3.0.CO;2-D. 4th edn. Wiley, Hoboken (2015). 9781118999400
15. Dahmann, J.S., Baldwin, K.J.: Understanding the current state of US defense systems of systems and the implications for systems engineering. In: 2nd Annual IEEE Systems Conference, 7–10 April 2008, pp. 1–7. IEEE (2008). https://doi.org/10.1109/SYSTEMS.2008.4518994
16. ISO/IEC/IEEE 15288: International Standard - Systems and software engineering – System life cycle processes, Piscataway, NJ, USA. IEEE (2015). https://doi.org/10.1109/IEEESTD.2015.7106435
17. VDI 2206:2004-06: Design methodology for mechatronic systems. Beuth Verlag (2004)
18. VDI/VDE 2206:2021-11: Development of mechatronic and cyber-physical systems (2021)
19. Fornari, G., de Santiago Júnior, V.A.: Dynamically reconfigurable systems: a systematic literature review. J. Intell. Rob. Syst. **95**(3–4), 829–849 (2018). https://doi.org/10.1007/s10846-018-0921-6
20. Haimes, Y.Y.: Modeling complex systems of systems with phantom system models. Syst. Eng. **15**, 333–346 (2012). https://doi.org/10.1002/sys.21205
21. Bodeau, D., Brtis, J., Graubart, R., Salwen, J.: Resiliency techniques for systems-of-systems extending and applying the Cyber Resiliency Engineering Framework to the space domain. In: 7th International Symposium on Resilient Control Systems, ISRCS 2014, pp. 1–6 (2014). https://doi.org/10.1109/ISRCS.2014.6900099
22. Kossmann, M., Samhan, A., Odeh, M., Qaddoumi, E., Tbakhi, A., Watts, S.: Extending the scope of configuration management for the development and life cycle support of systems of systems—an ontology-driven framework applied to the enceladus submarine exploration lander. Syst. Eng. **23**(3), 366–391 (2020). https://doi.org/10.1002/sys.21532
23. Männistö, T., Soininen, T., Tiihonen, J., Sulonen, R.: Framework and conceptual model for reconfiguration (1999)
24. Petitdemange, F., Borne, I., Buisson, J.: Assisting the evolutionary development of SoS with reconfiguration patterns. In: Bahsoon, R., Weinreich, R. (eds.) Proccedings of the 10th European Conference on Software Architecture Workshops, pp. 1–7. ACM, New York (2016). https://doi.org/10.1145/2993412.3004845
25. Eigner, M., Dickopf, T., Apostolov, H.: Interdisziplinäre Konstruktionsmethoden und -prozesse zur Entwicklung cybertronischer Produkte – Teil 2/Interdisciplinary Design Methods and Processes to Develop Cybertronic Products – Part 2. Konstruktion **71**, 69–75 (2019). https://doi.org/10.37544/0720-5953-2019-01-02-69
26. Ries, E.: What is the Minimum Viable Product (2009). http://venturehacks.com/articles/minimum-viable-product. Accessed 22 Feb 2022

Lifecycle Engineering in the Context of a Medical Device Company – Leveraging MBSE, PLM and AI

Gregor M. Schweitzer[1](✉), Michael Bitzer[1], and Michael Vielhaber[2]

[1] Fresenius Medical Care, Bad Homburg, Germany
[2] Saarland University, Saarbrücken, Germany

Abstract. Medical devices today often consist of complex mechatronic products which help to provide treatment services for patients. Beside safety and reliability also sustainability aspects need to be considered for such products and services. The ability to describe, analyze and predict products and services throughout all phases of its lifecycle is becoming a core competence of engineering departments (Lifecycle Engineering). In order to have a digital representation of the product and services along the lifecycle (digital model, digital twin), well-established engineering approaches need to be combined.

This paper builds on a stream of research, proposing to leverage and combine Model-based Systems Engineering (MBSE), Product Lifecycle Management (PLM) and Artificial Intelligence (AI) to strengthen the Lifecycle Engineering. The so called Engineering Graph is a key element of this research work to bridge those engineering disciplines and enable AI-driven engineering in the lifecycle context.

Keywords: Lifecycle Engineering (LCE) · Systems Engineering (SE) · Product Lifecycle Management (PLM) · Artificial Intelligence (AI) · Engineering Graph

1 Introduction

Mega-trends in society, economy, politics, regulatory and technology lead to increased volatility, uncertainty, complexity and ambiguity (VUCA) for companies in multiple industries [1]. Especially the trend of sustainability is getting more and more impact on products and services [2].

The ability to develop and assess products and services from a lifecycle perspective is a key success factor to operate in such a volatile and complex environment. Concepts such as Product Lifecycle Management (PLM) or Lifecycle Engineering (LCE) emerged to address this environment. The representation and assessment of the lifecycle in the virtual environment is the foundation.

By integrating and leveraging digital technologies – especially in early product development phases – innovative options and chances can be created. To represent products and services across the lifecycle, model-based engineering approaches can be used.

Published by Springer Nature Switzerland AG 2023
F. Noël et al. (Eds.): PLM 2022, IFIP AICT 667, pp. 557–566, 2023.
https://doi.org/10.1007/978-3-031-25182-5_54

These virtual product models can be analyzed by artificial intelligence (AI) and data science technologies to gain further information and support lifecycle spanning use cases such as Life Cycle Sustainability Assessments (LCSA).

This paper introduces the Engineering Graph as a concept that combines current engineering methods like Model-based Systems Engineering (MBSE) and PLM, modern theory from computer science for product modeling and machine learning to support LCE. This concept is applied at a leading medical device company.

2 Related Work for Engineering

Engineering as one of the core competences of manufacturing industries undergoes an evolution similar to the products and technologies developed by engineering itself. The digitalization is a main driver for this evolution in engineering: from geometry-oriented to behavior and meaning-oriented engineering and modelling approaches [3].

This section will point out disciplines and approaches of engineering which can be seen as a foundation for a lifecycle-oriented approach of engineering: LCE.

2.1 Model-Based Systems Engineering

The transdisciplinary and integrative approach of Systems Engineering (SE) enables the successful realization, use and retirement of engineered systems [4].

SE covers all processes of the system lifecycle: agreement and organizational project enabling processes; technical management processes and technical processes itself [5]. In the "Architecture Definition Process" the system architecture is developed. One core element of SE is the decomposition of a System of Interest in sub-systems. Those subsystems can be elaborated in System Elements. A System of Interest can also be described by its operational environment and enabling systems.

In order to describe those elements and relationships a model-based approach is used, typically supported by the modelling language System Modeling Language (SysML). SysML is a dialect of UML 2 that customizes the language via three mechanisms: Stereotypes, Tagged Values, and Constraints [6].

2.2 Product Lifecycle Management

PLM is a concept which enables representations, perspectives and validations of a product in its lifecycle phases. PLM is evolutionary based on Product Data Management (PDM). PDM was developed in the context of document management and Computer Aided Design (CAD). With the evolution towards PLM so called product models or virtual products were introduced [7].

PLM manages all data from development, production, warehouse and sales and supports single source of data through the entire lifecycle [8]. The whole product range is covered, from individual part to the entire portfolio of products [9].

2.3 Lifecycle Engineering

LCE is a sustainability-oriented engineering methodology that considers the comprehensive technical, environmental, and economic impacts of decisions within the product lifecycle [10]. In this context, the product lifecycle is formally defined by ISO 14040 as the "consecutive and interlinked stages of a product system, from raw material acquisition or generation from natural resources to final disposal." [11].

In recent years, modern concepts to LCE are emerging in the literature. These are leveraging different data sources and a high level of computational capabilities.

The Integrated Computational Life Cycle Engineering (IC-LCE) integrates data from the entire product lifecycle via coupled models [12]. The results of LCE can be visualized to be communicated to expert and non-expert users by combining LCE with Visual Analytics [13]. Also, knowledge-based engineering can be combined with LCE. A manual way to engineer knowledge and make it available for LCE is introduced [14]. Based on that, a framework to automatically collect data during a products lifecycle is developed [15].

Sakao et al. (2021) identify current challenges and opportunities of LCE and develop a vision for Adaptive and Intelligence LCE (AI-LCE) based on their findings [16]. Here, different engineering capabilities are supported by business intelligence tools based on a database called "memory" and external factors and requirements.

All studies identify issues of current LCE methodologies and therefore derive the need for a new concept. The issues identified are summarized in Table 1.

Table 1. Issues identified with current LCE

Issue #	Issue	Source
1	Lack of Speed	[12, 16]
2	Oversimplified models	[12]
3	Lack of comprehensiveness	[12]
4	Lack of transparency	[12]
5	Lack of integration between environments of core engineering disciplines	[13]

All concepts introduced to solve the issues identified require some sort of database, repository or memory. However, there is no concept of how to build that database and how existing methodologies like PLM and SE can be included into the database concept. Therefore, this paper proposes a new concept based on the Engineering Graph and combining the methodologies of PLM and SE.

3 Related Work for Computer Science

With increased digitalization in engineering and an increased amount of product data that needs to be stored and analyzed, theories from the field of computer science become important in engineering. This section introduces modeling languages that can be used to connect and store engineering data and technologies that are designed to derive information from large datasets such as AI and data science.

3.1 Modeling Languages

SysML has emerged as a machine and human understandable language which describes a product system through requirements, structure, and behavior [17]. The language was invented by the OMG in cooperation with the International Council of Systems Engineering (INCOSE) [6, 18], and was developed to support modeling and (re)-using of engineering information across the lifecycle [19].

Graphical databases focus on relationships between data points. They consist of nodes, which represent data points, and relationships, which connect them. Nodes and relationships can have properties that are used to filter and find data quickly [20]. Properties can be qualitative or quantitative information. The objects and their relationships are represented naturally and clearly by using abstraction concepts [21]. The schema of the graphical database is not fixed at its creation, contrary to relational databases [22]. This leads to their capability to include data from different sources without the need to match the schemas. Therefore, the graphical database can be extended with new and unexpected sources, which is especially useful in complex environments such as engineering [23].

The capability of graph databases to include data from different and unforeseen sources allows building a large and interconnected database from public sources. Thereby, public knowledge from semantic web sources such as Wikimedia [24] or the Google Knowledge Graph [25] can be harvested and linked to a company specific meta structure. That meta structure allows the connection of data from outside sources with company internal data, together building a dataset large enough to allow the application of AI technology.

3.2 Machine Learning and Graph Data Science

Machine Learning is a technology that is capable of making sense of large datasets and deriving information from them without explicit programming [26]. In recent years, several applications to engineering problems such as identification of new product ideas [27], requirements elicitation [28], creativity [29], configuration management [30] and decision support in early design phases [31] are explored.

Machine Learning works by showing a neural network a large dataset of training data. During the supervised training process, the internal weights in the neural network are adjusted automatically to create a model that achieves the desired outcome. This model is then tested on the verification dataset. If it passes, it can be applied to new data and moved to production [32].

Graph data science technology is especially developed to be used on graph databases [33]. The capabilities include community detection, centrality, link prediction and similarity, which will be described in the following.

Community Detection evaluates how a group is clustered or partitioned, as well as its tendency to strengthen or break apart [34]. The weakly connected components can analyze the graphs' structure and find not connected parts. Additionally, the number of communities within a graph can be identified, which brings an understanding of the number of subtopics a graph contains.

Centrality can be used to determine the importance of distinct nodes in a network [34]. One of the most widely applied algorithms is Pagerank [35].

Link prediction is done by using machine learning. A model is trained to learn where relationships between nodes in a graph should exist [34]. This model can then be used to predict further relationships.

Similarity algorithms compute the similarity of pairs of nodes [34]. The similarity between two nodes is calculated based on the nodes they are connected to.

4 Engineering Graph Enhancing MBSE for LCE

This section introduces the evolution in product modeling towards graph-based and summarizes the ongoing research on the Engineering Graph that brings this concept to live in the area of engineering. Lastly, a concept is introduced how the Engineering Graph can bridge PLM and SE and enable Data Science and AI to support LCE.

4.1 Evolution Towards Graph-Based Modeling

The scope of product development has increased. When a product was designed on its own, geometry-based models were sufficient. The increased scope and complexity to design product systems consisting of mechanic, electric and software parts and designing systems of systems where different product systems interact with each other to bring value to the end user needed a new modeling approach with higher abstraction: model-based [36].

Now that the scope is increasing again towards system environments, where product systems and systems of systems are viewed in their environment, e.g. by lifecycle assessments, there is a need for a new modeling approach with higher abstraction: graph-based. Figure 1 shows this evolution.

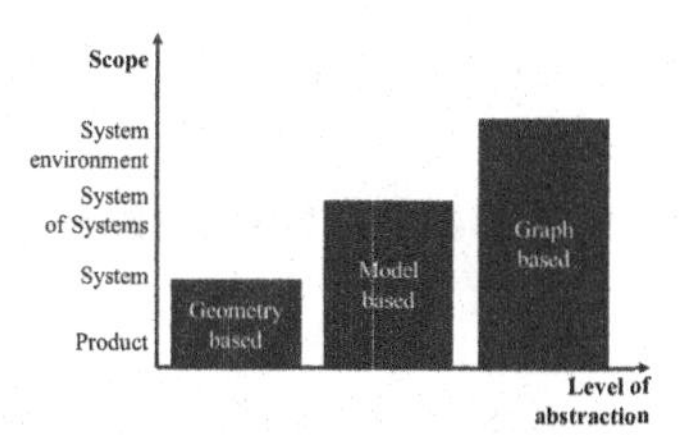

Fig. 1. Evolution of modeling

4.2 Engineering Graph

Previous research already explored the application of graphical databases in engineering. The Engineering Graph was introduced as a graph-based database to support engineering applications such as LCSA [37].

The Engineering Graph connects data from different sources within a company and from external sources such as suppliers, partners and public sources like the semantic web. Data is stored at a high level of abstraction, where the focus lies on the connections between data points and not the data itself. Already existing data and configuration information is not duplicated in the Engineering Graph.

4.3 Engineering Graph Bridging SE, PLM and AI to Support LCE

The Engineering Graph as a graphical database can be used to bridge different engineering methodologies and their underlying data and schemas. This results in a comprehensive and interconnected database.

SE offers the information of product breakdown, how parts are connected and how they work together using which interfaces. PLM offers the product data including configuration information. The information from both methods is connected and enriched with lifecycle data such as where and how a product is used, what norms and regulations it needs to comply with in its target market and additional information from non-government organizations such as the World Health Organization (WHO) or the United Nations (UN).

Graph database technology is able to support LCE and address the issues identified in Sect. 2.3 when the established engineering methodologies PLM and SE are combined. By leveraging already existing models, speed (Issue #1) is increased because there is no duplicate work. This also addresses the issue of oversimplified models (Issue #2) and lack of comprehensiveness (Issue #3) as the models from SE and PLM are very sophisticated. The integration between environments of core engineering disciplines (Issue #5) is increased because the graph can directly integrate the data from these different systems.

Many of the use cases of LCE such as LCSA or cost assessment are predefined in early design phases. The Engineering Graph can be one way to support design decisions in early phases to improve LCE measures by providing large amounts of data early. It contains all freely available LCE information and data from previous product generations. Data Science and AI technologies can be applied if the graph contains a large enough dataset for these technologies to be applicable.

5 Use Case in the Medical Device Industry

In this section the application of the Engineering Graph for LCE is demonstrated at a leading medical device company. First, it is described how the system is built and second, its application for LCE is shown.

5.1 Engineering Graph Connects PLM, SE and External Information

In order to move towards graph-based LCE and support the application of AI technologies, the Engineering Graph is created at a leading Medical Device Company using Neo4J software. The graph spans across different Systems of Interest that are defined as part of the companies SE activities. The "product system" data is stored in the PLM system and connected to the graph (see green nodes in Fig. 2). The brown nodes in Fig. 2 show norms such as the ISO 14044 that is relevant to LCSA. Additionally, it is important to include norms relevant to medical devices such as ISO 62304 as the medical device industry is highly regulated.

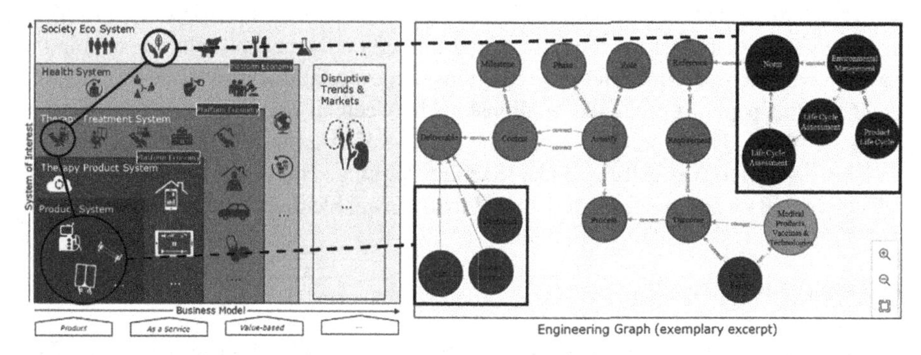

Fig. 2. Engineering Graph connecting data from SE and PLM

The data from the different sources is added to the graph via the import APIs of Neo4J and connected to each other by a predefined meta model (blue nodes). Company external sources such as the Google Knowledge Graph, Wikimedia Graph and information from the WHO as well as the UN are also added via the Neo4J APIs to further enrich the Engineering Graph. The connection to existing nodes is performed manually for the most obvious ones, other relationships can be proposed by the system as shown in the following section.

This database with many connections is made available in early phases of development. Here, it is leveraged to influence the sustainability impact of a products lifecycle. Having this information in early phases of development is important, as many decisions influencing its sustainability impact are made here.

5.2 Engineering Graph Supports LCE

After building the Engineering Graph, the following paragraphs will focus on the analyses that it enables to support LCE. These are the graph data science algorithms introduced in Sect. 3.2. Analyzing the graph across the entire product life cycle and considering relevant norms and regulations can lead to the detection of unknown and unexpected relationships. Discovering the impact of the elements in the graph on each other in an automated way can lead to decreased time to market due to less rework.

First, the graph is analyzed to ensure that it is well connected and that there are no unconnected nodes left. Therefore, the Weakly Connected Components is used. In this example, it could be shown that there exists only one component which means that the graph is well connected. Second, Label Propagation is used to identify communities of nodes in the graph. These can be an indication of how many subtopics the graph contains. In the graph for this paper, 7 communities could be detected.

After the graphs communities are analyzed, Pagerank is used to identify the most important nodes in the graph. In the case of the Engineering Graph in this paper, the "UN Sustainable Development Goals" is the most connected node in the graph.

Next, predictions for new relationships can be generated based on the existing graph. The predicted relationships can be used to complement the graph and to detect unexpected relationships that can represent cause-effect chains. This prediction is based on machine learning. Therefore, a model is first trained on the current relationships in the graph. Second, it is used to generate predictions for new relationships. In the Engineering Graph for this paper, it predicted relationships between all existing nodes with a probability of 49.9%. This result shows that the amount of data in the Engineering Graph is not large enough to successfully apply machine learning. Therefore, the database needs to be increased by adding product data and freely available data from the semantic web.

6 Discussion

The literature of LCE advances towards standardization, comparability and the adoption of new technologies. This paper proposes the application of the concept of the Engineering Graph to LCE to offer a standardized and easy to expand database that leverages existing models and concepts. The concept of the Engineering Graph was introduced in prior research and is here extended by adding data science and AI capabilities and showing its usefulness for LCE use cases.

Larger sample data in the graph will yield more exact data. Currently, the database is not large enough to yield robust results from machine learning technology.

Future research needs to be conducted to embed the creation and maintenance of the Engineering Graph into standard development processes such as the V-model or ISO 15288 for SE. Furthermore, it needs to be shown how the graph can be automatically extended leveraging Natural Language Processing technology. Additionally, the results based on a graph based on a larger dataset need to be reported. Lastly, the graph needs to be applied to further use cases to demonstrate general usefulness.

References

1. INCOSE. Systems Engineering Vision 2035. Accessed 02 Feb 2022
2. Dumitrescu, R., Albers, A., Riedel, O., Stark, R., Gausemeier, J.: Engineering in Deutschland – Status quo in Wirtschaft und Wissenschaft. Ein Beitrag zum Advanced Systems Engineering (2021)
3. Bitzer, M., Eigner, M., Faißt, K.-G., Muggeo, C., Eickhoff, T.: Framework of the evolution in virtual product modelling and model management towards digitized engineering. In: DS 87–6 Proceedings of the 21st International Conference on Engineering Design (ICED 17) Vol 6: Design Information and Knowledge, Vancouver, Canada, 21–25 August 2017 (2017)
4. Sillitto, H., et al.: Systems Engineering and System Definitions, p. 18 (2019)
5. ISO 15288. ISO/IEC/IEEE 15288:2015 (2015)
6. OMG. What is SysML? (2020). http://www.omgsysml.org/what-is-sysml.htm. Accessed 04 Mar 2020
7. Eigner, M., Stelzer, R.: Product Lifecycle Management: ein Leitfaden für Product Development und Life Cycle Management, 2, neu Bearb. Springer, Aufl. Dordrecht (2013)

8. Bracht, U., Geckler, D., Wenzel, S.: Digitale Fabrik: Methoden und Praxisbeispiele, 2. Aktualisierte und erweiterte Auflage. Springer, Berlin (2018). https://doi.org/10.1007/978-3-662-55783-
9. Terzi, S., Bouras, A., Dutta, D., Garetti, M., Kiritsis, D.: Product lifecycle management - From its history to its new role. Int. J. Product Lifecycle Manag. **4**, 360–389 (2010)
10. Hauschild, M.Z., Rosenbaum, R.K., Olsen, S.I.: Life Cycle Assessment: Theory and Practice (2018)
11. DIN EN ISO 14040. DIN EN ISO 14040:2009-11, Umweltmanagement_- Ökobilanz_- Grundsätze und Rahmenbedingungen (ISO_14040:2006); Deutsche und Englische Fassung EN_ISO_14040:2006. Beuth Verlag GmbH (2009)
12. Cerdas, F., Thiede, S., Herrmann, C.: Integrated computational life cycle engineering — application to the case of electric vehicles. CIRP Ann. **67**(1), 25–28 (2018)
13. Kaluza, A., Gellrich, S., Cerdas, F., Thiede, S., Herrmann, C.: Life cycle engineering based on visual analytics. Procedia CIRP **69**, 37–42 (2018)
14. von Drachenfels, N., Cerdas, F., Herrmann, C.: Towards knowledge based LCE of battery technologies. Procedia CIRP **90**, 683–688 (2020)
15. Dilger, N., et al.: Definition and reference framework for life cycle technologies in life cycle engineering - a case study on all solid state traction batteries. Procedia CIRP **98**, 217–222 (2021)
16. Sakao, T., Funk, P., Matschewsky, J., Bengtsson, M., Ahmed, M.U.: AI-LCE: adaptive and intelligent life cycle engineering by applying digitalization and AI methods – an emerging paradigm shift in life cycle engineering. Procedia CIRP **98**, 571–576 (2021)
17. Korthals, K., Auricht, M., Felten, M.: Systems engineering solution lab - experience model based systems engineering at CLAAS. In: Presented at the Prostep ivip Symposium 2020, 02 September 2020 (2020)
18. Incose. History of Systems Engineering (2020). https://www.incose.org/about-systems-engineering/history-of-systems-engineering. Accessed 04 Mar 2020
19. Estefan, J.A.: Survey of model-based systems engineering (MBSE) methodologies. Incose MBSE Focus Group **25**(8), 1–12 (2007)
20. Rawat, D.S., Kashyap, N.K.: Graph database: a complete GDBMS survey. Int. J **3**, 217–226 (2017)
21. Angles, R., Gutierrez, C.: Querying RDF data from a graph database perspective. In: Gómez-Pérez, A., Euzenat, J. (eds.) ESWC 2005. LNCS, vol. 3532, pp. 346–360. Springer, Heidelberg (2005). https://doi.org/10.1007/11431053_24
22. Angles, R., Gutierrez, C.: Survey of graph database models. ACM Comput. Surv. **40**(1), 1:1–1:39 (2008)
23. Vicknair, C., Macias, M., Zhao, Z., Nan, X., Chen, Y., Wilkins, D.: A comparison of a graph database and a relational database: a data provenance perspective. In: Proceedings of the 48th Annual Southeast Regional Conference, New York, NY, USA, pp. 1–6 (2010)
24. Wikimedia. Wikimedia Foundation (2021). https://wikimediafoundation.org/. Accessed 12 Nov 2021
25. Google. Google Knowledge Graph Search API. Google Developers (2021). https://developers.google.com/knowledge-graph?hl=de. Accessed 12 Nov 2021
26. Samuel, A.L.: Machine learning. Technol. Rev. **62**(1), 42–45 (1959)
27. Christensen, K., Nørskov, S., Frederiksen, L., Scholderer, J.: In search of new product ideas: identifying ideas in online communities by machine learning and text mining. Creat. Innov. Manag. **26**(1), 17–30 (2017)
28. Wang,Y., Zhang, J.: Bridging the semantic gap in customer needs elicitation: a machine learning perspective. In: DS 87–4 Proceedings of the 21st International Conference on Engineering Design (ICED 2017), vol 4: Design Methods and Tools, Vancouver, Canada, 21–25 August 2017, pp. 643–652 (2017)

29. Hein, A.M., Condat, H.: Can machines design? an artificial general intelligence approach. In: Artificial General Intelligence, Cham, pp. 87–99 (2018)
30. Liu, H., Huang, Y., Ng, W.K., Song, B., Li, X., Lu, W.F.: Deriving configuration knowledge and evaluating product variants through intelligent techniques. In: 2007 6th International Conference on Information, Communications Signal Processing, pp. 1–5 (2007)
31. Bertoni, A., Larsson, T., Larsson, J., Elfsberg, J.: Mining data to design value: a demonstrator in early design. In: DS 87–7 Proceedings of the 21st International Conference on Engineering Design (ICED 17), vol 7: Design Theory and Research Methodology, Vancouver, Canada, 21–25 August 2017, pp. 021–029 (2017)
32. Mahesh, B.: Machine learning algorithms-a review. Int. J. Sci. Res. (IJSR) **9**, 381–386 (2020)
33. Cook, D.J., Holder, L.B.: Mining Graph Data. John Wiley & Sons, Hoboken (2006)
34. The Neo4j Graph Data Science Library Manual v2.0 - Neo4j Graph Data Science. Neo4j Graph Data Platform. https://neo4j.com/docs/graph-data-science/2.0/. Accessed 30 Mar 2022
35. Brin, S., Page, L.: The anatomy of a large-scale hypertextual Web search engine. Comput. Netw. ISDN Syst. **30**(1), 107–117 (1998)
36. Schweitzer, G.M., Bitzer, M., Vielhaber, M.: Produktentwicklung: KI-ready? Zeitschrift für wirtschaftlichen Fabrikbetrieb **115**(12), 873–876 (2020)
37. Schweitzer, G.M., Mörsdorf, S., Bitzer, M., Vielhaber, M.: Detection of cause-effect relationships in life cycle sustainability assessment based on an engineering graph. In: DESIGN 2022 (2022)

Crowd Intelligence Driven Design Framework Based on Perception-Retrieval Cognitive Mechanism

Chen Zheng[1,2](✉), Kangning Wang[1], Tengfei Sun[1], and Jing Bai[1]

[1] School of Mechanical Engineering, Northwestern Polytechnical University, 127 West Youyi Road, Xi'an 710072, Shaanxi, People's Republic of China
chen.zheng@nwpu.edu.cn
[2] CSSC Huangpu Wenchong Shipbuilding Company Limited, Guangzhou 510715, People's Republic of China

Abstract. Currently, the use of crowd intelligence in which the knowledge from different disciplines is integrated for complex product design has attracted increasing attention from both academia and industry. However, the multi-modal, multi-temporal and multi-spatial characteristics of multi-disciplinary knowledge hinder its implementation. The perception-retrieval cognitive mechanism of human beings' brain shows its unique advantages in the cognitive process of multi-modal, multi-temporal and multi-spatial knowledge, and can quickly integrate external information and retrieve memory. In order to solve the problems of low efficiency and poor acquisition accuracy of multi-disciplinary knowledge, inspired by the brain's perception-retrieval cognitive mechanism, this paper adopts a crowd intelligence-drive to achieve efficient integration, dynamic storage and real-time acquisition of multi-disciplinary knowledge.

First, a deep survey relating to the current research studies on knowledge-based engineering approaches and the perception-retrieval cognitive mechanism is conducted. Second, the brain-inspired crowd intelligence-driven design approach for complex products and the techniques that can be used as the potential solutions to each step are presented. Finally, the authors draw the conclusion and point out the future research direction.

Keywords: Knowledge-based engineering · Systems engineering · Product design · Perception-retrieval cognitive mechanism · Crowd intelligence

1 Introduction

With the development of technologies such as the Internet of Things and big data in today's era, product has become more and more complex, and the use of crowd intelligence in which knowledge from various disciplines is integrated for the complex product design has become the current development trend [1]. However, if the multi-modal, multi-temporal and multi-spatial characteristics of multi-disciplinary design knowledge have not been considered during the knowledge-based engineering (KBE) process for

Published by Springer Nature Switzerland AG 2023
F. Noël et al. (Eds.): PLM 2022, IFIP AICT 667, pp. 567–576, 2023.
https://doi.org/10.1007/978-3-031-25182-5_55

complex products, the design efficiency and quality will be negatively affected, which may lead to the project delay or design failure [2].

Currently, the research field of neuroscience has brought great inspirations to the artificial intelligence, information engineering and other fields [3, 4]. According to cognitive psychology, biological cognition is a process of continuous perception of various information in the environment, and comparing and retrieving it with existing memory [5]. This perception-retrieval cognitive mechanism of human beings' brain has formed the unique natural advantage of organisms in the cognitive process of multi-modal, multi-temporal and multi-spatial information or knowledge [6, 7]. Inspired by brain's perception-retrieval cognitive mechanism, this paper proposes a human-like knowledge organization, integration and acquisition approach to support KBE process for complex products.

This paper is organized as follows. Section 2 presents current research studies on knowledge-based engineering approaches and perception-retrieval cognitive mechanism. Section 3 introduces the brain-inspired crowd intelligence-driven design approach for complex products and the techniques which can be used as the potential solutions to each step of the proposed design approach. Section 4 draws the conclusion and proposes the future research.

2 Literature Review

2.1 Knowledge Based Engineering Approaches for Complex Product Design

KBE is an automated process of identification, acquisition, and re-use based on design knowledge, and has been widely used to promote the rapid design of products [8]. Pokojski et al. proposed a KBE approach which tries to integrate the knowledge from designers, users, operators, etc. to achieve the multi-disciplinary integrated design for complex products [9]. Johansson et al. developed a KBE framework to combine the knowledge relating to information interaction, quality control and design evaluation [10]. Camarillo et al. presented a KBE approach which uses case-based inference to push similar cases to designers and resolve problems encountered by integrating multi-disciplinary knowledge of stakeholders during the entire product life cycle [11]. However, the multi-modal (i.e., design knowledge represented in different modalities such as natural language, video, image, etc.), multi-temporal (i.e., design knowledge proposed in different stage of the product lifecycle such as conceptual design, detailed design, manufacturing, maintenance, quality control stages, etc.) and multi-spatial (i.e., design knowledge proposed by different stakeholders, such as designers, users, operators, etc.) characteristics of design knowledge affect the crowd intelligence decision-making, because the information from different sources may conflict with each other and therefore negatively affect the reliability of crowd intelligence.

Therefore, the traditional KBE technology needs to be transformed to adapt to the new product design requirements.

2.2 Perception-Retrieval Cognitive Mechanism

Cognitive psychology believes that biological cognition is a process of continuous perception of various information in the environment through the senses, and retrieval of it compared with the existing memory. Under this perception-retrieval cognitive mechanism, the brain can quickly organize and integrate information with different modalities and spatiotemporal characteristics in an optimal way, and achieve a fast and accurate memory retrieval based on external information, thus demonstrating the inherent advantages in processing the multi-modal, multi-temporal and multi-spatial complex correlations and providing inspirations for the proposed research on the efficient organization and accurate acquisition of design knowledge for complex product design. The existing studies on perception-retrieval cognitive mechanism will be presented hereafter.

Perception Process Based on Multi-sensory Integration: McGurk and MacDonald first proposed the concept of biological multi-sensory integration, arguing that the information from different modalities such as image, text, sound, touch, etc. can be effectively integrated in certain areas of the brain to form unified, coherent and stable perceptual information [12]. In response to this phenomenon of sensory information integration, researchers have carried out research work in two directions, i.e., psychophysics and neuroanatomy.

Psychophysics focused on the relationship between stimulus information and sensation in the process of multi-sensory integration from the macroscopic behavior of biology. Tenenbaum et al. proposed a multi-sensory integration model based on Bayesian inference [13]. The visual and auditory integration experiment conducted by Battaglia et al. [14], the visual and haptic integration experiment by Ernst et al. [15], and the visual and vestibular signal integration experiment by Hou et al. proved the effectiveness of this model in the process of multi-sensory integration [16].

Neuroanatomy starts from the microscopic nerve cell level and studies the biological multi-sensory integration mechanism. Quiroga et al. reported for the first-time multi-modal nerve cells that can respond to both image and text modal information [17]; Stein and Meredith et al. reported multi-modal nerve cells that can simultaneously process vestibular and visual signals [18]. Based on these studies, Rowland et al. proposed a multi-sensory integration model at the level of individual nerve cells [19].

In order to establish a unified multi-sensory integration model at macro and micro levels, it was found that the process of multi-sensory integration was no longer regarded as the result of the action of single multi-modal nerve cells, but was realized by the collective action of nerve cell populations in specific regions of the brain [20–22]. Beck et al. proposed a centralized framework for the multi-sensory integration model [23]. Gu et al. proved through experiments that different brain regions can simultaneously participate in the same multi-sensory integration process. In addition to dorsolateral superior temporal, ventral parietal region [24], the frontal eye field [25] and the visual posterior sylvian area [26] can also participate in the integration of visual signals and vestibular signals. Based on this discovery, Zhang et al. proposed a distributed multi-sensory integration model (Fig. 1). In this model, different multi-sensory integration brain regions estimate the stimulus information according to the input they receive, and then send their estimates to other brain regions. Finally, each brain area can integrate multiple inputs, resulting in more accurate estimation of stimulus information [27].

Decision-making Process Based on Memorial Retrieval. The process of memorial retrieval is completed under the combined action of short-term and long-term memory.

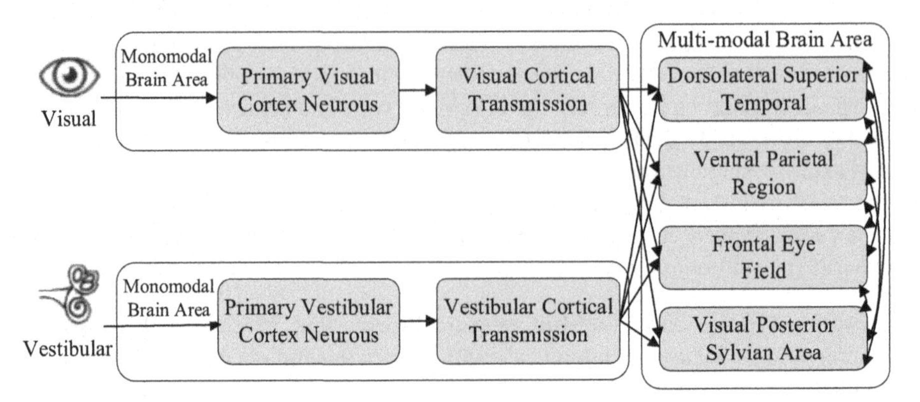

Fig. 1. Distributed multi-sensory integration model.

Short-term memory has little information and short storage time, but it is the main module to complete the complex reasoning and calculation process in the decision-making process. Long-term memory has a large amount of storage information and a long storage time [28]. Long-term memory can quickly recall the memory content associated with the received information [29] and can be updated and modified autonomously [30]. The Atkinson-Shiffrin memory model [31] proposed by Atkinson and Shiffrin and the Working memory model proposed by Baddeley [32] have both explained the memorial retrieval mechanism: after the process of multi-sensory integration, the stimulus of environmental information is stored in the form of short-term memory in the temporal lobe of the brain, and it is compared with the prior knowledge in long-term memory to infer the state of the outside world, so as to actively complete the final decision.

On the basis of the above mechanism, Damasio et al. further proposed that memory information is expressed and stored in the form of vectors after a large number of neuroanatomical experiments [33]. Shiffrin proposed the retrieving effective from memory (REM). When memory retrieval occurs, the received short-term memory feature vector matches the stored long-term memory feature vector, and Bayesian decision-making is used to calculate whether the received information has been learned. If the information has been learned, the decision is made using the existing experience; otherwise, it is considered as new information and stored in long-term memory. Jiang et al. proposed a method of learning, storing and extracting visual images based on memory recall mode [34].

2.3 Summary of Literature Review

In order to take advantage of the crowd intelligence for the design of complex products, the multi-modal, multi-temporal and multi-spatial knowledge needs to be integrated, stored and acquired. By using the perception-retrieval cognitive mechanism based on multi-sensory integration and REM, which enables the brain to continuously receive and rapidly organize and integrate external multi-modal, multi-temporal and multi-spatial information in an optimal way, and accurately achieve retrieval of memory according to the external information[35, 36], it is possible to realize the efficient integration, dynamic storage and real-time acquisition of design knowledge, thus achieving a crowd intelligence-driven design for complex products.

3 Crowd Intelligence Driven Design Framework Based on Perception-Retrieval Cognitive Mechanism

In order to solve the problems of low efficiency and poor accuracy in the organization and acquisition of the multi-modal, multi-temporal and multi-spatial knowledge during the design process of complex products, the framework of perception-retrieval cognitive mechanism was designed from the following three aspects.

3.1 Organization and Integration of Multi-modal, Multi-temporal and Multi-spatial Knowledge

The knowledge adopted for the design of complex products has the characteristics of multi-modal, so the first step to achieve the crowd intelligence-driven design should be the organization and integration of the different design knowledge.

According to the multi-sensory integration process in the perception-retrieval cognitive mechanism, the features of the design knowledge can be extracted. First, the structured knowledge represented in OWL, RDF and other formats can be directly extracted. Second, the semi-structured knowledge (represented by XML or JSON) which cannot express semantic information explicitly, should be extracted semantic information by analyzing the hidden in data tags and element structures, and an OWL ontology document and description can be constructed to represent the semi-structured knowledge structure. Third, the unstructured knowledge represented by natural language, pictures or videos is identified using methods such as Polyglot [37], Mask R-CNN [38] or LSTM [39], and feature extraction is carried out in combination with the semantic relationship between the identified entities. After extracting the semantic features of the design knowledge, a semantic network with different structures can be formed, which needs to be semantically aligned. Semantic alignment is accomplished through distance-based semantic similarity, which measures the location of knowledge entities in the design ontology database. The ontology library is based on OntoSTEP in the field of mechanical design, ORA in the field of robot design or SIARAS in the field of manufacturing. Once the semantic alignment is completed, a unified representation of the design knowledge should be provided. Knowledge graph is a semantic network with graph data, which uses nodes and edges in graph structure to express knowledge entities and their relations. However, knowledge graph can only be used to express static and data-oriented knowledge, and cannot express the multi-modal, multi-temporal and multi-spatial characteristics of knowledge in KBE. Therefore, it is necessary to construct a multi-modal dynamic knowledge graph based on the knowledge graph.

At the same time, knowledge provided by different sources may contain noise, redundancy or even conflicting knowledge, which must be processed and integrated to form a semantically unified and coherent knowledge representation (Fig. 2(a)). The attention mechanism in the multi-sensory integration model proposed by Tenenbaum is used to filter and sift through noisy, redundant or conflicting information.

3.2 Storage and Update of Multi-modal, Multi-temporal and Multi-spatial Knowledge

According to the REM model, human long-term memory is stored in the form of feature vectors. Therefore, the knowledge graph composed of knowledge in KBE should be represented in the form of vector. Once the multi-modal dynamic knowledge graph has been constructed, and the knowledge nodes and their relations in the graph should be expressed in the form of feature vectors. Knowledge graph embedding technology can directly map the knowledge entities and their relations in the multi-modal dynamic knowledge graph into the low-dimensional vector space to realize the feature coding of knowledge. Feature vector can be can be obtained by using the knowledge graph embedding technology based on TransR model.

After the feature vector is obtained, the feature vector should be stored. In order to reduce the computing and searching time, according to the distributed multi-sensory integration model of brain long-term memory, the knowledge in different disciplines is featured by clustering and stored in different knowledge modules (Fig. 2(c)). Graph attention networks is used to complete the clustering of design knowledge.

Large-scale knowledge in KBE shows a high dependence on time and space, so the knowledge stored in KBE needs to be in a dynamic form of continuous renewal. According to the perception-retrieval cognitive mechanism, the newly-received knowledge is retrieved using the knowledge subgraph coding algorithm that can be developed based on the TransR model, and then compared with the existing knowledge for storage and update. Since only the subgraphs with changes are encoded, the computational load of the encoding in the process of knowledge updating can be greatly reduced.

3.3 Push of Multi-modal, Multi-temporal and Multi-spatial Knowledge

Referring to the short-term memory mode in the REM, the existing knowledge is called and matched to complete the push of knowledge. The research on the praxeology shows that when design participants conduct the design work, their behavior patterns are not chaotic, but have their own rules. When analyzing the behavior patterns of operators to determine whether they need knowledge, the context aware computing is adopted to determine whether they need knowledge through real-time perceptual monitoring of their own behavior and software operation. At present, the context aware computing technology based on information communication, sensors and machine learning is very mature, which provides data and technical support for signal acquisition and processing in the process of behavior pattern recognition.

After confirming that operators need knowledge, it is necessary to further acquire their knowledge needs and judge what knowledge they need. However, the traversal method will consume a lot of computing time of the system in the mass and miscellaneous crowd knowledge, which seriously affects the real-time performance of knowledge push. Nowadays, inferential methods based on ontology and rules have been widely applied in the field of information, and various inference machines supporting ontology and rules have also been developed, such as Jena, Jess and Racer inference machines, which can provide support for the reasoning process required. Therefore, a semantic inferential model is used to locate the required knowledge quickly.

In order to reduce the time spent in knowledge search, the knowledge feature vector stored in the KBE is retrieved and compared with the knowledge requirement feature vector after the system is quickly positioned using probability-based reasoning. The retrieval and comparison between the knowledge feature vector in the knowledge module and the knowledge demand feature vector can be regarded as the process of calculating the likelihood of two equal-dimensional vectors. Therefore, Bayesian formula can be used to calculate their likelihood and complete the final matching of knowledge (Fig. 2(b)).

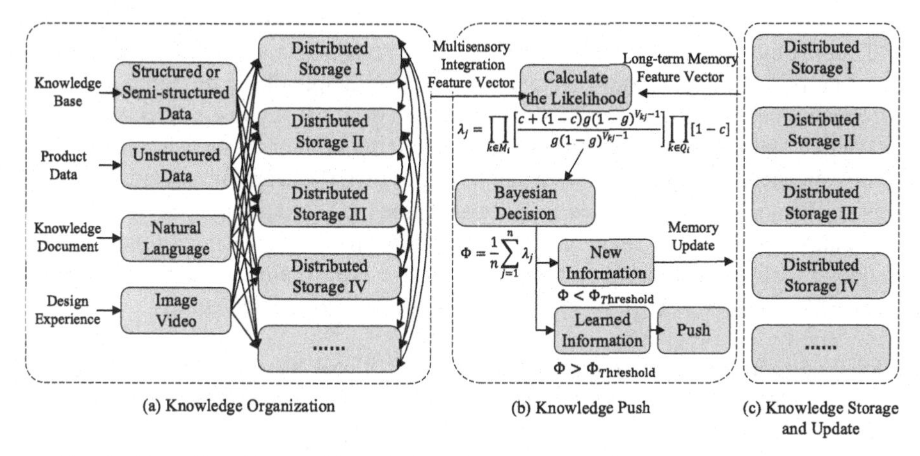

Fig. 2. Crowd intelligence driven design of perception-retrieval cognitive mechanism.

4 Conclusion

Considering the limitations of current KBE approaches, the authors propose a crowd intelligence-driven framework based on the brain's perception-retrieval cognitive mechanism for the organization, storage and push of multi-modal, multi-temporal and multi-spatial knowledge. This mechanism solves the problems of low organizational efficiency and poor acquisition accuracy of crowd design knowledge in different modalities at multiple time and space scales, and will provide theoretical basis and application prospect for promoting the intelligent design of advanced aviation manufacturing equipment and complex products such as aerospace, ships and automobiles in the future. Especially in the field of advanced aviation manufacturing equipment, the application in the field of manufacturing equipment is promoted by integrating process knowledge, product knowledge, equipment knowledge and program control knowledge commonly used in the development of manufacturing equipment.

Future research can be generally divided into two parts. First, knowledge graph has become a widely adopted knowledge representation method. The multi-modal dynamic knowledge graph proposed in this paper introduces new factors τ. On the one hand, on the basis of head node(h), relation node(r), and tail node(t), τ is integrated to realize the time correlation of the knowledge in the full-time domain. On the other hand, the modal and spatial features cannot only be integrated into the dynamic knowledge graph as the tail node of the knowledge graph, but also form a representational spatial relationship (hasProvider) and modal relationship (hasDescription or hasImage) with the entity head node. At the same time, τ represents the update cycle of crowd design knowledge. When a new integrated knowledge graph is received, the new knowledge graph is retrospectively compared with the previous knowledge graph.

Second, the existing TransE, TransH, TransR, etc. are used as plane distance models, and their corresponding algorithms are very mature. Compared with other plane distance models, the TransR model not only solves the complex one-to-many, many-to-one and many-to-many relationships between the head and tail nodes, which cannot be realized by TransE model, but also improves the semantic expression ability of TransH model for relationships between knowledge entities. The principle of the proposed algorithm can be summarized as follows. Firstly, the low-dimensional vectors is used to initialize knowledge entities (head node (h) and tail node (t)) and their relationship (r), and the positive and negative training samples consisting of (h, r, t) can be constructed. Secondly, the TransR model based on plane distance to define the scoring function fr(h, t) is used to calculate the total loss value of the feature vector for positive samples and negative samples. Finally, taking the minimum total loss value as the optimization goal, the feature vector of knowledge in KBE is obtained through continuous calculation.

Acknowledgement. This project is supported by the National Natural Science Foundation of China (Grant No. 51805437) and the Fundamental Research Funds for the Central Universities (Grant No. 31020210506005).

References

1. Jeong, Y., Pan, Y., Rathore, S., Kim, B., HyukPark, J.: A parallel team formation approach using crowd intelligence from social network. Comput. Hum. Behav. **101**, 429–434 (2019). https://doi.org/10.1016/j.chb.2018.07.018
2. Li, W., et al.: Crowd intelligence in AI 20 era. Front. Inf. Technol. Electron. Eng. **1**(18), 15–43 (2017). https://doi.org/10.1631/FITEE.1601859
3. Poo, M., Du, J., Ip, N.Y., Xiong, Z.-Q., Xu, B., Tan, T.: China brain project: basic neuroscience, brain diseases, and brain-inspired computing. NeuroView **92**(3), 591–596 (2016). https://doi.org/10.1016/j.neuron.2016.10.050
4. Pan, Y.: Heading toward artificial intelligence 2.0. Engineering **2**(4), 409–413 (2016). https://doi.org/10.1016/J.ENG.2016.04.018
5. Adolphs, R.: Cognitive neuroscience of human social behaviour. Nat. Rev. Neurosci. **4**(3), 165–178 (2003). https://doi.org/10.1038/nrn1056
6. Zhao, X., Chen, H., Xing, Z., Miao, C.: Brain-inspired search engine assistant based on knowledge graph. IEEE Trans. Neural Netw. Learn. Syst. (2021). https://doi.org/10.1109/TNNLS.2021.3113026
7. Wang, J., Cheng, R., Liao, P.-C.: Trends of multimodal neural engineering study: a bibliometric review. Arch. Comput. Methods Eng. **28**(7), 4487–4501 (2021). https://doi.org/10.1007/s11831-021-09557-y
8. Zheng, C., et al.: Knowledge-based program generation approach for robotic manufacturing systems. Rob. Comput.-Integrated Manuf. **73**, 102238 (2022). https://doi.org/10.1016/j.rcim.2021.102242

9. Pokojski, J., Szustakiewicz, K., Woźnicki, Ł., Oleksiński, K., Pruszyński, J.: Industrial application of knowledge-based engineering in commercial CAD/CAE systems. J. Ind. Inf. Integrat. **25** (2022). https://doi.org/10.1016/j.jii.2021.100255
10. Johansson, J., Contero, M., Company, P., Elgh, F.: Supporting connectivism in knowledge-based engineering with graph theory, filtering techniques and model quality assurance. Adv. Eng. Inf. **38**, 252–263 (2018). https://doi.org/10.1016/j.aei.2018.07.005
11. Camarillo, A., Ríos, J., Althoff, K.D.: Knowledge-based multi-agent system for manufacturing problem solving process in production plants. J. Manuf. Syst. **47**, 115–127 (2018). https://doi.org/10.1016/j.jmsy.2018.04.002
12. McGurk, H., MacDonald, J.: Hearing lips and seeing voices. Nature **264**, 746–748 (1976). https://doi.org/10.1038/264746a0
13. Tenenbaum, J.B., Griffiths, T.L., Kemp, C.: Theory-based Bayesian models of inductive learning and reasoning. Trends Cogn. Sci. **7**(10), 309–318 (2006). https://doi.org/10.1016/j.tics.2006.05.009
14. Battaglia, P.W., Jacobs, R.A., Aslin, R.N.: Bayesian integration of visual and auditory signals for spatial localization. J. Opt. Soc. Am. A **7**(20), 1391–1397 (2003). https://doi.org/10.1364/JOSAA.20.001391
15. Ernst, M.O., Banks, M.S.: Humans integrate visual and haptic information in a statistically optimal fashion. Nature **415**, 429–433 (2002). https://doi.org/10.1038/415429a
16. Hou, H., Zheng, Q., Zhao, Y., Pouget, A., Gu, Y.: Neural correlates of optimal multisensory decision making under time-varying reliabilities with an invariant linear probabilistic population code. Neuron **5**(104), 1010–1021 (2019). https://doi.org/10.1016/j.neuron.2019.08.038
17. Quiroga, R.Q., Reddy, L., Kreiman, G., Koch, C., Fried, I.: Invariant visual representation by single neurons in the human brain. Nature **435**, 1102–1107 (2005). https://doi.org/10.1038/nature03687
18. Stein, B.E., Meredith, M.A.: The Merging of the Senses. The MIT Press, Cambridge (1993)
19. Rowland, B.A., Stanford, T.R., Stein, B.E.: A model of the neural mechanisms underlying multisensory integration in the superior colliculus. Perception **10**(36), 1431–1443 (2007). https://doi.org/10.1068/p5842
20. Beck, J.M., et al.: Probabilistic population codes for Bayesian decision making. Neuron **6**(60), 1142–1152 (2008). https://doi.org/10.1016/j.neuron.2008.09.021
21. Gao, C., Weber, C.E., Wedell, D.H., Shinkareva, S.V.: An fMRI study of affective congruence across visual and auditory modalities. J. Cogn. Neurosci. **7**(32), 1251–1262 (2020). https://doi.org/10.1162/jocn_a_01553
22. Gilissen, S.R., Farrow, K., Bonin, V., Arckens, L.: Reconsidering the border between the visual and posterior parietal cortex of mice. Cereb. Cortex **3**(31), 1675–1692 (2021). https://doi.org/10.1093/cercor/bhaa318
23. Kuang, S., Deng, H., Zhang, T.: Adaptive heading performance during self-motion perception. PsyCh Journal **3**(9), 295–305 (2020). https://doi.org/10.1002/pchj.330
24. Gu, Y., Cheng, Z., Yang, L., DeAngelis, G.C., Angelaki, D.E.: Multisensory convergence of visual and vestibular heading cues in the pursuit area of the frontal eye field. Cereb. Cortex **9**(26), 3785–4380 (2016). https://doi.org/10.1093/cercor/bhv183
25. Chen, A., DeAngelis, G.C., Angelaki, D.E.: Convergence of vestibular and visual self-motion signals in an area of the posterior sylvian fissure. J. Neurosci. **32**(31), 11617–11627 (2011). https://doi.org/10.1523/JNEUROSCI
26. Zhang, W.H., Chen, A., Rasch, M.J., Wu, S.: Decentralized multisensory information integration in neural systems. J. Neurosci. **2**(36), 532–547 (2016). https://doi.org/10.1523/JNEUROSCI
27. McBride, D.M., Cutting, J.C.: Cognitive Psychology Theory, process, and Methodology. SAGE Publications, Thousand Oaks (2018)

28. Eysenck, M.W., Keane, M.T.: Cognitive Psychology. Psychology Press, London (2015). https://doi.org/10.4324/9781315778006
29. Lee, J.L., Nader, K., Schiller, D.: An update on memory reconsolidation updating. Trends Cogn. Sci. **7**(21), 531–545 (2017). https://doi.org/10.1016/j.tics.2017.04.006
30. Eysenck, M.W., Brysbaert, M.: Fundamentals of Cognition. Routledge, New York (2018)
31. Baddeley, A.: Working memory and conscious awareness. In: Theories of Memory, pp. 11–28(2019)
32. Damasio, A.R.: The brain binds entities and events by multiregional activation from convergence zones. Neural Comput. **1**(1), 123–132 (1989). https://doi.org/10.1162/neco.1989.1.1.123
33. Jiang, Y., Wang, Y.: Application of REM memory model in image recognition and classification. CAAI Trans. Intell. Syst. **3**, 310–317 (2017)
34. Seilheimer, R.L., Rosenberg, A., Angelaki, D.E.: Models and processes of multisensory cue combination. Curr. Opin. Neurobiol. **25**, 38–46 (2014). https://doi.org/10.1016/j.conb.2013.11.008
35. Bertin, R.J.V., Berthoz, A.: Visuo-vestibular interaction in the reconstruction of travelled trajectories. Exp. Brain Res. **1**(154), 11–21 (2004). https://doi.org/10.1007/s00221-003-1524-3
36. Gu, Y., Angelaki, D.E., DeAngelis, G.C.: Neural correlates of multisensory cue integration in macaque MSTd. Nat. Neurosci. **10**(11), 1201–1210 (2008). https://doi.org/10.1038/nn.2191
37. Nystrom, N., Clarkson, M.R., Myers, A.C.: Polyglot: an extensible compiler framework for java. In: Hedin, G. (ed.) CC 2003. LNCS, vol. 2622, pp. 138–152. Springer, Heidelberg (2003). https://doi.org/10.1007/3-540-36579-6_11
38. He K, Gkioxari G, Dollár P, et al.: Mask r-cnn. In: Proceedings of the IEEE International Conference on Computer Vision, pp. 2961–2969 (2017)
39. Greff, K., Srivastava, R.K., Koutník, J., et al.: LSTM: a search space odyssey. IEEE Trans. Neural Netw. Learn. Syst. **10**(28), 2222–2232 (2016). https://doi.org/10.1109/TNNLS.2016.2582924

Digital Transition and the Environment: Towards an Approach to Assess Environmental Impacts of Digital Services Using Life Cycle Assessment

Mohammed Ziani[1,2,3](✉), Nicolas Maranzana[3], Adélaïde Feraille[2], and Myriam Saade[2]

[1] Université Gustave Eiffel, MAST-EMGCU, Cité Descartes, Boulevard Newton, 77447 Marne-La-Vallée, France
m_ziani@gmx.fr
[2] Ecole des Ponts ParisTech, MSA NAVIER, 77455 Marne La Vallée, France
[3] Arts et Métiers Institute of Technology, LCPI, HESAM Université, 75013 Paris, France

Abstract. The adoption of digital technologies in the construction sector is seen by many as a catalyst for the ecological transition, through the optimization of resource and energy use. However, the environmental impact of the digital infrastructure supporting projects is not well known. In the example of life cycle management of linear infrastructures, the increased use of IoT (Internet of Things) sensors for maintenance/operation has been followed by a diversification of technologies for Building Information Modeling (BIM), Building Life cycle Management (BLM), equipment and exchange formats. The variety of computer terminals, hardware storage and communication equipment are extended by that of software tools and free or proprietary solutions. The main objective of this paper is to establish a methodology based on environmental Life Cycle Assessment (LCA) to quantify the environmental impacts of a digital service in Infrastructure Lifecycle Management (ILM). Relying on a case study of a bridge monitoring system implemented for the A71 near Orléans, France, we propose to model the assessed system as four interacting sub-systems: data acquisition, network and communication, consultation, and data storage. Based on such modelling and related data, a preliminary assessment highlights the contribution of data storage in the environmental impacts of the overall system.

Keywords: Life Cycle Assessment (LCA) · Ecological transition · Internet of Things (IoT) · Infrastructure Lifecycle Management (ILM)

1 Introduction

The Architecture, Engineering, and Construction (AEC) services sector has experienced a spectacular advance in Information and Communication Technologies (ICT), among other sectors [1]. Many examples in France demonstrate the importance of technological

Published by Springer Nature Switzerland AG 2023
F. Noël et al. (Eds.): PLM 2022, IFIP AICT 667, pp. 577–586, 2023.
https://doi.org/10.1007/978-3-031-25182-5_56

and digital change in the management of engineering structures, the specific case of bridges being for example studied in the EMGCU laboratory, with the implementation of demonstrators on real bridges (mixed or prestressed structures) with an integrated measurement system using Internet of Things (IoT).

The current ecological crisis is however pushing infrastructure researchers and stakeholders to assess the environmental impact of digital technologies [2]. Promoted as a mean to increase the energy efficiency of buildings, decrease construction time, optimize the use of natural resources [3], or improve the recovery of building materials [4], digital services supporting civil engineering projects consume resources and energy [5]. Their environmental footprint is currently not well known, and methodologies to assess such footprint poorly developed. This article aims to propose an approach for the evaluation of the environmental impact of digital services, based on environmental Life Cycle Assessment (LCA). After presenting the state of the art (Sect. 2), the methodological approach is presented (Sect. 3) and applied to a case study (Sect. 4), a bridge monitoring system implemented for the A71 near Orléans, France. Results are presented in Sect. 5, followed by a conclusion (Sect. 6).

2 Research Background

2.1 Literature Review

The bibliographic review was based on French and European scientific publications and grey literature, including reports from consulting firms, professional federations, think tanks and public authorities (Table 1).

Table 1. Public data

Knowledge base	Government digital services (GDS)	LCA data	Manufacturer data
ADEME	GreenIT.fr	Base Carbone	Dell
ARCEP	Bureau Veritas	Négaoctet	HP
CNNum	fing.org, EcoInfo	Ecoinvent v3.6	Lenovo
EPLCA	Boavizta, APL-datacenter	ILCD	Seagate
scorelca.org	The Shift Project		

Several studies and projects carried out over the last ten years have focused on specific issues, such as energy consumption in data centers, premature obsolescence of terminals or electronic waste management. Other studies have focused on the entire life cycle of digital equipment (Table 2), but with a single-criteria approach.

The LCA studies of the main equipment essential to the operation of networks and data transmission on the transport network are presented in the report EcoInfo - RENATER (EcoInfo - RENATER, 2021) and the study [13]. The institute “Efficacity”, has developed an estimation on the consumption of the IT room part.

Table 2. LCA studies

Datacenter	Whitehead et al., 2015	[6]
RENATER network	(EcoInfo - RENATER, 2021)	[7]
Software	(GREENSPECTOR, n.d.)	[8]
USB key	(Guern and Farrant, 2011)	[9]
Workstation	(Anders S.G. {Citation}, 2012)	[10]
cloud computing	(Itten et al., 2020)	[11]
Semiconductors	(Sarah B. Boyd, 2012)	[12]

2.2 Key Technologies

2.2.1 Digital Service

Definition from the Alliance GreenIT [14, 15]: "A digital service is constituted of a set of software, hardware, networks, infrastructures, and other digital services. It fulfills a functional unit such as "book a train ticket", "send an e-mail to friends", etc."

2.2.2 Internet of Things (IoT)

With the rapid development of IoT technology, it has become a promising technology that can make product management more flexible and efficient. According to ARCEP's [16, 17] definition, "The Internet of Things is a set of connected objects and network technologies that combine physical objects that have connected sensors, wired or wireless digital communication networks, remote storage spaces for the data collected and data processing applications that engage decision-making processes".

2.2.3 Building Information Modeling (BIM)

Is an approach to building construction and operation, whose development is being extended to infrastructures and, more broadly, to civil engineering works. It is at the heart of the work process around a digital model of the structure or a digital twin for the operation/maintenance of infrastructures. The BIM approach improves data management, information quality and cost control in the construction process.

2.2.4 Infrastructure Lifecycle Management (ILM))

The life cycle of data consists of four phases: The creation, the hot, warm and cold phase. Information Life cycle Management (ILM) is a comprehensive approach to managing the flow of an information system's data and associated metadata from creation and initial storage to the time when it becomes obsolete and is deleted [18].

2.2.5 Building Lifecycle Management Platform (BLMP)

As a methodology Based on BLM and BIM technology and Industry Foundation Classes (IFC) standard, The Building Lifecycle Information Management Platform Based on

BIM presents a practical and effective way to realize information creating, exchange, sharing and integration management of all participants of the construction project [19]. The BLMP framework divides Collaboration Workspace into three functions, which are cooperative service management platform, collaborative work information management platform, and cooperative design work platform [20, 21].

3 Research Methodology

As a methodology for assessing environmental impacts related to products and their potential impacts, Life Cycle Assessment (LCA) has been developed as an important tool for evaluating the energy and environmental performance of products over their life cycle [22]. Life cycle assessment is a widely recognized method for evaluating environmental impacts [23]. LCA is defined by the ISO 14040 as the compilation and evaluation of the inputs, outputs and the potential environmental impacts of a product system throughout its life cycle.

Our approach, adapted to digital services, was carried out following the general methodological recommendations of ISO 14040:2006 and 14044:2006. It necessarily takes into account the following points (Fig. 1):

1. The 4 main steps, respectively: the definition of the objectives and the system, the analysis of the inventory, the evaluation of the impacts and the interpretation of the results.
2. All stages of the life cycle of the service and associated equipment: manufacturing, installation, distribution, use, end of life;
3. The three thirds of the digital architecture: user terminals, communication networks and data centers.

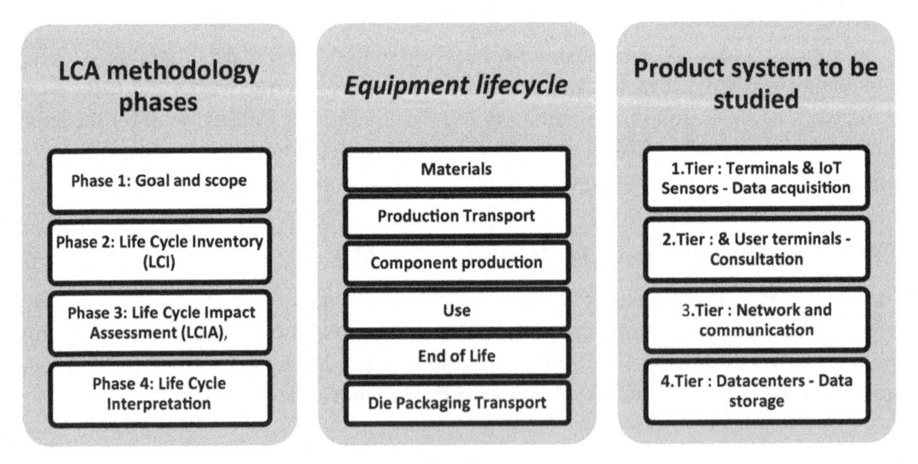

Fig. 1. Proposed LCA approach, adapted to digital services.

4 Case Study

4.1 Case Description

This study quantifies the environmental impacts of the digital equipment and infrastructures which support the monitoring system of a structure: ***A71 Viaduct on the Loire River***, according to the four bricks illustrated below: data acquisition system; network and Communication system; consultation system; data Storage system. The description of the global architecture of the system is illustrated in Fig. 2. The components of the system and communication networks are simplified to make the result readable and understandable. The equipment listed in Fig. 2, are the components included in the present study, based on the experimental protocol set up in the framework of EMGCU Lab activities, to evaluate the services of the monitoring structures.

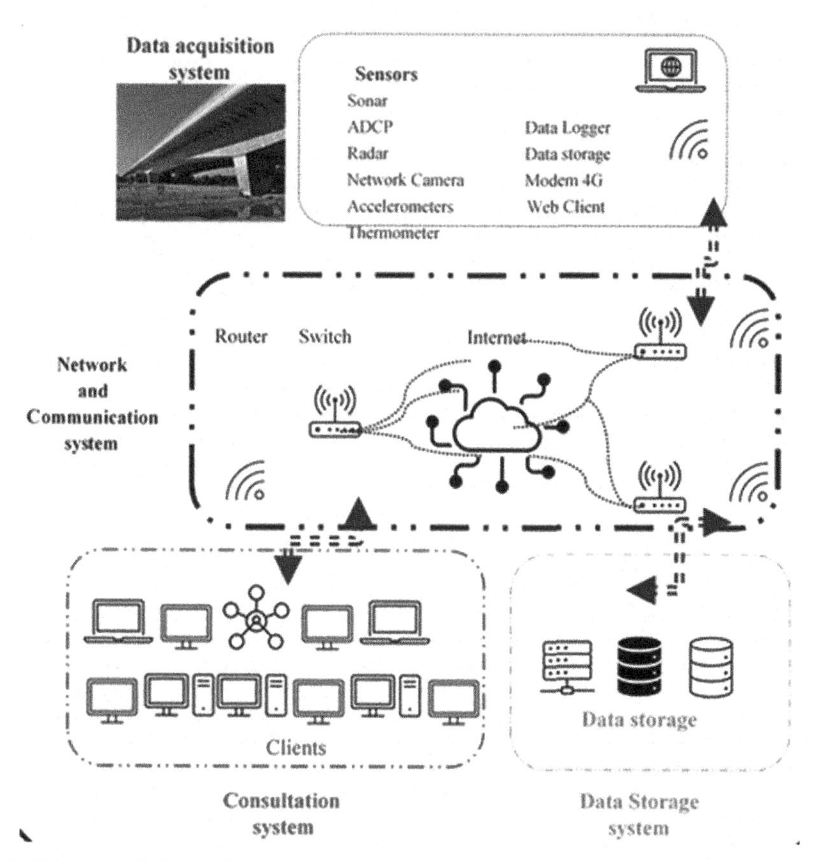

Fig. 2. Diagram of the equipment needed for bridge monitoring - Bridge Monitoring System Architecture

4.2 Function

The monitoring of structures implies to assess their structural health and to detect new developments early enough to avoid major problems. Sensor data is collected and read on a periodically basis by the nodes and sent to the gateway via the mesh network nodes. The gateway then sends the data to the EMGCU server, or to a third-party server. The data exchange flow consists of the following steps:

- Acquisition of input data from the sensors: bathymetry, water level, flow speed, camera, temperature, vibrations
- Transfer of input data to the calculation platform (local server)
- Calculation and centralization of the data on the local calculation platform,
- Transfer of the data on the transport network, on RENATER (Internet)
- Transfer of data to the storage platform (Data Center EMGCU)
- Analysis and exploitation of the data

The daily volume of data is estimated at 10 MB.

4.3 Functional Unit

Given the objective of the study, the Functional Unit (FU) is defined. The FU is the unit of reference used to link the inputs and outputs and the environmental performance of the system under study. The FU selected for this study is **one day of monitoring**.

4.4 System Boundary

The assessed system is modeled as 4 interacting sub-systems (Fig. 2).): data acquisition; network and communication; consultation; and data storage, and each subsystem is composed of several components. The system studied includes the following steps (Fig. 4):

- The manufacture of equipment: the manufacture of equipment takes into account the consumption of raw materials, the transport of raw materials and the energy consumption necessary for the production of equipment. The values of the coefficients used in the method come from measurements, from the Ecoinvent 3.7.1 database [24], from manufacturers or from the literature.
- The use phase: Operation and maintenance: For the reference scenario, the use phase requires the electricity consumption of the components of the monitoring system. As much as possible, we have relied on the real electrical consumption (especially for the sensors). When this information was missing, we relied on of the technical specifications of the equipment provided by the manufacturer and have reconstructed each equipment of the subsystems.

4.5 Inventory Data

Life Cycle Inventory emission factors, related to flows of raw material extracted and base material used, is extracted from the Ecoinvent 3.7 database. The energy mix considered is that for France. The model has been developed with OpenLCA [25] in line with the Ecoinvent database.

4.6 Life Cycle Impact Assessment

Impacts calculation was conducted using ReCiPe [26].

5 Results and Discussion

The following figure present the impact scores for different environmental categories related to the functional unit defined - one day of monitoring. Impact scores for eighteen impact categories are calculated considering the different life cycle stages of the monitoring system: production, assembly, transportation, use and end of life. The use phase includes all steps associated with the acquisition of measurement data transmitted to the storage system and the consultation of the data by EMGCU project beneficiaries. Specifically, the electricity consumption for the acquisition system is taken into account.

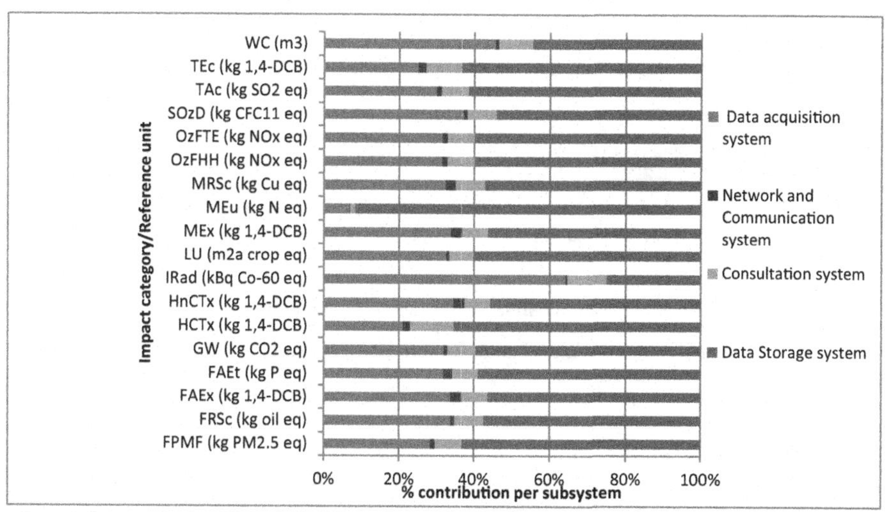

Impact characterization Method: ReCiPe midpoint H. FPMF : Fine particulate matter formation ,FRSc : Fossil resource scarcity ,FAEx : Freshwater ecotoxicity ,FAEt : Freshwater eutrophication ,GW : Global warming ,HCTx : Human carcinogenic toxicity ,HnCTx : Human non-carcinogenic toxicity ,IRad : Ionizing radiation ,LU : Land use ,MEx : Marine ecotoxicity ,MEu : Marine eutrophication ,MRSc : Mineral resource scarcity ,OzFHH : Ozone formation, Human health ,OzFTE : Ozone formation, Terrestrial ecosystems ,SOzD : Stratospheric ozone depletion ,TAc : Terrestrial acidification ,TEc : Terrestrial ecotoxicity ,WC : Water consumption.

Fig. 3. Environmental impact scores – contribution analysis to Monitoring System Impacts (by subsystem)

Results presented in Fig. 3 shows that for most impact categories, the consultation system is the most contributing subsystem, along with the data acquisition system. The most contributing processes are the production of screens for the consultation block, the production of electricity for the sensors and the production of the electronic components of the sensor system. The storage system has a low contribution, because we assumed that data are stored for a duration of one day.

5.1 Sensitivity Analysis Results

The sensitivity analysis focuses on the duration of data storage, considering a storage of 1 year and 10 years (the initial assumption being 1 day). The emissions table for the impact category/Global Warming (kg CO_2 eq) is given in Fig. 4. We find that the duration of the data storage has a strong influence on electricity consumption. By increasing the data storage duration, the overall impact scores increase.

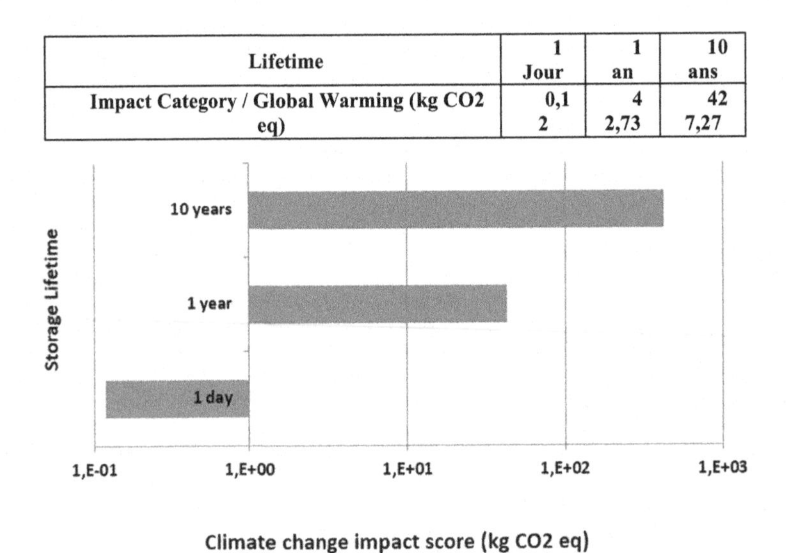

Lifetime	1 Jour	1 an	10 ans
Impact Category / Global Warming (kg CO2 eq)	0,12	42,73	427,27

Fig. 4. Sensitivity analysis - Climate Change for different data storage duration (kg CO_2 eq/FU)

5.2 Discussion

The results show that for the baseline scenario with 1 day of storage, the data storage system has a significantly lower impact than the other evaluated subsystems. The sensitivity results show that the impact increases significantly with increasing storage time. This demonstrates the importance of electricity consumption of data centers, which in our case is based on rudimentary assumptions. This parameter should be further explored.

While these preliminary results show the importance of data storage time, they also show the significant impact of the production of equipment, and in particular the consultation systems (screen). The hypotheses taken for the case study corresponding to an experimental system, it would be necessary to carry out a complementary sensitivity study on the duration of use of the consultation system per FU, in order to get closer to realistic hypotheses corresponding to an operational situation.

The case study presented here therefore demonstrates that LCA is a valuable tool to evaluate the environmental impacts of digital services. It allows to highlight the effect of each practice on the environmental performance of digital services.

6 Conclusions

This study quantified the environmental impacts of complex digital equipment and infrastructure in the construction and Engineering Construction (AEC) sector, using a multi-criteria, multi-stage LCA approach. This real-life case study made it possible to test the methodological and theoretical approach to evaluate the environmental impacts of a digital service. The model, at this stage, has allowed having an idea on the environmental performance of the digital service in a case of 'BIMP' for the Management-Operation-Maintenance step.

The use of life cycle inventories is derived from different data sources (Impacts Base, Ecoinvent, manufacturers, etc.). The large amount of data collected on the system satisfies our initial desire to do a simplified LCA study.

Nevertheless, these first results can be considered as encouraging with regard to the work carried out by the digital and Green IT community on this subject. This work is only a beginning; it can be completed within the framework of future analyses that will be carried out by the research teams and the national community of the Digital LCA.

6.1 Limitations of the Study

The accurate and comprehensive determination of the environmental impacts of digital equipment and infrastructure is a complex task that faces many limitations due to data access and associated uncertainties. In addition, this study has identified a number of complementary works that can be carried out to allow a better understanding and analysis of the environmental impacts of the digital sector, Particularly, the approach should be extended to other infrastructures project phases (BIM in design, deconstruction, renovation…) and not only monitoring.

Acknowledgments. This research was funded by IREX (https://irex.asso.fr/). The case study was proposed by the EMGCU lab (https://emgcu.univ-gustave-eiffel.fr/).

References

1. Sassanelli, C., Rossi, M., Terzi, S.: Evaluating the smart maturity of manufacturing companies along the product development process to set a PLM project roadmap. Int. J. Prod. Lifecycle Manag. **12**, 185–209 (2020). https://doi.org/10.1504/IJPLM.2020.109789
2. Aiouch, Y., et al.: Evaluation de l'impact environnemental du numérique en France et analyse prospective. **1**, 171 (2022)

3. Honic, M., Kovacic, I., Sibenik, G., Rechberger, H.: Data- and stakeholder management framework for the implementation of BIM-based Material Passports. J. Build. Eng. **23**, 341–350 (2019). https://doi.org/10.1016/J.JOBE.2019.01.017
4. Akanbi, L.A., et al.: Salvaging building materials in a circular economy: a BIM-based whole-life performance estimator. Res. Conserv. Recycl. **129**, 175–186 (2018). https://doi.org/10.1016/j.resconrec.2017.10.026
5. Pohl, J., et al.: Review how LCA contributes to the environmental assessment of higher order effects of ICT application: a review of different approaches. J. Clean. Prod. **219**, 698–712 (2019)
6. Whitehead, B., Andrews, D., Maidment, G., Dunn, A.: The screening life cycle assessment of a data center (2012)
7. EcoInfo - RENATER, Rapport: évaluation de l'empreinte carbone de la transmission d'un gigaoctet de données sur le réseau RENATER (2021)
8. Leboucq, T.: GREENSPECTOR. Guide_Methodologique : ACV des logiciels (2020)
9. Guern, Y.L., Farrant, L.: Contact Bio Intelligence Service S.A.S., p. 11 (2011)
10. Andrae, A.S.G.: Comparative micro life cycle assessment of physical and virtual desktops in a cloud computing network with consequential, efficiency, and rebound (2012)
11. Itten, R., et al.: Digital transformation—life cycle assessment of digital services, multifunctional devices and cloud computing. Int. J. Life Cycle Assess. **25**(10), 2093–2098 (2020). https://doi.org/10.1007/s11367-020-01801-0
12. Boyd, S.B.: Life-Cycle Assessment of Semiconductors. Springer, New York (2012). https://doi.org/10.1007/978-1-4419-9988-7
13. Damien, O.: Anaïs de l'équipe d'Objectif Carbone. Impact carbone de la connexion à internet (2020)
14. De Carlan, F., T. Guiot, J-C. Leonard, G. Castagna, Efficacity Data Center État des lieux. 2017
15. E.L.-P. SCORE LCA Workshop. Connected devices and related services envi-ronmental impacts: Order of magnitude and methodological recommendations (2021). https://seminaire2021.scorelca.org/storage/doc/11-scorelca-iot-210316.pdf
16. ClubGreenIT. Du Green IT au numérique responsable Lexique des termes de référence (2018). https://club.greenit.fr/doc/2018-05-ClubGreenIT-lexique-numerique_responsable-v1.8.3.pdf
17. Arcep. Préparer la révolution de l'internet des objets : Document n° 1 – Une cartographie des enjeux (7 novembre 2016), p. 46 (2016)
18. Kirchner, C.: Le monde de l'internet des objets: des dynamiques à maîtriser", p. 304 (2022)
19. Information Life Cycle Management (ILM) is a comprehensive approach to. Data Center Solutions (2017). https://en.vcenter.ir/storage/information-life-cycle-management-ilm/
20. An, J., Zou, Z., Chen, G., Sun, Y., Liu, R., Zheng, L.: An IoT-based life cycle assessment platform of wind turbines. Sensors **21**(4), 1233 (2021). https://doi.org/10.3390/s21041233
21. Akintoye, A., Goulding, J., Zawdie, G.: Construction Innovation and Process Improvement. John Wiley & Sons, Hoboken (2012)
22. Qing, L., Tao, G., Ping, W.-J.: Study on building lifecycle information management platform based on BIM. Res. J. Appl. Sci. Eng. Technol. **7**, 1–8 (2014). https://doi.org/10.19026/rjaset.7.212
23. Olivier Jolliet, P.C., Saadé-Sbeih, M., Jolliet-Gavin, N., Shaked, S.: Analyse du cycle de vie - Comprendre et réaliser un écobilan, 3eme édition. Librairie Eyrolles (2017). https://www.eyrolles.com/Sciences/Livre/analyse-du-cycle-de-vie-9782889151356/
24. Ecoinvent. Ecoinvent Database (Version 3.7). https://ecoinvent.org
25. OpenLCA. openLCA.org. https://www.openlca.org
26. ReCiPe2016. https://simapro.com/2017/updated-impact-assessment-methodology-recipe-2016/

Development of a Learning Ecosystem for Effective Learning in Socio-Technical Complex Systems

Maira Callupe(✉), Monica Rossi, Brendan Sullivan, and Sergio Terzi

Politecnico di Milano, Via Lambruschini 4B, 20156 Milan, Italy
maira.callupe@polimi.it

Abstract. This paper presents the development of a Learning Ecosystem (LE) to cope with the challenges brought by the characteristic complexity and uncertainty of research and innovation projects. These projects are made up of consortium partners who constitute a complex socio-technical system (STT), which works towards the development of cutting-edge technology by creating new and valuable knowledge. Through the creation of a LE grounded in complexity, learning and knowledge management theories, the project consortium aims to enable an efficient knowledge exchange and effective learning processes within the system, which are ultimately conducive to the achievement of the project goals. After a review of the literature about complex STTs and LEs, the development of the LE for a specific project taken as a pilot is discussed in detail, elaborating on its components, knowledge flows, learning processes, learning tools, and the methods to assess its effectiveness.

Keywords: Complex socio-technical system · Learning ecosystem · Effective learning · Knowledge gap · Research and innovation project

1 Introduction

Complexity in engineering has been studied from different perspectives and different foci due to the absence of a unique concept or an overarching complexity theory, with works discussing its source, typology, scope, concepts, etc. [1–3]. The study of complexity has been associated with challenges arising from uncertainty, iteration, multidisciplinary, and dynamic behavior, which are all characteristic of research and innovation (R&I) projects [4]. In this context, the complexity arises not only from the technology embedded in the subject of the project itself, but also the socio-technical interactions between actors involved in the project, as well as the social and technological pressures from external stakeholders which represent the target of the projects' output [3]. This degree of complexity is often found in R&I projects financed by the European Union, and calls for solutions that enable the management of complex socio-technical systems (STS) while allowing for effective innovation and supporting organizations in being more adaptive in their response to change and uncertainty [5]. Furthermore, modern views

Published by Springer Nature Switzerland AG 2023
F. Noël et al. (Eds.): PLM 2022, IFIP AICT 667, pp. 587–596, 2023.
https://doi.org/10.1007/978-3-031-25182-5_57

on STS aim to promote not only innovation, but also knowledge sharing and learning within, as creating these opportunities is conducive to an informed decision-making process along the system [6, 7].

The present work uses complexity theory to describe the organizational context of the R&I project "Ecosystemic knowledge in Standards for Hydrogen Implementation on Passenger Ship" (e-SHyIPS) as a complex STT, and to develop a methodology that aims to address, in particular, the efficient exchange of knowledge within the system. Section 2 discusses the characterization of the e-SHyIPS project as a complex STT, identifying its elements, their interactions, and highlighting the importance of knowledge and learning. Section 3 reviews the literature on LEs, and Sect. 4 builds on the previous sections to develop the "e-SHyIPS LE", detailing its components, learning tools, and measures of effectiveness. The work finishes with the conclusions and next steps presented in Sect. 5.

2 Complex Socio-technical Systems in Research and Innovation Projects: The e-SHyIPS Case

When it comes to R&I projects, adopting a socio-technical view is based on the notion that the actors involved in the project (project consortium members) interact and negotiate a solution that can be considered "satisfactory" in terms of the actors performance and the project outcomes, and that the activities conducted by the actors are not only affected by the technical properties of the project's subject, but also by the manner in which they act and interact, ultimately influencing their performance and the project outcomes [4]. Considering the external forces also exerting their influence, this perspective is more pragmatic and richer as it can provide a better understanding of the reality of a project, which is of great value for research and practice.

Several succinct definitions of complexity can be found in literature, but, due to the nature of the concept itself, they manage to capture its dimensions only partially. Thus, a number of studies are concerned, instead, with describing its characteristics as a way to grasp the complexity of a certain system [8]. Saurin and Sosa [8] describe a complex STS as having a particular set of characteristics which include: a large number of dynamically interacting elements, wide diversity of elements, unanticipated variability, and resilience.

The e-SHyIPS project is a Horizon Europe project that aims to define a pre-standardization plan for the update of the International Code of Safety for Ship Using Gases or other Low-flashpoint Fuels (IGF code) pertaining to hydrogen-based fuels passenger ships, and a roadmap for the boost of Hydrogen economy in the maritime sector. The scientific and technical contents addressed in the context of the project are divided into 4 categories called "experimental pillars". The project consortium is made up of 14 partners, each with a specific expertise that determines their role and participation in the project experimental pillars, which include: vessel designers, regulatory entities, fuel cells R&D centers, ferry and containership companies, computing centers, hydrogen suppliers, port operators, and engineering consultants. Several approaches and scientific research methodologies are used to gather relevant state of the art and to guide experimental activities, such as literature reviews, theoretical studies, surveys and interviews. The project also envisions the involvement of a wide list of external stakeholders in order to collect baseline information, best practices, expertise, feedback, and more.

These stakeholders include Advisory Board members and consortia of projects working in similar topics or "Cluster Projects". They are expected to be continuously involved in the project development and to provide constant and rigorous knowledge contributing to the activities of the consortium. Therefore, the interactions between all actors (external stakeholders and consortium partners) involve the flow of knowledge within and outside of the organization as well as the associated learning activities and processes. Table 1 gathers all elements of the organizational context and presents them as evidence to characterize the e-SHyIPS project as a complex STS as per Saurin and Sosa (2013) [8].

Table 1. Examples of evidence that characterizes the e-SHyIPS project as a complex STS.

Characteristics of complex STS [8]	Examples of evidence in the e-SHyIPS project
1. A large number of dynamically interacting elements	Number of consortium partners, number of Advisory Board members, number of Cluster Projects
2. Wide diversity of elements	Types of knowledge exchange, actors and stakeholders' types of expertise, types of learning processes
3. Unanticipated variability	Types of decisions taken by the consortium partners, doubts arising during decision-making, sources of uncertainty
4. Resilience	Adaptations of internal procedures on the basis of feedback received from external sources

3 Learning Ecosystems

Knowledge sharing and learning within the system are subjects of modern studies about complex STS due to their impact in the decision-making and the overall system output [6]. Furthermore, in recent years the learning environment has been approached as a system, with works highlighting the importance of mapping and understanding all complex relationships arising between the elements within [9]. The high complexity of modern learning setups calls for frameworks that can appropriately represent the dynamics between all involved stakeholders and the impact of external influences [10]. In this context, a LE is defined as consisting of stakeholders incorporating learning processes and learning utilities within specific environmental borders [11]. The individual behavior of stakeholders contributes positively or negatively to the success of the LE, while the relationships and interactions between them are related to the flow of information as well as the transfer and transformation of knowledge [12].

4 The e-SHyIPS Learning Ecosystem

The present work adopts the LE methodology to accurately represent the complexity identified in the e-SHyIPS project in order to address the challenges noted in Sect. 2 from a learning point of view, namely, the multiple knowledge flows and learning processes taking place between a variety of stakeholders. While the hydrogen maritime economy is an issue of direct application based off the project, the method described is intended to focus on areas where there is a high degree of uncertainty and multiple interconnected elements that are both known and unknown. The development of the e-SHyIPS LE is guided by 3 main principles: i) giving visibility to processes and outcomes, ii) encouraging the diversity of perspectives in the decision-making, and iii) having an efficient exchange of knowledge. A brief introduction to the e-SHyIPS LE is given in Sect. 4.1, while the theoretical foundation of the LE is laid out in Sect. 4.2. Finally, Sect. 4.3 describes the proposed methods to assess the effectiveness of the LE, this is, to assess the learning effectiveness.

4.1 Components

The e-SHyIPS LE (Fig. 1) is created with the involvement of the project consortium and stakeholders from maritime, technological, and hydrogen sectors. In the highly innovative and dynamic context of the project, this methodology will enable the ecosystem to react quickly to the changeability needs of hydrogen and fuel-cell fast developing technologies, as well as to bring together a diverse set of perspectives. In order to achieve the project objectives, the LE continuously monitors, controls, and pulls the knowledge gaps identification and resolution, integrates the results from experimental activities, and promotes the capture of new knowledge. Based on the contents described in Sect. 3, the components of the e-SHyIPS LE are as follows:

- *Learning stakeholders.* The 14 members of the consortium represent the learning stakeholders. Throughout the duration of the project they will engage with each other as well as external actors.
- *Learning contents.* The knowledge directly related to the 4 experimental pillars. This knowledge can be originated from within the ecosystem (internal sources) or from sources beyond the learning environmental borders (external sources).
- *Learning processes.* The activities facilitated through the use of specific tools for the acquisition and creation of knowledge and, as a consequence, the closure of knowledge gaps. There are 4 main learning processes taking place in the e-SHyIPS LE: i) learn by experimenting, ii) learn by searching, iii) learn by imitating, and iv) learn by interacting.
- *Learning environmental borders.* The boundaries of the ecosystem enclosing the learning stakeholders (i.e., the consortium members). The learning processes involving exclusively the learning stakeholders are referred to as endogenous learning, whereas the learning processes involving the learning stakeholders and interactions with external forces are referred to as exogenous learning.

Furthermore, the e-SHyIPS LE is affected by the following external forces:

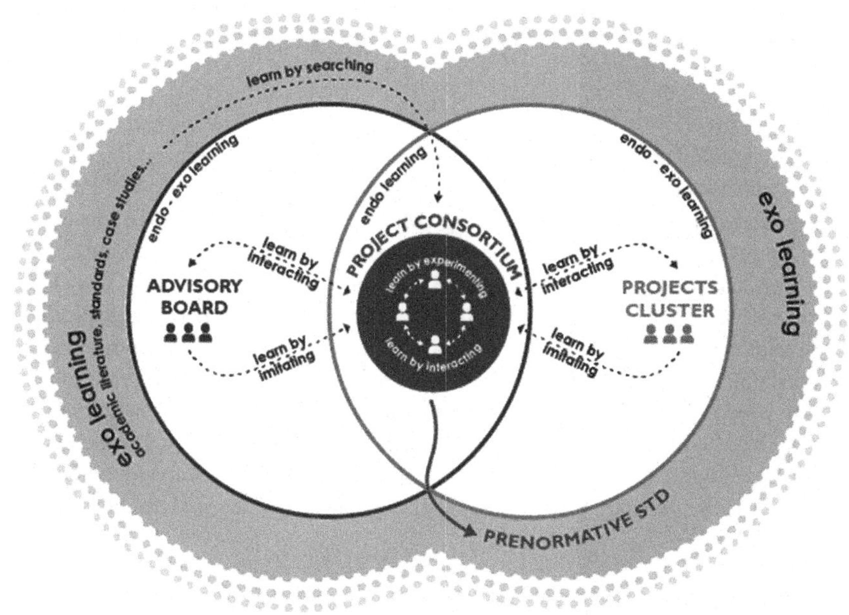

Fig. 1. The e-SHyIPS learning ecosystem

- *The Advisory Board and the Cluster Projects.* Actors with whom the learning stakeholders interact synchronously or asynchronously on a regular basis. They are considered as not fully external actors, since the consortium has a direct line of communication with them; but also not fully internal actors, since they are not part of the learning stakeholders, do not engage in their regular activities, and do not share the same objectives and attitudes.
- *External sources.* The external sources from which knowledge can be acquired. These include academic literature, standards and regulations currently in place, case studies, and the activities carried out by organizations of interest to the consortium (enterprises, similar projects' consortia, etc.) The feedback received from the European Commission about the project performance is also a valuable external source.

4.2 Learning Framework

The learning framework lays out the theoretical foundation of the LE as well as the learning tools developed on the basis of said theory. The interactions and relationships between the learning stakeholders are described in terms of knowledge flow and learning processes. The knowledge flow refers to the processes that the learning contents go through as the learning stakeholders make sense of it, while the learning processes refer to the activities that facilitate the flow of knowledge [13]. The developed tools are intended to support the learning stakeholders in the various learning processes. The main objective of the learning framework is to support the identification and closure of knowledge gaps.

Knowledge Flow

The knowledge flow refers to the processes that the learning contents go through as the learning stakeholders make sense of it. The processes involved in the flow of knowledge are encapsulated in the concept "Knowledge Management", which refers to the management of knowledge within an organization "by steering the strategy, structure, culture and systems and the capacities and attitudes of people with regard to their knowledge", ultimately achieving an organization's goals by making the factor knowledge productive [14]. Several models of Knowledge Management covering a wide spectrum of perspectives are abundantly described in literature. Therefore, there is some degree of diversity in the knowledge flows described by these models depending on the disciplinary context [15]. Those with relevance to context of the e-SHyIPS LE can be grouped into the following three main stages: i) Knowledge Gap Identification, ii) Knowledge Gap Resolution, and iii) Knowledge Capitalization. Each stage is made up of the phases shown and described in Fig. 2.

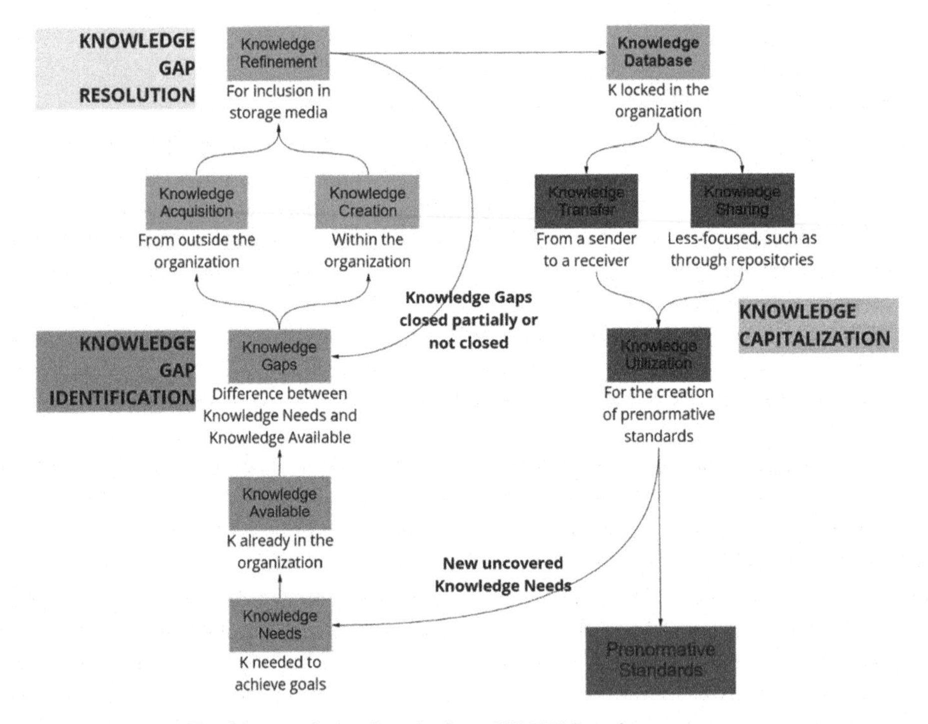

Fig. 2. Knowledge flow in the e-SHyIPS learning ecosystem

Learning Processes and Learning Tools

Once a knowledge gap is identified it can be bridged through the acquisition of existing knowledge, often from outside the organization, or the creation of new needed knowledge, often within the organization. In line with the boundaries described in Sect. 4.1, four learning processes, also referred to as "Learn by X", are put in place in order to acquire or create knowledge, each one with an associated tool and source of knowledge (Table 2) which are described as follows:

Table 2. Learning processes with their corresponding tool and source of knowledge

Learn by X	Learning tool	Source of knowledge
Learn by searching	Document library	Academic literature, state of practice and standards
Learn by imitating	Repository of observed actions	External actors, Cluster Projects, Advisory Board members
Learn by interacting	Repository of interactions	Consortium partners, Cluster Projects, Advisory Board members
Learn by experimenting	A3 Experiment cards	Experiments

1. *Learn by searching.* Exogenous learning process involving the comparison, integration, and synthetization of information from external sources [16]. To support the acquisition of knowledge from the academic literature, state of practice and standards, a document library has been created as the supporting learning tool. The documents are added by the consortium partners along with information meant to facilitate their consultation: general information, keywords, associated experimental pillars, and associated knowledge gaps.
2. *Learn by imitating.* Learning process involving the acquisition of knowledge through passive or active observation of an act followed by its imitation [17]. The dedicated learning tool is meant to ensure that all relevant observations are recorded in a single repository, storing those corresponding to external actors (exogenous learning) and Cluster Projects and Advisory Board members (exo/endogenous learning). These observations also include information about the associated experimental pillars and knowledge gaps.
3. *Learn by interacting.* Type of learning that relies on the experiences of the consortium partners and their interactions beyond the environmental boundary of the ecosystem [18]. Similarly to learn by imitating, the dedicated learning tool is meant to ensure that all relevant interactions are recorded in a single repository, primarily those with Cluster Projects and the Advisory Board (exo/endogenous learning). Internal interactions are already recorded during meetings. These interactions also include information about the associated experimental pillars and knowledge gaps.

4. *Learn by experimenting.* Learning process that involves the creation of new knowledge from the development of experiments within the consortium. In order to setup the experimental plans, the A3 problem-solving approach is used to document, communicate, and transfer the created knowledge [19]. The dedicated tool is called the "A3 experiment card", which has been adapted for the project and also includes information about the associated experimental pillars and knowledge gaps.

4.3 Assessing the Effectiveness of the Learning Ecosystem

Being able to assess the effectiveness of the overall LE is as important as laying the theoretical foundation and developing the corresponding learning tools. The attributes through which the effectiveness of the ecosystem can be measured are knowledge and learning; however, these are subjective and intangible concepts whose measurement is not simple as it depends on several contextual factors arising from the characteristics of the LE. The measurement of the effectiveness of knowledge management and learning are current topics of research focused not only on the LEs found in the educational system, but also in those in the scope of organizational learning.

In the context of the e-SHyIPS LE, the individual knowledge gap is selected as the unit of analysis to be used in the assessment of its effectiveness. Therefore, the effectiveness of the ecosystem will be determined primarily by the number of knowledge gaps that are successfully closed within a certain time period. Thus, effectiveness is defined as the degree to which something contributes towards the closure of knowledge gaps. Considering this, there are two manners in which the effectiveness of the LE can be measured: i) the performance of the learning tools, and ii) the feedback from the learning stakeholders making use of the learning tools.

Learning Tools Performance

The performance of the learning tools is measured through their contribution towards the resolution of knowledge gaps. Thus, a set of KPIs both absolute and relative were defined for each of the learning tools. These KPIs will be tracked throughout the development of the experimental pillars. A sample of the KPIs corresponding to the tool "Repository of observed actions" corresponding to the process "Learn by Imitating" is shown in Table 3 below. The KPIs will be measured and updated regularly, with eventual adjustments to better track the performance of the learning tools.

Table 3. KPIs to measure the effectiveness of the Learn by Imitating tool

KPIs	Description
Absolute	• Total number of organizations identified as Cluster Projects • Number of members belonging to each sector in the Advisory Board
Relative	• Number of organizations contacted/Total number of organizations in the cluster • Number of organizations contacted that contribute towards the closing of knowledge gaps/Total number of organizations in the cluster

Feedback from Learning Stakeholders

The effectiveness of the learning tools will also be measured as they are perceived by the learning stakeholders i.e., measuring the quality of the experience of learning stakeholders as they use the tools for the closing of knowledge gaps. The proposed media to collect the feedback are surveys, questionnaires and/or interviews scheduled every number of months.

- Surveys. The surveys aim to assess the following features:
  - Contents. Learning stakeholders' perception of the contents offered and whether it facilitates its expected purpose.
  - Ease of use. The difficulty of using the tools by the learning stakeholders in order to fulfill its purpose.
  - Availability. The guaranteed access to the tools at any time.
- Interviews. The interviews are meant to have immediate access to the experience of learning stakeholders as they are using the tools. The interviews also have the purpose of enabling the tracking of the status and evolution of knowledge gaps.

5 Conclusion and Next Steps

The objective of this work was to capitalize on complexity, learning, and knowledge management theories to develop a methodology to enable an efficient exchange of knowledge. The e-SHyIPS LE, thus, not only supports the project consortium to work towards achieving the objectives of the project, but also the overall objective of R&I projects, such as those funded by the European Union, to promote learning and to advance the scientific and technological community through the creation and sharing of knowledge. This methodology represents a pivotal work to create solid learning environments to guarantee achieving the project outcomes despite the inherent complexity and challenges presented by these type of R&D projects.

As for the next steps, the LE is planned to be deployed in the context of the project and real data is expected to be collected according to Sects. 4.2 (Learning Framework) and 4.3 (Assessment of the LE effectiveness). General findings will be exploited to implement the LE in R&I projects dealing with similar complexity or similar complex STS.

Acknowledgement. This project has received funding from the Clean Hydrogen Partnership under grant agreement No 101007226. This Joint Undertaking receives support from the European Union's Horizon 2020 Research and Innovation program, Hydrogen Europe and Hydrogen Europe Research.

References

1. Elmaraghy, W., Elmaraghy, H., Tomiyama, T., Monostori, L.: Complexity in engineering design and manufacturing. CIRP Ann. Manuf. Technol. **61**(2), 793–814 (2012). https://doi.org/10.1016/j.cirp.2012.05.001

2. Sheard, S.A., Mostashari, A.: A complexity typology for systems engineering. In: 20th Annual International Symposium of the International Council on Systems Engineering, INCOSE 2010, vol. 2 (2010). https://doi.org/10.1002/j.2334-5837.2010.tb01115.x
3. Sheard, S.A.: Systems engineering complexity in context. In: 23rd Annual International Symposium of the International Council on Systems Engineering, INCOSE 2013, vol. 2 (2013). https://doi.org/10.1002/j.2334-5837.2013.tb03077.x
4. Hassannezhad, M., Cantamessa, M., Montagna, F., Clarkson, P.J.: Managing sociotechnical complexity in engineering design projects. J. Mech. Des. Trans. ASME **141**(8) (2019). https://doi.org/10.1115/1.4042614
5. Kiridena, S., Sense, A.: Profiling project complexity: insights from complexity science and project management literature. Proj. Manag. J. **47**(6), 56–74 (2016). https://doi.org/10.1177/875697281604700605
6. Walker, G.H., Stanton, N.A., Salmon, P.M., Jenkins, D.P.: A review of sociotechnical systems theory: a classic concept for new command and control paradigms. Theor. Issues Ergon. Sci. **9**(6), 479–499 (2008). https://doi.org/10.1080/14639220701635470
7. Oosthuizen, R., Pretorius, L.: Establishing a methodology to develop complex sociotechnical systems (2013). https://doi.org/10.1109/ICIT.2013.6505890
8. Saurin, T.A., Gonzalez, S.S.: Assessing the compatibility of the management of standardized procedures with the complexity of a sociotechnical system: case study of a control room in an oil refinery. Appl. Ergon. **44**(5), 811–823 (2013). https://doi.org/10.1016/j.apergo.2013.02.003
9. Goodyear, P., Carvalho, L.: The analysis of complex learning environments. In: Rethinking Pedagogy for a Digital Age: Designing for 21st Century Learning (2013)
10. Gütl, C., Chang, V.: Ecosystem-based theoretical models for learning in environments of the 21st century. Int. J. Emerg. Technol. Learn. **3**(1) (2008). https://doi.org/10.3991/ijet.v3i1.742
11. Chang, V., Guetl, C.: E-learning ecosystem (ELES) - a holistic approach for the development of more effective learning environment for small-and-medium sized enterprises (SMEs) (2007). https://doi.org/10.1109/DEST.2007.372010
12. Giannakos, M.N., Krogstie, J., Aalberg, T.: Toward a learning ecosystem to support flipped classroom: a conceptual framework and early results. Lect. Notes Educ. Technol., 9789812878663 (2016).: https://doi.org/10.1007/978-981-287-868-7_12
13. Easterby-Smith, M., Lyles, M.: The Blackwell handbook of organizational learning and knowledge management. Oxford (2003)
14. Beijerse, R.P.U.: Knowledge management in small and medium-sized companies: knowledge management for entrepreneurs. J. Knowl. Manag. **4**(2), 162–179 (2000). https://doi.org/10.1108/13673270010372297
15. Mcadam, R., Mccreedy, S.: A critical review of knowledge management models. Learn. Organ. **6**(3), 91–101 (1999). https://doi.org/10.1108/09696479910270416
16. Kammerer, Y., Brand-Gruwel, S., Jarodzka, H.: The future of learning by searching the web: mobile, social, and multimodal. Front. Learn. Res. **6**(2), 81–91 (2018). https://doi.org/10.14786/flr.v6i2.343
17. Albino, V., Petruzzelli, A.M., Rotolo, D.: Exploring and exploiting through external sources: the effect of learning and technological proximity. Int. J. Innov. Learn. **8**(1), 1–11 (2010). https://doi.org/10.1504/IJIL.2010.034011
18. Bresman, H., Zellmer-Bruhn, M.: The structural context of team learning: effects of organizational and team structure on internal and external learning. Organ. Sci. **24**(4), 1120–1139 (2013). https://doi.org/10.1287/orsc.1120.0783
19. Morgan, J.M., Liker, J.K.: The toyota product development system: integrating people, process and technology. J. Prod. Innov. Manag. **24**(3), 276–278 (2006)

Manufacturing: Sustainable Manufacturing, Lean Manufacturing, Models for Manufacturing

Real-Time Allocation of Volatile Energy Related Emissions in Manufacturing

Dominik Leherbauer(✉) and Peter Hehenberger

Smart Mechatronics Engineering, University of Applied Sciences Upper Austria,
Stelzhamerstraße 23, 4600 Wels, Austria
{dominik.leherbauer,peter.hehenberger}@fh-wels.at

Abstract. International agreements target a reduction in greenhouse gas emissions. A major contributor to these greenhouse gas emissions is the generation and consumption of energy. By a varying supply and demand of different energy sources including renewables a varying energy mix results. A difficulty poses the determination of CO_2 equivalent emissions for volatile energy types, because different energy sources have type-specific emission amounts. Within the manufacturing environment, the challenge is to allocate the resulting energy flows, respectively emission flows through the existing hierarchical structure.

State of the art provides methods for the calculation of embodied energy. Life cycle assessment methods can be used to determine environmental impact, but are carried out mostly as static analysis. Within this publication an approach to determine the embodied emissions with the consideration of volatile CO_2 emissions is presented. The goal of the approach is to provide a path to map the resulting CO_2 equivalent emissions to produced goods in a manufacturing context.

Used methods include the analysis of existing methods for energy allocation in products and life cycle assessment methods. An analysis of the electricity grid has been conducted and a mathematical model for the calculation of inherent CO_2 equivalent emissions has been formulated. The paper provides a conceptual approach to map volatile equivalent CO_2 emissions to produced goods and can be used to minimize embodied energy in further applications.

Keywords: Embodied energy · Volatile emission allocation · Dynamic life cycle assessment · Sustainable manufacturing · CO_2 emissions

1 Introduction

International agreements e.g., the Paris Agreement target a reduction in emitted greenhouse gases [15]. Considering greenhouse gases, a distinction can be made between CO_2 and other gases. However, the indication of CO_2 equivalency converts the resulting effects of other gases to those of CO_2 [6]. Considering the share of energy consumed within the European Union, data shows approximately 26

F. Noël et al. (Eds.): PLM 2022, IFIP AICT 667, pp. 599–608, 2023.
https://doi.org/10.1007/978-3-031-25182-5_58

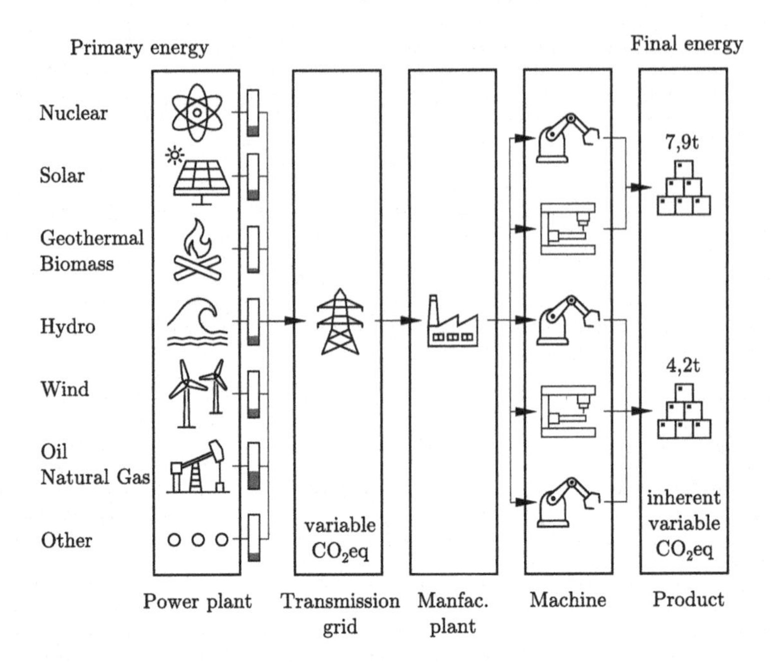

Fig. 1. Source to sink sequence for the mapping of volatile CO_2 equivalent emissions

percent of consumed final energy can be allocated in the manufacturing sector in 2020 [5].

A difficulty herein pose the temporal changes of composed energy sources e.g., electricity. While energy sources like oil or natural gas can be stored and vary only by small amounts in terms of inherent CO_2 equivalent emissions, through its non-storability the mixture of electricity in the electricity grid cannot be determined statically. By the usage of sustainable electricity sources as photovoltaic panels or wind, which vary in availability, a volatility in the electricity mix is introduced. This volatility is mainly caused by the condition, that the power fed in the energy grid must also be withdrawn at the same time. Missing energy must then be provided from control reserves or other power plants. However, these power plants may be less sustainable in terms of associated CO_2 equivalent. This publication aims to model this behavior by mapping the resulting CO_2 equivalent emissions to produced goods in the manufacturing context as visualized in Fig. 1. The publication highlights the need for a systematic approach to breakdown emissions and assess intense products and processes. A proposal is derived from the current state-of-the-art and shows a conceptual approach to carry out the respective mapping.

2 State of the Art

For the complete flow visualized in Fig. 1 multiple components are needed. For the individual electricity sources the variable CO_2 equivalent within the transmission network must be obtained. In the manufacturing environment an allocation algorithm is sought to distribute the resulting emissions. Though, no overall methodology is known, existing references indicates methodical approaches. Thus, in this section these existing approaches are presented.

2.1 Life Cycle Inventory Approach

A widespread calculation approach for the environmental impact of manufacturing goods is the life cycle assessment (LCA) approach. Following the definition of ISO14044:2006 LCA represents a 'compilation and evaluation of the inputs, outputs and the potential environmental impacts of a product system throughout its life cycle' [8]. As the name suggests LCA is applicable for the whole product life cycle and therefore, deals with unknown quality of input data. The gathering of the most applicable input data is covered by the life cycle inventory. In this approach within a manufacturing context, each individual manufacturing process is seen as an enclosed system. Each system features inputs for material and energy and outputs for resulting materials, emissions and other outcomes. For the usage of external resources LCA databases provide an indication of ecologic consequences per consumed unit. Though, a region specification should be taken into account [11], the provided values are static and averaged. Hence, inconsistencies can arise in case of a variable resource consumption for resources with a volatile composition. In general LCA is carried out as a static analysis and features known limitations [9]. An alternative would be a dynamic analysis i.e. taking into account the temporal behavior. In particular, a dynamic analysis should be considered if source and sink both feature a variable behavior. Within a manufacturing energy provision context, this may be caused by

- On-site energy generation
- Shift working times
- Variable energy consumption of products
- Variable CO_2 equivalent of energy sources

In this case a dynamic LCA (DLCA) as proposed by Collinge et al. may be used to better account for these characteristics [2]. In the given context considering emissions and the manufacturing phase the existing LCA respectively DLCA approaches can be considered as too exhausting. Instead, the proposed approach aims to extend the methodology of the DLCA for the manufacturing phase.

2.2 Electricity Generation and Characteristics

Considering the supply with electricity throughout the day, the CO_2 equivalent emissions vary. This is mainly caused by a varying composition of electricity generation as well as a continuous energy exchange between control zones. These can be distinguished by the representation of energy exchanges and are either based on trades or physical flow [7]. A distinction can be made between financial and ecologic costs. While plants aim to maximize financial profits, the ecologic costs result. However, due a correlation of these costs is implied by the fundamental characteristics of different generation types. While environmentally-dependant energy generation plants, e.g. photovoltaic panels or wind power plants create a low amount of operational costs, they are active most of the time. Commodity-dependant power plants aim to maximize profits based on the market price and operational costs and thus act differently to maximize profits. These aspects are covered by merit order models, which could be validated in various publications [10,13]. Likewise, the ecologic costs consist of initial costs embodied in infrastructure and running costs bound in operations [14]. The provision of data for these costs is covered by life cycle assessment (LCA) calculations as well as calculations by the Intergovernmental Panel on Climate Change (IPCC). A difference in these calculations exists in the granularity. While the IPCC provides calculation values per generation type, the LCA calculations can also be country specific.

A difficulty in the calculation of inherent emissions in the electricity grid is the network architecture allowing exchanges between the individual control areas. By the exchange of electricity between these areas, the emissions are also exchanged, surfacing a mathematical problem to solve. Hence, a differentiation must be made between occurring emissions by generation and consumption. To calculate the specific CO_2 equivalent emissions for the electricity grid network the calculation is based on the absolute released emissions. The mathematical relationship is trivial given the specific CO_2 equivalent emissions c, the amount of energy E and absolute released emissions C

$$C = c \cdot E \tag{1}$$

To calculate the absolute released emissions per control area by consumption, several steps are needed. In a first step, the released emissions per generation type must be calculated. The generated energy data per type and specific emissions for each generation type are needed as a basis. Thus, the control area specific emissions can be calculated by the sum of Eq. 1 for each generation type. To solve for emissions by consumption a linear set of equations is sought. Exemplary for control area A the absolute emissions by consumption $C_{A_{con}}$ result in consideration of electricity imports $C_{A_{imp}}$ and exports $C_{A_{exp}}$ with:

$$C_{A_{con}} = \sum C_{A_{gen}} + \sum C_{A_{imp}} - \sum C_{A_{exp}} \tag{2}$$

Expanding Eq. 2 by Eq. 1 and rewriting imports and exports per control area result in a new equation. Notably, control areas can simultaneously import and export electricity with each neighboring control zone. Declaring energy imports from B to A as E_{BA} and imports from C to A as E_{CA} yields:

$$\left(E_{A_{con}} + \sum E_{A_{exp}}\right) c_{A_{con}} = E_{A_{gen}} \cdot c_{A_{gen}} + E_{BA} \cdot c_{B_{con}} + E_{CA} \cdots \tag{3}$$

With the expansion and known parameters, a set of equations emerges and enables solving for the control area specific CO_2 emissions. The calculation of

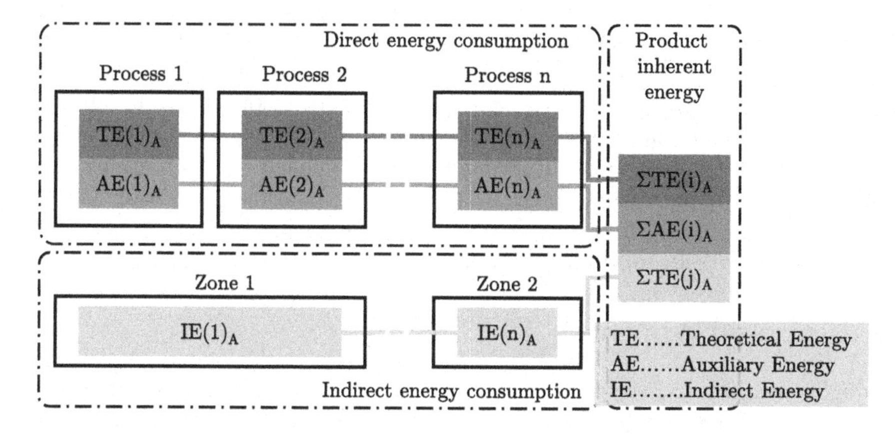

Fig. 2. Energy allocation in products by Rahimifard et al. [12]

the unknown specific CO_2 equivalent intensities c can be carried out by setting up Eq. 3 for each considered control area network node. By writing the equation set in matrix form and solving by matrix inversion the specific intensities can be obtained.

A limitation of this approach is the mathematical complexity and the data availability. Solving for the whole electricity grid network can be computationally expensive, in particular if c is only sought for a specific control area. An approximation is possible in this case by the assumption, that distant energy exchanges impact the considered control area less by more separating control areas. Hence, the approximation can be made by cutting off the calculation after a defined degree of neighboring control areas. By reaching the cutoff control area all imports are simply neglected, and the calculation complexity can be reduced. In practice, the integration of data sources can pose a problem. In case of unavailable data, the specific CO_2 equivalent can only be approximated by substitution of imported emissions with static LCA data.

2.3 Energy Allocation

The allocation of needed energy in goods originates from the economic input-output analysis as well as process analysis approaches. From these approaches the hybrid LCA approaches emerged, which combines these analysis methods as presented by Crawford et al. [4]. Based on the first documented hybrid analysis approach by Bullard et al. [1] the concept of embodied energy was introduced by Costanza [3] within an economic context. Since then, multiple approaches for specific use cases and product life cycle phases emerged. Rahimifard et al. [12] used the originally provided approach to allocate energy within a manufacturing environment. The framework presented by Rahimifard et al. aims to provide an indication for energy efficiency applications and is visualized in Fig. 2.

As displayed, the energy consumption is split into a part of direct and indirect energy consumption. The direct energy consumption results by manufacturing processes needed to manufacture a product. A distinction for energy efficiency related purposes can be made between the physical minimum to carry out the process, namely theoretical energy and all other direct consumption caused by inefficiencies or auxiliary consumption, namely auxiliary energy. The indirect energy consumption covers all other consumption needed by the manufacturing environment e.g., lighting. Caused by the missing causal relationship between fluctuations in indirect and direct consumption Rahimifard et al. suggested to average indirect consumption per hour.

3 Proposed Approach

Though, the discussed state-of-the-art approaches are complete in itself, integration poses several difficulties to consider. A core problem is the integration of the time intervals of the electricity with the manufacturing systems sensing. Hence, the occurring manufacturing environments' properties must be considered.

3.1 Hierarchical Classification

The manufacturing environment can be represented with a hierarchical model. In the hierarchy, multiple levels can be considered, depending on sensing equipment. The existing energy allocation approaches indicate the usage the need for manufacturing processes, zones, energy consumers, and products as pointed out in Fig. 2. Additional references for the hierarchy can be found in ISA-88 and IEC61512-1:2000 which, besides other, both mention manufacturing site, area and process cells. The hierarchical levels by Wiendahl et al. indicate, the manufacturing processes are the basement of the levels within a changeable factory [16]. An assignment of processes to jobs thus provides a transparent path from the manufacturing site level to individual manufacturing jobs. Extending the mentioned levels by a buffer for the averaging of indirect consumption the following hierarchy presented in Table 1 can be derived.

Table 1. Derived hierarchical levels for a multi-level allocation in a manufacturing environment

Name	Function	Example
Manufactory	Entry point of overall sensed energy times current CO_2 equivalent emissions	Power grid connected plant
Area	Distribution of indirect energy consumption and unassigned energy flows	Machining area
Energy consumer	Actual energy consuming entity with sensing capabilities	Machine
Indirect buffer	Averaging and allocation of indirect flows	Losses
Process	Manufacturing process caused by a product	Milling
Job	Container for one or more produced products	12 pcs of part no. 5

3.2 Energy Sensing

A requirement for the real-time allocation of emissions within the proposed hierarchy is an appropriate energy sensing system. To allocate the emissions the levels manufactory, area and energy consumers must feature a measurement system. The sampling rate of measurement must be adapted to the electricity grid and the processes. While the direct embodied energy relies on the difference of consumed energy at process start and end, the relative CO_2 equivalent by electricity composition changes in 15 min intervals. Hence, a requirement for the integration is a sampling rate of energy consumption sensing, which enables the mapping of CO_2 equivalent emissions in relation to the actual manufacturing process and changing electricity mix.

Another difficulty is the mapping of energy losses. These can occur by a multitude of reasons and are not explainable in every case. To obtain meaningful results in terms of emission allocation a representable coverage of consumption is required. The coverage of measurements can be determined by an energy transparency key performance indicator. Energy transparency γ in this context is defined as the total consumption of all energy consuming entities per the according sensing zones' consumption. In case of no untraceable losses, this factor γ would be 1. A factor $\gamma > 1$ implies bad matching of sampling rates or an erroneous hierarchical classification and requires adaption. A factor $\gamma < 1$ implies plausibility of the sensed values, should however, be ideally be close to 1. As a consequence of losses, an unassignable share also results. As a difference to the sensed energy consumption, the unassignable share can only be determined by calculation and shall be carried out for every sensed layer.

3.3 Allocation of Emissions

To integrate energy allocation and real-time emissions, the information output must be defined. Following the definition of Rahimifard et al. [12] the share of direct energy consumption is split into theoretical energy and auxiliary energy consumption. If the theoretical energy efficiency provides no viable information, this split can be neglected and only the measured direct energy share considered. The determination of theoretical and auxiliary energy depends also on the available information of the consuming entity. Considering a large scale manufacturing system, a substantial modeling effort would be necessary to obtain the classification for every consumer. By neglecting this split, the respective energy sensing can be the relevant information provider.

Combining the discussed indications, a multi-level mapping can be derived. The calculation of emissions can be carried out based on the discussed approaches. Combining the energy measurement system with the derived specific CO_2 equivalent intensities, the overall accountable emission can be calculated by Eq. 1. In order to distribute these emissions, the local zone-wise metering must be taken into account and an emission flow results.

Considering the allocation of direct and indirect energy consumption, there is a difference in the difficulty of the allocation. By means direct energy consumption is caused by a manufacturing process and the emission mapping to

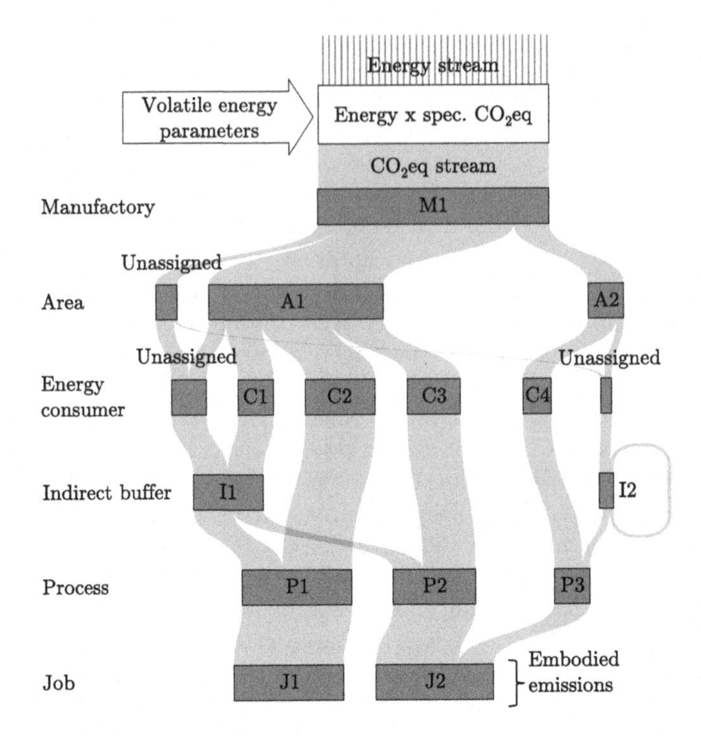

Fig. 3. Real-time emission flow from the manufacturing environment to products

the process is implied. The allocation of indirect emissions lacks the implication, thus an allocation strategy must be defined. Fundamentally, two allocation strategies are possible to split the unassignable emissions: (1) by share of allocable energy or (2) by share of other metrics. Other metrics can be chosen if they provide a better description to split the emissions. An example would be the allocation of other media consumption by tracing, like pressurized air or the split based on process times. In explanation the longer a process takes, the more indirect emissions from e.g., lighting shall be allocated. Otherwise, short-term processes with high energy respectively emission intensity would wrongly get penalized. Thus, analogous to LCA, the availability of additional data can significantly increase the validity of results.

Between the energy consumer and process level, an indirect buffer is located. If the sampling rates between the sensing equipment on the individual level is not in sync, discrepancies in the indirect consumption may happen. By averaging and buffering the indirect consumption, these effects can be mitigated. Visualizing the emission flows as Sankey diagram, the buffered emissions thus can be represented a loop The resulting approach to allocate the resulting emissions of the electricity grid is visualized in Fig. 3.

4 Discussion

The proposed conceptual approach enables the real-time allocation of emissions into manufacturing jobs respectively products. Especially in the context of LCA benefits in accuracy can result by consideration of real conditions. A limitation of the approach could be given by the need for smart-metered zone consumption. The availability of these data would however also increase the energy transparency and provide further value within data-driven manufacturing environments. Another synergetic relationship with cyber-physical production systems would be the availability of high quality process data needed for the indirect allocation. This would result in a wider range of applicable indicators, but would simultaneously introduce another problem. With a rising amount of available indicators, the possibilities of incomprehensible selections also rise. To maintain comprehensibility, a policy is needed defining the selection of indirect indicators.

Another possible application of the proposed approach are the usage as key performance indicator for embodied emissions of products. By also considering the indirect emissions and external parameters it aims to be a holistic indicator for environmental impacts. Thus, further usage in optimization problems would also be possible. However, a known problem in such applications are the needed predictions of the electricity markets composition.

The presented approach is conceptual, but built upon the intensively researched energy allocation. Additional validation is needed for industrial applications with real sensors and the matching of needed sensor sampling rates. The research contribution of this publication is the application of allocation methods in a temporal context.

5 Conclusion

In this paper, an approach for the dynamic allocation of CO_2 equivalent emissions in products was presented. A literature research revealed there was no consistent approach to allocate the emissions of volatile energy sources in products. Therefore, several components were researched including (1) state-of-the-art approaches to tackle this problem, (2) the real-time calculation of specific emissions of electricity and (3) energy allocation. A proposal was made to integrate these components into a real-time allocation of resulting CO_2 equivalent emissions. The proposed approach allows breaking down emissions to products and processes.

Potential for further research can be found in practical applications of approach. Possible applications are optimization problems to minimize embodied CO_2 equivalent emissions. The real-time allocation approach enables optimization on a global level by considering also implications on indirect energy consumers and energy sources' conditions. Further usages of the proposed approach can be made within a practical case study for validation. A major challenge in a practical application is the availability of data from electricity grid side and the hierarchical sensing data.

Acknowledgements. The research has been co-financed by the European Union's Horizon 2020 Program under Grant Agreement No. 958478 entitled 'ENERgy-efficient manufacturing system MANagement'.

References

1. Bullard, C.W., Herendeen, R.A.: The energy cost of goods and services. Energy Policy **3**(4), 268–278 (1975)
2. Collinge, W.O., Landis, A.E., Jones, A.K., Schaefer, L.A., Bilec, M.M.: Dynamic life cycle assessment: framework and application to an institutional building. Int. J. Life Cycle Assess. **18**(3), 538–552 (2013)
3. Costanza, R.: Embodied energy and economic valuation. Science **210**(4475), 1219–1224 (1980)
4. Crawford, R.H., Bontinck, P.A., Stephan, A., Wiedmann, T., Yu, M.: Hybrid life cycle inventory methods – a review. J. Clean. Prod. **172**, 1273–1288 (2018). https://doi.org/10.1016/j.jclepro.2017.10.176
5. Eurostat: Database - Energy - Eurostat. https://ec.europa.eu/eurostat/web/energy/data/database
6. Eurostat: Glossary: Carbon dioxide equivalent. https://ec.europa.eu/eurostat/statistics-explained/index.php?title=Glossary:Carbon_dioxide_equivalent
7. Fan, J.L., Hou, Y.B., Wang, Q., Wang, C., Wei, Y.M.: Exploring the characteristics of production-based and consumption-based carbon emissions of major economies: a multiple-dimension comparison. Appl. Energy **184**, 790–799 (2016)
8. International Organization for Standardization: Environmental Management – Life Cycle Assessment Requirements and Guidelines (ISO 14044:2006) (2006)
9. Lueddeckens, S., Saling, P., Guenther, E.: Temporal issues in life cycle assessment—a systematic review. Int. J. Life Cycle Assess. **25**(8), 1385–1401 (2020). https://doi.org/10.1007/s11367-020-01757-1
10. McConnell, D., et al.: Retrospective modeling of the merit-order effect on wholesale electricity prices from distributed photovoltaic generation in the Australian National Electricity Market. Energy Policy **58**, 17–27 (2013)
11. Patouillard, L., Bulle, C., Querleu, C., Maxime, D., Osset, P., Margni, M.: Critical review and practical recommendations to integrate the spatial dimension into life cycle assessment. J. Clean. Prod. **177**, 398–412 (2018). https://doi.org/10.1016/j.jclepro.2017.12.192
12. Rahimifard, S., Seow, Y., Childs, T.: Minimising embodied product energy to support energy efficient manufacturing. CIRP Ann. **59**(1), 25–28 (2010)
13. Rintamäki, T., Siddiqui, A.S., Salo, A.: Does renewable energy generation decrease the volatility of electricity prices? an analysis of Denmark and Germany. Energy Econ. **62**, 270–282 (2017). https://doi.org/10.1016/j.eneco.2016.12.019
14. Tranberg, B., Corradi, O., Lajoie, B., Gibon, T., Staffell, I., Andresen, G.B.: Real-time carbon accounting method for the European electricity markets. Energy Strat. Rev. **26**, 100367 (2019)
15. United Nations Framework Convention on Climate Change: The Paris Agreement—UNFCCC. https://unfccc.int/process-and-meetings/the-paris-agreement/the-paris-agreement
16. Wiendahl, H.P., et al.: Changeable manufacturing - classification, design and operation. CIRP Ann. **56**(2), 783–809 (2007)

Development of an Industry 4.0-oriented Tool Supporting Circular Manufacturing: A Systematic Literature Review

Marco Spaltini(✉), Federica Acerbi, and Marco Taisch

Department of Management, Economics and Industrial Engineering, Politecnico di Milano, via Lambruschini 4/B, 20156 Milan, Italy
{Marco.spaltini,federica.acerbi,marco.taisch}@polimi.it

Abstract. The traditional linear approach to economy has been recognized as not sustainable any longer. Among the different approaches identified to reverse the current socioeconomic system, Circular Economy (CE) appears a promising field. This approach is based on 3 pillars: preservation and enhancement of natural capital, optimization of resource yields components and materials and fostering system effectiveness. Together with CE, also the concept of Industry 4.0 (I4.0) is gaining traction among manufacturers, academia and policy makers. This manuscript merges these two concepts through a literature review. More specifically, this research has been focused on clarifying how manufacturers can be supported in understanding how to adopt digital technologies to re-design their processes under circular-driven perspectives. Indeed, from the literature emerged that I4.0 might benefit circularity both in terms of modalities to deliver value and rationalization of resources. However, case studies suggest that applications in this sense are still scarce and mostly oriented to service-based and analytics-related solutions. Three main gaps have been identified: lack of a structured and practical model to merge I4.0 and CE that consider the variables characterizing real manufacturing firms, lack of focus on multi-product and multi-asset when assessing the link between I4.0 and CE and lack of a quantitative evaluation of the impact of the application of I4.0 technologies and of the benefits, or pains, to support the transition toward CE. Hence, the objective of this paper is to set the basis for a tool able to support manufacturing firms in defining Circular Business practices enabled by Industry 4.0 technologies.

Keywords: Industry 4.0 · Best practices · Circular economy

1 Introduction

The call for overcoming the current take-make-dispose paradigm has now become a global imperative for Governments, Business and Academia [1]. Among the paths investigated in literature to achieve an environmentally sustainable sociotechnical system, Circular Economy (CE) appears as a promising solution, especially for manufacturing

Published by Springer Nature Switzerland AG 2023
F. Noël et al. (Eds.): PLM 2022, IFIP AICT 667, pp. 609–619, 2023.
https://doi.org/10.1007/978-3-031-25182-5_59

companies which can rely of several strategies starting from product design, circular industrial processes and circular networks within their supply chains (SCs) or districts [2]. In addition to CE, another relevant trend for both academia and practitioners is represented by Industry 4.0 (I4.0) paradigm. Despite I4.0 surely benefits firms by improving economic performances for plants and SCs [3], a structured and practical model enabling to merge I4.0 and CE that consider the variables characterizing real manufacturing firms is still missing [4]. Indeed, the implementation of I4.0 technologies might represents a key enabling factor for new intrinsically more circular Business Models (BMs) [5] and thus these technologies might be used to overcome the barriers highlighted by [6] such as: Difficulties in disassembly activities; SC complexity; Lack of information causing coordination problem, both internal and among stakeholders; Lack of CE culture in designing and production processes; Quality concerns about the material used; Lack of capital and high start-up cost threatening economic sustainability; Limited support from governments. Therefore, this research aims to investigate how the relationship between I4.0 and CE might be virtuous and how to practically link the two concepts together with no harm for economic profitability. Hence, the following Research Question (RQ) has been formulated: "How Industry 4.0 technologies could support manufacturing companies to embrace CE overcoming the challenges experienced?" This interpolation between CE and I4.0 is expected to be the key step to develop a framework able to support manufacturing companies in undertaking a digital transformation process oriented to reach their environmental sustainability objectives. The contribution is structured as follow. Section 2 describes the research methodology employed to set up the framework. Section 3 reports the results from the analysis of the state of the art about CE and I4.0 related studies. Section 4 discusses the results proposing a framework for the simultaneous adoption and Sect. 5 concludes the contribution by highlighting the main outcomes and the limitations leading to further research opportunities.

2 Methodology

To understand the relationship between I4.0 and CE, the authors developed a systematic literature review enriched by a search for industrial cases. In addition other relevant documents were collected according to the snowball effect [7, 8]. The search was performed on Scopus database. Three relevant keywords were selected: "Industry 4.0", "Circular" and "Manufacturing" followed by an analysis of the most adopted synonyms. "Circular Economy" has not been selected since the term economy is often substituted by other more focused word (e.g. Circular Manufacturing). Hence, the string chosen resulted in ("Industry 4.0" OR "Fourth Industrial Revolution" OR "4th Industrial Revolution") AND ("Circular" OR "Closed loop" OR "Cradle to Cradle") AND "Manufacturing". The filters chosen regarded: language, year of publication, subject area of reference and stage of publication [9]. Only English documents have included. The time horizon ranged from 2011 up to 2021. The selection of 2011 is motivated by the relative novelty of the concept of Industry 4.0 [10]. Concerning the subjects, the authors selected only those related to manufacturing and technology and business. The Literature Search resulted in 96 documents and 25 of them were accepted for the final review process. The sample of papers has been reviewed inspired by the classification done by [11] refining it through an iterative process.

3 Literature Review Results

In the extant literature, it has been highlighted that due to the higher relative percentage of activities manually performed and the averagely lower focus on sustainability issues, the benefits provided by I4.0 could have a greater impact for SMEs thus amplifying the potential competitive advantage in terms of costs, access to new markets and environmental performances [12]. As Shown in Table 2, except for Additive Manufacturing (AM), the greater contribution to the achievement of circularity in the manufacturing field is given by IT-related solutions. There are 3 main areas of application: Virtualization of processes (e.g. simulation) [13], optimization and resource reduction (e.g. AI-based decision making) [14]and Virtualization of resources (e.g. cloud solutions) [15] (Table 1).

Table 1. Number of papers linking I4.0 technology to a specific Circular approach

I4.0 technology	Rethink	Reduce	Repair	Remanufacture	Repurpose	Recycle	**Total**
CPS		4	1	3		1	**9**
IoT	1	4	2	3		3	**13**
Big data analytics	2	2		3		3	**10**
Cloud computing	3	4		2	1	1	**11**
Cloud manufacturing	1	3		2			**6**
AM	1	4	3	3	1	4	**16**
Cobotics				2			**2**
AR/VR	1	1		2			**4**
AI/ML		4		4		3	**11**
Advanced automation				1			**1**
Blockchain	3	4		1			**8**
Total	**12**	**30**	**6**	**26**	**2**	**15**	

As far as the findings enabled to highlight, Internet of Things (IoT) is considered, together with AM, the most mentioned technology able to support CE. In fact, the application of sensors to products and assets are essential to enable their monitoring in real-time and consequently condition-based or predictive maintenance [16]. According to [17], IoT technology is a key element for the development of new BMs like sharing mobility and so to increase the overall usage of assets along their lifecycle. Moreover, data collected from IoT devices, elaborated through Big Data Analytics models are key to support firms to define and optimise their sustainability performances [18]. Similarly,

Cloud-based solutions represent a viable element, on the one hand, to reduce the virtualise computing and production resources [19] and consequently increase the overall use of the assets along their Lifecycle [20]. On the other hand, [17] underlined that Cloud Manufacturing can also be an enabler of CE by supporting collaborative design, process resilience, waste reduction, re-use and recovery of materials. Also, Artificial Intelligence (AI) is considered a valuable technology to foster CE. In particular, modelling and simulation software are deemed useful to test and predict future behaviour of complex systems [21]. AI and Machine Learning (ML) have been also implemented in energy management to reduce the overall energy consumed by assets and achieve better efficiency [17]. CE enabled by I4.0 technologies could be an opportunity also for technology consumers. Virtualisation, through Virtual Reality (VR) and Augmented Reality (AR) systems, is a way to reducing the environmental footprint of manufacturing firms [13]. Finally, one of the main enablers of CM is determined by the Product-Service System (PSS) concept [22] which represents an opportunity for manufacturers to provide the functionality given by the product rather than the product itself and maintain the ownership [23]. This allows manufacturers to increase the usage of the good and reduce the physical resources to produce it and even extend the life cycle if supported by I4.0-based maintenance tools [24]. Literature suggests that also Blockchain technology has a role in facilitating the transition toward CE. Most of contribution in this direction are refer to it in terms of Track & Trace capabilities and control along the SCs [25–27]. Moving toward OT solutions, AM is currently considered one of the most disruptive technologies for business and society [8]. Regardless the purely economic benefits, this also has a wide range of applications to integrate CE principles. Firstly, [28] argued that AM can reduce the length of SCs by reducing the distance between consumers and producers with benefits for transportation emissions. In addition, most of the material wasted (e.g. metals) in AM processes can be recycled up to 95% of the total scraps [19, 29]. Regarding material and emission reduction, AM generates less scrap than extrusive methods [30] and allows to produce lightweight geometries without compromising resistance nor other functional proprieties [31]. [30] highlighted that AM has uses in repairing and remanufacturing activities too. Concerning collaborative robotics and automation, these are relevant to support disassembly and repair activities and to perform in unsafe working conditions [21]. Although literature identifies a variety of solutions supporting CE through the adoption of I4.0, few examples are available in Industry. Such mismatch between academia and business suggests that a set of blocking factors impedes firms to digitalize their processes by encompassing also environmental sustainability. Most of the case studies collected refer to firms applying I4.0 solutions to enable pay-per-use BMs. An example is provided by [17] that presented an electric scooter manufacturer who shaped its new BM by embedding IoT and cloud to enable a sharing service. Similar examples have also been analysed by [24] in which different pay-per-use/sharing-based models have been presented. [32] studied the modalities through high value manufacturers are moving to develop circular BMs. [33] collected 11 case studies involving large multinational firms from different sectors that implemented circular practices. Among them, several technologies were both embedded in the product offered (e.g. Michelin and EON ID) but also introduced in operations (e.g. Apple's disassembly robotic system) thus strengthening the dual benefit achievable by I4.0. Other common I4.0 application in defining sustainable practices refers to predictive maintenance solutions. Such solutions directly aim at reducing material by extending the lifecycle of products and assets. Evidence from manufacturing Industry is indeed available for both the cases. Several Pay-per-Use or sharing-based BM specifically included predictive or condition-based repair services with their strategy [17]. However, also pure predictive maintenance approaches are spreading as well [34]. Finally, [35] investigated the costs related to unsustainable approaches in manufacturing and the modalities through which I4.0 plays in supporting circular processes. In particular, the study introduced a quantitative methodology to convert in monetary terms environmental externalities thus supporting manufacturers in taking more rational and consistent evaluations of investments.

4 Discussion and Framework Proposal

The potentials represented by the I4.0 must be expressed in terms of improvement of economic performances for companies or SCs but also as elements to overcome the main barriers identified for the CE adoption [6] such as difficulties in disassembly activities, SC complexity, coordination problem (both internal and among stakeholders), lack of CE culture in designing and production processes, quality concerns about the material used, and high start-up cost threatening economic sustainability. Indeed, a proposed framework is reported in Fig. 1.

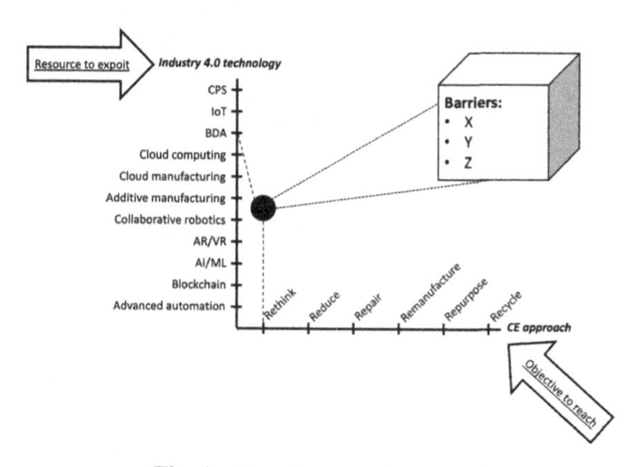

Fig. 1. New framework proposal

Among the major potentialities of I4.0 emerged that is possible to shorter time to market through visualization and simulation [36], to improve control over production processes through IoT and Big Data Analytics [37], to reduce the capital invested through Cloud Computing and Cloud Manufacturing [38], to design new product through AM [31], to establish new BMs and Revenue Streams like pay-per-use. [17], and many others.

Additionally, I4.0 represents an opportunity to foster circularity of products by extending their lifecycle [8] or by substituting them with equivalent services [23] creating new BMs [19]. In parallel, digital technologies enable CE adoption at process level too,

Table 2. Synergies between I4.0 and CE to overcome CE barriers

I4.0	CE approach	Opportunities	Barriers addressed
IoT	Rethink, repair, reduce	• Optimize sustainability performance • Increase assets usage along their lifecycle	• Lack of capital and high start-up cost • Quality concerns about the material used
Cloud-based solutions	Recycle, Repurpose, Remanufacture	• Virtualize production resources • Increase assets usage along their lifecycle • collaborative design waste reduction; re-use; materials recovery	• Lack of information causing coordination problem, both internal and external • Lack of capital and high start-up cost • Quality concerns about the material used
AI/ML		• Test and predict behavior of complex systems • assets lifecycle extension energy management	• SC complexity
Virtual Reality (VR) and Augmented Reality (AR)	Repair, Reduce	• Turn part of the costs from fixed to variables • Reduce environmental impacts	• Difficulties in disassembly activities • Lack of CE culture in designing and production processes
Sensors	Repair, Rethink	• increase the usage of the good • reduce the physical resources to produce it • extend the life cycle • improve maintenance activities	• Lack of capital and high start-up cost • Difficulties in disassembly activities • Quality concerns about the material used
Sensors	Repair, Recycling, Remanufacturing	• increase the usage of the good • extend the life cycle • improve maintenance activities	• Lack of capital and high start-up cost • Difficulties in disassembly activities • Quality concerns about the material used

(*continued*)

Table 2. (*continued*)

I4.0	CE approach	Opportunities	Barriers addressed
AM	Reduce, Recycling, Repair	• reduce the length of SCs • reduce transportation emission and costs • reduce the overall need for virgin material • reduce scrap generation lightweight products with special geometries high • functional proprieties • ease to repair and remanufacture products	• SC complexity • Quality concerns about the material used • Lack of capital and high start-up cost
Collaborative robotics and automation	Reduce, Repair, Remanufacturing	• support the disassembly • support repair activities • perform unsafe/dangerous working activities	• Difficulties in disassembly activities • Lack of information causing coordination problem, both internal and external

within factories' boundaries [39] and at SC or network level [40] supporting for instance the information sharing among several stakeholders [41]. Table 2 summarizes the outcomes of adopting I4.0 technologies to overcome the barriers emerged while embracing CE. Nevertheless, the limited availability of concrete robust case studies supporting the existence of a virtuous relationship between the two concepts suggests that practitioners still experience some challenges in putting into practice what has been conjectured so far by academia. It is also proven that such scarce population of empirical implementations is not due to a low interest toward the topic either from consumer or manufacturer side [20, 21, 42].

5 Conclusions

The present contribution aimed at investigating how to properly use I4.0 technologies to overcome the barriers experienced by manufacturing companies while embracing CE approaches. To address this goal, a systematic literature review has been conducted and once defined the barriers for adopting CE, the contributions were analysed by studying the potentialities of each I4.0 technology in this context. Therefore, as seen above in Table 2, most of the barriers could be solved or reduced through I4.0 technologies. In future works, the framework proposed in this contribution is supposed to be extended by exploring the different manufacturing processes opportunities in order to clarify the best practices to be implemented in order to use the right technology to overcome a specific CE barrier emerged in a certain manufacturing process, among the main characterizing and affecting operations [11], as depicted in Fig. 2.

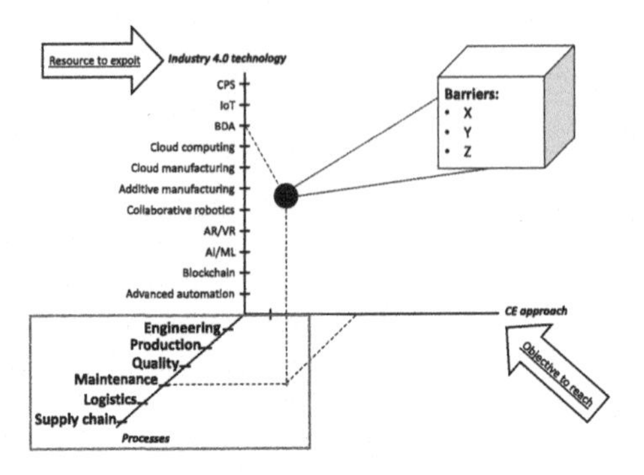

Fig. 2. Structure of the model conjectured.

Additionally, other two main gaps observed from the literature are still open and need to be addressed in future studies: i) Lack of focus on multi-product and multi-asset (e.g. machine, lines etc.) when assessing the link between I4.0 and CE [18]; ii) Lack of a quantitative evaluation of the impact of the application of I4.0 technologies and of the benefits, or pains, to support the transition toward CE [43].

References

1. World Economic Fourm, Ellen MacArthur Foundation, and Mckinsey & Company. Towards the Circular Economy : Accelerating the scale-up across global supply chains (2014). https://doi.org/10.1162/108819806775545321
2. Acerbi, F., Taisch, M.: A literature review on circular economy adoption in the manufacturing sector. J. Clean. Prod. **273**, 123086 (2020). https://doi.org/10.1016/j.jclepro.2020.123086
3. Raj, A., Dwivedi, G., Sharma, A., Lopes de Sousa Jabbour, A.B., Rajak, S.: Barriers to the adoption of industry 4.0 technologies in the manufacturing sector: An inter-country comparative perspective. Int. J. Prod. Econ. **224**, 107546 (2020). https://doi.org/10.1016/j.ijpe.2019.107546
4. Kumar, R., Singh, R.K., Dwivedi, Y.K.: Application of industry 4.0 technologies in SMEs for ethical and sustainable operations: analysis of challenges. J. Clean. Prod. **275**, 124063 (2020). https://doi.org/10.1016/j.jclepro.2020.124063
5. Piscitelli, G., Ferazzoli, A., Petrillo, A., Cioffi, R., Parmentola, A., Travaglioni, M.: Circular economy models in the industry 4.0 era: a review of the last decade. Procedia Manuf. **42**(2019), 227–234 (2020). https://doi.org/10.1016/j.promfg.2020.02.074
6. Halse, L.L., Jæger, B.: operationalizing industry 4.0: understanding barriers of industry 4.0 and circular economy. In: Ameri, F., Stecke, K.E., von Cieminski, G., Kiritsis, D. (eds.) APMS 2019. IAICT, vol. 567, pp. 135–142. Springer, Cham (2019). https://doi.org/10.1007/978-3-030-29996-5_16

7. Nobre, G.C., Tavares, E.: Scientific literature analysis on big data and internet of things applications on circular economy: a bibliometric study. Scientometrics **111**(1), 463–492 (2017). https://doi.org/10.1007/s11192-017-2281-6
8. Angioletti, C.M.: Achieving a circular economy through (more) circular products. Politecnico di Milano (2018)
9. Greenhalgh, T.: How to read a paper: Papers that summarise other papers (systematic reviews and meta-analyses). BMJ **315**(7109), 672–675 (1997). https://doi.org/10.1136/bmj.315.7109.672
10. Rojko, A.: Industry 4.0 concept: background and overview. Int. J. Interact. Mob. Technol. **11**(5), 77–90 (2017). http://online-journals.org/index.php/i-jim/article/view/7072/4532
11. Rosa, P., Sassanelli, C., Urbinati, A., Chiaroni, D., Terzi, S.: Assessing relations between Circular economy and industry 4.0: a systematic literature review. Int. J. Prod. Res. **58**(6), 1662–1687 (2020). https://doi.org/10.1080/00207543.2019.1680896
12. Mittal, S., Khan, M.A., Romero, D., Wuest, T.: A critical review of smart manufacturing & industry 4.0 maturity models: implications for small and medium-sized enterprises (SMEs). J. Manuf. Syst. **49**(October), 194–214 (2018). https://doi.org/10.1016/j.jmsy.2018.10.005
13. Charnley, F., Tiwari, D., Hutabarat, W., Moreno, M., Okorie, O., Tiwari, A.: Simulation to enable a data-driven circular economy. Sustain. **11**(12), 1–16 (2019). https://doi.org/10.3390/su10023379
14. Acerbi, F., Forterre, D.A., Taisch, M.: Role of artificial intelligence in circular manufacturing: a systematic literature review. IFAC-PapersOnLine **54**(1), 367–372 (2021). https://doi.org/10.1016/j.ifacol.2021.08.040
15. Lopes de Sousa Jabbour, A.B., Jabbour, C.J.C., Godinho Filho, M., Roubaud, D.: Industry 4.0 and the circular economy: a proposed research agenda and original roadmap for sustainable operations. Ann. Oper. Res. **270**(1–2), 273–286 (2018). https://doi.org/10.1007/s10479-018-2772-8
16. Navas, M.A., Sancho, C., Carpio, J.: Disruptive maintenance engineering 4.0. Int. J. Qual. Reliab. Manag. **37**(6–7), 853–871 (2020). https://doi.org/10.1108/IJQRM-09-2019-0304
17. Pham, T.T., et al.: Industry 4.0 to accelerate the circular economy: a case study of electric scooter sharing. Sustain **11**(23), 1–16 (2019). https://doi.org/10.3390/su11236661
18. Rajput, S., Singh, S.P.: Industry 4.0 Model for circular economy and cleaner production. J. Clean. Prod. **277**, 123853 (2020). https://doi.org/10.1016/j.jclepro.2020.123853
19. Spaltini, M., Poletti, A., Acerbi, F., Taisch, M.: A quantitative framework for Industry 4.0 enabled circular economy. Procedia CIRP **98**, 115–120 (2021). https://doi.org/10.1016/j.procir.2021.01.015
20. Kerin, M., Pham, D.T.: Smart remanufacturing: a review and research framework. J. Manuf. Technol. Manag. **31**(6), 1205–1235 (2020). https://doi.org/10.1108/JMTM-06-2019-0205
21. Blömeke, S., Rickert, J., Mennenga, M., Thiede, S., Spengler, T.S., Herrmann, C.: Recycling 4.0 - mapping smart manufacturing solutions to remanufacturing and recycling operations. Procedia CIRP **90**, 600–605 (2020). https://doi.org/10.1016/j.procir.2020.02.045
22. Khan, M.A., Wuest, T.: Upgradable product-service systems: implications for business model components. Procedia CIRP (2019). https://doi.org/10.1016/j.procir.2019.01.091
23. Isaksson, O., Hallstedt, S.I., Rönnbäck, A.Ö.: Digitalisation, sustainability and servitisation: consequences on product development capabilities in manufacturing firms. In: Proceedings of Nord Design Era Digital Nord 2018 (2018)
24. Jabbour, C.J.C., et al.: First-mover firms in the transition towards the sharing economy in metallic natural resource-intensive industries: Implications for the circular economy and emerging industry 4.0 technologies. Res. Policy **66**, 101596 (2020). https://doi.org/10.1016/j.resourpol.2020.101596

25. Upadhyay, A., Mukhuty, S., Kumar, V., Kazancoglu, Y.: Blockchain technology and the circular economy: implications for sustainability and social responsibility. J. Clean. Prod. **293**, 126130 (2021). https://doi.org/10.1016/j.jclepro.2021.126130
26. Esmaeilian, B., Sarkis, J., Lewis, K., Behdad, S.: Blockchain for the future of sustainable supply chain management in Industry 4.0. Res. Cons. Recycl. **163**, 105064 (2020). https://doi.org/10.1016/j.resconrec.2020.105064
27. Tozanlı, Ö., Kongar, E., Gupta, S.M.: Trade-in-to-upgrade as a marketing strategy in disassembly-to-order systems at the edge of blockchain technology. Int. J. Prod. Res. **58**(23), 7183–7200 (2020). https://doi.org/10.1080/00207543.2020.1712489
28. Sun, L., Wang, Y., Guowei Hua, T.C.E., Cheng, J.D.: Virgin or recycled? optimal pricing of 3D printing platform and material suppliers in a closed-loop competitive circular supply chain. Res. Cons. Recycl. **162**, 105035 (2020). https://doi.org/10.1016/j.resconrec.2020.105035
29. Vayre, B., Vignat, F., Villeneuve, F.: Metallic additive manufacturing: state-of-the-art review and prospects. Mech. Ind. **13**(2), 89–96 (2012). https://doi.org/10.1051/meca/2012003
30. Kravchenko, M., Pigosso, D.C.A., McAloone, T.C.: Circular economy enabled by additive manufacturing: potential opportunities and key sustainability aspects. In: Proceedings of Nord 2020 Conference Nord 2020 (2020). https://doi.org/10.35199/norddesign2020.4
31. Ford, S., Despeisse, M.: Additive manufacturing and sustainability: an exploratory study of the advantages and challenges. J. Clean. Prod. **137**, 1573–1587 (2016). https://doi.org/10.1016/j.jclepro.2016.04.150
32. Okorie, O., Charnley, F., Russell, J., Tiwari, A., Moreno, M.: Circular business models in high value manufacturing: five industry cases to bridge theory and practice. Bus. Strateg. Environ. **30**(4), 1780–1802 (2021). https://doi.org/10.1002/bse.2715
33. Hoosain, M.S., Paul, B.S., Ramakrishna, S.: The impact of 4ir digital technologies and circular thinking on the united nations sustainable development goals. Sustain **12**(23), 1–16 (2020). https://doi.org/10.3390/su122310143
34. Kristoffersen, E., Blomsma, F., Mikalef, P., Li, J.: The smart circular economy: a digital-enabled circular strategies framework for manufacturing companies. J. Bus. Res. **120**(August), 241–261 (2020). https://doi.org/10.1016/j.jbusres.2020.07.044
35. Dev, N.K., Shankar, R., Qaiser, F.H.: Industry 4.0 and circular economy: operational excellence for sustainable reverse supply chain performance. Res. Cons. Recycl. **153**, 104583 (2020). https://doi.org/10.1016/j.resconrec.2019.104583
36. Modoni, G.E., Caldarola, E.G., Sacco, M., Terkaj, W.: Synchronizing physical and digital factory: benefits and technical challenges. Procedia CIRP **79**, 472–477 (2019). https://doi.org/10.1016/j.procir.2019.02.125
37. Roy, R., Stark, R., Tracht, K., Takata, S., Mori, M.: Continuous maintenance and the future – foundations and technological challenges. CIRP Ann. **65**(2), 667–688 (2016). https://doi.org/10.1016/j.cirp.2016.06.006
38. Kumar, V., Vidhyalakshmi, P.: Cloud computing for business sustainability. Asia-Pacific J. Manag. Res. Innov. **8**(4), 461–474 (2012). https://doi.org/10.1177/2319510x13481905
39. Yang, S., Aravind, M.R., Kaminski, J., Pepin, H.: Opportunities for Industry 4.0 to support remanufacturing. Appl. Sci. **8**(7), 1177 (2018). https://doi.org/10.3390/app8071177
40. Manavalan, E., Jayakrishna, K.: A review of Internet of Things (IoT) embedded sustainable supply chain for industry 4.0 requirements. Comput. Ind. Eng. **127**, 925–953 (2019). https://doi.org/10.1016/j.cie.2018.11.030
41. Acerbi, F., Taisch, M.: Information flows supporting circular economy adoption in the manufacturing sector. In: Lalic, B., Majstorovic, V., Marjanovic, U., von Cieminski, G., Romero, D. (eds.) APMS 2020. IAICT, vol. 592, pp. 703–710. Springer, Cham (2020). https://doi.org/10.1007/978-3-030-57997-5_81

42. Kirchherr, J., Reike, D., Hekkert, M.: Conceptualizing the circular economy: an analysis of 114 definitions. Res. Cons. Recycl. **127**(April), 221–232 (2017). https://doi.org/10.1016/j.resconrec.2017.09.005
43. Rajput, S., Singh, S.P.: Connecting circular economy and industry 4.0. Int. J. Inf. Manag. **49**(March), 98–113 (2019). https://doi.org/10.1016/j.ijinfomgt.2019.03.002

First Approach to a Theoretical Framework for Carbon Footprint Management in the Aerospace Manufacturing Industry

Lucia Recio Rubio[1,2](✉), Amanda Martin-Mariscal[1], and Estela Peralta[1]

[1] MyM Group, Cadiz, Spain
lrecio@us.es
[2] University of Sevilla, Sevilla, Spain

Abstract. Reducing CO_2 emissions from aircraft and other artifacts, as well as minimizing the environmental footprint during their manufacture, is one of the current challenges facing the aerospace industry. This implies that all manufacturing companies related to the industry are faced with the need to implement strategies to manage and control emissions during their supply chain. Under this scenario arises the proposal of a conceptual framework integrated in a Product Lifecycle Management (PLM) environment based on the digital control of manufacturing, oriented to the fulfillment of carbon emission reduction objectives. The theoretical proposal is based on the use of an industrial Digital Twin (DT), combining the use of a digital model, a physical model, a CO_2 model, and a life cycle analysis (LCA) module. Those models work in an integrated way to estimate and control the carbon footprint with the goal of being able to reduce it. This study seeks to discern which are the appropriate measurement parameters in the CO_2 model. Starting from the proposed combined measures of manufacturing activities, support tasks, management tasks and industry suppliers. That will allow estimating the carbon footprint produced in the manufacture of each part through an LCA database.

Keywords: PLM · Carbon footprint · Digital twin · Aerospace industry · Sustainability

1 Introduction

Circular economy, cleaner production or green manufacturing are sustainable strategies on which research has been conducted in recent years, due to society's growing concern to be more careful with the planet.

In this research line, the aerospace industry included action plans to commit to sustainability in both final products and manufacturing processes. Among these commitments is a movement to decarbonize the entire aerospace industry that includes the objectives of carbon neutral growth from 2020 and a 50% reduction in CO_2 emissions by 2050, comparing with those recorded in 2005 [1]. This sector produces 2.5% of the world's energy-related CO_2 emissions [2, 3], reaching 3.8% in the EU [4]. In addition, it

Published by Springer Nature Switzerland AG 2023
F. Noël et al. (Eds.): PLM 2022, IFIP AICT 667, pp. 620–629, 2023.
https://doi.org/10.1007/978-3-031-25182-5_60

should be noted that emissions from the sector will increase annually by 5% until 2030, which may amount to as much as 22% by 2050 [5, 6], which in impact data amounts to 2.6 billion tons of CO_2 [7].

After reviewing the literature in search of application methods of these sustainable strategies, it was observed that although the subject is of interest, there are no frameworks that allow the implementation of carbon footprint management systems to the assembly and manufacturing processes industry.

The management of the carbon footprint is closely linked to the concept of "eco-efficiency". Eco-efficiency is understood as the ability to create more for less, as pointed out by Moreira [8], or as Clark [9] said, the ability to deliver goods and services at competitive prices that meet the needs of the industry and provide quality of life, progressively reducing the ecological impact and resource intensity throughout the life cycle.

Other authors, such as Oliva, wanted to apply this term eco-efficiency to the industrial domain and defined it as "reducing the consumption of resources, raw material, energy, waste and air emissions while maintaining or reducing the costs for manufacturing a product. For an assembly process, it is a question of balance of environ-mental and economic benefits in an integrated way" [10].

Currently, the latest advances in eco-efficiency have led to innovative circular economy strategies, which help to improve functionalities such as the management of efficiency and cost evaluation of assembly processes or the simulation and optimization of processes as shown in the 3-Layers Model [10, 11]. In the same way that the terms associated with the circular economy have been the subject of study in recent years, other concepts such as digital transformation or digital continuity of industrial companies have been core in the proposals and advances in the sector [12, 13].

With this knowledge, it is possible to understand how the management of an environmental impact is carried out in the industrial sector. First, the search for methods to assess the environmental impact is of interest, such as Life Cycle Assessment (LCA), standardized in ISO 14040: 2006 [14]. These methods make it possible to optimize the impacts through plan, do, act and check. Second, it is of great interest to carry out impact management by taking advantage of the industry's digital transformation trends [15]. This transformation enables sustainable and collaborative management in an agile and flexible way, one of the main challenges for organizations in the implementation of circular economy strategies.

To facilitate the implementation of carbon footprint management systems in the industry 4.0 context, the objective of this work is to propose a process-oriented framework based on circular economy strategies. The framework supports the implementation of a PLM system to carry out the management of the carbon footprint, specifically, for those dedicated to manufacturing processes. This is an innovative proposal that integrates the circular economy and takes advantage of the challenges of Industry 4.0 and of the tools of the Product Lifecycle Management (PLM) environments.

To this end, the paper is structured as follows: Sect. 2 reviews previous studies in the field of PLM, Industry 4.0, and its link to sustainability strategies. Section 3 presents the proposal of the framework for carbon footprint management in Industry 4.0 components highlighting the contribution and benefits of the research. Finally, Sect. 4 provides a discussion of the results and summarizes the conclusions of the work, establishing future lines of work. Throughout the study, the terms carbon footprint, CO_2 footprint, carbon contribution, CO_2 emissions, among others, will be used as synonyms to improve the richness and the readability of the paper.

2 Background

The large volume of data to be managed by companies in the product life cycle (quality, prevention, corporate social responsibility, costs, time, materials, human re-sources, environmental damage, etc.) makes the use of IT technologies indispensable to carry out their activities. In this field, multiple research projects have emerged to manage product and processes information through PLM environments [16]. Specifically, the aerospace industry devotes part of its interest and research efforts (time and investment) to green plans, integrating sustainability with the digitization of its processes, as for example Airbus does with its Digital, Design, Manufacturing & Services (DDMS) project [17].

PLM as enabler for Industry 4.0 has attracted much attention in the last decade, both from the industrial world and from academia and research. This recent upsurge is due to the potential benefits that PLM can offer in the face of today's design and manufacturing challenges, however, this is still a distant promise for most organizations [18]. PLM could be defined as a systematic concept of integrated management of product-related information and processes throughout the entire product lifecycle, from initial conception to end-of-life [19, 20], all managed through the methods and tools provided by the integrated collaborative platform or environment that is the PLM system itself. This collaborative management is one of the goals of Smart Factories where it is intended to achieve effective interoperability between all assets and information systems [21].

Product lifecycle management enables the development and control of information through the efficient storage, processing, collection, and distribution of data throughout the lifecycle of products (design, manufacturing, logistics, use and end-of-life), processes and resources (planning, execution, and disposal) in an organization. In this way, and with the goals of Industry 4.0, PLM systems become essential to achieve optimal interconnection and manage digital data together with other available enablers [22]: Internet of Things (IoT), cloud, big data, smart sensors, artificial intelligent, real-time data or robotics [23]. It requires agile, dynamic, and collaborative databases, whose functions can be covered by PLM systems. Furthermore, it makes it possible to handle the digital models of Cyber-physical systems (CPS), thanks to the integration of increasingly consolidated CAx techniques. In view of this integrated management, it could be said that the development of PLM has been highly dependent on the evolution and assimilation of computer-based software or product solutions, starting with computer-aided design or computer-aided manufacturing applications (CAD, CAM or CAE), passing through applications related to Enterprise Resource Planning (ERP), Customer Relationship Management (CRM) or Supply Chain Management (SCM) [20, 24].

Analyzing the studies on carbon footprint management in manufacturing and specifically in the aerospace industry, proposals are limited. Traditionally, they are aimed at managing energy use to minimize greenhouse gas (GHG) emissions with strategies for combustion efficiency, gearing ratios in engines, or redesigning wing aerodynamics [25].

Some PLM software (Siemens, Dassault Systèmes or Autodesk) do include carbon footprint calculation modules (such as Environmental Impact Calculator or Product Carbon Footprint), but they focus on calculating the system's contribution to the cli-mate change category as a product life cycle impact. It is necessary to incorporate a comprehensive system that considers, in addition to the calculation to minimize the impact, the management of holistic strategies and other key innovations for the minimization of CO_2 emissions. That can be implemented in the manufacturing process contributing to the other stages of the life cycle, such as the production of lighter, fuel efficient, less polluting and more aerodynamic next generation aircraft, closed-loop recycling, carbon-neutral technologies, waste management, hybrid-electric architectures, new manufacturing processes, new materials, sustainable alternative fuels, among others [26, 27]. From this review, it can be appreciated that the development of tools for environmental management is still under progress.

3 Materials and Methods

This section describes the framework for sustainability with the integration of a CO_2 model, managed through an LCA database, with the industrial Digital Twin of a manufacturing company and its application context. For the development of the conceptual proposal, the following steps were carried out and are explained in the following sub-sections:

- Development of the new carbon footprint estimation process oriented to Industry 4.0, composed of three interconnected models (physical model, digital model, and carbon model) (Sect. 3.1).
- Characterization of the carbon model differentiating into groups of activities that contribute to the carbon footprint and identification of sustainability indicators for each group of operations (Sect. 3.2).

3.1 Conceptual Framework Proposal

This conceptual framework is proposed for the estimation and management of the carbon footprint for a machining and metal forming company of aerospace parts.

Processes play a very important role when studying a machining plant, a proper characterization of them allows to know the production cost, processing time, product quality, production resources and environmental impact of the whole product. Focusing the interest on the carbon emission of the parts manufacturing process, the material flow, the energy flow and the environmental flow in the production and machining processes must be included in the study [28]. The characterization of manufacturing processes is strongly promoted to optimize the management of carbon emissions, showing special interest in the factors with the greatest influence on the carbon footprint in the machining process, such as materials, machinery, or personnel [29].

Derived from the commitments and objectives of circular economy and emission reduction set by the aerospace industry [30], the conceptual framework is proposed. Seeking, in the first place, to estimate the carbon footprint produced during the complete manufacturing life cycle of the components of the supplier companies. To be able to provide, together with the manufactured parts, a study of the CO_2 footprint generated during manufacturing, aided with the use of an industrial Digital Twin. In addition to this objective, it is intended that the model serves to facilitate the control and monitoring of carbon consumption throughout the production chain. This contributes to carried out periodic continuous improvement processes to bring the company closer to the commitments of circular economy with the help of advances in industry 4.0.

The model is based on the use of a Digital Twin of an industrial plant. Figure 1 shows the architecture of the base model. It combines the use of a physical model, a digital model, and a CO_2 model, as well as an LCA database to work together to estimate, control and optimize carbon consumption and emissions.

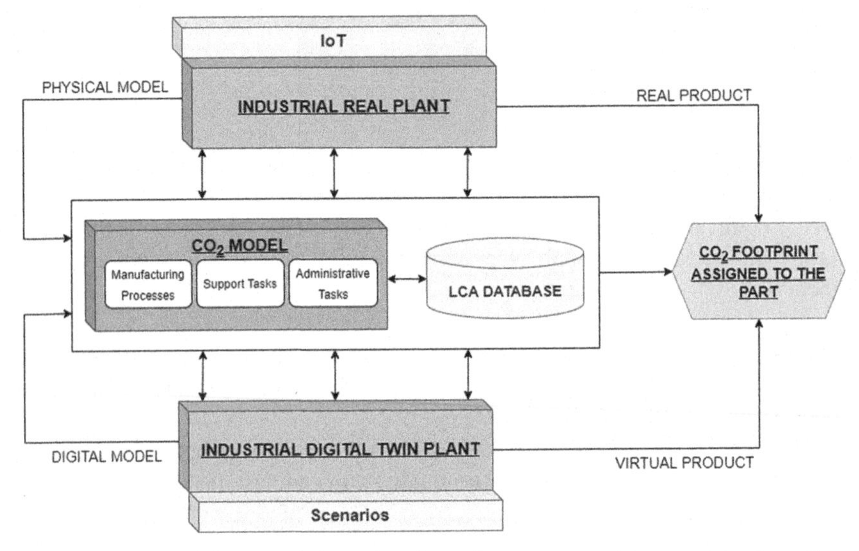

Fig. 1. A conceptual framework for carbon footprint management (prepared by the authors)

In the conceptual model shown in the figure above, you can see how the information flows, represented with arrows, derived from the real plant (physical model) and the digital plant (digital model) are managed and evaluated through the CO_2 model and the LCA database. The information collected from the actual situation and from the actual data history is collected and processed in the physical model. In addition, this information feeds the Digital Twin of the plant, where different scenarios are proposed, based on the information obtained, and process improvements are sought to reduce CO_2 emissions.

The next section is centered in the first approach to the components of the CO_2 model, which is the core of the framework, and to present the proposed parameters and indicators involved in the model, as well as the information expected to be obtained.

3.2 CO_2 Model

The next step of the characterization of the theoretical framework is to introduce the carbon model, which would be the engine for estimating and calculating the emissions generated during the production of each of the parts.

For the generation of this model, all operations related to the manufacture of the parts that may have a contribution to its carbon footprint are considered.

The first step to configure the model would be to study all the agents involved in the life cycle of the plant under study, identifying each of the tasks performed during manufacturing and the possible carbon contribution they may have.

In this way, aided by the job data obtained from the physical model and the digital model, it would be possible to map all the operations and processes involved in the CO_2 footprint of the plant.

Once the tasks involved have been identified, it is proposed to cluster them into groups differentiated by the type of activity and the relationship it has with the production process. So that it is easier to identify the sustainability indicators that can be measured in each of them.

The proposed division of groups by task is as follows:

- Manufacturing processes: all the carbon contribution generated by the machinery and manufacturing processes themselves, i.e., the direct contribution to the carbon footprint of the part.

  Within this grouping, sustainability indicators to be measured are identified, aided by Zhao's studies [31], such as:

  - Raw material consumption
  - Energy consumption (differentiating between renewable and non-renewable)
  - Water consumption
  - Consumption of additional materials

- Support tasks: All types of activities associated with production but not directly involved in the manufacture of the part (transport of materials, cleaning of machines, setting up the warehouse, etc.) but which do produce CO_2.

  Some of the sustainability indicators to be measured in this group would be:

  - Energy, materials, & water consumption in plant cleaning tasks.
  - Energy, materials, & water consumption in machinery maintenance tasks.
  - Energy consumption in transporting materials.
  - Plant air conditioning

- Administrative tasks: All the tasks necessary for the plant to be in operation and continue with production, from reception, human resources, suppliers, etc.

  The following sustainability parameters have been proposed for measurement:

  - Energy consumption by supplier management.
  - Energy consumption by human resources

- Energy consumption by communication personnel
- Energy consumption by management personnel.

To obtain, store and maintain all the data obtained from the sustainability indicators, both physical and digital models are needed as input, so that data can be measured in both cases, and the use of an LCA database that is able to cover all material flows such as inventory throughout the life cycle of all the processes carried out in the plant is necessary [32]. The required information flow is shown in Fig. 2 below.

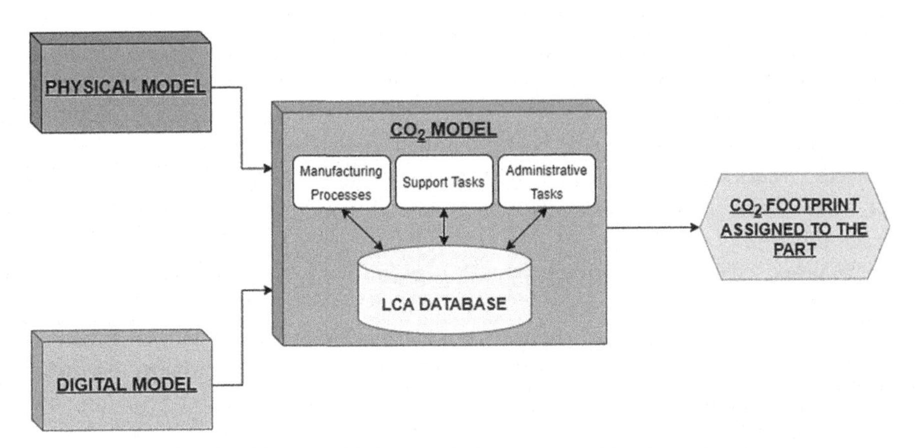

Fig. 2. CO_2 model information flow (prepared by the authors)

The expected output of the proposed carbon model is an estimate as accurate as possible of the carbon footprint of currently manufactured metal parts.

Once the actual footprint is obtained, the next objective is to take it as a starting point to, in conjunction with the other two models, realize periodic emission reductions thanks to the improvements generated in the Digital Twin scenarios that subsequently materialize in the actual industrial plant.

In this way, the conjunction of the three models, as shown in the conceptual framework, would allow organizations to meet their carbon footprint reduction commitments during their life cycle, aided by the implementation of a digital model sup-ported by the advances of Industry 4.0 and the implementation of a PLM.

4 Discussion

Through this work, it was proven that information management with a PLM and virtual representation through an industrial Digital Twin are an important improvement to address the sustainability parameters of companies.

Through the above study and the development of the circular economy framework, it has become evident that the need to develop tools to support sustainability strategies is a reality in the aerospace industry.

Without the inclusion of technology and innovation in the environmental management of the aerospace industry, it would not be possible to achieve the proposed sustainability and emission reduction commitments. Therefore, it is necessary to bring companies closer to Industry 4.0 and its benefits if merged with environmental strategies.

This study has provided a framework to work on for the estimation of the current carbon footprint in a parts manufacturing company. It also opens the door to the periodic management of these emissions through the joint use of the three models to achieve reduction targets.

The possibility of being able to count on the collaboration of a real company for its implementation provides this conceptual framework with an associated line of work with which to continue researching and refining the proposal to achieve the development of a functional tool. The next steps to follow in this line of action would be to launch an industrial project based on this research. This project would allow us to test the usefulness and reliability of the framework.

Being able to quantify and subsequently optimize the carbon footprint by using the Digital Twin of the plant, allows small companies in the aerospace industry to comply with the environmental requirements of the industry. This brings additional value to organizations to strengthen their position as suppliers to large multinationals in the sector.

In a future where digital transformation and continuity are indispensable and the environment is a constant concern, it is of vital importance for companies to possess tools that bring them closer to these goals in order not to be left behind compared to other companies with the same target.

The industry seeks to improve and innovate in production and assembly lines, in the manufacture of components and intermediate products, and thus favor the digital transformation of companies thanks to Industry 4.0 tools and PLM environments.

Acknowledgments. The authors would like to recognize M&M Group colleagues and Seville University colleagues support during the development of this work and thank them for their contribution.

References

1. Decarbonisation | Airbus. https://www.airbus.com/en/sustainability/environment/climate-change/decarbonisation. Accessed 11 Mar 2022
2. van Grootel, A., Chang, J., Wardle, B., Olivetti, E.: Manufacturing variability drives significant environmental and economic impact: the case of carbon fiber reinforced polymer composites in the aerospace industry. J. Clean. Prod. **261**, 121087 (2020). https://doi.org/10.1016/J.JCLEPRO.2020.121087
3. World Energy Outlook 2021 (2021). https://doi.org/10.1787/14FCB638-EN
4. Reducing emissions from aviation. https://ec.europa.eu/clima/eu-action/transport-emissions/reducing-emissions-aviation_en. Accessed 25 Mar 2022
5. Timmis, A.J., et al.: Environmental impact assessment of aviation emission reduction through the implementation of composite materials. Int. J. Life Cycle Assess. **20**(2), 233–243 (2014). https://doi.org/10.1007/s11367-014-0824-0

6. Global Market Forecast | Airbus. https://www.airbus.com/en/products-services/commercial-aircraft/market/global-market-forecast. Accessed 25 Mar 2022
7. Qiu, R., Hou, S., Meng, Z.: Low carbon air transport development trends and policy implications based on a scientometrics-based data analysis system. Transp. Policy **107**, 1–10 (2021). https://doi.org/10.1016/J.TRANPOL.2021.04.013
8. Moreira, F., Alves, A.C., Sousa, R.M.: Towards eco-efficient Lean Production Systems. In: Ortiz, Á., Franco, R.D., Gasquet, P.G. (eds.) BASYS 2010. IAICT, vol. 322, pp. 100–108. Springer, Heidelberg (2010). https://doi.org/10.1007/978-3-642-14341-0_12
9. Clark, G.L., Schmidheiny, S.: Changing course: a global perspective on development and the environment. Econ. Geogr. **69**, 436 (1993). https://doi.org/10.2307/143602
10. Oliva M., Mas F., Eguia I., et al: Product Lifecycle Management Enabling Smart X, vol. 594, pp. 448–459 (2020). https://doi.org/10.1007/978-3-030-62807-9
11. Mas, F., Racero, J., Oliva, M., Morales-Palma, D.: A preliminary methodological approach to models for manufacturing (MfM). In: Chiabert, P., Bouras, A., Noël, F., Ríos, J. (eds.) PLM 2018. IAICT, vol. 540, pp. 273–283. Springer, Cham (2018). https://doi.org/10.1007/978-3-030-01614-2_25
12. Mas, F., Luis Menéndez, J., Oliva, M., et al.: iDMU as the collaborative engineering engine research experiences in Airbus (2014)
13. Cantamessa, M., Montagna, F., Neirotti, P.: An empirical analysis of the PLM implementation effects in the aerospace industry. Comput. Ind. **63**, 243–251 (2012). https://doi.org/10.1016/J.COMPIND.2012.01.004
14. ISO 14040:2006(es), Gestión ambiental—Análisis del ciclo de vida—Principios y marco de referencia. https://www.iso.org/obp/ui#iso:std:iso:14040:ed-2:v1:es. Accessed 10 Mar 2022
15. Zhang, H., Zhu, B., Li, Y., et al.: Development and utilization of a Process-oriented Information Model for sustainable manufacturing. J. Manuf. Syst. **37**, 459–466 (2015). https://doi.org/10.1016/J.JMSY.2015.05.003
16. Singh, S., Misra, S.C., Kumar, S.: Institutionalization of product lifecycle management in manufacturing firms. IEEE Trans. Eng. Manag. (2021). https://doi.org/10.1109/TEM.2021.3097040
17. Digital Design, Manufacturing & Services | Airbus. https://www.airbus.com/en/innovation/disruptive-concepts/digital-design-manufacturing-services. Accessed 27 Mar 2022
18. Rangan, R.M., Rohde, S.M., Peak, R., et al.: Streamlining product lifecycle processes: a survey of product lifecycle management implementations, directions, and challenges. J. Comput. Inf. Sci. Eng. **5**, 227–237 (2005). https://doi.org/10.1115/1.2031270
19. Saaksvuori, A., Immonen, A.: Product lifecycle management (2008)
20. Cao, H., Folan, P.: Production Planning and Control Product Life Cycle: the evolution of a paradigm and literature review from 1950–2009 (2015)
21. Fraga, A.L., Vegetti, M., Leone, H.P.: Ontology-based solutions for interoperability among product lifecycle management systems: a systematic literature review. J. Ind. Inf. Integr. **20**, 100176 (2020). https://doi.org/10.1016/J.JII.2020.100176
22. Kulkarni, V.N., Gaitonde, V.N., Kotturshettar, B.B.: Product lifecycle management (PLM): a key enabler in implementation of industry 4.0. Handb. Smart Mater. Technol. Devices 1–32 (2021). https://doi.org/10.1007/978-3-030-58675-1_14-1
23. Martinelli, A., Mina, A., Moggi, M.: The enabling technologies of industry 4.0: examining the seeds of the fourth industrial revolution. Ind. Corp. Change **30**, 161–188 (2021). https://doi.org/10.1093/ICC/DTAA060
24. Ameri, F., Dutta, D.: Product lifecycle management: closing the knowledge loops. Comput. Aided Des. Appl. **2**, 577–590 (2005). https://doi.org/10.1080/16864360.2005.10738322
25. Eckelman, M.J., Ciacci, L., Kavlak, G., et al.: Life cycle carbon benefits of aerospace alloy recycling. J Clean Prod **80**, 38–45 (2014). https://doi.org/10.1016/J.JCLEPRO.2014.05.039

26. Gangi, F., Mustilli, M., Daniele, L.M., Coscia, M.: The sustainable development of the aerospace industry: drivers and impact of corporate environmental responsibility. Bus. Strateg. Environ. (2021). https://doi.org/10.1002/BSE.2883
27. Eisenhut, D., Moebs, N., Windels, E., et al.: Aircraft requirements for sustainable regional aviation. Aerospace **8**, 61 (2021). https://doi.org/10.3390/AEROSPACE8030061
28. Deng, Z., Lv, L., Huang, W., Shi, Y.: A high efficiency and low carbon oriented machining process route optimization model and its application. Int. J. Precis. Eng. Manuf. - Green Technol. **6**, 23–41 (2019). https://doi.org/10.1007/S40684-019-00029-0/FIGURES/10
29. Zhang, C., Liu, C., Zhao, X.: Optimization control method for carbon footprint of machining process. Int. J.Adv. Manuf. Technol. **92**(5–8), 1601–1607 (2017). https://doi.org/10.1007/s00170-017-0241-1
30. Destination 2050-A route to net zero European aviation Preface. https://www.destination2050.eu. Accessed 18 Mar 2022
31. Zhao, W.-B., Jeong, J.-W., Noh, S.D., Yee, J.T.: Energy simulation framework integrated with green manufacturing-enabled PLM information model. Int. J. Precis. Eng. Manuf.-Green Technol. **2**(3), 217–224 (2015). https://doi.org/10.1007/s40684-015-0025-8
32. Martínez-Rocamora, A., Solís-Guzmán, J., Marrero, M.: LCA databases focused on construction materials: a review. Renew. Sustain. Energy Rev. **58**, 565–573 (2016). https://doi.org/10.1016/J.RSER.2015.12.243

Improving the Sustainability of Manufactured Products: Literature Review and Challenges for Digitalization of Disassembly and Dismantling Processes

Chorouk Mouflih, Raoudha Gaha, Magali Bosch, and Alexandre Durupt(✉)

Université de Technologie de Compiègne, Roberval Lab., Rue du Dr Schweitzer, CS 60319, 60203 Compiègne Cedex, France

{chorouk.mouflih,alexandre.durupt}@utc.fr

Abstract. Digital technology used today within Industry 4.0 open up new opportunities for most areas of engineering, including the end of life and especially disassembly process, which was identified as the most critical and strategic step in the remanufacturing. The aim of this paper is to identify all I4.0 technologies that have been used in the process of disassembly (with a special focus on sustainability) in order to create a Digital Twin model for disassembly process based on sustainability indicators. In this research, firstly, we present a literature review on the related work based on existing I4.0 technologies in the disassembly process and the issues related to this process. Secondly, an overview of I4.0 technologies used in the optimization of the disassembly process research is presented. Thirdly, a discussion about the use of I4.0 technologies in order to have a sustainable disassembly process and a digital twin is presented. Finally, research questions about the use of I4.0 technologies in a digital twin for disassembly process based on sustainable indicators were conducted in the conclusion.

Keywords: Digital technology · Industry 4.0 · Disassembly · Remanufacturing · Digital twin

1 Introduction

The disassembly process is a critical step in the remanufacturing of end-of-life (EOL) products [1] that enables a circular economy (CE) [2, 3]. In the framework of sustainable development, from an environmental perspective, disassembly process reduces the use and waste of raw materials, increases the utilization of resources, and reduce pollution and emissions [4, 5]. From an economical point of view, it can reduce product life cycle costs [6], helps decrease product recovery value [7] as well as product production cost. Finally, from a society point of view, it creates new career opportunities and provides a safer, healthier and more ergonomic work environment [8]. However, due to the complexity and uncertainties of the disassembly products [9] this process faces several

Published by Springer Nature Switzerland AG 2023
F. Noël et al. (Eds.): PLM 2022, IFIP AICT 667, pp. 630–640, 2023.
https://doi.org/10.1007/978-3-031-25182-5_61

challenges (e.g. high cost, difficult sequence planning, operators safety and environmental impact). The disassembly process can benefit from the digital technologies within the Industry 4.0 (I4.0) such as Virtual Reality (VR), Augmented Reality (AR), Digital twin (DTW) or Robotics, in order to be optimized and be cost-efficient. I4.0 technologies can also be important enablers for industrial sustainability [1, 5] especially for disassembly process by reducing its energy consumption and its environmental impacts.

In this framework, the aim of this paper is to identify all I4.0 technologies that have been used in the process of disassembly (with a special focus on sustainability) in order to create a DTW for disassembly process based on sustainability indicators. In this research, firstly, we present a literature review on the related works based on existing I4.0 technologies in the disassembly process and issues related to this process. Secondly, an overview of I4.0 technologies used in the optimization of the disassembly process research is presented. Thirdly, a discussion about the use of I4.0 technologies in order to have a sustainable disassembly process and a digital twin is presented. Finally, research questions about the use of I4.0 technologies in a digital twin for disassembly process based on sustainable indicators were conducted in the conclusion.

2 Related Work

2.1 Disassembly Process

The disassembly process is a systematic approach separating a product into parts or removing its components, subsets or constructive parts [10], it plays a key role in recycling of materials and components [11].However, it is a very critical process due to its high degree of uncertainty caused by the quality and structure of EOL products [10]. Within the framework of this literature review, the issues concerning the disassembly are presented in three categories: Cost/Time efficiency, Operator/tasks/safety and Process information.

Cost/Time efficiency category addresses the issues regarding uncertainty related to the operating time, sequence planning and the process cost.

Operator/tasks/safety category addresses the issues that concern the disassembly tasks performed by the operators, their safety, the working environment and conditions.

Process information category deals with issues regarding the decision support information (data about design, materials, the usage, maintenance, and repair of a product throughout its lifecycle).

One of the most promising technologies in overcoming those issues and the sustainable challenges of the disassembly process is the DTW.

2.2 Digital Twin

The concept of DTW was firstly introduced by the National Aeronautics and Space Administration (NASA)'s Apollo program in 1962 [12]. Since then, numerous explanations and definitions of this concept have been proposed [12]. [13] defined the DTW as a virtual model that corresponds to physical entities in the real world. Based on information collection and sharing, it can simulate the behavior and performance of the entities in real time. It is used mainly for monitoring, diagnostics, prognostics as well as optimization [14]. Which makes it a very good solution for manufactures to improve the manufacturing process [15]. Despite the big potential gain that a DTW can offer to adopt a sustainable approach and to improve the sustainability of manufactured products, there is still a lack of research on this area.

2.3 I4.0 Enabling Technologies in the Disassembly Process

I4.0 relies on technologies that enables the connectivity and interaction between all stakeholders involved in manufacturing [16]. In the framework of this literature review, five main technologies were found mostly used in the optimization and problem solving of the disassembly process. Those I4.0 technologies can be classified into Virtual Reality (VR), Augmented Reality (AR), Human Robot Collaboration (HRC), Radio Frequency Identification (RFID) and Operational Research and Optimization-Based approaches.

VR technology is a three-dimensional technology. It is a computer-generated environment that provides to the users the feeling of being present through multiple synthetic feedback sent to sensorial channels like virtual, aural, haptic and others [17].

AR technology can be defined as a Human-machine technology that combines three-dimensional computer generated information and physical real scene, all in real time [18].

Human-robot collaboration (HRC) is a flexible semi-automated approach that aims to provide flexibility in order to reduce the uncertainties and have more adaptability [19].

Radio frequency identification (RFID) is a technology that uses radio waves to automatically read the digital data encoded in the RFID tags or smart labels [20].

Operational Research and Optimization-Based approaches are approaches based on analytical and optimizational methods such as algorithms, in order to optimize the disassembly process.

3 Overview of I4.0 Technologies Used in the Optimization of the Disassembly Research

The set of 18 documents selected have been analyzed. All the papers were focused on exploring the roles of I4.0 technologies to support, optimize and resolve issues regarding the process of disassembly.

Five main I4.0 technologies have been considered: VR, AR, RFID, HRC and Operational research and optimization-based approaches in the resolution of the disassembly process issues presented in the related work section.

3.1 Cost/Time Efficiency

Many researchers studied the use of I4.0 technologies in the optimization of the cost/time efficiency of the disassembly process. In the VR field [21] developed a system that integrates a virtual assembly and disassembly model into a Computer-aided design/computer-aided manufacturing (CAD/CAM) environment in order to evaluate the products in their design phase for virtual manufacturing. As a result, the cost and time of assembly/disassembly (A/D) operations were optimized. [22] represented a real-time driving model of the virtual human (VH) and analyzed corresponding VH driving error coming from the Motion Capture System. The simulation results can be used for process and operation sequence planning, also to optimize operator comfort, visibility, and access to controls.

Regarding the studies that have addressed the use of AR in order to optimize the cost/time efficiency, [23] presented an AR-guided product disassembly (ARDIS) framework with an automatic content generation module that translates a disassembly sequence into AR-based instructions for the human operators.

In the HRC technology, [24] proposed a framework and enabling technologies of cloud disassembly in combination of cloud manufacturing and robotic disassembly in order to improve the disassembly efficiency and generate the optimal disassembly sequence.

The RFID technology was also used in the optimization of the cost/time efficiency of the disassembly process. The concept of [25] proposed system is that it allows total control over material flow (fixtures, and their elements) which increased the degree of flexibility of the process. [26] presented Advanced Repair-to-Order and Disassembly-to-Order (ARTODTO) model that allowed a lower total cost of the disassembly by determining how to process every EOL product on hand to meet used product and component demands and recycled material demand. [27] proposed a method that aims to facilitate the disassembly strategy and to evaluate the expected profit after product disassembling. It enables to detect and exclude the elements with hazardous substances and exclude them from other materials and minimalize the cost.

In the Operational research and optimization based approaches area, [28] proposed a model based on triple bottom line (TBL) dimensions (i.e. human safety, environmental safety and business criteria) that included 22 disassembly criteria. The results of the study implemented for computer disassembly show an improvement in cycle time, a minimization of the amount of wastes and maximization of recovered materials from EOL. [6] proposed a development of a disassembly sequence planning (DSP) model based on a constraint satisfaction problem (CSP) and its solution method, based on a backtracking algorithm. In order to find all of the feasible disassembly sequences as well as the optimum sequence for recycling. [29] suggested an ant colony algorithm in order to obtain a set of optimization program of the disassembly and assembly sequence and other information in the disassembly process.

3.2 Operator Tasks/Safety

Three I4.0 technologies were used in the category of operator tasks/safety. First, the VR technology where [30] presented a disassembly task evaluation method. It aims to quantify and analyze fatigue induced during disassembly sequence simulation by using the expenditure volume of metabolic and muscle fatigue. [31] developed an operators training system in A/D that exploits an immersive mixed reality system. The application allows operating on a 3D model of a machine, defining the A/D constraints that the operators will respect during the training. Therefore, the operations can be performed and practiced safely, the experience can be repeated several times and incidents can be avoided.

Second I4.0 technology used is the AR technology where [32] presented a model based on AR (self-disassembly) that can be viewed and followed by the operator, based on the laboratory tests that have been performed, this model helps improve the time and costs for industrial maintenance. As well as the development of the operators' technical skills and a good evaluation of their working conditions' safety.

Last I4.0 technology used is the HRC technology. The concept and initial investigations regarding the implementation of a hybrid disassembly work station with a compliant robot assistant for the task of unscrewing in disassembly with only simple tools. Presented by [33] reduces human contact with potentially harmful situations or substances, adds more flexibility to the process, reduces setup, tool replacement time and costs.

3.3 Process Information

Regarding the process information category, the VR technology was used by[34] with the aim of the prevention and reduction of negative environmental impacts of all activities related to the production of a new product. [34] proposed a method of ecological-oriented product assessment performed at early stage of design process in order to improve the skills and knowledge of future designer about environmental aspects of the designed product.

Concerning the AR technology, in order to establish an educational environment for producing robots on line and to deliver enough information for the products assembly and disassembly [35] studied a real space robot disassembly and assembly processes and the movement of the robot work in animation.

Regarding the HRC technology, [36] developed a strategy for HRC disassembly using active compliance monitoring of collaborative robots, that delivers enough information for the products assembly and disassembly to assist the human operator with complex disassembly tasks, the collaborative robot compliance capability is used.

Table 1 is inspired by the tables of the eight defined groups of the roles of simulation to support disassembly under a circular economy perspective by [37] in their study about how disassembly process simulation tools have been exploited to foster the adoption of Circular Economy in the Printed circuit board domain through a symmetric literature review. However, Table 1 summarizes and categorizes all the papers studied according to three main issues of the disassembly process in terms of: a) the main proposition of the article and its type (method, approach, system, etc.); b) the lifecycle phase improved through the concerned technology (Design; Assembly; Maintenance; EOL); c) I4.0 technology used and d) the environmental improvement.

4 Discussion

4.1 Sustainable Disassembly Process

Up to now, the sustainable aspect of disassembly process have been conducted with special focus on reducing the process time and cost [38]. However, the optimization of disassembly plan, sequence and time may lead to an important reduction in energy consumption and environmental impact improvement [8].

According to the literature review above, [6, 7, 22–24, 30, 35, 36] used different I4.0 technologies in order to resolve issues regarding the disassembly process and to optimize the disassembly sequence time and cost. The use of those I4.0 technologies can lead to a more sustainable disassembly process by improving its environmental impact through the optimization of the disassembly process. However, other authors have addressed the environmental aspect in the disassembly in a more direct way in their studies. For instance, the PCDEE-Circle (perception, cognition, decision, execution and evolution – circle) framework developed by [4] towards HRC disassembly meet sustainable manufacturing requirements such as minimizing energy consumption under the premise of ensuring safety. The sustainable disassembly line presented by [28] allows minimizing the amount of wastes and maximizing recovered materials from EOL. In the aim of the prevention and reduction of negative environmental impacts of all activities related to the production of a new product during its lifecycle. Also, [34] proposed a method of ecological-oriented product assessment.

Table 1. Literature summary.

Disassembly issues	Ref	Paper content	Lifecycle phase concerned by the I4.0 technology for disassembly	I4.0 technology for disassembly	Environmental improvement
Cost/time efficiency	[21]	A system that integrates a virtual A/D model into a CAD/CAM environment for evaluating products in design phase	Design/Assembly/EOL	VR	X
	[22]	A model of a virtual real-time human for assembly and disassembly operation in virtual reality environment	Assembly/maintenance/EOL	VR	X
	[23]	AR-guided product disassembly (ARDIS) framework with an automatic content generation module	maintenance/EOL	AR	X
	[4]	Perception, cognition, decision, execution and evolution – circle framework and an approach for HRC disassembly	EOL	HRC	X

(*continued*)

Table 1. (*continued*)

Disassembly issues	Ref	Paper content	Lifecycle phase concerned by the I4.0 technology for disassembly	I4.0 technology for disassembly	Environmental improvement
	[24]	A study of the framework and enabling technologies of cloud disassembly	EOL	HRC	X
	[25]	A system for identification of machining fixtures, and their elements in an assembly/disassembly	EOL	RFID	X
	[26]	Advanced Repair-to-Order and Disassembly-to-Order model determining how to process each and every EOL product	EOL	RFID	
	[27]	A method for facilitating the disassembly strategy and evaluation of expected profit after product disassembling	EOL	RFID	
	[28]	A model based on triple bottom line (TBL) dimensions, i.e. human safety, environmental safety and business criteria	EOL	OR & OBA	X
	[6]	A method of a disassembly sequence planning based on a backtracking algorithm	EOL	OR & OBA	X
	[29]	Ant colony algorithm-based disassembly to obtain a set of optimization program of the disassembly and assembly	Assembly/EOL	OR & OBA	X
Operator/tasks/safety	[30]	A disassembly task evaluation method That aims to quantify fatigue by using the expenditure volume of metabolic	EOL	VR	

(*continued*)

Table 1. (*continued*)

Disassembly issues	Ref	Paper content	Lifecycle phase concerned by the I4.0 technology for disassembly	I4.0 technology for disassembly	Environmental improvement
	[31]	An operator training system in A/D, by means of a 3D camera placed on the Human Machine Disassembly	Design/EOL	AR	
	[32]	AR model (self-disassembly) that can be viewed and followed by the operator	Design/maintenance/EOL	AR	X
	[33]	A concept and initial investigations regarding the implementation of a hybrid disassembly work station	EOL	HRC	
Process information	[34]	A method of ecological-oriented product assessment performed at early stage of design process	Design/EOL	VR	X
	[35]	A study in real space robot disassembly and assembly processes and the movement of the robot work in animation	Assembly/EOL	AR	X
	[36]	A strategy for HRC disassembly using active compliance monitoring of collaborative robots	EOL	HRC	

4.2 Digital Twin for Disassembly Process

Based on the literature review above, decision-making is one of the most important factors in disassembly optimization. A decision-making based on reliable and up-to-date information allows for the optimization of disassembly sequences and a proper planning of the process. Which leads directly to a less costly and more sustainable process (if environmental issues are taken into consideration). The potential of using DTW is very high in terms of decision-making. However, at the end of life step there is a lot of uncertainty regarding the products [9], their traceability and their information (e.g. information about their recyclability, used materials, management of hazardous substances, etc.). To face this problem, a DTW based on knowledge can be very interesting. Which means that the DTW will rely on knowledge-based technologies that we define as the technologies based on reference database that includes all the products information throughout all its life cycle. The knowledge-based technology will provide the entire product's information despite its unknown traceability, which will allow us to avoid the majority of the disassembly process issues mentioned in the literature review and a good decision-making.

5 Conclusion

This paper presents a literature review about the use of I4.0 technologies in the disassembly and dismantling processes. After analyzing the set of 18 documents selected, that focused on exploring the roles of I4.0 technologies to support the process of disassembly and resolving the issues regarding the cost/time efficiency, operator/tasks/safety and the process information. Five main I4.0 technologies have been considered: VR; AR; RFID; HRC; Operational research & optimization based approaches.

In order to construct a DTW for disassembly and dismantling processes, it is very crucial to implement a variety of enabling technologies such as I4.0 technologies conducted from the literature review. However, in order to have a sustainable disassembly and dismantling process, it is necessary to rely on indicators of sustainability. Leading us to the following research questions: Are the I4.0 technologies studied above exploitable in a framework of the creation of a DTW model based on sustainability indicators in order to have a sustainable disassembly and dismantling processes? And which sustainable indicators can be beneficial in the case of a sustainable disassembly and dismantling processes?

References

1. Jovane, F., et al.: A key issue in product life cycle: disassembly. CIRP Ann. Manuf. Technol. **42**, 651–658 (1993). https://doi.org/10.1016/S0007-8506(07)62530-X
2. Vongbunyong, S., Kara, S., Pagnucco, M.: Application of cognitive robotics in disassembly of products. CIRP Ann. Manuf. Technol. **62**, 31–34 (2013). https://doi.org/10.1016/j.cirp.2013.03.037
3. Duflou, J.R., Seliger, G., Kara, S., Umeda, Y., Ometto, A., Willems, B.: Efficiency and feasibility of product disassembly: a case-based study. CIRP Ann. Manuf. Technol. **57**, 583–600 (2008). https://doi.org/10.1016/j.cirp.2008.09.009
4. Liu, Q., Liu, Z., Xu, W., Tang, Q., Zhou, Z., Pham, D.T.: Human-robot collaboration in disassembly for sustainable manufacturing. Int. J. Prod. Res. **57**, 4027–4044 (2019). https://doi.org/10.1080/00207543.2019.1578906
5. Bentaha, M.L., Dolgui, A., Battaïa, O., Riggs, R.J., Hu, J.: Profit-oriented partial disassembly line design: dealing with hazardous parts and task processing times uncertainty. Int. J. Prod. Res. **56**, 7220–7242 (2018). https://doi.org/10.1080/00207543.2017.1418987
6. Li, B., Ding, L., Hu, D., Zheng, S.: Backtracking algorithm-based disassembly sequence planning. Procedia CIRP **69**, 932–937 (2018). https://doi.org/10.1016/j.procir.2017.12.007
7. Wan, H.D., Gonnuru, V.K.: Disassembly planning and sequencing for end-of-life products with RFID enriched information. Robot. Comput.-Integr. Manuf. **29**, 112–118 (2013). https://doi.org/10.1016/j.rcim.2012.05.001
8. Andriankaja, H., Le Duigou, J., Danjou, C., Eynard, B.: Sustainable machining approach for CAD/CAM/CNC systems based on a dynamic environmental assessment. Proc. Inst. Mech. Eng. Part B: J. Eng. Manuf. **231**, 2416–2429 (2017). https://doi.org/10.1177/0954405416629104

9. Abdullah, T.A., Popplewell, K., Page, C.J.: A review of the support tools for the process of assembly method selection and assembly planning. Int. J. Prod. Res. **41**, 2391–2410 (2003). https://doi.org/10.1080/0020754310000087265
10. Gungor, A., Gupta, S.M.: Disassembly sequence planning for products with defective parts in product recovery. Comput. Ind. Eng. **35**, 161–164 (1998). https://doi.org/10.1016/j.rcim.2012.05.001
11. Mok, S.M., Wu, C.H., Lee, D.T.: A hierarchical workcell model for intelligent assembly and disassembly. In: Proceedings - 1999 IEEE International Symposium on Computational Intelligence in Robotics and Automation, CIRA 1999, pp. 125–130. Institute of Electrical and Electronics Engineers Inc. (1999). https://doi.org/10.1109/3468.925662
12. Boschert, S., Rosen, R.: Digital twin—the simulation aspect. In: Hehenberger, P., Bradley, D. (eds.) Mechatronic Futures, pp. 59–74. Springer, Cham (2016). https://doi.org/10.1007/978-3-319-32156-1_5
13. Grieves, M., Vickers, J.: Digital twin: mitigating unpredictable, undesirable emergent behavior in complex systems. In: Kahlen, F.-J., Flumerfelt, S., Alves, A. (eds.) Transdisciplinary Perspectives on Complex Systems, pp. 85–113. Springer, Cham (2017). https://doi.org/10.1007/978-3-319-38756-7_4
14. Haag, S., Anderl, R.: Automated Generation of as-manufactured geometric representations for digital twins using STEP. Procedia CIRP **84**, 1082–1087 (2019). https://doi.org/10.1016/J.PROCIR.2019.04.305
15. Gaha, R., Durupt, A., Eynard, B.: Towards the implementation of the digital twin in CMM inspection process: opportunities, challenges and proposals. Procedia Manuf. **54**, 216–221 (2020). https://doi.org/10.1016/j.promfg.2021.07.033
16. Dantas, T.E.T., de Souza, E.D., Destro, I.R., Hammes, G., Rodriguez, C.M.T., Soares, S.R.: How the combination of circular economy and industry 4.0 can contribute towards achieving the sustainable development goals. Sustain. Prod. Consum. **26**, 213–227 (2021). https://doi.org/10.1016/j.spc.2020.10.005
17. Bamodu, O., Ye, X.: Virtual reality and virtual reality system components. Presented at the (2013). https://doi.org/10.2991/icsem.2013.192
18. Stock, T., Seliger, G.: Opportunities of sustainable manufacturing in industry 4.0. Procedia CIRP **40**, 536–541 (2016). https://doi.org/10.1016/j.procir.2016.01.129
19. Huang, J., et al.: An experimental human-robot collaborative disassembly cell. Comput. Ind. Eng. **155**, 107189 (2021). https://doi.org/10.1016/j.cie.2021.107189
20. Temjanovski, R.: How IT can add Value to logistic sector: Bar code systems and RFID (radio frequency identification) in logistics services. J. Econ. **5**, 47–56 (2020). https://doi.org/10.46763/joe205.20047t
21. Mok, S.M., Wu, C.H., Lee, D.T.: Modeling automatic assembly and disassembly operations for virtual manufacturing. IEEE Trans. Syst. Man Cybern. Part A Syst. Hum. **31**, 223–232 (2001). https://doi.org/10.1109/3468.925662
22. Qiu, S., Fan, X., Wu, D., He, Q., Zhou, D.: Virtual human modeling for interactive assembly and disassembly operation in virtual reality environment. Int. J. Adv. Manuf. Technol. **69**(9–12), 2355–2372 (2013). https://doi.org/10.1007/s00170-013-5207-3
23. Chang, M.M.L., Ong, S.K., Nee, A.Y.C.: AR-guided product disassembly for maintenance and remanufacturing. Procedia CIRP **61**, 299–304 (2017). https://doi.org/10.1016/j.procir.2016.11.194
24. Yuan, L., Cui, J., Zhang, X., Liu, J.: Framework and enabling technologies of cloud robotic disassembly. Procedia Comput. Sci. **176**, 3673–3681 (2020). https://doi.org/10.1016/j.procir.2016.11.194
25. Vukelic, D., et al.: Machining fixture assembly/disassembly in RFID environment. Assem. Autom. **31**, 62–68 (2011). https://doi.org/10.1108/01445151111104182

26. Ondemir, O., Ilgin, M.A., Gupta, S.M.: Optimal end-of-life management in closed-loop supply chains using RFID and sensors. IEEE Trans. Industr. Inf. **8**, 719–728 (2012). https://doi.org/10.1109/TII.2011.2166767
27. Nowakowski, P.: A novel, cost efficient identification method for disassembly planning of waste electrical and electronic equipment. J. Clean. Prod. **172**, 2695–2707 (2018). https://doi.org/10.1016/j.jclepro.2017.11.142
28. Kazancoglu, Y., Ozkan-Ozen, Y.D.: Sustainable disassembly line balancing model based on triple bottom line. Int. J. Prod. Res. **58**, 4246–4266 (2020). https://doi.org/10.1080/00207543.2019.1651456
29. Shan, H., Li, S., Huang, J., Gao, Z., Li, W.: Ant colony optimization algorithm-based disassembly sequence planning. In: Proceedings of the 2007 IEEE International Conference on Mechatronics and Automation, ICMA 2007, pp. 867–872 (2007). https://doi.org/10.1109/ICMA.2007.4303659
30. Chen, J., Mitrouchev, P., Coquillart, S., Quaine, F.: Disassembly task evaluation by muscle fatigue estimation in a virtual reality environment. Int. J. Adv. Manuf. Technol. **88**(5–8), 1523–1533 (2016). https://doi.org/10.1007/s00170-016-8827-6
31. Sportillo, D., Avveduto, G., Tecchia, F., Carrozzino, M.: Training in VR: a preliminary study on learning assembly/disassembly sequences. In: De Paolis, L.T., Mongelli, A. (eds.) AVR 2015. LNCS, vol. 9254, pp. 332–343. Springer, Cham (2015). https://doi.org/10.1007/978-3-319-22888-4_24
32. Freddi, M., Frizziero, L.: Design for disassembly and augmented reality applied to a tailstock. Actuators **9**, 1–14 (2020). https://doi.org/10.3390/act9040102
33. Chen, W.H., Wegener, K., Dietrich, F.: A robot assistant for unscrewing in hybrid human-robot disassembly. In: 2014 IEEE International Conference on Robotics and Biomimetics, IEEE ROBIO 2014, pp. 536–541. Institute of Electrical and Electronics Engineers Inc. (2014). https://doi.org/10.1109/ROBIO.2014.7090386
34. Grajewski, D., et al.: Improving the skills and knowledge of future designers in the field of ecodesign using virtual reality technologies. Procedia Comput. Sci. **75**, 348–358 (2015). https://doi.org/10.1016/j.procs.2015.12.257
35. Jung, H.-J., Park, D.-W.: A study of user-controlled robot simulation on augmented reality. In: Lee, G., Howard, D., Ślęzak, D., Hong, Y.S. (eds.) ICHIT 2012. CCIS, vol. 310, pp. 582–587. Springer, Heidelberg (2012). https://doi.org/10.1007/978-3-642-32692-9_73
36. Huang, J., et al.: A strategy for human-robot collaboration in taking products apart for remanufacture. FME Trans. **47**, 731–738 (2019). https://doi.org/10.5937/fmet1904731H
37. Sassanelli, C., Rosa, P., Terzi, S.: Supporting disassembly processes through simulation tools: a systematic literature review with a focus on printed circuit boards. J. Manuf. Syst. **60**, 429–448 (2021). https://doi.org/10.1016/j.jmsy.2021.07.009
38. Zhou, Z., et al.: Disassembly sequence planning: Recent developments and future trends. Proc. Inst. Mech. Eng. Part B: J. Eng. Manuf. **233**, 1450–1471 (2019). https://doi.org/10.1016/j.rcim.2019.101860

Characterization of Integrated Lean Tools, Adapted to the Upstream Phase of the Product Life Cycle of Stäubli Electrical Connectors SAS

Sébastien Maranzana[1,2(✉)], Nedjwa Elafri[2,4(✉)], and Rose Bertrand[2,3(✉)]

[1] Stäubli Electrical Connectors SAS, Hésingue, France
s.maranzana@staubli.com
[2] Laboratoire ICUBE, Strasbourg, France
elafrinedjwa@gmail.com, bertrand.rose@unistra.fr
[3] Université de Strasbourg, Strasbourg, France
[4] Université de Constantine, Constantine, Algeria

Abstract. Evolution of markets induced by strong competitive and technological pressure that forces companies to constantly evolve to gain in performance and remain attractive. Customer needs are also evolving and become more complex with ever shorter deadlines. In this context, the launch of a new product represents a challenge for companies.

Lean methodology appears as an interesting way for companies to improve their performance. Native from the manufacturing industry, Lean Manufacturing has widely demonstrated its effectiveness in improving the performance of production workshops. Today it is derived for a large number of activities that are grouped under the methodology of Lean Management and are adapted to different environments such as services, information technologies, software development, ...

Key period for the success of a new product, the upstream phase of the product life cycle relies mainly on the efficiency of operational and strategic processes as well as the information systems of the company.

Deploying a Lean approach to improve the performance of the organization in the launch phase of a new product means intervening in different interconnected environments: product development, administration and information technology. An integrated Lean approach that compiles the tools of Lean IT, Lean Office and Lean Product Development is interesting.

In this article we will first try to establish the characteristics of integrated Lean adapted to the first stages of the product life cycle of the company Stäubli Electrical Connectors SAS by a study case approach. In a second step we will define by a bibliographic analyze tools of Integrated Lean and their fields of application. The objective of this article is to show the interest of a global Lean approach in the upstream phase of the product life cycle rather than a spitted vision by environment.

Keywords: Lean Management · Product life cycle · Lean tools · Integrated Lean

F. Noël et al. (Eds.): PLM 2022, IFIP AICT 667, pp. 641–651, 2023.
https://doi.org/10.1007/978-3-031-25182-5_62

1 Research Methodology

This research work use a Bottom Up research strategy based on the Case Study approach [1]. To introduce this study, we will start by presenting the industrial context and issues based on the Company Stäubli Electrical Connectors SAS. And then, we will continue by presenting the framework of this study: the upstream stage of the product life cycle.

1.1 Stäubli Electrical Connectors SAS

Stäubli Electrical Connectors SAS is a company specialized in the manufacture of electrical connectors. These connectors are based on the Multilam® contact technology, which provide to Stäubli products high performance even in strict environments. Based in Hésingue (France), the company belongs to the Electrical Connectors division of the Stäubli Group. This division is composed of six sites, three of them have been historically located in the Three Border Region (Germany, Switzerland and France) around the city of Basel. This location is the result of a desire to promote exchanges between the three sites and to ensure national representation. Stäubli EC SAS is the competence center of the Group for railways and E-mobility Markets.

Hésingue factory has a complete industrial tool: project and sales teams, design office, purchasing and quality department, a modern production tool (assembly) and a complete test laboratory. These resources allow us to manage strong R&D activities.

The range of products developed and manufactured by Stäubli EC SAS is very wide and varied, from standard products (in the catalog) to specific solutions for customer applications, including a large part of R&D solution development to follow the markets. This important product mix implies a great variability in terms of production mean, volumes and required deadlines. This imposes a great complexity of the Supply Chain management modes.

1.2 Industrial Issues

Stäubli EC SAS is present on markets that are constantly changing and subject to strong competitive and technological pressure. The company must continually evolve to improve its performance by optimizing the cost-quality-delivery triptych. One of the factor of success is the ability to develop and produce complex and varied solutions to meet the needs of customers within short deadlines.

Since 2000, the electrical connector market has seen the establishment of new actors based in low production costs countries, and the opening up of markets has given customers access to a wide choice of suppliers. Historical and leading players are particularly sensitive to this competitive pressure.

Among the key cost-quality-delay factors taken into account by customers, product quality and price are no more the only differentiating factor. Through decisive elements: delivery time and customer care, are taken into consideration by the customer in the choice of suppliers. In addition, the market and needs are changing. Now customers are more looking for global solutions that provide a complete function rather than a supplier of components to be integrated.

This change of consumption model is leading to increased product complexity and changes in manufacturers (specialization and/or change of core business).

The electrical connector market is characterized by a rapid evolution of technologies, norms and standards as well as the constant arrival of new applications. This evolution offers interesting opportunities, however to take advantage of these, it is essential to develop the ability of design more complex products in a shorter time frame.

The company have to adapt to the needs of the Market and their customers: Offer the right products in the right timeframe. To achieve this objective, Stäubli EC SAS want to improve its industrial performance and reactivity during the upstream phase of the product life cycle. Starting with the definition of the specifications before the design of the product until the first deliveries to the customer.

The launch of a new product represents a key period. The whole success depends particularly on the performance of the operational processes in place and the ability of the various stakeholders in the company (Project, Design Office, Purchasing, Planning, Production, etc.) to communicate effectively each other.

The industrial performance of a company is therefore based on its ability to adapt its organization, its operational processes and its communication system to meet the needs of its customers. To perform on the market, companies have to improve their industrial performance in the upstream phase of the product life cycle.

1.3 Upstream Stage of Product Life Cycle

The concept of product life cycle appeared for the first time in the 1950s and 1960s, following the analysis of the characteristic evolution of the markets and products, since their appearance until their disappearance.

The product life cycle is commonly represented by a "life curve": this bell-shaped curve represents the three characteristic phases of the evolution of the product during its life. The first phase represents an expansion phase corresponding to the marketing of the product and its possible adaptations to needs. The second phase takes the form of a flat line representing the period of use of the product by users. The last and third phase: the recession which corresponds to the obsolescence of the product mainly linked to the technological evolution or to the modification of the consumption practices. The usual representation shows the evolution over time of the "satisfaction index" translated into accounting units: sales, turnover, etc. [2].

The upstream phase of the product cycle is particularly interesting for us, because it condenses key factors of the product (quality/cost/delay) and processes used will depend the industrial performance of companies.

The "birth" of a new product is closely linked to the innovation process used by the company. The purpose of representing the innovation process is to describe the different functions and trades mobilized as well as the mechanism and the different steps/milestones. The innovation process allows decisions to be made to arrive at the development of a new product [3].

The innovation process is intended to define the organization of activities throughout the process and to enable decision making. Each stage becomes more complex and costly, as the commitment of the company to product development increases.

The proper conduct of this process ensures the maturity and viability of the project and its product.

Nevertheless, this project management-oriented vision remains distant from the company operational activities. The link between the "project" activity and the "operational activity" is very important, because often the same actors act on several projects simultaneously in addition to their operational activities outside the project phases. It is also necessary for project managers to be aware of the operational tasks at all levels of the project life cycle [4].

We often forget that it is on the operational processes that depend a part of the industrial performance. Indeed, in the innovation process, the actions allowing the launch of a product are based on the operational process, its actors and how they communicate together. In the upstream stage of product life cycle the industrial performance is linked to the integration of companies' operational process in their innovation process.

2 Lean Method

In this part, we will first introduce the Lean method as a global approach. In a second time we will make a focus on its use in the upstream phase of the product life cycle and try to characterized the Integrated Lean notion.

2.1 Lean Basics

Since its appearance at the end of the 1970s, Lean has demonstrated its effectiveness in improving the industrial performance of manufacturing companies.

The definition of Lean varies according to the authors. However, it can be defined as a philosophy or a set of practices aimed at achieving improvements by following the most economical paths while focusing on reducing waste [5].

The book "Lean Thinking" [6], is a retrospective study of Lean practice, synthesizes the Lean philosophy into five principles:

- Accurately identify the added value of products, from a customer perspective.
- Determine the value chain of products.
- Establish continuous value flows.
- Pull the flows through the customer.
- Aim for excellence.

Lean has its origin in practices developed by the automotive company TOYOTA and its production system: The Toyota Production System (TPS). These practices were analyzed and transposed in the book "The Toyota Way" [7]. The TPS is based on the elimination of non-value added. Seven types of waste (Muda) have been identified: overproduction, overstock, unnecessary transport, unnecessary processes, unnecessary movements, errors and rejects and waiting time. Two other types of non-value added were also defined: variability (Mura) and work overload (Muri).

In the early 2000s, Lean leaves production workshops and its principles are declined to different environments. First, in the world of management (Lean Management) and

Services (Lean Service) and then adapted to many activities as: administrative services (Lean Office), product development (Lean Product Development) or Software development (Lean Software Development) [8].

2.2 Integrated Lean

We have seen that the success of a company is linked to its industrial performance, its ability to "deliver the right product at the right time at the right price". All these key characteristics for the success of a product are defined during the upstream phase of the product life cycle.

The Lean methodology has now proved its worth in improving industrial performance, and its use has been democratized and adapted to different environments.

The use of Lean in the upstream phase of the product life cycle is therefore an interesting way to improve the industrial performance of Stäubli EC SAS.

When we schematize the upstream phase of the product life cycle (Fig. 1), two processes appear. The innovation process, which groups together all activities linked to the project and the operational process, which groups together all the functional activities of the company. These two processes are interconnected and linked by an ecosystem of information systems.

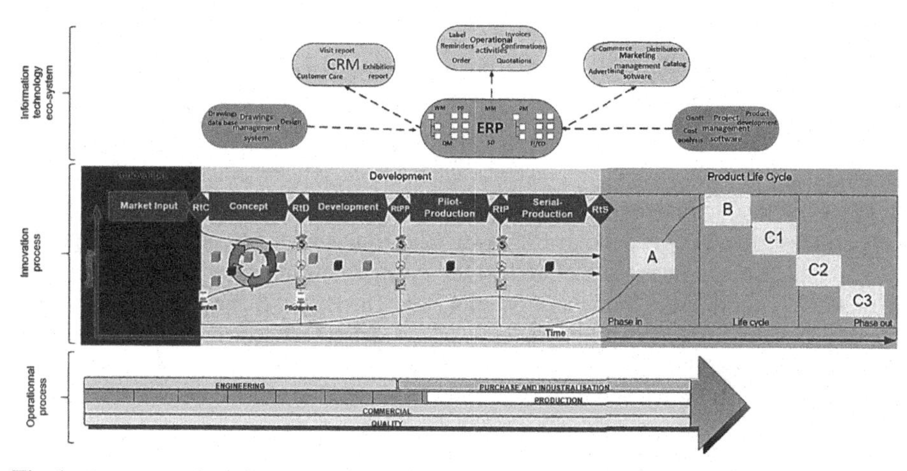

Fig. 1. Representation of the different processes that compose the upstream phase of the Stäubli EC SAS product cycle (Source: Stäubli EC SAS)

The upstream phase of the product life cycle is therefore based on three distinct but linked environments:

- "Project" environment, which consists of developing new products.
- Company operational environment, which is common to all projects, in the upstream phase of the product life cycle corresponds mainly to administrative activities.
- Information systems ecosystem: all of the company's internal communication resources.

For each of these environments an adaptation of the Lean methodology exists: Lean Product Development for new product development [9], Lean Office for administrative activities [10] and Lean IT for information systems [11].

These different environments are interconnected and inseparable. Each action has a direct or indirect impact on the other activities in the upstream phase of the product life cycle. It is therefore difficult and limited effectiveness to conduct a focused and unilateral Lean approach on a separate environment in this product life cycle phase.

A holistic and integrated approach of Lean in the upstream phase of the product life cycle and focused to eliminating Lean waste (Muda) seems interesting. A new approach to improving industrial performance named Integrated Lean. Who integrates components of Lean IT, Office and Product Development characterized by a global approach in the upstream stage of product life cycle [11].

3 Integrated Lean Tools

In this part we are going to present our bibliographical analysis work. At first, define major Integrated Lean Tools and then make the link between these tools and their uses.

3.1 Definition of Integrated Lean Tools

Goal

The aim of this study is to define the major Integrated Lean tools available for the upstream stage of product life cycle by using the knowledge already present in the Lean literature.

Methodology

To define the major Integrated Lean tools, we performed a literature review based on forty articles from Google Scholar Data base (show in annex). The notion of Integrated Lean in the upstream phase of the product life cycle is new and poorly described in the literature. We get interested by these components: The Lean IT, Product Development, Office. For each component of Integrated Lean, we identified the different tools and their recurrences in the bibliography.

Result

Figure 2 shows the result of the review. Forty-three Lean tools have been identified in overall. Six tools are found in both three components of Integrated Lean: VSM, Standard Work, KPI's, A3 problem solving, Heijunka (reduce variability) and DMAIC. Some other tools are available in two components: Kaizen, 5S, Just in Time, Kanban and Gamba Walk. Other tools appear only in one components. We also note that the tool VSM come the most offen in this review.

Analyze

The first observation we can make concerns the representation in the different components of the integrated Lean of each tool.

Tools common for the three components IT, Product Development and Office are the most versatile and able to act on the whole upstream stage of product Life cycle. These six tools defined during our analysis appear as the major tools of Integrated Lean.

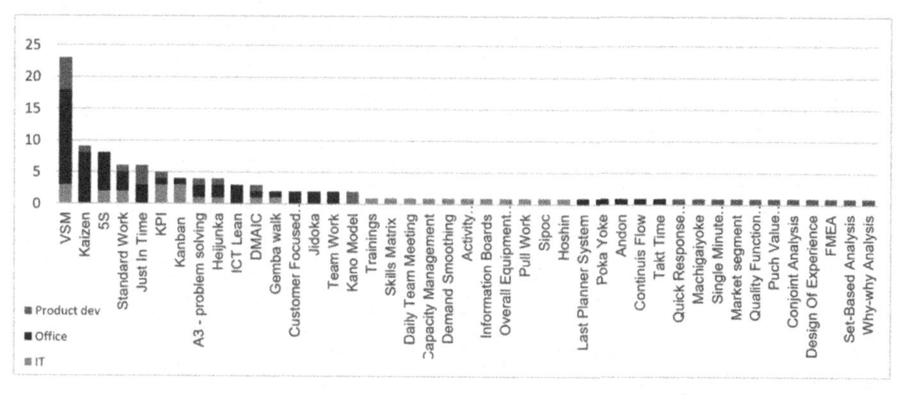

Fig. 2. Recurrence of lean IT/Product development/Office tools

Several tools are present in a recurrent number on two of the three components of Integrated Lean. Versatile enough to act on several part of the upstream phase of the product life cycle, these secondary tools of Integrated Lean are useful to improve numerous activities but not on the whole upstream phase.

The last group of tools that appears from this analysis seems to act only on one of the components of Integrated Lean. They are specific tools dedicated to specialized activities and environments. These tools are not part of the Integrated Lean toolbox and remain dedicated to their components.

We also notice some frequency variations between the principal tools of Integrated Lean: VSM tool is much more frequent than the DMAIC tool for example. This difference is perhaps related to the ease of implementation or a particular interest of the authors for this tool and its value. The last statement concerns the literature review itself. Not all components of Integrated Lean are at the same level of maturity. Lean Office seems to be more explored and described than Lean Product Development or IT which promise interesting opportunities.

3.2 Integrated Lean Tools and Associate Muda

Goal

Now that we have defined the major tools of Integrated Lean, we tried to establish the link between these tools and their use against Lean waste, the seven Muda.

Methodology

For the six major Integrated Lean tools identified earlier, we reviewed twenty-two article from Google Scholar database to identify the impacts of each tool on the different Lean wastes. Specifically, for each integrated Lean tool, we examined the literature to identify the link between the use of this tool and its efficacy on each Muda.

Results

Figure 3 shows the impact of the six selected Integrated Lean tools on the Lean Muda.

We observe that some tools impact almost all Muda: VSM and A3 problem solving. The others tools like Standard Work, Heijunka, DMAIC, KPI impact only some Muda. As in the previous analysis, the tool VSM appear the most often in this review.

We saw also that the six tools selected as Integrated Lean tools cover all seven Muda: Inventory and Waiting come back in a more recurrent way in the field of action of identified tools. Muda Motion seems to be the least impacted.

Tools / Muda	Over productivity	Waiting	Inventory	Transpor-tation	Over processing	Motion	Defects	Authors
VSM	X	X	X		X	X	X	(Perdana and Tiara, 2020), (Garza-Reyes, Romero, Govindan, Cherrafi and Ramanathan, 2018), (Kosasih, Sriwana, Sari and Doaly, 2019), (Guo, Jiang, Xu and Peng, 2019), (Jamil and Al., 2020)(Parab and Shirodkar, 2019), (Maalouf and Zaduminska, 2019)
Standard Work	X	X	X					(Vajna and Tangl, 2017), (Okpala, 2014)
KPI		X					X	(Leksic, Stefanic and Veza, 2020), (Maulina, 2021)
A3	X	X	X	X	X	X	X	(Anouar and Al., 2022), (Flug and Nagy, 2016)
Heijunka			X	X	X			(Gerger, 2019), (Aoki and Katayama, 2017), (Bebersdorf and Huchzermeier, 2022)
DMAIC		X	X	X				(Ferreira and Al., 2019), (Arafeh, 2015), (Guo and Al., 2019), (De Barros and Al., 2021), (Rifqi, Zamma, Souda and Hansali, 2021), (Jamil and Al., 2020)

Fig. 3. Impacts of integrated Lean tools on the seven Lean Muda.

Analyze

We can draw the following interpretations. VSM et A3 problem solving impact almost all Muda, they are therefore versatile in treating all Lean wastes. Other Integrated Lean tools are more specialized against specific Muda. As evoked by the distribution of the impacts on the Muda: the joint use of these tools would make it possible to reduce all type of waste define in the Lean literature.

In addition to targeting Muda, some tools are known to act against other forms of non-value added. Heijunka is a tool used to limit the variability (Mura) and Standard Work is an efficient way to reduce overburden processes (Muri). The more specific tools can therefore also have effects on other non-value added than Muda.

Knowledge of the impact of Integrated lean tools on the Muda will allow for a better choice of tools according to the needs and priority of the company.

4 Conclusion and Opportunities

We are going to conclude on this bibliographical analysis work and open on the limits and opportunities that it offers.

4.1 Conclusion

The upstream phase of the product life cycle appears to be a key moment in the success of a product and the stage that defines the industrial performance of the company. Lean, is a well-known way to improve industrial performance and is now widely used and adapted to many environments and activities. This upstream stage composed of three linked and interconnected environments: operational and innovation processes support by information ecosystem. A global and common Lean approach on the upstream phase of the product lifecycle is necessary and interesting. The Integrated Lean, that includes the components of Lean IT, Office and Product Development. This study based on the Lean literature allowed to define the major tools of Integrated Lean: VSM, Standard Work, KPI's, A3 problem solving, Heijunka and DMAIC. Secondly, we analyzed the impact of these tools on the Muda of Lean. This enables us to describe that some tools can cover all the Muda and that others seem to be more specialized.

These elements highlighted by this study will help Stäubli Electrical Connectors SAS in its deployment of an Integrated Lean strategy.

4.2 Opportunities

This study based on the review of Lean literature shows the limits of the bibliographic approach: data's does not come from the field but from the existing state of the art. To establish and validate conclusions, a field assessment is required. To confirm elements highlighted in this study. It would be necessary to set up a field study with companies implementing a Lean approach in the upstream phase of the product life cycle. The scientific methodology we want to apply is a crossed approach based on "Case study" method, benchmarking and analysis industrial practices and the "Grounded theory" a bibliographic approach [12].

The final objective would be to set up a roadmap for the deployment of Integrated Lean based on the knowledge of the Lean tools and their impacts on Muda associated with methodology crossing theory of constraints and influence modeling [13].

Annex: Integrated Lean Tools Definition

Category	Reference	Tools
Lean IT tools	Kobus, (2016)	Standard Work, KPI, Trainings, Skills Matrix , Daily Team Meeting, VSM, Capacity Management, Demand Smoothing, Activity Implementation Plan
	Kobus et al., (2018)	5S, Kanban, KPI
	Kobus, Westner, & Strahringer, (2017)	Information Boards, 5S, Kanban, Overall Equipment Effectiveness, Pull Work, VSM
	Ferreira, Almeida, & Grilo, (2018)	VSM, SIPOC, A3 problem solving, DMAIC, Gemba Walk, Kanban, Standard Work, KPI, Heijunka, Hoshin
Lean Office tools	J. C. Chen & Cox, (2012)	VSM
	(Sabur & Simatupang, (2015)	Customer Focused Approach
	Tapping & Shuker, (2003)	VSM
	Chiarini & Gabberi, (2020)	VSM
	Tonkin, (2019)	Customer Focused Approach
	Power, Sinnott, & Mullin, (2021)	Last Planner System
	Magalhães, Alves, Costa, & Rodrigues, (2019)	KPI, 5S, Standard Work
	J. Monteiro, Alves, & do Sameiro Carvalho, (2017)	Standard Work, Kaizen, VSM, 5S, Poka Yoke
	F. Costa, Kassem, & Staudacher, (2021)	A3 – Problem solving
	Rossiti, Serra, & Lorenzon, (2016)	VSM
	Rüttimann, (2019)	Andon, VSM, Kaizen, DMAIC, Kanban
	Freitas & Freitas, (2020)	ICT Lean, 5S, VSM
	Pontes, Silva, Martins, Santos, & Siqueira Silveira, (2019)	Just In time ,Jidoka, Team Work
	Gauze, Souza, & Vaccaro, (2017)	Kaizen
	Macedo, Silva, & Freitas, (2021)	Just in Time, Team Work, Jidoka, Heijunka, Kaizen
	Santos, Freitag, & Quelhas, (2020)	VSM
	Sastre, Saurin, Echeveste, de Paula, & Lucena, (2018)	Continuous Flow, VSM, 5S, Standard Work, Takt Time, Heijunka, Kaizen
	M. F. J. R. Monteiro, Pacheco, Dinis-Carvalho, &Paiva,(2015)	Team work, Gemba Walk ,5S, A3–problem solving
	NĂFTĂNĂILĂ & MOCANU, (2014)	5S
	Antoniolli, Lima, Argoud, & Batista de Camargo Jr, (2015)	VSM, Just In Time, Kaizen
	Danielsson, (2013)	ICT Lean
	Rossiti et al., (2016)	Kaizen
	Rüttimann, Fischer, & Stöckli, (2014)	Kaizen
	Rüttimann & Stöckli, (2016)	Kaizen, Just In Time
	Sukma, Amrina, & Hasan,(2018)	VSM
	L. De Souza, Tortorella, Cauchick-Miguel, & Nascimento, (2018)	VSM
	De Siqueira & Da Silva, (2020)	VSM
	Strandhagen, Vallandingham, Alfnes, & Strandhagen, (2018)	VSM
	Cavdur, Yagmahan, Oguzcan, Arslan, & Sahan, (2018)	VSM
	Jiménez Calzado, Domínguez Somonte, Espinosa, Awad Parada, & Romero Cuadrado, (2021)	ICT Lean
Lean Product Dev. tools	Gershenson & Pavnaskar, (2003)	VSM, Just In Time, Quick Response Product Development, Heijunka, Machigaiyoke, Single Minute Exchange Of Projects, Kaizen
	León & Farris, (2011)	VSM, A3, Just In Time
	Wang, Ming, Kong, Li, & Wang, (2012)	Market Segment, Kano Model, Quality Function Deployment Analysis, Puch Value Engineering, Conjoint Analysis, Design Of Experience, FMEA, Set-Based Concurrent Engineering, Standard Work, Why-Why Analysis, VSM
	Marodin, Frank, Tortorella, & Netland, (2018)	Just In Time
	Albuquerque, Torres, & Berssaneti, (2020)	VSM
	Cooper & Appell, (2021)	VSM, DMAIC, KPI, Kano Model

References

1. Yin, R.K.: Case Study Research Design and Methods, 5th edn. Sage Publication, Thousand Oaks (2014)
2. Bernat, J.P., Marcon, C.: Cycle de vie et courbe d'apprentissage de produits complexes: le cas des outils coopératifs de management des connaissances. In: VSST 2001, Barcelone, Espagne (2001)
3. Ferioli, M.: Phases amont du processus d'innovation: proposition d'une méthode d'aide à l'évaluation d'idées. Institut National Polytechnique de Lorraine, France (2010)
4. Legardeur, J., Boujut, J.F., Tiger, H.: Lessons learned from an empirical study of the early design phases of an unfulfilled innovation. Res. Eng. Design **21**, 249–262 (2010). https://doi.org/10.1007/s00163-010-0090-5
5. Elrhanimi, S., El Abbadi, L., Abouabdellah, A.: Proposition d'un tableau de bord pour l'évaluation de l'impact du Lean manufacturing sur la performance globale de l'entreprise. In: Xème Conférence Internationale: Conception et Production Intégrées, Tanger, Maroc (2015)

6. Womack, J.P., Jones, D.T.: Lean thinking—Banish waste and create wealth in your corporation. J. Oper. Res. Soc. **48**(11), 1148 (1996)
7. Liker, J.K.: The Toyota Way: 14 Management Principles from the World's Greatest Manufacturer. McGraw-Hill Edition (2004)
8. Ughetto, P.: Le lean: pensée et impensé d'une activité sans relâchement. Activités **9**(2) (2012)
9. Hoppmann, J.: The lean innovation roadmap – a systematic approach to introducing lean in product development process and establishing a learning organization. Institute of Automotive Management and industrial Production Technical University of Braunschweig (2009)
10. Reinhard, E.: Contribution méthodologique à l'introduction du lean office dans un service support de gestion des approvisionnements: analyse longitudinale par étude de cas dans une entreprise fournisseur du secteur de la santé. Université de Strasbourg (2017)
11. Ignace, M.P., Ignace, C., Médina, R., Contal, A.: La pratique du Lean management dans l'IT. Pearson (2012)
12. Maranzana, S., Rose, B., Leiritz, J.M.: Une démarche Lean dans la phase amont su cycle de vie produit – Application au processus opérationnel de l'Entreprise Stäubli Electrical Connectors SAS. In: CIGI Qualita 2021, Grenoble, France (2021)
13. Seleem, S.N., Attia, E.A., Karam, A., El-Assal, A.: A lean manufacturing road map using fuzzy-DEMATEL with case-based analysis. Int. J. Lean Six Sigma (2020)

The Use of the Toyota Kata Approach: A Literature Review and SWOT Analysis

Raphael Odebrecht de Souza[1](✉), Danilo Ribamar Sá Ribeiro[2], Helio Aisenberg Ferenhof[3], and Fernando Antônio Forcellini[1]

[1] Department of Mechanical Engineering, Federal University of Santa Catarina, Florianópolis, Brazil
raphael.odebrecht@posgrad.ufsc.br

[2] Graduate Program in Production Engineering, Federal University of Santa Catarina, Florianópolis, Brazil

[3] Post-Graduate Program in Knowledge Engineering and Management, Federal University of Santa Catarina, Florianópolis, Brazil

Abstract. Toyota Kata (TK) approach emerged to how Toyota can continuously improve its processes over time. Thus, several western companies started using this approach to improve their production processes. However, there is a lack of studies that present and summarize the main opportunities and challenges when implementing and using TK in practice. Based on this, we present the result of a Structured Bibliographic Research (SBR) and a conducted SWOT analysis (strengths, weaknesses, opportunities, and threats) of the TK in a manufacturing environment. As result, it was possible to analyze eighteen works that discussed the implementation of the TK in industries. Several common aspects of the success, as well as the main barrier, could be identified. In the SWOT analysis, we demonstrated the main strengths to achieve the benefits of implementing TK. The weaknesses and threats identified may challenge some organizations. We also present a first conceptual model that aims to help organizations that want to use the TK approach in conjunction with others management approaches to improve its processes. As an implication, this manuscript presents the important steps organizations have taken to improve its processes. This way opens a new perspective for managers to lead in process optimization.

Keywords: Toyota Kata · Lean · Process improvement · Structured Bibliographic Research · SWOT analysis

1 Introduction

Since the publication of the book "The Machine that Changed the World" by [1], the Toyota Production System titled "Lean Manufacturing" has become a popular method within western manufacturing companies. However, despite all the efforts of copying and studying the Toyota improvement method, no organization has ever matched the performance of its Japanese precursor [2]. Until today, most companies' efforts were

Published by Springer Nature Switzerland AG 2023
F. Noël et al. (Eds.): PLM 2022, IFIP AICT 667, pp. 652–662, 2023.
https://doi.org/10.1007/978-3-031-25182-5_63

directed toward implementing Lean tools to improve operational processes while placing little, if none, attention to the people who improve these processes [3].

To answer why and how Toyota can continuously improve its processes, [2] went to the company and studied the employees' routines. The answers found by the author were published in the book "Toyota Kata: Managing People for Improvement, Adaptiveness and Superior Results" in 2010. In short, the routines, patterns, and their resulting behavior conducted daily within Toyota are called Kata. The Toyota Kata (TK) approach consists of two complementary routines called Coaching Kata and Improvement Kata. When these routines are repeated constantly, they can make the continuous improvement part of the organization's culture.

In his work, [2] describes how the routine for process improvement is carried out by Toyota and presents some brief examples. However, it does not show a detailed case study on the use of the TK with metrics and results. In addition, the book is a description of a routine already practiced by Toyota. Thus, an example of its implementation from scratch is not presented.

Given the importance of the topic, the TK has become the theme of research by several universities and is used by companies worldwide to improve their processes Based on this scenario, to complement existing studies and understanding of the TK, this paper aims to present the result of a Structured Bibliographic Research (SBR) and SWOT analysis to identify success cases and the main opportunities and barrier when using the TK in the manufacturing environment. We also present conceptual model that aims to help organizations that want to use the TK approach to improve its processes.

2 Theoretical Background

2.1 Toyota Kata

Kata means routine or "way of doing" and is described as a repeated pattern and thus develops skills and a new mindset for continuous improvement. This pattern directly interferes with the way employees think and behave, as it is a continuous improvement routine that provides people training based on applying the scientific method for sustainably solving problems [2]. The Toyota Kata approach aims to incorporate continuous improvement thinking in employees [4]. According to [2], in a process improvement situation, there is an unknown path between the desired performance (Target Condition) and the current state of the process (Current Condition). Therefore, continuous improvement is the ability to walk to the desired condition through a murky territory, be sensitive to it, and respond to the terrain conditions. The Toyota Kata approach consists of two complementary routines: The Improvement Kata and the Coaching Kata. The Improvement Kata is performed by leaner and can be used at all organization levels. It is divided into four steps, as shown in Fig. 1.

The Coaching Kata routine is practiced by all organizational levels at Toyota. Each employee is assigned a more experienced mentor (or coach) who provides operational guidance by generating real improvements or dealing with work-related situations. The coach conducts cycles with the learner next to a Storyboard, where information and the PDCA cycle are recorded.

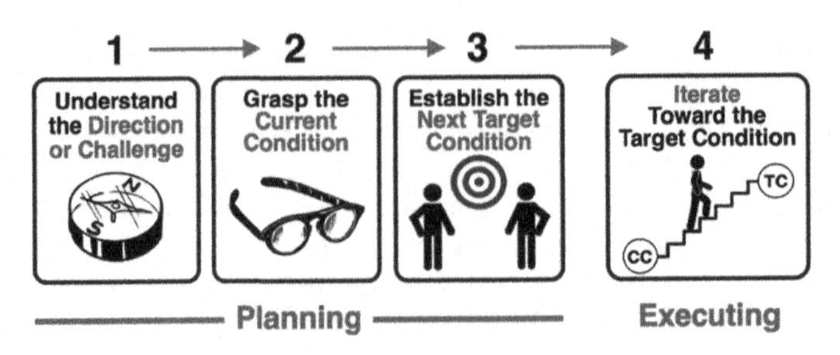

Fig. 1. The Improvement Kata routine. Source: [2].

3 Methodology

An extended Structures Bibliography Research (SRB) following the Systematic Search Flow (SSF) method from [5] was conducted to gather information about the utilization of the TK in manufacturing companies. Leading journals, conference proceedings, master dissertations, and Ph.D. thesis were searched using the keywords that the online databases Scopus, Web of Knowledge, Engineering Village, and ProQuest offer. The search was also complemented by exploratory research on Google Scholar. In addition, texts not available in their entirety were excluded, considering works in English, Portuguese and Spanish, without limitation of date, and containing the respective words in their abstracts, titles, or keywords. The query used was ("Toyota Kata"); the research was carried out with the respective query in each of the bases. Table 1 shows the inclusion and exclusion criteria used in this research.

Table 1. Inclusion and exclusion criteria.

Criteria	Explanation
Inclusion	The research efforts related to studies that focused on TK, i.e., studies have some kind of practical implication;
	Document type: Articles; Master's dissertations/PhD thesis;
	Language: English, Portuguese and Spanish;
	Time: Until March 25, 2022
Exclusion	Duplicate in databases;
	Documents that are not available online;
	It is used only as an example/part of research direction and future perspective/a quoted expression/in keywords and references;
	Non-adequacy to the scope of this study

The searches returned 115 works that were evaluated according to the reading of their titles, abstracts, and keywords. Of these, 39 were in line with the work context and were read in full. In the end, 18 works were fully aligned with the research context.

The synthesis of the results was divided into thematic and SWOT analysis. Thematic analysis is a common technique for qualitative data, which comprises the organization of the studies into guiding axes for the research [6]. At last, a SWOT analysis was conducted. This tool is characterized by the analysis of strengths (s), weaknesses (w), opportunities (o), and threats (t) [7]. Recently, it has been used to provide comprehensive synopsis in several areas [8–12]. Its use in this research was because it is considered a robust approach to summarize the results [13] and allows to present all the information collected in the literature review also with a visual approach that is useful to better frame the different instruments [10].

4 Results and Discussion

4.1 Content Analysis

This section presents the main results concerning the bibliographic portfolio. Of the analyzed works, seven were master's dissertations/Ph.D. thesis, and eleven were articles. Table 2 summarizes the analyzed works.

Table 2. Resulting bibliographic portfolio.

Classification	Authors
Master's/Ph.D. thesis	[14–20]
Articles	[21–31]

In the following paragraphs, the main aspects found in the literature review will be discussed. First, an overall overview of the prerequisites (e.g. team training) needed before implementing the TK is presented, followed by combining TK with other tools/methods. Next, aspects related to continuous improvement, implementation, and the sustainability of improvements are presented. Then, some works that discuss qualitative results will be presented.

Before starting to optimize the production systems using the TK, several studies such as [14, 16, 17, 27] discussed the importance of conducting a training phase. To make the process more interactive, [14] suggest using a gamification method so that training takes place in a simple way and with the involvement of all employees. Generally, the gamification process simulates a production system so that employees are gradually familiarized with aspects of Lean and TK. [16] work went one step further, and the improvement team conducted what was called "dry run training" for four weeks to familiarize them with TK before use. Only after this time did the improvement team begin using TK to improvise the entire production system. It was also evident that in the training phases, trainers should encourage students to think and develop new solutions using the PDCA cycle.

After employees are familiar with the TK approach, most works emphasize that the transformation process should start small at first, using a pilot group containing only a few participants. Routines must be practiced daily so that once the participants learn the routine, they can become coaches and the process can be expanded. Also important is the process of sharing the improvements that happen over time. As shown by [20], not only does this process develops the awareness of other co-workers who will recognize the benefits of the TK, but managers will also have the opportunity to visualize the results and recognize the efforts of workers.

Regarding combining the TK with other tools/methods, [14–17, 23, 25] discussed using the Value Stream Method (VSM) to better characterize the current state, list the problems, and characterize the states to be reached by the improvement team. [25] argued that continuous improvement is only achieved when the TK is combined with the VSM. This way, the improvement team can clearly define the organization's vision and the strategy to achieve it. This is in agreement with [4], which says that "trying to implement continuous improvement in an organization without having a clear and defined vision is like driving your car without knowing your destination: you are only wasting gas".

[21] and [22] jointly used the TK with TRIZ and the Agile methodology, respectively. According to [21], by combining the two concepts, the focus of the TK shifts from minor incremental improvements to continuous innovation. Similarly, [22] combines the TK with the Agile methodology to reduce uncertainties in organizations' annual goal planning. Thus, the TK allows different work teams to be aligned to achieve the same objective. In addition, short PDCA cycles break the work into small batches, which provide quick feedback, and rework is reduced to small actions.

[28] applied a Project-based Learning theoretical model to teach Process Design through TK patterns in a Computer Numerical Control (CNC) Laboratory by Mechanical Engineering undergraduate students. The students use a systematized routine with small experiments and reflections on the actions performed. [29] combined Failure Mode and Effects Analysis (FMEA) with TK. The proposed method was validated in an electric grill product company. As result, TK changed the way of thinking and it was perceived by the multifunctional team. [26] used the TK to re-design business processes in a manufacturing company. According to the authors, adopting the TK to complex innovation projects can achieve remarkable results and might be an alternative to the implementation of software-related frameworks, like, e.g., SCRUM, particularly in manufacturing companies that already have successfully applied problem-solving routines.

Most authors cited that TK is often seen as synonymous with a continuous improvement routine. [14] state that routines of improving processes are needed to create clear, realistic, and reachable targets. Learning from previous experience for improvement is of great importance in this matter. Similarly, [23] argued that the level of learning and the continuous improvement process drops to zero if the work team does not establish challenging target conditions. Thus, it is necessary for everyone involved to change their way of thinking from the traditional way, in which the more assertive or predictable the steps, the better, to how it is necessary to use the PDCA method and experiments to reach the intended results.

The case study of [24], demonstrated that using the TK to improve industrial processes allowed the team of workers to end up assuming their routines of continuous improvement and finding solutions to reach the target condition. In contrast, [19] argues that management participation is necessary for the continuous improvement routine to happen efficiently. It is because the goal of the TK is to run small incremental improvement cycles, and managers often expect a significant change overnight. Thus, everyone must participate in the process so that expectations are aligned.

Regarding implementation and sustainability of the TK, [20] identified that companies often pay much attention to the tools instead of the people and how they think. The author also makes it clear that the focus of implementing the TK must be on people. In this sense, the author even says that people who are not aligned with the intended objectives should be disconnected from the organization.

About qualitative data, [15] used the TK in a company to reduce the Work in Progress (WIP) of a specific production process. As a result, the author highlights that it was possible to group processes and form a work cell, thus significantly increasing productivity. [23] used the Improvement Kata approach to analyze and solve a problem in the area of Occupational Safety and Heath in a unit of the Social Service of Industries. As result, the error in filling out documents decreased and there was an improvement in lead time. [24] presents a case study of applying the TK in a wood industry to optimize a work cell. As a result, the authors highlight that the work team developed a new routine of continuous improvement and developed their solutions to reach each established target condition. Thus, they ably reduced the lead time. [17] developed a systematic for implementing Lean Maintenance by associating the five Lean principles with the TK. The systematic was implemented in a thermoplastics company in the area of road signs. As a result, it was possible to implement the main lean tools in the maintenance sector in shorts cycles. [18] used the TK to improve the fuel tank painting process in the motorcycle manufacturing industry. In this study, the use of the TK met expected expectations with a reduction in paint usage, reduction of failures, and development of a routine of continuous improvement of the employees. [30] applied the TK along with Kaizen events to improve a restaurant's processes. This methodology reduced the main impact of the problem's effects (such as customer complaints, process reworking, extra-cost, and delays, among others).

[19] compared and evaluated three methods (TK, daily Kaizen events, and monthly meetings) to carry out continuous improvement in industries. For this, three sectors of the wood artifact manufacturing industry were selected. As result, the improvement in teams' performance over time was better when using methods in which monitoring took place more frequently, thus obtaining better performance.

4.2 SWOT Analysis

Table 3 highlights the main results found in the studied publications by a SWOT (strengths, weaknesses, opportunities, and threats) analysis of the TK approach. The SWOT analysis is focused on using the Toyota Kata approach in manufacturing systems as a whole and not focused on solving a specific problem. Many factors influence (facilitate or inhibit) the use of this approach, such as the creation of an operational improvement routine, and employee engagement, among others [31]. However, the use of this approach can be elevated by using strengths, opportunities, and potential obstacles. Based on [32], the following classification was adopted for the SWOT dimensions: benefits obtained by use of TK (Strengths); potential benefits that can be achieved by the

Table 3. SWOT analysis results.

Strengths	Weakness
- Reduce processing time, WIP, stocks, and the implementation of a one-piece flow system [14, 15] - Generating innovative solutions, decreasing the effects of psychological inertia, and offering guidance on where reasonable solutions can be found [21] - TK provides focus and organizational alignment without replanning, teams work in different obstacles aligned to achieve the same goal, and short PDCA cycles divide the work into small batches [22] - Develop a vision of what is expected [16] - It's a new routine of continuous improvement and developed solutions to reach each established target condition [24] - short experiments stimulated improvement as a routine among those involved in the process, as it required the participation and involvement of people to think of innovative solutions and not just apply pre changes [17, 20] - To learning by experience and by trial and error under guidance is most effective in conveying organizational culture and developing specific patterns of behavior [31] - Improves processes and communication among collaborators [26]	- Lack of direction, non-use of the value stream mapping tool, and lack of routine meetings with the work team [25] - Management often expects a significant change. However, the approach aims to run small incremental improvement cycles [19] - leaders often think that improvements only happen on the shop floor and do not feel they need to participate in the process actively - It is hard for everyone involved to set clear target conditions at the beginning of the process [14] - leaders feel that the benefits of implementing the TK will bring quick and fast results, when in fact, the results take some time to appear [16]
Opportunities	**Threats**
- A form of learning based on constructivism must be used [20] - Start small with a pilot group and daily routines. Employees must be encouraged and recognized when using the TK routine. Leaders must participate in the process [20] - It should be disseminated to all organizational sectors with training on lessons learned to ensure its sustainability [20] - For complete immersion, gamification can be used to create a type of Lean Game [14] - can provide benefits in the implementation of lean service for companies [31]	- Improvements tend to slow down with more complex challenges [14] - Accompanied by an increase in the daily workload and stress [16]

use of TK (Opportunities); difficulties and/or barriers in the use of TK (Weaknesses); potential failures/problems caused by the use of TK (Threats). The benefits perceived in the strengths section, in general, are already known and studied in articles in the literature. In the opportunities dimension, future research can focus on analyzing the potential advantages obtained in some areas, such as the implementation of Lean Game/Service. Moreover, in the weaknesses and threats dimensions, barriers, difficulties, and problems encountered in TK. These dimensions open a range of possibilities for future research.

5 Conceptual Model

From the SBR and SWOT analysis, it was possible to develop a first conceptual model that connect manufacturing problems with the best combination between the TK approach and others management approaches. This conceptual model aims to help companies that want to improve its processes using the TK approach. Figure 2 briefly presents the training phase, TK patterns, the combination of the TK approach with other tools and the expected results on the manufacturing environment.

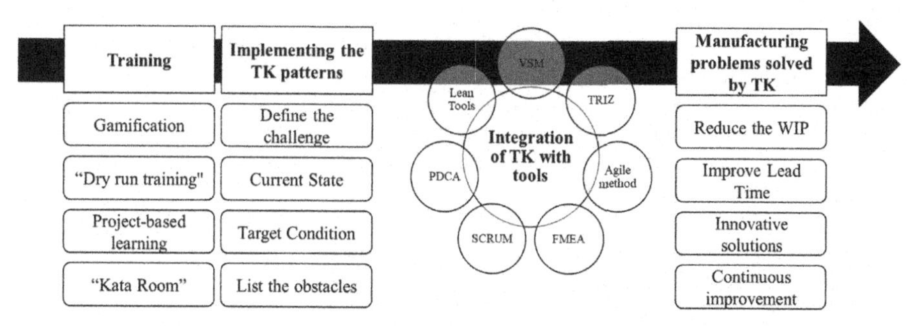

Fig. 2. Conceptual model for implementing the TK approach. Source: Authors

6 Conclusion

This paper aimed to perform a literature review and a SWOT analysis about the use of the TK to better understand how these concepts are used in practice by organizations other than Toyota. Following the Systematic Search Flow (SSF) method, it was possible to understand better how the TK is used in the manufacturing environment. From the analysis of the studies, it is evident that the TK can be used to improve the efficiency of production systems. However, to get there, some steps must be taken. It must be clear to managers that improvements will not happen overnight. It takes time for those involved to learn and understand the concepts of the Toyota Kata approach. Once the approach becomes part of the employees' routine and personal daily lives, the process will become natural.

The SWOT analysis demonstrated the strengths that allow the achievements of the benefits provided by the implementation of the TK. The main aspects are processes improvements, development of innovative solutions, focus and alignment, new routines to solve problems, using short experiments, and improvement of communication among workers. In contrast, the weakness and threats can be a challenge for industries, especially for those who expect quick results. The conceptual model presented can be used by companies that want to start using the TK approach combined with other methodologies to improve its processes.

With this study, it was possible to identify that the TK can and should be used by organizations that aim to develop a new mentality of problem-solving by workers to continuously improve its processes. However, the authors recognize that this study has some limitations, such as: (i) the keyword employed for the literature search phase may have left the research spectrum too open, especially in the exploratory research using google scholar, which brought up numerous pages of results, (ii) some works that only presented conceptual models were left out, but they may also contribute to the understanding of the subject, (iv) few studies presented discussions about western companies, and (v) content analysis may generate interpretation bias.

References

1. Womack, J.P., Jones, D.T., Roos, D.: The Machine that Changed the World. Free Press, Simon and Schuster, Inc., New York (1990)
2. Rother, M.: Toyota Kata: Managing People for Improvement, Adaptiveness, and Superior Results. McGraw-Hill Companies Inc., New York (2010)
3. Hines, P., et al.: Staying Lean: Thriving not Just Surviving. Cardiff University, Lean Enterprise Research Centre, Cardiff (2008)
4. Reverol, J.: Creating an adaptable workforce: using the Coaching Kata for enhanced environmental performance. Environ. Qual. Manag. **22**(2), 19–31 (2012)
5. Ferenhof, H.A., Fernandes, R.F.: Desmistificando a revisão de literatura como base para redação científica: método SSF. Revista ACB **21**(3), 550–563 (2016)
6. Munaro, M.R., Tavares, S.F., Bragança, L.: Towards circular and more sustainable buildings: a systematic literature review on the circular economy in the built environment. J. Clean. Prod. **260**, 121134 (2020). https://doi.org/10.1016/j.jclepro.2020.121134
7. Humphrey, A.: SWOT analysis for management consulting. SRI Alumni Newslett. **1**, 7–8 (2005)
8. Ribeiro, D.R.S., Mendes, L.G., Forcellini, F.A., Frazzon, E.M.: Maintenance 4.0: a literature review and SWOT analysis. In: Freitag, M., Kinra, A., Kotzab, H., Megow, N. (eds) Dynamics in Logistics. LDIC 2022. Lecture Notes in Logistics, pp. 409–422. Springer, Cham (2022). https://doi.org/10.1007/978-3-031-05359-7_33
9. Bonfante, M.C., Raspini, J.P., Fernandes, I.B., Fernandes, S., Campos, L.M., Alarcon, O.E.: Achieving sustainable development goals in rare earth magnets production: a review on state of the art and SWOT analysis. Renew. Sust. Energ. Rev. **137**, 110616 (2021). https://doi.org/10.1016/j.rser.2020.110616
10. Olivieri, M., Andreoli, M., Vergamini, D., Bartolini, F.: Innovative contract solutions for the provision of agri-environmental climatic public goods: a literature review. Sustainability **13**(12), 6936 (2021). https://doi.org/10.3390/su13126936
11. Paes, L.A.B., Bezerra, B.S., Deus, R.M., Jugend, D., Battistelle, R.A.G.: Organic solid waste management in a circular economy perspective – A systematic review and SWOT analysis. J. Clean. Prod. **239**, 118086 (2019). https://doi.org/10.1016/j.jclepro.2019.118086

12. Zonta, T., da Costa, C.A., da Rosa Righi, R., de Lima, M.J., da Trindade, E.S., Li, G.P.: Predictive maintenance in the Industry 4.0: a systematic literature review. Comput. Ind. Eng. **150**(2019), 106889 (2020). https://doi.org/10.1016/j.cie.2020.106889
13. Helms, M.M., Nixon, J.: Exploring SWOT analysis – where are we now? J. Strategy Manag. **3**, 215–251 (2010). https://doi.org/10.1108/17554251011064837
14. Bouwkamp, J.: Toyota Kata's: the missing link for success of Lean within SME's. Masters Dissertation. Master Technology Management, Rijksuniversiteit Groningen (2012)
15. Mendes, P.M.T.: Lean manufacturing na eliminação do uso do empilhador para a manipulação do WIP de golas. Masters Dissertation. Faculdade de Ciências e Tecnologia – Universidade de Coimbra (2015)
16. Pogliaghi, A.: Can Toyota Kata revitalize the Lean Journey of Hilti Italian warehouse in Carpiano? Master Dissertation. Copenhagen Business School (2016)
17. Ribeiro, D.: Sistemática para implementação de Lean maintenance em processos de manufatura com base na abordagem Toyota Kata. Masters Dissertation. Universidade Federal de Santa Catarina (2017)
18. Silva, T.F.: Aplicação do Toyota KATA para reduzir o consumo de tinta num processo de pintura. Masters Dissertation. Universidade do Minho (2018)
19. Ferreira, A.F.C.: Análise e melhoria de um sistema de melhoria contínua. Masters Dissertation. Universidade do Minho (2019)
20. Harvey, W.: Examining Toyota Kata's best demonstrated practices from implementation to sustainment. Doctoral Thesis. Northern Kentucky University (2019)
21. Toivonen, T.: Continuous innovation: combining Toyota Kata and TRIZ for sustained innovation. Procedia Eng. **131**, 963–974 (2015)
22. Iberle, K.: The improvement Kata: annual improvement planning meets agile. In: Pacific NW Software Quality Conference, pp 1–14 (2015)
23. Martins, C.F., Röse, A.S., Brognoli, A.C.S., Lima, M.B.B.P., Barddal, R.: Kata de melhoria: desenvolvendo habilidades para resolver problemas e aprender de forma sistemática no SESI Santa Catarina: uma aplicação Lean na área de Segurança e Saúde do Trabalho. J. Lean Syst. **1**(2), 107–121 (2016)
24. Dinis-Carvalho, J., Ratnayake, R.M.C., Stadnicka, D., Sousa, R.M., Isoherranen, J.V., Kumar, M.: Performance-enhancing in the manufacturing industry: an improvement KATA application. In: 2016 IEEE International Conference on Industrial Engineering and Engineering Management (IEEM), pp. 1250–1254 (2016). https://doi.org/10.1109/IEEM.2016.7798078
25. Michels, E., Forcellini, F.A., Fumagali, A.E.C.: Opportunities and barriers in the use of Toyota Kata: a bibliographic analysis. Revista Gestão da Produção Operações e Sistemas **15**(5), 262–285 (2019)
26. Brandl, F.J., Ridolfi, K.S., Reinhart, G.: Can we adopt the Toyota Kata for the (re-)design of business processes in the complex environment of a manufacturing company? In: 53rd CIRP Conference on Manufacturing Systems, vol. 93, pp. 838–843 (2020)
27. Odebrecht de Souza, R., Ferenhof, H.A., Forcellini, F.A.: Systematic for process improvement using cyber-physical systems and Toyota Kata. In: Canciglieri Junior, O., Noël, F., Rivest, L., Bouras, A. (eds) Product Lifecycle Management. Green and Blue Technologies to Support Smart and Sustainable Organizations. PLM 2021. IFIP Advances in Information and Communication Technology, vol. 639, pp. 447–460. Springer, Cham (2022). https://doi.org/10.1007/978-3-030-94335-6_32
28. Ribeiro, D.R.S., Forcellini, F.A., Pereira, M.: Toyota Kata patterns to help teach process design: applying a project-based learning model. In: Canciglieri Junior, O., Noël, F., Rivest, L., Bouras, A. (eds) Product Lifecycle Management. Green and Blue Technologies to Support Smart and Sustainable Organizations. PLM 2021. IFIP Advances in Information and Communication Technology, vol. 640, pp. 55–67. Springer, Cham (2022). https://doi.org/10.1007/978-3-030-94399-8_5

29. Zanatta, F.E., Ferreira, J.C.E.: Deploying strategies with integrated events in a home appliance manufacturer: application of FMEA through Toyota Kata. In: Canciglieri Junior, O., Noël, F., Rivest, L., Bouras, A. (eds) Product Lifecycle Management. Green and Blue Technologies to Support Smart and Sustainable Organizations. PLM 2021. IFIP Advances in Information and Communication Technology, vol. 640, pp. 438–455. Springer, Cham (2022). https://doi.org/10.1007/978-3-030-94399-8_32
30. Suárez-Barraza, M.F., Miguel-Dávila, J.A., Morales-Contreras, M.F.: Application of Kaizen-Kata methodology to improve operational problem processes. A case study in a service organization. Int. J. Qual. Serv. Sci. **13**(1) (2021). https://doi.org/10.1108/IJQSS-07-2020-0113
31. Ferenhof, H.A., Cunha, A.H., Bonamigo, A., Forcellini, F.A.: Toyota Kata as a KM solution to the inhibitors of implementing lean service in service companies. VINE J. Inf. Knowl. Manag. Syst. **48**, 404–426 (2018)
32. Wijesooriya, N., Brambilla, A.: Bridging biophilic design and environmentally sustainable design: a critical review. J. Clean. Prod. **283**, 124591 (2021). https://doi.org/10.1016/j.jclepro.2020.124591

From Ontologies to Operative Data Models: A Data Model Development Supporting Zero Defect Manufacturing

Claudio Turrin[1], Federica Acerbi[2(✉)], Antonio Avai[3], Arnaldo Pagani[4], Manfredi Giuseppe Pistone[5], Angelo Marguglio[5], and Pierluigi Petrali[4]

[1] SAM Software, Via Muller 37, 28921 Verbania, Italy
[2] Department of Management, Economics and Industrial Engineering, Politecnico di Milano, via Lambruschini 4/b, 20156 Milan, Italy
Federica.acerbi@polimi.it
[3] TTS – Technology Transfer System, Via Francesco d'Ovidio, 3, 20131 Milan, Italy
avai@ttsnetwork.com
[4] Whirlpool EMEA spa, Via Aldo Moro, 5, 21024 Biandronno, VA, Italy
Arnaldo_Pagani@whirlpool.com
[5] Engineering Ingegneria Informatica, Piazzale dell'Agricoltura, 24, 00144 Rome, Italy
{ManfrediGiuseppe.Pistone,angelo.marguglio}@eng.it

Abstract. Resource consumption is expected to reach 167 Gigatonnes by 2060 doubling the amount accounted in 2017. More than ever, manufacturing companies are asked to rapidly face this issue to limit their negative impacts on the entire society. Factories are supposed to act on their internal operating activities to enhance the quality of both products and processes reducing as much as possible the industrial waste generated. Among all, the zero-defect manufacturing principles represent an opportunity towards this direction, but companies necessitate to rely on their existing assets, both digital and physical, to create value based on that. Therefore, this contribution aims to develop an operating database based on a structured data model enabling to link, and reason over, three main areas (i.e. material, process, function) influencing the final product quality. These areas were identified in an already validated ontology, GRACE, considering them as the key elements to be kept under consideration to evaluate the impacts on product quality facilitating to make zero-defect oriented decisions. Indeed, the developed model, to be useable by a manufacturing company, was aimed to ensure to be data source independent, to embed these four areas, to include all the internal processes (i.e., pre-assembly, assembly) of a factory, and to transform abstracts ideas into operative actions. The data model development procedure is described, and its application was performed in a case study (i.e., an assembly line). This application enabled to validate it highlighting the key requirements for a discrete manufacturing company to embrace zero-defect manufacturing.

Keywords: Zero-defect manufacturing · Assembly line · Data model

Published by Springer Nature Switzerland AG 2023
F. Noël et al. (Eds.): PLM 2022, IFIP AICT 667, pp. 663–672, 2023.
https://doi.org/10.1007/978-3-031-25182-5_64

1 Introduction

Resource consumption is increasingly rising, supposing to reach 167 Gigatones by 2060 doubling the amount accounted in 2017 [1]. This problem requires to be rapidly faced by the entire society, and especially by manufacturing companies due to their high resource intensity characteristic [2] which leads also towards an augment of waste and pollution generation [3]. Regarding the waste, the enhancement of process quality can be a great driver towards waste reduction and the concept of "zero defect manufacturing" is gaining momentum representing a cornerstone in this domain [4]. Actually, the "zero defect manufacturing" (ZDM) is easily adoptable in discrete manufacturing companies, advancing the traditional quality management techniques like Six Sigma and Lean [5]. In this context, the ZDM concept relies on a set of inspection activities covering the different parts of the products to be manufactured [4]. Data collection and management represent a fundamental element to monitor product and process quality enhancing the company's sustainability attitude [6]. Indeed, data mining technique was adopted to investigate and analyse the causes (e.g. the equipment status) affecting the product quality during the production process to strengthen the ZDM benefits obtainable by discrete manufacturing companies [7]. Nevertheless, especially in discrete manufacturing companies, data sources are heterogeneous and complex to be managed in a coordinated way. For this reason, it is required to reduce the complexity and create interoperable systems enabling the exchange of the right data, at the right time to provide useful information to end users based on an appropriate data analysis.

Data collection, analysis and sharing are easily achievable in the contemporary era thanks to the diffusion of the key enabling technologies characterizing the Industry 4.0 (I4.0) paradigm. Among them, in the context of product quality management relying on a ZDM approach, a digital twin has been developed based on the R-MPFQ model (Resource, Material, Production Process, Function, Quality) [8]. This model has been linked to an ontology developed in a European project "inteGration of pRocess and quAlity Control using multi-agent technologies (GRACE)", which was aimed at clarifying the main entities to be kept into account in integrating process and quality control. The main gap observed was the difficulty in translating this standard methodology into concrete actions implementable by a factory by relying on the data gathered from its information systems. In the European Project QU4LITY https://qu4lity-project.eu/, focused on ZDM in the I4.0 era, an operative data model has been developed by relying on the already validated ontological model. This model aims at supporting the decision-making process of both managers and operators facilitating them in performing the proper corrective actions to improve product quality. This contribution will explain how a holistic semantic data model has been developed and how it can be used to monitor the product quality and systematically detect the causes fostering ZDM.

The remainder of the paper is structured as follows. Section 2 presents the theoretical background explaining the MPFQ model in its detail and how it is linked with the concept of zero-defect manufacturing. Section 3 elucidates the data model development and application in a concrete case. Section 4 concludes the contribution by highlighting the main implications to practice and theory underlining the open opportunities based on the limitations of this research.

2 Research Context

2.1 Quality Management and Zero-Defect Manufacturing

The concept of zero-defect manufacturing is gaining momentum being it considered an opportunity for factories to create more eco-friendly and more efficient production lines with zero defects [9]. This is extremely important, especially for discrete manufacturing companies which are forced to increase the customization level of products ensuring the quality of both products and processes, thus limiting the products defects [10]. Indeed, process control and monitoring may empower product quality reducing defected parts since process degradation and malfunctions can cause defects on both parts and products [11]. Actually, zero-defect manufacturing strategies can be of two types: (i) zero-defect prevention, according to which knowing in depth the production process conditions and the machine state it is possible to prevent a defect, and (ii) zero defect compensation, according to which a downstream change of the process is made to avoid reworking activities [12]. In both cases, information sharing through integrated information and communication solutions may benefit the efficiency improvement by limiting processes and products defects, thus enhancing their quality [13].

2.2 MPFQ Ontological Model

As just mentioned, quality management can be highly enhanced by the proper use and sharing of information and knowledge which can be facilitated by the availability in factories of integrated and/or interoperable systems [14]. Nevertheless, interoperability among systems still represents an issue for several manufacturing companies causing semantic misalignment which indirectly is linked to a limited capability to reuse and share knowledge in a factory [15]. To tackle this issue, ontologies have been developed with the final aim to create a common and shared view over knowledge representation that facilitates the communication among distributed entities [16].

In the context of quality management, an ontology (i.e. GRACE ontology) has been developed, that aimed at covering the knowledge about the resources in production lines, the production process, and the testing of the produced products to link this knowledge with the distributed quality controls [17]. Based on this ontology, to furtherly focus the attention on product quality, the MPFQ model has been created, which stands for Material, Process, Functions and Quality, covering the key areas influencing product quality [18]. Recently, it has been added the R of Resources to the MPFQ model, considered as the industrial assets required to produce the products, to differentiate the industrial assets from the other elements included in the model [8].

Although the ontology generated represents a comprehensive picture supporting knowledge sharing for quality management issues, its concrete operationalization is still missing. Therefore, based on this generated and validated scientific knowledge, a relational data model has been elaborated in this contribution to overcome the key limitations emerged so far. This model has been developed by operationalizing the ontology so far created with the final goal to enable the exploitation of data coming from the factory information systems to take informed decisions.

3 Data Model Development and Application

The R-MPFQ modelling approach has been used in the QU4LITY project with the goal to model assembly and quality processes by using, as pilot, the Lodz Dryer factory of Whirlpool (WHR). The first main action of the QU4LITY WHR team was to map the WHR Lodz Dryer factory assembly processes and the final product to understand where and how the R-MPFQ model could be generated. Starting from the bill of material of a dryer and the experience of the quality team of the Lodz factory, the dryer was decomposed into six sub-units, and then the relations between components of each sub-unit were identified. The assembly process mapping aimed not only to model the organization (sequential or parallel) of the final product as well as the sub-units assembly processes but also to organize how the processes and related operating resources are harmoniously structured within the assembly sequences and how the "Function" element is generated. This aspect is very important for the QU4LITY project because it represents what a customer mainly perceives from a dryer, either in a positive or negative sense. The outcome of this activity has been the MFQ model of the product itself, which represents the key element on which the entire operational modelling is based.

Figure 1 represents the MFQ model of a dryer where rectangular boxes are the main functional units of a dryer (M), the orange hexagons show some Quality indicators (Q), the light blue hexagons are the external elements and the arrows are the main functions (F) generated by each functional unit. To report an example explaining the functional model, the Quality element "noise" is generated by the air flow, mechanical and hydraulic sub-units of the dryer, while the chassis has the function of stopping or reducing the noise. If "noise" is perceived as a critical aspect for customers, it means the previously mentioned sub-units should be investigated deeper to find out the key elements in terms of material, process or function that allow reducing noise and achieving zero-defect manufacturing. A detailed graph was implemented for each sub-unit of the dryer.

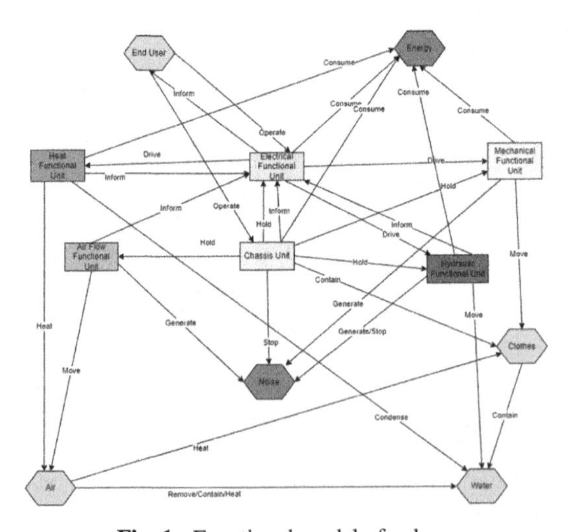

Fig. 1. Functional model of a dryer

The next step was the creation of the R-MPFQ model for the dryer and all its sub-units. Figure 2 depicts a small part of the developed model referring to the heat pump sub-unit where the different colours of the boxes represent the four elements of the model. For instance, the compressor (M) is placed on the silent block (M) by the pick and place process (P). The function (F) of the silent block serves to hold the compressor and to reduce the noise (Q) of the compressor that, on the contrary, generates noise (Q) and consumes energy (Q). The pick and place process may impact on the noise if the compressor is not properly assembled on the silent block. All these Quality elements (noise, energy consumption, leakage) are measurable during the assembly process.

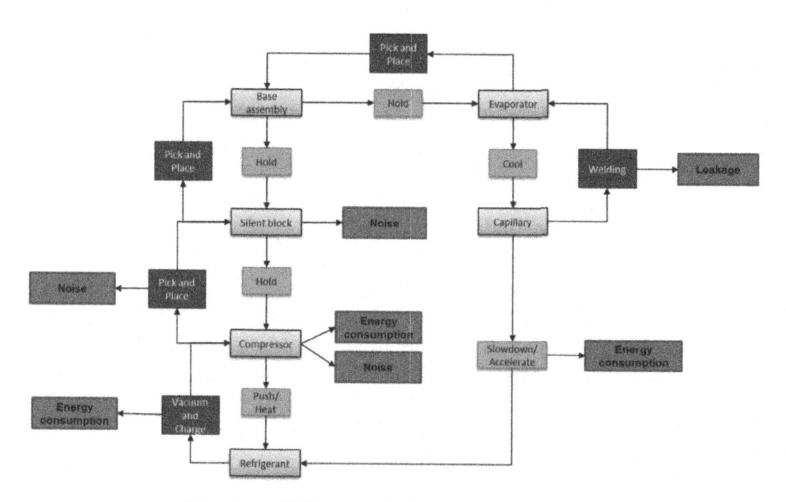

Fig. 2. MPFQ model of heat pump sub-unit

Along with the creation of this model, it was needed to define the data ontology model able: (i) to represent the R-MPFQ model; (ii) to represent any type of assembly process, pre-assembly, testing, typology of resources, type of operations, etc.; (iii) to turn in an operational object capable of transforming logical abstraction concepts into concatenations of events and data; (iv) to transform the information in a "standard" format. Therefore, to ensure all these issues, the team started to create the relational database from the ontological model of manufacturing developed under the project GRACE [17]. This model was deeply analyzed to check if it was able to represent all the requirements listed before. A revised ontology model was defined integrating new concepts in the initial one, such as "Measure", "Resource" or "Function", based on other existing IOF ontologies (link). In parallel, based on this ontology, a relational database operative for the factory was developed. The result was the development of an ontological model represented by the graph in Fig. 3 which has been "transformed" into relational entities [19], representative of reality. This "transformation" means giving each entity an operational value and transforming abstract entities into physical entities clarifying the relations among them based on normalized information.

Fig. 3. The final ontological data model

Once the relational data structure and its alignment with the ontological model have been finalised, the description of all the parts that compose the assembly process started and, in particular, the following entities were defined and related:

- Journal: it describes the production of a product instance belonging to a production order executed in the assembly line.
- Journal Details: it describes the execution of an operation by a single station.
- Operation: a job executed by one resource.
- Process: a single process composing an operation.
- Resource: an asset that can perform a certain range of processes.
- Function: a feature of the product or the process that can be measured.

The relationships were defined based on the ontological model and this composition of the entities led to the development of two types of representations: the "Static" one, where the system is described by the relations between the entities, and the "Dynamic" one, which enables to correlate and associate the events that occurred during the execution of the single process along a timeline. This correlation is possible thanks to the information about the connection with the production tracking. The product tracking hence is easily linked to process and product measures.

Once the series of tables representing ontological entities were developed and their links were defined, the relational process was then translated into a relational database system. In particular, by introducing the R-MPFQ model fused with Autonomous Quality (AQ) control loops in production, it was addressed unresolved problems in the vertical integration of data management (from data gathering to visualization and decision making), enabling a holistic vision to be achieved. To this end, a set of solutions (see Fig. 4), namely QU4LITY Cloud Solutions, were built to provide a seamless solution to exchange data using the R-MPFQ ontology model, enabling a semantic enriched data exchange from on-premises data lakes to enable the Cloud Data Storage using a time-based approach: QU4LITY Cloud Bridge and Q-Ontology Enabler. The former, QU4LITY Cloud Bridge, built on top of Node.js and Sequelize (an open-source ORM Object/Relational Mapper), offers a REST API layer to ease the interfaces between R-MPFQ relational database and other processing and visualization components within the pilot, taking care of any data decoding/encoding needs (i.e. IEEE754 data encoding). The latter, the so-called Q-Ontology Enabler, consisting of a set of scripts built with Python, delivers a powerful tool to enable semantic interoperability between legacy data and newly engineered R-MPFQ Ontology with extreme ease. Indeed, thanks to the ontology built, only input data format has to be mapped for the scripts to work.

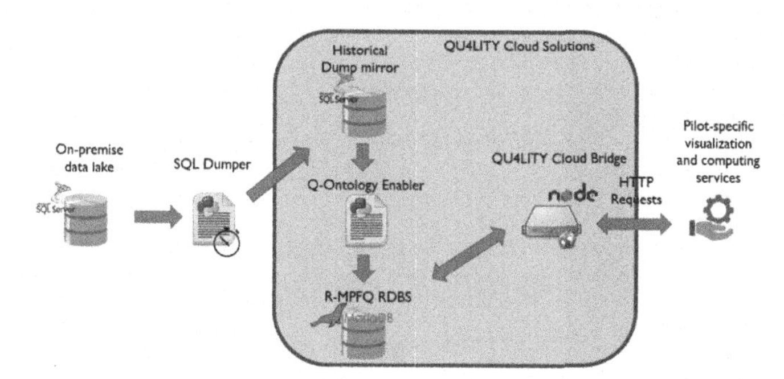

Fig. 4. QU4LITY Cloud Bridge and Q-Ontology Enabler positioning

Subsequently, to support the decision-making on the shopfloor, a dashboard QU4LITY trend cockpit was developed as a Web-based application to query the QU4LITY cloud bridge and report the measurements and the results according to the relations defined with the R-MPFQ model. Thanks to this implementation, it is possible to carry on some R-MPFQ analysis starting from the selection of a Quality indicator, then of a material or a process looking at a certain timeframe as visible in Fig. 5. In the example, the end-user is interested in investigating the energy consumption of the drum. The application, leveraging on the R-MPFQ model and the related database, can show all the materials, functions and processes/resources involved in the energy consumption of the drum.

Then, the end-user can select an item and can analyze the data collected as shown in Fig. 6, where the data about drum concentricity are plotted and the relative CPK (Process Capability Index) is assessed. In the same way, the end-user can retrieve data about processes or functions and find out trends or critical values impacting the Quality of the final product negatively.

The Qu4lity trend cockpit, integrated with the R-MPFQ model, the ontology data model and the operative relational database, allows the autonomous ZDM which entails humans in the loop. Therefore, different stakeholders' profiles such as Quality manager, product development team and industrial engineer can exploit this Quality system to make fast and effective analysis on production and process quality, to correlate consumers' needs with products and processes quality metrics and to take decisions based on data and trends. This advancement was possible thanks to the transformation of the

Fig. 5. QU4LITY trend cockpit interface

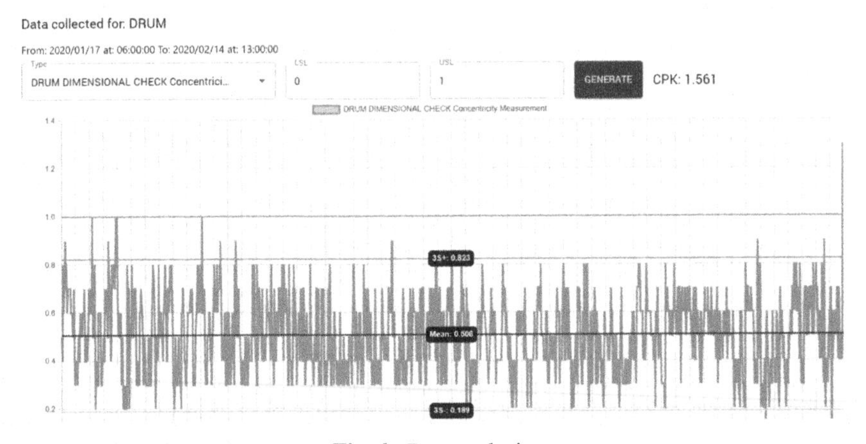

Fig. 6. Data analysis

R-MPFQ model from an abstract model to an operating one as integrated into the ontological model. Usually, for companies, the measurements are associated "vertically" to specific parameters of specific objects. In an ontological model, the measurements may be considered as enlarged covering a broader area, thanks to the correlations established among entities both directly and indirectly. Indeed, the measurement of an element means to measure a chain of process elements and functional elements that enable to reach a certain result, which is possible thanks to the new developed model.

Additionally, on one side, the manufacturing sector does not use the functional model of the product represented as composed by elements of varied nature (mechanical, hydraulic, electric, etc.). This functional model hence has never been used as a lever, in the manufacturing world, to understand how good the product production is and thus, this operational model facilitates the opening towards this direction. On the other side, also quality management is still limited to only part of the parameters which might be explored thanks to this model. Indeed, sophisticated data analysis tools such as neural networks should be used for issues like the monitoring of those parameters enabling to make the manufacturing company knows what makes the final customer be satisfied or dissatisfied referring to the production process.

4 Conclusions

This contribution enabled to highlight the main challenges that might be faced when an ontological model needs to be concretely adopted by an industrial company and thus, it needs to be transferred into an operational data model highlighting also the key benefits. More specifically, considering that ontologies are created to develop standards aiming at facilitating systems interoperability and thus, knowledge sharing within the factory boundaries, these must be easily adoptable by companies and need to be transferred into their industrial contexts. Therefore, the transformation process, required to generate a data model starting from an ontological model, is deeply described in this contribution with the aim to share the knowledge on how to develop a solution enabling to exploit the data collected from heterogeneous data sources characterizing many manufacturing companies. Such a process required the involvement of a team with different knowledge on products and processes of Lodz Dryer factory, on ontologies definition, and on data models and relational database development. The close collaboration of the team was mandatory to achieve this result.

Last, the final operational data model is under validation by WHP in the Lodz Dryer factory. As highlighted in the application part of this contribution, it should facilitate the analysis of the data collected from the assembly lines with the ultimate goal to take corrective actions on the shop floor by acting directly on those, often hidden, elements influencing customers' perspectives. The first results are very promising and based on these tests, WHP will decide to extend this methodology into other factories of its group. Moreover, an additional user-friendly application is under development with the aim to support operators in their daily activities reaching zero-defect manufacturing.

This research was funded by the European Project QU4LITY (GA. 825030) https://qu4lity-project.eu/.

References

1. OECD: Global material resources outlook to 2060. OECD (2019). https://doi.org/10.1787/9789264307452-en
2. Acerbi, F., Taisch, M.: A literature review on circular economy adoption in the manufacturing sector. J. Clean. Prod. **273**, 123086 (2020). https://doi.org/10.1016/j.jclepro.2020.123086
3. Despeisse, M., Evans, S.: Improving factory resource and energy efficiency: the FREE toolkit. In: Umeda, S., Nakano, M., Mizuyama, H., Hibino, H., Kiritsis, D., von Cieminski, G. (eds.) APMS 2015. IAICT, vol. 459, pp. 640–646. Springer, Cham (2015). https://doi.org/10.1007/978-3-319-22756-6_78
4. Psarommatis, F., Kiritsis, D.: Identification of the inspection specifications for achieving zero defect manufacturing. In: Ameri, F., Stecke, K.E., von Cieminski, G., Kiritsis, D. (eds.) APMS 2019. IAICT, vol. 566, pp. 267–273. Springer, Cham (2019). https://doi.org/10.1007/978-3-030-30000-5_34
5. Psarommatis, F., Kiritsis, D.: Comparison between product and process oriented zero-defect manufacturing (ZDM) approaches. In: Dolgui, A., Bernard, A., Lemoine, D., von Cieminski, G., Romero, D. (eds.) APMS 2021. IAICT, vol. 630, pp. 105–112. Springer, Cham (2021). https://doi.org/10.1007/978-3-030-85874-2_11

6. Acerbi, F., Taisch, M.: Towards a data classification model for circular product life cycle management. In: Nyffenegger, F., Ríos, J., Rivest, L., Bouras, A. (eds.) PLM 2020. IAICT, vol. 594, pp. 473–486. Springer, Cham (2020). https://doi.org/10.1007/978-3-030-62807-9_38
7. Wang, K.S.: Towards zero-defect manufacturing (ZDM)—a data mining approach. Adv. Manuf. **1**, 62–74 (2013). https://doi.org/10.1007/s40436-013-0010-9
8. Zheng, X., Psarommatis, F., Petrali, P., Turrin, C., Lu, J., Kiritsis, D.: A quality-oriented digital twin modelling method for manufacturing processes based on a multi-agent architecture. Procedia Manuf. **51**, 309–315 (2020). https://doi.org/10.1016/j.promfg.2020.10.044
9. Psarommatis, F., May, G., Dreyfus, P.A., Kiritsis, D.: Zero defect manufacturing: state-of-the-art review, shortcomings and future directions in research. Int. J. Prod. Res. **58**(1), 1–17 (2020). https://doi.org/10.1080/00207543.2019.1605228
10. Eger, F., et al.: Zero defect manufacturing strategies for reduction of scrap and inspection effort in multi-stage production systems. Procedia CIRP **67**, 368–373 (2018). https://doi.org/10.1016/j.procir.2017.12.228
11. Ferretti, S., Caputo, D., Penza, M., D'Addona, D.M.: Monitoring systems for zero defect manufacturing. Procedia CIRP **12**, 258–263 (2013). https://doi.org/10.1016/j.procir.2013.09.045
12. Powell, D., Magnanini, M.C., Colledani, M., Myklebust, O.: Advancing zero defect manufacturing: a state-of-the-art perspective and future research directions. Comput. Ind. **136**, 103596 (2022). https://doi.org/10.1016/j.compind.2021.103596
13. Eleftheriadis, R.J., Myklebust, O.: A guideline of quality steps towards zero defect manufacturing in industry. In: Proceedings of the International Conference on Industrial Engineering and Operations Management, pp. 332–340 (2016)
14. Polenghi, A., Acerbi, F., Roda, I., Macchi, M., Taisch, M.: Enterprise information systems interoperability for asset lifecycle management to enhance circular manufacturing. IFAC-PapersOnLine **54**(1), 361–366 (2021). https://doi.org/10.1016/j.ifacol.2021.08.162
15. Polenghi, A., Roda, I., Macchi, M., Pozzetti, A., Panetto, H.: Knowledge reuse for ontology modelling in maintenance and industrial asset management. J. Ind. Inf. Integr. **27**, 100298 (2022). https://doi.org/10.1016/j.jii.2021.100298
16. Negri, E., Fumagalli, L., Garetti, M., Tanca, L.: Requirements and languages for the semantic representation of manufacturing systems. Comput. Ind. **81**, 55–66 (2016)
17. Leitão, P., Rodrigues, N., Turrin, C., Pagani, A., Petrali, P.: GRACE ontology integrating process and quality control. In: IECON 2012-38th Annual Conference on IEEE Industrial Electronics Society, pp. 4348–4353. IEEE (2012). https://doi.org/10.1109/IECON.2012.6389189
18. Foehr, M., Jäger, T., Turrin, C., Petrali, P., Pagani, A., Leitão, P.: Implementation of a methodology for consideration of product quality within discrete manufacturing. IFAC Proc. **46**(9), 863–868 (2013). https://doi.org/10.3182/20130619-3-RU-3018.00181
19. Trinkunas, J., Vasilecas, O.: A graph oriented model for ontology transformation into conceptual data model. Inf. Technol. Control **36**(1), 126–132 (2007)

Evaluation of a Commercial Model Lifecycle Management (MLM) Tool to Support Models for Manufacturing (MfM) Methodology

Rebeca Arista[1,2](✉), Fernando Mas[1,3], Domingo Morales-Palma[1], Dominique Ernadote[4], Manuel Oliva[5], and Carpoforo Vallellano[1]

[1] Universidad de Sevilla, Sevilla, Spain
rebeca.arista@airbus.com
[2] Airbus SAS, Blagnac, France
[3] M&M Group, Cadiz, Spain
[4] Airbus D&S, Elancourt, France
[5] Airbus, Sevilla, Spain

Abstract. Model-Based System Engineering (MBSE) supports the conceptualization of systems, proposes independence between knowledge conceptualization and tools, and fosters reusability and interoperability. There are several research works oriented to define ontologies in the manufacturing domain, but only a few consider their use in an Ontology-based Engineering (OBE) perspective, or consider a framework to manage the whole lifecycle of the models in supporting software tools.

Models for Manufacturing (MfM) is an OBE methodology which helps to define and manage ontologies to capture not only explicit domain knowledge, but also implicit and tacit knowledge from domain experts, which are crucial for design and decision support tools. The ontology models lifecycle is managed through a Model Lifecycle Management (MLM) system where the tool agnostic MfM metamodels are implemented.

This paper shows the implementation of one of the MfM metamodels, the Scope metamodel, in a commercial software tool which covers one of the MLM main functions, the editing function, and demonstrates the ontology scope model definition with an example. The purpose of this evaluation is to demonstrate that the MfM methodology can be applied to a largely used standard language implemented in COTS (Commercial Off-the-Shelf) tools.

Keywords: Models for Manufacturing (MfM) · Model Lifecycle Management (MLM) · 3-Layer Model (3LM) · Ontologies · Manufacturing modelling

1 Introduction

In order to secure a short time-to-market, products and manufacturing systems need to be designed collaboratively in a shorter time, often reusing existing product designs

Published by Springer Nature Switzerland AG 2023
F. Noël et al. (Eds.): PLM 2022, IFIP AICT 667, pp. 673–682, 2023.
https://doi.org/10.1007/978-3-031-25182-5_65

and industrial resources. Ontology-Based Engineering (OBE) emerges as a new approach from earlier Knowledge-based Engineering techniques thanks to the technological advancements in computer science. This allows to enhance multidisciplinary design, realistic simulations and trade space exploration of complex systems.

Models for Manufacturing (MfM) is a recent OBE methodology [1] for defining and replicating data, functions and behaviors of complex manufacturing systems using computer-aided graphical modelling authoring tools. It is derived from the concept of MBSE methodology, using models and their associated behavioral abstraction for enabling a more robust engineering definition [2].

Different tools, languages and formalisms can be used to design and develop ontologies [3]. MfM methodology fosters the design of ontologies not only to capture explicit domain knowledge, but also implicit and tacit knowledge from domain experts. This crucial part of the conceptualization process needs to be well addressed, in order to enable intelligent design support tools development in its full potential, and use them in complex system design processes.

MfM methodology proposes a 3-Layers Model (3LM) architecture, which is described in detail in Sect. 2.1. It consists of a Data layer, an Ontology layer, and a Service layer, where the Ontology layer is the core, and where knowledge is captured and managed through four model types: Scope model, Data model, Behavior model and Semantic model. The authors have published several research papers for modelling manufacturing systems in the aerospace industry using MfM methodology [4–6], including functional and data models deployed using data structures available in commercial PLM systems [7]. Nevertheless, MBSE implementation challenges in lifecycle management systems still need to be covered [8].

The lifecycle of the ontology models created using MfM methodology are proposed to be managed through a Model Lifecycle Management (MLM) system, implementing the MfM metamodels defined for each type of model independently of any commercial software tool and modelling language [9]. Further details of MfM metamodels can be seen in Sect. 2.2.

The five basic functions of an MLM system, based on a PLM system, are [10, 11]: editing or visualizing, vaulting, workflow management, configuration management and interfacing to other systems. This paper demonstrates the Scope metamodel implementation in UML standard and in Dassault Systems commercial software MagicDraw, to evaluate it regarding MLM editing functionality. The purpose of this evaluation is to demonstrate that the MfM methodology can be applied to a largely used standard language implemented in COTS (Commercial Off-the-Shelf) tools.

The Scope model of the Industrial System Design (ISD) ontology, which was created to support the industrial system reconfiguration in concept phase [12], will be reused as an example to demonstrate the MLM toolset implementation in MagicDraw. A similar implementation can be done for the rest of the metamodels supporting the MfM methodology, and will be part of the future work. The next sections are structured as follows: Sect. 2 provides an overview of MfM methodology and the details of the Scope metamodel; Sect. 3 describes the Scope metamodel implementation in UML standard and MagicDraw tool; Sect. 4 shows the ISD scope model as example model in MagicDraw; and Sect. 5 discusses the commercial software solution used as MLM and outlines further work to be done.

2 Review of Models for Manufacturing (MfM) Methodology

2.1 3-Layers Model (3LM) as Framework

Models for Manufacturing methodology described in [1] is based in a 3-Layers Model framework as shown in Fig. 1. The 3LM guarantees the independence between the three layers, holding Service Layer, Data Layer and Ontology Layer isolated from each other.

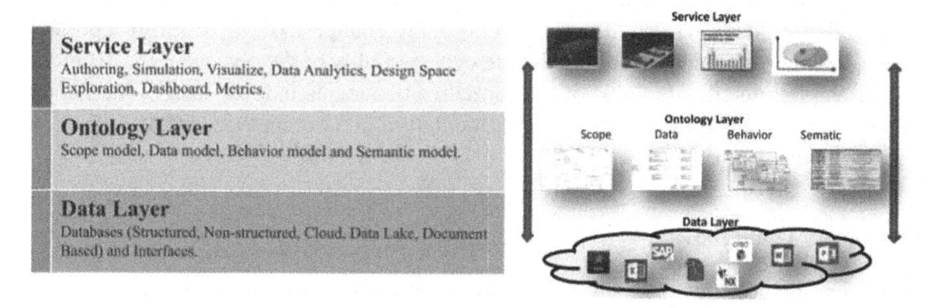

Fig. 1. 3-Layers Model (3LM) framework.

Each layer collects the same type of items. The Data layer collects all the information in terms of databases and interfaces: legacy databases related to legacy software, commercial software databases, clouds, data lakes and many others. The Ontology layer holds the Company knowledge in the form of scope, data behavior and semantic models, and will be described in detail in Sect. 2.2. The Service layer holds the software services, legacy or commercial, like authoring tools, simulation tools, visualizers, data analytics services, dashboards, and space design exploration tools.

The ontologies in the Ontology layer are created using four model types: Scope model, Data model, Behavior model and Semantics model. The first step is to create the Scope model, which defines the limits where the ontology is valid and contains all main objects of the Data model and system behavior.

The next step is to create the Data model, which defines the information managed in the selected scope. The Behavior model comes after, defining the system behavior for simulation. Finally, a Semantic model is created to support the Data model instantiation with real data coming from different databases, different languages or formats.

2.2 MfM Metamodels

The MfM methodology follows a set of golden rules:

- MfM methodology is agnostic against tools, and does not define any preferred tool.

- MfM methodology encourages tools with easy to write models and very easy to read, understand, share, and discuss by skilled engineers on the topic.
- MfM methodology encourages a mechanism to manage the model's lifecycle, configuration and effectivity using PLM tools to fulfil this requirement.

To fulfil these rules, all the concepts of the ontology layer models have been formalized in metamodels using the Meta-Object Facility (MOF) standard from the Object Management Group [9]. The Scope Metamodel, Data Meta-model, Behavior Metamodel and Semantic Metamodel have been defined in a first version, as well as the relationships between them [13]. The metamodels do not include information related to the diagrams visualization (positions of elements, size, etc.).

The formalization of all the ontology layer concepts became evident when a Model Lifecycle Management (MLM) was proposed to support MfM method implementation. The MLM would allow each instantiated concept to change and evolve during a project or use case development, managing the life cycle not only of the objects of the different use cases (the models), but also of the formalized concepts independently. The Scope Metamodel, object of this paper for implementation in a commercial MLM toolset, is detailed in the next subsection.

2.3 Scope Metamodel

The Scope model contains the definition of the main system functions and main data objects. Figure 2 shows the metamodel for the Scope model ('scope' package). The root class of the 'scope' package is "ScopeModel" which is composed of a main 'root' activity that is decomposed into sub-activities.

The "Activity" concept defines the system functions or activities performed; "Means" concept defines the resources or means needed and used for realizing an "Activity"; "DataObject" concept (from the 'data' package) contains the data objects required ('input') or produced ('output') by activities. An "Activity" class can be a "ComposedActivity" (see 'parent' and 'child' roles) or an "ElementaryActivity" (an activity without children; from the 'behavior' package). The latter is a key concept in the MfM Behavior metamodel.

An Activity may require different "Means" for its realization. Sometimes, the different Means that are used in sub-activities can be grouped in a single Means, easier to associate with the parent activity. For example, "“CAx" (Computer-aided technologies)" can be used to group "“CAD" (Computer-aided design)", "“CAD/CAM" (Computer-aided design/manufacturing)" and "“CAPP" (Computer-aided process planning)" systems. On the other hand, a DataObject can belong to an aggregate set of objects ('parent'-'child' relationship) or have some other types of relationship with other objects (role 'related' in Fig. 2).

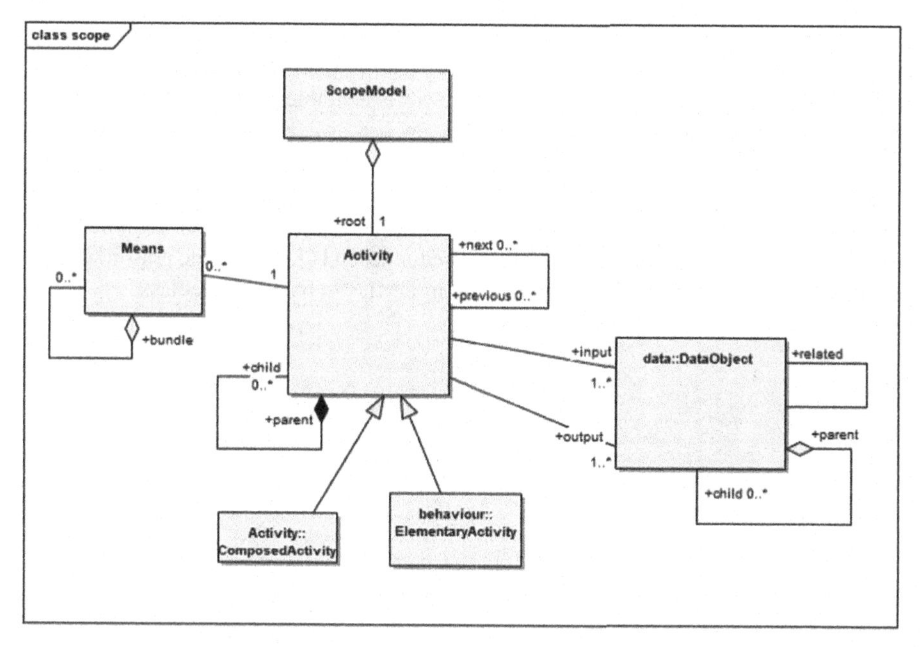

Fig. 2. Metamodel of the Scope model.

3 Scope Metamodel Implementation in MagicDraw

3.1 Context

UML is a modelling language of general purpose defined by OMG group [14] used to specify and document the elements of systems to be used across application domains and platforms. It is widely used in industry and academia as the standard language for describing software systems.

As described in Sect. 2.2, MfM metamodels are described in MOF (MetaObject Facility), to ensure a language and tool-agnostic application of MfM methodology. MOF is a standardized description of metamodels, foundation of OMG industry-standard environment [15]. As UML language is MOF-based, we can implement MfM MOF-metamodels in UML language.

MagicDraw commercial solution implements UML standard profile, and provides means to adapt domain-specific profiles using a DSL engine, making it possible to create custom diagrams and specification elements [16]. In projects created by the modelling tools developed by Dassault Systèmes, the UML model is an implementation of the OMG UML 2.5.1 metamodel, covering the whole UML specification and also extends it by introducing some additional Elements (like Diagram) [17].

3.2 Implementation

The implementation proposed for the Scope Metamodel in UML 2.5.1 specification is described in Fig. 3. In UML 2.5.1 metamodel, an "Activity" is a kind of Behavior that is specified as a graph of nodes interconnected by edges. A subset of the nodes are executable nodes that embody lower-level steps in the overall "Activity". "Object nodes" hold data that is input to and output from executable nodes, and moves across object flow edges. Control nodes specify sequencing of executable nodes via control flow edges. Activities are essentially what are commonly called "control and data flow" models. Such models of computation are inherently concurrent, as any sequencing of activity node execution is modelled explicitly by activity edges, and no ordering is mandated for any computation not explicitly sequenced [14]. This is the reason why the Scope Metamodel "Activity" is implemented in UML "Activity" Metaclass.

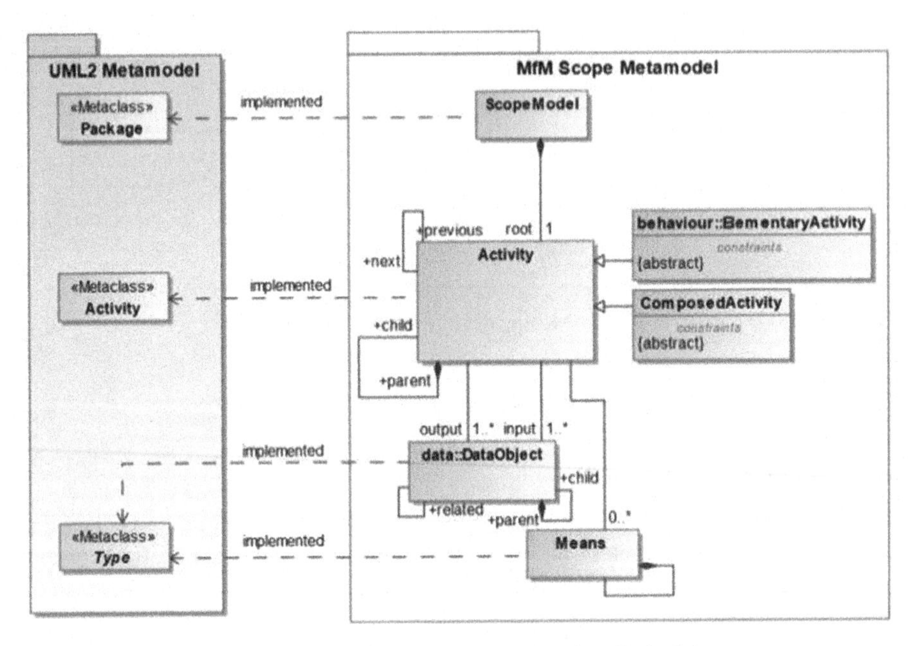

Fig. 3. Mapping between Scope metamodel and UML Activity syntax.

The "Means" concept in the Scope Metamodel defines the resources or means used for realizing an "Activity", and can be represented by any "Type" which is a generic UML metaclass grouping the classifiers used as type for a parameter defined as input or output of an Activity. The same rationale can be applied to the "DataObjects" in the Scope Metamodel, therefore we propose the same representation. Finally, the Scope Metamodel "ScopeModel" is proposed to be implemented as a UML "Package" Metaclass, in order to contain all the scope model knowledge and information.

4 Example of Scope Model in MagicDraw

To illustrate an MfM ontology model in MagicDraw, we will use the Industrial System Design ontology (ISD ontology) which has been presented by the authors in previous work to support the Industrial System Design in the concept phase [13]. In this previous work, the scope models were described using RAMUS software and IDEF0 modelling language [18] implementing MfM Scope meta-model in the scope models shown in Fig. 4.

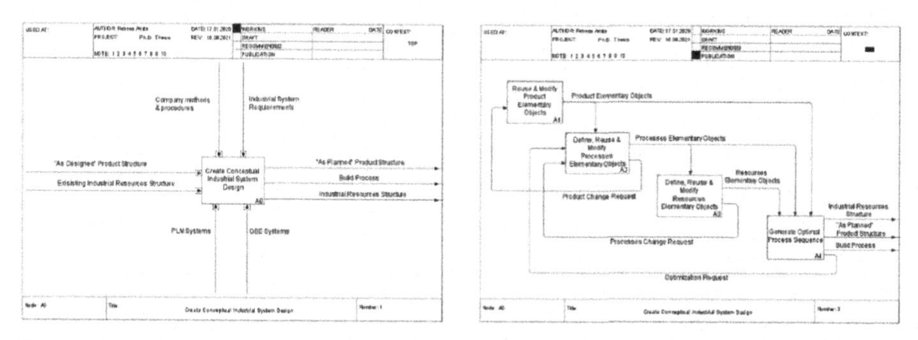

Fig. 4. ISD ontology scope models from previous work, with knowledge represented in IDEF0.

The knowledge in the scope model of the ISD ontology is now modelled in MagicDraw with UML, following the Scope Metamodel implementation defined in Sect. 3.2. In the ISD ontology, the scope is the activity "Create Conceptual Industrial System Design" of the industrial system design process. Figure 5 shows this top level activity, which is a Composed Activity of MfM Scope Metamodel.

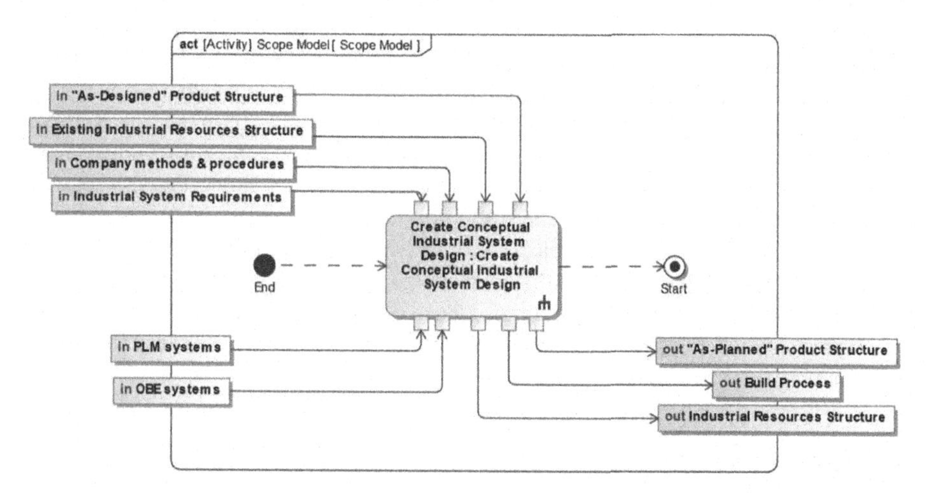

Fig. 5. Top-level Activity diagram "Create Conceptual Industrial System Design".

The activity receives as input the "As-Designed Product Structure", "Existing Industrial Resources Structure", "Company methods & procedures" and the "Industrial System Requirements", which are DataObjects of MfM Scope Metamodel. It also receives as input the "PLM systems" and "OBE systems" supporting the activity, which are Means in the MfM Scope Metamodel. The activity provides as output the "As-Planned Product Structure", the "Build Process" and the "Industrial Resources Structure", all of them DataObjects of MfM Scope Metamodel.

The white-box activity diagram of "Create Conceptual Industrial System Design" activity is shown in Fig. 6, with the following Elementary Activities of MfM ScopeMetamodel: "Reuse & Modify Product Elementary Objects", "Define, Reuse & Modify Process Elementary Objects", "Define, Reuse & Modify Resource Elementary Objects" and "Generate Optimal Process Sequence".

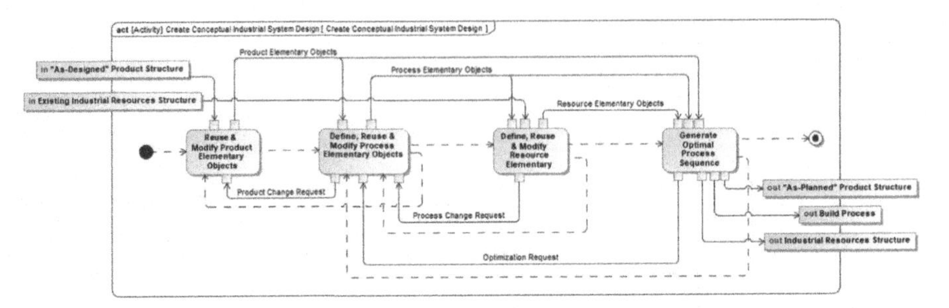

Fig. 6. Activity diagram "Create Conceptual Industrial System Design".

The activity "Reuse & Modify Product Elementary Objects" uses the "As-Designed Product Structure" input to define the Product Elementary Objects which are the first product split including interfaces. They are received as input in the activity "Define, Reuse & Modify Process Elementary Objects" which provide the Process Elementary Objects needed to manufacture, assemble or transport each Product. A Product Change Request can be sent back if no process can be defined for a given product.

The Process Elementary Objects are received as input in the activity "Define, Reuse & Modify Resource Elementary Objects" defining the Resource Elementary Objects from the existing industrial system capability to fulfill the processes. A Process Change Request can be sent if no resource can fulfill the process.

The three sets of elementary objects are received as input in the activity "Generate Optimal Process Sequence" where different process sequences are generated and optimized considering business rules and precedence constraints. This activity provides the final outputs "As-Planned product structure", "Build Process" and "Industrial Resources Structure".

5 Discussion and Further Work

Models for Manufacturing methodology aims to be an agnostic methodology successfully applied in different industrial and non-industrial cases for ontology-based engineering. The MfM metamodels define an independent link with any modelling software covering as minimum MfM metamodels.

To support the lifecycle of the MfM models and to establish a collaborative environment, an implementation in UML 2.5.1 standard enabled by MagicDraw software was shown covering all concepts of the Scope Metamodel. An example has been made with the ISD ontology, showing the knowledge captured in previous scope models now captured using MagicDraw. As a general conclusion MagicDraw may be used to cover MLM function of editing or visualizing to support the MfM method of the knowledge contained in a scope model. Nevertheless, the UML implementation enabled by MagicDraw can only be effectively used to model the Scope, complementing the implementation with a specific extension that allows to effectively capture the different semantics of the DataObjects and Means.

Further work will focus on this extension though a stereotype of UML to improve the Scope metamodel implementation. This will allow the use of MagicDraw as editing software for MfM scope models in different use cases. The same exercise will be made for the rest of the metamodels supporting the MfM methodology, in order to enable a wider implementation of MfM methodology in different industrial cases.

Acknowledgement. The authors would like to thank Sevilla University, Airbus and M&M Group colleagues for their support and contribution during the development of this work.

References

1. Mas, F., Racero, J., Oliva, M., Morales-Palma, D.: A preliminary methodological approach to models for manufacturing (MfM). In: Chiabert, P., Bouras, A., Noël, F., Ríos, J. (eds.) PLM 2018. IAICT, vol. 540, pp. 273–283. Springer, Cham (2018). https://doi.org/10.1007/978-3-030-01614-2_25
2. Holland, O.: Model-based systems engineering. In: Loper, M.L. (ed.) Modeling and Simulation in the Systems Engineering Life Cycle. SFMA, pp. 299–306. Springer, London (2015). https://doi.org/10.1007/978-1-4471-5634-5_23
3. Slimani, T.: Ontology development: a comparing study on tools, languages and formalisms. Indian J. Sci. Technol. **8**(24), 1–12 (2015)
4. Morales-Palma, D., Mas, F., Racero, J., Vallellano, C.: A preliminary study of models for manufacturing (MfM) applied to incremental sheet forming. In: Chiabert, P., Bouras, A., Noël, F., Ríos, J. (eds.) PLM 2018. IAICT, vol. 540, pp. 284–293. Springer, Cham (2018). https://doi.org/10.1007/978-3-030-01614-2_26
5. Mas, F., Racero, J., Oliva, M., Morales-Palma, D.: Preliminary ontology definition for aerospace assembly lines in Airbus using models for manufacturing methodology. In: Proceedings of 7th International Conference on Changeable, Agile, Reconfigurable and Virtual Production CARV18 (2018)
6. Arista, R., Mas, F., Oliva, M., Morales-Palma, D.: Applied ontologies for assembly system design and management within the aerospace industry. In: Proceedings of the Joint Ontology Workshops JOWO 2019, 10th International Workshop on Formal Ontologies Meet Industry 2518 (2019). http://ceur-ws.org/Vol-2518/paper-FOMI1.pdf
7. Gómez, A., Ríos, J., Mas, F., Vizán, A.: Method and software application to assist in the conceptual design of aircraft final assembly lines. J. Manuf. Syst. **40**, 37–53 (2016). https://doi.org/10.1016/j.jmsy.2016.04.002
8. CIMData: Model-based systems engineering: business opportunities and overcoming implementation challenges. A CIMData Report (2014). https://www.cimdata.com/en/component/docman/doc_download/3491-model-based-systems-engineering-business-opportunities-and-overcoming-implementation-challenges

9. Morales-Palma, D., Oliva, M., Racero, J., Eguia, I., Arista, R., Mas, F.: Metamodels approach supporting models for manufacturing (MfM) methodology. In: Canciglieri Junior, O., Noël, F., Rivest, L., Bouras, A. (eds) Product Lifecycle Management. Green and Blue Technologies to Support Smart and Sustainable Organizations. PLM 2021. IFIP Advances in Information and Communication Technology, vol. 640, pp. 398–409. Springer, Cham (2022). https://doi.org/10.1007/978-3-030-94399-8_29
10. CIMData: Product lifecycle management: empowering the future of business. A CIMData Report (2002). https://www.cimdata.com/en/component/docman/doc_download/2523-product-lifecycle-management-plm-definition
11. Arista, R., Mas, F., Oliva, M., Racero, J., Morales-Palma, D.: Framework to support models for manufacturing (MfM) methodology. IFAC-PapersOnLine **52**(13), 1584–1589 (2019). https://doi.org/10.1016/j.ifacol.2019.11.426
12. Arista, R., Mas, F., Morales-Palma, D., Oliva, M., Vallellano, C.: A preliminary ontology-based engineering application to industrial system reconfiguration in conceptual phase. In: Proceedings of the Joint Ontology Work-shops JOWO 2021, International Workshop on Formal Ontologies meet Industry (2021). http://ceur-ws.org/Vol-2969/paper20-FOMI.pdf
13. MfM research group: Metamodels. https://github.com/ModelsforManufacturing/Metamodel. Accessed 29 Mar 2022
14. UML OMG standard: Object management group. https://www.omg.org/spec/UML/2.5.1/About-UML/. Accessed 6 Feb 2022
15. Meta Object Facility: Object management group. https://www.omg.org/spec/MOF/. Accessed 6 Feb 2022
16. UML profiling: Dassault Systèmes CATIA No Magic documentation. https://docs.nomagic.com/display/MD190/UML+Profiling+and+DSL+Guide. Accessed 6 Feb 2022
17. Standard profile implementation: Dassault Systèmes CATIA No Magic documentation. https://docs.nomagic.com/display/MD190/Standard+profiles+implementation. Accessed 6 Feb 2022
18. NIST: Integration Definition for Function Modeling (IDEF0). Computer Systems Laboratory of the National Institute of Standards and Technology (1993). http://www.idef.com/wp-content/uploads/2016/02/idef0.pdf

An Approach for the Assessment of Water Use in Batik Production Processes

Ida Nursanti[1,2](✉), Alison McKay[1], and Richard Chittenden[1]

[1] University of Leeds, Leeds LS2 9JT, UK
Ida.Nursanti@ums.ac.id
[2] Universitas Muhammadiyah Surakarta, Surakarta, Central Java 57162, Indonesia

Abstract. Batik production processes use large quantities of water that are currently not evaluated. The Indonesian government has set the direct water use standard in the batik production process, but a method for assessing this water use indicator is needed. In this paper, we report work on developing an approach for assessing water used in the batik production process as an integral part of the batik fabric design process. The input parameters used in the approach are available in the design process, while the output parameter is the total direct water used in the production process of batik fabric made using a batik design. Three batik fabrics with different designs were selected as a case study. The production process model of the three designs was developed and implemented in a discrete event simulation software package. To validate the simulation results, data related to water use in batik production from batik factories were obtained using an online survey. The output of the proposed approach can be used as data for water use analysis to carry out process improvement in batik production processes.

Keywords: Water use assessment · Batik production process · Product design

1 Introduction

Batik production is a wet textile process. Over 47,000 manufacturers undertake batik production in Indonesia, spread across 101 industrial centers [1]. Although most are small and medium sized enterprises, they make a significant contribution to Indonesia's economic development [2]. However, the batik industry uses large quantities of water that currently are not evaluated [3]. In response, the Indonesian government proposed green industry standards for the batik industry in 2019. These include two indicators of water use: direct water use and wastewater reuse ratio [4]. The unit used for the indicator of direct water use in the production process is liters of water per square meter of fabric per color (l/m^2/color). Therefore, the mechanism used to assess the direct water used in batik production must consider the fabric design.

The literature on sustainable water use highlights several methods for assessing water used in manufacturing processes [5]. The goal of these methods is to identify potential improvements for stakeholders and decision-makers [6]. However, current methods support neither the operational assessment of direct water use in batik production nor the

Published by Springer Nature Switzerland AG 2023
F. Noël et al. (Eds.): PLM 2022, IFIP AICT 667, pp. 683–692, 2023.
https://doi.org/10.1007/978-3-031-25182-5_66

evaluation of alternative scenarios to reduce water use considering product design, and production order and planning issues such as production volumes. This paper reports research on developing an approach to assess water used in batik production that is specific to each batik fabric design based on operating conditions at the design stage. The remainder of this paper proceeds as follows. Section 2 provides a literature review concerning the characteristics of batik production processes and previous research on water use in the textile processing industries. The proposed approach, which is evaluated through a single case study, is introduced in Sect. 3, while Sect. 4 presents a discussion of the results. The conclusion is provided in Sect. 5.

2 Literature Review

2.1 Characteristic of Batik Production Processes

The batik production process produces colored and patterned fabric called "batik fabric". The design and manufacture of this fabric is both an art and a craft, and part of an ancient tradition [7]. Batik fabrics are made with the wax-resist dyeing technique which forms pattern motifs by applying wax and dyes to the fabric [8]. A batik design is characterized by its color, composition, and pattern. Batik patterns have hundreds of variations and unique features [9] and can be applied to various types of fabric. However, cotton fabric is more commonly used [10]. The main auxiliary materials used in the batik production process are wax and dyes. Batik wax is made of ingredients including residues of pine-gum distillation, resin from Shorea Javanica, paraffin, micro wax, animal fat such as Kendal which is fat from cows, coconut oil, beeswax, and lancing wax [11]. Based on the tools used for applying hot wax, batik fabrics are divided into three types: written, stamped, and combination. As regards dyes, two types of dye are used: synthetic and natural. Most batik manufacturers prefer synthetic dyes because natural dyes do not have as much color variation and have longer processing times [8]. Data from several studies suggest that the most widely used synthetic dyes in the batik production process are Remazol, Naphthol, Indigo Soluble, Direct, and Reactive [2, 12]. Other auxiliary materials for producing batik fabrics are soda ash (sodium carbonate), water glass (sodium silicate), sulphur, and chlorine.

Batik production consists of three main processes (dyeing, waxing, and dewaxing) and three additional processes (fabric preparation, bleaching, and finishing). Dyeing is a process in which color is transferred to the fabric to add permanent and long-lasting color. The dyeing process can be carried out with various dyeing techniques such as dipping, smocking, dabbing, spraying, and painting. More than one color can be applied to the fabric in each dyeing process, depending on the color combination in the product design. Waxing is a process in which hot wax is applied to fabric to prevent dye from penetrating, and so color change in, areas of fabric which are covered by wax. Dewaxing is the process for removing wax from the fabric by dipping it in a vat of boiling water that melts the wax. The rinsing process is carried out in the process of dyeing and dewaxing. This process is also done in fabric preparation, bleaching, and finishing processes. The batik production process can be started from either the dyeing or waxing process. In addition, the dyeing, waxing, and dewaxing processes can be repeated several times to form a finished batik design [13]. The source of water used by most Indonesian batik manufacturers is groundwater. Water is used in every stage of the batik production process except waxing. Most water used in the batik production processes is discharged to rivers as wastewater, while water absorbed by the fabric evaporates in the drying process. Visually, wastewater produced in the batik production process has color and is highly polluting due to synthetic dyes and other chemical materials [14].

2.2 Water Use in the Textile Processing Industries

A large and growing body of literature has investigated water use in the textile processing industry because it not only uses a large amount of water but also is one of the main producers of industrial wastewater in many countries [15]. Hussain and Wahab [16] categorized research on water uses in the textile industry into wastewater treatment and reuse, production equipment, process and chemical innovations, and advanced water analysis and saving tools. A considerable amount of literature has been published on wastewater treatment and reuse in textile processes, including the batik production process [13]. These studies propose several possible treatment processes for wastewater including physical, oxidation, and biological methods [17]. Further work reports methods for managing water use in textile processing through innovations; however, capital investment is identified as a central barrier for applying these methods [16].

Meanwhile, methods used for analyzing water use in textile processes are life cycle assessment (LCA) and water footprint assessment [16]. LCA focuses on identifying potential environmental impacts due to water use, while the water footprint assessment method is used to quantify the volume of water needed to produce a product [18, 19]. In the water footprint assessment method, a product's water footprint is divided into three components: green water, blue water, and grey water footprints. The green water footprint is the volume of rainwater consumed during the production process [20]. In the Indonesian batik industry this is not considered unless rainwater is used as a water source. The blue water footprint is the amount of surface and ground water used in the production of the product and disposed of as wastewater. Finally, the grey water footprint is the theoretical volume of water required to dilute a critical pollutant load to meet water quality standards [19]. The total water footprint of a product includes its direct and indirect water footprints [19]. The direct water footprint is the volume of water consumed or polluted when producing a product, while the indirect water footprint refers to water consumed to produce raw materials and energy for the production process, sometimes referred to as virtual water [19, 20]. However, the grey and indirect blue water footprints were out of scope in this research.

The water footprint method has been applied to a number of textile applications. For example, Chico et al. [21] calculated the total water footprint of a pair of jeans based on the types of textiles and production methods. Zhang et al. [22] used it to evaluate the total water footprint of three kinds of zipper. Handayani et al. [10] examined the total water footprint of a natural-colored batik fabric. Other authors investigated the total water footprint in the ready-made garment (RMG) production in Bangladesh [23]. In addition, Wang et al. [24] introduced a methodology of water footprint assessment for the Chinese textile manufacturing sector. They calculated the blue water footprint by summing the volume of freshwater appropriated in all processes. In contrast, Lévová and Hauschild [25] developed a method to calculate an impact score for water use based on LCA. In quantifying the amount of water used in the production process, this method considers the negative effects of removing the water from the water source, changing water quality, and causing a time delay between extraction and discharge. This model was applied in the textile industry to select the production site and the type of chemicals and was also employed in a study by Nursanti et al. [26] to assess the impact of water use in a batik factory. However, these studies did not consider the use of water in the production process and the characteristics of the production process system in detail based on operating conditions. These studies also did not evaluate different what-if scenarios for reducing water use that considers product design, production order and production planning, such as production volumes. Adjusting production orders and production plans are practiced to reduce water use [22].

3 Proposed Approach

3.1 Research Methodology

An inductive methodology was used to identify requirements (Sect. 3.2) for the proposed approach. These informed the development of a conceptual model (Sect. 3.3) of the production process for the three batik designs that were selected as case studies. The WITNESS discrete event simulation system was used to implement the conceptual model (Sect. 3.4) and enable the quantification and visualization of direct water use. For validation, outputs of the approach were compared with data on water use in batik production from factories in Central Java, Indonesia, obtained with an online survey.

3.2 Requirement Definition

The purpose of the proposed approach was to assess direct water used in batik production, based on operating conditions, at the design stage. This allows water to be incorporated into the design assessment process and provides opportunities for identifying process improvements. Using the approach, experiments with alternative production scenarios can be carried out to explore how they affect water use. The input parameters used in this approach were the information available during the design process of a batik pattern and its production process, including the type of batik, production process flow, techniques used in each process, and the dye used in each dye application process. Three batik fabrics with different designs were selected as case studies, as shown in Fig. 1. The first design is a written batik fabric [27], the second is stamped [28], and the third is a combination fabric [29]. The first design was a traditional one, the second modern and the third a combination of the two. Data to support the model, including water used in each process and the size of each batik fabric, was gathered from a previous study by Nursanti et al. [26]. Each piece of batik fabric was 2.75 m long and 1.15 m wide. The volume of water used in the dye application and color fixation was assumed to be the same for all types of dyes and dye application techniques and used to process one piece of fabric. This assumption was made due to data availability. In addition, the number of pieces of equipment used in each process was assumed to be one except for the rinsing process in the dyeing and dewaxing processes which use two water tubs.

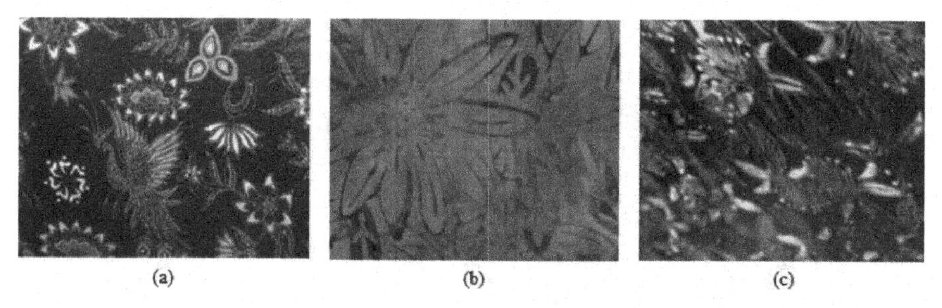

Fig. 1. (a) First batik design; (b) Second batik design; (c) Third batik design

3.3 Conceptual Model

The conceptual model presents the batik production processes as described in the literature review; the production process of batik fabric varies depending on the pattern made. The process flow for each dyeing process also varies depending on the technique and dye used. In addition, the model shows in detail the processes in batik production that use water. The conceptual model is shown in Fig. 2. There are three production process flows in the model: one for each of the three batik designs. Waxing, dyeing, and dewaxing processes are carried out several times. The bleaching process is carried out to produce the second batik design. In processes that use water, water was added and discharged each operation. The volume of water for wetting, rinsing, bleaching, and wax removal is the capacity of equipment used, which can be used to process more than one piece of fabric. The number of fabrics processed or the batch size in each operation varies. The water used in each operation is discharged if it is concentrated due to dyes, waxes, and other substances that dissolve.

3.4 Implementation

The model of the batik production was implemented in the Witness discrete event simulation package. The production flow for each design was defined by the routing through the different processes. The routing of the different fabrics as they proceed through the various processes was controlled by the input/output logic at each production stage. The model was run until all pieces of fabric were processed into the final products. It was run 100 times for each case study with different production volumes and batch sizes. Together with the production process flow of the case study, the production volume and batch size were defined and set up each time the model was run. The production volumes used ranged from five to one hundred pieces of fabric. One of a range of batch sizes (10, 20, 30, 40, and 50 pieces of fabric), which is the number of pieces of fabric processed per operation at the workstation for wet, rinse, bleach, and remove wax, was used in each simulation. The volume of water for each process used was from Nursanti et al. [23].

The simulation results for all three batik designs are shown in Fig. 3. Figure 3(a) presents water use to produce one piece of batik fabric for each specified production volume and batch size per operation. It can be seen that water used to produce a piece of batik fabric decreases if the production volume increases and the production volume

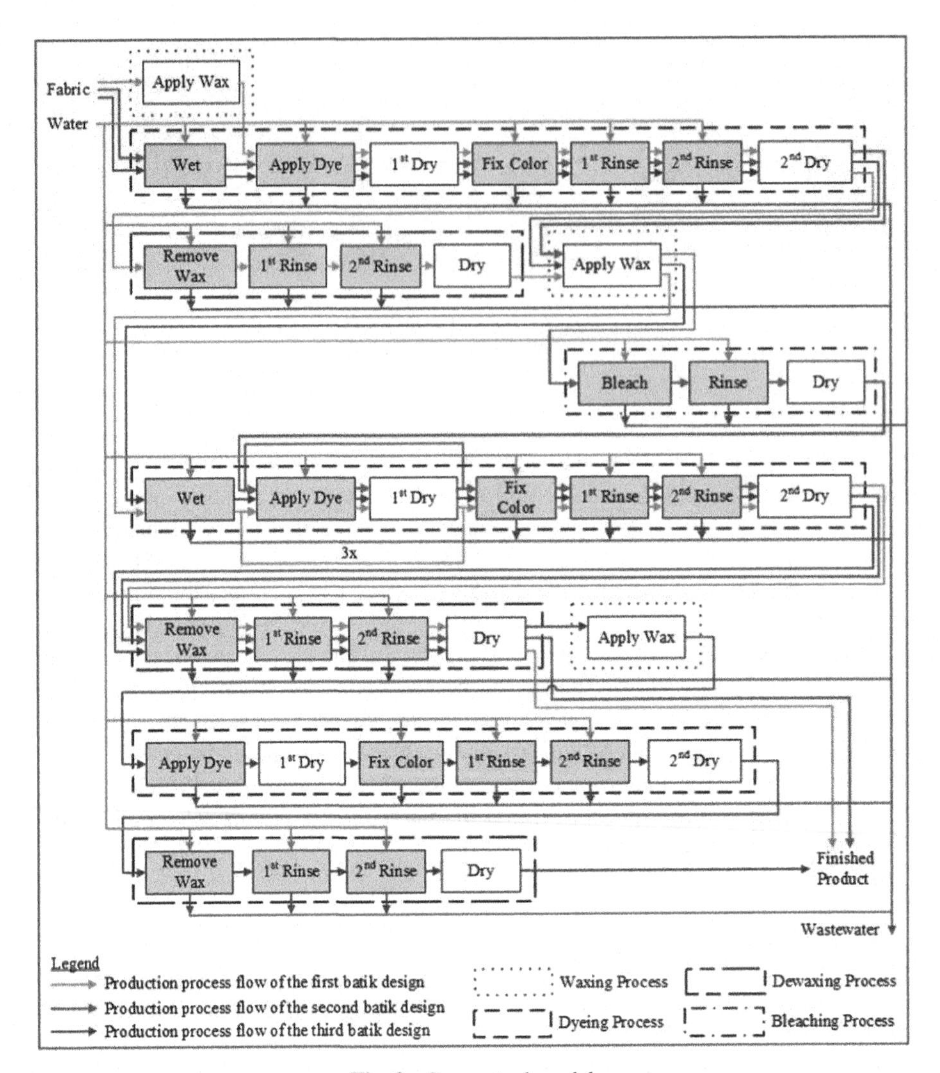

Fig. 2. Conceptual model

exceeds the number of pieces of fabric processed per operation at each workstation. In addition, the total water used graphs that are presented in Fig. 3(b) indicate that there are fixed, semi-variable, and variable water uses. Fixed water usage is the volume of water used for a given operation regardless of production volume. Semi-variable water use is fixed for a given production volume but increases when that production volume is exceeded. In contrast, the variable water use is water that fluctuates with production volume. The total water use from this assessment can be used as input data to analyze the water use of a batik design.

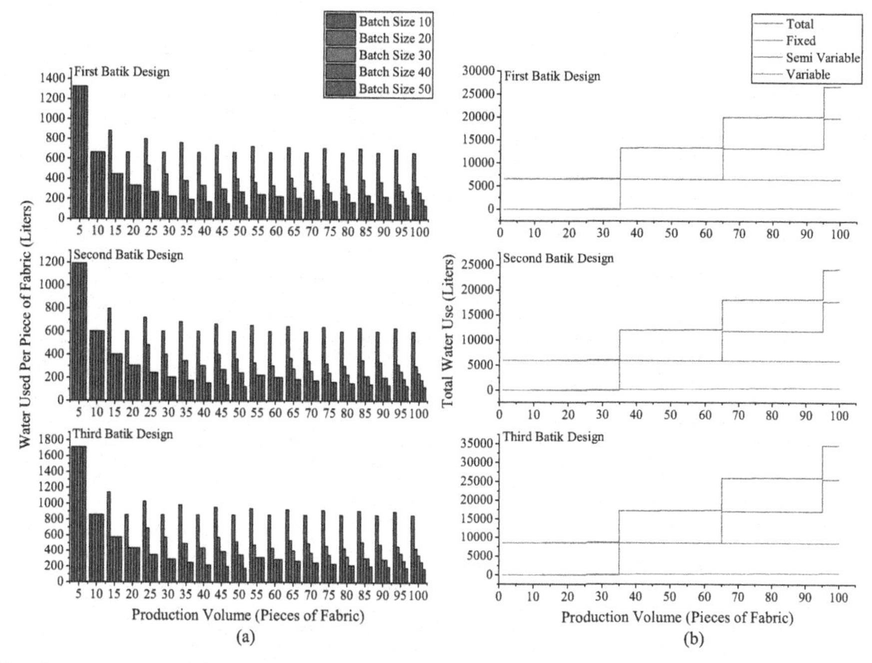

Fig. 3. (a) Water use per piece of fabric; (b) Total water use if the batch size per operation is 30 pieces of fabric (the fabric size is 2,75 m long and 1,15 m wide)

4 Discussion

Figure 4 presents the water used to produce one square meter of batik fabric based on the simulation and survey results. Since the size of the fabric used in batik factories varies, the unit of fabric used to present the data in the figure was the square meter. The data obtained from the survey was the average water use per day, the average production volume per day, and the type of batik produced in several batik factories. However, the batik patterns and the number of batik designs made in these factories were unknown. Thus, in calculating water use per square meter of fabric, all batik fabric made in a batik factory were considered to have the same production process. The batik production process varies depending on the design of the pattern, as shown in this paper.

According to the survey results, water used in the production of written batik fabric is high because the production volume is low. However, as in the simulation results, the water used to produce written batik fabric is greater than the water used for the stamped batik fabric production process. In addition, if the fabric produced about 270 square meters, the water used to make combination batik fabric in the simulation results is close to the value of water used in the production process of combination batik fabric in the survey results. The water use per square meter in the simulation results is higher because the production equipment used in batik factories varies such as the size of the tubs, and the batch size per operation at each workstation also varies.

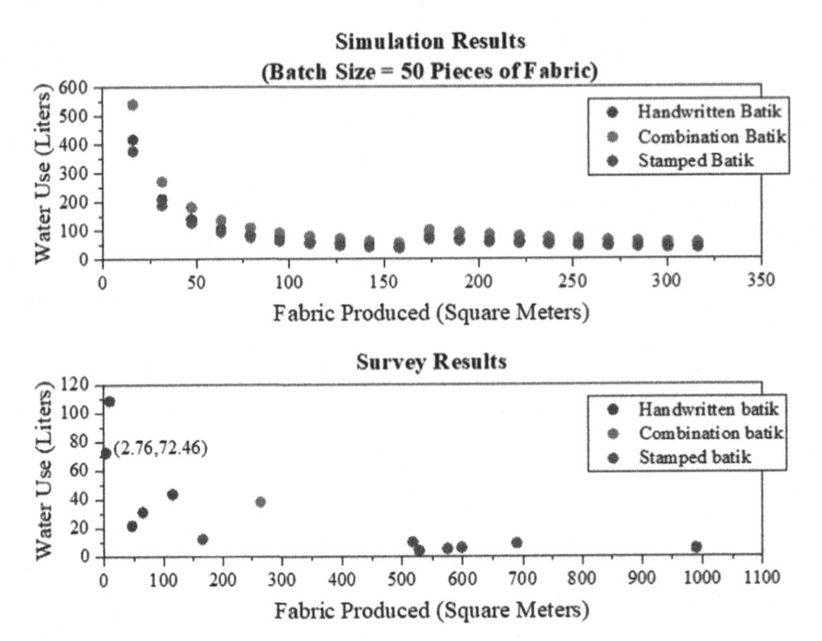

Fig. 4. Water use per square meter of fabric based on simulation results and survey results

5 Conclusions

An approach for evaluating water use in batik production processes based on information available in the process of designing a batik pattern and its production process has been proposed in this paper. The data determined when designing a batik pattern and its production process are the batik production processes carried out and the sequence of the processes, the technique used in each process, and the type of dye used in each dye application process. Production volumes are determined later in production planning. In addition, to calculate water consumption in the production process, historical data is needed, such as the number of products processed, and the volume of water used per operation at each workstation. The output of this approach is the total direct water used to produce batik fabric made using a batik design. Two limitations of the research are (i) only three production processes of batik designs were used as a case study, and (ii) limited water use data was obtained from batik factories. However, this study's findings provide much room for further progress in determining the strategies to improve water sustainability in the batik production process. Results from the proposed approach can be used as data for water use analysis of each batik design produced in a batik factory so that process improvement can be carried out. In addition, the proposed approach can facilitate the evaluation of water use based on green industry standards for the batik industry since the batik production process is not generalized. In future work, other factors related to water used in batik production will be considered in the approach. The approach also will be applied to assess water use in batik production processes in several batik companies.

Acknowledgements. This work is supported by the Directorate General of Higher Education, Ministry of Education and Culture, Republic of Indonesia.

References

1. Marifa, P.C., et al.: Production waste analysis using value stream mapping and waste assessment model in a handwritten batik industry. In: MATEC Web of Conferences, ICET4SD 2017, vol. 154, p. 01076. EDP Sciences (2018)
2. Handayani, W., Kristijanto, A.I., Hunga, A.I.R.: Are natural dyes eco-friendly? A case study on water usage and wastewater characteristics of batik production by natural dyes application. Sustain. Water Resour. Manag. **4**(4), 1011–1021 (2018). https://doi.org/10.1007/s40899-018-0217-9
3. Handayani, W., Widianarko, B., Pratiwi, A.R.: The water use for batik production by batik SMEs in Jarum Village, Klaten Regency, Indonesia: what are the key factors? In: IOP Conference Series: Earth and Environmental Science, vol. 716, No. 1, p. 012004. IOP Publishing (2021)
4. Kemenperin: Regulation of the minister of industry of the republic of Indonesia number 39 year 2019 regarding green industry standard for the Batik industry. Kementerian Perindustrian Republik Indonesia, Jakarta (2019)
5. Willet, J., Wetser, K., Vreeburg, J., Rijnaarts, H.H.M.: Review of methods to assess sustainability of industrial water use. Water Resour. Ind. **21**, 100110 (2019)
6. Angelakoglou, K., Gaidajis, G.: A review of methods contributing to the assessment of the environmental sustainability of industrial systems. J. Clean. Prod. **108**, 725–747 (2015). https://doi.org/10.1016/j.jclepro.2015.06.094
7. Susanty, A., et al.: Achieving cleaner production in SMEs batik toward innovation in production process. In: ICE & IEEE International Technology Management Conference, pp. 1–11. IEEE (2013). https://doi.org/10.1109/ITMC.2013.7352704
8. Rinawati, D.I., et al.: Natural dyes product design using green quality function deployment II method to support batik sustainable production. In: E3S Web of Conferences, ICENIS 2018, vol. 73, p. 04014. EDP Sciences (2018)
9. Nurhaida, I., et al.: Automatic Indonesian's batik pattern recognition using SIFT approach. Procedia Comput. Sci. **59**, 567–576 (2015)
10. Handayani, W., Kristijanto, A.I., Hunga, A.I.R.: A water footprint case study in Jarum village, Klaten, Indonesia: the production of natural-colored batik. Environ. Dev. Sustain. **21**(4), 1919–1932 (2019)
11. Malik, A., et al.: The effect of microwax composition on the staining quality of Klowong Batik Wax. In: MATEC Web of Conferences, ICET4SD 2017, vol. 154, p. 01118. EDP Sciences (2018). https://doi.org/10.1051/matecconf/201815401118
12. Rashidi, H.R., Sulaiman, N.N., Hashim, N.A.: Batik industry synthetic wastewater treatment using nanofiltration membrane. Procedia Eng. **44**, 2010–2012 (2012)
13. Rahayu, I.A.T., Peng, L.H.: Sustainable batik production: review and research framework. In: Advances in Social Science, Education and Humanities Research, ICRACOS 2019, vol. 390, pp. 66–72. Atlantis Press (2020)
14. Lestari, S., et al.: Effect of batik wastewater on Kali Wangan water quality in different seasons. In: E3S Web of Conferences, ICENIS 2017, vol. 31, p. 04010. EDP Sciences (2018). https://doi.org/10.1051/e3sconf/20183104010
15. Hasanbeigi, A., Price, L.: A technical review of emerging technologies for energy and water efficiency and pollution reduction in the textile industry. J. Clean. Prod. **95**, 30–44 (2015). https://doi.org/10.1016/j.jclepro.2015.02.079

16. Hussain, T., Wahab, A.: A critical review of the current water conservation practices in textile wet processing. J. Clean. Prod. **198**, 806–819 (2018)
17. Holkar, C.R., et al.: A critical review on textile wastewater treatments: possible approaches. J. Environ. Manage. **182**, 351–366 (2016)
18. Pfister, S.: Understanding the LCA and ISO water footprint: a response to Hoekstra (2016) "A critique on the water-scarcity weighted water footprint in LCA." Ecol. Indic. **72**, 352 (2017). https://doi.org/10.1016/j.ecolind.2016.07.051
19. Hoekstra, A.Y., et al.: The Water Footprint Assessment Manual: Setting the Global Standard. Earthscan, London (2011). https://doi.org/10.4324/9781849775526
20. Gu, Y., et al.: Calculation of water footprint of the iron and steel industry: a case study in Eastern China. J. Clean. Prod. **92**, 274–281 (2015)
21. Chico, D., Aldaya, M.M., Garrido, A.: A water footprint assessment of a pair of jeans: the influence of agricultural policies on the sustainability of consumer products. J. Clean. Prod. **57**, 238–248 (2013). https://doi.org/10.1016/j.jclepro.2013.06.001
22. Zhang, Y., et al.: The industrial water footprint of zippers. Water Sci. Technol. **70**(6), 1025–1031 (2014). https://doi.org/10.2166/wst.2014.323
23. Hossain, L., Khan, M.S.: Water footprint management for sustainable growth in the Bangladesh apparel sector. Water **12**, 2760 (2020)
24. Wang, L.-L., et al.: The introduction of water footprint methodology into the textile industry. Industria Textilă **65**(1), 33–36 (2014)
25. Lévová, T., Hauschild, M.Z.: Assessing the impacts of industrial water use in life cycle assessment. CIRP Ann. Manuf. Technol. **60**(1), 29–32 (2011)
26. Nursanti, I., et al.: Water footprint assessment of Indonesian Batik production. In: AIP Conference Proceedings 1977, ICETIA 2017, p. 050008. AIP Publishing (2018)
27. Khuri Vidy: https://www.youtube.com/watch?v = TgHk0MP1Fmc&t = 616 s. Accessed 22 Feb 2022
28. AvlynFabrics: https://www.youtube.com/watch?v=-nsMFIZfBiE. Accessed 22 Feb 2022
29. Danang Pamungkas: https://www.youtube.com/watch?v = FITBZB3py1Q&t = 897 s. Accessed 22 Feb 2022

Correction to: MBSE-PLM Integration: Initiatives and Future Outlook

Detlef Gerhard, Sophia Salas Cordero, Rob Vingerhoeds, Brendan P. Sullivan, Monica Rossi, Yana Brovar, Yaroslav Menshenin, Clement Fortin, and Benoit Eynard

Correction to:
Chapter "MBSE-PLM Integration: Initiatives and Future Outlook" in: F. Noël et al. (Eds.): ***Product Lifecycle Management. PLM in Transition Times: The Place of Humans and Transformative Technologies*****, IFIP AICT 667,**
https://doi.org/10.1007/978-3-031-25182-5_17

In the originally published version of chapter 17 the last name of one of the authors has been tagged incorrectly. The last name of the author has been corrected as "Salas Cordero".

The updated original version of this chapter can be found at
https://doi.org/10.1007/978-3-031-25182-5_17

Published by Springer Nature Switzerland AG 2023
F. Noël et al. (Eds.): PLM 2022, IFIP AICT 667, p. C1, 2023.
https://doi.org/10.1007/978-3-031-25182-5_67

Author Index

Printed in the United States
by Baker & Taylor Publisher Services